MW00343040

DATE DUE

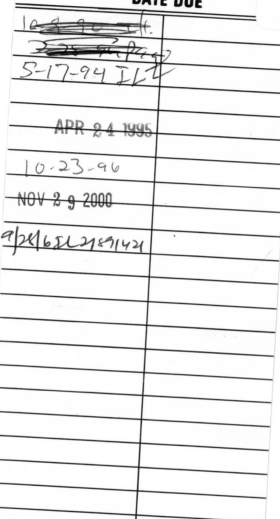

10-8-90	
~~illegible~~	
5-17-94 ILL	
APR 24 1995	
10-23-96	
NOV 2 9 2000	
9/24/6 ILL 2/89/421	

Handbook of
Clinical Psychology in
Medical Settings

Handbook of
Clinical Psychology in
Medical Settings

Edited by

Jerry J. Sweet

Ronald H. Rozensky

and

Steven M. Tovian

Evanston Hospital
and Northwestern University
Evanston, Illinois
and Northwestern University Medical School
Chicago, Illinois

UNIVERSITY LIBRARY
GOVERNORS STATE UNIVERSITY
UNIVERSITY PARK, IL. 60466

Plenum Press • New York and London

Library of Congress Cataloging-in-Publication Data

Handbook of clinical psychology in medical settings / edited by Jerry
 J. Sweet, Ronald H. Rozensky, and Steven M. Tovian.
      p.   cm.
   Includes index.
   ISBN 0-306-43550-0
   1. Clinical psychology.   I. Sweet, Jerry J.   II. Rozensky, Ronald
H. III. Tovian, Steven M.
   [DNLM: 1. Psychology, Clinical--handbooks.   WM 34 H236]
   RC467.H269 1991
   616'.001'9--dc20
   DNLM/DLC
   for Library of Congress   91-2606
                                                        CIP

RC 467 .H269 1991

Handbook of clinical
  psychology in medical
        279297

ISBN 0-306-43550-0

© 1991 Plenum Press, New York
A Division of Plenum Publishing Corporation
233 Spring Street, New York, N.Y. 10013

All rights reserved

No part of this book may be reproduced, stored in a retrieval system, or transmitted
in any form or by any means, electronic, mechanical, photocopying, microfilming,
recording, or otherwise, without written permission from the Publisher

Printed in the United States of America

To Nancy, Christopher, and Jamie

—JJS

To Patti, Sarah, and Jordyn

—RHR

To Merle, Avra, Zev, Rachel, and Ayal

—SMT

Contributors

Roger T. Anderson
Behavioral Medicine Branch
National Heart, Lung, and Blood Institute
Bethesda, Maryland 20892

Cynthia D. Belar
Department of Clinical and Health
  Psychology
University of Florida Health Science Center
Gainesville, Florida 32610

Linas A. Bieliauskas
Veterans Administration Medical Center
  and University of Michigan
Ann Arbor, Michigan 48105

Joseph Bleiberg
National Rehabilitation Hospital
102 Irving Street N.W.
Washington, DC 20010

Bruce Bonecutter
Department of Psychiatry
Cook County Hospital
Chicago, Illinois 60612

Rosalind D. Cartwright
Sleep Disorder Service and Research Center
Rush–Presbyterian–St. Luke's Medical
  Center
1653 West Congress Parkway
Chicago, Illinois 60612

Stanley L. Chapman
Pain Control and Rehabilitation Institute
  of Georgia
350 Winn Way
Decatur, Georgia 30030

James P. Choca
Department of Psychiatry and Behavioral
  Sciences, Northwestern University Medical
  School, and Psychology Service, Lakeside
  Veterans Administration Hospital
Chicago, Illinois 60611

Robert Ciulla
Islip Mental Health Center
1747 Veterans Highway
Islandia, New York 11769

Daniel J. Cox
Behavioral Diabetes Research
Department of Behavioral Medicine and
  Psychiatry
University of Virginia Health Sciences Center
Charlottesville, Virginia 22901

Thomas L. Creer
Department of Psychology
Ohio University
Athens, Ohio 45701

Nicholas A. Cummings
American Biodyne, Inc.
400 Oyster Point Boulevard, Suite 218
South San Francisco, California 94080

Patrick H. DeLeon
U.S. Senate Staff
Washington, DC 20510

Jean C. Elbert
Department of Pediatrics
University of Oklahoma Health Sciences
   Center
Oklahoma City, Oklahoma 73117

Kevin Franke
Department of Psychiatry
Northwestern University Medical School
Chicago, Illinois 60611

Alice G. Friedman
Department of Psychology
State University of New York at Binghamton
Binghamton, New York 13901

Linda Gonder-Frederick
Behavioral Diabetes Research
Department of Behavioral Medicine and
   Psychiatry
University of Virginia Health Sciences Center
Charlottesville, Virginia 22901

Martin Harrow
Department of Psychiatry
Michael Reese Hospital and Medical Center
Chicago, Illinois 60616

David E. Hartman
Department of Psychiatry
Cook County Hospital
Chicago, Illinois 60612

Heather C. Huszti
Department of Psychiatry and Behavioral
   Sciences
University of Oklahoma Health Sciences
   Center
Oklahoma City, Oklahoma 73190

Craig Johnson
Laureate Psychiatric Clinic and Hospital
Tulsa, Oklahoma 74147-0207, and
Department of Psychiatry
Northwestern University Medical School
Chicago, Illinois 60611

Michael Jospe
California School of Professional Psychology
1000 South Fremont Avenue
Alhambra, California 91803-1360

Bonnie L. Katz
National Rehabilitation Hospital
102 Irving Street N.W.
Washington, DC 20010

Daniel S. Kirschenbaum
Department of Psychiatry and Behavioral
   Sciences
Northwestern University Medical School
Chicago, Illinois 60611

Benjamin Kleinmuntz
Department of Psychology
University of Illinois at Chicago
Chicago, Illinois 60680

Mary P. Koss
Department of Psychiatry
University of Arizona
Tucson, Arizona 85724

Harry Kotses
Department of Psychology
Ohio University
Athens, Ohio 45701

Asenath La Rue
Department of Psychiatry and Biobehavioral
  Sciences
University of California at Los Angeles
Los Angeles, California 90024-1759

Kris R. Ludwigsen
Kaiser Foundation Hospital
200 Muir Road
Martinez, California 94553

Charles McCreary
Department of Psychiatry and Biobehavioral
  Sciences
University of California at Los Angeles
Los Angeles, California 90024-1759

James Malec
Department of Psychiatry and Psychology
Mayo Clinic and Foundation
Rochester, Minnesota 55905

Donald Meichenbaum
Department of Psychology
University of Waterloo
Waterloo, Ontario N2L 3G1, Canada

Barbara G. Melamed
Ferkauf Graduate School of Psychology and
  Albert Einstein College of Medicine
Yeshiva University
1300 Morris Park Avenue
Bronx, New York 10461

Russ V. Reynolds
Department of Psychology
Ohio University
Athens, Ohio 45701

Ronald H. Rozensky
The Evanston Hospital and Northwestern
  University and Medical School
Evanston, Illinois 60201

J. Terry Saunders
Behavioral Diabetes Research
Department of Behavioral Medicine and
  Psychiatry
University of Virginia Health Sciences Center
Charlottesville, Virginia 22901

Harry S. Shabsin
Department of Psychiatry and Behavioral
  Sciences
Johns Hopkins University School of
  Medicine, and
Division of Digestive Diseases
Francis Scott Key Medical Center
Baltimore, Maryland 21224

Edward P. Sheridan
Department of Psychology
University of Central Florida
Orlando, Florida 32816

Kathleen Sheridan
Department of Psychology
University of Central Florida
Orlando, Florida 32816

Sharon A. Shueman
Shueman, Troy & Associates
246 North Orange Grove Boulevard
Pasadena, California 91103

George C. Stone
Department of Psychiatry
University of California
San Francisco, California 94143

Jerry J. Sweet
The Evanston Hospital and Northwestern
  University and Medical School
Evanston, Illinois 60201

Robert J. Thompson, Jr.
Duke University Medical Center
Durham, North Carolina 27710

David L. Tobin
Department of Psychiatry
University of Chicago
Chicago, Illinois 60637

Steven M. Tovian
The Evanston Hospital and Northwestern
    University and Medical School
Evanston, Illinois 60201

Warwick G. Troy
Shueman, Troy & Associates
246 North Orange Grove Boulevard
Pasadena, California 91103

Dennis C. Turk
Pain Evaluation and Treatment Institute
University of Pittsburgh School of Medicine
Baum Boulevard at Craig Street
Pittsburgh, Pennsylvania 15213

C. Eugene Walker
Department of Psychiatry and Behavioral
    Sciences
University of Oklahoma Health Sciences
    Center
Oklahoma City, Oklahoma 73190

Sharlene M. Weiss
Health Promotion and Disease Prevention
    Branch
National Center for Nursing Research
Bethesda, Maryland 20892

Stephen M. Weiss
Behavioral Medicine Branch
National Heart, Lung, and Blood Institute
Bethesda, Maryland 20892

William E. Whitehead
Department of Psychiatry and Behavioral
    Sciences
Johns Hopkins University School of
    Medicine, and
Division of Digestive Diseases
Francis Scott Key Medical Center
Baltimore, Maryland 21224

Jack G. Wiggins
Psychological Development Center
7057 West 130th Street
Cleveland, Ohio 44130

David J. Williamson
Department of Clinical and Health
    Psychology
University of Florida Health Science Center
Gainesville, Florida 32610

Diane J. Willis
Department of Pediatrics
University of Oklahoma Health Sciences
    Center
Oklahoma City, Oklahoma 73117

W. Joy Woodruff
Department of Psychiatry
University of Arizona
Tucson, Arizona 85724

Logan Wright
Department of Psychology
University of Oklahoma
Norman, Oklahoma 73072

Foreword

For two decades, I have been responding to questions about the nature of health psychology and how it differs from medical psychology, behavioral medicine, and clinical psychology. From the beginning, I have taken the position that any application of psychological theory or practice to problems and issues of the health system is health psychology. I have repeatedly used an analogy to Newell and Simon's "General Problem Solver" program of the late 1950s and early 1960s, which had two major functional parts, in addition to the "executive" component. One was the "problem-solving core" (the procedural competence); the other was the representation of the "problem environment." In the analogy, the concepts, knowledge, and techniques of psychology constitute the core competence; the health system in all its complexity is the problem environment. A health psychologist is one whose basic competence in psychology is augmented by a working knowledge of some aspect of the health system.

Quite apparently, there are functionally distinct aspects of health psychology to the degree that there are meaningful subdivisions in psychological competence and significantly different microenvironments within the health system. I hesitate to refer to them as areas of specialization, as the man who gave health psychology its formal definition, Joseph Matarazzo, has said that there are no specialties in psychology (cited in the editors' preface to this book). With all respect to Matarazzo, whose knowledge of the history of psychology is as immense as his experience with present-day psychology, I would suggest that there are, at least, proto-specializations. We can recognize such functional specializations on the basis of the time required for a person trained to the level of journeyperson in one or more areas of psychology to achieve an equal level of competence in some other specific area.

As one considers that statement, it is also apparent that the time required depends on the areas in which competence was previously attained. There are various bodies of theory, of technique, and of knowledge of environments (tacit and otherwise) that are essential to competent performance in many different settings where psychologists are to be found. There is a loosely defined disciplinary core of theory and knowledge that defines a psychologist. Beyond that, there is a vast amount of specialized theory and factual knowledge far too great to be mastered in total by any individual. So there is selective mastery of sections of the whole that produces, perhaps not specialties, but at least fuzzy sets of psychologists who communicate differently within and between their groupings.

There appears to be a growing consensus that there is a core of knowledge and technique common to all *practice* of psychology, that is, the professional core. Evidence of such a core would be found, I suppose, in the demonstration that it takes less time for someone who has become a competent professional in one type of application to become competent in another than it takes for persons who have never practiced. Surely, that is a plausible expectation. Are there differing degrees of overlapping subsets of professional competence between industrial-organizational, school, counseling, forensic, and clinical psychology? That seems less certain.

In the final analysis, the demonstration of competence must always occur in a particular environment and problem setting. It is at least an implicit assumption of this book that medical environments and problems have more in common with each other than they do with nonmedical settings, including mental health settings. The book offers a powerful test of this assumption. For the most part, it takes for granted at least an adequate preparation in a generic clinical training program, and it undertakes to demonstrate how the competence acquired elsewhere must be modified and extended if proficiency in medical settings is to be achieved. The first 16 chapters are concerned with issues common to all practices in a medical environment. Of these, the first 6 chapters focus upon types of hospital structure, education and training, and professional and political issues. Chapters 7 through 10 examine quality assurance, finances, marketing, and computers in practice. Chapters 11 through 16 cover broad clinical and research topics that cross diagnostic categories.

The second half of the book consists of 15 chapters that discuss in detail the clinical considerations that arise as the general principles are put to work in specific work environments. These chapters were written by persons actually experienced in the delivery of such services, and they are rich in the practicalities that must be faced day to day. A special quality emerges from the authors' adherence to the principle of presenting the assessment and treatment of particular psychomedical problems in the context of developing and operating a successful service program. This reader found himself frequently surprised by the ingenious way that some practical, micropolitical, or administrative problems had been handled rather than avoided, as has often been the case in other presentations, in which common problems have been glossed over or simply ignored. As one reads about program after program, one has a chance to see how and to what degree the general principles advanced in the introductory sections are realized in actual medical settings.

The descriptions of each of the kinds of programs discussed here could be expanded into books in their own right. There are, in fact, books about a number of the chapter topics. Yet the presentations here are informative and comprehensive, and they are short enough so that they can be read for an overall impression. Thus, this book can serve as an orientation for postdoctoral trainees preparing for careers in medical settings, or for psychologists about to start new jobs in such settings, where the specific tasks to be undertaken are not fully defined, or for experienced clinical health psychologists who propose to expand their work into new settings. At this stage of the development of clinical psychology in medical settings, the psychologist associated with a medical center is likely to be functioning in a somewhat exploratory and entrepreneurial manner. Here is an excellent guidebook for such an exploration.

GEORGE C. STONE

San Francisco, California

Preface

The strength of clinical psychology is its scientific basis and its application of scientific methodology to the practice of psychology. It is that psychology of empiricism and that style of problem definition and hypothesis testing that we believe has uniquely fitted the practice of clinical psychology into the medical setting. The expansion of roles and the numbers of psychologists in medical settings set forth in this volume support this position. It was our intent as editors to highlight this growth and the wide range of clinical and nonclinical services provided by psychologists in medical settings. In order to present the information on clinical practice within the proper context, we felt that it was necessary to set the scene by reviewing the characteristics of the medical environment in general, the hospital itself, the politics and professional issues of hospital practice, finances, marketing, quality assurance, research, and the educational and training issues that prepare psychologists for this milieu. In order to truly underscore the strength of psychological practice within medical settings, we asked the chapter authors to stress the scientist-practitioner focus of the field. Clinical programs are described and program development is offered to the reader based on the scientific foundations of the practice of clinical psychology.

Throughout the text, the reader will find various generic labels for the activities of clinical psychologists in medical settings. These include *health psychology*, *medical psychology*, *clinical health psychology*, *neuropsychology*, *rehabilitation psychology*, *behavioral medicine*, and *clinical psychology*. The title, *Handbook of Clinical Psychology in Medical Settings*, reflects our view that the success of psychology in medical settings is rooted in the core knowledge of the field of psychology as discussed by Joseph Matarazzo (1987; there is only one psychology, no specialties, but many applications—*American Psychologist*).

A brief personal digression by the editors is in order. Each of us entered psychology with an interest in research and practice in "mental health." Undergraduate psychology backgrounds and graduate courses and seminars prepared us for a Boulder model outlook on the mental health field. None of us considered *a priori* that a medical facility would be a place to carry out the scientific or clinical life of a psychologist. Serendipity found each of us, as part of our training, assigned to a "medical facility." The graduate-student peers of one of us thought that he had received the least desirable practicum spot, as there were "real" mental health facilities within which he could have trained. As they say, the rest is history. The applicability of the general skills we learned in clinical training fit so well and so successfully within the medical setting and with medical patients that each of us received strong reinforcement, and we have remained within, and in combination, now have experienced well over 40 years of professional time in, that environment. We hope that this text reflects our commitment to and excitement about the medical setting as a place for research on and the practice of clinical psychology.

Given our enthusiasm, we necessarily had to restrict the size and range of the topics so that they could be covered reasonably within this one book. Therefore, the reader will no doubt notice that not all topics, disease entities, diagnoses, or problems found in the medical setting are covered here. Again, the range of possible topics attests to the scope of clinical psychology in medical settings.

Throughout the text, the reader will find mention of guild, turf, or political issues that can and do influence the daily practice, academic, and financial freedoms of clinical psychologists in medical settings. The cartoon shown below appeared in the *New Yorker* and hangs in the office of Bryant Welch, Executive Director of the American Psychological Association's Practice Directorate. Little interpretation is needed to see the irony of the heretofore unwanted dragon being asked to address his foes. However, it is time that clinical psychologists in medical settings not see themselves as dragons (or the outsiders we once were) but begin to see themselves as sitting in the audience with the rest of the knights. Our armor may be less well heeled or just as shiny as the next knight's; our armor may be more-or-less tarnished; our helmets may sport blue rather than purple plumes; but we are now and will continue to be members of the modern-day health-care alliance. It is

"To begin with, I would like to express my sincere thanks and deep appreciation for the opportunity to meet with you. While there are still profound differences between us, I think the very fact of my presence here today is a major breakthrough."

Drawing by W. Miller; © 1983 The New Yorker Magazine, Inc. Reprinted by permission.

the theme of science and service and the politics of competence that support our place in the medical setting, not the appearance of our knightly accoutrements.

We would like to thank our chapter authors for their energy and enthusiasm. Their forbearance through the editorial process and their gracious acceptance of our comments are gratefully appreciated. We thank Eliot Werner, Executive Editor at Plenum, for his interest and guidance in bringing a dream and an outline to press. We also want to thank our secretaries, Vanessa Schaeffer, Pat Barnes, and Elaine Richardson, for their hours of typing, photocopying, phone calling, and patience.

We owe a debt of gratitude to our professors in psychology, who helped breathe life into the Boulder model, and to our colleagues in psychology, medicine, and psychiatry, who help reinforce the vitality of our professional lives. Thanks are also due to Edward Sheridan for his encouragement and helpful suggestions throughout this project. And a final, special thank you to Ira H. Sloan, Chairman, Department of Psychiatry, Evanston Hospital, for his support in providing a truly collegial, multidisciplinary environment.

Glenwick *et al.* (1987) have offered a discussion of helpful suggestions and encouragements for, and the pitfalls encountered in, editing a book. Thus far, all of their insights have rung true to the present editors. They suggested that working with coeditors who have a similar value system is a strength. The present editorial product, for us, represents that strength and reflects a process of equal input, equal energy, equal commitment, and equal responsibility of the editors for the whole of this project.

Time spent on this endeavor was personally and professionally enjoyable. Relationships renewed or established with colleagues across the continent reflect for us the vitality of clinical psychology in medical settings.

RONALD H. ROZENSKY
JERRY J. SWEET
STEVEN M. TOVIAN

Evanston, Illinois

References

Glenwick, D., Brodsky, S., Franks, C., Hess, A., Balch, K., Frank, J., Garfield, S., & Jason, L. (1987). Issues in preparing an edited volume in psychology. *American Psychologist*, 4, 405–407.
Matarazzo, J. (1987). There is only one psychology, no specialties, but many applications. *American Psychologist*, 42, 893–903.

Contents

PART VI. TEMPLATES FOR PROGRAM DEVELOPMENT

PART VII. FUTURE DIRECTIONS

PART I

Introduction

Since the inception of clinical psychology in medical settings, much has changed and is changing with regard to its practice. Before one can fully appreciate the current diversity and complexity of professional and practice issues and program development concepts, it is essential to garner an understanding of the scope of health-related clinical psychological practice and the nature of the medical environment itself. Therefore, in this introductory section, the reader is provided with the foundations from which a full appreciation of later sections can be derived. The editors (Chapter 1) provide a brief overview of the historical events, the trends of growth, and the ever-changing economic factors and professional roles impinging on clinical psychologists in medical settings. Robert J. Thompson, Jr. (Chapter 2) details the characteristics and interactions of psychology with the health care system, with an emphasis on the implications of ongoing changes for psychologists. This chapter concludes with a critique of psychology's current "developmental tasks." Bruce Bonecutter and Martin Harrow (Chapter 3) delineate the goals, the driving forces, and the internal workings of hospitals. Extending to virtually all medical settings, the information provided by these authors is central to understanding how and why medical systems operate the way they do.

CHAPTER 1

Clinical Psychology in Medical Settings

Past and Present

Jerry J. Sweet, Ronald H. Rozensky, and Steven M. Tovian

Introduction

The history of psychology has been traced through the behavior of early civilizations covering a span of 4,000 years (Kimble & Schlesinger, 1985a,b). The history of *modern* scientific psychology is relatively brief, encompassing a little more than a century. Nevertheless, Pion (in press) noted that, since 1977, the demand for doctoral psychologists, as reflected by employment figures, has actually increased more rapidly than the employment of all other doctoral scientists. Further, Pion cited National Science Foundation statistics indicating that, in 1986, 94% of doctoral psychologists were actually working in psychology. This figure represents a higher percentage than in any other scientific field.

It has been generally agreed that Lightner Witmer's founding of the first American psychological clinic in 1896 marked the beginning of clinical psychology (McReynolds, 1987). However, clinical psychology did not become well organized and recognized as a discipline, with guidelines for formal training programs and significant identification with the vocational title *clinical psychologist*, until the 1940s (Strickland, 1988). Recent data gathered by Norcross, Prochaska, and Gallagher (1989) support the "veritable explosion" of clinical psychology in numbers, activities, and knowledge since World War II.

Within clinical psychology, there has been a further specialization, referred to generically by various authors since the early 1970s as *health psychology, clinical health psychology, medical psychology,* or *behavioral medicine*. Rehabilitation psychology, described in hallmark texts such as that by Neff (1971), preceded clinical psychology into the medical setting. With its own brief 40-year history, but now sharing many of the same goals and practices as health and medical psychology, modern American clinical neuropsychology has developed largely since the 1970s. Matarazzo (1987) argued that there is *de facto* recognition of subspecialties in clinical psychology, although formal or quasi-legal recognition is yet to occur. Further, he

Jerry J. Sweet, Ronald H. Rozensky, and Steven M. Tovian • The Evanston Hospital and Northwestern University and Medical School, Evanston, Illinois 60201.

stated that, although formal specialization is evolving, today there is only one psychology with many applications.

Current Trends

Regardless of whether one prefers "health psychologist," "medical psychologist," or some other title, the rapid growth and the strong presence today of clinical psychologists in medical settings are easy to observe. The number and quality of professional journals, professional organizations, and professional texts devoted to the interests and activities of clinical psychologists in medical settings support this point. Wedding and Williams (1983) demonstrated convincingly the proliferation of new graduate-psychology training-courses in behavioral medicine and clinical neuropsychology, as well as the dramatically increasing number of neuropsychology citations in psychology and medical journals during the 1970s and the early 1980s.

Howard, Pion, Gottfriedson, Flattau, Oskamp, Pfafflin, Bray, and Burstein (1986) documented a dramatic shift in doctorate production favoring those in the health-service-provider subfields of psychology. These authors noted that, in 1984, the majority (53.2%); approximately 1,750 of new psychology doctorates were in the health-service-provider subfields. These authors also noted that, whereas the number of new health service providers in psy-

chology has continued to rise dramatically over the years, the number of traditional academic and research providers peaked in the mid-1970s and had returned by 1984 to its 1968 level (approximately 675 new doctorates).

Equally impressive are data from the National Library of Medicine indicating an enormous increase in the number of books published yearly on neuropsychology (from 3 in 1979 to approximately 24 in 1986; Wedding, Franzen, & Hartlage, 1987). Clinical and research interest in biofeedback during its brief history, has generated more than 2,000 journal citations since 1970 (Hatch & Riley, 1985) and has resulted in 3,300 applications and 1,727 successful candidates for biofeedback certification (Biofeedback Certification Institute of America, 1989). As can be seen in Table 1, the relative percentage of growth of membership in Divisions of the American Psychological Association (APA) with particular relevance to medical settings (e.g., Division 38—Health Psychology; Division 40—Neuropsychology) has kept pace with and, in the case of Division 40, has greatly outdistanced the growth of the general clinical division (Division 12) and of the APA as a whole. In addition to the growth of selective divisions of the APA, Pion (in press) documented an increase of approximately 32% among psychologists working in "hospitals and clinics" from 1977 ($n = 5,393$) to 1987 ($n = 7,155$). The growth of Division 42 (Independent Practice) seen in Table 1 may reflect the switching of more experienced members in the health-service-pro-

Table 1. Membership of Selected Divisions of the American Psychological Association

Year	APA total	Division				
		Clinical (12)	Rehabilitation (22)	Health (38)	Neuropsychology (40)	Independent practice (42)
1983	56,402	4,944	881	2,303	1,009	4,614
1984	58,222	5,055	931	2,419	1,469	5,057
1985	60,131	5,418	974	2,507	1,785	5,020
1986	63,146	5,678	1,001	2,704	2,117	5,090
1987	65,144	5,920	939	2,702	2,334	5,140
1988	66,996	5,765	963	2,608	2,463	5,413
1989	69,366	5,832	932	2,749	2,610	5,293
7-year growth (%)	23	18	6	19	159	15

vider subfields in psychology from organized human service settings (i.e., medical settings) to independent practice (Howard *et al.*, 1986), which of course may continue to involve medical patients. Division 42 (Independent Practice) has its own "Hospital Practice Committee," and Division 29 (Psychotherapy), published a special issue of its journal, *Psychotherapy*, in Fall 1988 focusing on special issues of psychotherapy in the new health care system.

When examining Table 1, one may wonder why there has been so little growth in Division 22 (Rehabilitation), as the number of ads in the *APA Monitor* and the evaluation of employment trends for psychologists (e.g., Pion, in press) strongly suggest an increasing demand for clinicians to work in rehabilitation settings. Because many of these rehabilitation positions involve working with head injury and stroke patients, it is likely that neuropsychologists, strongly identified with Division 40 rather than with Division 22, are filling many of these positions.

Overall, the available data strongly suggest that the rate of growth of the *de facto* specialty areas most associated and identified with medical settings is sizable and, in some cases is unparalleled by other areas within clinical psychology, either past or present.

It remains historically important in demonstrating this continuing trend to note that the number of identified clinical psychologists employed as faculty in medical schools rose from a total of 255 in 1953 (Mensch, 1953) to 2,336 in 1976 (Lubin, Nathan, & Matarazzo, 1978) and recently estimated to be approximately 3,000 (Clayson & Mensh, 1987). In 1953, only 2.5% of the APA membership were identified as working in medical schools, as compared to 6.0% in 1976 (Gentry & Matarazzo, 1981). Gentry and Matarazzo further reported that only 47% of medical center psychologists were employed full-time in 1955, whereas 71% were employed full-time in 1976. In addition, these authors noted that, in 1981, professional psychologists were employed in virtually all (98%) medical schools in the United States as compared to only 73% of such institutions in 1953. The ratio of the available psychologists in medical set-

tings to the number of medical students decreased significantly from 1:88 in 1955 to 1:24 in 1976 (Gentry & Matarazzo, 1981).

Impact of Education and Training

Strong emphasis has been placed on the scientist-practitioner model for clinical psychologists working in medical settings (e.g., Miller, 1983; Stone, 1983). Interestingly, and perhaps not well known to many psychologists, some of our medical colleagues, particularly those within physician training programs, also adhere to a scientist-practitioner model. In fact, the American Society for Clinical Investigation, founded in 1909, is an organization of physicians that promotes the scientist-practitioner model within the medical community (Kelley, 1984). Chapter 16 by Malec and Chapter 17 by Rozensky, Sweet, and Tovian in this book discuss the rationale, growth, and necessity of scientifically based practice in the medical setting.

Conferences on education and training have resulted in guidelines for health psychologists and clinical neuropsychologists at both the predoctoral and the postdoctoral levels (American Psychological Association, Division of Health Psychology National Working Conference on Education and Training in Health Psychology, 1983; International Neuropsychological Society Task Force Report on Education, Accreditation, and Credentialing, 1981). The result has been that more clinicians have been better prepared for work in medical settings and have been more interested in seeking work within medical settings. Position openings in these areas have also expanded rapidly, as exemplified by the increase in the percentage of *APA Monitor* advertisements for clinical neuropsychology positions from 1% in 1976 to 9% in 1986 (D'Amato, Dean, & Holloway, 1987). Howard *et al.* (1986) found that nearly one half of new doctorate recipients were taking full-time positions in health care settings. A prior study by Grzesiak (1984) of *APA Monitor* advertisements suggested that approximately 40% of 330 health-related positions in a one-year period were for health psychology and behavioral medicine positions, whereas 24% were in neuropsychology, 14% in rehabilitation, and 11% in pediatrics. Hospitals and clinics, colleges and universities, and medical schools were the highest ranking employers of these health-related professionals. A quick tabulation of recent ads in the *Monitor* will show these proportions to have varied in recent years (e.g., rehabilitation positions appear to have increased dramatically since 1986). However, the conclusion is the same: Clinical psychology in medical settings continues to grow.

In its biennial survey of psychologists' salaries, the American Psychological Association (1988) found that 15% of the respondents in the human services areas were in hospital settings. The survey offers comparisons of salaries of those employed in general hospitals, psychiatric facilities, independent practice, academia, and other settings and may be of interest to the reader. In their practice survey of 270 Division 38 (Health Psychology) practitioners, Piotrowski and Lubin (1989) found that, of the respondents' primary occupational settings, 25% were in a medical center or hospital, 17% in a medical school, 17% in a university or college, 5% in the Veterans Administration, 4% in an outpatient clinic, and 28% in private practice. Norcross *et al.* (1989) reported that 5% of Division 12 (Clinical) members worked primarily in a general hospital. According to data published by the National Science Foundation, the number of doctoral psychologists employed in hospitals and clinics was 7,155 in 1987 (Pion, in press). Even this figure undoubtedly underestimates the total number of psychologists working with medical patients by not including all those doing so as private practitioners (not employees).

Financial Imperatives

The rapidly changing economics of American health-care practice today have fostered an increasingly competitive environment. For example, Light (1986) highlighted the compet-

itiveness between for-profit hospital chains and traditional teaching hospitals. Within this context of competition and change, the need for documentation of cost effectiveness seems to have served to support explicitly those programs and practices that are empirically based and highly accountable, whether they are staffed by clinical psychologists or by other health-care providers. As discussed by Schneider (1987), Genest and Genest (1987), and Cummings in this book (Chapter 8), the available data on the cost effectiveness of behavioral medicine interventions is impressive in arguing that they be funded. For example, von Baeyer (1986) noted that the majority of behavioral medicine intervention studies have found a reduced use of medical resources (and therefore cost savings) following treatment. Boudewyns and Nolan (1985) argued that the framers of health care policies such as Medicare, based on diagnosis-related groups (DRGs), should recognize and support clinical psychologists because many of their techniques can reduce the cost and consumption of health care through programs that reduce behavioral risk factors, increase compliance with medical regimens, and prepare patients for stressful medical procedures.

The already research-based practices of clinical psychologists in medical settings have, for the most part, continued to grow, despite general economic setbacks in many areas of health care, including mental health. Nevertheless, the seemingly unpredictable and increasingly hostile economic climate of health service delivery in all disciplines has served to create increased competitiveness, both between and within disciplines. This tense climate has caused some concern among clinical psychologists that we be careful to respond to the increased competitiveness in a manner that is not harmful to relationships with our colleagues in medicine, or in psychology, whether applied or academic (Sweet & Rozensky, 1991). Unfortunately, an obvious outgrowth of the sometimes painful evolution of our discipline and the larger political-economic systems with which we interact has been increasing friction between academic and clinical psychologists. The recent turmoil regarding the failed attempts to reorganize the American Psychological Association (Rodgers, 1988) reflect that friction (see Wright and Friedman, Chapter 32 in this book).

Roles of the Psychologist

From the overview of the historian, Kimble (1985) noted that the history of psychology has been one of evolution, rather than revolution. Instead of "Kuhnian breakthroughs," there has been "a gradual accumulation of knowledge that produces a change in atmosphere, a change in the way in which we think about psychological problems" (p. 18). Kimble also noted that "progress in psychology appears to come from the application of objective and quantitative methods" (p. 18). As the application of psychological methods in medical settings becomes more specialized, the diagnostic and treatment roles of clinical psychologists will also become more subspecialized, as has already been evident for some time within medicine (e.g., radiologists who only diagnose spinal disorders, neurologists who only diagnose and treat brain tumors, cardiologists who only perform angioplasty, and oncologists who only engage in radiotherapy of cancer patients). Diekstra and Jansen (1988) described aptly the need for a natural evolution of psychology's role in the "new" health care systems:

> Of decisive importance for the future of psychology as a discipline within the health sector will be its ability to establish an accepted role within primary health care. This necessarily implies within the discipline itself a well-defined and visible distinction between psychologists as general or primary mental health care providers and psychologists as specialty mental health workers. One point to be emphasized is that simply repackaging an old product (specialized care) rather than changing the product to meet the changing needs of people and the changing ways in which health care will be offered in the future, is self-destructive. Appropriate research and training must be an ongoing part of psychology's activities or else patients and policymaking bodies will simply move on to other professional groups who may be offering services which seem more helpful, relevant, appropriate, and cost efficient. (p. 350)

The specific roles filled by clinical psychologists working in medical settings are diverse and have changed across time. These roles include diagnostician, therapist, teacher, researcher, and administrator (e.g., Gentry & Matarazzo, 1981). If we look only at psychologists working in medical schools, it appears that the roles of therapist, teacher, and administrator increased substantially from 1955 to 1964 to 1977 (Nathan, Lubin, Matarazzo, & Persely, 1979). For example, among psychologists working in medical schools, Nathan *et al.* found that the percentage of time spent performing administrative activities increased from 11% in 1955 to 22% in 1977. As pointed out by Gentry and Matarazzo (1981), it is extremely important that the roles of psychologists in medical settings evolve and diversify. Matarazzo's discussion (1987) of the inherent strength of the application of the principles of psychology to the medical or health care arena illustrates this diversity and adaptability of clinical practice in medical settings. This diversification and adaptability are perhaps the best indicators of prosperity and survivability within a politically and economically complex system that is constantly evolving at what appears to be an increasing rate.

Future Directions

Pion (in press) described the major American societal transformations as being driven by a shift toward the delivery of services and a growing reliance on specialized scientific and technical knowledge. Pion noted that all disciplines have felt the impact of these changes, and that psychology's strong scientific framework has permitted gains during these times. It is interesting that the same strategies that Goldsmith (1989) pointed out as necessary for hospitals to survive the "next wave of economic pressures" can also be recommended to psychologists who seek to survive the changing health care economy while working in medical settings. With regard to hospitals, Goldsmith suggested that a successful strategy will require renewed and deeper collaboration with physicians; solutions to problems of productivity; refocusing ambulatory and chronic care services; and managing for medical value (p. 107). Goldsmith expected that (1) hospitals will offer more ambulatory services and will continue to move services now housed within the hospital out into the community; (2) acute care will be concentrated in a relatively small number of high-tech regional centers; (3) community hospitals will decentralize to the point that they primarily provide diagnosis and treatment of the chronically ill; (4) as technology makes advances over disease, most illness will be associated with aging (i.e., hospitalized patients will be older); and (5) by the mid-1990s, quality indicators will be a major factor among consumers seeking health care.

Within the context of Goldsmith's observations, it is noteworthy that a survey by Stabler and Mesibov (1984) found that both pediatric psychologists and health psychologists reported physicians' lack of knowledge about the availability of psychological services as the number one perceived roadblock (by a large margin) to working with physicians. By comparison, lack of professional preparation, mental-health-professional overpopulation, and insufficient patients were reported to be *negligible* roadblocks to providing services in medical environments. The clinical chapters in this book address how to acquaint physicians with the psychological services available in medical settings.

Summary

Although there are numerous professional and clinical problems, significant bodies of knowledge now exist for and measurable progress has been made in providing psychological services in the medical setting. In their national survey of psychologists regarding medical staff membership and clinical privileges, Boswell, Litwin, and Kraft (1988) documented that progress in this area is coming, although slowly. Two publications by the American Psychological Association (*A Hospital Practice Primer for Psychologists*, 1985, and *Hospital Practice: Advocacy Issues*, 1988) and a recent special issue of the APA's *Journal of Consulting and Clinical Psychology* (57, 3, 1989) on coping with medical illnesses and procedures underscore the importance to psychologists of practice in medical settings. The chapters that follow elucidate and define the professional, economic, political, and clinical status of psychological practice in medical settings. The breadth and diversity of the content in the following pages are a testament to the scope of clinical psychology in medical settings. Whether referred to in terms of the newer applications of health psychology or of the more traditional psychological practices, the science and practice of clinical psychology in the medical setting have become a significant part of the evolving history of psychology as a whole. Finally, the continued successful evolution, survival, and growth of clinical psychology in medical settings will depend on the scientifically based clinical practice of psychology and the flexibility and adpatability of that practice in the changing context of health care. As pointed out earlier, Matarazzo (1987) eloquently reminded us all that we remain "one psychology" with many applications.

ACKNOWLEDGMENTS

The authors wish to thank the American Psychological Association and the Biofeedback Certification Institute of America for their assistance in providing membership data.

References

American Psychological Association, Committee on Professional Practice of the Board of Professional Affairs. (1985). *A hospital practice primer for psychologists*. Washington, DC: Author.

American Psychological Association, Committee on Professional Practice of the Board of Professional Affairs. (1988). *Hospital practice: Advocacy issues*. Washington, DC: Author.

American Psychological Association, Division of Health Psychology. (1983). Proceedings of the National Working Conference on Education and Training in Health Psychology. *Health Psychology, 2*. (Supplement)

American Psychological Association, Office of Demographic, Employment and Educational Research. (1988). *Salaries in psychology: 1987.* Washington, DC: Author

Biofeedback Certification Institute of America. (1989). *Register of the Biofeedback Institute of America.* Wheat Ridge, CO: Author.

Boswell, D., Litwin, W., & Kraft, W. (1988, August). *Medical staff membership and clinical privileges: A national survey.* Paper presented at the annual meeting of the American Psychological Association, Atlanta, Georgia.

Boudewyns, P., & Nolan, W. (1985). Prospective payment: Its impact on psychology's role in health care. *Health Psychology, 4,* 489–498.

Clayson, D., & Mensh, I. (1987). Psychologists in medical schools: The trials of emerging political activism. *American Psychologist, 42,* 859–862.

D'Amato, R., Dean, R., & Holloway, A. (1987). A decade of employment trends in neuropsychology. *Professional Psychology: Research and Practice, 18,* 653–655.

Diekstra, R., & Jansen, M. (1988). Psychology's role in the new health care systems: The importance of psychological interventions in primary health care. *Psychotherapy, 25,* 344–351.

Genest, M., & Genest, S. (1987). *Psychology and health.* Champaign, IL: Research Press.

Gentry, W., & Matarazzo, J. (1981). Medical psychology: Three decades of growth and development. In C. Prokop & L. Bradley (Eds.), *Medical psychology: Contributions to behavioral medicine.* New York: Academic Press.

Goldsmith, J. (1989). A radical prescription for hospitals. *Harvard Business Review, 67,* 104–111.

Grzesiak, R. (1984). Employment opportunities in health psychology: Four years of "Monitor" advertisements. *The Health Psychology Newsletter, 6,* 6–8.

Hatch, J., & Riley, R. (1985). Growth and development of biofeedback: A biliographic analysis. *Biofeedback and Self-Regulation, 10,* 289–299.

Howard, A., Pion, G., Gottfriedson, G., Flattau, P., Oskamp, S., Pfafflin, S., Bray, D., & Burstein, A. (1986). The changing face of American psychology: A report from the Committee on Employment and Human Resources. *American Psychologist, 41,* 1311–1327.

International Neuropsychological Society. (1981, September). Report of the Task Force on Education, Accreditation, and Credentialing. *The INS Bulletin.*

Kelley, W. (1984). Clinical investigation and the clinical investigator: The past, present, and future. *Journal of Clinical Investigation, 74,* 1117–1122.

Kimble, G. (1985). Overview: The chronology. In G. Kimble & K. Schlesinger (Eds.) *Topics in the history of psychology,* Vol. 1. Hillsdale, NJ: Erlbaum.

Kimble, G., & Schlesinger, K. (1985a). *Topics in the history of psychology,* Vol 1. Hillsdale, NJ: Erlbaum.

Kimble, G., & Schlesinger. K. (1985b). *Topics in the history of psychology,* Vol. 2. Hillsdale, NJ: Erlbaum.

Light, D. (1986). Corporate medicine for profit. *Scientific American, 255,* 38–45.

Lubin, B., Nathan, R., & Matarazzo, J. (1978). Psychologists in medical education. *American Psychologist, 33,* 339–343.

Matarazzo, J. (1987). There is only one psychology, no specialties, but many applications. *American Psychologist, 42,* 893–903.

McReynolds, P. (1987). Lightner Witmer: Little-known founder of clinical psychology. *American Psychologist, 42,* 849–858.

Mensh, I. (1953). Psychology in medical education. *American Psychologist, 8,* 83–85.

Miller, N. (1983). Some main themes and highlights of the conference. *Health Psychology, 2,* 11–14.

Nathan, R., Lubin, B., Matarazzo, J., & Persely, G. (1979). Psychologists in schools of medicine: 1955, 1964, and 1977. *American Psychologist, 34,* 622–627.

Neff, W. (1971). *Rehabilitation psychology.* Washington, DC: American Psychological Association.

Norcross, J., Prochaska, J., & Gallagher, K. (1989). Clinical psychologists in the 1990's: 1. Demographics, affiliations, and satisfactions. *The Clinical Psychologist, 42,* 29–39.

Pion, G. (in press). Psychologists wanted: Employment trends over the last decade. In R. Kilburg (Ed.), *Managing your psychological career.* Washington, DC: American Psychological Association.

Piotrowski, C., & Lubin, B. (1989). Assessment practices of Division 38 practitioners. *The Health Psychologist: The Official Publication of Division 38 of the American Psychological Association, 11,* 1–2.

Rodgers, J. (1988). Structural models of the American Psychological Association in 1986. *American Psychologist, 43,* 372–382.

Schneider, C. (1987). Cost effectiveness of biofeedback and behavioral medicine treatments: A review of the literature. *Biofeedback and Self-Regulation, 12,* 71–92.

Stabler, B., & Mesibov, G. (1984). Role functions of pediatric and health psychologists in health care settings. *Professional Practice: Research and Practice, 15,* 142–151.

Stone, G. (1983). Summary of recommendations. *Health Psychology, 2,* 15–18.

Strickland, B. (1988). Clinical psychology comes of age. *American Psychologist, 43,* 104–107.

Sweet, J., & Rozensky, R. (1991). Professional relationships. In M. Hersen, A. Kazdin, & A. Bellack (Eds.), *The clinical psychology handbook* (2nd ed.). New York: Pergamon Press.

von Baeyer, C. (1986). *Do psychological services reduce health care costs? An introduction and research summaries in nontechnical language.* Ottawa, Canada: Applied Division, Canadian Psychological Association.

Wedding, D., & Williams, J. (1983). Training options in behavioral medicine and clinical neuropsychology. *Clinical Neuropsychology, 5,* 100–102.

Wedding, D., Franzen, M., & Hartlage, L. (1987). Milestones, assessment models, and emerging issues in clinical neuropsychology. *Bulletin of the National Academy of Neuropsychologists, 4,* 6–12.

CHAPTER 2

Psychology and the Health Care System

Characteristics and Transactions

Robert J. Thompson, Jr.

Introduction

The purpose of this chapter is to delineate the challenges and opportunities confronting psychology as a health care science and profession that arise from the transactions of psychologists with medical settings. These transactions are considered from the perspective of the unique characteristics and contributions of psychologists and the characteristics of a health care system that is currently undergoing unprecedented changes.

Two premises affect this perspective. One is that psychology as a behavioral science and profession has much to offer the field of health care and, in turn, has much to gain from its involvement in that field. The second premise is that the development of psychology as a health care science and profession is related to how psychologists handle the developmental

This chapter is an expanded version of an invited address, "Psychology in hospitals: Challenges and opportunities," presented at the American Psychological Association Annual Convention, August, 1988, Atlanta, Georgia.

Robert J. Thompson, Jr. • Duke University Medical Center, Durham, North Carolina 27710.

tasks of identity, autonomy, and competency that arise out of the transactions of the substantive advances in psychology with the needs and expectations of society.

Historical Perspective

One way to appreciate the developmental course is to examine the formative processes of psychology as a health care science and profession. What have been the essential, characteristic, and defining features of psychology's transaction with societal health needs and expectations, and what do these features portend for psychology's future development?

Dennis (1950) pointed out that, in the history of medicine and in the history of psychology, one frequently encounters the same names: Galen, Descartes, Helmholtz, Locke, Hartley, Fechner, Wundt, Janet, McDougall, James, and Freud. Not only have medicine and psychology been historically intertwined, but since psychology emerged as a separate discipline a little more than 100 years ago, their interrelationship has been periodically addressed.

In 1911, a symposium on the topic "The Relations of Psychology and Medical Education" was sponsored by a joint session of the American Psychological Association (APA) and the Southern Society for Philosophy and Psychology. The papers presented at this symposium by psychologists (Shepard Ivory Franz and John Broadus Watson) and physicians (Adolf Meyer, E. E. Southard, and Morton Prince) (see the *Journal of the American Medical Association*, 1912, *58*, 909–921) reflect the perception 75 years ago that the new science of psychology had something to contribute to medical education, research, and practice. However, to realize its potential contribution, psychology needed to become more valuable or useful by addressing the problems confronted by the medical practitioner. Through this process, psychology in turn would be enhanced.

Almost four decades after the 1911 symposium, another symposium was held, and the papers were subsequently published in a book, *Current Trends in the Relation of Psychology to Medicine* (Dennis, 1950). The major questions addressed were: "What have we [psychologists] offered, what do we now have to offer, what do we purpose to offer to medicine?" (p. 5). Today, 40 years later, these remain the salient questions, with one important modification. Today, the question is what psychology has to offer not to medicine, but to health care.

In addressing the questions of 1950, Dennis pointed out the contributions that psychology had already made in the areas of intelligence, child development, and clinical psychology. He maintained that "the application of learning principles to the prevention and cure of disease remains primarily for the future" (p. 6). Dennis also made the following observation: "It may be that our most basic contribution will not be a specific discovery, but rather a promulgation of an objective approach to human behavior" (p. 7).

Robert Felix (1950), a physician and director of the Institute of Mental Health of the U.S. Public Health Service, portrayed psychology as one of the basic sciences for public health and made the point that "public health makes use of the kinds of personnel it needs" (p. 14). Felix outlined the five major objectives of the public health program: the measurement of the nature and extent of health problems; the develop-

ment of inexpensive, rapid, and valid methods for the identification of those who need care; rapid and economical treatment; prevention of illness; and health promotion. Felix maintained that, although the field of mental health was the public health area in which psychology had been most active, pertinent knowledge existed in many fields of psychology that were germane to the task confronting public health more generally. However, in addition to advancing knowledge through research, he contended that psychologists must be willing to work in public health organizations "translating the findings of research into the life patterns of people" (p. 22). Felix also maintained that psychology could deal with these important public health problems with "no loss of vigor and with every expectation of maintaining a thoroughly health scientific selfhood" (p. 24).

Carlyle Jacobsen (1950), a psychologist and Executive Dean for Medical Education at the State University of New York in Brooklyn, noted that

> the psychologist is distinguished from the psychiatrist in interest and training, in the kinds of service which he is prepared to bring to the patient, and by training in research that most closely resembles that of the investigator in physiology and the biological and natural sciences. (p. 33)

Jacobsen maintained that psychologists are appointed to the hospital staff or to the teaching and research staff of the medical school because they have contributions to make to patient care and to the education of medical students. However, Jacobsen saw the increase in medical knowledge and the mobilization of this knowledge for patient service as coming about because medicine, as a university discipline, accepts responsibility for the advancement of knowledge. Although acknowledging that psychology could contribute to patient care and medical education, Jacobsen urged that psychology not neglect its responsibility for basic research.

In retrospect, several themes can be discerned that would prove to be formative in the development of psychology in the years to follow. By 1950, psychology had grown to be useful particularly in the field of mental health in response, in large part, to the influence of Veterans Administration programs after World War II. However, the potential for psychology to contribute to the broad societal mission of public health had not yet been realized. Increasing demands from society for psychology to contribute could be expected. The unique contributions of psychology to advancing knowledge, to teaching, and to service would stem from psychology as a behavioral science.

Psychology as a Health Science and Profession

Additional initiatives by the federal government in the 1960s provided new opportunities for psychology in the mental health field and facilitated the interface of psychology with physical health. For example, in 1962, the National Institute of Child Health and Development was founded to support biomedical and psychological research related to child development. Another example was the enactment of federal legislation in 1963, the Mental Retardation Facilities and Community Mental Health Centers Construction Act of 1963 (Public Law 88-164) (Thompson & O'Quinn, 1979). This legislation reflected societal concerns about providing services, and about training service providers, for citizens with mental illness and mental retardation and was a stimulus to the further development of clinical and community psychology. In addition, this 1963 legislation established a network of University-Affiliated Facilities for training students from a number of disciplines in the provision of interdisciplinary services to children with developmental disorders.

Shortly thereafter, within the disciplines of psychology and pediatrics, there was recognition of the utility of collaboration. Julius Richmond, U.S. Assistant Secretary for Health and Surgeon General, stressed that pediatrics needed to emphasize all aspects of the child, not just biological development, and also that child development rather than child psychiatry was an appropriate basic science for pediatrics (Richmond, 1967).

Kagan (1965) sensed "a new liaison between the behavioral sciences and pediatrics" (p. 272), and he foresaw the marriage of pediatrics and psychology as leading to "a corpus of cognitive products that would not have occurred had each remained single for much longer" (p. 272). Wright (1967) saw the "pediatric psychologist" as one of the offspring of this marriage. He defined a pediatric psychologist as one who worked primarily with children in a nonpsychiatric medical setting and who was ideally "competently trained in both child development and in the child clinical area" (p. 323). Wright cautioned that it would be important to develop unique knowledge and competencies from this interface of psychology and pediatrics and not just to rely on applications of current knowledge. Wright wrote that "no behavioral discipline or subspecialty can justify its existence in the present world unless it has something unique to contribute to knowledge" (p. 325).

The recognition of the potential offered by the involvement of psychology in health problems was not limited to the area of pediatrics. In a 1969 article that has proved to be seminal, William Schofield provided the vision of a broader role for psychology expanding out from the confines of mental health to the broader purview of health. Schofield developed the view of psychology as a life science and wrote:

> if we accept psychology as a life science, we need only to acknowledge the possibility that its discoveries may have health fostering applications at every level and in every dimension of its total endeavors. (p. 567)

Schofield perceived that there existed in the late 1960s new stimuli to a broader role for psychology as one of the health sciences. These new stimuli were the goals and objectives of federal agencies such as the National Institute of Mental Health, the National Institute of Child Health and Development, the Bureau of the Education of the Handicapped, and the Office of Education. These federal agencies, charged with responsibilities for the health and welfare of our society, wanted more from psychology than they were then getting and had the funds to support research, training, pilot programs, and program evaluation.

Schofield also perceived that the most valuable and unique contributions of psychologists were those "evolving from our expertise in the study of complex behavior and from our fundamental commitment to critical evaluation" (p. 58). He also recognized that, although supportive, society would also be increasingly demanding, and he commented on psychology's obligation as a profession to be socially responsive:

> it is no longer a question of fighting for a place at the table. We are accepted there. But our continued presence will demand justification in terms of our day to day contributions. If we wish to eat at society's table we must be able and willing to till society's fields. (p. 581)

Biopsychosocial Model

By the mid-1970s, there was an increasing recognition within medicine of the need for a new model. George Engel (1977), a physician, traced the historical origins of the reductionistic biomedical model to the emergence of mind–body dualism, which had resulted in the application of the scientific approach to biological processes and the ignoring of the behavioral and psychosocial processes. He advocated a biopsychosocial model as a way to "broaden the approach to disease to include the psychosocial without sacrificing the enormous advantages of the biomedical approach" (p. 131).

In 1977, behavioral medicine emerged as the interdisciplinary field concerned with the development, integration, and application of behavioral and biomedical scientific knowledge and techniques to health and illness (Schwartz & Weiss, 1978). Psychology's research, education, and service contribution to the field spawned the new area of health psychology. The new APA Division of Health Psychology emerged in 1978.

In these developments, Weiss (1987) called our attention to the critically important role of underlying models. Weiss maintained that it was the emergence of multifactorial approaches to the pathogenesis of disease that enabled the links between the behavioral and the biomedical sciences and that facilitated conceptual development in behavioral medicine.

Again, the federal government played a major role in the development of these new bio-behavioral initiatives. The National Institute of Health was most helpful through the National Heart, Lung, and Blood Institute support of research training and the establishment of a Behavioral Medicine Study Section (Matarazzo, 1980).

It was clear that multifactorial approaches had become necessary to confront the complexity of health problems. The concern now is to determine how biological and psychosocial processes act together in health and illness across the life span.

The Changing Health Care System

Having examined the characteristics that psychology, as a behavioral science and health care profession, brings to the transaction with societal needs and expectations, we now need to examine the current environment of major importance, that is, the characteristics of today's health care system. The system is experiencing major changes fueled by increasing demands for health services and an increasing determination to contain costs. There are four major issues to discuss: (1) the increasing number of health care providers; (2) corporatization and bureaucratization; (3) administrative waste; and (4) the crisis in medical education.

Health Care Providers

Starr (1982) provided figures reflecting the physician surplus in this country. From 1965 to 1980, the number of medical schools increased from 88 to 127. The number of annual physician graduates rose from 7,409 to 17,000. Physicians in active practice increased from 337,000 in 1975 to 450,000 in 1980 and will reach 600,000 by the end of the decade. Starr wrote that, for every 100,000 people, we had 148 doctors in 1960, 177 in 1975, and 202 in 1980, and we will have 245 by 1990.

Starr (1982) also pointed out one of the consequences of the physician surplus: "In 1975, there were 565 Americans per doctor; by 1990 there will be 404—a reduction of nearly 30 percent in the potential clientele for the average physician" (p. 424).

In addition to the increase in physicians' numbers, there also has been growth in other health care disciplines. Dorken and Bennett (1986) reported that the number of licensed psychologists was 20,000 in 1974 and increased 128% to 45,600 in 1985. During the same period, there was a 46% increase in psychiatrists to about 38,000. Based on data from several national surveys, Matarazzo, Carmody, and Gentry (1981) chronicled the almost tenfold increase in the number of psychologists employed in medical schools from 255 in 1953 (Mensh, 1953) to 2,336 in 1976 (Lubin, Nathan, & Matarazzo, 1978). These increasing numbers of physicians, psychologists, and other health care providers have resulted in increasing competition for the health care dollar.

Corporatization and Bureaucratization

The United States is also in the midst of what has been called the corporatization and bureaucratization of health care. Although there have been concerted efforts to keep socialized medicine from coming in the front door, corporate medicine has come in the back. We are now experiencing rampant entrepreneurialism, which is fueled by cost containment pressures on one hand and profit motives on the other.

Starr (1982) provided the following data. In 1970, the Hospital Corporation of America (HCA) controlled 23 hospitals. In 1981, HCA owned or managed about 300 hospitals. In 1976, profit-making chains owned or managed hospitals with 72,282 beds. By 1981, this figure had increased 68%, to 121,741 beds.

There has also been a proliferation of health care models, particularly prepaid programs, such as health maintenance organizations (HMOs) and preferred-provider organizations (PPOs), efforts at horizontal and vertical integration, managed care, and sophisticated marketing. At the same time, there is increasing external regulation, such as DRGs (diagnosis-related groups), to control costs by focusing on the clinical services provided. Cost containment measures increase competition among health care providers for health dollars and also foster bureaucratization. We have seen the rise of the managers with a perceived mandate to "rationalize" the health care system. We can anticipate increasing tension between managers and health care professionals, as well as an erosion of the previous prerogatives of professionals to manage their professional activities. Rationality is allegedly to be achieved through better organization and control over providers in the name of efficiency and quality. Strategic decisions, the "what" and "how much" regarding services, will be the purview of the manager, and the professional will be left with the "how." Managers can be expected to provide the services that are adequately reimbursed and profitable and to curtail those that are not. Furthermore, Scott (1985) warned us that "managers are increasingly likely to promote deprofessionalization, if not proletarianization, of allied health workers" (p. 119). One consequence of corporatization and bureaucratization is likely to be increasingly inadequate health care for the poor. Various estimates suggest that there are already 35 million persons in this country who are inadequately insured.

Administrative Waste: Cost without Benefit

Recently, Himmelstein and Woolhandler (1986) called attention to the fact that the focus of cost containment has been on curtailing the volume of clinical services, whereas the costs of health administration are usually regarded as fixed. There needs to be a focus on administrative waste, which is cost without benefit. They added that the bureaucratization of medical care in the United States is reflected in the rising cost of health insurance overhead, hospital and nursing-home administration, and doctors' office expenses. Whereas from 1970 to 1982, the number of physicians and total health care personnel increased 48% and 57%, respectively, the number of health care administrators increased 171%. The total cost for health care *administration* in the year 1983 was estimated to be $77.7 billion, which amounted to 22% of all spending for health care in the United States. Himmelstein and Woolhandler (1986) maintained that much of the administrative expense is attributable to the reimbursement system, which requires that charges for each service be attributed to a specific patient. Also contributing are the costs of the regulation and enforcement bureaucracies and of marketing services.

Cost containment efforts must focus on administration as well as on clinical practice. There are those who believe that a national health insurance program may offer substantially reduced health care administrative costs through a more efficient reimbursement system. In terms of providing access to health care for all our citizens and in terms of reducing administrative cost, national health insurance warrants fuller consideration.

Medical Education

The crisis in medical education involves doing a better job of training students in a biopsychosocial model and at the same time surviving in the increasingly competitive health care marketplace that presents a very real threat to the educational viability and fiscal solvency of academic medical centers and their teaching hospitals.

The Flexner report of 1910 contributed to the decision to locate medical schools within universities and facilitated the development of the modern medical university, with its basic science and clinical curricular components. At a recent conference on medical education marking the 75th anniversary of the Flexner report, Charles Vevier (1987), a historian, maintained that, since the Flexner report, there has been a tension in medical schools between the educational mission and the press to function as a public service agency taking in social, business, and applied-science concerns. Vevier noted that, although medical schools became university-based and adopted biological science as the knowledge base for professional medical training, they did so on a guarded and autonomous basis and have been able to maintain an operational relationship almost independent of the university. Over the years, universities and academic health science centers have expanded to encompass educational, service, and research activity. Vevier also commented that the combination of economic stringency and government oversight has placed more emphasis on accountability and has exacerbated existing "separatist tensions involving operational control and academic integration between the university and medical education" (p. 10).

Robert Petersdorf (1987), president of the Association of American Medical Colleges, was a conference participant. He perceived that the surplus of physicians, which has resulted in a decrease in patients in general and in full-pay patients in particular for teaching hospitals, and the changing methods of reimbursement for both hospitals and physicians have contributed to an emphasis on cost containment and competition. These factors have made a profound impact on medical education by affecting medical school revenues. Petersdorf maintained that teaching hospitals "have become too costly in an era of competition and cost cutting" (p. 26). Their cost is influenced by having to maintain a high-quality paid faculty; to provide multiple, special, and intensive services; and to provide services for the poor. Discounted reimbursement practices such as DRGs make cost shifting more difficult. Because the hospital overhead is passed on, charges for ambulatory medicine are noncompetitive and also less efficient. Academic clinical departments have had to become very large to meet their teaching, research, and patient care missions; to earn enough income through the practice plans to keep faculty salaries competitive; and to subsidize the educational and research missions. These demands have placed many academic departments on treadmills, and the effect on the quality of the teaching of medical students has been adverse. Changes in the medical marketplace are such that price has become a major determinant as reflected in the rise of preferred-provider organizations. The shift toward "marketed" medicine and the vertical integration of new forms of health care delivery that incorporate sophisticated marketing and sales operations present a very real threat to the educational viability and the fiscal solvency of academic medical centers and their teaching hospitals.

At the same time that medical schools are confronting these economic realities, there is increased pressure to incorporate behavioral and social science perspectives into the training of physicians and to reconcile humanism and technology in medical education. One solution is a closer integration of the medical school and the university (Fein, 1987, p. 73), which would enable scholars from the university to address medical care issues. Edmond Pellegrino (1987), a physician who has been a medical school dean and president of Catholic University, argued that medicine, to be taught liberally rather than vocationally, must take advantage of its location in the university. He views the university as the locus for the dialogue between medicine and the humanities and social sciences that is necessary for the reconciliation of technology and humanism.

To meet these substantial educational and economic challenges, integration is necessary on two fronts. Medical schools and teaching hospitals will need to cultivate collaborative relationships with community-based components of the continuum of health care services, not only to maintain economic viability but also to provide quality education that utilizes a broad spectrum of patients and services. In addition, universities with vision and leadership will promote integration of the medical school and the arts-and-science components to achieve both cost effectiveness and a higher quality in educational programs. Integration will enable an infusion into the medical school of university resources in the behavioral and social sciences and can reduce the duplication and competition of limited basic-science resources. Because of their long-standing separatist posture, medical schools are as yet undeveloped with regard to what they, in turn, could offer that would enhance undergraduate and graduate education in the arts and sciences. There are educational and economic benefits to be had by institutions that are able to actualize the concept of one university.

Quality Assurance and Cost-Effectiveness

In addition to these changes in the health care system, involving increasing numbers of providers, corporatization and bureaucratization, administrative waste, and medical education, there is increasing awareness of the variability in the utilization rates, the patterns of practice, and the clinical outcome of health care services. Consumers and decision makers—individual patients, insurers, members of business and industry, and government officials—are increasingly concerned not only about cost containment but also about reconciling cost and quality considerations. These concerns will result in increasing demands to demonstrate the effectiveness of health care services. Quality assurance will be the major issue of the 1990s.

Scientific Bases of Clinical Practice

Recently, John Wennberg (1988), a physician, addressed the crisis regarding the scientific basis of clinical practice. He said that "over the next decade or so, the issue of what is appropriate practice will dominate the health policy debate" (p. 100) and "it is clear that the scientific basis of clinical practice is much less well-developed than previously assumed" (p. 101).

In contrast to the well-established scientific paradigm for the evaluation of the efficacy of new drugs, the outcomes of diagnostic and treatment procedures, including major and minor surgery, and the efficient utilization of hospitals are not routinely evaluated. To illustrate the situation, Wennberg noted the very different likelihoods of hospitalization and of specific surgical procedures for the same conditions depending on the city in which one resides. For example, there was a twofold increase in the likelihood of having coronary bypass surgery in one city over another.

The Omnibus Budget Reconciliation Act of 1986 contains legislation that established a national program for the assessment of patient outcomes, administred by the National Center for Health Services Research and Health Care Technology Assessment (NCHSR). This legislation targets funds from the Medicare trust fund to further develop the methodology and personnel to assess the outcome significance of the differences in risks and costs of care revealed through geographic variation studies. Importantly, the legislation requires peer review mechanisms that ensure that the research will proceed as part of regular science (Wennberg, 1988).

We also see that the Joint Commission on the Accreditation of Hospitals (JCAH) and the Health Care Financing Administration (HCFA) are intent on improving the quality of care through the assessment of outcome.

JCAH: Agenda for Change

The JCAH has recently embarked on a major research and development project called the Agenda for Change. For more than 30 years, the JCAH has endeavored to improve the quality of care through its accreditation process. At present, accreditation addresses the question: "Can this organization provide quality health care?" (JCAH, 1987, p. 2). An essential component of the Agenda for Change is to move beyond evaluating capability and into assessing actual clinical and organizational performance. The question to be addressed currently is "Does this organization provide quality health care?" (JCAH, 1987, p. 2).

In the future, the JCAH will assess an institution's performance in relation to outcome criteria for diagnostic and treatment services, in addition to assessing compliance with accepted standards of structure and process.

The assessment of structure involves factors such as the existence of monitoring equipment in operating rooms, the hospital's procedures for delineating clinical privileges, and care-monitoring mechanisms such as surgical morbidity and mortality conferences (Schroeder, 1987).

The assessment of processes involves factors such as whether the appropriate laboratory tests have been used, whether there is documentation in the medical record of the performance of the relevant parts of the physical examination, and whether medical records are signed in a timely manner (Schroeder, 1987).

The assessment of outcome is to be accomplished through the development and use of carefully selected valid and reliable clinical in-

dicators, which "describe measurable care processes, clinical events, complications, or outcomes" (JCAH, 1987, p. 4). Indicators are to serve as "flags" of instances of apparently substandard care or outcomes, which can then be scrutinized by the institutions' quality assurance system. "The Joint Commission's interest lies principally in the quality of the organization's response to the potential problem highlighted by apparently aberrant indicator data" (JCAH, 1987, p. 4).

HCFA's Agenda for Promoting High-Quality Care

The Health Care Financing Administration (HCFA) is responsible for managing the Medicare program and also has an agenda for promoting high-quality care. The Medicare program has 31 million elderly and disabled beneficiaries and in 1988 spent roughly $85 billion, which represented almost 20% of health care spending in the United States (Roper & Hackbarth, 1988).

Recently, the administrator and the deputy administrator of the HCFA, William Roper and Glen Hackbarth (1988, p. 91), described the HCFA's agenda to go one step beyond quality assurance to quality promotion. During its first decade, Medicare focused on improving access to health care. During the second, Medicare focused on cost containment. In the third decade, Medicare is focusing on the quality of care and the closely related concept of value.

The current quality-assurance system in Medicare is two-tiered. One tier involves monitoring by the HCFA of the compliance of state agencies with the minimum requirements for facilities and personnel who are providing care for Medicare patients. The second tier comprises review, primarily by peer review organizations (PROs), of the care provided to individual medical care beneficiaries to make sure that the services provided meet professionally recognized standards.

Like the JCAH, the HCFA contends that the focus needs to be more on the outcome of care and less on the process. The HCFA administrators see the current focus on monitoring discrete encounters with the health care systems—for example, hospital admissions—as reflecting Medicare's fee-for-service payment system and the fragmented nature of today's health care delivery system, which "makes little medical or financial sense" (Roper & Hackbarth, 1988, p. 93). Thus, an increasing emphasis on outcome is seen as resulting in less emphasis on monitoring discrete encounters and in more focus on how illnesses are treated across a full range of inpatient and outpatient services.

The HCFA is endeavoring to add a third tier to the system of quality assurance: the dissemination of quality-related information (Roper & Hackbarth, 1988). The HCFA is currently engaged in studying how to measure quality objectively.

In one effort, the HCFA released the information for 1986 on deaths within 30 days of admission to the approximately 6,000 hospitals caring for Medicare patients. Although acknowledging that it was not a direct measure of a given hospital's quality, the HCFA contended that the mortality information could be reviewed by hospital administrators, physicians, PROs, and patients as a screening tool to identify situations requiring more detailed examination. The "release of such information is now institutionalized in HCFA" (Roper & Hackbarth, 1988, p. 95) and will eventually include information on nursing home, HMOs, physicians, and other providers.

There are opportunities and challenges for psychologists to respond to increased societal demands for quality health care services with demonstrated effectiveness and cost-effectiveness. First, within each institution, psychologists need to be involved in the process of helping to establish the quality assurance monitors pertinent to their own clinical activities. Second, as is fitting in a behavioral science, our clinical practice is based on documentation of effectiveness; psychologists have made and will continue to make valued contributions to the development of effective diagnostic and treatment procedures. Third, the evaluation skills of psychologists are particularly germane to the overall mandates for effective quality-assurance programs at the departmental, hospital, and national levels.

Psychology's Developmental Tasks

The developmental tasks of the maturing science and profession of psychology arise from transactions with the changing demands and expectations of the health care system. With its considerable competencies, psychology now confronts the developmental tasks of autonomy and identity.

The maturity of an emerging discipline must be responsibly asserted by itself and must be concurrently acknowledged by significant others (Matarazzo et al., 1981). Psychology is increasingly recognized as a health care science and profession. However, it is still necessary that psychologists advocate for an appropriate role in the health care system. Exercising appropriate professional autonomy is fundamentally essential to fulfilling psychology's responsibility to society for having been granted the privileged status of a profession. In our science and service, we fulfill that responsibility by ensuring that quality psychological services will be available and will be provided. Thus, the primary rationale for our advocacy efforts to exercise appropriate professional autonomy within hospital and medical settings is to ensure that quality psychological services will be provided.

One function of advocacy is to achieve representation, "at the table," along with our colleagues from medicine and the other health care disciplines. That is, psychology must participate in the structures and processes that affect health care service, education, and science. Often, this participation will take the form of membership in professional organizations such as the Association of American Medical Colleges (AAMC) and regulatory bodies such as the JCAH and organizations that govern reimbursement. For example, the AAMC is a major forum in which to promote psychology's role in, and contributions to, medical education. Psychology also needs representation, along with medicine, in the Blue Cross/Blue Shield specification of the Current Procedures

Terminology (CPT) coding system. A case in point is that the 1987 revision of the CPT "medicalized" the section on psychiatry procedures. Thus, psychological testing is now referred to as "psychological testing by a physician."

Another function of advocacy is to remove barriers to our autonomous functioning. These barriers have consisted primarily of federal and/or state statutes, laws, or regulations that restrict the range of practice or limit access to psychological services. Two of the areas in which barriers exists are Medicare and membership on the hospital medical staff.

Medicare

Psychology must obtain professional autonomy and parity under Medicare, which Patrick DeLeon (1987) described as "the federal government's national health program" (p. 81). Senator Daniel Inouye and Patrick DeLeon have been tireless in their efforts to obtain the necessary legislation to have psychology appropriately recognized in Medicare to enable "reimbursement for psychological services in the same manner that physician services are currently reimbursed" (p. 81). The recent efforts by Bryant Welch of the Office of Professional Practice, Senator Jay Rockefeller, John Linton, and others (through the West Virginia State Psychological Association) made some headway in legislating psychologists' participation in Medicare with regard to rural health clinics and community mental health centers. However, to ensure that the elderly, the disabled, and the poor will have direct access to quality psychological health care services, psychologists must engage in a concerted and organized effort to have psychology appropriately recognized in Medicare and Medicaid, which account for 90% of the federal health expenditures (DeLeon, 1988).

Hospital Staff Membership

Professional autonomy requires that psychologists become members of the medical staff of the hospital in which they practice (Thompson, 1987; Thompson & Matarazzo,

1984). This is essential because of the two self-governance responsibilities afforded to the hospital medical staff. One is to monitor and evaluate the quality of patient care. The second is to develop rules, regulations, and policies regarding the granting, delineating, and renewal of clinical privileges, which are patient care responsibilities of individual practitioners. Psychologists, as members of an autonomous profession, must be active participants in the hospital processes that govern who may provide which patient care services, and in those hospital processes through which the responsibility for quality of care is exercised.

It is necessary to recognize that the functioning of psychologists in hospitals depends not only on federal statutes but also on state laws and individual hospital bylaws (see Chapter 5 for further discussion). State statutes specify which practitioners are allowed to provide services independently. State statutes also specify the composition of the medical staff of hospitals. The statutes can require, enable, or prohibit psychologists from being members of the active medical staff. Currently only three states— North Carolina, Georgia, and California—plus the District of Columbia have statutes specifically prohibiting hospitals from discriminating against psychologists regarding membership on the hospital medical staff. In eight other states, the definition of medical staff is broad enough to include psychologists, and in eight additional states, psychologists may work in conjunction with physicians. There are 27 state statutes that limit membership to physicians and dentists, and another 4 statutes are "silent" on this issue.

Hospital bylaws also affect the functioning of psychologists in hospitals. The hospital bylaws are not allowed to be in conflict with state statutes and are also subject to other regulatory bodies, such as the JCAH. A 1978 survey revealed that only 6 of the then 115 schools of medicine had bylaws that enabled psychologists to be full voting members of the medical staff of their associated teaching hospitals (Matarazzo, Lubin, & Nathan, 1978). The JCAH regulations at that time were viewed as a potential threat to the role of psychologists in medical schools because medical staff membership was limited, unless otherwise provided by law, to licensed physicians and dentists (Matarazzo *et al.*, 1978). The findings of a 1983 survey (Thompson & Matarazzo, 1984) showed that the efforts of psychologists to influence JCAH policy and to revise the bylaws of medical schools had resulted in an almost threefold increase, with 16 of the then 123 schools reported to have hospital bylaws enabling psychologists to be full voting members of the medical staff.

The JCAH *Accreditation Manual for Hospitals* (1986) requires that there be "a single organized medical staff that has overall responsibility for the quality of the professional services provided by individuals with clinical privileges, as well as the responsibility of accounting therefore to the governing body" (p. 109). The medical staff includes fully licensed physicians and may include other licensed individuals who are permitted by law and by the hospital to provide patient care services independently in the hospital. The JCAH regards as an independent provider any individual who is permitted by law and also permitted by the hospital to provide patient care services without direction or supervision within the scope of his or her license and in accordance with individually granted clinical privileges. All members of the medical staff have delineated clinical privileges.

Clinical Privileges

The JCAH requires that the processes of the hospital for the delineation of clinical privileges be described in the medical staff bylaws, rules, and regulations and be implemented by the medical staff. Whatever method is used, the JCAH (1986) requires that there be evidence that "the granting of clinical privileges is based on the individual's demonstrated current competence" (p. 118). When the system involves categories of privileges, the scope of each level of privileges is well defined, and the standards to be met by the applicant are stated clearly for each category. Clinical privileges are granted for an interval that is not longer than two years.

Individuals are granted the specific privilege to admit patients to inpatient services in accordance with state laws and the criteria for standards of medical care established by the medical staff of each hospital. The JCAH requires that, when nonphysician members of the medical staff are granted privileges to admit patients to inpatient services, provision be made for prompt medical evaluations of these patients by a qualified physician.

The renewal of clinical privileges is based on a reappraisal of the individual at the time of reappointment:

> The reappraisal includes information concerning the individual's current licensure, health status, professional performance, judgment, and clinical and technical skills as indicated by the results of quality assurance activities and other reasonable indicators of continuing qualifications. (p. 121)

Peer and departmental recommendations are part of the basis for the renewal of privileges. In addition, documentation of participation in continuing education activities is also considered.

To protect the public from incompetent physicians, Congress passed the Health Care Quality Improvement Act of 1986, which provides legal protection for peer review. Effective October 14, 1986 the act generally provides immunity from liability under federal law ("New Legal Protection," 1988). This legislation also includes the establishment of a central data bank to contain all disciplinary actions taken against physicians and all settlements and verdicts in medical malpractice cases ("New Legal Protection," 1988). Professional societies, health care organizations, and insurance companies must report at least once a month to the data bank ("New Legal Protection," 1988). Although there have been delays in funding the data bank, it is an indication of future directions for peer review. Forty states already require health care providers and insurance companies to report medical liability claims filed against physicians, and most states require the reporting of settlements and judgments to the state's insurance department ("New Legal Protection," 1988).

There is variability across hospitals in the specific organizational structures involved in clinical privileges. What is essential is that psychologists actively participate with their colleagues from medicine and other disciplines in these processes as they pertain to psychologists.

In summary, these processes include six related functions: (1) delineating which clinical patient care responsibilities psychologists as independent health care professionals will have in the hospital; (2) specifying the qualifications and criteria necessary for such privileges; (3) reviewing credentials; (4) recommending specific privileges for each psychologist functioning within the hospital; (5) specifying the criteria for the renewal of privileges; and (6) participating in the process of the renewal of privileges for psychologists.

Quality Assurance

Although clinical privileges have been the focus of recent initiatives by psychologists, the second area of self-governance responsibility of the medical staff—monitoring and evaluating the quality of patient care—has received much less attention. Quality assurance will be the next step in psychology's professional maturation and the major issue of importance in the next decade for psychology as a profession providing health care. This importance stems from increasing competition and increasing efforts at cost containment, with a resulting emphasis on the demonstrated effectiveness of the diagnostic and treatment services provided. Accrediting bodies such as the JCAH and reimbursement agencies such as the HCFA are using their leverage to promote quality assurance.

JCAH standards require hospitals to have quality assurance programs through which the medical staff strives to ensure the provision of high-quality patient care. This requirement is fulfilled through the monitoring and evaluation of the quality and the appropriateness of patient care and of the clinical performance of all individuals with clinical privileges. Chapter 7 offers a further discussion of quality assurance.

It is essential that psychologists, as members of an autonomous profession, be involved in the quality assurance structures and processes of their hospital as these relate to the provision of psychological assessment, consultation, and treatment services to patients. It is clear that major regulatory bodies such as the JCAH and the HCFA will use their accreditation and reimbursement authorities to promote quality assurance and health promotion by hospitals. It is also clear that the focus of the quality assurance programs will have to move from structure and process to outcome, i.e., to a determination of the actual effectiveness of the services provided.

Prescribing Medications

Psychology is facing what amounts to a "life course" policy decision. The question of whether psychologists should seek to prescribe psychotropic medication is basic to our developmental tasks of autonomy, competency, and identity. Clearly, psychology must continue to define its functions. However, prescription privileges are not a mark of psychology's professional autonomy, and to the extent that these privileges are portrayed in this way, the emphasis is misplaced. It is evident that psychologists could develop the necessary competencies for the responsible exercise of this clinical care function. However, it is also clear that current training programs and experiences do not provide the necessary level of expertise in this area. Consequently, it is not an area of practice with which our profession is associated. The quest for these privileges must come, as with any other clinical privilege, only after a demonstration of the requisite skills. Furthermore, the rationale for moving in this direction must be based on more than "me-too-ism." That is, psychology's advocacy must be based on a fundamentally sound rationale other than that, if physicians do it, psychologists should be able to do it also. Can we next anticipate advocacy for psychologists to provide electroconvusive therapy?

The point being made here is that the rationale for what professional psychologists do must emanate from what is fundamental to the science and profession of psychology. Medication prescription is not fundamentally associated with the behavioral science of psychology.

Prescribing medication is a new venture rather than an extension of the psychologist's behavioral science expertise to mental and physical problems. Furthermore, the case has not yet been made that it is in the public interest for psychologists to have prescription privileges. Is this a responsive step for psychology to take in a period requiring that all health care professionals address cost-containment and quality-of-care issues? With the physician surplus, are more prescribers needed? What about our advocacy of team approaches to accomplish the efficient provision of behavioral and biomedical services? Do we really want to advocate that prescription privileges be "generic" to psychology, so that, as a discipline and a profession, we must take on the responsiblities for training, credentialing, and monitoring, as well as underwriting liability insurance?

This is not to say that individual psychologists, depending on the nature of their interest and practice, should not develop the skills and competencies requisite necessary to prescribing privileges. An alternative to the generic approach is to advocate for the establishment of competency-based criteria for this privilege that are not discipline-specific, but subspecialty-specific. There are a number of functions and services already recognized as not being specific to a discipline, such as "counseling" and "hypnosis." Psychologists also obtain the specific credentials necessary to other practices, such as the interpretation of sleep EEG's. Because of the lack of consensus among psychologists about the prescription issue, further dialogue and consideration are necessary.

Conclusion

Psychology is now recognized one of the health care professions and is now challenged by the transition from professional adolescence to an exciting and productive young adulthood (Thompson, 1987). Psychology has proved to be resilient in dealing with internal and external threats and to be extraordinarily competent in meeting the expectations and needs of society when allowed to compete fairly. Medicine can be viewed as an older sibling, which, in the past, was the only child and now needs to adjust to increasingly competent, valued, and autonomous siblings. The task for medicine is to accept its role as only one of the disciplines in the health care family, along with dentistry, psychology, social work, nursing, and other health care disciplines.

There will inevitably be some overlap in functions and contributions in this family of disciplines. Psychology's self-actualization will depend on its unique contributions to improving the human condition. Psychology's uniqueness lies in its behavioral science perspective, which is what is common to the various subspecialties and interest areas within psychology. Cost-effective high-quality health care will increasingly require the blending, in proper sequence, of an array of both behavioral and biomedical services. The last several decades have shown that advances in biomedical science and care increase, rather than decrease, the need for advances in the behavioral areas. Furthermore AIDS, as a behaviorally transmitted disease, is presenting psychologists with unpresented challenges and responsibilities (see Chapter 31). In the coming years, psychology will have to continue to develop not only unique disciplinary competencies but also expertise with, and commitment to, interdisciplinary collaboration. Moreover, psychologists need to demonstrate compassion, along with their competencies, to avoid the perception that they are motivated by self-interest in just another way to earn a living.

It is very important for us to recognize that, as psychologists, we have been afforded a unique opportunity to influence how our society views health and illness in general and the contribution of the behavioral sciences, in interaction with the biomedical sciences, in particular. This opportunity has arisen from the integral role of psychology in undergraduate education in the approximately 2,100 baccalaureate-granting colleges and universities in this country, and from the developing role of psychology in high school education. This is another advantage of being both a basic science and a profession. In this situation, psychologists have more than an equal opportunity to argue their case with those who will define society's needs and expectations in the years ahead. We should see to it that health psychology is a visible component of psychology's educational offerings. Further, psychology has a unique opportunity to serve as a "behavioral science bridge" for the integration of the medical school and the arts-and-sciences components of the university.

To foster the continued development of psychology as a health care science and profession, it is necessary to confront the challenges and opportunities that lie before us by being responsive to societal needs. Psychology has much to offer and to gain in this process.

ACKNOWLEDGMENT

Joseph Matarazzo's input into and critique of this chapter are gratefully acknowledged.

References

DeLeon, P. H. (1987). The leaves of fall. *The Clinical Psychologist, 40,* 81–84.

DeLeon, P. H. (1988). Public policy and public service: Our professional duty. *American Psychologist, 43,* 309–315.

Dennis, W. (1950). Interrelations of psychology and medicine. In W. Dennis (Ed.), *Current trends in the relation of psychology to medicine* (pp. 1–10). Pittsburgh: University of Pittsburgh Press.

Dorken, H., & Bennett, B. E. (1986). How professional psychology can shape its future. In H. Dorken & Associates (Eds.), *Professional psychology in transition* (pp. 350– 387). San Francisco: Jossey-Bass.

Engel, G. L. (1977). The need for a new medical model: A challenge for biomedicine. *Science, 196,* 129–136.

Fein, R. (1987). Medical education: The impact of the social sciences on a changing delivery system. In C. Vevier (Ed.), *Flexner: 75 years later. A current commentary on medical education* (pp. 63–75). Lanham, MD: University Press of America.

Felix, R. H. (1950). Psychology and public health. In W. Dennis (Ed.), *Current trends in the relation of psychology to medicine* (pp. 11–27). Pittsburgh: University of Pittsburgh Press.

Himmelstein, D. V., & Woolhandler, S. (1986). Cost without benefit: Administrative waste in U.S. health care. *The New England Journal of Medicine, 314,* 441–445.

Jacobsen, C. (1950). Psychology in medical education. In W. Dennis (Ed.), *Current trends in the relation of psychology to medicine* (pp. 28–59). Pittsburgh: University of Pittsburgh Press.

Joint Commission on the Accreditation of Hospitals. (1986). *Accreditation manual for hospitals.* Chicago: Author.

Joint Commission on the Accreditation of Hospitals. (1987, August). *Overview of the Joint Commission's "Agenda for Change."* Chicago: Author.

Kagan, J. (1965). The new marriage: Pediatrics and psychology. *American Journal of Diseases of Childhood, 110,* 272–278.

Lubin, B., Nathan, R. G., & Matarazzo, J. D. (1978). Psychologists in medical education: 1976. *American Psychologist, 33,* 339–343.

Matarazzo, J. D. (1980). Behavioral health and behavioral medicine: Frontiers for a new health psychology. *American Psychologist, 35,* 807–817.

Matarazzo, J. D., Lubin, B., & Nathan, R. G. (1978). Psychologists' membership on the medical staffs of university teaching hospitals. *American Psychologist, 33,* 23–29.

Matarazzo, J. D., Carmody, T. P., & Gentry, W. D. (1981). Psychologists on the faculties of United States schools of medicine: Past, present and possible future. *Clinical Psychology Review, 1,* 293–317.

Mensh, I. N. (1953). Psychology in medical education. *American Psychologist, 8,* 83–85.

New legal protection paves way for peer review. (1988, Winter). *The Digest: A Medical Liability and Risk Management Newsletter,* pp. 1–2.

Pellegrino, E. D. (1987). The reconciliation of technology and humanism: A Flexnerian task 75 years later. In C. Vevier (Ed.), *Flexner: 75 years later, A current commentary on medical education* (pp. 77–111). Lanham, MD: University Press of America.

Petersdorf, R. G. (1987). Medical education: The process, students, teachers and patients. In C. Vevier (Ed.), *Flexner: 75 years later, A current commentary on medical education* (pp. 17–33). Lanham, MD: University Press of America.

Richmond, J. B. (1967). Child development: A basic science for pediatrics. *Pediatrics, 39,* 649–658.

Roper, W. L., & Hackbarth, G. M. (1988). HCFA's agenda for promoting high-quality care. *Health Affairs,* Spring, 91–106.

Schofield, W. (1969). The role of psychology in the delivery of health services. *American Psychologists, 24,* 565–584.

Schroeder, S. A. (1987). Outcome assessment 70 years later: Are we ready? *The New England Journal of Medicine, 316,* 3, 160–162.

Schwartz, G. E., & Weiss, S. M. (1978). Behavioral medicine revisited: An amended definition. *Journal of Behavioral Medicine, 1,* 249–251.

Scott, W. R. (1985). Conflicting levels of rationality: Regulators, managers, and professionals in the medical care sector. *The Journal of Health Administration and Education, 3,* 113–131.

Starr, P. (1982). *The social transformation of American medicine.* New York: Basic Books.

Thompson, R. J., Jr. (1987). Psychologists in medical schools: Medical staff status and clinical privileges. *American Psychologist, 42,* 866–868.

Thompson, R. J., Jr., & Matarazzo, J. D. (1984). Psychology in United States medical schools: 1983. *American Psychologist, 39,* 988–995.

Thompson, R. J., Jr., & O'Quinn, A. N. (1979). *Developmental disabilities: Etiologies, manifestations, diagnoses, and treatments.* New York: Oxford University Press.

Vevier, C. (1987). The Flexner Report and change in medical education. In C. Vevier (Ed.), *Flexner: 75 years later. A current commentary on medical education* (pp. 1–15). Lanham, MD: University Press of America.

Weiss, S. M. (1987). Behavioral medicine in the trenches. In J. Blumenthal & D. McKee (Eds.), *Applications in behavioral medicine and health psychology: A clinician's source book* (pp. xvii–xxiii). Sarasota, FL: Professional Resource Exchange.

Wennberg, J. E. (1988). Improving the medical decision-making process. *Health Affairs*, Spring, pp. 99–106.

Wright, L. (1967). The pediatric psychologist: A role model. *American Psychologist, 22,* 323–325.

CHAPTER 3

The Structure and Authority of Hospitals

Bruce Bonecutter and Martin Harrow

> The morbidity and mortality rates of Americans are no longer related to infectious diseases prevalent at the turn of the century; instead, they are related to chronic disorders related to our life-styles.
>
> —Thomas J. Stachnik (1980)

Introduction

The relationship between mind and body has been the object of great curiosity, discussion, and scientific study for a very long time. It is important both to understanding psychology as currently practiced in health center settings and to a proper sense of humility for one to acknowledge that, in the scheme of history, the science and practice of psychology are recent entrants into the study of the relationship between mind and body.

Historically, clinical psychologists first entered the hospital with test kit in hand, the Stanford–Binet and later the Rorschach. Thanks to psychological research and consumer demand, the scope of psychologists' hospital practice has greatly expanded over the past 25 years. The psychologist now enters the hospital with numerous screening and focused tests, clinical and health psychology treatment techniques, and the knowledge needed to tailor diagnostic assessment and treatment to the individual and the setting. No longer, as in the 1960s, does psychology's role in hospitals focus exclusively on the mentally ill. The applied and basic research in areas such as neuropsychology, biofeedback, cognitive-behavioral treatments for health-damaging lifestyles, and other health psychology and behavioral medicine foci have radically boosted psychology's value to the general health of the human species. Psychology is entering the general hospital in a very big way (American Psychological Association, 1985, p. 88).

Elsewhere in this book, the clinical, legal, political, economic, and informal aspects of the practice of psychology in medical settings are elucidated. This chapter provides the reader with a map of the terrain. As in any adventure, knowing how to read the map helps psychologists avoid harm and increases the efficiency of their professional travels.

Bruce Bonecutter • Department of Psychiatry, Cook County Hospital, Chicago, Illinois 60612. Martin Harrow • Department of Psychiatry, Michael Reese Hospital and Medical Center, Chicago, Illinois 60616.

Historical Definition and Purpose of a Hospital

The word *hospital* comes from the Latin word *hospitium*, originally used to designate a place of hospitality and shelter for pilgrims and so-journers. Early Christian governing counsels added the function of caring for the sick and established the tradition of providing hospitals in every city large enough to have a cathedral. Hernán Cortés established the first North American hospital in Mexico City in 1524, and the Jesus of Nazareth Hospital is still functioning over four and a half centuries later. William Penn established the first almshouse in the United States with a hospital function in 1713 in Philadelphia. The American voluntary hospital, the current concept of a hospital, was established in the North American colonies by Benjamin Franklin and Dr. Thomas Bond (Pennsylvania Hospital, 1751), also in Philadelphia. In the late 1800s, with the progress of the health sciences, especially anesthesiology and bacteriology, resulting in techniques in antiseptic surgery, hospitals became more treatment centers than shelters for the sick. Sick people could actually expect to live after going to the hospital. The perception that hospitals are places where various scientist practitioners make people healthier is barely three generations old (Snook, 1981).

Hospitals have specific, legally stated mission or purpose statements. For example, Cook County Hospital's (Cook County Hospital, bylaws 1984) say:

> WHEREAS, Cook County Hospital is a public hospital organized under the laws of the state of Illinois; and
>
> WHEREAS, its purpose is to serve as a general hospital providing care to all patients regardless of race, color, religion, age, sex or national origin, providing education and undertaking research, and
>
> WHEREAS, it is recognized that the medical staff has the overall responsibility for the quality of medical care provided to patients, and for the professional practice and ethical conduct of its members, as well as accounting therefore to the Board of Commissioners of Cook County and that the cooperative efforts of the medical staff, the administration and the Board are necessary to fulfill the hospital's goals in providing patient care.

> THEREFORE, the physicians and dentists [other hospitals add podiatrists and psychologists] practicing in this hospital hereby organize themselves into a medical staff [or if the others are included "organized professional staff"; Matarazzo, 1978; Dorken, 1982; Tanney, 1983] in conformity with these bylaws.

Psychologists are expected to link their services and treatment results to these definitions and purposes for the benefit of the patients, the professionals, and the economic needs of a hospital. The psychologist's authority and permission to practice in hospital and health center settings vary.

A frequently cited test of professional independent practice is to ask the professional, "From how many people (or committees) must you ask permission, in order to practice and get reimbursed?" As evidenced by the definition and purpose statements above, hospitals themselves must have permission to practice and get reimbursed. In the modern professional world, psychologists should know the formal and informal ecosystem of their practice and the financial interests in their hospital (American Psychological Association, 1985, 1988).

Psychologists practice in a hospital in three main contractual ways: as *staff employees* paid by the hospital for a broad range of time and services; as a *consultants* in a narrow, often educative, range of services; and as *private practitioners* permitted or "privileged" to provide psychological services to patients (credentialed to do specific practice) in the hospital, but paid by the patient. All psychologists practicing in a hospital are under the legal, administrative, and clinical supervision of that hospital.

The authority to "treat" or "practice on" a patient is extremely important from the legal, ethical, and financial points of view (Bule, 1988–1989). It is therefore wise for psychologists to know the legal, ethical, and financial road map of the hospital.

The Authority of the State

Each of the 50 states in the United States and similar government structures internationally have the legislative authority to license a hospital to open and practice. With this authority, the state may close the hospital entirely or in part. The legal philosophy for this authority is the protection of the public in the state.

However, the legislators of any state are far too busy to monitor hospitals directly. Therefore, they rely on state inspectors, blue-ribbon governors' committees, independent authorities such as the Joint Commission on Hospital Accreditation, and professional organizations such as the American Medical Association and the American Psychological Association to (1) establish the criteria for hospital practice and (2) monitor, report, and discipline hospitals not in compliance with these criteria.

The psychologist should begin to understand the great importance of legislators, especially health, hospital, and health insurance subcommittees; governors' hospital regulatory boards; the Joint Commission on Hospital Accreditation, and the American Psychological Association's Office of Professional Affairs. These authorities write or advise legislators on criteria that grant the psychologist permission to treat, conduct research, and get paid in a hospital setting. The right to research and practice in the "centerpiece of the health care system"—the hospital—strongly determines the worth of a specific health science to society in general. Therefore, the time and money invested in working toward protecting public accessibility to psychology in the medical center are time and money well spent (American Psychological Association, 1985, 1988; Dorken, 1982; Tanney, 1983).

In most hospital practice, the state regulation of reimbursement by health insurance, Medicare, Medicaid, and other third-party payers is also of significant interest to the psychologist practicing in a medical setting. The psychologist may have completed a scientifically respectable psychodiagnostic write-up or an effective episode of psychotherapy, but the psychologist or the hospital will not be reimbursed for the skill and time in doing quality work if the state's third-party payment criteria for psychology are not influenced by the psychologist and his or her professional association. The state legislators and the insurance industry are quite understandably highly motivated to keep the costs of health care down. This motivation also adds incentive for the psychologist to use rigorous, sound social science methodology in operationally defining assessment and treatment procedures and fair reimbursement rates.

Both in countries where there are national health insurance plans and in the variety of health-care cost-management systems in the United States, the legislators and the administrators value solid scientific information on treatment effectiveness and a fair cost–benefit analysis of psychological services.

Hospital and medical setting legislation is the foundation for other applied health care research and practice legislation. The legal status of psychologists practicing in a hospital is often used to define their worth for public health research grants and for treatment reimbursement from government or private third-party payers. Legislation and guidelines affecting hospital and health service practice affects nearly all psychology practice (American Psychological Association, 1985, 1988; Tanney, 1983).

The Lines of Authority in the Hospital Itself

The structures of authority follow the functional goals (purpose and mission statement above) of the hospital. That is, the structure is altered if the hospital is organized as (1) a private or public hospital; (2) a for-profit or not-for-profit hospital; (3) part of a chain, either corporate, religious, or state-owned; (4) a stand-alone facility; (5) a specialized or general hospital; (6) university-affiliated or research-affiliated training hospital, or a nontraining research hospital; or (7) "other." Knowing the hospital's combination of these legally defined goals will help the psychologist to draw a working mental map of the lines of authority.

The health care world is currently moving more toward a business model. This model is in conflict with the original charity and service models, but it is here to stay. This newer orientation affects the structure of a hospital. In some cases, hospitals close, or they are taken over by other authorities and are then restructured to become business-efficient or to accomplish some other mission. The psychologist then has to learn new routes of gaining permission to practice, to do research, and/or to be reimbursed.

The American Psychological Association's *Hospital Practice Primer for Psychologists* (1985) offers a sample organizational chart of a predominantly mental-health-focused hospital (see Figure 1). A general hospital is much more diverse and complicated. Most psychologists working in hospital settings are situated organizationally either within a division of the department of psychiatry or within a small department of their own as depicted in the sample organizational chart. Increasingly, psychologists are also budgeted directly to medical, surgical, pediatric, or family practice departments and are often "credentialed and privileged" by the larger group of their peers in the psychology or psychiatry departments in the hospital. It is wise for psychologists to be responsible to members of their own profession within a hospital, even if they are in another budget unit of the hospital. Other psychologists can best understand the psychologist's practice and research and support him or her when an internal or external question of practice, research, or reimbursement for practice or research arises.

The Board of Directors

Regardless of the functional goal or purpose of the hospital, the board of directors is the highest ranking body responsible for the hospital's operation. The U.S. courts hold hospital boards legally responsible for the activities of the staff practicing in their hospital (Snook, 1981).

In a federal, state, or local hospital, the board is comprised of officials elected to a general board or is a separate board (appointed or elected) empowered to govern the hospital. In religious and private-sector hospitals, the board may be substantial investors, owners of the hospital, members of the clergy or the religious laity, and/or leaders in the community where the hospital is located.

The board of directors has a structural map for itself, usually with a president, other officers, and committees specializing in areas of hospital oversight, such as the finance committee and the personnel committee. There are often "interface" committees or subcommittees designed to communicate between the board and the hospital. It is very wise for psychologists in responsible positions to spend time on

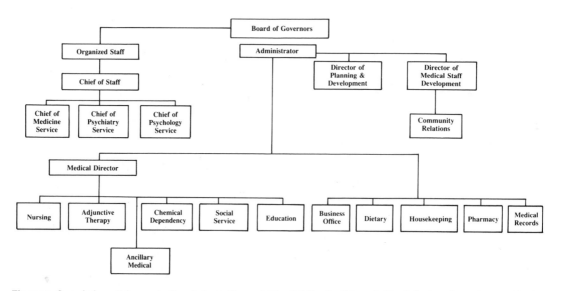

Figure 1. Sample hospital organizational chart. (From *A Hospital Practice Primer for Psychologists*, Committee on Professional Practice of the Board of Professional Affairs of the American Psychological Association, 1985.)

these committees in order to establish mental health and health psychology programs in the budget and administrative structure of the hospital. Additionally, this committee work serves the important function of fostering better communication with the administration and the board of the hospital. If the psychologist who is not able to spend time on these committees should discover who the hospital staff liaisons with these joint committees are. Through contact with these liaisons, psychologists can communicating their needs to the organized professional staff and the appropriate hospital board members.

Typically, the main work and societal roles of board members lie outside the health care field. For the most part, they are busy and dedicated people who expect clear systems theory thinking and presentations. A psychologist's skills in measuring need, treatment effectiveness, and cost-effectiveness are of great benefit in relating to the board of directors.

Organized Professional Staff, Chief of Staff, Hospital Director, and Nursing Staff

The next level of authority is shared by three general administrative structures and the nursing staff, again each having its own internal structures.

The *organized professional staff* focuses on the science and art of practice in the hospital. The medical and/or organized professional staff is an organized body of physicians and other licensed professionals, often called *attending staff*, who diagnose, treat patients, and participate in related duties (e.g., research, program development, and evaluation). Without an active structural link to the medical or organized professional staff, it is becoming very difficult for psychology to be practiced or reimbursed as an independent and respected scientific profession.

The professional staff may have levels of membership. It is best for psychologists to have a level of membership that permits them to vote on hospital professional issues affecting all professionals, and to sit, vote, and hold office on committees of the organized profes-

sional staff. It is essential to the modern practice of psychology that psychologists fully participate on the hospital staff committees that affect the practice of psychology, such as the credentials committee, which decides who is a trained and licensed psychologist and what specific privileges or psychological specialties this person is to be able to practice on the patients of the hospital; the bylaws committee, which determines the framework for professional self-government and the means of accountability to the board of directors; and the quality assurance committee, which decides how to efficiently assess the process and outcome of psychological services offered in the hospital. Specialized *ad hoc* committees may be important as well, such as the *ad hoc* committees on substance abuse treatment or on child abuse and neglect. Other organized professional staff committees include the medical records committee, the drug and formulary committee, the tissue review committee, the peer review committee, and the medical audit committee. Psychology's failure to win a voting presence on the general organized professional staff and the focused committees shuts psychology out of most of hospital practice. Participating on these committees (e.g., the research committee or the ethics committee) also has the informal and very potent effect of displaying the usefulness of psychology in the care of patients and in the collegial governance of the practice, research, and business concerns of the hospital. Psychologists are usually very busy and understandably dislike the time and paperwork involved in committee work. However, the advantages of putting in time and effort are significant.

Unfortunately, some hospitals try to designate psychologists as "technicians" or "allied health care staff," who are not useful to or interested in the broader problems of patient care and professional and scientific interaction. Thus, it is essential for psychologists to work with their state association to gain an established and functioning role in hospital practice.

The *chief of staff*, or *medical services director*, is the treatment staff administrator, nearly always a respected physician, who oversees the medical-legal aspects of all treatment delivered to patients in the hospital. The chief of staff often has an administrative staff to assist her or him this process. The heads of the clinical departments report to the chief of staff and form an executive committee for the chief of staff. The usual departments in a general hospital include medicine, surgery, pediatrics, obstetrics and gynecology, family practice, neurology, radiology, anesthesiology, psychiatry, pathology, and emergency medicine, with specialized divisions reporting to these departments. The smaller the hospital, the smaller is the number of departments, and the more likely it is that specialties will be divisions reporting to a department.

A few hospitals have separate departments of psychology, but most have divisions (or even smaller "sections" or "services") of psychology within the department of psychiatry, with a psychologist as official or unofficial chief. The unit structurally below the division is the section. Frequently, psychologists are section directors in their specialty, that is, the neuropsychology and rehabilitation section, the section for consultation to or liaison with inpatient pediatrics, and the pain clinic and biofeedback section. All these clinical supervisors or section chiefs report upward to their division and then department head, who reports to the chief of staff.

Nearly all budget and program proposals are worked out in this part of the structural map. That is, the department chief huddles with the division, section, and service chiefs to gather and display data indicating the need, efficacy, and cost of the existing and proposed treatment programs. The organized professional staff then reviews the scientific and clinical merit of these programs. The department chief then reviews the proposals and asks for the chief of staff's support in forwarding these requests to the director of the hospital and to the hospital board of directors.

The *director of the hospital* (sometimes titled the *administrator*, the *chief executive officer*, the *president*, or the *executive vice-president*) is appointed by the board of directors as the chief executive

responsible for the performance of all functions of the institution and accountable to the governing authority. The chief executive, as head of the organization, is responsible for all functions including a medical staff, nursing division, technical division, and general services division which will be necessary to assure the quality of patient care. (American College of Hospital Administrators, 1973)

The director is the day-to-day agent of the board. In addition to the chief of staff (or medical service director), the director of the hospital has a group of assistant directors working in areas such as billing and finance; personnel, payroll, and benefits; building and grounds; dietary; and transportation. In larger hospitals, the hospital director may have a deputy director just under the director in authority. The director and the deputy and assistant directors have major responsibility for financial and day-to-day operations. Historically, they were nurses with administrative skills, but with the need for tight business management, the hospital director is now a physician or a high-ranking nurse with additional administrative degrees, or a person who holds a single degree such as a master's degree in business administration (M.B.A.), a master's degree in public health (M.P.H.), a master's degree in social work (M.S.W.), or a master's degree in social service administration (M.S.S.A.). Usually, these people are all well trained and have extensive hospital experience.

Just like the hospital board members, the hospital director and the hospital administration staff are usually busy and dedicated people who respect clear, well-conceived systems theory communications. They focus on getting the best care at the most reasonable cost. That is, they are motivated to please the patients who use the hospital and to please the board, who entrust the management of the hospital to them. The psychologist's skills in research, both library literature review and quasiexperimental design research, are respected by administrative directors. It is important for psychologists to be independent professionals who provide expertise to the administration of a hospital, as the administration has a great deal to say about programs, facilities, and benefits.

The numerically largest single category of professionals in almost any hospital is the nursing staff. The nursing department sometimes reports to the chief of staff, sometimes to the administrative director, and sometimes to both. Because the nursing profession is absolutely vital to a hospital, and because a psychologist is very likely to share patient care with a bachelor's or master's level nurse, it is essential to consider the nursing department in a view of the lines of authority in a hospital. The psychologist will do well to plan programs and to discuss cases with the clinical and administrative nurses. Often, in fact, the nurses have structural authority in clinical management and quality assurance review. Again, the psychologist's scientist-practitioner skills can be used to assist the nursing staff in these areas.

The internal organizational structure of a hospital's nursing staff somewhat parallels that of the medical director and the department, division, section, and service chiefs. In addition, the nursing staff has the responsibility for coverage 24 hours every day. Obviously, this requires more supervisors and direct service staff. It is a very good practice to know the organizational hierarchy of the nursing staff, especially those in the psychologist's areas of practice and those responsible for providing care and treatment during each of the three eight-hour shifts.

Additionally, the profession of nursing may have certain highly trained and qualified specialty practitioners, such as nurse midwives, visiting nurses, and psychiatric nurse practitioners, who have a great deal of independent practice permission or "privilege" within the hospital and the health care community. For the best care of the clients, the psychologist does well to know the role, the range of services, and the system for requesting the services provided by these nurse specialists.

We have presented the organized professional staff, the chief of staff, and the administrative director as being parallel and as possibly reporting separately to the board of directors. Usually, there is a balance of power between these three units of authority and the powerful

nursing department, even though they may at times be hierarchically under each other. It helps to view a hospital as a matrix organizational system with connected lines of authority. In fact, many hospitals explicitly state that the simple hierarchical organizational chart is so oversimplified as to be somewhat misleading (Snook, 1981).

Departmental Authority

In discussing the chief of staff, or medical services director, we have seen that departments, divisions, and sections report to him or her. On the department level, there are structures that organize and regulate a psychologist's day-to-day practice and reimbursement. The department heads hire and fire; monitoring and staff appeals are handled by the credentials and peer review committees of the organized professional staff, or medical staff. Often, the psychologist's direct supervisor initiates the hiring or firing procedure, but it does not even begin to become final until the department chair signs the forms. Many psychologists are not aware of the many levels of permission necessary in order to be hired or permitted to practice within a hospital. In most cases, no one may come in to practice privately or to begin work as a staff psychologist until all levels of the hospital hierarchy, up through the board, sign the permission forms, as the state government authority focused on providing for and protecting the public will hold all these levels of authority responsible for the psychologist's practice.

After the psychologist is hired, the next most crucial aspect of practice in a hospital is determining what privileges the psychologist is to have in that hospital. The adequacy of the training and background experience of the psychologist, or of any other hospital professional, is paramount. For the protection of the patient, and for the protection of the hospital against a lawsuit for conducting clinical activities outside the bounds of the psychologist's training and expertise, the psychologist, or any other professional, should not see a patient at all, even after being hired, until he or she has defined privileges. Hospitals try to be very serious about this. Therefore, the department heads and the organized professional staff carefully scrutinize every professional's privileges. Usually, the bylaws provide for ongoing review by a peer review committee of any incident that puts the competence and ethics of a professional's practice into question, and for a routine, often biennial, review of every department's professional staff's privileges. Such reviews try to answer the question: Did the psychologist demonstrate competence doing what the psychologist was permitted to do? If it is discovered that the psychologist did not demonstrate competence, he or she will lose privileges or will be placed on supervised probation in a manner very similar to the American Psychological Association's system of ethical review (American Psychological Association, 1985, 1988; Beresoff, 1983; Bule, 1988–1989; Tanney, 1983).

Figure 2 provides a sample of the privileges of psychologists and psychiatrists with "attending" status in a mythical department of psychiatry, psychology, and behavioral sciences (see American Psychological Association, 1985). There are other systems of listing privileges in a hospital. At present, these vary by hospital. Each hospital is responsible for establishing and maintaining its own system. In general, the psychologist will be asked to document her or his competence in treating specific age groups, gender groups, and cultural groups for specific diagnoses and/or dynamic problems with specific diagnostic and treatment modalities. The psychologist may add to or subtract from his or her privilege list. It is a sign of responsible practice if the psychologist who has become rusty in some skill requests to have a privilege lessened or deleted. It is a sign of responsible practice if, in the course of receiving continuing education (hospitals expect vigorous continuing education), the psychologist gains another skill and requests that it be added to his or her privilege list.

The psychologist who has become privileged to practice in a hospital needs to receive credit (both professional and financial) for psychological services delivered to patients in the hospital. "Privileging" is tied to both professional status and financial worth (Bersoff, 1983; Tanney, 1983).

For a psychologist practicing in a hospital as a staff member or as a private practitioner the privileging process presumes a high degree of training. Division 42 of the American Psychological Association (APA) has drafted the following "Proposed Guidelines for Evaluating Psychologists' Application for Hospital Practice":

1. Doctoral degree in psychology and internship/residency, preferably from an APA approved clinical setting.

2. State licensure at the independent practice level, as required by JCAH and state law.

3. Completion of a course of not less than four (4) contact hours on staff relations, hospital procedures, hospital administrative practices, charting and other appropriate content for the successful passage of an examination covering these matters.

4. Demonstrated knowledge of the "Diagnostic and Statistical Manual-III R" and the "International Classification of Diseases-9," nosology including proficiency in differential diagnosis. Familiarity with CPT-Procedures Terminology codes and previous graduate coursework in psychopathology. This requirement can be met through: Specific coursework of at least 12 contact hours, i.e., APA or state approved continuing education workshops or courses or as part of a graduate school, and/or internship course or seminar series.

5. Coursework in clinical psychopharmacology of at least 16 contact hours or a hospital inservice training course, to include psychoactive drugs, abused drugs and related laboratory tests, is recommended since hospital psychologists should be able to distinguish drug symptomatology from behavioral pathology. Previous graduate coursework in psychophysiology is desirable.

6. Additional competencies would include Cardio-Pulmonary Resuscitation Certification, required by most hospitals, and familiarity with relevant medical tests. In addition to the usual neurological diagnostic and treatment techniques, neuropsychologists will want to understand EEG, CT-Scan, PET-Scan and X-Ray findings. Health psychologists may find it desirable to be acquainted with EKG and other lab tests.

7. Practicum experience should include at least 2000 hours (one year, full time) of supervised practice in a hospital setting. Of this, 1000

hours should be post-doctoral and post-internship/residency so that the psychologist has at least six months of functioning as a fully accountable practitioner without the internship/residency umbrella. For experienced clinicians, e.g., those board certified or board eligible for the American Board of Examiners in Professional Psychology, who want to develop competence in hospital practice, successful management of 12 hospitalized cases, supervised by a hospital based preceptor, within a 12–24 month period would be accepted as an alternate means of meeting this requirement. (Enright, 1987, pp. 1–2)

The *Hospital Practice Primer for Psychologists* (APA, 1985) sets more general standards than those proposed by APA Division 42. Being approved for staff membership and being privileged at a hospital involves many of these APA guidelines as "requirements" and some as "desired." In general, hospitals and their overseers strive to ensure high standards of competency.

The Structure and Authority of Documentation in Medical Settings

"If it isn't written (properly), it didn't happen" is the universal anthem of hospital practice from the president of the board on down (Snook, 1981).

Even before the increase in health service litigation, a hospital operated on documentation: patient records, prescriptions, laboratory test orders, nursing orders, orders for treatment, billing slips, admission and discharge records, procedural notes, summaries of treatment, physical histories, psychosocial histories, and on and on. The reason for this very high volume of documentation is still primarily the benefit of the patient. A team of professionals must treat the patient three shifts a day. They must consider not only the episode of care in this hospitalization, but also the meaning of the records from prior episodes of care for that patient (and even for that patient's family and support system members) in refining the diagnosis and in developing current treatment and disposition plans. The professional and the hospital must also know the billable services delivered to patients in order to make

money and/or to contain financial losses (Hall, 1988; Snook, 1981; Soisson *et al.*, 1987).

If psychologists wish other professionals to take their scientific practitioner findings of treatment interventions seriously, and if psychologists wish to be paid, then they will thoroughly learn and punctiliously practice the hospital's methods of documentation.

The relatively recent introduction of computer network systems in hospitals has both facilitated and complicated the documentation process. Because many psychologists are computer-wise, this is an opportunity to be valuable to the hospital. However, at present, most of the documentation must start with paper entry, must be entered into the computer data bank, and is later produced as a paper report when it is needed.

So, even with a good computer system, if it isn't officially written, it never happened or never will happen. The good news about the better hospital-information-management systems is that multiple entries of the same information can be reduced. However, the psychologist delivering a service in a hospital must be aware of the numerous kinds of documentation needed for each instance of psychological service (i.e., productivity, billing, and needs assessment data). Computerization of patient records also makes efficient archival research more possible.

In a simple 45-minute individual psychotherapy session, many documents need to be generated. The *treatment plan* in the patient's chart must include the treatment modality of individual psychotherapy and a why this treatment is scientifically recommended. The psychologist needs to document the fit of the patient's diagnosis, dynamics, and other system factors with the specific mode of individual psychotherapy implemented. As in research, the psychologist needs to be able to rule out threats to the validity of accepting this treatment approach as being efficacious for this patient at this time. Can the psychologist reject the null hypothesis that no treatment or that other treatments will also be effective?

The session itself needs to be summarized relative to the treatment goals in the *progress*

```
"HIGH QUALITY HOSPITAL"  DEPARTMENT OF PSYCHIATRY, PSYCHOLOGY
AND BEHAVIORAL SCIENCES

                      PRIVILEGES GRANTED

All Departmental privileges are granted pursuant to the "High
Quality Hospital" Organized Professional Staff Bylaws and are to
be administratively supervised by the appropriate Division,
and Section Chief.

The two categories of privileges are: "Full" = F, for an
staff member who is licensed or registered in the specific
privileges and who demonstrates competency, and "Supervised" = S,
for a staff member who is not licensed or registered in the
specific privileges and/or does not demonstrate adequate
competency.

I.   General Adult Psychiatry

_____  Supportive and/or Brief Psychotherapy
_____  Focused/Time-Limited Psychotherapy
_____  Intensive Psychotherapy
_____  Consultation Procedures on Patient Care re:
       Treatment and Disposition
_____  Administration and Evaluation of Psychological
       Tests - Specify Types:

_____  Psychopharmacotherapy - Specify:


II.  Child (C) , Adolescent (A) or Geriatric (G) Psychiatry
       Please Specify

_____  Direct Diagnostic and Treatment Procedures
_____  Indirect Diagnostic and Treatment Procedures
_____  Consultation Procedures Resulting in Disposition
_____  Consultation  to  Staff,  Services,  Departments  or
       Institutions
_____  Administration and Evaluation of Psychological
       Tests - Specify Types:
_____  Psychopharmacotherapy - Specify:


III. Specific Forms of Treatment

_____  Individual Psychotherapy      _____  Rehabilitation Tx.
_____  Group Psychotherapy           _____  Sexual Disfunction Tx.
_____  Family Psychotherapy          _____  Narcosynthesis
_____  Couples Psychotherapy         _____  Electroconvulsive Tx.
_____  Cognitive/Behavioral          _____  Milieu Psychotherapy
       Psychotherapy                 _____  Community Education
```

Figure 2. Sample privilege sheet for prototypical hospital practice.

notes section of the patient's chart. This summary has multiple purposes. Primarily, it functions to remind the psychologist of the process of psychotherapy in that session. Additionally, it informs the other treatment staff of the progress toward the treatment goals, serves as a backup to the bill generated for a 45-minute individual psychotherapy session, and provides very acceptable documentation for legal and ethical questions that may arise about the psychologist's competence during that session. Most hospitals and medical settings require problem-oriented and "SOAP" charting of sessions. The problem-oriented progress system refers each note by title to one or more of the problems listed on the treatment plan in the patient's chart (e.g., "aggressive verbalization toward authority" and "denial of substance abuse—cocaine"), along with the date and the session number at the beginning of the note. Each progress note is further outlined according to the SOAP categories: Subjective—close paraphrase

```
_____   Hypnotherapy                        _____   Geriatric Psychotherapy
_____   Biofeedback                         _____   Mental Retardation
_____   Pain Management                     _____   Other, Specify:
_____   E.R. / Crisis Intervention
_____   Substance Abuse Tx.: Acute___, Extended___

IV. Patient Management Privileges

_____   Admit Patients
_____   Co-Admit Patient with a Credentialed Physician
_____   Discharge Patients
_____   Coordinate Psychotherapy with Hospital Care
_____   Write and Sign Treatment Formal Plans
_____   Write Orders for Assessment and Treatment Procedures
_____   Write Orders for Medication
_____   Write Orders for Consultation and Other Professional
        Services
_____   Supervise Staff and Trainees
_____   Write Consultation Notes in Patient Records
_____   Treatment Program Planning and Evaluation
_____   Staff Development and Training
_____   Other, Specify:

V.  Clinical Testing (T) and Assessment (A) Privileges
    Please Specify

_____   Behavioral                  _____   Vocational/Educational
_____   Psychophysiological         _____   School Psychological
_____   Neuropsychological          _____   Psychosocial
_____   Mental Status               _____   Support System
_____   Intellectual/Cognitive      _____   Organizational/Management
_____   Personality                 _____   Substance Abuse
_____   Forensic                    _____   Other, Specify:

VI.  Scientific and Research Privileges

_____   Design Scientific and Clinical Research
_____   Direct the Conduct of Research
_____   Administer Research Protocol to patients


I,_____, attest to the accuracy of   this
list of privileges indicating my training and competency to
practice these privileges. _____, Date:_____
                                    Signature
```

Figure 2. (*Continued*)

and exact quote of statements the patient made related to the treatment problems; *Objective*—behavioral, physiological, and measurable data related to that session; *Assessment*—minimal deductive inference interpretations of what the client said and what the psychologist observed; and *Plan*—the methodology for testing hypotheses generated in listening and observing the patient and/or helping the client to progress in his or her psychotherapy (Hall, 1988; Soisson *et al.*, 1987).

Some health care systems have developed a system of using progress notes to generate *billing* and to measure the *productivity* of the psychologist. However, most hospitals and most private-practice psychologists must also generate a bill for the patient and/or the third-party payer. There are very nice computer or self-carboning-paper business accounting systems for such billing and accounting. As mentioned above, the practice and business authorities of the hospital need documentation that the psychologist is using his or her time efficiently and is practicing competently. In private practice, the open market takes care of the productivity aspects and some aspects of quality evaluation. That is, the psychologist who is slow or imprecise in documenting bills is paid less per hour, and the psychologist who bills for an unrealistic rate of service and number of services will be audited by any third-party payer. Third-party payers are experimenting with various means of documenting quality.

In a hospital practice, the business-wise hospital managers will want to know what the psychologist's services cost in comparison to what the hospital receives for these services during the salaried or hourly contract with the hospital. Obviously, the psychologist needs to document that she or he is performing a reasonable amount of high-value services. Most psychologists, therefore, document services in a log sheet designed by the hospital's management information system (MIS) department and in their own professional appointment calendar. This is very similar to the documentation done by lawyers in billing for their time by type and rate of service.

The fact of the service is documented in the progress note; the volume and reimbursement rate is documented in the log and the billing system. Some of the quality of the service can be measured by productivity rates, but *quality assurance* concerning the services delivered to patients requires more structure and documentation. The state government, the citizens of the state, and third-party payers all expect productivity and service of good quality. This fact has given rise to numerous attempts to measure quality in the hospital and other health service sectors. For the psychologist practicing in the hospital and other health care settings, the following are of significant importance in surveying, sampling, and correcting faults in the quality of psychological services: the rules, regulations, and guidelines of the American Psychological Association, the National Institute of Mental Health, and the appropriate sections of the Joint Commission on the Accreditation of Hospitals. Most practitioners admit anxiety and annoyance with having their practice monitored by quality assurance programs. Psychologists have a decided advantage with such programs if they choose to use it. Quality assurance monitoring, documentation, and interpretation are very close to "process and outcome evaluation" in the psychotherapy treatment outcome literature. The social science methodology skills that make good psychologists are very close to the skills needed in working with a good quality assurance program. Operationally defining the variables from existing or easily obtainable data that distinguish good from unacceptable treatment process and treatment outcome is the object of quality assurance. Sampling technique, systems theory, solid logic, and basic statistics can be valuable in a hospital's quality assurance program for both mental health services and other health services (for psychology's role in quality assurance see Chapter 7).

Other Structures and Authorities in the Hospital: Formal, Cultural, and Informal

Successful professional travels through the world of hospital practice will intersect with other paths of structure and authority.

Unions

Even before hospitals and medical settings became increasingly operated on the "big business" model, there were certain trades and professions that were organized into labor, trade, and professional unions. And after Congress passed Public Law 93-360 in 1974, removing the nonprofit hospital's exemption from the Taft–Hartley Act, the union movement grew rapidly in hospitals. Unions are important to a psychologist practicing in a hospital in many ways. Trade unions and union contracts often determine the quality of the psychologist's physical environment. Installing new phone lines for voice or computer modem, soundproofing an office, rewiring an office for biofeedback or supervisory videotaping purposes, and so on bring the psychologist into contact with both the hospital administrators in charge of buildings and grounds and the trade unions. Learning the general rules under which these tradespersons work will greatly assist the psychologist and his or her patients in managing the stress of construction or repairs (Snook, 1981).

In some public and a few private hospitals,

there are professional and support staff unions. The receptionists, clerks, master's level therapists, nurses, and pharmacists may be members of a union or unions with specific job descriptions and the right to negotiate the routine work load of the hospital. The psychologist who takes the time to understand union members' view of the hospital will have better results and successes in dealing with them. In a few hospitals non-management-level psychologists are themselves in a union, either with physicians or with other doctoral-level staff. This membership has the potential for further informing the public and hospital management of the effectiveness and value of psychological services through interaction with fellow union members and through union public relations efforts.

Other Ph.D. and Master's Level Professionals

As a Ph.D., the psychologist may be grouped with the other Ph.D.'s for purposes of governing the hospital, as discussed above, through the organized professional staff. Often, the psychologist's clinical and scientific research interests put him or her in contact with these other Ph.D.'s in the hospital. For example, the emerging interest in psychoneuroimmunology may put a psychologist in touch with immunologists and biochemists in the laboratories of the hospital. For political and governing reasons, and for clinical and research reasons, it is wise for a psychologist to become acquainted with the organizational structure of the laboratories, audiology, speech pathology, and other areas of the hospital where Ph.D.'s practice and conduct research. Sometimes, psychologists or other Ph.D.'s act as representatives on committees or as scientific staff representatives on the executive committee of the organized professional staff. Knowing some of each other's professional skills and needs greatly benefits psychologists and the other Ph.D.'s.

Many master's level professionals work in the modern hospital: licensed social workers, physical therapists, occupational therapists, and others in the treatment regions of the hospital, as well as masters in business administration, public health, marketing, and so on in the business regions of the hospital.

On the treatment side, it is wise to know each other's skills in order to maximize good patient care and to minimize interprofessional conflict. Some of these master's level professionals and the profession of psychology have had a varied history of collaborative and conflictual relationships in both research and practice areas. A wise psychologist learns the specific history of professional interaction in the hospital in which she or he intends to practice. The guidelines emerging from the Joint Commission on Interprofessional Affairs (Bule, 1989a) will be helpful in the interactions of psychology, psychiatry, social work, and master's level nursing.

The psychologist will encounter the pastoral counselor or chaplain in most hospital settings, and also in the business and governance of church- or religious-affiliated hospitals. In hospitals owned and administered by explicitly religious groups, the psychologist is well advised to become very familiar with the purpose or mission statement of the hospital in order to negotiate any potential interprofessional "philosophical" conflicts before settling into a full practice in that hospital. A good person with whom to discuss this subject is another psychologist who has been practicing there for years, either at a supervising level or with a vigorous private practice.

Religious treatment personnel include persons with titles such as *chaplain* and *pastoral counselor*. Most of these religious treatment professionals have at least a master's degree in divinity and pastoral counseling, considerable community and hospital experience, and possibly a good deal of clinical skill as well. Most hospitals, not just the religious affiliates, have such professionals privileged to practice in the hospital. Their training and organizational affiliation are important on the psychologist's hospital road map, as they sometimes provide a quick understanding of the clinical philosophies of these personnel. Division 36 of the American Psychological Association ("Psychologists Interested in Religious Issues") can be of assistance in helping a psychologist interact with religious professionals.

Hospital Business and Marketing Functions

On the business side, many psychologists fail to notice that, by presenting their skills in a scientifically defensible way and by describing their referral pattern needs relative to the hospital's service mission and marketing plan, they can both build their practice and improve the service that their clients receive. However, because certain psychologists have overstated the efficacy of their science and/or have ridden a pop psychology wave for economic gain, psychologists need to use ethical professional judgment in such communications. Psychologists often avoid the business areas of the hospital because business personnel may be seen by psychologists and other treatment staff as misattributional stereotypes such as "inhumane capitalists" or "heartless Scrooges," often because of the business professional's focus on economic growth and cost efficiency. If the psychologist knows the diagnostic and treatment efficacy of the science of psychology as applied to modern health care and community preventive medicine, the psychologist's practice and the health of the community served by the hospital can be enhanced through working with business and marketing professionals in a mutually scientific and ethical manner. Snook's extended definition (1981) of hospital marketing is worth considering:

> Marketing is a total system of interacting management activities that are designed to plan, price, promote and distribute need- and want-satisfying services to a hospital's present and potential patients. . . . According to another definition, "managing or planning simple exchange relationships" is the essence of marketing. . . . The purchase behavior can be viewed as simple exchange of resources, that is, a certain individual gives another individual something in order to receive in exchange a privilege, a good or a service. (p. 217)

Marketing should also be looked at by psychologists in terms of the struggle between hospitals for a share of the patient market. Historically, competition between hospitals has always existed, and marketing has always played some role in it. However, in the current fiercely competitive atmosphere, this struggle to increase the share of the market and to gain more patients has intensified for both private for-profit hospitals and not-for-profit hospitals. In many cases, losing patients, and thus losing some of the share of the market, has become more critical than ever, particularly as a greater number of hospitals are faced with the threat of closing. Marketing efforts have become increasingly important to the survival of many hospitals. Sometimes, the marketing people make legitimate claims about the hospital's desirability, value, efficiency, or specialized programs (e.g., for substance abuse), and sometimes, these claims are questionable.

Professional staff, including psychologists, have always been looked at partly in terms of whether a staff member's presence in a hospital will enhance the hospital's share of the patient market. This tendency has increased in the current, fiercely competitive atmosphere. Nevertheless, from the marketing perspective, psychologists who can demonstrate that they have the right product or service at the right place and price, which can be promoted in a scientifically supportable way, can gain entry into the modern hospital system. The marketing executive will confront the psychologist's understanding of the true efficacy and worth of the research and practice of the science of psychology: Is what psychology studies valid enough clinically to be worth something to the health of the public? If so, tell the public exactly how, and bring them in for psychological help and services. If the psychologist does not think that what he or she does helps and serves people, the marketing executive will certainly suspect that psychologist's ethical principles.

Support Staff Structure and Authority

There are many other systems of structure and authority in the modern hospital. Some of these systems are discussed in following chapters dealing with specific research, diagnostic, and treatment skills in the practice of modern psychology.

The systems of structure and authority that affect a psychologist's day-in and day-out practice in a hospital are worth mentioning here. They are the same as those encountered in any bureaucratically run institution.

The first system of authority to be mentioned may seem relatively inconsequential, it is a practical archetype of the several others in the hospital setting. Almost every hospital has some person who assigns parking privileges. If one is a full member of the organized professional staff and is polite, responsible about the rules, and so on, then and only then will one get a parking space near one's office. This seems simple, but in many medical centers with thousands of staff, patients, patients' families and friends, salespersons, and so on vying for a few parking spaces, the staff member's comfort and safety depend on the person who assigns these parking spaces. Going through departmental and organized-professional-staff channels of authority to settle difficulties here is highly advised.

The scenarios are similar with hospital security, transportation, dictation pools, petty cash reimbursement, tradespersons, housekeeping, and so on. A psychologist's life can be very pleasant or very miserable, depending on how well she or he understands and follows the structure and authority of these support services. Frankly, disrespecting these hospital team members' authority smacks of prejudice toward working-class persons. Sublimating one's frustration with the inevitable slowness of bureaucracy in order to turn "bitching into bettering" is very wise. Psychologists have the advantage of self-supervising themselves by using the concepts of anger management, learned helplessness, empathy, negotiating skills, and self-efficacy in dealing with these authorities. As in the parking-space example, an understanding of the contingencies placed on a person or system is vital to minimizing hassles and smoothing out routines.

Summary and Conclusion

The modern hospital is a very potent arena for psychological practice. Private practice has many positive features, but many find it a lonely occupation. Hospitals offer the oppor-tunity of collegial group interaction, consultation, and continuing education.

Many psychologists rule out hospital practice to avoid a setting where the interprofessional and socioeconomic politics are likely to be complex. However, it is becoming increasingly clear that this avoidance is not working, largely because all health care is becoming managed health care. The individual psychologist's practice and the professional and research value of psychology as a whole are necessarily connected to the centerpiece of the health care system, the modern hospital and clinic. Not being involved in the "politics" of the hospital and the health care system is becoming functionally equivalent to concluding that psychological research and practice are scientifically worthless and clinically useless. Ethical psychologists advocate for the proven worth of their profession, or they resign from research and practice.

In the hospital and health clinic setting, clinical skills are used, sharpened, and kept up to date. The research potential is great. The potential for the legitimate exchange effective treatment for a professional fee or salary is also great. This all starts with an understanding of the hospital's structural and authoritative terrain. Failure to interact with the complex map of the hospital terrain leaves a psychologist practicing in an isolated region of the hospital, and eventually losing the property rights needed to practice or engage in research in the hospital at all.

References

American College of Hospital Administrators. (1973). *Principles of appointment and tenure of chief executive officers.* Chicago: Author.

American Hospital Association. (1988). *Hospital Statistics,* Evanston, IL: Author.

American Psychological Association. (1985). *A hospital practice primer for psychologists.* Washington DC: Author.

American Psychological Association. (1988). *Hospital Practice Advocacy Issues.* Washington DC: Author.

Berg, M. R. (1984). Teaching psychological testing to psychiatric residents. *Professional Psychology: Research and Practice, 15*(3), 343–352.

Bersoff, D. N. (1983, November). Hospital privileges and the antitrust laws. *American Psychologist, 38*(11), 1238–1242.

Bule, J. (1988a, December). Brief attacks decision in Calif. hospital case. *American Psychological Association Monitor, 19*(12), 22.

Bule, J. (1988b, December). Enlist Marketing Director to Build Hosp. Referrals. *American Psychological Association Monitor, 19*(12), 23.

Bule, J. (1988c, November). Hospital privileges survey finds barriers still exist. *American Psychological Association Monitor, 19*(11), 17.

Bule, J. (1988d, December). Privileges Bill on Hold in New York Until 1989. *American Psychological Association Monitor, 19*(12), 23.

Bule, J. (1988e, December). Privileges Issues Divides Psychiatrists. *American Psychological Association Monitor, 19*(12),.

Bule, J. (1989a, January). Joint Commission on Interprofessional Affairs: Cutting conflict to nurture cooperation. *American Psychological Association Monitor, 20*(1), 20.

Bule, J. (1989b, January). Senate panel adopts psychiatrists' terms. *American Psychological Association Monitor, 20* (1), 18.

Cattell, R. B. (1983, July). Let's end the duel. *American Psychologist, 38*(7), 769–776.

Cook County Hospital, Bylaws Committee of the Executive Medical Staff (1984). *Cook County Hospital Medical Staff Bylaws,* adopted by the Cook County Board of Commissioners.

Dorken, H., Webb, J. T., & Zaro, J. S. (1982, December). Hospital practice of psychology resurveyed: 1980. *Professional Psychology, 13*(6), 814–829.

Enright, M. F. (1987). *Proposed guidelines for evaluating psychologists' application for hospital practice approved in principal by Division 42, Psychologists in Independent Practice* P.O. Box 1483, Jackson, WY 83001, March 2.

Hall, J. E. (1988, June). Records for psychologists. *National Register of Health Service Providers in Psychology, Register Report, 14*(3), 1, 4.

Matarazzo, J. D., Lubin, B., & Nathan, R. G. (1988, January). Psychologist's membership on the medical staffs of university teaching hospitals. *American Psychologist, 33*(1), 23–29.

Presser, N. R., & Pfrost, K. S. (1985). A format for individual psychotherapy session notes. *Professional Psychology: Research and Practice, 16*(1), 11–16.

Snook, I. D., Jr. (1981). *Hospitals, what they are and how they work.* Rockville, NY: Aspen Systems Corporation.

Soisson, E. L., VanderCreek, L., & Knapp, S. (1987). Thorough record keeping: A good defense in a litigious era. *Professional Psychology: Research and Practice, 18*(5), 498–502.

Stachnik, T. J. (1980, January). Priorities for psychology in medical education and health care delivery. *American Psychologist, 35*(1), 8–15.

Tanney, F. (1983, November). Hospital privileges for psychologists: A legislative model. *American Psychologist, 38*(11), 1232–1237.

PART II

Professional Issues

As a health care profession, clinical psychology has always had a major emphasis on professional issues. These issues span a wide range of topics and professional behavior, including the establishment of a professional identity through training programs and credentialing, the ethics and competence of practice, the interaction of individuals with each other and other professionals, and, ultimately, the interaction of organized clinical psychology with other organized disciplines. At this point in the history of the development of professional psychology, and of the evolution of the health care system, these issues are of tantamount importance to psychologists practicing in medical settings. Therefore, they are placed before the chapters on practical issues and program development. Edward P. Sheridan and James P. Choca (Chapter 4) delineate the salient issues in predoctoral, internship, and postdoctoral training. Ronald H. Rozensky (Chapter 5) presents macropolitical issues such as standards of practice, third-party reimbursement, the politics of health care economics, hospital bylaws, and hospital privileging, and micropolitical issues such as how to seek practice privileges within a hospital. Concluding the section, Cynthia D. Belar (Chapter 6) discusses important aspects of appropriate professional behavior within a medical setting.

CHAPTER 4

Educational Preparation and Clinical Training within a Medical Setting

Edward P. Sheridan and James P. Choca

Introduction

The diversity and depth of knowledge required to educate a professionally competent clinical psychologist have expanded significantly since the early 1980s. Although the Boulder model (Raimy, 1950) of the scientist-practitioner is still favored in many programs (Perry, 1979, 1983), there is disagreement on how much emphasis to place on training in science or practice (Peterson, 1987). New models, such as the scientist-professional or scholar-professional, enjoy popularity and compete significantly with the traditional model. The reality that the majority of clinical psychologists will spend most of their lives as practitioners sometimes conflicts with the aspirations of academicians to educate persons who will devote their careers to research. It is no small task to weave an educational experience that recognizes these

diverse forces and purposes into an integration of the current richness found in psychology. Some of this difficult integration was attempted at the National Conference on Graduate Education in Psychology (Bickman & Ellis, 1989). A more sophisticated statement was developed at the 1990 National Conference on Scientist-Practitioner Education and Training for the Professional Practice of Psychology. This chapter focuses on one piece of this education: what clinical training possibilities exist in a hospital setting and how such training helps fulfill psychology's larger goal, the education of scientifically minded clinicians.

Regardless of the model or the philosophical position that one takes toward the training of clinical psychologists, hospital settings provide an ideal base for education in many forms of practice. There are few educational experiences that create the sense of vitality and excitement that one finds in such an environment. The extremes and varieties of real-life suffering, both psychological and physical, are everyday occurrences. Whether conducting research or performing clinical services, individuals realize that their work is of great importance. This setting provides the inspiration to

Edward P. Sheridan • Department of Psychology, University of Central Florida, Orlando, Florida 32816-0340. James P. Choca • Department of Psychiatry and Behavioral Sciences, Northwestern University Medical School, and Psychology Service, Lakeside Veterans Administration Hospital, Chicago, Illinois 60611.

intervene, and to assist the most troubled of persons. At the same time, only the most insensitive are not humbled by the enormity of some of the tasks presented, coupled with the limited knowledge and skill we have in ameliorating psychological and physical pain.

On entering training in a hospital setting, students are usually excited and bewildered. The intensity of activity, the importance of the work, and the complexity of the requirements can be extremely stimulating and, at the same time, overwhelming. It is the purpose of this chapter to suggest a path that may help an individual move through the various stages of training from being a novice to being prepared for independent practice. This sequence is divided into three traditional sections: preinternship or practica, internship or residency, and postdoctoral training. Within these sections, some themes (e.g., rotations and supervision) are repeated to emphasize a specific intensity of training. Attention is given to the recent National Conference on Internship Training in Psychology (Belar, Bieliauskas, Larsen, Mensh, Poey, & Roehlke, 1987), which proposed a two-year internship: one year predoctoral and one postdoctoral. In addition, because this emerging model suggests that internship and postdoctoral education should occur after the completion of a doctoral dissertation, special focus is given to the possibility of increased practicum training during the preinternship years.

The proposal outlined in these pages attempts to anticipate the training demands of the 21st century (Sheridan, 1987). Both the National Conference on Graduate Education (Bickman & Ellis, 1989) and the National Conference on Internship Training in Psychology (Belar *et al.*, 1987) attempted to review the vast changes that have occurred in psychology in the past 30–40 years. They proposed that the knowledge base of the profession now requires a more complex education before an individual emerges as a journeyman. This same position was taken earlier, at the National Conference on Education and Training in Health Psychology (Stone, 1983).

The following presentation is not wedded to a specific theory of personality, psychopathology, or psychotherapy. Rather, it intends to acknowledge that expertise in clinical psychology demands an understanding of several theoretical approaches, as well as assessment and intervention strategies. Thus, the chapter proposes that training in service delivery strategies be tied as closely to research results as possible, and that students be taught the theoretical and scientific assets and limitations of their endeavors.

Preinternship

The 1947 Shakow Committee Report (American Psychological Association, 1947) and the Boulder Conference of 1949 (Raimy, 1950) introduced the need for required clinical practicums. The Boulder Conference specifically established an internship requirement and strongly recommended the use of preinternship practicums (Raimy, 1950). Since then, the consensus of professionals in the field has been that, in psychological specialties that require an internship, the student should be exposed to a certain amount of preparatory practicum training before the internship (Hoch, Ross, & Winder, 1965; Roe, Gustad, Moore, Ross, & Skodak, 1958).

The function of preinternship training in a medical center is to introduce psychology students to the work of professionals. At this stage in the development of psychologists, there is frequently a gap between students' knowledge and the practical skills they need to see clients (Autor & Zide, 1974; Choca, 1988). Often, students arrive at training centers with few, if any, supervised clinical experiences; their ideas about how psychologists function may be vague, inaccurate, or even distorted. Typically apprehensive about the work he or she is about to do, the average beginner is capable of doing little more for the patient than a layperson with some common sense. The goal of training at this level is to develop the initial skills that students will need for the internship or residency.

There are no convincing data to guide us in determining what preliminary knowledge or experiences a trainee needs before going into an internship or residency (Malouf, Haas, & Farah, 1983). However, specific theoretical objectives are available. The Boulder Conference offered the following:

1. Developing a feeling of responsibility for the client and a sensitivity to the clinician-client relationship.
2. Developing of minimal competence in the use of psychological techniques in the clinical setting.
3. Familiarization of students with a wider range of techniques.
4. Teaching of the nature and meaning of service.
5. Beginning of integration of university course content with the clinical viewpoint and with procedures in a service setting.
6. Introducing the interdisciplinary approach to clinical problems; learning to cooperate with colleagues of other disciplines.
7. Applying professional ethics.
8. Learning to communicate by the writing of case reports for clinical use.
9. Providing a wide range of clinical contacts at a relatively superficial level through a variety of clerkships. (Raimy, 1950, p. 105)

For the present discussion, the objectives of the practica are grouped into two areas. Under professional training, we examine what progress can be expected in the way that trainees relate to clients or other professionals. Under the rubric of specific skills, we discuss the capacities that the typical internship site would expect entering trainees to have developed. Before addressing those objectives in greater depth, however, it is necessary to consider the requirements that the student can be expected to have fulfilled before arriving at the practicum training site.

Responsibilities of the Academic Institution

Perhaps the first responsibility of the graduate or professional school regarding the practicum training of its students is attitudinal, that is, seeing clinical training as an important and

necessary contribution to students' development. Professionals in the field have repeatedly expressed dissatisfaction with the degree of cooperation that is prevalent between academia and the training facilities (Roe *et al.*, 1958; Strother, 1956; Zolik, Bogat, & Jason, 1983). The Miami Conference report offers the concern that academic centers may not view the practicum experience as "broadening the student's horizons and drawing his attention to promising new areas of research" but as taking the student away from "academic work and interfering with . . . dissertation research" (Roe *et al.*, 1958, p. 59). Needless to say, it is most useful for each setting to see the other as having a complementary and equally important function in students' development.

The delineation of the responsibilities for training of the academic and the practicum setting has been left notably ambiguous by the existing guidelines (e.g., Raimy, 1950; Roe *et al.*, 1958). Medical centers accepting trainees at a preinternship level must be well aware that these trainees are not "finished clinicians" and are limited in their capacity to render psychological services (Raimy, 1950, p. 115). Importantly, even when trainees are not able to make a great contribution to the evaluation or treatment of a client, they should be prepared enough so that they "do no harm."

In the typical medical center, a student is exposed to clients almost from the start. As a result, the university or professional school must share some responsibility for "certifying" that the student being sent to a training site is ready to accept that experience responsibly (Raimy, 1950, p. 115). This duty involves an assessment both of the student's personal and emotional capacity and of the student's training in certain skills.

Academic centers training clinical psychologists should not assume that the mere fact that the student has fulfilled academic requirements qualifies the person to become a trainee at a medical center. This does not imply that the academic center is responsible for thoroughly evaluating the emotional stability of its students in the way that one would a client. In the course of the first year of graduate studies, however, the academic faculty often discover that a student is having significant problems. It is their duty in those cases to assess how impaired the student is and what action is appropriate.

In instances where there is significant psychopathology that would interfere with the individual's ability to function, the faculty must have the courage to counsel the student to seek help. That student should never be certified as ready for a practicum experience. In cases where emotional problems are judged to be less severe, it may be feasible to have the student enter a training site, but only when appropriate steps have been taken. Such steps may include apprising the training director at the medical center of whatever concerns there have been and ascertaining that the setting is one that can monitor and address the student's problem areas and difficulties as they arise. It is most helpful, especially in these cases, to emphasize the need for supervisors and the faculty to work in a well-coordinated manner to help the trainee overcome emotional problems as the training proceeds. This need cannot be overstressed, especially in light of the "well-known massive gaps in communication" between training facilities and academic training programs (Zolick *et al.*, 1983).

Regardless of their level of training and professional sophistication, trainees must behave ethically and morally the moment they enter the medical center. Malouf *et al.* (1983) observed that, if a trainee violates the client's right to confidentiality, a later correction by the supervisor cannot repair the client–therapist relationship or undo the harm done to the client. The trainee's behavior must also be in compliance with the law, particularly with regard to patients' rights.

Before assigning clinical tasks, medical center supervisors often do not take the time to review the specific education in ethics that trainees have received. Although almost 80% of the internship programs that Newmark and Hutchins (1981) surveyed claimed to offer some training in the area of ethics, only 45% of the responders provided a formal, systematic, and comprehensive learning experience. All but

two of the programs that did not offer training in ethical standards were of the opinion that this was the responsibility of the professional or graduate school. Consequently, it appears that practicum directors assume that the academic setting has provided an education in ethics before clinical training is begun. The requirement of teaching professional ethics in graduate or professional schools has been well recognized (e.g., Raimy, 1950; Roe *et al.*, 1958).

Psychology trainees must come into the clinical setting with an academic background that is sufficient to allow them to benefit from the practicum experience. Centers expect students to have some knowledge of the main therapeutic modalities and the theories behind these modalities (Raimy, 1950). In addition, students are assumed to have the ability to administer and score the better known psychological tests or inventories (Raimy, 1950) and to conduct an initial diagnostic interview.

Finally, the graduate or professional program should make an effort to help students choose the practicum centers that best fit their needs. Variables such as the quality of training offered, the skills that may be taught, the way in which they are taught, and the kind of patient population that the center serves are all important considerations. The recommendations to the student should be based on what serves the student's interest best rather than other "expediencies," such as "politics" or the presence of "remunerative consultantships for university faculty" (Roe *et al.*, 1958, p. 58). The recommendations should also follow a comprehensive plan, designed to give students breadth of experience rather than specialization this early in their career.

Training in Professionalism

An important objective of practicum-level training is developing a professional attitude in the student. Many beginners feel uncomfortable with patients. This discomfort may surface as anxiety or nervousness, or it may be defended against in a number of ways, such as a tendency to under- or overestimate the emotional problems that clients present. The predilection of beginners to becoming "patient advocates" may make them less effective and may interfere with their ability to consult with other professionals (Gabinet & Schubert, 1981). Thus, one important function of the practicum experience is to have trainees work with enough patients so that they come to feel comfortable with various expressions of psychopathology.

Beginners typically need to develop their own identity as professionals. To complicate this task, for years to come they will be functioning in a gray area in that they are given professional responsibilities but do not have the full authority of a seasoned professional. They need to become comfortable with developing their own ideas or treatment plans while accepting the guidance of a supervisor. They must also develop a way of interacting with professionals of other disciplines that is neither conflictual, disdainful, nor submissively compliant. As Gabinet and Schubert (1981) suggested, until students develop confidence about what they, as professionals, can contribute to patients' care, they will not be able to participate effectively as caregivers.

Trainees must leave behind the "student syndrome," that is, the mental set that the practicum experience is a school requirement following scholastic traditions. They need to accept, for instance, that the patient remains in the clinical setting even after school has closed for the holidays or for the summer. This is not to say that a trainee does not have a right to a vacation, but that the responsibility must be accepted for arranging coverage at times when one is not available. By the time students move on to the internship, the development of their image of themselves as psychologists should be sufficient so that they can project a professional appearance, even if their self-confidence or their security in the newly found role needs further development.

Skills Training

To help with the task of identifying and teaching the beginner the functions that psychologists perform, the second author has written a manual that presents the information needed simply and comprehensively (Choca, 1988). The topics covered include psychopathology, interviewing, assessment, psychological testing, interventions, ethics, and the hospital system.

In the area of interviewing, preinternship-level training may take a student from talking to the patient as an interested and reasonable layperson might to being able to conduct a fairly professional inquiry. In the process, the student must learn the important indicators of psychopathology and must develop the ability to elicit the needed information unobtrusively and effectively. The talkativeness and personal disclosures that may have characterized the style of relating at the beginning (Johnson, 1981) should be all but extinguished, and the trainee should become comfortable asking questions about delicate or personal areas when they are warranted.

The literature available regarding practicum training (e.g., Zolik et al., 1983) suggests that training in diagnostic techniques tends to be stressed during the preinternship period. By the end of the practicum, psychologists in training should be able to administer and accurately score any of the popular psychological tests without constantly referring to the test manual. The student must have gained some experience in interpreting results and writing testing reports, even if the reports still lack the maturity and sophistication that may be expected from a psychologist at a more advanced level of development.

Another objective of preinternship-level training is the development of the student's ability to generate reasonable treatment plans for a client after an evaluation. To accomplish this task, the trainee must have developed an understanding of appropriate treatment goals and what interventions lead to these goals.

Perhaps the most difficult developmental objective of preinternship training involves designating what skills the student should have mastered as a psychotherapist. Modalities differ to such a vast degree, and the kinds of intervention that one may learn even within one modality are so varied, that it is almost impossible to design a system of general objectives, a fact that was noted in the Boulder Conference report (Raimy, 1950, p. 96). Nevertheless, the objectives developed by the Chicago Conference seem useful:

1. Develop therapeutic competence.
2. Help the student develop greater ability to enter a meaningful relationship with others.
3. Increase self-awareness, sensitivity, and understanding of themselves and others.
4. Develop the ability to recognize and conceptualize human problems. (Hoch *et al.*, 1965, p. 90)

Focusing exclusively on one therapeutic modality at this stage in the person's training was "disapproved" by the Boulder Conference (Raimy, 1950, p. 97).

Trainees should also be expected to have a good command of the of the revised third edition of the American Psychiatric Association's *Diagnostic and Statistical Manual of Mental Disorders* (DSM-III-R; 1987) and should be comfortable with the concept of differential diagnosis. Many beginners are unaccustomed to this way of thinking and may be particularly unprepared to understand the effect of medical problems (Gabinet & Schubert, 1981). In this training stage, expertise in making complex diagnoses begins.

Finally, by the end of pre-internship level practica, the trainee should have some ability to read and write in a hospital chart (Gabinet & Schubert, 1981). This skill involves developing enough familiarity to understand medical nomenclature and what information is contained in the different parts of charts.

Training Program Specifications

There are many ways of achieving the objectives detailed above. No "ideal" way of preparing trainees for their internship experience has been determined. Nevertheless, a number of recommendations regarding practicums seem reasonable.

For instance, because of the lack of sophistication of typical preinternship trainees, it is necessary to have them observe how other psychologists perform their duties. Programs that offer a milieu in which the student can observe how experienced practitioners handle different situations may be most beneficial.

An apprenticeship model, in which the trainee works very closely with the supervisor, meets both the student's needs and the moral and ethical requirement, so that the client is assured competent treatment. Students "should not be expected to render services beyond their level of training" (Raimy, 1950, p. 100). The supervisor, therefore, has a responsibility to both the student and the client (Malouf *et al.*, 1983). In a quality apprenticeship, care is taken to chose competent supervisors (Raimy, 1950), and the ratio of students to supervisors is monitored. Although no figures are available for practicum centers, research apprenticeships at universities have been found to have a modal ratio of about three or four students per professor (Clark & Moore, 1958), a number that would appear to be workable in a clinical setting if the students are working part time.

The training program must also invest a reasonable amount of time in supervision. The somewhat dated American Psychological Association (APA) guidelines emphasize that a center should be willing to offer a "learning experience" rather than use the student to perform a "service function" (Roe *et al.*, 1958, p. 58), and concern is voiced about the need to "guard" the student against "irresponsible and exploitative behaviors" on the part of the centers (p. 59). Although exploitation is totally unacceptable, our discipline must recognize that these are times of fiscal restraint for health centers, and that some facilities would not be able to offer training if the student did not provide services to their clients, a fact that has been made clear by articles on the cost efficiency of training programs (e.g., Loucks, Burstein, Schoenfeld, & Stedman, 1980; Rosenberg, Bernstein, & Murray, 1985). Although the ratio of supervisory time to the amount of time that the student spends in face-to-face contact with a client may be arbitrary and should remain flexible, the guideline of one hour of supervision for each two to four hours of care delivered seems reasonable. Especially with a beginner, the amount of supervision available needs to be flexible because, at times, the student will require a considerable amount of time in order to perform a task with which he or she is unfamiliar. Gysbers and Johnson (1965) found that, in the earlier stages of the practicum, trainees wanted more supervision and guidance than the supervisor thought appropriate. Although limits may have to be established for particular students, it is often best if some of the supervision is available on demand. Frequently, such supervision deals with the student's anxiety about or emotional reactions to providing services to the client, in addition to monitoring how the work is performed.

Finally, the ideal preinternship program includes a system of rotations. Especially in the area of health psychology, the different applications of our profession vary tremendously when applied to different clients. For instance, learning how to handle the emotional problems presented by patients who have intractable chronic lower back pain is very different from learning how to help clients facing a terminal illness, or from learning how to help smokers quit their habit. In order to begin developing a knowledge of the spectrum of problems that psychologists treat, students need to be exposed to more than one area of application. As already suggested, such rotations also prevent premature specialization.

Internship

Internship training has represented the capstone of clinical experience for over four decades. Psychology as a discipline has done much to highlight this training. A national organization, the Association of Psychology Internship Centers (APIC), was formed to develop guidelines for announcing internship opportunities and developing a uniform system for selection (Klepac & Reynes, 1989). The APA has been accrediting internships for over 30 years. Many doctoral programs in clinical psychology require their students to complete APA-approved internships. Currently, the Veterans Administration, one of the leading employers of psy-

chologists, will not hire persons who have not completed an APA-approved internship.

In spite of the central and pervasive role that internship training plays in the development of clinical psychologists, with rare exceptions such training occurs independently of the student's doctoral program. Thus, individuals wishing to enter clinical psychology must meet two standards, one set by their university (which includes completing an internship) and a second developed by the internship they choose. In the latter case, hospitals have become the premier training sites, offering myriad opportunities for the development of clinical expertise. And as the knowledge base of psychology expands, hospitals, coupled with expanded outpatient systems, will continue to be highly desired for internship training.

At the same time, a number of variables affecting training are changing. The internship in clinical psychology now tends to occur later in training than it did in the early 1980s. A number of factors have influenced this change. First, leading internship centers want students who are truly prepared for an advanced experience. Thus, some students are expanding their practicum training so that they will be better prepared to compete for these internship opportunities. In addition, as already noted, the National Conference on Internship Training (Belar *et al.*, 1987) proposed that internships should extend over two years, the second year being postdoctoral. The same conference proposed that internships should not begin until the doctoral dissertation is completed. Additional factors have influenced the decision to place the internship at the end of the doctoral requirement sequence. For students completing their internship but not having a Ph.D., finding employment is difficult. Students who have left their university and must return to complete dissertations often find this sequence arduous and unappealing. State laws require that individuals have a year of supervised postdoctoral experience before they are eligible for licensure. As a result, clinical training performed after the internship but without a Ph.D. degree does not count toward the experience required for licensure.

Structure

Quality clinical training demands an identifiable psychology service with a clear commitment to education. A specific psychologist must be responsible for the training. That person may chair a committee or may have assistants who serve as directors of practicum training, internship training, and/or postdoctoral education. The director of training must establish a philosophy of education that is communicated to all applicants to the internship program. This written statement informs applicants of the various clinical training opportunities or rotations (including which opportunities are mandatory and which are elective), the type of responsibility given to interns, the amount of clinical work and supervision required, and the salary and benefits associated with internship status.

Although neither the APA's Committee on Accreditation nor the Association of Psychology Internship Centers has mandated that internships have a number of rotations, two models seem to be the more common. One model permits students to treat a specified number of patients throughout their internship to develop familiarity with serious and complex psychopathologies; at the same time, interns may rotate for a period of three or four months to various sections within a hospital as an introduction to different clinical problems. A second model permits students to have two or more rotations that they continue throughout their internship year.

Within both of these models, other opportunities can greatly enhance clinical training. For example, it is imperative that interns receive some extensively supervised experience in doing two types of evaluations. First, interns need to increase their skill in examining persons with complex disorders presenting for initial evaluations. Such opportunities permit the intern to develop a key talent needed for independent professional practice. It is during this experience that interns often work with other professionals in developing comprehensive treatment plans or making referrrals to formulate complex diagnoses. The hospital provides

a marvelous opportunity for the intern to avoid the insulation of a training site where only pscyhologists work and diagnoses tend to be limited to their expertise.

The second type of training assessment occurs through the sophisticated use of psychological tests. Unlike in university training, and sometimes in practicum training, interns must be able to identify problems that require psychological testing and the test(s) to be used and must be prepared to examine the patient shortly after referral and to provide feedback quickly. The importance of the last skill cannot be overemphasized. Too often, interns (especially if they did not have practicum experience in a hospital) still have the idea that they are taking a course and that "papers" can be written at any time. Thus, students may control their anxieties by not writing "reports" until they are more comfortable. Hospitals simply do not permit such an inappropriate approach to diagnosis.

An additional experience that is mandated of all physicians in training and that is currently expected in many psychology internship programs is being on call. It is well known that teaching hospitals are run by the faculty and staff from approximately 7 A.M. to 7 P.M., Monday through Friday. At other times, hospitals are staffed primarily by interns and residents who are on call. These trainees are required to be present in the hospital and available for all consultations or new patients who arrive during evenings and weekends. This experience is both extremely challenging and anxiety-producing (Sheridan, 1981a; Zimet & Weissberg, 1979). The challenge makes it invaluable, and the anxiety can be controlled by the availability of supervisors. If psychologists are to maintain status and responsibility equal to those of other professionals within a hospital, performing call is a necessity. In addition, the result of call—that is, learning to make independent professional judgments in emergency situations—is a most appropriate internship goal. In two-year internship programs, first-year interns can be coupled with a second-year intern to learn some of the skills and responsibilities of call. When this arrangement is not possible, as in one-year internships, faculty must teach these critical skills. In all cases, it is imperative that a faculty member be available by phone to assure interns that they are not alone in difficult situations.

Rotations

The work in hospital psychology internships has traditionally centered on psychiatric patients. Although some internships continue to have this focus, a number of innovations have broadened training perspectives. For example, internships may exist in independent departments of psychology, in pediatrics, in a mental health center that is part of the hospital, or in a department of psychiatry. The patient focus may be as specific as children under 12, adults over the age of 18, or the developmental spectrum from infancy to old age. Rotations may include cardiology, gerontology, neuropsychology, oncology, chemical dependence, neurology, consultation and liaison, pediatrics, rehabilitation medicine, specialized units (e.g., the burn center, the child abuse clinic, the AIDS center, or the smoking-cessation clinic), or psychiatry. In psychiatry, the rotations may be to programs for the seriously mentally ill, to crisis intervention, to work with inpatients, to partial hospitalization, or to special clinics (e.g., eating disorders, phobias, or intensive psychotherapy). This listing is only a modest suggestion of the training possibilities currently existing within many hospitals.

The major question is how a psychology staff develops a quality internship program that meets the needs of students and the hospital, while anticipating the future job market and the needs of society. This is a difficult and controversial task. However, some suggestions are offered here.

The first premise is that interns need some system of rotation. As mentioned earlier, rotations can either be for a few months or last throughout the internship. The important factor is that interns be exposed to a broad rather than a narrow range of supervised experiences. If the training is to be primarily within a mental illness model, the following experiences appear to be necessary: intake evaluations; diagnostic assessments involving the use of psychological tests; short-term crisis therapy; longer term therapy with seriously mentally ill persons; milieu, family, and group psychotherapies; and consultations with other health care providers.

When the internship is devoted primarily to developing expertise in clinical health psychology, there is both overlap and uniqueness in the required training. For example, intake interviews, crisis intervention skills, diagnostic evaluations using psychological tests, and consultation with other health care providers are essential. However, psychological testing would need to be expanded to include both neuropsychological expertise and familiarity with tests that are specifically sensitive to medical problems (e.g., headaches and lower back pain). Rotations to specific medical services are also necessary if the intern is to understand the unique manifestations of psychopathology in various physical disorders. Importantly, an internship cannot take place in one specific unit, such as oncology or spinal cord injury.

Within rotations, it is important that interns spend sufficient time to become a part of the service milieu. Whether the rotation is primarily inpatient or outpatient, there are many factors related to patient care that cannot be learned without regular contact with the entire treatment team. Thus, interns need to be an integral part of service staff meetings, diagnostic conferences, and rounds. In addition, their presence must be felt by other health care workers who will share information on patient contacts or unique cases that may present when the intern is not available.

Supervision

Quality clinical training has always been associated with excellent supervision. As the world of applied psychology becomes more complex, it becomes imperative that interns, like practicum students, be exposed to a range of supervisors. A system that meets this need can parallel the traditional systems for rotations. That is, on long-term cases, it may be helpful to have a supervisor who guides the intern through the entire process of a case. At the same time, as an intern moves from rotation to rotation, it is likely that the best education will occur through supervision from individuals with specific knowledge of the unique patient problems being presented within a given clinic. Ideally, one or two supervisors for long-term cases, separate supervisors for group and family therapy, and separate supervisors for rotations would help interns develop an appreciation of the complex role they will eventually fill as independent practitioners, hospital staff members, and/or faculty members.

Supervisors who are not psychologists also play a significant role. Although it is important to have psychologists available on a service as models for interns, supervision by talented persons in other professions can sensitize interns to new ways of approaching problems. This system is especially helpful in guaranteeing that interns have a culturally diverse set of supervisors. Such a model may be demanding when the number of staff members available for intern supervision are limited. However, it is imperative that, in the growing, sophisticated world of psychology, we educate our new members as thoroughly and knowledgeably as we possibly can.

Seminars

Internship faculty assume that a student arrives with a basic knowledge of personality theory, psychopathology, psychological testing, interviewing, and intervention techniques acquired at the university and in practica. To develop this learning to a sophisticated level of application demands both direct practice and further didactic learning. The National Conference on Internship Training in Psychology proposed that seminars on the following topics are necessary (Carrington & Stone, 1987): professional issues, including discussions of ethics and specific laws that influence practice; cross-cultural and ethnic differences; services to special populations; and research with a clinical application. A hospital must add seminars in psychodiagnosis and intervention strategies.

Special-topics seminars that enrich the offerings of a specific facility also greatly add to the intern's learning.

An important conflict may occur in an intern's life between the demands of clinical service and the need of further education. In developing responsibility for patient care, interns sometimes give lessened (or a very low) priority to didactic learning. They may schedule an extra psychotherapy appointment that conflicts with a seminar or may frequently be paged from didactic experiences. Such emphasis on patient care is inappropriate in that it fails to acknowledge that interns are in training and must devote the same quality of energy to education as they do to direct service. It is important that internship directors assist trainees in learning the necessary balance. This philosophy is particularly meaningful in hospitals where other disciplines may fail to encourage balance, placing direct service as the first priority in all circumstances.

Funding

In general, psychology has enjoyed modest success in funding internship positions. The range typically runs from salaries equal to those of medical house staff to unfunded positions. The former situation is appropriate but difficult to achieve in many facilities because of a tradition of lower funding for psychology interns. The latter situation was voted as unacceptable at the National Conference on Internship Training (Belar et al., 1987). In our current cost-conscious hospitals, psychologists have an important task in justifying trainee salaries and educating administrators regarding the economic value of trainees, especially interns and postdoctoral fellows (Loucks et al., 1980; Rosenberg et al., 1985). Data on the revenue generated by assessments and intervention need to be presented, accompanied by the economic advantage of having interns on call. In most hospitals, these figures impressively demonstrate the value of persons in training in psychology and highlight the fact that they are considerably underpaid.

A related issue not to be overlooked is how

psychology identifies its levels of training in a hospital, an issue particularly relevant to paying trainees. In medicine, the discipline with the largest number of trainees, externs are often paid little if any stipend. However, interns, residents, and postdoctoral fellows are always paid. It has been argued for some time (Matarazzo, 1965; Sheridan, 1981a,b) that psychology can appropriately use the title *resident* to identify its senior trainees, thus informing colleagues in other disciplines what level of expertise should be expected. Few psychologists in hospitals are unaware of the discrepancy between the more advanced training of psychology interns and the beginning qualifications of first-year psychiatry residents. By virtue of the established hierarchy in medical settings, we continue to promote a junior status for interns, although often, they could more appropriately be called residents. Although this latter title is generally appropriate for psychology interns, it is certainly a fact that *resident* is a far more appropriate term for any clinical or health psychologist in postdoctoral training (Sheridan, Matarazzo, Boll, Perry, Weiss, & Belar, 1988). This identification issue is clearly a prerogative of psychology as a discipline. In settings where trainee salaries are unacceptable, it may be important to combine the case for increased funding with titles that accurately inform hospital administrations what they can expect from our students.

Postdoctoral Training

Postdoctoral training has a long history in psychology, but it has received relatively little organized attention. Over 30 years ago, the Stanford Conference on Training for Psychological Contributions to Mental Health (Strother, 1956) discussed postdoctoral education. It suggested a predoctoral core curriculum, including practicum experience in diagnosis and therapy (but no internship) leading to a Ph.D. degree in clinical psychology. After the awarding of the degree, students would enter a two-year postdoctoral program from which they would emerge prepared for independent prac-

tice. A similar model has been proposed by Matarazzo (1965, 1987).

More than 20 years ago, Alexander (1965) summarized the status of postdoctoral training in clinical psychology. He particularly focused on the 1949 Boulder conference suggestion that, although doctoral-level clinical training was extensive, it was not intensive enough to produce independent practitioners at the time the Ph.D. degree was awarded. Thus, postdoctoral training was recommended for practitioners. Although Alexander's comments were important, they did not receive wide recognition.

The 1965 Chicago Conference on the Professional Preparation of Clinical Psychologists concluded that postdoctoral training was desirable for all clinicians and essential for those who anticipated a career as clinician-teachers or independent practitioners (Hoch *et al.*, 1965). As specific as this recommendation was, there was little evidence 20 years later that it had led to a significant change in clinical training.

Perhaps the strongest argument for mandated postdoctoral education came from the 1983 National Working Conference on Education and Training in Health Psychology (Stone, 1983). This conference proposed that there be a continuum for health psychology training from an organized Ph.D. curriculum (Boll, Thoresen, Adler, Hall, Millon, Moore, Olbrisch, Perry, Weiss, Woodring, & Wortman, 1983) through a formal one-year internship (Strickland, Follick, Altman, Cahn, Dingus, Kurz, Temoshek, & Trickett, 1983), culminating in a two-year postdoctoral residency (Matarazzo, Best, Belar, Clayman, Jansen, Jones, Russo, & Sheridan, 1983). Such a program in health psychology should offer an integrated, sequential experience that requires specific educational and clinical training at the predoctoral, internship, and postdoctoral levels. The conferees believed that the field of clinical health psychology had expanded so greatly that two years rather than one year of postdoctoral training was necessary to prepare an independent practitioner.

The National Conference on Internship Training (Belar *et al.*, 1987) developed a system that expands current expectations of clinical training and meets the standards of the licensing laws. Although this conference did not extend postdoctoral education to two years, the spirit of this conference emphasized the strong need to mandate more comprehensive clinical training for individuals currently emerging as independent practitioners of psychology.

The addition of a postdoctoral year to the current practice of one-year internships should not prove difficult in teaching hospitals. First, all branches of medicine have been doing this for many years. Psychology would simply be joining a long-standing system that recognizes that quality practitioners require extensive education and training. Second, as previously noted, in a highly cost-conscious health care environment, it is clear that both internship and postdoctoral trainees are excellent hospital investments. In some cases, convincing hospital administrators of this fact will require considerable work. Third, with the emergence of health maintenance organizations (HMOs), psychologists will need extensive interaction with internists, pediatricians, and other primary-health-care providers to develop a culture in which psychologists are seen as active and needed participants in the emerging cost-driven health-delivery-service models.

Postdoctoral Curriculum

Should the two-year internship program simply be an extension of the one-year internship, or should there be a significant difference between these two years? Strong opinions (Belar *et al.*, 1987; Beutler, 1981) favor a model in which first-year internship training is generic and postdoctoral training intensifies this generic education and/or provides specialization. In particular, the postdoctoral year or years should provide significant latitude for the individual trainee to choose experiences that are of high personal interest.

At the same time, it has been suggested (Sheridan *et al.*, 1988) that some structure exist during postdoctoral education. Valuable diagnostic and clinical skills that may have been learned either only at an apprentice level or not

at all may receive significant focus in postdoctoral education. In the area of assessment, neuropsychological testing and an increased familiarity with specific behavioral assessment techniques are prominent examples. In the area of intervention, developing further skills in hypnosis, biofeedback, relaxation therapies, group and family interventions, psychoanalysis, and specific-symptom-focused treatment strategies may all receive particular attention, depending on the expertise of the teaching hospital and the interests of the postdoctoral fellows. In addition, postdoctoral education should include skills in supervising younger members of the profession and in providing consultation to a variety of allied health workers. For those so inclined, this is also an appropriate time to develop initial skills in administration, particularly if a postdoctoral resident aspires to certain careers. For example, a person who intends to enter a community mental health system is very likely to have significant administrative responsibilities in the very near future, as may an individual who accepts a position in a community hospital or a small counseling center.

This discussion of training opportunities in a postdoctoral year deviates from the traditional apprentice model found in many hospitals. Because of grants and other specific funding mechanisms, the postdoctoral fellow frequently works with a single faculty member. Although this arrangement has worked well in research, it must be expanded to meet the needs of contemporary professional psychologists. Considerable time may be devoted to specializing with a specific supervisor but it is essential that postdoctoral programs not overlook the opportunity to complete the educational spectrum of the young professional.

Although the tasks already outlined are demanding, if psychology is to be true to its unique foundation, time must also be allotted to those new doctoral-level psychologists wishing to incorporate research into their careers. Psychology's proposed mission is to develop individuals not only knowledgeable in the most recent interventions and research, but also educated and motivated to extend the frontiers of new knowledge. Unfortunately, although significant effort is given to developing this latter talent during doctoral training, it is unfortunately lost in many cases during postdoctoral education. Although part of the reason is individual personal preferences for the role of practitioner, the role of researcher is also lost because of economic demands and the lack of research opportunities in many systems. Training directors need to come to grips with this perplexing problem not only because many psychologists are prepared to conduct research that will advance our science, but because psychologists are also prepared to be excellent collaborators with health service providers in all specialties who wish to examine the psychological, sociological, and systems variables in their treatment worlds, but who lack the background to propose and carry out such investigations. Because well-trained psychologists understand both the treatment and the research environments in hospitals, they are uniquely qualified to be expert collaborators. To leave research to nonclinical psychologists or to experts in other disciplines is to forfeit an opportunity to play a leading role in the future development of our health care system.

Summary

Hospitals present an exciting and special opportunity for the training of clinical psychologists. This education can span practicum, internship, and postdoctoral training. Although initial skills in administering and scoring psychological tests, forming a theoretical understanding of psychopathology and psychotherapy, and developing a beginning ability in interviewing should all be learned in university programs, hospital experiences can also lead psychologists through graded skills in evaluation, testing, psychotherapy, milieu interventions, and the acquiring of supervisory, consultative, and administrative skills. Such education should prepare individuals for the responsibilities of independent practice.

It is the joint responsibility of university and hospital faculties to develop a philosophy of education for clinical psychology that leads the field into the twenty-first century. Major conferences focused on graduate education (Bickman & Ellis, 1989), health psychology (Stone, 1983), clinical child psychology (Tuma, 1985), and internship training (Belar et al., 1987) have all attempted to address the complexity of this mission. With the emerging establishment of an Education Directorate within the American Psychological Association, it is time for leaders in universities and hospitals to join together to formulate creative training programs for the future of clinical psychology. Although arguments may abound about the conflict between the freedom needed for a creative university education and the responsibilities felt by hospital educators and licensing boards, it is time for fruitful dialogue. The chasm between the university education and the hospital education of the clinical psychologist must be bridged so that the fine education that can be produced only by a marriage of the two is more thoughtfully discussed and planned for the next generation.

References

Alexander, I. E. (1965). Postdoctoral training in clinical psychology. In B. B. Wolman (Ed.), *Handbook of clinical psychology* (pp. 1415–1426). New York: McGraw-Hill.

American Psychiatric Association. (1987). *Diagnostic and statistical manual of mental disorders.* (3rd ed., DSM-III-R). Washington, DC: Author.

American Psychological Association, Committee on Training in Clinical Psychology. (1947). Recommended graduate training program in clinical psychology. *American Psychologist, 2* 539–558.

Autor, S. B., & Zide, E. D. (1974). Master's level professional training in clinical psychology and community mental health. *Professional Psychology, 5,* 115–121.

Belar, C. D., Bieliauskas, L. A., Larsen, K. G., Mensh, I. N., Poey, K., & Roehlke, H. J. (Eds.). (1987). *Proceedings: National Conference on Internship Training in Psychology.* Baton Rouge, LA: Land and Land Printers.

Beutler, L. E. (1981). Should internships be extended? *APIC Newsletter, 6,* 12–14.

Bickman, L., & Ellis, H. C. (Eds.). (1989). *Preparing psychologists for the twenty-first century.* Hillsdale, NJ: Erlbaum.

Boll, T., Thoresen, C., Adler, N., Hall, J., Millon, T., Moore, D., Olbrisch, M. E., Perry, N., Weiss, L., Woodring, J., & Wortman, C. (1983). Working Group of Predoctoral Education/Doctoral Training. *Health Psychology, 2*(Suppl.), 123–130.

Carrington, C., & Stone, G. (1987). What is the core content of internship training? In C. C. Belar, L. A. Bieliauskas, K. G. Larsen, I. N. Mensh, K. Poey & H. J. Roehlke (Eds.), *Proceedings: National Conference on Internship Training in Psychology*. Baton Rouge, LA: Land and Land Printers.

Choca, J. (1988). *Manual for clinical psychology trainees* (2nd ed.). New York: Brunner/Mazel.

Clark, K. E., & Moore, B. V. (1958). Doctoral programs in psychology. *American Psychologist, 13*, 631–633.

Gabinet, L., & Schubert, D. S. (1981). Teaching hospital inpatient consultation-liaison to psychology trainees and interns. *Teaching of Psychology, 8*, 85–88.

Gysbers, N. G., & Johnson, J. (1965). Expectations of a practicum supervisor's role. *Counselor Education and Supervision, 4*, 68–74.

Hoch, E. L., Ross, A. O., & Winder, C. L. (1965). *Professional preparation of clinical psychologists*. Washington, DC: American Psychological Association.

Johnson, W. R. (1981). Basic interviewing skills. In C. E. Walker (Ed.), *Clinical practice of psychology: A guide for mental health professionals*. Elmsford, NY: Pergamon Press.

Klepac, R. K., & Reynes, R. L. (1989). *Directory: Internship programs in professional psychology*. Washington, DC: Association of Psychology Internship Centers.

Loucks, S., Burstein, A. G., Schoenfeld, L. S., & Stedman, J. M. (1980). The real cost of psychology intern services: Are they a good buy? *Professional Psychology, 11* 898–900.

Malouf, J. L., Haas, L. J., & Farah, M. J. (1983). Issues in the preparation of interns: Views of trainers and trainees. *Professional Psychology: Research and Practice, 14*, 624–631.

Matarazzo, J. D. (1965). A postdoctoral residency program in clinical psychology. *American Psychologist, 20*, 432–439.

Matarazzo, J. (1987). Postdoctoral education and training of service providers in health psychology. In G. C. Stone, S. M. Weiss, J. D. Matarazzo, N. E. Miller, J. Rodin, C. D. Belar, M. J. Follick, & J. E. Singer (Eds.), *Health psychology: A discipline and a profession* (pp. 371–388). Chicago: University of Chicago Press.

Matarazzo, J., Best, J. A., Belar, C., Clayman, D., Jansen, M., Jones, P., Russo, D., & Sheridan, E. (1983). Working group on postdoctoral training for the health psychology service provider. *Health Psychology, 2*(Suppl.), 141–145.

Newmark, C. S., & Hutchins, T. C. (1981). Survey of professional education in ethics in clinical psychology internship programs. *Journal of Clinical Psychology, 37*, 681–683.

Perry, N. J., Jr. (1979). Why clinical psychology does not need alternate training models. *American Psychologist, 34*, 603–611.

Perry, N. J., Jr. (1983). Majority position in favor of health service delivery by fully-trained psychologists. *Health Psychology, 2*(Suppl.), 115–116.

Peterson, D. (1987). Education for practice. *Clinical Psychologist, 40*, 7–9.

Raimy, V. C. (1950). *Training in clinical psychology*. New York: Prentice-Hall.

Roe, A., Gustad, J. W., Moore, B. V., Ross, S., & Skodak, M. (1958). *Graduate education in psychology*. Washington, DC: American Psychological Association.

Rosenberg, H., Bernstein, A. D., & Murray, L. (1985). Cost-efficiency of psychology internship programs: Another look at the monetary and non-monetary considerations. *Professional Psychology: Research and Practice, 16*, 17–21.

Sheridan, E. P. (1981a). Advantages of a clinical psychology residency program in a medical center. *Professional Psychology, 12*, 456–460.

Sheridan, E. P. (1981b). A resident by any other name . . . is short-changed: Clinical psychology training in medical schools and hospitals. *Clinical Psychologist, 34*, 5–6.

Sheridan, E. P. (Ed.). (1987). The twenty-first century: The challenge to educate clinical psychologists. *Clinical Psychologist, 40*, 3–14.

Sheridan, E. P., Matarazzo, J. D., Boll, T. J., Perry, N. W., Weiss, S. M., & Belar, C. D. (1988). Postdoctoral education and training for clinical service providers in health psychology. *Health Psychology, 7*, 1–17.

Stone, G. C. (Ed.). (1983). National Working Conference on Education and Training in Health Psychology. *Health Psychology, 2*(Suppl.).

Strickland, B. R. Follick, M., Altman, D., Cahn, J., Dingus, C. M., Kurz, R., Temoshok, L., & Trickett, E. (1983). Working group on apprenticeship. *Health Psychology, 2*(Suppl.), 131–134.

Strother, C. R. (Ed.). (1956). *Psychology and mental health*. Washington, DC: American Psychological Association.

Tuma, J. M. (1985). *Proceedings: Conference on Training Clinical Child Psychologists*. Washington, DC: American Psychological Association.

Zimet, C. N., & Weissberg, M. P. (1979). The emergency service: A setting for internship training. *Psychotherapy: Theory, Research and Practice, 16*, 334–336.

Zolik, E. S., Bogat, G. A., & Jason, L. A. (1983). Training of interns and practicum students at community mental health centers. *American Journal of Community Psychology, 11*, 673–686.

CHAPTER 5

Psychologists, Politics, and Hospitals

Ronald H. Rozensky

Introduction

Understanding the rapidly changing political machinations of professional psychology in its evolving clinical roles within hospitals is a little like trying to read the poker hand of a card player in the club car of a train that is speedily traveling west while you are sitting on the station platform. Unless you have very sharp eyes, and excellent timing and direct a lot of energy in the right direction, you will miss the whole picture.

The intertwining of the history of professional clinical psychology and medicine (see Chapter 2), the increasing number of psychologists working within medical settings and the multiplicity of their roles in hospitals (see Chapter 1), and hospitals and their administrative organization (see Chapter 3) are defined and discussed elsewhere in this book. The chapters that detail the wide range of innovative programs provided by clinical psychologists in medical settings support the need for and the strength of those services as offered by psychologists. This expansion in actual clinical responsibilities and the increase in the formal-

ized hospital privileges that support these clinical roles have come about not only because of excellence in patient care services but because of the aggressive activities of organized psychology within the political arena.

The concepts surrounding "politics" may have derogatory definitions, definitions that focus on the use of political schemes or strategies that are devices toward small ends or selfish advantage (*Random House Dictionary*, 1966). However, politics may be used to obtain a position of power or control. The *self-control* of one's own professional behavior and, with it, one's own clinical, scientific, and financial independence seems to be an admirable (political) goal. Politics may be carried out by a statesperson who uses foresight and unselfish devotion for the betterment of the state. In this chapter, the "state" is the practice of clinical psychology in medical settings. The welfare of our patients and the right to practice as an autonomous health care profession are the political goals sought.

The issues affecting the activities of clinical psychologists in medical settings can be divided into macro- and micropolitical issues. The macroissues include legal, political, and organizational activities at the national and state level; licensing, sunset issues, federally mandated health insurance regulations, free-

Ronald H. Rozensky • The Evanston Hospital and Northwestern University and Medical School, Evanston, Illinois 60201.

dom-of-choice legislation, and ethical and professional guidelines, to name a few, are macroissues that transcend the individual hospital but also influence daily practice in the hospital. Micropolitical issues occur within one's own hospital milieu and include the structure of each hospital's bylaws, privileging and credentialing (including hospitalization privileges), medical or professional staff voting rights, academic freedom, supervision requirements versus autonomy, the building of professional relationships, and the politics of competence.

The politics of competence are based on the clinical training (see Chapter 4), the clinical skills, the ethical standards, the scientific foundations (see Chapters 6 and 16), and the personal style and professionalism of the practicing psychologist in the medical setting (see Chapter 6; Sweet & Rozensky, 1991). It is precisely that competence that forms the data base on which those psychologists who use the political arena make their case, thus furthering the cause of the practicing psychologist. The broadening of the roles, responsibilities, and privileges to be discussed within this chapter is a direct reflection of the field's ability to let it be known through political channels that clinical psychologists in medical settings provide invaluable services to their patients.

Macropolitical Issues

Standards of Practice

Ethics and Guidelines

The ethical standards of and guidelines for psychologists address the expectations, scope, professional requirements, and competencies that one must satisfy before seeking a clinical role as a specialist practicing in a medical setting. The American Psychological Association (APA) stated that psychologists should "not put themselves forward as *specialists* in a given area of practice unless they meet the qualifications noted in the Guidelines" (APA, 1981, p. 640). Before referring to themselves as clinical psychologists, for example, psychologists must

meet the generic standards of training and experience stated in the guidelines (APA, 1972). Doctoral training and internships that meet APA accreditation standards certify that these basic requirements have been fulfilled. Within the specialty of clinical psychology, guideline 1.6 (APA, 1981) states that clinical psychologists must limit their practice to demonstrated areas of competence "as defined by verifiable training and experience" (p. 644).

Before one can ask to be regarded as a practitioner who works within a *medical* setting, both the training and the experience that prepare one to function within that *specialized* environment must be sought, successfully accomplished, and documented. Only then can a clinical psychologist ethically ask to be regarded as a member of a hospital's professional staff. Adherence to these guidelines offers some assurance that a given individual is prepared both to work in and to have a reasonable chance of succeeding in a medical environment. Further, adherence to the guidelines increases the likelihood of maintaining good professional relationships while protecting the good name of the profession and the expectation of success of other clinical psychologists who follow one into that medical setting. In discussing the need to work toward specifying the credentials of those seeking hospital privileges, Enright (1987) suggested that there is an overwhelming need for specific training and credentialing of psychologists in hospital practice. Further, he saw the hospital environment as fraught with pitfalls and traps that may make well-intentioned, but untrained and unaware, psychologists entering hospital practice a liability both to themselves and to the profession. "Entering the hospital without knowing the customs, language, protocol and procedures is like entering the Amazon without a map, provisions or guide," Enright wrote, and "unfortunately, in both cases, the head hunters still roam free and remain quite hungry" (p. 11).

Clearly, then, these ethical guidelines must be met before embarking on hospital practice. Dorken (1981) wrote that "the development of personal accord and collegial interdisciplinary relations on which mutual trust and respect are based—and which are critical to quality and continuity of patient care—must of necessity evolve at the facility level" (p. 604). The chance to develop professionally within a hospital begins with training (see Chapter 4) and is delimited by the macropolitical environment. These macroissues must be addressed before the in-house or micropolitical issues. Licensing, recognition by third-party payers, and hospital privileging form the external, politically defined legs on which the day-to-day competence of psychology must stand.

Licensing

Every state and the District of Columbia regulate the professional practice of psychology through some type of licensure or certification statute designed to inform and protect the public (Stromberg, Haggarty, Leibenluft, McMillian, Mishkin, Rubin, & Trilling, 1988). This regulation is established and maintained through the political process. Regulation serves to define and recognize psychology as an independent professional group, while identifying and certifying competence, and protects the consuming public (Hess, 1977). Bennett (1988) further stated that the need to exercise this regulatory control over a profession is directly proportional to the need to protect the consumer from the acts and practices of untrained or unqualified practitioners who may claim to offer those same services.

By providing regulation, the state can set minimal standards and qualifications for entry into psychological practice. Regulation also allows for the removal of unethical or incompetent providers. Although there is academic debate over the relevance or utility of the regulation and licensing of psychologists (Gross, 1978; Phillips, 1982; Wiens & Menne, 1981), Bennett (1988) stated that, with the limited number of investigators available and the high costs of pursuing complaints against already licensed practitioners, the state is more effective in enforcing entry-level requirements than in enforcing discipline after entry. Thus, regulations that maintain high entry standards of education and training are "essential for the protec-

tion of the public health, safety and welfare" (Bennett, 1988, p. 2). Once the psychologist is licensed, Koocher's concept (1979) of "competence validity" might be served through his suggestion of required peer review, continuing education requirements, and assessment center methodologies (actual observations of professional behavior).

At present, there exist three levels of regulation of practitioners. *Registration* is simply an eligibility listing that is based on minimal standards and that provides for a minimal review of credentials. The title *psychologist* is restricted by the *certification* process, which reviews minimal educational requirements, training, and experience. Often, a written exam is required. Certification does not restrict practice; anyone may do the work of a psychologist in a state with a certification law. He or she may not, however, use the title *psychologist* without certification. In one state with a certification statute, however, Beach and Goebel (1983) found that a full 40% of advertisements in the yellow pages under the heading "Psychologist" were illegal; certification apparently did not prevent those not certified from calling themselves psychologists. *Licensing* is the most restrictive form of regulation. It restricts both the use of the title and the actual practice of the professional and limits those not licensed from acting as if they were (Bennett, 1988). According to Stromberg *et al.* (1988), about 25 states have true licensure laws, and 15 to 20 have certification statutes for psychologists.

The American Psychological Association (1987) prepared a model act for state licensure that addresses the profession's general concerns regarding the proposed wording of any laws designed to specify the regulation of practice. Stromberg *et al.* (1988) offered a detailed discussion of the psychological and legal interplay in licensing.

The extent to which a state's regulatory laws define, permit, or limit the practice of clinical psychology has a direct effect on the boundaries of practice within the medical setting. Each practicing psychologist seeking privileges in a hospital should know these regulations.

Sunset Laws

The political realities of "sunset" legislation began to threaten the legal recognition and regulation of psychology in 1979, only two years after the field could claim that it now was a regulated profession nationwide. Sunset laws were originally designed to reassert citizen control over runaway governmental costs and programs. Sunset laws call for state-run programs to be evaluated periodically for their effectiveness and necessity. This evaluation includes questioning whether the funds needed to implement programs are being well spent and if the initial legislative intent of the programs is being followed (Kilburg & Ginsberg, 1983). Briefly, if no action is taken to reinstate an agency during the sunset process, it goes out of existence on a given date. If, after review, it is recommended that an agency be retained, the legislation must enact a law to maintain that agency. There are no guarantees that a legislative body will follow recommendations. Political pressures can be brought to bear either to follow or not to follow a recommendation to maintain or terminate a given agency. If maintained, an agency will be reviewed again before another sunset date. Sunset activities, however, began to focus on professional and occupational regulatory statutes. This focus was due to the sunset concepts' being experimental. The professions, being well defined and generating the data necessary for evaluation more readily than other governmentally defined entities, easily became an ideal group on which to test the new sunset concepts (Bennett, 1988). Bennett (1988) wrote that "although designed to be an objective review process, in most cases the sunset approach has turned into a *political-legislative* problem for the affected agency" (p. 5; italics added).

The sunset problem has permitted groups hostile to the independent practice of professional psychology to attempt to use the process to their own ends (Kilburg & Ginsberg, 1983). An unregulated health care profession ceases to be seen as an important force in health care planning, third-party reimbursement is unavailable to the unlicensed, and professional autonomy is threatened. Kilburg and Ginsberg

(1983) detailed the nationwide response of organized psychology to the sunset crisis. In highlighting how psychology "learned to love this crisis," Kilburg and Ginsberg pointed out that sunset laws have resulted in a strengthening in many states of the "definition-of-practice" section of their licensure laws. Finally, they stated that organized psychology has learned that it "is required to establish and maintain a strong and permanent legal foundation for the discipline" (p. 1231). Licensing, it would appear, is a professional responsibility, not a right. A constant professional vigil is needed to protect it.

Third-Party Reimbursement

Insurance reimbursement for mental health services has been a central issue in the politics of organized clinical psychology and a natural second step after the assurance of legal recognition via licensure and regulation. Meltzer (1975) noted that psychologists could not have parity with psychiatry as long as psychologists had to "go through" or "collaborate" with a psychiatrist to receive third-party reimbursement for clinical services rendered. Freedom-of-choice legislation began to be introduced at the state and federal level and became a rallying point for psychology and an "emotional cause of the 1970s" (Meltzer, 1975, p. 1150).

Freedom of Choice

Freedom of choice contains two issues: (1) independent insurance reimbursement for psychologists can occur without authorization or monitoring by another health care professional, and (2) patients are free to choose their mental-health-care provider without the restriction of being told that their insurance company will financially support their treatment only if it is provided by a physician. The intensity of the initial struggles is suggested by one of many articles on the early freedom-of-choice battles: "We are engaged in a hard nosed political and economic struggle with physicians for a piece of the insurance pie. The survival of our profession is at stake; and if we win, we will probably survive" (*Illinois Psychologist*, 1977, p. 10). Within this political environment, psychology's professional advocacy groups were formed. One of the early, high-priority issues for these groups was making psychologists (financially) independent mental-health-care providers (Meltzer, 1975).

Although third-party reimbursement provides patients with freedom of choice and the psychologist with increased practice and financial freedom, some authors, reflecting political differences within professional psychology itself, have questioned the general effects of this financial issue on the field. Albee (1975) warned of medicalizing "problems of living" as illness or defects via a system that reimburses health care services. Additionally, it has been argued that insurance reimbursement for psychotherapy may well become a subsidy to those who are more able to afford and use it at the expense of those who are less able to afford psychotherapy and who also tend not to understand or accept mental health treatment anyway (Albee, 1977; McSweeny, 1977). Finally, the profit motive of those drawn to health care by the dollar has been questioned (Meltzer, 1975). The cost savings in medical care afforded by psychological intervention (see Chapter 8), however, may well counter most arguments against the use of insurance dollars to support psychologists' work. Cost-effectiveness may be particularly important for psychologists who work within medical settings.

ERISA

Although freedom-of-choice laws exist across the country, a political-economic issue has arisen to threaten this freedom. The Employee Retirement Income Security Act (ERISA) was implemented by Congress in 1974 to protect the retirement plans of employees. This Federal law preempted state insurance laws by setting minimal national standards for retirement plans. ERISA was later broadened to include many employee welfare plans that included health care benefits and life and disability insurance, as well as pensions. In many cases, companies can establish independent trusts to administer these benefit packages, including the purchase of insurance (Peres, 1986). Although ERISA

stipulates that it does not exempt state laws that regulate insurance, many companies have interpreted ERISA as giving them a means to circumvent individual states' freedom-of-choice laws and thus to deny insurance claims submitted by psychologists. What was won via the legislative process, freedom of choice, had to be defended in the courts.

In 1985, a decision by the U.S. Supreme Court held that Massachusetts had a right to regulate insurance companies and that state laws requiring insurance companies to cover mental disorders were not preempted by ERISA (Peres, 1986). In that same year, court decisions held that ERISA interpretations did not preempt Michigan state laws mandating that substance abuse benefits must apply to employee benefit trust funds relying on reinsurance or stop-loss policies. In Illinois in 1985, the courts held that an insurance company was in violation of state law when it denied coverage to a state resident who had received treatment from a clinical psychologist and not a physician. There was a similar court ruling in Pennsylvania in 1986 regarding ERISA interpretations and psychological testing (Turkington, 1986), and finally, a federal appeals court ruled that self-insured benefits plans are covered by state laws rather than ERISA (Bales, 1988).

The APA's Office of Professional Practice offers an extensive listing entitled "Freedom of Choice and Mandated Mental Health Insurance Legislation as Applicable to Psychologists." This state-by-state listing displays the extent of freedom-of-choice legislation as accomplished by professional psychology via the political process. Additionally, the listing provides practical information on the definition of qualified providers, mandated benefits for mental health care, and physician referral requirements by state.

The Politics of Finance

Federal Legislation

Federal legislation can set national trends in the health care and insurance industries. These, in turn, can affect the independence and scope of clinical practice and the financial viability of clinical psychology in medical settings. Of our gross national product, 10.5% is being spent on health care, more than in any other nation (DeLeon, VandenBos, & Kraut, 1984). With more than 10% of the federal budget alone going toward health care funding and only 1% of the members of Congress having any health care background, it has been suggested that psychologists need to educate the lawmakers as to the scope and cost-effectiveness of psychological services so as to be assured inclusion as providers in any federally funded program (Folen, 1985).

At present, there are several major federally mandated health care initiatives and programs that have an impact on the practice of psychology: the Civilian Health and Medical Program of the Uniformed Services (CHAMPUS), the Federal Employees Health Benefit Program (FEHBP), the Medical Expense Deduction provision of the Internal Revenue Code, Medicare and Medicaid (Title XVIII and Title XIX of the Social Security Act; DeLeon et al., 1984), and the largest health-care-delivery system in the United States, the Department of Medicine and Surgery of the Veterans Administration (West & Lips, 1986).

CHAMPUS. CHAMPUS health care (insurance) coverage provides for 6.295 million dependents of active-duty, deceased, or retired military personnel. It is the single largest health care plan in the nation (Dorken, 1988). The appropriations bill that funds CHAMPUS includes the phrase "medically or psychologically necessary," and with a few political ups and downs (DeLeon & VandenBos, 1983), practicing psychologists have been assured their place as qualified providers to this large population. Dorken (1988) reported that, for the year 1986, psychologists provided some 33,761 *inpatient* mental health procedures to eligible CHAMPUS patients, compared to 192,502 inpatient procedures provided by psychiatrists.

FEHBP. DeLeon et al. (1984) noted that almost another 10 million federal employees, annuitants, and their dependents participate in the FEHBP. Under this federally funded program, psychologists have been accorded completely independent status as health care providers. The language of the law authorizing the inclusion of psychologists as providers in FEHBP can be viewed as a federal freedom-of-choice law.

Medicare and Medicaid. DeLeon *et al.* (1984) offered the following summary of Medicare and Medicaid issues.

Medicare is a federally funded health insurance program that covers the health care for the nation's elderly and disabled, over 29 million people. Part A of Medicare covers mandatory hospitalization and extended-care insurance. Part B covers physicians' fees and other services not related to hospital care. Part A is 100% federally funded, and Part B is voluntary and is cofinanced by the government and the participant. The Medicaid program is about 56% funded by the federal government, is administered by each state, and is designed to provide medical assistance to the needy and low-income population, about 19 million persons. Although not directly recognized as providers in the supporting legislation, psychologists can receive reimbursement for Medicare Part A services to inpatients if the psychologist either is an employee of a hospital or offers services under arrangement with the hospital. The hospital bills the government for the services of the psychologist as a hospital service. Part B services for nonhospitalized patients may be billed by psychologists if the patient is referred for these services by a physician. Under Medicaid, services by psychologists are reimbursable at the discretion of each state. Twenty-five states allow reimbursements to psychologists under Medicaid. A recent guideline provided by the Health Care Financing Administration in September 1988 (National Council of Mental Health Centers, 1988) approved a ruling that affords Medicare B patients the right to the reimbursement of services provided by psychologists working within community mental health centers (CMHCs). No physician supervision or involvement in is necessary in cases within CMHCs. On December 19, 1989, President

Bush signed into law the Medicare bill that now includes direct reimbursement of psychologists for services provided to the elderly and disabled. This will cover about half of all inpatient and outpatient psychological services and culminates a 25-year-long political struggle to make psychological services finanically available to the growing elderly population.

National Health Insurance. As far back as January 10, 1977, Senator Daniel Inouye (Democrat of Hawaii) stated that, unless psychology is able to obtain independent recognition under Medicare, it is doubtful that, once some type of national health insurance is enacted, psychology will be included in such a plan ("In or Out?" 1977). In discussing the importance of the Medicare issue to psychologists, DeLeon (1987), urged "active participation in the political process" (p. 9) by contacting Congress to ensure psychology's inclusion in health care legislation. The concept of national health insurance and clinical psychology's role in such a program have been the focus of much discussion over the years (Cummings, 1977; Julius & Handal, 1980; Kiesler, Cummings, & Vanden-Bos, 1979).

More recently, the concept of health insurance for citizens across the nation was reflected in a bill sponsored in the Senate (bill 1265) by Senator Edward Kennedy (Democrat of Massachusetts) and in the House by Representative Henry Waxman (Democrat of California), H.R. 2508. Usually referred to as the Kennedy–Waxman Bill, this initiative would have required businesses to provide for all their employees a minimum level of health care insurance (Buie, 1987b). This approach has been seen as a political move away from a too costly and too bureaucratic tax-supported national health insurance to a privately funded, but federally mandated, system. Initially, to attain the support of business, Kennedy agreed to seek to override state laws requiring the coverage of mental health, optometry, dentistry, and podiatry (Buie, 1987b). However, because of the political intervention of the APA, including the hiring of an attorney and a lobbyist, and because of the close day-to-day scrutiny of the

legislative process by the APA's Health Policy Directorate (Buie, 1987a), mental health coverage, including psychologists' services, were included in the bill. The APA's Practice Directorate now "strongly supports the bill, both as a professional issue and as an issue of broader public interest. As a professional issue for psychologists, the bill would achieve a national freedom-of-choice law" (Buie, 1988d, p. 19). This measure would, if passed, add some 37 million Americans to those with health insurance. The grass-roots politicking of some 40,000 psychologists was noted as the "clincher" in seeing mental health coverage and psychologists included in the bill (Buie, 1988e). However, a bill can be amended during the process of becoming a law, and psychologists have been encouraged to let their legislators know of their support for such a bill so that they will continue to be included in the final version.

HMOs. Health maintenance organizations stemmed from political and social changes in American society and were designed to offer prepaid health care to their enrollees (Kisch & Austad, 1988). According to DeLeon, Uyeda, and Welch (1985), there were some 15 million HMO enrollees in 323 HMOs nationwide with the number increasing to 700 HMOs and 29 million enrolles by 1988, (Buie, 1988f). Kisch and Austad (1988) reported that, during 1986, 13.3% of HMO participants received some type of mental health services. Physicians remain as gatekeepers in most HMOs, and psychologists are employed in relatively few plans, even though services of the sort provided by psychologists are offered by most plans (Cheifetz & Salloway, 1984). Given the existence of HMOs, psychologists have been encouraged not only to provide direct clinical services but also to offer health education programs (Tulkin & Frank, 1985) and, in a *zen*-like manner, to join HMOs (in "one reality") (Budman, 1985). Be that as it may, the political reality of recognizing the autonomy of psychologists practicing within risk-sharing HMOs was ensured by Senator Daniel Inouye's 1984 amendment to the Deficit Reduction Act. According to DeLeon *et al.* (1985), the wording of the law stipulates that "the services furnished . . . by a clinical psychologist (as defined by the Secretary) . . . and such services and supplies furnished as (an incident) to [these shall be considered] as would otherwise be covered under this part if furnished by a physician or an incident to a physician's service" (p. 1123). Subsequent regulation has defined the "clinical psychologist" as one who holds a doctoral degree and is licensed by the state to practice autonomously. Thus, according to DeLeon et al. (1985), some 400 psychologists have established the HMO as their primary work environment. Politics have defined that medical setting as a work place and have ensured the availability of doctoral-level practitioners in HMOs for HMO subscribers.

Research and Training. Within the area of federal recognition of and funding for the research activities and training of psychologists, the APA's Science Directorate has taken the political lead. As a result of the directorate's advocating for psychology in Congress, the National Institutes of Health (NIH) are now required to include a behavioral scientist on each of the 15 national advisory councils: "The additional power psychology will now wield in the NIH could translate into a larger share of the research funds netted out by the huge $7.2 billion agency" (Landers, 1988, p. 1). Also according to Landers, the Directorate was able to influence a bill that requires that a clinical psychologist serve on the health professions advisory board of the Health Resources and Services Administration. This move ensures the recognition of psychologists as a health service professionals along with physicians, nurses, dentists, and optometrists and permits psychology students access to a large amount of scholarship and loan money to support training. Finally, the APA was able to get scientific research concerning psychological issues vis-à-vis AIDS included in the AIDS Amendments of the Omnibus Health Act ("On Behalf of Science," 1988).

Although some internal political struggles have occurred between clinical service providers and researchers within psychology (Fisher, 1988; Salziner, 1988), it is clear that, when psychologists go up Capitol Hill *together* to meet and politic with Congress, all of those in psychology prosper.

IRS Regulations. DeLeon (1981) and DeLeon et al. (1984) have pointed out that, as psychologists become more active in the health care arena in general, such issues as the extent to which the Internal Revenue Service's rules allow medical deductions may affect professional psychology. Just the inclusion of psychologists' fees as medical deductions, or the change of the term *medical expenses* to *health care expenses*, has an effect on those seeking care. The political arena is wide-ranging.

Veterans Administration. The Veterans Administration (VA) is the single largest employer of psychologists in the United States, with over 1,400 full-time doctoral-level psychologists (DeLeon et al., 1984). The VA was the first employer to establish the doctoral degree as the entry level for training of practicing psychologists and has been a primary source of predoctoral and internship training in clinical psychology. Traditionally, psychologists have had a wide range of administrative and clinical responsibilities in over 155 VA medical centers (VAMCs) nationwide. Psychologists serve on the executive committees of 72% of the VAMCs and have negotiated membership on the medical staff in many VA hospitals (West & Lips, 1986). The relatively autonomous functioning of psychologists in the VA has lead to the professional freedom to produce a large amount of published research (Boudewyns, 1986), as well as to the development or improvement of clinical techniques in medical psychology (Stenger, 1979). However, even with the long history of growth in professional psychology in the VA, federal budgetary and administrative changes have threatened psychology's voice in the VA central office and thus its autonomy in the VAMCs. As a federally funded health care system, the VA reflects in many ways the political winds that move the health care industry and affect the practice of clinical psychology in medical settings. West and Lips (1986) reminded us that "the VA is a *politically* driven system that must rely on Congress for its budget as well as general policy guidelines" (p. 999; italics added). The hospital practice of clinical psychology within the VA system is one of those policy areas vulnerable to the political winds of change.

Clinical privilege and financial policy as established at the federal level set standards for the health care industry that can directly affect a large segment of the population. Such policies, as set in Washington, D.C., come from the very heart of American politics.

Macro- and Micropolitical Interface: Hospital Bylaws

Those individuals who are members of the hospital's medical or professional staff are authorized by the process of privileging to treat that hospital's patients. When discussing the general topic of hospital privileges, it must be remembered that hospitalization itself is only one of a wide range of clinical services that clinical psychologists may be privileged to provide their patients. For many practicing clinical psychologists, the rights to "attend" their patients as outpatient psychotherapists, to be diagnosticians, to act as emergency-room decision-makers, to be consultants on medical units, to be voting members of the professional staff, and to act as independent, principal research investigators are as important as, if not more important, than, hospitalization rights.

A hospital's bylaws define the organizational structure of the professional staff and prescribe the process for applying for and obtaining hospital privileges (Stromberg *et al.*, 1988). Although the bylaws vary from institution to institution, they routinely spell out the qualifications for membership on the staff, the conditions for and the duration of appointments, the categories of professional staff membership, and privileging for the various types of clinical activities and procedures that each staff member may perform (Committee on Professional Practice of the Board of Professional Affairs, 1985). Bylaws also establish quality assurance expectations, governance of the professional staff, the com-

mittee structure, and the voting rights of members of the staff.

Additionally, hospitals may have a written set of rules, often referred to as "policies and procedures," that standardize clinical documentation expectations (charting), admission and discharge practices, emergency or disaster plans, and personnel or human services regulations. These rules vary greatly across medical centers and may be more flexible and changeable than bylaws. The individual psychologist bears the burden of understanding and following these expectations of his or her institution (Committee on Professional Practice of the Board of Professional Affairs, 1985).

Hospital Privileges and Psychology

Staff Membership Categories

Hospitals generally have several categories of professional staff membership, and although the credentials required for membership in these categories may vary from institution to institution, they include the following:

The *active staff* is the highest level of privileging and responsibility. These staff members can admit patients to the hospital, can provide continuous care to their patients, can serve on committees, are required to attend staff meetings, and can vote and hold office on the professional staff.

The *associate staff* is usually made up of newly appointed health care providers who work in outpatient departments or as research fellows but who do not have admitting privileges. Occasionally, psychologists are assigned this classification.

The *courtesy staff* are those who actually admit only a few patients per year and may well be on the staff of another hospital.

The *consulting staff* are those who act only as consultants in a specific specialty. They, like the courtesy staff, may not hold office or have voting privileges.

Honorary or *emeritus staff* members are not active in the hospital but are given such honors as are set forth in the bylaws.

The *affiliate* or *adjunct staff* consists of allied

health care professions, nonphysicians who have been granted privileges by the organized professional staff. Affiliates may participate in patient care under the direct and continued supervision of a member of the active professional staff. Affiliates usually do not have voting privileges pertaining to professional staff matters and are not assigned committee responsibilities. Usually, psychologists can seek professional staff membership only at the affiliate level. In their definitions and discussion of staff membership, from which the above were taken, both Stromberg *et al.* (1988) and the Committee on Professional Practice of the Board of Professional Affairs (1985), have suggested that psychologists are more and more able to seek membership on professional staffs at the active staff level.

The Political Struggle for Hospital Privileges

Historically, psychologists were precluded from becoming full members of the active medical staff in most hospitals. Before 1983, the Joint Commission on Accreditation of Hospitals (JCAH; now the JCAHO, Joint Commission of Accreditation of Healthcare Organizations) limited medical staff membership to physicians and dentists. In recent years, and in response to continued concerns about its potential antitrust liability, the JCAHO has modified its stance to permit, but not require, hospitals to offer staff appointments to other licensed professionals as permitted by the law and the individual hospital (Stromberg *et al.*, 1988). The JCAHO (JCAHO, 1987) now permits, but does not require, hospitals to offer professional staff appointments to "other licensed individuals permitted by the law and the hospital to provide patient care services independently . . . within the scope of [their] license and in accordance with individually granted clinical privileges" (Stromberg *et al.*, 1988, p. 327). These changes reflect the results of a lengthy struggle with the Joint Commission to include professional psychology as one of the formal voices in hospital health-care-accreditation policy during a time when physicians were defining and limiting the roles of psychologists in medical

and psychiatric settings (Zaro, Batchelor, Ginsberg, & Pallak, 1982). Finally, the recent appointment of a psychologist to the JCAHO's Professional and Technical Advisory Committee of the Accreditation Program for Psychiatric Facilities (APA, 1988b) affords organized psychology its first voice from within this largest of hospital accreditation bodies.

The Reason for All the Fuss about Privileges. Stromberg (1986) stated that access to hospitals helps in the acknowledgment of psychology as an independent health care profession. Clinically, to be able to follow patients in need of hospital care is important to some clinicians. Financially, Stromberg sees privileges as easing the process of selection as HMO and Preferred Provider Organization (PPO) providers. Finally, networking with other health care providers is helpful in conveying that psychologists can provide a full range of services. This can happen most readily within the structure of an organized professional staff.

Thompson (1987) focused on the importance of professional autonomy as reflected in the obtaining of clinical privileges. The medical staff is afforded two important responsibilities: (1) monitoring and evaluating quality of care and (2) developing rules, regulations, and policy regarding staff membership, including the granting, delineating, and renewal of clinical privileges. The seeking and obtaining of privileges ensures that psychologists will, as members of an autonomous profession, participate "in the hospital processes that govern who can provide which patient care services" (p. 868). This is the center of the micropolitical environment that defines the role of clinical psychology within each hospital.

Contained in a packet of information concerning hospital privileges for psychologists (Plunkett & Morgan, 1987) is the statement that

> to a large extent, psychology's limited access to privileges in hospitals appears to be related to a prevailing climate of adverse regulation and medical institution barriers, coupled with restrictive or sometimes prohibitive hospital bylaws.

The sample of headlines from professional newspapers and newsletters and the popular press displayed in Table 1 illustrates the extent of the lengthy political battles that have been waged thus far over the topic of hospital privileges for psychologists. These headlines reflect a roller-coaster struggle, with undulations of increasing frequency, that has lasted for well over a decade in venues from coast to coast.

JCAHO. In 1983, the standards of the JCAH stated that medical staff privileges were limited to "individuals who are currently fully licensed to practice medicine and in addition, to licensed dentists" (p. 93). A hospital that violated this mandate would have risked losing its JCAH accreditation. Bersoff (1983) stated that the JCAH standards that limited psychologists from being members of a hospital's medical staff "unjustifiably and unlawfully deprive the public of access to an entire body of qualified, trained, and licensed providers of psychological services and absolutely exclude competent psychologists from independent hospital practice" (p. 1238).

Antitrust. The practice of limiting or barring the hospital practice of psychologists can be seen as violating the federal Sherman Antitrust Act (Bersoff, 1983). Two types of illegal boycotts of psychologists exist:

> In cases where physicians (through their staff privileges committees) refuse to deal with their competitors, they are engaging in a horizontal boycott. In cases where a hospital collaborates with physicians to eliminate physicians' competitors, the hospital (through its board of trustees) and the physicians are engaging in a vertical boycott. Both forms are per se violation of the antitrust laws. (Bersoff, 1983, p. 1240)

By establishing that psychologists are competitors with physicians in the field of the diagnosis and treatment of mental disorders, by stating that psychologists' ethics prohibit them from seeking privileges beyond their competence and training, and by arguing that "the rights, privileges, health, and pocketbooks of patients and the public are at stake" (Bersoff, 1983, p. 1241), organized psychology has been able to establish legal precedents for the granting of hospital privileges to psychologists. Cal-

Table 1. Hospitals, Politics, and Psychologists in the News

Psychiatrists oppose key bill. *APA Monitor*, July 1974.

"Medical psychotherapy"—Psychiatry's trump? *APA Monitor*, June 1975.

War between the shrinks. *New York*, May 1979.

Characteristics of health services, hospital privileges and issues with third party reimbursements." *The Register Report*, May, 1982.

Hospital privileges. *Advance*, August 1983.

Psychologists get hospital admitting privileges in D.C. *Psychiatric News*, November 1983.

Successful passage of the hospital privileges for psychologists bill in the District of Columbia. *Independent Practitioner*, January 1984.

Political action committee plays crucial role in hospital privileges legislation in D.C. *Illinois Psychologist*, November 1984.

Nonphysician practitioners make slow headway on staff privileges. *Hospitals*, December 1984.

Illinois: Battles on two fronts. *APA Monitor*, August 1986.

Court upholds barring of psychologists on medical staffs. *Psychiatric News*, August 1987.

The rocky road to hospital privileges. *APA Monitor*, September 1987.

Appeal dismissal affirms Calif. hospital privileges. *APA Monitor*, October 1987.

Illinois hospital battle rages on. *APA Monitor*, October 1987.

N.Y. may join states with hospital privilege laws. *APA Monitor*, March 1988.

CA: Hospital battle raging. *APA Monitor*, April 1988.

IL: Settlement may end [medical staff] dispute. *APA Monitor*, May 1988.

OH: Hospital bill passed by house. *APA Monitor*, May 1988.

MO: Psychologists thrive in med center. *APA Monitor*, June 1988.

NY: Hearing fuels hopes for hospital privilege law. *APA Monitor*, June 1988.

Court reverses psychologists' hospital rights. *Psychiatric News*, July 1988.

IPA [Illinois Psychological Association] hospital lawsuit settled. *Make the Future, the Health Service Advisory Newsletter*, July 1988.

CA: Group readies to fight hospital privileges ruling. *APA Monitor*, August 1988.

AMA and psychiatry join forces to oppose psychologists. *Practitioner Focus*, Summer 1988.

California case could have dire impact, lawyers say. *The APA Monitor*, November 1988.

Hospital survey finds barriers still exist. *APA Monitor*, November 1988.

Florida psychiatrists challenge psychology's scope of practice. *Practitioner Focus*, Fall 1988.

President signs Medicare bill—Victory caps uphill trek. *APA Monitor*, January 1990.

ifornia, the District of Columbia, Georgia, and North Carolina now have laws that prohibit discrimination against psychologists when medical staff appointments are made. Stromberg *et al.* (1988) offered the District of Columbia's act as the most sweeping statement that psychologists are "accorded clinical privileges and appointed to all categories of staff membership at those facilities and agencies that offer the kinds of services than can be performed by either members of these health professions or physicians" (p. 328).

In the Courts. Progress on the professional staff issue for psychologists has been slow. At times, the macropolitics of organized medicine and issues of turf and financial incentives have had actual or potentially deleterious effects on the practice of professional psychology. An example is the Illinois Department of Public Health's attempts in 1988 to ban psychologists from hospital medical staffs by threatening those hospitals *with* psychologists on their medical staff with loss of their license to operate as a hospital. A newsletter to Illinois psychologists is offered *in toto* (see the Appendix to this chapter) because it not only clearly illustrates the issues but emphasizes the macropolitical realities that can influence daily hospital practice anywhere.

In a similar vein, the travails of California's psychologists reflect first the gaining of hospital privileges over a decade ago and then increased attacks by organized (political) medicine and psychiatry. As a result of that attack, the court reversed the interpretation of the original California law (APA, 1988a). In 1978, after nine years of political struggles by psychologists, California had passed S.B. 259, a bill that authorized each hospital to make inpatient care available to "psychological patients" by "clinical psychologists with appropriate training and clinical experience" (Dorken, 1981, p. 600). After several years of practice (within the scope of their training and licensure), psychologists faced a regulation set forth by the California Department of Health Services that "a psychiatrist shall be responsible for the diagnostic formulation for each patient and the develop-

ment and implementation of the individual patient's treatment plan" (APA, 1988a). In a trial held in 1986, a California court held that psychologists were indeed allowed to admit and discharge their own patients. However, on appeal by the California Medical Association and the California Psychiatric associations in 1988, the appeals court ruled that only a physician is authorized to render a diagnosis pertaining to a mental disorder that is organic in origin or nature, thus restricting a psychologist's access to hospital practice. In 1988, the California Supreme Court agreed to review the appeals court's decision to limit psychologists' practice after the psychologists' attorneys warned that "the rights, privileges, health and well-being of patients and the public will be sacrificed if the lower court's decision is upheld" (Buie, 1988a, b,c). On June 29, 1990, the California Supreme Court decided in favor of a patient's right to be treated by a psychologist upon admission to a hospital. This decision, in the case that came to be known as *CAPP v. Rank*, is not appealable and ends the six-year legal battle. According to an APA press release, as a result of this ruling, psychologists in California will be permitted full responsibility for their hospitalized patients, including admitting and discharging patients, if the psychologist is licensed by the state, is providing services legally defined by the license, and is complying with the rules of the facility.

In this case, it was a legal-political process that assured psychologists their rights to practice within the scope of their license, and it was that same process that assured the patients of psychologists their rights to continuity of care and freedom of choice. It was a legal-political process championed by the opposition, organized medicine and psychiatry, that attempted to reverse those rights. Dorken (1981) had felt at the time of the implementation of the California law that "selected amendments and close consultation [and] reasonable accommodations" had brought about "a quite cordial relationship" between organized psychology and medicine (p. 604). However, the ever-changing financial and regulatory milieu of present-day health care can bring about challenges to hitherto "cordial relationships" and can thus affect the day-to-day practice of hospital psychology. It would appear that, buoyed by the laws and with antitrust thoughts in mind, this struggle will continue. In reviewing the success of organized psychology's struggle for hospital privileges in the District of Columbia, Mikesell (1984) noted the words of one observer: "The hospital privileges bill is where it is today because of the political activity of psychologists in the District of Columbia over the past two years" (p. 19).

Micropolitical Issues

Practice Patterns of Psychologists in Hospitals

Boswell, Litwin, and Kraft (1988) carried out a national survey of 1,061 practicing psychologists in order to study hospital privilege issues. The 72.2% return rate was interpreted as reflecting practicing psychologists' high interest in the topic. Of the questionnaires returned, 582 indicated some type of hospital affiliation: 8% of the respondents indicated a primary affiliation with a public general hospital, 17% with a private general hospital, 19% with a public psychiatric hospital, 21% with a private psychiatric hospital, 20% with a VA hospital, 11% with a medical school, and 5% with some other type of hospital. Of those, 71% were affiliated full time and 28% were affiliated part time or had a specific arrangement with a hospital. *Only 16.3% of the hospital-affiliated psychologists were full professional staff members with voting privileges;* 22.1% were associate members without a vote, 50.1% were not members of the staff of their hospital, and 11.4% were special members of the staff. In general medical hospitals, psychologists were less likely to be full members of the staff (public, 8.7%; private 7.4%) and were more likely to be associate members (public, 32.6%; private, 38.9%). When looking at satisfaction with 18 individual practice privileges, Boswell *et al.* (1988) concluded that "the only level of clinical privilege with which hospital affiliated psychologists appear to be satisfied

is the full, independent level" (p. 12). This study concluded that psychologists clearly desire expanded privileges and believe that organized psychology should advocate for this expansion. However, "this overwhelming support breaks down when the issue of expanding privileges to include prescribing certain medications is included" (p. 16).

It is clear, though, from the Boswell *et al.* (1988) study that the political struggle for autonomy within hospitals still has a way to go: "For instance, 49.2% cannot diagnose independently, 13.5% cannot do psychological testing independently, 19.6% cannot provide therapy without the approval of a physician, and 25.4% cannot conduct research independently" (p. 17). Certainly these are areas in which psychologists are highly trained and licensed to practice without supervision. Finally, Boswell *et al.* found that 94.8% of hospital psychologists were not privileged to admit, 96.3% were not allowed to discharge, and 83.4% were not permitted to order a medical referral. These limitations affect not only professional freedom but the quality of care that hospital psychologists are formally privileged to provide their patients in a hospital setting.

In further discussion of these issues, Litwin, Kraft, and Boswell (1988) reported that only 4% of hospital psychologists surveyed believed that they had had difficulty obtaining medical staff (voting) memberships because of lack of competence, education, training, and experience. A full 71% either disagreed or strongly disagreed that staff membership problems are related to issues of competence or qualification. However, of those, 83% agreed or strongly agreed that their difficulties in obtaining staff privileges had been primarily because of "turf" and financial factors. Only 6% of those surveyed did not agree that turf and finances were the source of staff membership struggles. At least as hospital psychologists view themselves and their practice milieu, they are competent clinicians who see the financial threat to organized medicine of their independent practice as a major motivator of those who wish to restrict their independence.

Although these views reflect the source of

political struggles that occur nationally or at formal bylaws committee meetings from hospital to hospital, the reality of the politics of clinical competence can be seen "in the trenches." Litwin *et al.* (1988) found that 80% of those surveyed reported that they either agreed or agreed strongly that they often are led to "informally" perform duties (e.g., discharge decisions) for which they have no *formal* privilege. In other words, what we might define as these "hidden or secret privileges" show that formal bylaws often do not represent *actual* practice or competency.

Seeking Privileges, or Applied Micropolitics

While political forces work to shape legal statutes and hospital bylaws, the practicing psychologist seeking privileges in a medical settings should seek those privileges in an orderly, ethical, and professional manner. "Just because I want referrals from the local HMO" and "just because I might want to follow my patient into the hospital" and "just because I want to consult or do testing in the hospital" are not reason or rationale enough to justify seeking privileges. Proper training and experience within the medical setting, to ensure success in that environment, are not only the necessary but the ethical basis for seeking privileges. Once those requirements are met, the privileging process can begin.

The acquisition of medical staff membership, according to Stromberg *et al.* (1988), does not ensure an adequate range of clinical roles for the psychologist in a hospital. "Medical staff membership is only an empty vessel of eligibility" (Stromberg *et al.*, 1988, p. 328) into which the delineation of one's clinical privileges is poured. Such privileges are based on an individual's training, experience, and demonstrated competence, according to the APA's Committee on Professional Practice of the Board of Professional Affairs (1985). The committee said that "psychologists should attempt to gain explicit approval for the services they provide rather than permitting the institution to allow them to function in an informal fashion" (p. 19).

A Hospital Practice Primer for Psychologists (1985) and *Hospital Practice: Advocacy Issues* (1988), each prepared by the APA's Committee on Professional Practice, offer forms for documenting and approving the lists of clinical, consulting, and scientific privileges commonly sought by hospital-based psychologists.

Criteria for Privileging

Permissible Criteria. Hospitals can lawfully grant or decline staff membership and privileges to a psychologist "on any basis which is rationally related to his or her competence" (Stromberg *et al.*, 1988, p. 329). The following are criteria that Stromberg *et al.* stated should be permissible as part of the evaluation process for membership and privileging: Requiring Information that includes education, internship, fellowships or other training, experience, current competence and licensure, and health status is acceptable. Requiring information on pending or completed professional liability actions or loss of membership or privileges at other institutions is also acceptable. Hospitals should be permitted to set academic or educational criteria if they are clearly related to professional competence. For example, the requirement that psychologists have a doctoral degree from an accredited educational institution and a documented amount of training and experience is not seen as an unfair, exclusionary criterion in the application process. Similarly, hospitals should be able to ask applicants to carry a certain amount of malpractice insurance in order to protect the hospital and members of the professional staff from "having to bear a disproportionate share of any costs stemming from a malpractice claim involving an inadequately insured practitioner" (Stromberg *et al.*, 1988, p. 331). Hospitals should be able to set a geographic limit on staff members' residences to ensure that patients will be attended adequately in an emergency. Hospitals should be able to deny privileges to those whose ability to work with others or whose personal style can be documented as jeopardizing quality of care. Hospitals should be allowed to have a "closed-staff policy" in which the entire hospital staff or the staff of a given department is closed to membership for the reason that "the hospital's facilities are not adequate to treat the patients that are likely to be admitted by additional staff members, or that the existing staff is sufficient to meet the hospital's needs" (Stromberg *et al.*, 1988, p. 332). In order to avoid antitrust issues, hospitals must be able to document the rationale behind such exclusionary privileging procedures. Finally, observed and documented competency to use clinical privileges appropriately is a legitimate requirement to ensure quality of care.

Nonpermissible Criteria. According to Stromberg *et al.* (1988), several criteria should *not* be permissible for a hospital to use during the application or privileging process. These include membership in professional societies, the recommendation of a member of the present professional staff, or an affiliation with a particular HMO. Discriminatory criteria such age, sex, race, or handicap must not be used to exclude someone from staff membership:

> Finally, a key question is whether hospitals are free to deny staff membership and privileges to all *psychologists* as a category. State law varies on this question, with some states allowing such exclusion, others mandating nondiscrimination against psychologists, and still others silent on the issue. (Stromberg *et al.*, 1988, p. 334)

That issue is the source of a key political battle already discussed in this chapter. Stromberg *et al.* recommended that, "in areas where state law does not expressly allow the blanket exclusion of psychologists, a challenge to such a policy may be brought on antitrust or other grounds" (p. 334).

The Privileging Process

According to Stromberg *et al.* (1988), the application process should afford the psychologist reasonable written notice of the results of action on the application. Those making the staff privilege decisions should be objective, and those with a personal bias or prejudice should not be among them. Decisions should be based on stated credentialing criteria, privi-

leging criteria, and the information presented by the applicant. There should be a right to a hearing and an appeal process, should the applicant question the outcome of the appointment and privileging committee(s). Certainly, a committee of peers—that is, other psychologists working in the medical setting—should adjudicate applications for their own discipline and should be advisers to the hospital credentialing committee(s). This arrangement should ensure the most credible review of a psychologist's credentials.

Should a psychologist receive an adverse decision, Stromberg *et al.* (1988) suggested that "the first step should be to examine the basis for the proposed decision and to consider (with counsel) what procedural steps are available to oppose or appeal the decision" (p. 338). The first appeal should be carried out within the hospital's own appeal mechanisms. Should that appeal fail, legal action is possible. Stromberg *et al.* noted that there have been a large number of legal actions in which health care practitioners have challenged staff privileging decisions. Practitioners have lost most (but not all) of these legal appeals.

> In order for the psychologist to succeed in court, it will be necessary for a psychologist to show that the hospital violated its own rules or the law, or that the decision was arbitrary, capricious, or entirely lacking in evidentiary support, or that the decision was part of an effort to stifle competition or otherwise contrary to law. (p. 338)

In order to avoid a negative experience, those seeking privileges should carry out the following two steps to ensure that they will approach the credentialing and privileging process professionally.

Using Appropriate Resources. If there is an organized department, division, or section of psychology within the medical setting in which the psychologist applying for privileges, she or he should seek out the chairperson, director, or chief of that program and ask for a meeting to discuss the role of the psychology program in that institution. If there is no formal program, the applicant should seek out either the formal or the informal senior psychologists in order to discuss the role of psychology in that institution and how the applicant and her or his particular expertise and experience will complement the existing program. If there is no such formal or informal structure, the applicant should learn what is the existing psychiatry or social work structure. Working within the mental health disciplines, at least initially, will help the applicant to discover the range and level of the services that exist.

If there is a formal mental health structure within the medical setting, it is politically unwise to go outside that organization to seek privileges. Seeking privileges in the department of medicine or pediatrics, for example, when psychologists already have primary appointments within their own department or a psychiatry department only causes a division of forces and may serve to weaken the political and clinical liaisons already established within a given medical setting. As the data from Boswell *et al.* (1988) suggest, psychologists as a rule do not yet have a strong *formal* base within most recognized medical staffs, and any entrance of a newcomer into the system may threaten the possibly delicate balance of the informal privileges noted by Litwin *et al.* (1988). Respect for an existing hierarchy assures the applicant support from those already established and theoretically able to help newcomers establish their role within the medical setting. If that hierarchy does not exist, direct communication with the hospital president, administrative director, or hospital attorney (Enright, 1986) is appropriate.

Once the applicant has identified the appropriate initial contact person, then a meeting is useful to review the hospital's actual requirements for a psychologist to become a member of the professional staff. Discussing the criteria and one's credentials informally will help acquaint the applicant with the hospital and the hospital professional staff with the applicant. If the applicant perceives at this point any difficulty with particular credentials meeting the criteria, they can be discussed so that the application will be in order before the formal process begins.

The applicant should be aware of any re-

quirements put on the members of the professional staff and should determine his or her willingness to meet those expectations. For example, dues, voluntary teaching time, or *pro bono* services in a clinic may be required of all professional staff members. If the medical facility has a teaching program for psychologists or psychiatrists, applicants may be required to meet the standards of the parent medical school or university before they can be accepted on the staff. The applicant should become acquainted with those academic requirements, if they exist, to determine *a priori* both his or her qualifications for an academic appointment and the institution's or training program's need for his or her academic services. Finally, the applicant should make an effort to understand the extent of the privileges for psychologists actually available within the medical setting in question. If a physician's order or cosignature is necessary for psychological practice, the applicant must decide if he or she wishes to practice with those restrictions on autonomy. For example, the applicant who expects to hospitalize his or her patients where this is not a privilege available to psychologists should discuss how flexible that rule is and to what extent a challenge to that rule will be accepted by the psychology department or division. It is certainly easier to change rules from within an organization than to enter new and challenge the existing establishment without a full appreciation of the institution's history as it pertains to the role and function of clinical psychology.

Defining One's Role in the Medical Setting. Once the applicant has been accepted on the professional staff and is seeking a role within the medical setting with a specific skill to offer (e.g., biofeedback, neuropsychological assessment, or thanatology), the senior psychologists helping with the credentialing process can direct the applicant within the political and clinical boundaries of the hospital. If no psychology service exists or no other psychology specialist in that field practices in that setting, then approaching the director of the relevant medical service is appropriate.

Approaching other psychologists or medical

personnel aggressively to prove that one's services are a necessity is not only unprofessional, but certain to be met with poor results. Clinical psychologists in medical settings should pride themselves on coming from a data-based profession (see Chapter 17) and should realize that physicians in medical settings "respect critical thinking much more than exotic intuition" (Wright, 1982, p. 3). Therefore, providing one's *vita* as a means of introduction and any materials one has written or published in one's area of specialization, as well as other supporting literature documenting the usefulness of psychological intervention in one's specialty, will help in establishing both one's own and psychology's scientific and clinical credibility. Meyer, Fink, and Carey (1988) found that practicing physicians tend to view psychologists' consultations as generally available and helpful but are concerned about the adequacy of psychologists' training in providing such consultations. Liese (1986) reported that, for medical problems that physicians perceived as having a psychological component, they were as likely to refer the patient to a psychologist as to a psychiatrist or a social worker. Therefore, the clinical psychologist in the medical setting can expect that the medical establishment will find his or her services interesting and useful if the credibility and competency issues are handled matter-of-factly. It is up to individual psychologists to practice their own politics of competence day-to-day in order to market their services within the medical setting.

Becoming personally acquainted with primary referral sources, nurses, physicians, and other psychologists is of great help. Offering physicians who are potential referral sources collaboration on research with a shared patient population or offering to add a psychosocial component to their ongoing clinical research is an excellent way to enter a system. Presenting a paper at medical grand rounds that supports one's type of clinical service or research can be useful. Discussing a trial case that one is sure of handling well, and not simply for the sake of making a new contact (Wright, 1982), can build a new bridge. Offering one's skills as an educator to teach residents or medical students about the psychological components of a particular medical problem is another means of showcasing the range of clinical skills of psychologists. Offering to carry a case or two *pro bono* or on a sliding scale in a hospital clinic is a means of establishing oneself within the medical setting, as well as a means of carrying out one of psychologists' ethical responsibilities. Finally, one's availability to serve on the credentialing, research, human subjects, child abuse, or quality assurance committee(s) illustrates one's commitment to serving the greater goals of the medical setting and promotes acquaintance with a wider circle of practitioners.

Summary

Bismarck was purported to have said that one should avoid watching the making of both "policy and sausage." From the above examples, it is clear that that view, if carried out by professional psychology over the last few decades, would have greatly limited research funding, clinical training grants, professional practice both in the office and in hospitals, and ultimately, service of the public good (Sweet & Rozensky, 1991). Kelly, Garrison, and DeLeon (1987) noted that a large number of psychologists have become involved in legislative and public policy issues. The APA's Office of Legislative Affairs, Office of Professional Practice, and Science Directorate, among other APA offices, reflect organized psychology's professional relationship with public policy and the legislative system. Activities carried out by individual psychologists reflect the impact each one of us can have on these issues (Kelly *et al.*, 1987). Further, Ebert-Flattau (1980) prepared a legislative guide that can help psychologists understand the legal and political policy processes.

The *Monitor*, *The American Psychologist*, *The Clinical Psychologist*, and *The Register Report*, among other publications, can keep the reader abreast of political changes on the national level. Membership in and subscription to the state's psychological association or society's newsletters can keep the psychologist current on local politics, policy, and legislative issues. Regulation and licensing, freedom of choice, eligibility for third-party and governmental reimbursement for services provided, and clinical or hospital privileges are legally defined rights and responsibilities that are born in the political area and can be redefined, limited, or even removed in that same arena.

Appendix

The following piece, written by M. R. Levinson and B. E. Bennett and entitled "IPA Lawsuit Settled," appeared in the July 1988 issue of *Make the Future: The Health Service Advisory Board Letter*.

A Case Example of an Attempt to Restrict Privileges

The issue came about two years ago when the IDPH (Illinois Department of Public Health) issued an "interpretation" of an old regulation. That interpretation was that only physicians, osteopaths, dentists and podiatrists were eligible under the law to be members of hospital medical staffs in Illinois. Since psychologists had been serving successfully on hospital medical staffs for years, the Illinois Psychological Association (IPA) filed suit.

Medical staff membership is important because under Illinois Rules members of the medical staff are the only practitioners who may 1) vote on hospital policy, 2) write treatment orders, and 3) independently admit patients to the hospital. Dentists and podiatrists are permitted to function in the hospital within the scope of their license provided that they co-admit their patients with a physician who maintains responsibility for the medical treatment of the patient. Admission—or co-admission—privileges are important for psychology to assure continuity of care for the patient who seeks outpatient psychotherapy from a psychologist and who may need to be treated in the hospital setting. It is well recognized that psychologists independently diagnose their patients as well as develop and implement treatment plans. It is a matter of common practice that psychologists determine a patient's treatment plan and write the treatment orders for their hospitalized patients. Finally, unless psychologists are recognized as voting on hospital policy, the expertise of psychologists

and the advantages of the psychological model as opposed to the drug model of treatment will be overshadowed.

At first glance, this issue seems like standard fare, with the courts left to battle out the legalities and interpretation of a regulation. But closer examination shows the matter to be much more controversial. The Department's restrictive interpretation was made at the behest of the Hospital Licensing Board, a group of people appointed by the governor to advise the Director of the Department of Public Health. At the time, the Board had two physician members. Both of those physicians were psychiatrists and were outspoken critics of the increasing competition faced by psychiatrists from other providers such as psychologists, social workers and even non-psychiatrist physicians. They first expressed the view that the Department's regulation limited medical staffs to the above-mentioned providers. But before they formalized their view, they asked the Department's legal staff to provide its interpretation.

The Department's legal staff responded with a memorandum which concluded that "traditionally" the regulation was interpreted to mean that hospitals were free to appoint any licensed health care provider, such as psychologists, to their medical staffs. Apparently not satisfied with this response, the Board went directly to the recently appointed Director of the Department for his interpretation. The Director, a physician, wrote a letter setting forth his view that the medical staff was limited to physicians, osteopaths, dentists and podiatrists. And though the Director explained that while the Department had not enforced the regulation this way in the past, it would start to do so now. The Director even proposed a new regulation to make his restrictive interpretation "clear." It was when that new regulation was about to go into effect that the IPA sued.

The IPA's lawsuit was never finally decided on the merits. In fact the IPA never tried the merits of the case. The consent judgment requiring the Department to consider changing the rule to list psychologists among those eligible for medical staff membership ended the case. But some of the facts discovered through the lawsuit lend support to the view that this whole episode was the result of professional and economic protectionism.

First, the Director admitted under oath in the case that he did not consider the public health in making his restrictive interpretation of the regulation. He did not consider the training or capabilities of psychologists, the impact of excluding psychologists on the quality of care in general or on the continuity of care to particular patients, or the fact that excluding psychologists would reduce competition and patient choice. Nor did he consider that psychologists had served on medical staffs in Illinois hospitals for years and that the Department had never taken action against either the psychologists or the hospitals.

Second, what the Director said he did consider the regulation itself, some legislative history on the regulation and the views of the Department's legal staff. But even then, the Director testified that he did *not* review the memo prepared by his legal staff which concluded that psychologists traditionally *were* eligible for medical staff membership. And the legislative history contained only two comments from Board members who originally approved the regulation in the early 1970's. Both of those comments support psychologist membership on medical staff. One member said that "we are not trying to force any hospital by having decided who shall make up its staff membership, but to regulate and not prohibit appointment of (licensed) individuals." The other Board member agreed, stating that "it is not the intent to require that persons licensed by the Department be appointed to Medical staffs "but rather it is the intent *not to prohibit* their appointment" [italics added].

Third, though the Department denied that the state Medical Society (or any other medical organization) played any part in its interpretation, the Illinois Psychiatric Society filed a friend of the court brief in support of the Department's position. Their brief was paid for in part by the American Psychiatric Association. The Illinois Psychiatric Society is the professional association of psychiatrists and its brief was, in effect, an *ad hominem* attack on psychology. The psychiatrists argued that psychologists were unqualified to practice psychiatry.

But the IPA made its position clear from the outset. This case was not about psychologists practicing medicine or prescribing drugs. Psychologists are not qualified or licensed to do that. They are, however, qualified and licensed to practice psychology. And only as members of a hospital medical staff can psychologists practice to the full extent of their licenses, the same way physicians, osteopaths, podiatrists and dentists practice to the full extent of theirs.

Indeed, the Department's own witness admitted that psychology offers proven and well recognized methods of treatment for mental illnesses. Psychological treatment often provides a better alternative to medical treatment. In many circumstances, psychological treatment is a valuable complement to medical treatment. Clinical psychologists are among the highest trained health care professionals, having earned a doctoral degree from an approved program, having completed two years of supervised experience, and having passed a comprehensive national examination. Perhaps most significantly, many psychiatrists work closely with psychologists in many situations, in and out of hospitals.

In light of all this, it is just too bad that the Department apparently bowed to special interests. It is too bad for the psychologists who find their stature, professionalism and pocket books, demeaned. It is too bad for the hospitals who have lost the right to decide without interference from the state who can and cannot be on their medical staffs. And it is too bad for the public which cannot count on the Department of Public Health having the public health in mind when it makes rules and regulations.

The Department has sent letters to all hospitals in Illinois specifying the restriction on medical staff membership. IDPH policy currently prohibits psychologists from membership on the medical staff. We should point out that non-membership on the medical staff will not further restrict any privilege granted by the hospital to a psychologist other than independent admission to a hospital and voting on medical staff issues. The new interpretation does not prohibit psychologists from having voting rights on the hospital professional or scientific staff—a category of staff often reserved for doctoral level or independently licensed practitioners.

Finally, the Department may be faced with an enforcement problem regarding this issue. The only power the Department has for non-compliance is to revoke the hospital's license. It is hardly likely that such a scenario would occur. We hope that this interpretation does not lead to a curtailment of the role and services psychologists provide in the hospital setting.

References

Albee, G. W. (1975). To thine own self be true. *American Psychologist, 30,* 1156–1158.

Albee, G. W. (1977). Does including psychotherapy in health care insurance represent a subsidy to the rich from the poor? *American Psychologist, 32,* 719–721.

American Psychological Association. (1972). Guidelines for conditions of employment of psychologists. *American Psychologist, 27,* 331–334.

American Psychological Association. (1981). Specialty guidelines for the delivery of services by clinical psychologists. *American Psychologist, 36,* 640–651.

American Psychological Association. (1987). Model act for state licensure of psychologists. *American Psychologist, 42,* 696–703.

American Psychological Association. (1988a). Court rules in favor of psychiatrists in hospital privilege case. *Practitioner Focus, 2,* 4.

American Psychological Association. (1988b). Tom Cooke appointed to JCAHO. *Practitioner Focus, 2,* 18.

Bales, J. (1988; April). ERISA: Another victory. *The APA Monitor.* 20.

Beach, D. A., & Goebel, J. B. (1983). Who is a psychologists? A survey of Illinois Yellow Page directories. *Professional Psychology, 14,* 797–802.

Bennett, B. E. (1988). Professional regulation and sunset. In L. M. Foster, B. E. Bennett, W. K. Carrol, M. B. Epstein, A. R. Howard, & M. C. Weber (Eds.), *The Illinois mental health professional's law handbook.* Chicago: Illinois Mental Health Professional's Law Handbook Editorial Committee.

Bersoff, D. N. (1983). Hospital privileges and antitrust laws. *American Psychologist, 38,* 1238–1242.

Boswell, D. L., Litwin, W. J., & Kraft, W. A. (1988, August). *Medical staff membership and clinical privileges: A national survey.* Paper presented at the meeting of the American Psychological Association, Atlanta.

Boudewyns, P. A. (1986). *Psychological research in the Veterans Administration: The Boulder model works here.* Unpublished manuscript. (Available from P. A. Boudewyns, Chief of Psychology Section, VA Medical Center, Augusta, GA 30910).

Budman, S. H. (1985). Psychotherapeutic services in the HMO: Zen and the art of mental health maintenance. *Professional Psychology: Research and Practice, 16,* 798–809.

Buie, J. (1987a; August). Field put off by Kennedy trade-off. *APA Monitor,* pp. 32–33.

Buie, J. (1987b; October). Kennedy health bill stalled in committee. *APA Monitor,* p. 25.

Buie, J. (1988a, August). CA: Group readies to fight hospital privileges ruling. *APA Monitor,* p. 19.

Buie, J. (1988b, April). CA: Hospital battle raging. *APA Monitor,* p. 14.

Buie, J. (1988c, October). Court to review ruling on hospital privileges. *APA Monitor,* p. 27.

Buie, J. (1988d, April). Health insurance leaps first hurdle. *APA Monitor,* p. 19.

Buie, J. (1988e, January). Psychology wins place in health insurance bill. *APA Monitor,* p. 12.

Buie, J. (1988f, March). New data show HMO dream unfulfilled. *APA Monitor,* p. 18.

Buie, J. (1990, February). President signs Medicare bill: Victory caps uphill trek. *APA Monitor,* pp. 1, 17.

Cheifetz, D. I., & Salloway, J. C. (1984). Patterns of mental health services provided by HMOs. *American Psychologist, 39,* 495–502.

Committee on Professional Practice of the Board of Professional Affairs. (1985). *A hospital practice primer for psychologists.* Washington, DC: American Psychological Association.

Committee on Professional Practice of the Board of Professional Affairs. (1988). *Hospital practice: Advocacy issues.* Washington, DC: American Psychological Association.

Cummings, N. A. (1977). The anatomy of psychotherapy under national health insurance. *American Psychologist, 32,* 711–718.

DeLeon, P. H. (1981). The medical expense deduction provision: Public policy in a vacuum? *Professional Psychology, 12,* 707–716.

DeLeon, P. H. (1987, November). The importance of S123-Medicare. *Independent Practice, 7,* 9.

DeLeon, P. H., & VandenBos, G. R. (1983). The new federal health care frontiers: Cost containment and wellness. *Psychotherapy in private practice, 1,* 17–32.

DeLeon, P. H., VandenBos, G. R., & Kraut, A. G. (1984). Federal legislation recognizing psychologists. *The American Psychologist, 39,* 933–946.

DeLeon, P. H., Uyeda, M. K., & Welch, B. L. (1985). Psychology and HMOs: A new partnership or new adversary? *American Psychologist, 40,* 1122–1124.

Dorken, H. (1981). The hospital practice of psychology. *Professional Psychology, 12,* 599–605.

Dorken, H. (1988). Psychotherapy in the market place: CHAMPUS 1986. *Psychotherapy, 25,* 387–392.

Ebert-Flattau, P. (1980). *A legislative guide.* Washington, DC: American Psychological Association.

Enright, M. F. (1986; July). Hospital rounds. *Independent Practitioner, 5,* 14.

Enright, M. F. (1987; January). Hospital rounds. *Independent Practitioner, 7,* 20.

Fisher, K. (1988; April). A new APA? *APA Monitor,* p. 1.

Folen, R. (1985). Interview with Representative Cecil Heftel. *American Psychologist, 40,* 1131–1136.

Gross, S. J. (1978). The myth of professional licensing. *American Psychologist, 33,* 1009–1016.

Hess, H. (1977). Entry requirements for professional practice of psychology. *American Psychologist, 32,* 365–368.

Joint Commission on Accreditation of Hospitals. (1983). *Accreditation manual for hospitals.* Chicago: Author.

Joint Commission on Accreditation of Hospitals. (1987). *Accreditation: Manual for hospitals—1988.* Chicago: Author.

Julius, S. M., & Handal, P. J. (1980). Third-party payment and national health insurance: An update on psychology's efforts toward inclusion. *Professional Psychologist, 11,* 955–964.

Kelly, D. M., Garrison, E. G., & DeLeon, P. (1987, Spring). Psychology and public policy: Why get involved? *Clinical Psychologist, 40.*

Kiesler, C. A., Cummings, N. A., & VandenBos, G. R. (1979). *Psychology and national health insurance: A source book.* Washington, DC: APA.

Kilburg, R. R., & Ginsberg, M. R. (1983). Sunset and psychology or how we learned to love a crisis. *American Psychologist, 38,* 1227–1231.

Kisch, J., & Austad, C. S. (1988). The health maintenance organization: 1. Historical perspective and current status. *Psychotherapy, 25,* 441–448.

Koocher, G. P. (1979). Credentialing in psychology? Close encounters with competence? *American Psychologist, 34*, 696–702.

Landers, S. (1988, December). APA scores hat trick in Congress. *APA Monitor*, p. 1.

Levinson, M. R., & Bennett, B. E. (1988, July). IPA hospital lawsuit settled. *Make the Future: The Health Service Advisory Board Letter.* Available from the Illinois Psychological Association, 203 N. Wabash, Chicago, IL 60601.

Liese, B. S. (1986). Physicians' perceptions of the role of psychology in medicine. *Professional Psychology, 17*, 276–277.

Litwin, W. J., Kraft, W. A., & Boswell, D. L. (1988, October). *The hospital practice of psychology: Current trends and issues.* Paper presented at the annual meeting of the Louisiana Psychological Association, New Orleans.

McSweeny, A. J. (1977). Including psychotherapy in national health insurance: Insurance guidelines and other proposed solutions. *American Psychologist, 32*, 722–730.

Meltzer, M. L. (1975). Insurance reimbursement: A mixed blessing. *American Psychologist, 30* 1150–1156.

Meyer, J. D., Fink, C. M., & Carey, P. F. (1988). Medical views of psychological consultation. *Professional Psychology, 19*, 356–358.

Mikesell, R. H. (1984, January). Successful passage of the hospital privileges for psychologists bill in the District of Columbia. *Independent Practitioner, 4*, 18–19.

National Council of Mental Health Centers. (1988, November). Legislative advisory, Medicare reimbursement for CMHC psychologists: Instructions to intermediaries issued. Available from National Office, 12300 Twinbrook Pkwy., Rockville, MD 20852.

On behalf of science. (1988, Fall). *Science Agenda, 1*, 1.

Peres, K. E. (1986, Summer). ERISA and self-insured trusts. *Psychotherapy Bulletin, 21*, 15–17.

Phillips, B. N. (1982). Regulation and control in psychology—Close encounters with competence? *American Psychologist, 37*, 919–926.

Plunket, S., & Morgan, B. (1987). Memorandum—Hospital privileges for psychologists. Available from APA, 1200 17th St., Washington, DC 20036.

Random House. (1966). *The Random House Dictionary of the English Language.* New York: Author.

Salzinger, K. (1988, Fall). APA reorganization defeated: Now What? *Science Agenda, 1*, 8.

Social issues section looks at "freedom of choice." (1977, March). *Illinois Psychologist*, 10–14.

Staff. (1977, June). In or out? Psychology holds its own fate. *Advance*, p. 1.

Stenger, C. A. (1979). The role of the Veterans Administration in the emergence and development—and future—of clinical psychology as a profession. *Clinical Psychologist, 32*, 4.

Stromberg, C. D. (1986, August). Hospital staff privileges. *Register Report No. 25.*

Stromberg, C. D., Haggarty, D. J., Leinbenluft, R. F., McMillian, M. H., Mishkin, B., Rubin, B. L., & Trilling, H. R. (1988). *The psychologist's legal handbook.* Washington, DC: The Council for the National Register of Health Service Providers in Psychology.

Sweet, J. J., & Rozensky, R. H. (1991). Professional relationships. In M. Hersen, A. Kazdin, A. Bellack (Eds.), *The clinical psychology handbook (2nd ed.).* New York: Pergamon Press.

Thompson, R. J. (1987). Psychologists in medical schools: Medical staff status and clinical privileges. *American Psychologist, 42*, 866–868.

Tulkin, S. R., & Frank, G. W. (1985). The changing role of psychologists in health maintenance organizations. *American Psychologist, 40*, 1125–1130.

Turkington, C. (1986, July). ERISA exemptions widened. *APA Monitor*, p. 21.

West, P. R., & Lips, O. J. (1986). Veterans Administration psychology: A professional challenge for the 1980s. *American Psychologist, 41*, 996–1000.

Wiens, A. N., & Menne, J. W. (1981). On disposing of "straw people" or an attempt to clarify statutory recognition and educational requirements for psychologists. *American Psychologist, 36*, 390–395.

Wright, L. (1982). Incorporating health care psychology into independent practice. *The Independent Practitioner, 2*, 1–4.

Zaro, J. S., Batchelor, W. F., Ginsberg, M. R., & Pallak, M. S. (1982). Psychology and the JCAH: Reflections on a decade of struggle. *American Psychologist, 37*, 1342–1349.

CHAPTER 6

Professionalism in Medical Settings

Cynthia D. Belar

Introduction

It is well recognized that there is more to being a competent professional psychologist than specific knowledge and skills in the discipline of psychology and the delivery of psychological services. According to Sales (1983), who edited a monumental handbook on this topic, professionalism requires knowledge and skills related to (1) standards of professional practice; (2) professional organizations; (3) professional developments; (4) laws and regulatory processes affecting professional practice; (5) management and business; and (6) values and interests affecting professional decision-making. Other chapters in this handbook deal with many of these topics as they pertain to clinical psychologists working in medical settings. These chapters address professional issues through the delineation of the characteristics and politics of the medical environment, special educational and training requirements, accountability, and economic concerns.

This chapter also addresses professional issues for psychologists within medical settings, especially with respect to the roles that psychologists occupy and the functions that they serve. These roles and functions include education and training, clinical service (consultation, diagnostic assessment, and intervention), administration, and research. This chapter also includes a special focus on the psychologist's behavior. As noted in a primer on the practice of clinical health psychology, the personal conduct and attitude of the practitioner determines the difference between success and failure in the medical setting (Belar, Deardorff, & Kelly, 1987). So it is for all clinical psychology. Because issues in professionalism arise in relation to intradisciplinary behavior, interdisciplinary behavior, and relationships to the consumers of professional services, it is within these contexts that professionalism is discussed here. The underpinnings for these discussions are the "Ethical Principles of Psychologists" of the American Psychological Association (APA, 1990), the "Specialty Guidelines for Delivery of Services by Clinical Psychologists" (APA, 1981), and the author's observations during some 15 years' experience in medical settings.

Cynthia D. Belar • Department of Clinical and Health Psychology, University of Florida Health Science Center, Gainesville, Florida 32610.

General Issues of Professionalism in Medical Settings

A general issue related to professionalism in the medical setting concerns Ethical Principle 2f:

> Psychologists recognize that personal problems and conflicts may interfere with professional effectiveness. Accordingly they refrain from undertaking any activity in which their personal problems are likely to lead to inadequate performance or harm to a client, colleague, student or research participant. (APA, 1990, p. 391)

In the medical setting, psychologists are often confronted with situations that involve death, disability, and disfigurement. Personal concerns triggered by these experiences may result in an inability to work effectively with certain populations (e.g., cancer patients, burn unit patients, or head and neck surgery patients). There is also a risk that actual harm may be done to some patients, for example, if the psychologist were to communicate revulsion in the presence of a postmastectomy patient or were to refuse to look at the stoma of a colostomy patient. Clinical psychologists new to medical settings need to work through personal issues related to body image and physical vulnerability and must develop some awareness of their own concerns regarding death and dying before undertaking unsupervised research or practice. At a minimum, a period of acclimatization is required.

It is also important that psychologists possess a high frustration tolerance of problems that arise from being a Ph.D. in the M.D.-controlled medical setting. Despite the many opportunities for professional development found there, significant clashes in values can occur, and the psychologist is usually decidedly less powerful. The attending physician usually has the ultimate authority over the treatment of his or her patient; physicians also usually control hospital policymaking committees. Although there has been increasing attention to relationships among behavior, emotions, and health, sometimes only lip service is given to the role of the psychologist. Patients may be discharged before the consultation is completed; advice on patient management may be completely disregarded (as may be the advice of any consultant). Some physicians appear more interested in the disease and its related medical technology than in the patient who has the disease, and overtly eschew the "soft" behavioral sciences. Many psychologists have reported feeling as if they were treated as second-class citizens although this situation is by no means universal. Ideally, the psychologist is persevering and patient and can get along well on a thin schedule of external reinforcement. In the author's experience, psychologists with a strong need for validation by others and positive recognition by broad groups of physician referral sources may not do well in the long haul unless well supported by a strong psychology group. All psychologists need to attend to their own needs related to burnout prevention.

It is also noteworthy that M.D. control of medical settings has weakened substantially since the early 1980s, as corporate health care has become more pervasive. Thus, the psychologist will increasingly need to develop skills in dealing with business-minded, bottom-line-oriented administrators as well. Although there may be significant differences in the values of these groups, psychology shares with business a commitment to data, and this can provide fruitful avenues for communication as well as opportunities for collaboration in the development of mutually interesting projects.

Despite the tensions noted above, psychologists must be careful to avoid being overly defensive about the M.D.–Ph.D. issue. Shows (1976) pointed out that a readiness to project conflict into professional interactions can result in a defensive or aggressive stance that makes collaboration difficult. With time and experience, the psychologist learns that significant conflicts also occur among various subspecialties within medicine (e.g., medicine and surgery), and that some physicians actually *prefer* to consult with psychologists because of their expertise in the measurement of behavior and its change.

Another general professional issue in the medical setting is the nature of the setting itself. Within medical settings, disease is treated or prevented, and the goal is good health. Although recognizing that personal behavior is a private matter, Ethical Principle 3 clearly states that "psychologists are sensitive to prevailing community standards and to the possible impact that conformity or deviation from these standards may have upon the quality of their performance as psychologists" (APA, 1990, p. 396). Thus, psychologists in medical settings need to be aware of personal health habits (e.g., smoking and weight) and to make decisions about acceptable, ethical public behavior. Social learning theory underscores this need as well, especially as it relates to therapeutic effectiveness (Bandura, 1969). Should the psychologist offer a smoking cessation program with a pack of cigarettes visible in his or her pocket? Appropriate role modeling is, of course, relevant to all areas of professional activity within the medical setting.

Related to health behavior modeling is the stimulus value of the individual psychologist. In engaging in any professional activity, it is important for the psychologist to understand whether he or she has any peculiarities in manner that could interfere with the establishment of rapport. Mental health professionals have not always had good press in medical settings; stereotypical "shrink" behavior is often ridiculed. Although psychologists tend to be inquisitive and cerebral in their approach to problems (e.g., puzzling about the why's), physicians are often action-oriented individuals in search of concrete solutions and how-to formulas. Schenkenberg, Peterson, Wood, and DaBell (1981) found that physicians valued the following qualities in a psychological consultant: pleasant, personable, friendly, compassionate, empathic, sensitive, interested, available, able to communicate effectively, cooperative, intelligent, open, perceptive, and displaying common sense. Common sense dictates that psychologists need to behave in ways that do not alienate other professionals.

Professionalism in Education and Training

Clinical psychologists in medical settings may be engaged in a variety of educational and training activities with members of their own discipline, with those in other disciplines, and with consumers of professional services. For example, a psychologist may be a supervisor of a clinical psychology intern or fellow, a trainer in interviewing skills for family practice residents, or a provider of psychoeducational programs on stress management to hospital staff and patient groups. There are a number of intradisciplinary, interdisciplinary, and consumer-oriented issues in the provision of educational services.

Intradisciplinary Issues

The psychologist needs to be aware of the ethical and legal issues involved in the supervision of trainees. Newman (1981) articulated a number of ethical issues related to supervision, perhaps the most important being competence in the specific area of practice supervised and the avoidance of exploitation in the supervisory relationship.

There are also legal issues associated with trainee supervision in that "the relationship of an assistant to a licensed professional is, legally, akin to an 'extension' of the professional himself" (Cohen, 1979, p. 237). Supervisors in medical settings need to be aware of their increased malpractice liability because of this extension. Supervision should be based on the needs of the supervisee *and* the patient; a record of supervisory activity should be kept. Although there are explicit rules in medical settings regarding the cosignatures required for inpatient chart notes, detailed regulations for outpatient charts have not as yet been mandated by the Joint Commission for the Accreditation of Healthcare Organizations (JCAHO). In addition to conforming to the recordkeeping guidelines found in the "Specialty Guidelines for Delivery of Services by Clinical Psychologists" (APA, 1981), it behooves the clinical supervisor to periodically document agreement with the trainee's diagnostic formulation and treatment plan in the patient's chart.

Although supervisory activities are crucial to the development of future psychologists and are frequent among clinical psychologists, few are actually trained in supervision (Hess & Hess, 1983). Moskowitz and Rupert (1983) reported that 39% of clinical psychology graduate students had experienced a major conflict with a supervisor that made it difficult for them to learn; 20% of these conflicts were attributed to theoretical orientation, 30% to the supervisor's style, and 50% to a personality conflict. Discussions between supervisor and student helped resolve issues related to style and theoretical orientation but were somewhat less effective when personality issues were involved. Nearly a quarter of students having major conflicts failed to discuss these with their supervisor, and of these, a significant proportion handled their concerns by censoring their verbal reports to supervisors and their progress notes. Twenty five percent indicated that they *appeared* to comply with the supervisor, while doing what they wanted to in therapy sessions. Because this kind of solution not only affects learning but can have direct effects on the quality of professional care provided, supervisors need to be especially alert to, and to solicit, feedback regarding supervisory processes to determine whether problems exist.

Carifio and Hess (1987) suggested that good supervisors demonstrate flexibility, concern, attention, investment, curiosity, and openness. Although exhibiting characteristics valued in psychotherapeutic relationships (respect, empathy, concreteness, and appropriate self-disclosure), these authors asserted, the ideal supervisor avoids conducting psychotherapy with supervisees. Freeman (1985) indicated that supervision is maximized by giving feedback in an objective, timely, and clear manner, and by providing problem-solving alternatives.

In addition to training in specific knowledge and skills, supervisors also provide role models for professional behavior. While modeling the attributes described above, psychologists should also demonstrate "due regard for the needs, special competencies and obligations of their colleagues in psychology and other professions" (APA, 1990, p. 393). Within psychology, the modeling of intradisciplinary relationships has not always been positive. Since the mid-1970s, this author has worked with over 100 interns and fellows representing a variety of graduate departments. She has repeatedly heard from these trainees how graduate faculty openly

demean clinical course work and denigrate professional practice as a legitimate activity. In fact, students have viewed such attitudes and behaviors as actually being sanctioned by psychology department authorities. More recently, arguments for the separation of professional and scientific training at the 1987 National Conference on Graduate Education in Psychology reflected the common assertion that these areas of psychology are "basically incompatible" and that, in fact, "familiarity breeds contempt."

Despite these forces, the national conference delegates adopted resolutions reflecting a continuum of training emphases and reasserted that training in the *conduct* of research was fundamental to the *entire* discipline of psychology (Resolution 1.3, 1987). They also asserted that academic departments should discourage attitudes and behavior that disparage work in nonacademic settings (Resolution 5.3c). Both of these resolutions reinforce the centripetal as opposed to the previous centrifugal trends in American psychology, which were more fully developed by Altman (1987) in his excellent analysis of converging and diverging forces within disciplinary, educational, and societal contexts.

Tensions in psychology exist not only between models of professional training, but among different specialty areas as well (e.g., clinical and counseling psychology). Sweet and Rozensky (in press) noted that psychology must improve its intradisciplinary relationships, especially as they affect the public image of psychology. A failure in this regard will have negative consequences in the health care marketplace. These authors offered some excellent suggestions for professional behavior.

When interacting with other psychologists, *don't*:

1. assume your training background is superior to that of your colleagues, regardless of the type of specialty training completed or the type of degree granted,
2. be "narrow minded" with regard to the type of theoretical orientation, clinical techniques, or modes of practice which you publicly state are acceptable,
3. be unduly or overly critical with regard to peer review of research or clinical activities of your colleagues,
4. procrastinate or otherwise passively obstruct

the work of colleagues when asked to perform some type of peer review function, or

5. resist or obstruct the attempts of patients or representatives of the patient as they seek second opinions in either clinical or forensic situations.

Sweet and Rozensky strongly encouraged psychologists to "provide reasoned and thoughtful public statements regarding alternative theoretical orientations, competence, clinical techniques and modes of practice."

Interdisciplinary Issues

Psychologists in medical settings find it necessary to continually educate other professions regarding the discipline of psychology. Intradisciplinary conflicts that are poorly handled can result in conflicting messages that confuse other groups. Psychologists in medical settings often need to remind physicians that medication recommendations are beyond their scope of practice, that the appropriate method of consultation is through a referral question as opposed to a request for the administration of a specific test, and that, despite overlap among the mental health professions, psychologists have unique skills that are valuable in health care delivery. Medical center administrative personnel often need to be educated in the latter point as well. Psychologists in medical settings should not take it for granted that others understand the nature of their training or their areas of competency; they need to provide this education in a manner that is neither arrogant nor defensive.

Another important educational activity is helping the physician learn to prepare the patient for a psychological consultation. Bagheri, Lane, Kline, and Araugo (1981) reported that 68% of patients at one medical center had not been informed by their physicians that a psychiatric consultation had been requested, often because the physician feared that the patient would view such a referral as an insult. Indeed, such is frequently the case, and the psychologist is then presented with a hostile patient whose attitudinal set makes an adequate assessment more difficult. Educating physicians about how to refer, and when to refer, is an important professional activity that can be facilitated by the development of written materials, such as easily accessible brochures or service descriptions. Seeing patients along with medical personnel within the hospital or medical clinic can also serve to model ways to introduce psychological services.

Issues Related to Consumers

On many occasions, psychologists are called on to provide specific educational materials to patients and other professional staff (e.g., on parenting, stress, burnout, and behavioral health). In addition, psychologists are increasingly called on by the media for presentations of research findings. Professionalism requires that the psychologist keep abreast of current developments in the field so as to ensure that public statements will be based on "scientifically acceptable psychological findings and techniques with full recognition of the limits and uncertainties of such evidence" (APA, 1990, p. 392). Since the mushrooming of information concerning relationships among health and behavior, some patients are actually reporting guilt feelings because they cannot cure their own illnesses (e.g., thwart tumor growth through imagery or prevent exacerbations of systemic lupus erythematosus through stress management). Psychologists must be especially careful about sensational claims that go beyond what has been validated through careful about sensational claims that go beyond what has been validated through careful scientific inquiry. Many consumers of psychological information cannot distinguish between causal and correlational models or understand that the usefulness of psychological treatments does not mean that the problems have been caused psychologically. Professional responsibility requires clear articulation of these concepts to both professional and nonprofessional consumer groups.

Educating patients also requires special attention to their health belief model, which in turn requires sensitivity to Ethical Principle 2d ("Psychologists recognize differences among people, such as those that may be associated with age, sex, socioeconomic, and ethnic backgrounds"; (APA, 1990, p. 391). Given the reluctance of some ethnic minorities to seek mental health services, it is possible that psychologists working in medical settings will be exposed to a wider cross section of society than those working in mental health facilities. Indeed, the psychologist is often called on to help negotiate between the physician's health model and that of the patient so as to arrive at a mutually satisfactory model related to treatment and outcome. In this regard, special care needs to be taken to avoid imposing the values of the health care provider or system on the patient. Because the application of expertise in behavior change to the health care system has meant more opportunities for coercive control of patients, this issue requires special attention.

Professionalism in Clinical Service

Psychologists perform a number of clinical service activities within medical settings, including consultation, diagnostic assessments, and intervention. As in other areas, effective functioning requires clinical knowledge and skills, adherence to professional standards, and an understanding of the sociopolitical aspects of the medical setting.

Intradisciplinary Issues

The psychologist working in a medical setting needs to learn another culture and language, not in an effort to become a "junior M.D.," but so that he or she can communicate effectively and can behave in a manner considered acceptable. In this author's experience, failures of this kind (e.g., "irrelevant" reports, gross misinterpretation of medical abbreviations, and inappropriate charting) have spoiled professional opportunities for subsequent psychologists. Significant deviations in areas such as dress or customary referral patterns can also lead to outright rejection (e.g., referring a patient to a subspecialist for additional evaluation without first going back to the referral source to seek agreement).

However, overidentification with medicine is to be avoided. Elfant (1985) warned against the inappropriate "medical socialization" of psychologists, noting that there are strong pressures in the health care system to come to bottom-line decisions and to "fix" people in a rather heroic and imperious manner. The psychologist needs to maintain the psychological treatment model that insists on autonomy and freedom of choice for both patient and therapist. The area of compliance interventions offers special problems in this regard; psychologists must carefully evaluate who the client is (the health care system or the patient) and must clarify "the nature and direction of their loyalties and responsibilities and keep all parties informed of their commitments" (APA, 1990, p. 393).

Another issue that the psychologist must consider is the adequacy of his or her prior training with respect to clinical practice in the medical setting. Although many clinical psychologists in medical settings practice primarily with "psychiatric" populations, they may be called on as well to see patients coming from "medical-surgical" populations, or they may specialize in this area of practice. In the latter case, education and training standards have been developed by the National Working Conference on Education and Training in Health Psychology (Stone, 1983). One cannot obtain such training in weekend workshops, as supervised clinical experience is viewed as essential to the training of the clinical health psychologist. Indeed, a specialty diplomate is currently under development for clinical health psychology by the American Board of Health Psychology. In the former case, clinical psychologists must use their own judgment about their competency to deal with a specific clinical problem, making use of consultation or supervision as appropriate.

Especially noteworthy are issues related to the use of standardized psychological tests in medical settings. Ethical Principle 8 clearly states that psychologists "strive to ensure the appropriate use of assessment techniques by others" (APA, 1990, p. 394). Many psychological tests have been normed on psychiatric populations; their generalization to medical-surgical patients is questionable. Psychologists need to be aware of possible differences in test interpretation, the availability of other normative data, the increased risk of the inappropriate use of test results by other disciplines, and the language used to convey results. No psychological test can absolutely "rule out" organic problems and provide the conclusion that a problem is "functional" in origin. In addition, one might conclude that it is unethical to include in the charts of medical-surgical patients the computer-generated intrepretations of psychological tests normed on psychiatric patients without addressing the issue of the validity of the report. Chapters in this book by Hartman and Kleinmuntz (Chapter 10) and by Sweet (Chapter 18) elaborate on these topics. An increased risk of successful malpractice suits is present in this area if the psychologist is not attentive to these issues.

Knapp and Vandecreek (1981) detailed some of the other malpractice risks that psychologists working in the area of behavioral medicine face. First, there is the danger of inadvertently practicing medicine without a license (e.g., providing medical diagnoses, recommending medication increase or decrease, or recommending withdrawal from standard medical treatment). Psychologists in medical settings must take the *initiative* in clarifying their areas of professional expertise. For patients with physical problems, collaboration with a physician in the initial diagnosis is crucial. In addition, the psychologist needs to inform the physician of changes in patient behavior that may influence the patient's physical status (e.g., changes in relaxation or arousal levels that may influence a diabetic's insulin needs). These consultations should also be documented.

Knapp and Vandecreek (1981) believe that malpractice risks are expanded for clinical health psychologists because the nature of the clinician–patient relationship does not militate against such suits as it does in more traditional areas of practice. In these more traditional areas, the therapist–patient bond is often emotionally close; the patient often suffers from low self-esteem and is quite dependent on the therapist. In addition, suing for emotional harm would require a discussion of personal problems in court and the possibility of being socially stigmatized as a "mental health" patient. The medical patient, who has often had

only a consulting or short-term therapy relationship with the clinician, may not be as reluctant to initiate a suit. In addition, it seems to be much easier to prove "physical" as opposed to "emotional" harm.

Other areas of potential malpractice suits are patient confidentiality and informed consent, topics addressed later in this chapter. Finally, the psychologist in the medical setting is often called in for consultation regarding a patient's danger to himself or herself or to others. The psychologist often has to work rapidly and without prior knowledge of the patient. Unfortunate outcomes in these cases can elicit blaming behavior on the part of significant others, with whom the clinician has often had no previous relationship.

Interdisciplinary Issues

In interdisciplinary functioning, an important issue to consider when receiving a request for clinical service is the reason for the referral. Sometimes, the problem is more staff-centered or family-related and is not a problem of the identified patient. Empathy for the perspective of professional staff, plus an in-depth understanding of the thinking styles, roles, functions, and stressors relevant to various medical units, facilitates understanding and is perhaps best obtained via naturalistic observation.

It is also noteworthy that there are a number of physicians who feel threatened by having to call for psychological consultation. The request for help often reflects a breakdown in human relations and is thus a blow to self-esteem in a culture where everyone is expected to have expertise in human relationships. It may actually be more threatening to call for psychological consultation concerning behavioral problems than to call for the medical expertise of another subspecialty. Tact is required to handle these situations.

It is hypothesized that professional arrogance in psychologists is relatively more damaging to collaborative relationships with physicians, in part due to the nature of the problems being addressed, than would be arrogance displayed by another medical specialist (e.g., a cardiologist to a family practitioner. (Belar *et al.*, 1987, p. 23)

Another difficulty in interdisciplinary relationships occurs when dealing with hostility and arrogance on the part of another discipline, a problem encountered in medicine as it is in any profession. In these cases, it is not always possible to understand, or to empathize with, the physician's perspective. But it is always possible to use a task-oriented focus on the patient's needs, highlighting these as the mutual goal. Emphasizing the benefits to the physician in changing her or his behavior toward the patient can lay the groundwork for change. Other coping strategies include consistent assertiveness, confrontation as necessary, and, most of all, a wealth of good humor.

In order to provide adequate services, the professional psychologist must have a working knowledge of the roles of other professionals in the medical setting and must call on them as appropriate. On many occasions, psychologists find themselves in the role of patient advocate to obtain needed services. It has also been suggested that patient advocacy is even more important for health professionals who work in HMOs, where the autonomy of the patient is less than in the fee-for-service model.

Good communication among professionals is also required. This is greatly facilitated by succinct, relevant, and direct verbal and written communications that contain concrete suggestions and are devoid of "psychobabble." Even seemingly innocuous terms may be somewhat alarming to, and thus may be misinterpreted by, other groups (as this author recently learned when discussing "cognitive restructuring" with neurological physicians). Whenever possible, the implications of findings for the behavior of health care providers should be addressed.

Methods of communicating in hospital charts are specified in the rules and regulations sections accompanying hospital bylaws, or in policies adopted by medical records committees. Psychologists working in medical settings must seek out these policies. For example, it is important to note that (1) only certain approved abbreviations can be used; (2) errors need to be corrected with a single line and must be initialed; (3) lines should not be skipped; (4) there must be some indication that the medical chart has been reviewed; and (5) in some hospitals, only black ink may be used. Special committees are charged with monitoring compliance with these regulations. Psychologists' notes that deviate from accepted procedure are regarded as violating accepted standards, which, at a maximum, may have implications for staff privileges and, at a minimum, will be seen as less than professional.

Another risk to professionalism in medical settings is the diffusion of responsibility that can occur when multiple providers are involved in the care of a patient—a not uncommon model in multidisciplinary care. Because so many aspects of care are occurring simultaneously, often with segmented areas of responsibility, there may be a tendency to see the patient as being cared for by the "team" or the "hospital," with less accountability for the individual practitioner. The psychologist must conform to ethical principles with respect to responsibility in patient care—ensuring quality of care, appropriate communication with other professionals, timeliness, and follow-up as needed.

In providing clinical services, the psychologist should pay special attention to the need for prompt responses and feedback to the referral source. In general, consultations for inpatients should be provided within 24 hours of the request; emergencies require a more immediate response. Although psychological solutions cannot be rushed inappropriately, there must be some sensitivity to the exigencies of the hospital setting and its cost containment mechanisms.

Issues Related to Consumers

An important consumer-oriented issue relevant to patient care is the confidentiality provided the patient. Special problems arise in medical settings because medical records may be widely circulated, cases are often discussed within the context of a multidisciplinary team, and family members may be involved. There are also setting-related issues in that hospital rooms do not always provide the privacy desirable for conducting a psychological evaluation. The limits of confidentiality need to be specifically stated to the patient in order to obtain informed consent, and the patient (especially the hospitalized one who did not initiate the appointment) should be given a very explicit option of declining the services.

The psychologist in the medical setting may also find that he or she has more responsibility for the physical health of the patient than anticipated. In fact, a change in health outcome may be the goal of treatment. Mismanagement of this responsibility may be grounds for malpractice, especially if one is viewed as practicing outside the scope of psychology licensure or as practicing medicine without a license. For example, difficulties can arise if efforts are made to reduce the need for blood pressure medication without appropriate (and documented) consultation with the prescribing physician. Likewise it is outside the boundaries of competence for psychologists to diagnose "tension headache." Treatment for headache should never be undertaken without a prior medical evaluation. In general, it seems prudent to record diagnoses of medical problems in a manner reflecting the source of the information (e.g., migraine headache per patient or per medical record).

Dilemmas for psychologists occur when, despite their lack of competency in the area, there is a need to judge whether a previous medical work-up has been adequate. Depending on areas of practice, psychologists should develop relationships with relevant medical specialists in whom they have confidence with respect to standards of practice, and to whom the psychologists can turn for informal, as well as formal, opinions.

On occasion, psychologists working in medical settings are likely to find themselves consulting on a patient who they then learn is receiving services from another mental health professional. Professional ethics requires extreme caution in such cases, proceeding with due regard to the therapeutic issues for the client: "If a person is receiving similar services from another professional, psychologists do not offer their own services directly to such a person" (APA, 1990, p. 393). However, such situations do present opportunities to enhance the information available to the treating professional, and psychologists should seek to collaborate in such endeavors for the welfare of the patient (Sweet & Rozensky, in press).

Professionalism in Administration

Professionalism in administrative activities includes accepting responsibility for the professional development of others within one's employ (Ethical Principle 7c, APA, 1990, p. 393). As noted by Sweet and Rozensky (in press), the nature of the working relationship should be made explicit with respect to hours, pay, duties, and the nature and amount of supervision. In performance evaluations, feedback should be timely and constructive and should bear on the behavior of the employee, as opposed to his or her personality features. Sensitivity and tact are often needed to ensure that feedback can be used by the employee, and that a working relationship will still be possible.

Adherence to professional ethics also prohibits psychologists from exploiting subordinates and engaging in dual relationships that could increase the risk of exploitation (Ethical Principles 6a and 7d, APA, 1990, p. 393). These principles cover behavior in educational, clinical service, and research activities as well as administrative activities, but it is perhaps in the administrative area that such influences might be less well recognized. For example, a chief of service could unduly influence subordinates to practice within a particular theoretical orientation by repeated criticism of other viewpoints, or by giving seemingly favorable performance evaluations to those with similar models. The administrator must make clear the differences between administrative issues, performance criteria, and his or her own personal professional viewpoint, while communicating due respect for alternative views. Many medical settings tend to operate in a more autocratic manner; thus, this culture of psychology is not always well understood.

It is important for the psychologist administrator in the medical setting to advocate for hiring and practice standards for psychologists. Silence in the face of events such as the inappropriate designation of an employee as a psychologist is not acceptable. Such designations actually happen, especially in state-affiliated medical schools or in state or federal hospitals that may be exempt from state psychology licensure laws.

An organization with a long history of involvement in administrative issues related to professional psychology is the Association of Medical School Professors of Psychology. How the profession of psychology is to be organized is a topic of continuing interest. Whereas many find homes outside departments of psychiatry to be conducive to professional growth, where then is the locus of professional responsibility for psychology, for example, certification for staff privileges (see Rozensky, Chapter 5) and quality assurance (see Jospe, Shueman, & Troy, Chapter 7)? Medical settings would rarely, if ever, permit the hiring of an anesthesiologist by a department of surgery without professional management by the department of anesthesiology. Why should psychology be any less an integrated, self-regulated discipline? A rare, but very successful, model of administrative organization is that at the University of Florida's Health Science Center, where the Department of Clinical and Health Psychology is an independent department, and the department chair exercises control over hospital staff privileges for all psychology practitioners within the teaching hospital.

A final area of advocacy for professional psychologists is organizational policies that are either discriminatory or not in keeping with the welfare of patients (Principle 3d, APA, 1990, p. 391). Once again, psychologists in managed health care systems, where patients may have less autonomy to seek services elsewhere, must actively pursue policies that contribute to patient welfare. The HMO Special Interest Group of the APA Division of Health Psychology offers opportunities for collaboration with respect to these issues.

Professionalism in Research

A number of professional issues arise with respect to research activities in the medical setting. Guidelines for behavior are found within the previously mentioned "Ethical Principles of Psychologists" (APA, 1990) and are fully addressed in the publication *Ethical Principles in the Conduct of Research with Human Participants* (APA, 1983). The introduction to the second document describes the complexity of the ethical considerations in research and makes a statement reflecting a fundamental role of clinical psychologists in medical settings: "For psychologists . . . the decision not to do research is in itself a matter of ethical concern since one of their obligations is to use their research skills to extend knowledge for the sake of ultimate human betterment" (p. 15).

One of the most frequent areas of professional conflict in the area of research is publication authorship. Principle 7f (APA, 1990, p. 393) clearly indicates that publication credit is to be assigned in proportion to the individual's professional contribution; however, this guideline is sometimes at odds with the practices extant in medical settings, where the chief of the department or the service gets authorship on all papers produced by his or her faculty. In a related matter, during the course of research grant reviews, this author has often seen a physician's name listed as principle investigator on a grant proposal that is clearly the work of the collaborating psychologist. In some of these cases, the physician obviously does not have the competence to chair the project adequately. Psychologists need to work toward changing such unprofessional and potentially exploitive practices, even if hospital bylaws need to be revised to allow psychologists to be principle investigators.

In the design of studies, the psychologist needs to minimize the chance that the findings will be misleading (Principle 1a, APA, 1990, p. 390). A common error in this regard, as it is in clinical practice, is the use of psychological measurements with patient populations for whom they have not been validated. For example, measures of depression developed for psychiatric patients are not necessarily appropriate when applied to arthritis patients, as some of the items may reflect the arthritis disease process itself rather than a clinical psychological depression. If this possibility is not considered, the levels of "depression" in arthritis sufferers will be reported as spuriously high. Inattention to design issues and the consequent misinterpretation of data have led to many false conclusions regarding the "personality types" of various disease groups; these conclusions, in turn, have often had negative consequences for certain groups of medical patients.

In the context of medical setting research, it has been noted that the subjects in a psychological study may be confused about whether it is relevant to their medical treatment. It is imperative that subjects be reassured that ongoing medical treatment will not be jeopardized by their refusal to participate in a psychological study, and that the investigator refrain from using the physician–patient relationship to coerce patients into participation. Psychologists conducting such research need to be especially sensitive to these potential problems for medical patients.

The competencies of the psychological investigator also need to be taken into account. Consider the following example. A psychologist advertises for a population of "headache sufferers interested in obtaining treatment for headache." These subjects are classified as having migraine, tension, or mixed headache on the basis of descriptive criteria obtained in interviews by the psychologist, and they are then assigned to either an electromyographic biofeedback treatment or a thermal biofeedback treatment. Has the investigator obtained adequate medical consultation regarding the diagnosis of these subjects? The kind of research performed by psychologists in medical settings often requires consultation with physicians in order to be conducted ethically.

Finally, although it is well known that all research must have institutional approval before the investigator proceeds (Principle 7e, APA, 1990, p. 393), psychologists need to consider the impact of their research on that of other investigators who may subsequently wish to conduct research in the institution. It also behooves psychologists to seek positions on research committees and institutional review boards in order to facilitate adequate reviews of other behavioral studies within the organization, and to share some of the responsibilities for professional conduct within the medical setting.

Summary

In summary, the thesis of this chapter is that clinical psychologists who work in medical settings must have more than just knowledge and skills in the discipline of psychology and its application. Professionalism requires sensitivity to sociopolitical issues, as well as adherence to established ethical principles and standards of practice. The issues regarding professionalism may be intradisciplinary, interdisciplinary, or consumer-oriented. They pervade psychologists' roles as educators, service providers, administrators, and researchers in medical settings.

References

Altman, I. (1987). Centrifugal versus centripetal trends in psychology. *American Psychologist, 42,* 1058–1069.

American Psychological Association. (1990). Ethical principles of psychologists. *American Psychologist, 45,* 390–395.

American Psychological Association. (1981). Specialty guidelines for delivery of services by clinical psychologists. *American Psychologist, 36,* 640–651.

American Psychological Association. (1983). *Ethical principles in the conduct of research with human participants.* Washington, DC: Author.

Bagheri, A. S., Lane, L. S., Kline, F. M., & Araujo, D. M. (1981). Why physicians fail to tell patients a psychiatrist is coming. *Psychosomatics, 22,* 407–419.

Bandura, A. (1969). *Principles of behavior modification.* New York: Holt, Rinehart & Winston.

Belar, C. D., Deardorff, W. W., & Kelly K. E. (1987). *The practice of clinical health psychology.* New York: Pergamon Press.

Carifio, M. S., & Hess, A. K. (1987). Who is the ideal supervisor? *Professional Psychology: Research and Practice, 18,* 244–250.

Cohen, R. J. (1979). *Malpractice: A guide for mental health professionals*. New York: Free Press.

Elfant, A. B. (1985). Psychotherapy and assessment in hospital settings: Ideological and professional conflicts. *Professional Psychology: Research and Practice, 16*, 55–63.

Freeman, E. (1985). The importance of feedback in clinical supervision: Implications for direct practice. *The Clinical Supervisor, 1*, 5–26.

Hess, A. K., & Hess, K. A. (1983) Psychotherapy supervision: A survey of internship training practices. *Professional Psychology: Research and Practice, 14*, 504–513.

Knapp, S., & Vandecreek, L. (1981). Behavioral medicine: Its malpractice risks for psychologists. *Professional Psychology, 12*, 677–683.

Moskowitz, S. A., & Rupert, P. A. (1983). Conflict resolution within the supervisory relationship. *Professional Psychology: Research and Practice, 14*, 632–641.

Newman, A. S. (1981). Ethical issues in the supervision of psychotherapy. *Professional Psychology, 12*, 690–695.

Sales, B. D. (Ed.). (1983). *The professional psychologist's handbook*. New York: Plenum Press.

Schenkenberg, T., Peterson, L., Wood, D., & DaBell, R. (1981). Psychological consultation/liaison in a medical and neurological setting: Physicians' appraisals. *Professional Psychology, 12*, 309–317.

Shows, W. D. (1976). Problem of training psychology interns in medical schools: A case of trying to change the leopard's spots. *Professional Psychology, 7*, 393–395.

Stone, G. C. (1983). Proceedings of the National Working Conference on Education and Training in Health Psychology. *Health Psychology, 2*(5) Supplement, 1–153.

Sweet, J. J., & Rozensky, R. H. (in press). Professional relationships. In M. Hersen, A. Kazdin, & A. Bellack (Eds.), *The Clinical Psychology Handbook* (2nd ed.). New York: Pergamon Press.

PART III

Practical Issues

Practice Management

With increasing demands for accountability, there is an increasing need to develop, implement, and maintain internal and collegial quality assurance procedures for psychological services in medical settings. Michael Jospe, Sharon A. Shueman, and Warwick G. Troy (Chapter 7) outline the specific quality assurance challenges facing psychologists. Clinical psychologists in the medical setting must demonstrate cost-effectiveness, as well as clinical effectiveness, as third-party payers continue to require accountability. Nicholas A. Cummings (Chapter 8), writing on financial efficacy, provides empirical evidence that psychological services can reduce the inappropriate use of expensive medical care among specific populations, as well as improve medical management and behavioral outcomes among the chronically ill. Jack G. Wiggins and Kris R. Ludwigsen (Chapter 9) outline numerous marketing strategies for psychologists in the medical setting and also offer practical guidelines for specific programs in psychiatric settings and nursing homes. Finally, although computers can simplify practice management, psychologists need to be sensitive to both the professional benefits and the ethical, legal, and scientific pitfalls of computers in clinical practice. David E. Hartman and Benjamin Kleinmuntz (Chapter 10) discuss the history, uses, ethics, advantages, and disadvantages of computers in psychological practice.

CHAPTER 7

Quality Assurance and the Clinical Health Psychologist

A Programmatic Approach

Michael Jospe, Sharon A. Shueman, and Warwick G. Troy

Introduction

The major concern of this chapter is the increasing need for the development, implementation, and maintenance of internal, collegial quality assurance (QA) procedures for mental health services in organized care settings. This need is reflected in the highly visible pattern of external demands for accountability, demands that have grown consistently in number, scope, and urgency since the early 1980s (Rodriguez, 1988a,b). These pressures include externally imposed regulatory mechanisms such as certification and accreditation, as well as pressures for QA from consumers, government, and other sanctioners. They include, as well, unrelenting calls for cost containment of health services from payers and purchasers of these services. Calls for cost containment are often, if

not invariably, linked with demands for improved quality of care (Brook & Kosecoff, 1988).

In hospitals, which continue to consume the largest portion of public and private health care dollars (Waldo, Levit, & Lazenby, 1986), demands for cost containment and quality assurance have been felt more acutely than in any other health care setting. With economic survival at stake in an entrepreneurial and competitive marketplace, pressures to maintain budgets have posed significant difficulties in guaranteeing acceptable levels of care (Fuchs, 1988). For the hospital-based health psychologist, there is a significant challenge: not only to provide high-quality psychological consultation services, but also to devise and adhere to mechanisms for monitoring the delivery of services in order to demonstrate that they are necessary, appropriate, and cost-effective.

In our experience, health care psychologists working in hospital settings, although having a general familiarity with the elements of the medical-chart-based review system, are often significantly less comfortable with the rationale and theoretical bases of QA in its generic

Michael Jospe • California School of Professional Psychology, 1000 South Fremont Avenue, Alhambra, California 91803-1360. **Sharon A. Shueman and Warwick G. Troy** • Shueman, Troy & Associates, 246 North Orange Grove Boulevard, Pasadena, California 91103.

sense. More particularly, the psychologist tends not to assume ownership of the medical model of services review, even though he or she continually contributes to it. Thus, one of the prime messages to be developed in this chapter is that *psychologists must themselves assume formal responsibility for initiating and maintaining a quality assurance system for their own consulting services*, a QA system that is concurrent with and contributes to that used to "assure" (ensure) the quality of the medical and surgical services provided at the hospital. The accountability pressures mentioned above strengthen the need for this kind of discipline-centered review; the forces of economic competition make it the more urgent.

In this chapter, we try to inform the reader about quality assurance in general and to identify a number of specific challenges facing the clinical health psychologist who wishes to participate in implementing effective quality assurance activities in medical settings. The focus is on what the clinical health psychologist, as part of the interdisciplinary team, can contribute from within the service delivery setting. We distinguish this kind of QA from the external imposition of standards by regulatory agencies and other sanctioners. We also attempt to provide the reader with a concise model for developing and implementing QA activities in this setting.

In striving to be pragmatic, we emphasize applications of the principles of QA and provide examples that illustrate the circumstances prevailing in the working lives of clinical health psychologists in hospital settings. There is a significant amount of rigor implied in our definition of the QA process, including formal procedures for establishing and evaluating adherence to explicit standards. What we wish to emphasize, however, is that many of the activities that best serve the QA function in many service delivery settings are not explicitly identified as QA activities per se. Rather, they are viewed as integral parts of service provision that are implemented to satisfy the professional expectations of those persons working in the delivery system.

We find that professional psychologists tend to be unfamiliar with QA precepts and practices. We attribute this unfamiliarity both to the nature of the field and, probably more importantly, to inadequacies in professional psychology training programs (Shueman & Troy, 1982). Because of this perceived lack of knowledge, we provide some formal, though elementary, information about QA.

Absent in this chapter is a specific treatment of managed care (Henderson & Collard, 1988; Shueman, 1987) and issues related to viewing inpatient care as one component of a comprehensive service continuum (Evashwick & Weiss, 1987). Managed approaches are believed to offer significant hope for ensuring a high quality of care as well as cost containment. Although questions have inevitably been raised concerning the quality of the health and mental health services offered through managed programs, the authors are among those who believe in the promise of such models. The managed-care model and its QA correlates are omitted from discussion here only because the designated focus is on inpatient treatment settings rather than on the continuum of care.

A final caveat is that any QA plan that is developed specifically for a hospital unit or department such as psychological services needs to be functionally linked to the hospital-wide comprehensive QA plan, the existence of which is mandated by the Joint Commission for Accreditation of Healthcare Organizations (JCAHO). It might be expected that the standards promulgated by the JCAHO (JCAH, 1984a,b) would determine the nature of quality assurance efforts in hospitals. In fact, however, the scope and focus of the approach embodied in the standards are such that they serve mainly as a broad-based set of normative guidelines, particularly for ancillary services such as those provided by psychological service units.

Because the authors' intent is to examine in some detail the practicalities of the structure and function of a QA system that is appropriate to the psychological services unit, our treatment of the JCAHO is essentially a cursory one. The reader is, however, encouraged to become familiar with the sets of standards for hospitals and psychiatric facilities. These standards designate the context in which all quality assurance activities take place.

Chapter Organization

This chapter consists of five sections. The first section introduces working definitions of quality and QA, exemplified by a series of practical questions, each of which provides a different perspective on service quality as it relates to the planning and delivery of psychological care in hospitals. The second section provides a standard, more formal definition of the process of quality assurance and includes examples that demonstrate how the process may be implemented in hospital settings.

In the third section of the chapter, we suggest guidelines for avoiding the difficulties commonly encountered in clinical settings when implementing QA activities. The fourth section presents a model for implementing a quality assurance program in the multidisciplinary health care setting. Finally, the fifth section offers a brief analysis of the political and administrative considerations associated with gaining support for psychological quality assurance activities in medical settings.

What Is Quality Assurance?

We have elsewhere defined *quality assurance* as "any formal activity implemented within the service delivery system to improve the outcome of care" (Shueman & Troy, 1988, p. 267). Obviously, this definition covers a broad range of activities, as an almost endless number of variables could conceivably affect treatment outcome. These variables include the clinician's technical and interpersonal skills, the appropriateness of the technology applied by the clinician, the patient's understanding of and compliance with the treatment regimes, factors outside of the treatment context such as the patient's family or social support, and the adequacy of the communication among professional treatment team members. The variables are generally viewed as reflecting two major components of the process of health care: "tech-

nical care and the management of the interpersonal relationship between the practitioner and the client" (Donabedian, 1982, p. 4).

Dimensions of Quality

In practice, quality assurance studies focus on any of five dimensions of quality: *accessibility*, or the extent to which the consumer has the potential to use the health care system; *necessity*, the extent to which formal evaluation determines that professional care is indicated; *appropriateness*, the extent to which both the level and the type of care provided have been indicated for the given condition; *efficacy*, the extent to which the treatment does what it is intended to do; and *cost-effectiveness*, the extent to which a particular service program yields effective care commensurate with costs.

QA systems developed for hospital-based psychological consultation and liaison services tend not to involve all five of these dimensions. The responsive QA program does, however, need formally to consider necessity, appropriateness, and efficacy. Cost-effectiveness evaluation requires a technical expertise beyond that of many psychologists. Furthermore, if the system focuses effectively on the necessity and the appropriateness of services, cost considerations tend to be accounted for.

Whereas the benefits to any QA system that directly confronts problems of efficacy are well documented, there are very few examples within the psychological services literature of creative approaches to the assessment of necessity and appropriateness of care. This is particularly unfortunate because criterion development within these two domains is probably less exacting a task than it is for efficacy. A QA program that can thoroughly examine whether professional care is required at all and whether, if required, it ought to be delivered at that particular level has significant consequences for patient welfare, staff and professional time, and program costs.

QA program development involves identifying appropriate indicators for each of the dimensions and reconstituting them in operational terms in the form of functional criteria.

The process of criterion development and use is described in a later section of this chapter.

Consider now an inpatient medical setting where a psychological consultation team provides services to patients who are admitted for the treatment of medical and surgical problems. In such a setting, one may initially attempt an assessment of the quality of the services provided by the consultation team by asking any of a series of questions, the answers to which would, presumably, reflect aspects of quality of care such as we have considered above. A set of such questions is found in Table 1.

These questions address a range of normative issues with respect to an organized system of psychological services. Questions 1–3 refer to the adequacy and sufficiency of the resources necessary for the implementation of the psychological services and to the patient's access to these services. Questions 4–6 denote the actual activities of service implementation. Finally, Question 7 refers to the consequences of the psychological intervention. These three categories of questions exemplify the three foci or domains of QA:

1. *Input.* Input activities include resources such as staff, facilities, funding, and access to services, which are prerequisites for effective service delivery.
2. *Process.* Process activities consist of the actual delivery of services and embrace both the technical aspects of the services delivered by a clinician and interpersonal factors such as the clinician's warmth and empathy.
3. *Outcome.* Objective and subjective changes in patients that may be attributed to the delivery of the psychological services are referred to as *outcomes of treatment.* QA activities focused on outcome may assess, for example, whether patient functioning is improved as a result of the interventions, and whether patients are satisfied with the services.

The ultimate goal of quality assurance activities and the focus of QA have traditionally been viewed as the attainment of valued outcomes (Donabedian, 1982), and it is assumed that ensuring high standards for input and process increases the probability of attaining valued outcomes (i.e., improved efficacy). Most of the QA activities with which the authors are familiar focus on the input and process aspects of service delivery. In recent years, however, there has been a significant increase in the emphasis on outcome in quality assurance, even by traditionally input- and process-focused organizations such as the JCAHO (McAninch, 1988; JCAH, 1986). This increased emphasis is primarily a response to demands for accountability from consumers and purchasers of health care.

Although we applaud the move toward an outcome orientation, we give a significant role in this chapter to process. One reason is that QA activities focusing on process are relatively easy to implement as part of the routine professional activities of the clinical health psychologist. We discuss this subject in a later section. A second reason is that, in medical settings, many variables affect clinical outcome (the health of the patient), only some of which are within the span of control of the psychologist.

Table 1. Examples of Questions Reflecting Quality of Input, Process, and Outcome of Service Delivery

Quality questions	Domain
1. Are all psychologists who provide the consultations qualified for their role?	Input
2. Do all patients who could benefit from such a consultation have access to the service?	Input
3. Are staff and other resources adequate in number and training to assess situations and to implement intervention plans?	Input
4. Are consultations timely?	Process
5. Are appropriate intervention plans developed for all patients who are determined by the psychologist to need an intervention?	Process
6. Are all intervention plans implemented appropriately?	Process
7. Do all patients benefit from the psychological intervention?	Outcome

In such settings, and from the perspective of the members of the psychological consultation team, an exclusive focus on outcomes may be self-defeating.

Consider an example of a situation in which the psychologist's lack of control over treatment results in a poor psychological, but an adequate medical, outcome.

A 10-year-old boy is admitted to the hospital emergency room suffering from an acute asthma attack brought on (as far as can be determined) by his failure to comply with his medication regimen. According to the record, he has been admitted under similar circumstances four times in the previous six months. It is learned from talking with the boy that he stops taking his medicine during periods of intense fighting by his parents because his medical emergency is the one thing that appears to subdue the parents and stop their fights for some period of time. The boy is medically stabilized, kept in the hospital for two days, and then discharged with strict instructions about taking his medication regularly. The psychologist is not consulted, and no intervention with the parents is attempted.

The child is about to return to the same problem environment, and the probability of his being readmitted to the emergency room under similar circumstances remains high. Although the medical outcome may have been adequate (the boy was medically stabilized and discharged), from a QA perspective the outcome was inappropriate because there was no input from psychological professionals.

Refocusing the Quality Questions

Because the generality of the questions in Table 1 begs other questions, precision in delimiting the scope of each quality question is crucial. For example, to determine whether all persons who might benefit from psychological consultation have access to such services (Question 2), we would need to define precisely the types of patients who we believe could be helped by the addition of psychological services to their medical treatment plan. In other words, who are the targets of the interventions?

One option is to identify patients by diagnosis or disorder. Pediatric cancer patients and their families, for example, are generally considered appropriate candidates for assessment and intervention by a clinical health psychologist. Another appropriate group consists of patients for whom posthospitalization compliance with the treatment regimen is judged to be critical.

A second option would be to identify patients with particular diagnoses who exhibit maladaptive behaviors that may negatively affect their medical outcome and that may also be positively affected by psychological interventions. For example, current practice in cardiac rehabilitation includes the consideration of significant psychological factors in the development of the comprehensive rehabilitation plan. Indeed, the psychological dimensions in such treatment may be viewed, first, as modifiers of the situation being addressed (e.g., what sort of personality style does the patient have?) and, second, as variables that must be explicitly addressed in the treatment plan (e.g., does the patient need certain types of emotional support that he or she is not currently getting?).

To continue our exploration of Question 2, in addition to identifying patients who may receive interventions, we would need to define *access*. At the extreme, access means that all patients in the target categories are seen by a member of the consultation team. More parsimoniously, it could mean that any nonpsychological staff member having responsibility for such patients is knowledgeable about the types of interventions made by the psychologist and is capable of making judgments about the necessity of such a consultation.

Once we have identified the target groups and defined access, we need to establish standards or criteria that describe the expectations for patient access. In conducting a formal QA investigation of this particular quality assurance question, we may compare the actual percentage of patients given access to the service with a previously established minimum proportion deemed to reflect adequacy. If this aspect of the service system meets the established standard, it is considered of adequate quality.

Formal QA programs also include rules that guide decision making about when action for improvement must be initiated. Referred to by terms like *threshold for action*, these rules have the power to define much more adequate standards of performance. Table 2 presents a rationale and principles for the establishment of thresholds. Figures 1 and 2 provide examples illustrating the use of criteria and thresholds that are easily adaptable to any service delivery setting.

The final phase of QA would be to report the results of the evaluation to those involved in service delivery. If the standard has not been met, procedures such as continuing education activities may be implemented to improve the system. If the standard has been met, feedback serves the purpose of providing a reward for those involved in the delivery of services, as well as of consolidating appropriate and effective practices.

The Formal Definition of Quality Assurance

The kind of conceptualization reflected in our consideration of Question 2, above, relates to what is called the *process of quality assurance.* This three-stage process includes the *establishment of criteria* defining the acceptable levels of performance or outcome; *assessment* of the extent to which specific aspects of the service delivery system satisfy the criteria; and the implementation of *feedback and corrective actions* intended to improve those aspects of the system that are shown by the assessment to be deficient.

We depart slightly from the usual definition of the QA process (Donabedian, 1982) by including criterion development as an explicit component. We do this to emphasize the importance that we ascribe to the process of developing standards for quality.

Establishing Criteria

Criterion development is one of the most critical activities in QA, because it is in this activity that the constituency groups (e.g., service providers, program managers, advisory boards, and consumers) within the service system reach a consensus on how they want the system to operate and what they want it to accomplish. For psychological services in medical settings, the three critical constituency groups are, typically, psychologists, physicians, and nurses, each contributing various types of technical expertise and assuming unique areas of responsibility for treatment.

Criteria are generally established *a priori* and

Table 2. Developing the Threshold-for-Action Level[a]

Threshold: Minimum percentage of charts that should meet criterion; performance below that level requires implementation of action to improve.

Reasons for setting a threshold for action:
1. To state how often the criterion must be met to determine that high quality of care has been achieved for this patient sample.
2. To look for obstacles affecting group performance.
3. To have an objective way of determining when action must be taken to improve patient care.
4. To indicate the value of the criterion to those doing the audit. If the value of the criterion is set below 85%, the group should consider whether this criterion is an indicator of critical care.
5. To commit the group to examining why the criterion's actual performance level is lower than its threshold-for-action level.

To determine the threshold for each criterion, ask the following:
1. Assuming that 100 patients in the same category were to be reviewed, would you be satisfied with the quality of service if 98%, 95%, or 90% of the patient records documented that the criterion has been met?
2. If fewer than 98%, 95%, or 90% of the records contained such documentation, would you be concerned about the quality of patient care?
3. What, if anything, does the psychological literature reveal about the norm for this group of patients on this variable?
4. If the data indicate that this threshold was met or exceeded, would you be willing to defend the quality of patient care in the hospital?
5. If the data indicate that this threshold was not met, would you be willing to defend the quality of patient care in the hospital?

[a]Adapted from *Health Services Review Training Manual,* Veterans Administration, Regional Medical Education Centers, October 1975. Washington, DC: United States Veterans Administration.

PSYCHOLOGICAL CONSULTATION/LIAISON SERVICES UNIT

QUALITY OF CARE/SERVICE REPORT

PSYCHOLOGICAL TESTING

	JA	FE	MA	AP	MA	JU	JU	AU	SE	OC	NO	DE
						MONTH						

INDICATORS

<u>Volume</u>

	JA	FE	MA	AP	MA	JU	JU	AU	SE	OC	NO	DE
# referrals (30)	50	45	43	39	30	37	37	44	32	41	41	43
# referral sources (8)	27	23	19	21	13	25	24	26	15	22	23	24

<u>Quality</u> (Speed of response in days)

	JA	FE	MA	AP	MA	JU	JU	AU	SE	OC	NO	DE
Referral to assessment (1.5)	1.3	1.4	1.1	1.1	1.0	1.4	1.0	1.0	1.2	1.2	1.2	1.1
Initiation to completion (2.5)	1.6	1.7	1.8	2.4	2.0	2.5	2.3	1.8	1.4	2.2	2.9	2.0
Completion of report (4.0)	3.4	4.4	3.2	3.4	2.7	4.1	3.6	2.7	2.0	2.9	5.0	3.4

<u>Volume</u> indicators are reviewed if standard in parentheses is <u>not exceeded</u>.

<u>Quality</u> indicators are to be reviewed if standard in parentheses is <u>exceeded</u>.

Figure 1. Summary report of monthly statistics for a psychological testing service showing performance standards for selected quality indicators.

describe the conditions that the constituency groups believe should prevail within the service system. Because there is no absolute definition of quality, the criteria or standards are usually defined *normatively*, and what is considered "acceptable" depends on any number of context-specific factors.

It is important to set criteria at realistic levels. If they are set too high, the result may be that problems are left unresolved; they may frustrate those involved in the delivery of services and may act to undermine the QA program. Criteria set too low result in lost opportunities to improve patient care.

The case example at the bottom of page 102 demonstrates that there can be different acceptable as well as unacceptable levels of intervention for the same quality issue. This example illustrates Question 5 in Table 1: "Are appropriate intervention plans developed for all patients determined by the psychologist to need an intervention?"

PSYCHOLOGICAL CONSULTATION/LIAISON SERVICE

REVIEW OF RANDOMLY SELECTED CHARTS

Service Unit/Month: Emergency Room/December 1989

Criterion	Observed %	Standard %	Threshold %
[**Consultations requested:**]			
1. Pt. medical needs attended to	100*	100	98
2. Pt. conscious/ able to communicate	95	100	98
[**Consultations not requested:**]			
1. Pt. sustained amputation or other serious loss.	0**	0	0
2. Pt. involved in event resulting in death of other	5	0	0
3. Pt. exhibits psychotic behaviors	0	0	0
4. Pt. injury intentionally self-inflicted	0	0	0
5. Pt. seen in ER prior to current episode	8	0	10

*Percent of cases in which consultation was requested and criterion was satisfied.

**Percent of reviewed cases which satisfied criteria but did not result in consultation-request.

Figure 2. Monthly report for emergency room showing, for each service criterion, observed frequency, performance standard, and threshold for action.

A 76-year-old widow who lives in a city-run retired persons' housing complex near the hospital is brought into the emergency room after being found wandering around the perimeter of the facility. She is tearful and depressed and has not eaten anything for five days. After administration of fluids and electrolytes, she is transferred to a medical ward for observation.

The following day, the psychologist receives a consultation request from the patient's primary nurse and discovers that the patient, distraught over the death of her sister three weeks earlier, is moderately depressed but otherwise psychologically in order. The patient reveals that she hates growing older and has been feeling abandoned by the world as more and more relatives and friends die each year. She feels lonely and has no support systems; she has no children.

Possible interventions by the psychologist in this case (in increasing order of quality) might be:

1. No intervention.
2. Brief crisis intervention, focusing only on the immediate situation in the hospital.
3. Brief crisis intervention in the hospital and a consultation with the medical social work department regarding an assessment of the patient's living situation and her needs on discharge.
4. Numbers 2 and 3, together with securing a follow-up outpatient appointment with a geriatric psychology specialist in the medical center's psychiatry department, and making arrangements for the patient's transport to the appointment.
5. Numbers 2–4, together with a referral to a bereavement group run by a nearby church that the patient occasionally attends. A special feature of the bereavement group is that initial contact will be made with the patient by one of the couselors, who will visit her while she is still in the hospital, and who will attempt to give her meaningful support both immediately and following discharge.

One of the factors that needs to be considered in making judgments of quality is the availability of both personnel or service resources and the financial resources necessary to pay for them. Each of the five intervention options listed above has resource implications. If the services judged most appropriate (e.g., the bereavement group or the geriatric psychologist) are not available, or if there is no way to pay for them, then one could hardly criticize the clinician in the above example for not making such a referral. Realistic financial and resource constraints need to be recognized as issues in their own right rather than as reflections of unwillingness to expand the range of high-quality service options. On the other hand, from an external perspective, one might say that the quality of the service system (from an input perspective) is compromised as a result of the unavailability of the resources.

Ensuring the Consequences of Quality Assessment

An effective QA program must result in program improvement activities, or consequences, when quality assessment activities determine that deficiencies exist. Consequences are most commonly manifested in two forms: as administrative changes in procedures, personnel, or resource allocation, and as formal or informal educational activities targeted at individuals or groups within the service system.

To guarantee consequences, QA activities need to have the sanction and the unqualified support of those individuals who have the authority within the organization to ensure that changes will be made. Such authority issues present particular problems for psychologists trying to assure quality in medical settings because, by virtue of their professional affiliation, they do not have ultimate control. In addition, psychological considerations are not likely to be given priority in those medical settings in which patient problems are construed as purely medical rather than as interactions between medical and psychological factors.

Guidelines for Implementation of QA Activities

We next offer some guidelines for the development of quality assurance programs and activities in any setting where mental health services are provided.

Guideline 1: Define quality by consensus. We conjecture that much of the contention surrounding the development of QA criteria and associated measures stems from organizations' or individual service providers' desire to avoid doing QA. It is important to remember that one does not have to discover the "true" definition of quality before initiating the development of a QA program. It is necessary only to obtain consensus about the focus of the quality assessment and the level of the standards. Although consensus may prove difficult to establish, the process of seeking and consolidating agreement among professional peers is, in and of itself, a valuable exercise for the service unit and the organization.

Decisions arrived at through consensus are not invariant, and it is not unusual for an agency to discover that its original decisions about foci or standards need to be modified as more is learned about factors such as site-specific patterns of practice and resource limitations. For example, the ideal in adding significant psychological dimensions to the treatment of certain types of patients in a particular service might be to have every such patient seen by both a physician and a health psychologist. At the same time, adding such psychological dimensions to every patient contact might increase the amount of time required for each contact, might result in a patient backlog, and might subvert what was, judged in retrospect, a too-optimistic goal.

Guideline 2: Assess quality using multiple measures. Obviously, generalizations about the quality of services delivered by a system cannot be based on a single indicator such as the length of time a patient must wait for a psychological consultation or the inclusion of parents in the development of a treatment plan for a child. It is only by focusing on such specific aspects of the service system, however, that the "weak spots" in the system can be identified and illuminated, and that strategies can be developed for their remediation.

Global indicators of "quality," although intuitively appealing, serve little useful purpose if the ultimate goal of quality assurance is (as it should be) system improvement. (Global indicators are probably most appropriately construed as a kind of weighted average of many discrete indicators, each of which provides some perspective on how the service system is operating and performing.) It behooves the developers of quality assurance programs, therefore, to focus on multiple attributes of the service system, each of which is considered critical to the delivery of clinical services, and each of which can be operationalized as an objective criterion.

Guideline 3: Use QA for improvement, not punishment. The poor reception that QA programs tend to get from service providers has been

attributed to an aversion among clinicians to being evaluated, or to the fact that many QA programs are externally and abruptly imposed requirements whose perceived value to clinicians may be minimal. Whatever the explanation, QA systems tend to be viewed as punitive. And indeed, in some cases, their consequences as well as their process may be negative. For example, hospitals may lose JCAHO accreditation or Medicare certification because of failure to satisfy certain standards, and staff may or may not be promoted or given salary increments based on their performance as assessed by QA indicators.

If the organization focuses on the ultimate purpose of quality assurance—improvement in patient care—it may be easier for participants to view QA as constructive and useful and the consequences of quality assessment as helpful. This is even more likely if the QA activities are seen to be integral aspects of appropriate professional practice as defined by the practitioners themselves.

Guideline 4: Minimize disruption to the service system. One of the best ways to ensure the failure of a quality assurance program is to package it in a plethora of new forms and procedures. The typical clinician's apprehension regarding QA is often legitimized in the allegation that new responsibilities introduced by the QA process will reduce the amount of time he or she can devote to patient care. This negativism is more likely to occur if the QA process is perceived as the product of nonclinicians who, it may be believed, have no true understanding of what it is like to be directly involved in patient care.

It is crucial, therefore, that QA activities be integrated as much as possible into prevailing, valued professional activities such as treatment planning, service monitoring, and records maintenance. The need to integrate QA activities into the everyday working activities of medical service points up the significance of the medical chart. In hospitals, for example, the chart may be the sole repository of the aggregate of records that relates to both past and present medical history and treatment, and in which the various caregivers involved with the patient are able to communicate with each other through such routine activities as progress notes.

Guidelines 3 and 4 point to the importance of QA activities' being presented as a direct means to service improvement and being endorsed as such by providers and administrators alike. Furthermore, after the foci of assessment are determined, the responsibilities for QA activity development should be centrally vested in the unit's service providers.

Guideline 5: The care that is reviewed should be representative of the care provided. Probably the most common type of QA activity in organized care settings is chart review. In this activity, the entries found in the medical record are viewed as a reflection of the process and the outcome of the services provided, and judgments about the quality of care are based on the contents of the record. What often happens in chart review is that clinicians, if allowed, choose their "best" cases to be reviewed. Clearly, effective QA can not be built on a biased selection of samples of the work of clinicians in any organization. Work samples, whether charts or other reflections of clinical care, need to be representative of the range of quality of the services provided within the service setting. Thus, the charts to be reviewed should be selected randomly and from the categories being focused on in the particular review.

A Model Process for Implementing QA Activities

We now outline a model for the structure and the process of QA of psychological services in a hospital setting. This model should include activities that are both concurrent (during the time period that services are being provided) and retrospective (after the episode of care has concluded). In addition, the foci of the activities should be episodes of care looked at both individually and collectively. The former focus is often referred to as *case review*; the latter, as an *audit* or *focused study*.

Our model is comprehensive, and resource limitations or other deficiencies may not allow implementation of all parts of the model in

many hospitals. We believe, however, that implementing only one or some of the components could contribute significantly to the quality of the services provided.

Structural Entities for QA of Psychological Services

Figure 3 presents a suggested structure for a QA program that includes four functional entities: a multidisciplinary hospital QA committee; a psychological QA committee; a multidisciplinary integrative review team; and multidisciplinary treatment teams. Our emphasis in the model is on multidisciplinary organization because the context in which services are provided and interventions are developed depends irrevocably on the cooperation of all disciplines.

In this model, the day-to-day responsibility for QA at the individual case level is assumed by multidisciplinary treatment teams that serve a case management function. A typical team might include a physician, a nurse, a psychologist, and a social worker. Membership would vary, depending on who has patient care responsibilities for the patient being treated. Each team is responsible for the development, implementation, and monitoring of a treatment plan (of which the psychological intervention

Figure 3. Structure for a hospital QA program for clinical health psychology services.

is one component), for discharge planning, and for postdischarge follow-up. *It is in the process of plan development and implementation by the team that the QA is done.*

The multidisciplinary integrative review team (MIRT) is a standing committee, representing medicine, nursing, social work, and psychology. The responsibility of this group is to conduct document-based case review of the work of the treatment teams, to provide feedback to the teams on their performance, and to provide summaries of the findings to the psychological QA committee.

The psychological QA committee (PQAC) is composed of those psychologists who have formal consultation and liaison responsibilities for a range of services (e.g., emergency room, pediatrics, and cardiology) within the institution. The primary charges of the PQAC are to monitor the work of the MIRT, to oversee the development of QA criteria for psychological services, to design and ensure the implementation of focused studies and audits, and to summarize and report to the staff and the QA committee the results of MIRT reviews and other QA activities.

The final entity is the QA committee, the hospitalwide committee representing the hospital's administration and medical and professional staffs. This entity, which exists in all hospitals, has as its general charge the evaluation and monitoring of the quality and the appropriateness of patient care and clinical performance (JCAH, 1984a).

The Multidisciplinary Treatment Team Process

Before an effective team treatment including psychological services can be established, the hospital needs to ensure effective liaison between the psychological services unit and each service unit that uses the consultations of the psychologists. The basis of these relationships should be a written contract outlining the circumstances (criteria) under which a particular service unit will request psychological consultations. This contract should be developed as a cooperative effort of the psychological services unit and the other service unit. Its implementa-

tion should be preceded by in-service training conducted by the staff of the psychological services unit. The training would include the appropriate implementation of the criteria describing requests for consultation, the procedures for dealing with patients for whom consultations are requested, and the procedures for dealing with the emotional concerns of patients for whom consultations may not have to be requested.

Examples of criteria that might be part of a contract between the psychological services unit and the emergency room (ER) are:

1. Consultations should be sought and referrals made for the following types of patients presenting in the ER:
   a. Patients whose injuries are a result of intentional self-injury.
   b. Patients exhibiting psychotic behaviors such as hallucinations, delusions, and other common aberrant behaviors that can be easily described as such.
   c. Persons surviving a trauma that resulted in the death of another person or in which significant losses (such as amputations) were sustained.
   d. Patients who have been repeatedly treated in the emergency room.
2. Consultations should be sought only after the immediate medical needs of the patient have been attended to.

The Case Management Process

Figure 4 is a flowchart of the activities involved in the initiation of a consultation, the development and implementation of the psychological treatment plan, the discharge, and the follow-up. These activities, implemented by the individual treatment teams, constitute the primary day-to-day QA for psychological services within the medical setting.

The consultation is generally initiated, according to the criteria, by a member of the staff in the service unit responsible for the patient. The psychologist conducts an assessment, makes a decision about the appropriateness of an intervention, and develops the likely range of inter-

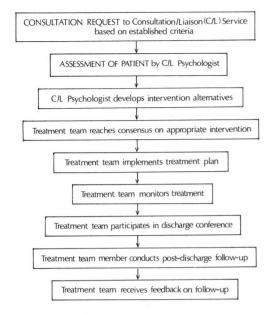

Figure 4. Flowchart of the case management activities of treatment teams. A necessary assumption of the process is that all meetings and discussions at each step *must be documented* in the patient's chart.

vention alternatives. The professionals providing the patient care need to concur on an intervention that will meet the needs of the patient, to define a realistic role for each of the treatment team members, and to make appropriate use of the available resources.

Once a treatment plan has been developed, the team members responsible for its implementation meet regularly and monitor its implementation, making modifications in the plan as necessary. They meet, finally, for the discharge conference and to determine the extent of postdischarge follow-up. Each patient who has received a psychological intervention is contacted at least once postdischarge to determine his or her status, and this information is shared with the treatment team members. All aspects of the intervention, including this follow-up information, are documented in the medical record.

One of the most widely used systems for documenting and organizing the types of issues dealt with in this chapter is the problem-oriented medical record, or POMR (Weed, 1964). At the same time, the development of interven-

tion alternatives as reflected in the examples in this chapter is based loosely on the goal attainment scaling (GAS) process (Kiresuk & Sherman, 1968). The value of the POMR in delineating and following the progress of problems cannot be disputed. The present authors, however, view GAS as a superior mechanism in that it avoids the thorny issue of distinguishing between the "subjective" and the "objective," a potentially confusing distinction in clinical psychology, particularly in relation to problems that are clearly interactions between medical and psychological signs and symptoms. Although the work of Kiresuk and Sherman (1968) involves quantitative outcome information, we do not believe it essential that outcome measurement be used here. Rather, the model is useful in providing a simple, yet sophisticated, approach to operationalizing the dimensions of input, process, and outcome in such a way that all team members may cooperate in the process and that the various levels of quality are clear, specific, and even, occasionally, self-evident. The GAS model is based on the identification of the problem being addressed; the specification of the components of each of the three foci (input, process, and outcome), based on the normative (expected) standard of practice for the identified problem and the projected levels of performance above and below those expected; and a comparison of actual input, process, or outcome with these specifications.

The specification of input, process, and outcome criteria is simply a matter of stating the expected or normative criterion and then stating criteria that would reasonably fall above and below it:

1. Much less than the expected attainment of criterion.
2. Less than the expected criterion attainment.
3. The expected criterion attainment.
4. More than the expected criterion attainment.
5. Much more than the expected criterion attainment.

The appendix contains an example of a QA exercise that uses a GAS strategy for establishing criteria describing the quality of input, process, and outcome.

Activities of the MIRT

The multidisciplinary integrative review team is charged with the concurrent and retrospective monitoring of the quality of treatment plans and the work of treatment teams. Although concurrent monitoring may be done by a MIRT representative meeting with each treatment team as they discuss their cases (facilitated by the use of GAS or POMR), it is probably more cost-effective for the MIRT to base its routine reviews on the medical record. Evaluations of the quality of interventions should be based both on *explicit* written criteria, such as those required by the JCAHO, and on more *implicit*, context-specific (*ad hoc*) criteria (Donabedian, 1982).

The MIRT may select cases for review on both random and targeted bases: random groups of patients, cases of particular providers or treatment teams, patients with particular problems or diagnoses, or cases that are believed to exemplify particular quality problems. The MIRT provides written feedback to the treatment teams on the particular reviews, summarizing the results of its reviews for reporting to the psychological QA committee.

The MIRT may also be responsible for conducting occasional focused studies to determine, among other things, the validity of the existing criteria or the need for new criteria. This process is akin to that of establishing the construct validity of the criteria. For example, there is evidence that a sibling of a child being treated for cancer may have psychosocial difficulties requiring intervention by a mental health professional (Spinetta, 1981). The MIRT may examine a set of cases in which siblings were identified, assessed, and treated as required and may compare them with another set in which such identification, assessment, and treatment did not occur. The aims of the comparison would be to examine the effects on process and outcome of including a previously neglected population (siblings of pediatric oncology patients) in the overall treatment plan and giving them access to treatment.

The Psychological QA Committee

In order to ensure the systematic and comprehensive implementation of discipline-specific QA activities, the responsibility for QA should be vested in a formally recognized group of professionals within the psychological services unit. This committee would have overlapping membership with the hospital QA committee and would be responsible for the "meta" issues of psychological QA.

We recommend a separate psychological QA committee because psychologists are required to be the final arbiters of the criteria that define the quality of psychological services (American Psychological Association, 1987) and are therefore the ones who should make judgments about the extent to which those psychological criteria are satisfied in the service system. Just as psychologists should not be responsible for determining what is good technical quality with respect to medical care, so, too, should physicians not be held responsible for establishing criteria for the quality of psychological interventions. At the same time, physicians and other stakeholders should have input into what the criteria will be. Because there is ultimately an interaction between the medical and psychological factors, the judgment about the overall quality of care should be made by multidisciplinary groups: the MIRT and the hospital QA Committee.

Problems of Implementation

Our predominant focus on the basic elements, structures, and processes defining a QA program appropriate for application to psychological consultation services in hospitals should not lead the reader to minimize the problems inherent in developing such a program. Nor should it be concluded that QA activities should be restricted to those discussed in this chapter.

All successful QA programs must meet the unique needs of the service system. In particular, they must explicitly acknowledge and incorporate the realities of the resources available in the immediate context in which they are embedded. Further, there are many accountability-oriented activities in organized-care settings that, although they relate to QA, are not incorporated into a formal QA program. This may be the rule rather than the exception. Naturalistic and *ad hoc* kinds of QA activities abound and have abounded since the advent of organized care in institutional settings. Psychologists in medical settings have long been familiar with such activities as chart maintenance and postintake disposition reviews, prospective individualized treatment plans, and multidisciplinary team consultations. All such professional activities are part of the complex of events that are QA. Although the authors are committed to the importance of the formal QA program, they also consider it unfortunate that, from the viewpoint of bodies that regulate and certify, such activities are not seen as QA unless they are formally documented as such.

What this chapter has tried to do is to signal the value of and some directions for a programmatic approach to QA. It has not been the concern in this chapter to discount what currently goes on in the name of quality assessment and review. Rather, it has sought to underscore and conserve such existing practices by highlighting the merits of a more formal, balanced, and integrated team approach to QA. Programmatic approaches, if properly handled, lend both efficacy and acceptability to the endeavor. On the other hand, as has been implied, implementing programmatic approaches is easier said than done, and QA-related conceptual awareness and technical sophistication alone will not ensure an effective and enduring QA program.

Administrative and Political Considerations

It seems axiomatic that the successful development of programs for the QA of psychological consultation services depends as much on the kind and degree of administrative support available as on the levels of acceptance and technical expertise existing among the providers of psychological services. Administrative support for QA must be available at both the level of the service unit and the level of the institution. Without this local and institutional administrative support, the resources, including the time, money, and leadership necessary for QA program development and maintenance, will not be forthcoming.

To emphasize the role of administrative support within the setting is to acknowledge the fact that, for most psychological service entities, the development of a formal QA program closely resembles the diffusion of an innovation. This is a gradual process through which the rationale, goals, objectives, content, and procedures of the nascent QA program become sanctioned, implemented, and consolidated over time. Thus, administrative support is most crucial when the program is most vulnerable: in its formative stages.

These considerations aside, it is the providers working as a team who constitute the core of any QA program. *The success of the program requires both an acceptance and an ownership of the program by the psychologists, who, in surrendering some individual autonomy, acknowledge the greater benefits of a well-articulated QA program for patients, for interprofessional linkages, for the institution, and, ultimately, for the profession.* The characteristics, dynamics, and history of group decision-making are all central to how the members of a psychological services unit approach what is (for most of the team) an innovation. And the authors cannot assert too strongly that the development of a QA program necessarily involves the development of the professional team itself. If team-building issues are not confronted directly by the unit head, the future of the innovation will be in jeopardy (Troy, 1988).

One aspect of QA program implementation that deserves discussion is interprofessional linkages. Because the development and maintenance of effective and collegial relationships among professional groups are central to service delivery in medical settings, the quality of communication and the structural linkages between professions help to determine the quality of the care delivered. Because it is in the interdisciplinary team that the responsibility for patient care is vested, there is an equivalent responsibility to nurture and sustain these ties.

At the same time, QA program development in, for example, the psychology unit cannot possibly proceed in isolation from other professional groups. In the presentation of the model, the authors have endeavored to acknowledge the interdependence of professional functioning within the medical setting.

It should also be emphasized that the strategic exploitation of interprofessional linkages in the cause of patient welfare requires, potentially, an even greater degree of surrender of autonomy than has already been noted for individual members of the intraprofessional service unit. Given the long history of the role of the interdisciplinary team in inpatient medical settings, the potential of rivalry, guild issues, and relative status considerations to disrupt collegial relationships is considerable. The conservation of effective interprofessional relationships is very likely due to the communality historically prevailing among the service disciplines with respect to professional goals and concern for patients. The current, rapid flux in the topology of health care service delivery and provider roles may significantly threaten this relative harmony. For the consultant health psychologist or behavioral health specialist, issues such as admitting privileges, prescription writing, and leadership roles in case management decisions clearly challenge the professional status quo and, particularly, the traditional leadership of the treatment team.

If both levels of quality assurance, interdisciplinary and intraprofessional, are to be developed in the greater interest of the patient, one can only hope that the general thrust of interdisciplinary problem-solving will proceed along functional, and not political, lines. If not, the unique and heretofore enduring role of the interdisciplinary team in case management in health care settings will be significantly, and interestingly, challenged.

Appendix

The following is an expanded example of the use of goal attainment scaling (GAS) in the development of alternatives that may be adopted by a treatment team and that focus on the input, process, and outcome of treatment.

Ms. B is a 38-year-old woman of Nordic extraction who has lived in the same city all her life. When she was 12, she developed juvenile onset diabetes, and she has been insulin-dependent ever since the condition was first diagnosed. She has been married to Mr. B since she was 22, and the marriage is described as a good one by both partners. Because of her medical condition, she could never carry a child to term, and after three miscarriages, she and her husband were advised not to attempt another pregnancy.

Ms. B maintained good control of her illness until three years ago, when vascular problems were experienced for the first time in her feet. The problems continued plaguing her, despite excellent compliance with both diet and insulin dosage regimens, and about six months ago, she developed gangrene in the toes of her left foot. At that time, she became mildly depressed, and her previously excellent compliance was compromised. Her endocrinologist, although concerned about her mood and compliance problems, felt that giving her "a good talking to" would suffice, thought that such problems were normal, and did not think it necessary to refer her to a mental health professional.

Three weeks ago, Ms. B had to undergo a below-the-knee amputation after developing uncontrollable gangrene. After the surgery, she began withdrawing, was having problems sleeping, cried a lot, lost her appetite, and felt ashamed that she was having any problems. She explained to the surgeon and the endocrinologist that she was usually a stoic person, that people of her background did not display negative emotion because it indicated that they were out of control, and that nobody should take any notice of her condition. The surgeon, coming from a similar background, experienced some identification with Ms. B and did not respond to his endocrinologist colleague's request to call for a psychological consultation. The patient's primary nurse, however, went ahead and called for the consultation anyway, following which the clinical health psychologist on call that day came up to the surgical unit and assessed the patient.

Based on a goal-attainment-scaling model, the intervention alternatives that might be proposed for the treatment of Ms. B are as follows. Note that Alternative 3 would be the expected (most likely) occurrence, Alternative 1 would be the worst that would be expected, and Alternative 5 would be the best expected of the service system.

Input

1. Ms. B is told that she can tough it out, that things will get better, and that she should not worry.

2. Ms. B is told that she should understand that having severe psychological symptoms is normal following an amputation, and that if the problem continues for another week, she should agree to a psychological consultation.

3. (Expected) The reason for the psychological consultation is carefully explained to Ms. B, and an agreement is reached that she will at least talk to the psychologist for about 15 minutes, but that she will not be forced to undergo treatment. The consultation is explained only in terms of the mood problem itself, however, and not in relation to her particular situation.

4. The reason for the consultation is carefully explained to Ms. B, particularly in relation to issues such as loss and the mood changes attendant on both diabetes and procedures such as amputation.

5. The reason for the consultation is carefully explained to Ms. B in relation to the factors mentioned in Number 4, above. In addition, the consultation itself is carefully explained, including a request to rule out major depression, the formulation of a treatment plan, and the inclusion of compliance issues in the overall picture.

Process

1. Medical and nursing staff pay no attention to Ms. B's psychological needs, agreeing with her that people of her background are tough, and attending only to her medical needs.

2. Medical and nursing staff attempt to find out whether Ms. B will admit that her psychological status is compromised, attempt to be somewhat supportive, but still focus primarily on her medical status.

3. (Expected) Medical and nursing staff, having succeeded in getting Ms. B to accept and undergo the evaluation, continue attending primarily to her medical needs and see the locus of all treatment as being exclusively in the psychologist's domain.

4. Medical and nursing staff, although focusing on Ms. B's medical needs to the expected degree, attempt to be supportive and understanding, to include extensive processing of Ms. B's resistance, and to work toward her accepting the psychological consultation.

5. An integrative treatment plan, is formulated, involving medical and nursing staff, the psy-

chologist, and a psychiatric consultation regarding psychoactive medication. Due attention is paid to the patient's medical and psychological issues, including compliance, depression, and loss.

Outcome

1. Ms. B's psychological condition is not attended to and becomes worse after discharge. She is admitted to a psychiatric hospital for the inpatient treatment of major depression three months after the amputation.

2. Ms. B sees a psychologist three times but refuses psychoactive medication.

3. (Expected) Ms. B sees a psychologist three times, accepts the psychiatrist's prescribed tricyclics, and agrees to one follow-up visit after discharge.

4. Ms. B is seen daily by both the psychologist and the psychiatrist while in the hospital, and a treatment plan is worked out in relation to several problems: depression, loss, changes in body image and self-esteem, and compliance.

5. In addition to the outcome described in Number 4, above, careful follow-up plans are developed, including one home visit four days after discharge, the inclusion of Mr. B in the treatment, and arrangements for transportation to the hospital for amputee-clinic and rehabilitation appointments. Outpatient appointments with the psychologist are scheduled following Ms. B's rehabilitation treatments, thereby ensuring maximum follow-through by the patient herself, as she need make the arduous trip to the medical center only once a week.

References

American Psychological Association. (1987). *General guidelines for providers of psychological services.* Washington, DC: Author.

Brook, R. H., & Kosecoff, J. (1988). Competition and quality. *Health Affairs, 7*(3), 150–161.

Donabedian, A. (1982). *Explorations in quality assessment and monitoring: Vol. 2. Criteria and standards of quality.* Ann Arbor, MI: Health Administration Press.

Evashwick, C., & Weiss, L. (1987). *Managing the continuum of care: A practical guide to organization and operations.* Rockville, MD: Aspen.

Fuchs, V. (1988). The "competition revolution" in health care. *Health Affairs, 7*(3), 5–24.

Henderson, M. G., & Collard, A. (1988). Measuring quality in medical case management programs. In K. Fisher & E. Weissman (Eds.), *Case management: Guiding patients through the health care maze*. Chicago: Joint Commission on Accreditation of Healthcare Organizations.

Joint Commission on Accreditation of Hospitals. (1984a). *1985 Accreditation Manual for Hospitals*. Chicago: Author.

Joint Commission on Accreditation of Hospitals. (1984b). *1985 consolidated standards manual for child, adolescent, and adult psychiatric, alcoholism, and drug abuse facilities serving the mentally retarded developmentally disabled*. Chicago: Author.

Kiresuk, T. J., & Sherman, R. E. (1968). Goal attainment scaling: A general method for evaluating comprehensive community mental health programs. *Community Mental Health Journal, 4*, 443–453.

McAninch, M. (1988). Accrediting agencies and the search for quality care. In G. Stricker & A. R. Rodriguez (Eds.), *Handbook of quality assurance in mental health*. New York: Plenum Press.

Rodriguez, A. R. (1988a). Effects of contemporary economic conditions on availability and quality of mental health services. In G. Stricker & A. R. Rodriguez (Eds.), *Handbook of quality assurance in mental health*. New York: Plenum Press.

Rodriguez, A. R. (1988b). Introduction to quality assurance in mental health. In G. Stricker & A. R. Rodriguez (Eds.),

Handbook of Quality Assurance in Mental Health. New York: Plenum.

Shueman, S. A. (1987). A model of case management for mental health services. *Quality Review Bulletin, 13*, 314–317.

Shueman, S. A., & Troy, W. G. (1982). Education and peer review. *Professional Psychology, 13*, 58–65.

Shueman, S. A., & Troy, W. G. (1988). Quality assurance in outpatient psychotherapy. In G. Stricker & A. R. Rodriguez (Eds.), *Handbook of quality assurance in mental health*. New York: Plenum Press.

Spinetta, J. J. (1981). The sibling of the child with cancer. In J. J. Spinetta & P. Deasy-Spinetta (Eds.), *Living with childhood cancer*. St. Louis: Mosby.

Troy, W. G. (1988). Quality assurance in a university counseling center: There is always more than meets the eye. In G. Stricker & A. R. Rodriguez (Eds.), *Handbook of quality assurance in mental health*. New York: Plenum Press.

Veterans Administration, Regional Medical Education Centers. (1975, October). *Health Services Review Training Manual*. Washington, DC: United States Veteran Administration.

Waldo, D. R., Levit, E., & Lazenby, H. (1986). National health expenditures, 1985. *Health Care Financing Review, 8*, 1–21.

Weed, L. L. (1964). Medical records, patient care, and medical education. *Irish Journal of Medical Sciences, 6*, 271–282.

CHAPTER 8

Arguments for the Financial Efficacy of Psychological Services in Health Care Settings

Nicholas A. Cummings

Introduction

By the beginning of the 1980s, it was generally conceded that psychologists had won their hard fought struggle to be included in third-party payment for mental health services. The notable exception was in Medicare and Medicaid, where the inclusion of psychological services was, at best, sparse and spotty and indicated a job yet to be concluded. The history of this movement, which began in the late 1950s, was detailed by Cummings (1979) in an article that fell just short of signaling victory.

The effort to achieve parity with psychiatry in the recognition of psychology by the insurance industry and the government is a case history in the struggle for the autonomy of the psychological profession, most often bitter, sometimes comical, but always colorful. Professional psychologists had to overcome the resistance not only of the insurance industry, which was reluctant to add another class of practitioners, but also of their own American Psychological Association (APA), which was

Nicholas A. Cummings • American Biodyne, Inc., 400 Oyster Point Boulevard, Suite 218, South San Francisco, California 94080.

then academically dominated and was indifferent to the professional psychologists' struggle for survival. This phase of the struggle concluded with psychology's being recognized along with psychiatry as one of the dominant practitioner forces in the field of mental health. It also concluded with a thoroughly professionalized APA, which, in contrast to the prior era, now spends a very significant portion of its resources on professional issues.

Now all of this is rapidly changing. Those who would plead for the financial efficacy of psychological services in health care settings are once again on the defensive. This threat has been brought about by what has been termed the *health care revolution* (Kiesler & Morton, 1988; Kramon, 1989), which has created an array of new health and mental health delivery systems to which psychology must adapt in its effort to be included. In the original struggle for the inclusion of psychologists in third-party reimbursement in private practice, failure to be included would have spelled economic extinction. Again, in the present situation, failure to be included in the new delivery systems may result in the demise of professional psychology. The arguments for the financial efficacy of psychological services in health settings have never been more crucial.

This chapter delineates the original arguments for the inclusion of psychology in third-party payment, for they are as valid now as they were then. Second, the health care revolution, with its implied threat to professional psychology as it is now constituted, is described. Finally, the chapter addresses how the arguments for the financial efficacy of the inclusion of psychologists in health care must be modified in the face of the health care revolution.

The Historical Perspective

Immediately following World War II, the Kaiser-Permanente Health Plan on the West Coast began offering total health benefits to millions of enrollees without the usual limitations, copayments, first-dollar deductibles, and other such restrictions that were customary in all other health plans at that time. In that era, mental health and substance abuse benefits were totally excluded, as they were also at Kaiser-Permanente. The research that led Kaiser-Permanente to include mental health and substance abuse as covered services became the basis of over two dozen research replications and constituted the arguments for the inclusion of psychologists in health care. The 20-year experience at Kaiser-Permanente, which demonstrated the financial efficacy of psychological services and the public policy implications in favor of the inclusion of psychologists, was summarized by Cummings and Vanden-Bos (1981).

The Beginning of the Kaiser-Permanente Mental Health Benefit

Kaiser-Permanente soon found, to its dismay, that, once a health system makes it easy and free to see a physician, there occurs an alarming inundation of medical utilization by seemingly physically healthy persons. In private practice, the physician's fee has served as a partial deterrent to overutilization, until the recent growth of third-party payment for health care services. The financial base at Kaiser-Permanente is one of capitation, and neither the physician nor the health plan derive an additional fee for seeing the patient. Rather than becoming wealthy from imagined physical ills, the system could have been bankrupted by what was regarded as abuse by the hypochondriac.

Early in its history, Kaiser-Permanente added psychotherapy to its list of services, first on a courtesy reduced fee of five dollars per visit and eventually as a prepaid benefit. This addition was initially motivated not by a belief in the efficacy of psychotherapy, but by the urgent need to get the so-called hypochondriac out of the physician's office. From this initial perception of mental health as a dumping ground for bothersome patients, 25 years of research has led to the conclusion that no comprehensive prepaid health system can survive if it does not provide a psychotherapy benefit.

The Patient with "No Significant Abnormality"

Early investigations (Follette & Cummings, 1967) confirmed physicians' fears they were being inundated, for it was found that 60% of all visits were by patients who had nothing physically wrong with them. Add to this the medical visits by patients whose physical illnesses were stress-related (e.g., peptic ulcer, ulcerative colitis, and hypertension), and the total approached a staggering 80%–90% of all physicians visits. Surprising as these findings were 30 years ago, nationally accepted estimates today of stress-related visits range from 50% to 80% (Shapiro, 1971). Interestingly, over 2,000 years ago, Galen pointed out that 60% of all persons visiting a doctor suffered from symptoms that were caused emotionally, rather than physically (Shapiro, 1971).

The experience at Kaiser-Permanente subsequently demonstrated that it is not merely the removal of all access barriers to physicians that fosters somatization. The customary manner in which health care is delivered inadvertently promotes somatization (Cummings & Vanden-Bos, 1979). When a patient who has not been feeling up to par attempts to discuss a problem in living (e.g., job stress or marital difficulty) during the course of a consultation with a physician, that patient is usually either politely dismissed by an overworked physician or given a tranquilizer. This reaction unintentionally implies criticism of the patient, which, when repeated on subsequent visits, fosters the translation of this emotional problem into something toward which the physician will respond. For example, in a purely psychogenic pain patient, the complaint that "My boss is on my back" may become at some point a lower back pain, and neither the patient nor the physician may associate the symptom with the original complaint. Suddenly, the patient is "rewarded" with X rays, laboratory tests, return visits, referrals to specialists, and, finally, even temporary disability, which removes the patient from the original job stress and tends to reinforce protraction and even permanence of the disability.

Estimates of stress-related physical illness are subjectively determined, whereas the number of physician visits by persons demonstrating no physical illness can be objectively verified through random samplings of all visits to the doctor. After more-or-less exhaustive examination, the physician arrives at a diagnosis of "no significant abnormality," noted by the simple entry of *NSA* in the patient's medical chart. Repeated tabulations of the NSA entries, along with such straightforward notations as "tension syndrome" and similar designations, consistently yielded the average figure of 60%. This figure is now generally recognized in the medical profession, which refers to these patients as *somaticizers*, and in behavioral health, which has named them the *worried well*.

During the early years of Kaiser-Permanente, there was considerable resistance to accepting such estimates because it was reasoned that if 60%–90% of physician visits reflected emotional distress, 60%–90% of the doctors should be psychotherapists! This concern, as will be demonstrated below, was unfounded because subsequent research indicated that a relatively small number of psychotherapists can effectively treat these patients.

In an effort to help the physician recognize and cope with the distress–somatization cycle, Follette and Cummings (1967) developed a scale of 38 Criteria of Distress. These criteria do not use psychological jargon; rather, they are derived from typical physicians' entries in the medical charts of their patients. The researchers worked back from patients seen in psychotherapy to their medical charts, on which the diagnosis NSA had been made. They gathered extensive samplings of typical entries that connoted distress and validated these into the 38 criteria shown in Table 1. Physicians were urged to refer patients for psychotherapy who scored 3 points or more on this scale as attested by the physician's own medical chart entries.

After expending considerable effort and time validating this scale, it was discovered that emotional distress could be just as effectively predicted by weighing the patient's medical chart. The reason is that patients with chronic illness (or those involved in prenatal care) tend to see a physician at more-or-less scheduled appointments, whereas a patient suffering from emotional distress tends to use drop-in services, night visits, and the emergency room. In the instance of the chronically ill patient, the physician makes each entry in the chart immediately under the one bearing the date of the previous visit; thus, several visits are recorded on one sheet, front and back, in the medical chart. By comparison, when emotionally distressed persons make nonscheduled visits, the medical chart is not available, and the physician makes the entry on a new and separate sheet, which is later filed in the chart by medical records librarians. Repeating this practice through months and years builds up enormous medical charts, sometimes into the second and third volume.

Once the patient enters the somatization cycle, there is an ever-burgeoning symptomatology because the original stress problem still exists in spite of all the physician's good efforts to treat the physical complaints. The patient's investment in his or her own symptom is only temporarily threatened by the physician's eventual exasperation, often accompanied by that unfortunate phrase, "It's all in your head." A

Table 1. Criteria of Psychological Distress with Assigned Weights

1 point	2 points	3 points
1. Tranquilizer or sedative requested	23. Fear of cancer, brain tumor, venereal disease, heart disease, leukemia, diabetes, etc.	34. Unsubstantiated complaint that there is something wrong with genitals
2. Doctor's statement patient is tense, chronically tired, was reassured, etc.	24. Health questionnaire: yes on 3 or more psychological questions	35. Psychiatric referral made or requested
3. Patient's statement as in No. 2	25. Two or more accidents (bone fractures, etc.) within 1 year; patient may be alcoholic	36. Suicidal attempt, threat, or preoccupation
4. Lump in throat		37. Fear of homosexuals or of homosexuality
5. Health questionnaire: yes on 1 or 2 psychological questions[a]	26. Alcoholism or its complications: delirium tremens, peripheral neuropathy, cirrhosis	38. Nonorganic delusions and/or hallucinations; paranoid ideation; psychotic thinking or psychotic behavior
6. Alopecia areata	27. Spouse is angry at doctor and demands different treatment for patient	
7. Vague, unsubstantiated pain		
8. Tranquilizer or sedative given		
9. Vitamin B_{12} shots (except for pernicious anemia)	28. Seen by hypnotist or seeks referral to hypnotist	
10. Negative EEG	29. Requests surgery that is refused	
11. Migraine or psychogenic headache	30. Vasectomy: requested or performed	
12. More than 4 upper-respiratory infections per year	31. Hyperventilation syndrome	
13. Menstrual or premenstrual tension; menopausal sex	32. Repetitive movements noted by doctor: tics, grimaces, mannerisms, torticollis, hysterical seizures	
14. Consults doctor about difficulty in child bearing	33. Weight lifting and/or health faddism	
15. Chronic allergic state		
16. Compulsive eating (or overeating)		
17. Chronic gastrointestinal upset; aerophagia		
18. Chronic skin disease		
19. Anal pruritus		
20. Excessive scratching		
21. Use of emergency room; twice or more per year		
22. Brings written list of symptoms or complaints to doctor		

[a]Refers to the last four questions (relating to emotional distress) on a Modified Cornell Medical questionnaire given to patients undergoing the Multiphasic Health Check in the years 1962–1964.

new physician within the care system is found, one whose sympathy and eagerness to determine the physical basis for the symptom have not been worn down by this particular patient. The inadvertent reward system continues, as does the growth of the medical chart. In a similar fashion, stress can impact on an existing physical illness, exacerbating its symptomatology and increasing its duration. The baffled and frustrated physician uses such terminology as "failure to respond" to account for the ineffectiveness of the treatment and often silently suspects noncompliance or malingering.

The Effect of Psychotherapy on Medical Utilization

In the first of a series of investigations into the relationship between psychological services and medical utilization in a prepaid health plan setting, Follette and Cummings (1967) compared the number and type of medical services sought before and after the intervention of psychotherapy for a large group of randomly selected patients. The outpatient and inpatient medical utilization by these patients for the year immediately before their initial interview in the Kaiser-Permanente Department of Psychotherapy, as well as for the five years following that intervention, was studied for three groups of psychotherapy patients (1 interview only, brief therapy with a mean of 6.2 interviews, and long-term therapy with a mean of 33.9 interviews) and a "control" group of matched patients who demonstrated similar criteria of distress but who were not, in the six years under study, seen in psychotherapy.

The findings indicated that (1) persons in emotional distress were significantly higher users of both inpatient facilities (hospitalization) and outpatient medical facilities than the health plan average; (2) there were significant declines in medical utilization by those emotionally distressed individuals who received psychotherapy, compared to that of the "control" group of matched patients; (3) these declines remained constant during the five years following the termination of psychotherapy; (4) the most significant declines occurred in the second year after the initial interview, and those patients receiving one session only or brief psychotherapy (two to eight sessions) did not require additional psychotherapy to maintain the lower level of medical utilization for five years; and (5) patients seen two years or more in continuous psychotherapy demonstrated no overall decline in total outpatient utilization (inasmuch as psychotherapy visits tended to supplant medical visits). However, even for this group of long-term therapy patients, there was a significant decline in inpatient utilization (hospitalization), from an initial rate several times that of the health plan average to a level comparable to that of the general adult health plan population. Thus, even long-term therapy is cost-effective in reducing medical utilization if it is applied only to those patients that need and should receive long-term therapy.

In a subsequent study, Cummings and Follette (1968) found that intensive efforts to increase the number of referrals to psychotherapy by computerizing psychological screening with early detection and alerting the attending physicians did not significantly increase the number of patients seeking psychotherapy. The authors concluded that, in a prepaid health plan that already maximally uses educative techniques for both patients and physicians, and that provides a range of psychological services, the number of subscribers seeking psychotherapy at any given time reaches an optimal level and remains constant thereafter.

In another study, Cummings and Follette (1976) sought to answer, in an eighth-year telephone follow-up, whether the results described previously were a therapeutic effect, were the consequences of extraneous factors, or were a deleterious effect. It was hypothesized that, if better understanding of the problem had occurred in the psychotherapeutic sessions, the patient would recall the actual problem rather than the presenting symptom and would have lost the presenting symptom and coped more effectively with the real problem. The results suggest that the reduction in medical utilization was the consequence of resolving the emotional distress that was being reflected in the symptoms and in the doctor's visits. The modal patient in this eighth-year follow-up may be described as follows: She or he denied ever having consulted a physician for the symptoms for which the referral was originally made. Rather, the actual problem discussed with the psychotherapist was recalled as the reason for the psychotherapy visit, and although the problem had been resolved, this resolution was attributed to the patient's own efforts, and no credit was given the psychotherapist. These results confirm that the reduction in medical utilization reflected a diminution in the emotional distress that had been expressed in symptoms presented to the physician.

Although they demonstrated in this study, as they did in their earlier work, that savings in medical services do offset the cost of providing psychotherapy, Cummings and Follette insisted that the services provided must also be therapeutic in that they reduce the patient's emotional distress. Both the cost savings and the therapeutic effectiveness demonstrated in the Kaiser-Permanente studies were attributed by the authors to the therapists' expectations that emotional distress could be alleviated by brief, active psychotherapy. Such therapy, as Malan (1976) pointed out, involves the analysis of transference and resistance and the uncovering of unconscious conflicts and has all the characteristics of long-term therapy, except length. Given this orientation, it was found over a five-year period that 84.6% of the patients seen in psychotherapy chose to come for 15 sessions or fewer (with a mean of 8.6). Rather than regarding these patients as "dropouts" from treatment, it was found on follow-up that they had achieved a satisfactory state of emotional well-being that had continued into the eighth year after the termination of therapy. Another 10.1% of the patients were in moderate-term therapy with a mean of 19.2 sessions, a figure that would probably be regarded as short-term in many traditional clinics. Finally, 5.3% of the patients were found to be "interminable," in that, once they had begun psychotherapy, they had continued, seemingly with no indication of termination.

In another study, Cummings (1977) addressed the problem of the "interminable" patient, for whom treatment is neither cost-effective nor therapeutically effective. The concept that some persons are so emotionally crippled that they may have to be maintained for many years or for life was not satisfactory, for if 5% of all patients entering psychotherapy are "interminable," within a few years a program will be hampered by a monolithic caseload, a possibility that has become a fact in many public clinics where psychotherapy is offered at nominal or no cost. It was originally hypothesized that these patients required more intensive intervention, and the frequency of psychotherapy visits was doubled for one experimental group, tripled for another experimental group, and held constant for the control group. Surprisingly, the cost–therapeutic-effectiveness ratios deteriorated in direct proportion to the increased intensity; that is, medical utilization increased, and the patients manifested greater emotional distress. It was only by reversing the process and seeing these patients at spaced intervals of once every two or three months that the desired cost–therapeutic-effect was obtained. These results are surprising in that they are contrary to traditionally held notions that more therapy is better, but they demonstrate the need for ongoing research, program eval-

uation, and innovation if psychotherapy is going to be made available to everyone as needed.

The Kaiser-Permanente findings regarding the offsetting of medical-cost savings by providing psychological services have been replicated by others (Goldberg, Krantz, & Locke, 1970; Rosen & Wiens, 1979). In fact, such findings have been replicated in over 20 widely varied health care delivery systems (Jones & Vischi, 1978). Even in the most methodologically rigorous review of the literature on the relationship between the provision of psychotherapy and medical utilization (Mumford, Schlesinger, & Glass, 1978), the "best estimate" of cost savings is seen to range between 0% and 24%, with the cost savings increasing as the interventions are tailored to the effective treatment of stress.

The Effects of Behavioral Medicine on Medical Utilization

The foregoing addresses interventions with the patients who comprise 60% of all physician visits: somaticizers who have no physical disease but are replicating physical symptoms as a result of stress, and who are commonly referred to as the *worried well*. There is also the worried sick patient whose physical illness is a source of stress (secondary stress attendant on physical illness, e.g., fear of death following a myocardial infarct), or whose stress has contributed to succumbing to a physical illness or complicates a physical illness (e.g., tension-induced peptic ulcers or ulcerative colitis, failure-to-thrive syndrome). Finally, there is the asymptomatically sick patient who experiences no discomfort and for whom a medical evaluation is necessary to establish the existence of the disease (e.g., essential hypertension). Mechanic (1966) estimated that, if one looks at all three of the preceding categories, 95% of all medical-surgical patients could profit from psychotherapy or behavioral medicine interventions. Even with many supposedly biologically based physical health disorders, psychotherapy and behavioral medicine work and are cost-effective in that they reduce medical utilization (VandenBos & DeLeon, 1988; Yates, 1984).

Although physicians are becoming increasingly cognizant of the somaticizer, there still is resistance to referring to a mental health professional in cases of actual illness. This resistance prompted an editorial in *Newsweek* by a journalist with breast cancer (Kaufman, 1989):

> Curiously, while I was advised to see an internist, a surgeon, cosmetic surgeon, an oncologist and radiation therapist, at no point did anyone in the medical fraternity recommend that I see a mental health professional to help me cope with the emotional impact of breast cancer. Perhaps they didn't realize that breast cancer had an emotional impact. But I did. So, I went to see a psychologist, ironically the one specialist not covered by my insurance. It was worth the cash out of pocket. (p. 32)

Patients like this journalist, report beneficial effects from counseling and behavioral medicine. That this benefit translates into a medical offset for the physically ill is demonstrated by a growing body of research, a few studies of which will serve as examples.

Schlesinger, Mumford, and Glass (1980) found that the greatest medical offset was obtained in the chronic diseases of diabetes, ischemic heart disease, airways diseases (e.g., emphysema), and hypertension. This finding was corroborated by Shelleberger, Turner, Green, and Cooney (1986), who reported a 70% reduction in physician visits in a chronically ill population following a 10-week biofeedback and stress management program. Fahrion, Norris, Green, and Schnar (1987) were able to alter dramatically, through behavioral medicine interventions, including biofeedback, a group of hypertensives' reliance on medication. A 33-month follow-up revealed that 51% had been well controlled off medication, an additional 41% had been partially controlled, and only 8% had been unsuccessful in lowering their blood pressure without medication. Assuming a five-year medication cost of $1,338, the authors demonstrated significant cost savings.

Olbrisch (1981) found a savings of 1.2 hospital days on average in surgical patients who received preoperative interventions. Similarly, Jacobs (1988), using biofeedback training before surgery, reduced hospital days by 72% and postoperative outpatient visits by 63%. Friedman, Ury, Klatsky, and Siegelaub (1974) found that they could predict through an automated screening the recovery rate following myocardial infarct and could influence that recovery rate through behavioral medicine interventions. This recovery period varied more than six months, which reflects high potential savings through behavioral medicine. Flor, Haag, Turk, and Koehler (1983) reported a significant reduction in physician visits and medication rates in rheumatological back pain patients after EMG biofeedback.

Impressive as the savings to the medical system can be through behavioral medicine, the potential savings in workers' compensation costs can even be greater. Steig and Williams (1983) calculated, for both treatment costs and disability payments, the estimated lifetime medical savings per patient as a result of a behavioral outpatient pain treatment program. Gonick, Farrow, Meier, Ostmand, and Frolick (1981) studied hospital costs five years pre- and posttreatment for 235 consecutive patients referred to behavioral medicine. The cost of providing the behavioral interventions, related to the savings in medical offset, yielded a cost–benefit ratio of $5 to $1.

Cummings and VandenBos (1981) described in detail the public policy implications that resulted in the eventual inclusion of psychologists as mental health providers in health care settings, as did DeLeon, VandenBos, and Cummings (1983). These conclusions indicated that any comprehensive health system that did not include a mental health or behavioral health benefit would pay for that lack of benefit in its medical-surgical benefit. Also, that cost would amount to far more than the cost of providing a psychological benefit. Insurers became convinced. Then came a whole new ballgame (Duhl & Cummings, 1987).

The Health Care Revolution

Actually, the health care revolution has been occurring since the early 1980s (Bevan, 1982), but it did not impact on the field of mental health until a few years ago because the initial cost containment efforts focused largely on reducing medical and surgical costs. An early alert was sounded by Cummings and Fernandez (1983) and three years later by Cummings (1986). By the time Duhl and Cummings (1987) described it extensively and Kiesler and Morton (1988) sought to inform all of psychology, the mental health part of the health care revolution was well under way, with over 31 million Americans covered under *managed* mental health rather than traditional fee-for-service. The figure is growing at 25% a year, and it is predicted that, by 1995, at least half of all Americans will receive their mental health benefits under managed mental health care and that 50% of all present fee-for-service mental health practitioners will be out of business (Cummings, 1986; Cummings & Duhl, 1986, 1987). At the same time, psychiatry is undergoing what it terms *remedicalization*, a euphemism for the position that only the *medical* aspects of mental health should be covered, and is fiercely opposing the extension of hospital privileges to psychologists ("Supreme Court to Review," 1988; see Chapter 5). In the absence of such privileges, psychiatry would have no competitors, as it is the only mental health profession licensed to perform medical services. Because federally chartered health maintenance organizations (HMOs; to be described further below) are largely exempt from state statutes, they are under no duress to recognize and employ psychologists. Many are seduced into "going on the cheap" in mental health and employing less expensive providers, not only social workers who are qualified, but also mental health "counselors" with as little as one or two years of community college psychology training (Cummings & Duhl, 1987). This is a crisis for psychology of enormous proportions.

The fuel for the health care revolution came from spiraling health care costs, which were exceeding twice the rate of inflation for the rest of the economy, and the thrust came from the entry into the health care arena of the new heavy hitters: American corporations. Where, in the previous struggle, professional psychologists had to persuade the insurance industry and their own APA, now they are confronted by those who pay the bills and who have cried, "Enough!" The new drive for health care cost containment not only has produced such unlikely bedfellows as industry and labor but has been joined by farmers and consumers as well. In 1965, health care accounted for 6% of the gross national product (GNP). The projections made from the accelerating costs in 1979 predicted a doubling to 12% of the GNP. The beginning of 1989 saw it at just over 11% of the GNP, attesting to the success of cost containment efforts to slow it down. The exception has been the cost of mental health, which is running away.

Whereas the current inflation in the health care field is about 9% a year, in 1988 mental health care was increasing at almost twice that rate, at 16.5% per year (Mullen, 1988). Aside from the fact that the health care industry was not confronting mental health care costs because of the overriding priority of medical and surgical costs, how did this happen? The efforts to control health care costs caused mental health care to balloon like an aneurism in a blocked artery. Primary among these efforts was the introduction of diagnosis-related groups (DRGs) by Medicare and Medicaid. DRGs imposed on hospitals lengths of stay limited by the diagnosis for each patient (or category of patients, over 300 in all). This limitation resulted in thousands of empty hospital beds throughout the nation, threatening the financial stability of the American hospital system. Hospitals were quick to note that DRGs did not apply to psychiatry and substance abuse, and they began a rapid conversion of their excess beds to adult psychiatry, substance abuse, and the new phenomenon of adolscent psychiatric hospitalization. They embraced huckstering, and marketed these beds in slick television commercials that were guaranteed to frighten any spouse or parent into hospitalizing a husband, wife, or chlid. General hospitals, which never had psychiatric beds, soon had 50% of their beds converted to mental health and substance abuse. Something that was never predicted became commonplace. Psychiatrists were lured into lucrative hospital-based practices and began to fill these beds. In 1986, the last year for which statistics are now available, psychiatric hospital beds increased by 37%, and expenditures for psychiatric hospitalization increased by 44% in the United States (Mullen, 1988). Preliminary data for 1988 and 1989 suggest similar increases for each of those years (Mullen, 1989). There is now a saying in the psychiatric units of private hospitals: "A built bed is a billed bed."

Outpatient psychotherapy still accounts for a relatively small portion of the increase in mental health costs, and hospitalization is responsible for the runaway costs. Nonetheless, the health care industry is turning its attention to aggressively reducing the cost of mental health. Some insurers, regarding mental health services as unimportant, are severely reducing this benefit. Others are aggressively turning to managed mental health. This reaction has created an industry where none existed before, with companies such as American Biodyne, American PsychManagement, Metropolitan Clinics of Counseling (MCC), Preferred Healthcare, Plymouth, United Clinics of Counseling (UCC), and U.S. Behavioral Health, to name only a few, are suddenly having an impact on the manner in which mental health care is dispensed.

Necessary Modifications

Any health system that does not include a comprehensive mental health service will pay for stress-related conditions through the over-utilization of its medical services. This fact was learned by the insurance industry in the 1980s, and resulted in the inclusion of mental health services (and the subsequent inclusion of psychologists as providers). Now there is an entirely new set of players that have to be persuaded: the giant health corporations that are rapidly gaining control of our health system and instituting managed care. The foregoing arguments are all still valid, but they will have to be reiterated.

There is an array of new delivery systems, sometimes called the alphabet soup of health-care, which psychologists and other mental health practitioners must learn (Cummings & Duhl, 1987). These managed-care systems include the HMO (health maintenance organization, which is capitated and closed-panel), the PPO (the preferred-provider organization, the purpose of which is to compete with existing providers), the EPO (the exclusive provider organization, in which all health enrollees must seek the benefit services), and the IPA (the independent provider association, where capitated providers practice in their own office), to name only the dominant few. Psychologists will have to adapt to and be willing to assume the risk for mental health services, which means that, if a prospective reimbursement is not sufficient because of provider inefficiencies, the provider sustains the financial loss. Psychologists will need a great deal of training, as well as encouragement from the APA and the leadership of the profession. Unfortunately, most of our resources at the present time are expended in attempting to preserve the status quo and to stave off the rapid emergence of managed care.

Psychologists are in an excellent position to innovate delivery systems. Psychiatry has all but abandoned psychotherapy, and because the psychologist can not prescribe medication, our profession has developed an impressive number of targeted, brief interventions. These targeted interventions, focused on specific psychological conditions, can bring rapid relief from pain, anxiety, and depression that is a change of behavior, rather than the masking of behavior that is accomplished by most chemotherapies (Cummings, 1985, 1988a). The managed health care industry must be made aware of our expertise in this regard.

We must abandon the concept of cure (Cummings & VandenBos, 1979). This concept has held back psychotherapy more than any other. First of all, we are dealing with psychological conditions, not an illness. Furthermore, behavioral health has shown that stress derives from the way we live: what we eat or do not eat; how we eschew exercise; how we smoke, drink, and pollute; and an array of other lifestyle variables. Psychologists have developed wellness programs and need to demonstrate their importance in any comprehensive health system. We are on the defensive here, because many in the health industry remember psychologists as those ethereal beings who were committed, for all their patients, to the nirvana of self-actualization and human potential, the so-called happiness variables that no one has ever been able to measure adequately in psychotherapy outcome studies. Psychology has innovated brief, intermittent therapy throughout the life cycle, which is focused, problem-solving therapy at stress points in a person's life (Cummings, 1986, 1988; Cummings & VandenBos, 1979). The ultimate cure of anxiety is never the focus, as anxiety is a normal accompaniment of life. Rather, the person is encouraged to seek brief therapy at various stress points throughout the life cycle.

Because most increases in mental health expenditures in the past several years have resulted from unnecessary psychiatric hospitalization, psychology is in an excellent position to demonstrate that it has proven outpatient alternatives to the overhospitalization of emotionally disturbed adults and adolescents, as well as of substance abusers. The average for nonmanged mental health plans currently exceeds 100 hospital days per year per 1000 enrollees, and we have seen it approach 300. Contrast these numbers with those for a well-run HMO, which average between 40 and 50 hospital days per year per 1,000 enrollees. American Biodyne, a psychology-driven mental health maintenance organization (MHMO) using a wide array of aggressive psychotherapy protocols, has achieved what is regarded as the lowest psychiatric and substance abuse hospitalization in the nation. On its entering one market, the 178,000 enrollees averaged 114 days of hospitalization per year per 1,000 enrollees. Within 60 days, Biodyne reduced the yearly hospital days to 4 per 1,000 enrollees per year and demonstrated what can be accomplished with the appropriate application of current psychological services.

Psychology is now engaged in a national struggle to obtain hospital privileges for psychologists. It will one day succeed, but at that time, it would be a tragedy if psychologists were to succumb to the temptation of the temporary, lucrative, hospital-filling practices that have attracted many psychiatrists. Rather, psychologists need to continue to demonstrate that outpatient psychotherapy can reduce unnecessary psychiatric hospitalization. Care outside the mental hospital is likely to be the wave of the future because it is more effective and can be less expensive (Kiesler, 1982; Kiesler & Sibulkin, 1987).

Finally, psychologists are uniquely prepared to render program evaluation and outcome measures of the effectiveness of all of health care, not just mental health care. In an era of health care rationing, public policy concerns center on the adequate distribution of our health care resources, the elimination of waste and duplication, the quality assurance, and the strengthening of our limited resources through efficacy of treatment and efficiency of delivery (Reinhardt, 1987). The profession of psychology, with its scientific base, is integral to the design, delivery, and outcome evaluation of *all* health care (Cummings, 1987).

Outreach: Physician Cooperation and Consumer Education

It has been demonstrated that physician referral is the most effective way to triage a patient into a behavioral health system (Friedman *et al.*, 1974). Patients respect their physicians and will generally accept such a referral. Unfortunately, in the case of the somaticizer, the exasperated physician often refers in a manner not conducive to compliance: "It's all in your head." Various methods have been used to help the physician identify and refer the somaticizer early in the cycle, by far the most frequent of which has been screening through computer-based test instruments (CBTI). Cummings (1985) issued a note of caution. In a study of the practice patterns over a two-year period of 34 primary-care physicians who received regular CBTI printouts identifying somaticizing patients, it was found that the rate of missed diagnoses of actual physical illness increased dramatically. The physicians began to rely overly on the results of CBTI screening and did not look further into the symptomatology of patients identified as somaticizers, thus failing to heed the age-old adage, "Hypochondriacs can get sick, too."

The economics of practice influence whether a physician will refer a somaticizing patient. Capitated physicians readily refer such patients , as there is no ecomomic incentive to hang on to them, whereas fee-for-service physicians regard the high utilizing patient as a source of revenue (Rand Corporation, 1987). It becomes useful, in such a setting, to access the somaticizer directly through outreach and consumer education.

One of the most successful triaging methods to directly address the overutilizer has been in operation for several years at American Biodyne and was reported on by Cummings and Bragman (1988). Founded in 1985, American Biodyne is a for-profit behavioral health maintenance organization (BHMO) that services the mental health and behavioral health needs of 2.1 million enrollees of several health insurers (e.g., Blue Cross and Blue Shield, CIGNA, Humana, and SelectCare) in eight states. Where its triage is in operation, Biodyne receives a monthly computer printout of the highest of 10% utilizers of medical facilities and resources as identified by frequency of service, not cost. The somaticizer is characterized by excessive visits to a physician, whereas high dollar amounts identify supercostly interventions such as open-heart surgery, organ transplants, and other medical heroics. Of these 10%, more than half are either somaticizers or persons suffering from physical illness whose treatment may be enhanced by behavioral health intervention. The outreach program is directed toward getting that 5%–6% of the patients into mental health or behavioral health treatment.

Because of their extensive knowledge of physical illness, and their ability to be conversant with patients about physical illness, coupled with their psychotherapeutic skills, psychiatric nurses are usually employed by Biodyne to conduct the telephone outreach. A medical so-

cial worker having similar knowledge and skills is an acceptable substitute for the psychiatric nurse. However, as a new center is being implemented, our procedure is to use the initial free time of psychologists, at least until the therapeutic load builds to the level where their time is not available. Therefore, it is necessary for each therapist on the staff to learn the outreach procedure.

The nurse, the social worker, or the psychologist is responsible for calling a predetermined number of these high utilizers. From the outset, it is important that the patient's belief in the somatic nature of his or her complaints not be challenged, even to the slightest degree. The patient's interest can usually be aroused by the statement, "Someone who has had as much illness as you have had certainly must be upset about it." This statement usually elicits an immediate reaction, ranging from an exposition of symptoms to the complaint that physicians don't seem to understand or to be sympathetic to the patient's plight. After patient has been heard out sufficiently to permit the development of some initial trust, the patient is invited to come in to explore how Biodyne can investigate the possibilities of an alternative to the treatments that have not worked, or perhaps, the patient, once the difficulty is better appraised, may be put in touch with a more sympathetic physician. Then, an initial appointment for psychotherapy is made. If the psychologist is doing the outreach, there is the immediate advantage that an appointment can be made with that therapist.

The telephone outreach is only one method used in an attempt to bring somaticizing patients into therapy. At Biodyne, there are periodic mailings of brochures or newsletters to remind these high medical utilizers of the services offered. In addition, each issue of the monthly newsletter features an article about a specific somatic complaint. The condition is discussed, and suggestions for change are made, along with the suggestion that an appointment at Biodyne may be appropriate. Psychologists and outreach personnel are also encouraged to take part in community presentations and presentations to local industries, in an attempt to further identify the high utilizers and to encourage their participation in psychotherapy.

Once the patient comes to Biodyne, it is vital that the therapist continue meticulously not to challenge the somaticization. The therapist's interviewing skills are marshaled to detect the problem that is being somaticized. Once this problem or set of problems is determined, the therapist treats these without ever relating them to the physical symptoms. In fact, most patients conclude rather brief therapy with a relief of somatic symptoms without every consciously relating psychological discoveries to the previous physical complaints.

It is important to note that the Biodyne model of triaging somaticizers out of the medical system and into a psychological system was not developed as a cost containment procedure. Rather, it was developed first to bring therapeutic effectiveness and relief of pain, anxiety, and depression to the patient in psychological distress. The model became an integral part of a therapeutically effective, comprehensive mental health treatment system, and only then was it discovered that it was also cost-effective.

In cases of actual physical illness, the psychologist accepts the illness as a given. At that point, the therapist concentrates on the patient's reaction to the condition (e.g., depression, rage, or despair), and also to any neurotic conflicts that may be impeding or showing recovery. These issues are then addressed in the course of the psychotherapy or the behavioral health intervention with the patient.

Perhaps one of the most effective methods developed for triaging the worried well into a behavioral health system and the asymptomatic sick into the medical system has been automated multiphasic health screening, which includes psychological screening (Friedman *et al.*, 1974). In the early 1980s, it was by far the approach of choice in most comprehensive prepaid health plans, and some had as many as 30 to 35 laboratory and other health checks computer on-line and all within a two-hour period. Eventually, such elaborate automated systems proved too costly, and they have given way to smaller, less ambitious health-screening systems, most of which can be quite effective when there is an awareness of physicians' propensity to miss physical diagnoses (as noted above).

Summary

Like all health care disciplines, clinical psychology is increasingly confronted with the need to prove not only its clinical effectiveness, but also its *cost*-effectiveness. Numerous powerful forces, whether in the form of DRGs or HMOs or some new form that has yet to appear on the health care scene, will continue to require accountability and justification for the expenditure of health care dollars. In part because of the scientific motivation of clinical psychologists to study what they do, data relevant to the clinical *and* the financial efficacy of services in health care settings are available to address these issues. The evidence thus far suggests that psychological services can reduce the inappropriate utilization of expensive medical care among the "worried well" and can improve medical management and behavioral outcomes among the chronically ill.

References

Bevan, W. (1982). Human welfare and national policy: A conversation with Stuart Eizenstat. *American Psychologist, 37,* 1128–1135.

Cummings, N. A. (1977). Prolonged or "ideal" versus short-term or "realistic" psychotherapy. *Professional Psychology, 8,* 491–501.

Cummings, N. A. (1979). Mental health and national health insurance: A case study of the struggle for professional autonomy. In C. A. Kiesler, N. A. Cummings, & G. R. VandenBos (Eds.), *Psychology and national health insurance: A sourcebook* (pp. 5–16). Washington, DC: American Psychological Association.

Cummings, N. A. (1985). Assessing the computer's impact: Professional concerns. *Computers in Human Behavior, 1,* 293–300.

Cummings, N. A. (1986). The dismantling of our health system: Strategies for the survival of psychological practice. *American Psychologist, 41,* 426–431.

Cummings, N. A. (1987). The future of psychotherapy: One psychologist's perspective. *American Journal of Psychotherapy, 61,* 349–360.

Cummings, N. A. (1988a). Brief, intermittent psychotherapy throughout the life cycle. *News from EFPPA* (European Federation of Professional Psychologists Association), 2(3), 4–11.

Cummings, N. A. (1988b). Emergence of the mental health complex: Adaptive and maladaptive responses. *Professional Psychology: Research and Practice*, 19(3), 308–315.

Cummings, N. A., & Bragman, J. I. (1988). Triaging the "somaticizer" out of the medical system into a psychological system. In E. M. & V. F. Stern (Eds.), *The psychotherapy patient* (pp. 109–112). Binghamton, NY: Syracuse University Press.

Cummings, N. A., & Duhl, L. J. (1986). Mental health: A whole new ballgame. *Psychiatric Annals*, 16, 93–100.

Cummings, N. A., & Duhl, L. J. (1987). The new delivery system. In L. J. Duhl & N. A. Cummings (Eds.), *The future of mental health services: Coping with crisis* (pp. 85–88). New York: Springer.

Cummings, N. A., & Fernandez, L. E. (1985, March). Exciting future possibilities for psychologists in the marketplace. *Independent Practitioner*, 3, 38–42.

Cummings, N. A., & Follette, W. T. (1968). Psychiatric services and medical utilization in a prepaid health plan setting: Part 2. *Medical Care*, 6, 31–41.

Cummings, N. A., & Follette, W. T. (1976). Psychotherapy and medical utilization: An eight-year follow-up. In H. Dorken (Ed.), *Professional psychology today* (pp. 176–197). San Francisco: Jossey-Bass.

Cummings, N. A., & VandenBos, G. R. (1979). The general practice of psychology. *Professional Psychology*, 10, 430–440.

Cummings, N. A., & VandenBos, G. R. (1981). The twenty year Kaiser-Permanente experience with psychotherapy and medical utilization: Implications for national health policy and National Health Insurance. *Health Policy Quarterly*, 1(2), 159–175.

DeLeon, P. H., VandenBos, G. R., & Cummings, N. A. (1983). Psychotherapy—Is it safe, effective and appropriate? The beginning of an evolutionary dialogue. *American Psychologist*, 38, 907–911.

Duhl, L. J., & Cummings, N. A. (1987). The emergence of the mental health complex. In L. J. Duhl & N. A. Cummings (Eds.), *The future of mental health services: Coping with crisis* (pp. 1–13). New York: Springer.

Fahrion, S. (1990). Cost effectiveness in biobehavioral treatment of hypertension. *Biofeedback and Self-Regulation*, 14, 131–152.

Fahrion, S., Norris, P., Green, E., & Schnar, R. (1987). Behavioral treatment of hypertension: A group outcome study. *Biofeedback and Self-Regulation*, 11, 257–278.

Flor, H., Haag, G., Turk, D., & Koehler, H. (1983). Efficacy of biofeedback, pseudotherapy, and conventional medical treatment for chronic rheumatic back pain. *Pain*, 17, 21–31.

Follette, W. T., & Cummings, N. A. (1967). Psychiatric services and medical utilization in a prepaid health plan setting. *Medical Care*, 5, 25–35.

Friedman, G. D., Ury, H. K., Klatsky, A. L., & Siegelaub, A. B. (1974). A psychological questionnaire predictive of myocardial infarction: Results of the Kaiser-Permanente epidemiologic study of myocardial infarction. *Psychosomatic Medicine*, 36, 71–97.

Goldberg, I. D., Krantz, G., & Locke, B. Z. (1970). Effect of a short-term outpatient psychiatric therapy benefit on the utilization of medical services in a prepaid group practice medical program. *Medical Care*, 8, 419–428.

Gonick,U., Farrow, I., Meier, M., Ostmand, G., & Frolick, L. (1981). Cost effectiveness of behavioral medicine procedures in the treatment of stress-related disorders. *American Journal of Clinical Biofeedback*, 4, 16–24.

Jacobs, D. (1988). Cost-effectiveness of specialized psychological programs for reducing hospital stays and outpatient visits. *Journal of Clinical Psychology*, 21, 23–49.

Kaufman, M. (1989). Cancer: Facts vs. feelings. *Newsweek* (April 24), 10.

Kiesler, C. A. (1982). Mental hospitals and alternative care. Non-institutionalization as potential public policy for mental patients. *American Psychologist*, 37, 349–360.

Kiesler, C. A., & Morton, T. L. (1988). Psychology and public policy in the "health care revolution." *American Psychologist*, 43(12), 993–1003.

Kiesler, C. A., & Sibulkin, A. (1987). *Mental hospitalization: Myths and facts about a national crisis*. Newbury Park, CA: Sage.

Kramon, G. (1989). Taking a scalpel to health costs. *New York Times* (January 8), sect. 3, pp. 1, 9–10.

Malan, D. H. (1976). *The frontier of brief psychotherapy*. New York: Plenum.

Mechanic, D. (1966). Response factors in illness: The study of illness behavior. *Social Psychiatry*, 1, 106–115.

Mullen, P. (1988). Big increases in health premiums. *Healthweek*, 2, 25, (December 27), pp. 1, 26.

Mullen, P. (1989). Increases in health premiums continue. *Healthweek*, 3, 22 (November).

Mumford, E., Schlesinger, H. J., & Glass, G. V. (1978). A critical review and indexed bibliography of the literature up to 1978 on the effects of psychotherapy on medical utilization. NIMH: Report to NIMH under Contract No. 278-77-0049-M.H.

Olbrisch, M. (1981). Evaluation of a stress management program. *Medical Care*, 19, 153–159.

Rand Corporation. (1987, July). *A report on the changing practice patterns of primary care physicians in geographical areas with too many physicians*. Santa Monica, CA: Author.

Reinhardt, U. E. (1987). Resource allocation in health care: The allocation of lifestyles to providers. *The Milbank Quarterly*, 65, 153–176.

Rosen, J. C., & Wiens, A. N. (1979). Changes in medical problems and use of medical services following psychological intervention. *American Psychologist*, 34, 420–431.

Schlesinger, H. J., Mumford, E., & Glass, G. V. (1980). Mental health services and medical utilization. In G. R. VandenBos (Ed.), *Psychotherapy: Practice, research, policy.* Beverly Hills, CA: Sage.

Shapiro, A. K. (1971). Placebo effects in medicine, psychotherapy and psychoanalysis. In S. L. Garfield & A. E.

Bergin (Eds.), *Handbook of psychotherapy and behavioral change: An empirical analysis.* New York: Wiley.

Shellenberger, R., Turner, J., Green, J., & Cooney, J. (1986). Health changes in a biofeedback and stress management program. *Clinical Biofeedback and Health, 9,* 23–24.

Steig, R., & Williams, P. (1983). Cost effectiveness study of multidisciplinary pain treatment of industrial-injured workers. *Seminars in Neurology, 3,* 375.

Supreme Court to review psychology's arguments in CAPP v. Rank Case. (1988). *California Psychologist,* Special Edition: CAPP v. Rank, December 1.

VandenBos, G. R., & DeLeon, P. H. (1988). The use of psychotherapy to improve physical health. *Psychotherapy, 25*(3), 335–342.

Yates, B. T. (1984). How psychology can improve effectiveness and reduce costs of health services. *Psychotherapy, 21*(3), 439–451.

CHAPTER 9

Marketing Psychological Services in Hospitals

Jack G. Wiggins and Kris R. Ludwigsen

Introduction

Hospitals are a new and evolving market for psychological services. Changes in health care delivery are now creating a climate in which hospitals are much more receptive to and interested in the diagnostic, consultative, and treatment services of psychologists. These changes include changing health care economics, the need for innovative programs that meet patient needs, and the evolving concept of the hospital as being responsive to the needs of the community. Because these factors closely coincide with the interests of psychologists in health care delivery, there appears to be a significant though underdeveloped mutual interest between psychologists and hospitals in serving the community. Finally, hospitals are an important market for psychologists because they are a central coordinating element in health care delivery and consume the lion's share of the health care dollar.

Jack G. Wiggins • Psychological Development Center, 7057 West 130th Street, Cleveland, Ohio 44130. Kris R. Ludwigsen • Kaiser Foundation Hospital, 200 Muir Road, Martinez, California 94553.

Changes in Health Care Delivery

Both marketing and hospital practice are relatively new concepts for psychologists to integrate into their practices. Cummings (1986) pointed out that psychologists have traditionally eschewed anything to do with business, marketing, or merchandising their services. However, the growing revolution in health care delivery as a response to much needed cost-containment reform indicates that both hospitals and psychologists must develop innovative models of health care delivery and then learn to market these models competitively to the patient-consumer.

Although the need for health care has continued to outstrip the availability of services, health care costs consume approximately 11% of the U.S. gross national product (GNP), a figure that many view with an alarm similar to rises in the prime lending rate. Health care is the third largest industry in our nation (DeLeon, 1986). Medical costs are increasing at twice the rate of inflation, and mental health costs are rising at four times the inflation rate. Clearly, the health care industry is ripe for modification. It might be said that the mission of third-party payers in the 1990s is to decrease the total amount spent on health care through controlling their costs and thus decreasing the percentage of the GNP now consumed by health care. The winds of change affect psychologists no less than hospitals. There are some 6,900 hospitals, and there are over 45,000 psychologists in health care (Dörken & Associates, 1986). If hospitals and psychologists are to survive the health care revolution, both must develop treatment strategies responsive to the current cost-containment climate as well as to the needs of the public.

The Role of Hospitals in Health Care

To understand the current position of hospitals in health care delivery, it is useful to understand the developmental history of the model. The evolution of the hospital as a health care facility has been intimately tied to the development of medical science and technology, as well as to the development of the nursing profession (Rosenberg, 1988). In the first half of the 19th century, there were few hospitals; many towns cared for their sick poor in infirmaries, while more fortunate patients received better care at home. The nursing profession, inspired by the leadership of Florence Nightingale, constantly worked to upgrade the quality of patient care and to improve sanitary conditions in inpatient facilities. Nursing schools were largely responsible for transforming the hospital routine and for enhancing the quality of care. Physicians returning from study abroad began to view the hospital as the most appropriate place for medical research and education because patients and technological resources could be brought together in the hospital. Thus, the interests of the medical profession began to shape the institution of the hospital while medical practice was being increasingly influenced by the developing knowledge of infectious diseases.

Until recently, hospitals relied on physicians to refer patients for health care services. They depended on physicians and courted their business through such means as offering outpatient office space on hospital grounds. In describing the rise of America's hospital system, Rosenberg (1988) reported that, until the

early 20th century, hospitals still tended to be the professional resource of a privileged few physicians who used them to advance their careers. Physicians often owned hospitals until Medicare abolished this practice as a conflict of interest. For the middle-class patient, hospital care became increasingly attractive as it was perceived to be effective. Today, virtually all American physicians in private practice have hospital privileges, and hospitals have continued to serve physicians as the most efficient source of income generation.

Nevertheless, hospitals have been responsible for a major share, approximately 70%, of the upward spiral of medical costs. Hospital costs have escalated even more rapidly than physicians' fees because of advances in technology, and have become the special target of regulators in the government and the private sector. Although hospitals are a convenient site for providing care, they are more costly, but not necessarily more effective, than other alternatives. Government efforts to control hospital costs include certificates of need, reimbursement by DRGs (diagnostic related groups), and other prospective payment plans. Efforts in the private sector include preadmission screening and indemnity contracting. As a result of these fiscal pressures, hospital bed occupancy in the 1980s has fallen substantially. With the increasing trend toward outpatient and adjunctive care, hospitals have had to develop new sources of referral.

The Development of Hospital Practice in Psychology

The role of psychologists in hospitals evolved from World War II with the development of clinical psychology as a profession. Initially, the psychologist's predominant role was that of psychodiagnostician, for example, evaluating the cognitive-intellectual, neuropsychological, and emotional impact of war injuries in the Veterans Administration hospital system. In this capacity, psychologists were initially viewed as consultants or technicians. The primary clinical decision-making responsibility rested with physicians, while the nursing staff were responsible for the regimen of care. Since the early 1950s, psychologists have served on the staffs of state mental hospitals, Veterans Administration hospitals, mental retardation facilities, and, more recently, private psychiatric hospitals, usually on a salaried basis. Despite indications that 60% of visits to physicians are for somatized emotional problems and that psychological intervention reduces overutilization of medical services as much as 79% (Cummings, 1986), there has been little involvement of psychology in general medical hospitals. Although the number of psychologists in hospital practice has been increasing, as noted in Chapter 1, hospital practice by psychologists has been quite limited in comparison with the need for inpatient psychological services.

Legal and Regulatory Changes Impacting Hospital Practice

Until recent years, many restraints were placed on the practice of psychologists in inpatient facilities, partly as a result of the philosophy that hospitals were the exclusive province of physicians. The Joint Commission on Accreditation of Healthcare Organizations (JCAHO; formerly known as the Joint Commission on Accreditation of Hospitals), the primary organization that establishes standards for and awards accreditation to hospitals, was authorized by the U.S. Department of Health, Education, and Welfare to regulate hospitals and thus was, in effect, authorized to promulgate rules and enforce them as an extension of the federal government. The JCAHO standards reinforced the institutional power base of physicians, as supported through federal statutes and state legislation, by explicitly requiring a physician to take "medical responsibility" for and be in charge of patients in hospital settings. With this power, the JCAHO created a medical monopoly in hospitals through such auspices as the medical executive committee.

In 1975, hospitals and their medical staffs became subject to antitrust laws following the U.S. Supreme Court decision in *Goldfarb* v. *The Virginia State Bar* (421 U.S. 773; Sciara, 1989). Prior to the Supreme Court decision, health

care and other professions were viewed by the legal community as having immunity to antitrust statutes (Overcast, Sales, & Pollard, 1982). After the *Goldfarb* decision, the Ohio Attorney General, at the request of psychologists in the state, filed a complaint that forced the JCAHO to conform to state law or be subject to antitrust action. The result was a consent agreement in which the JCAHO modified its rules and regulations to permit a broader scope of practice for nonmedical health care professionals and to limit the power of the medical executive committee.

In the late 1970s and early 1980s, a series of antitrust suits, collectively known as the Virginia Blues, was filed in Virginia to enforce the state's freedom-of-choice statute (Resnick, 1985). The ultimate holding by the courts stressed that psychologists and psychiatrists were economic competitors and that policies or institutional regulations that served to decrease their competition were subject to the scrutiny of state and federal antitrust statutes (Enright, Resnick, Sciara, DeLeon, & Tanney, in press). In 1984, the JCAHO modified its rules barring psychologists and other nonphysicians from hospital staff membership. In response to considerable pressure from the American Psychological Association (APA) and other nonphysician health care professions over a number of years (see Zaro, Batchelor, Ginsberg, & Pallak, 1982), the revised rules allowed psychologists to be members of the medical staff where permitted by state law and the individual hospital. Throughout the states, psychology and physician practice acts define *scope of practice* but place no restraints on *locus of care*. A suicidally depressed patient may need the containment and intensive treatment of a hospital environment but also needs the continuity of care of his or her outpatient psychologist in collaboration with the hospital treatment team. Any constraints on practice disrupt continuity of care for the patient as well as hindering the practitioner's ability to compete in the health care marketplace (Dörken, 1989).

As a result of this series of advances toward parity with the medical profession, hospital practice has become one of the primary legislative priorities for psychology. Currently,

California, Georgia, the District of Columbia, North Carolina, and Florida have state laws enabling psychologists to have hospital privileges, and approximately 14 other state psychological associations are pursuing hospital practice legislation. The hospital practice movement has developed slowly because of intense political pressure and litigation, as in the *CAPP* v. *Rank* suit (1988), now settled in favor of psychologists. However, the window of opportunity has been created, and the practice of psychology in hospitals has begun to flourish regardless of state regulations.

The JCAHO policy did not immediately result in substantial numbers of psychologists and other nonphysician practitioners applying for staff privileges. Nevertheless, Dorken *et al.* (1986) reported that over 95% of procedures billed to CHAMPUS (Civilian Health and Medical Program of the Uniformed Services) by psychiatrists in hospitals were within the scope of psychologists' recognized practice. With over 6 million beneficiaries, CHAMPUS is the single largest health plan in the country. Since some 80% of CHAMPUS expenditures are for inpatient care, this plan represents a significant potential market for psychologists. In addition, psychologists who treat depressives, anorexics, and others requiring long-term or intensive psychotherapy benefit from being able to follow their patients through hospitalization. Psychologists have been serving as consultants to physicians in their communities for many years and naturally want hospital affiliation to provide these services to patients more efficiently.

Hospital privileges offer opportunities for greater professional autonomy and recognition for the profession at large. In addition, Dorken (1989) pointed out that some 60% of inpatients receive outpatient follow-up care. Thus, hospitals function as gatekeepers, creating a quasi monopoly where psychologists who do not have staff privileges lose their outpatients when admitted, and where inpatients are discharged to outpatient care by a psychiatrist. Because hospitals consume the lion's share of the health care dollar, psychologists without inpatient privileges are economically handicapped in the marketplace.

Opportunities for Hospital Practice

Several factors favor psychologists' entry into hospitals at this time. First, many hospitals are adding outpatient programs and partial care to develop their referral bases. In addition to inpatient consultation and liaison, assessment, and psychotherapy, psychologists can offer outpatient services within a hospital setting. Psychological treatment in a hospital can include child, adolescent, and family therapy; parenting groups; biofeedback; pain and stress control; treatment of eating disorders; and burnout syndrome. Other hospital services to which psychologists can make contributions include birthing centers, women's health programs, substance abuse programs, senior citizens' programs, rehabilitation, and sports medicine. With the development of health psychology as a specialty, psychologists have demonstrated expertise in addressing the needs of stroke victims; patients with colostomies; and families experiencing labor, delivery, and ob-gyn complications and other medical disorders (Ludwigsen & Enright, 1988). Many physicians have long recognized the benefit of psychological intervention in internal medicine, oncology, pediatrics, and neurology, as well as psychiatry. Such interventions can shorten hospital stays and can improve treatment compliance, lessening costs to patients, hospitals, and third-party payers. Psychologists can also provide expertise in program development and evaluation, research, staff training, patient education, quality assurance, marketing and organizational development. Part IV of this book illustrates the broad scope of psychologists' involvement in the hospital arena.

The promotion of psychological services in hospitals is a three-pronged effort involving advocacy, marketing, and education. Advocacy involves legislative action and initiatives, regulatory modification, and, when necessary, litigation. While such efforts are essential within organized psychology, they are also time-consuming and costly, and can create adversarial roles. Marketing is the promotion of psychological services through private contractual arrangements or "contract services" for the benefit of the hospital, its patient-consumers, and the providing psychologist. As such, it creates mutually advantageous relationships. Marketing by individual psychologists or groups of practitioners can open doors to hospital practice at the local level. Finally, educating both psychologists and hospitals on the benefits of developing a hospital-based psychology practice and training psychologists for hospital practice ensure that psychologists will be systematically prepared for the opportunities that are developing.

Economic Pressures on Hospitals

In the 1980s, hospitals became subject to fiscal accountability and cost containment pressures. Medicare initiated a prospective payment plan through DRGs, in which the hospital was reimbursed for a particular condition—for example, an appendectomy—rather than for length of stay (Binner, 1986). Commercial insurers and Blue Cross designed a similar payment system using indemnity contracting. These changes in reimbursement exerted financial pressures on hospitals to control costs through limiting lengths of stay. Bed occupancy dropped from 90% to less than 65%. As a result, smaller hospitals were forced to merge with larger hospitals or go bankrupt. Changes in reimbursement resulted in radical changes in treatment policies. To be cost-effective, hospitals now had to move patients in and out more rapidly while maintaining a patient census to meet expenses and provide a basis for growth. Hospitals have generally adopted two courses of action: one is to increase the number of admissions; the other is to integrate services on an outpatient, partial hospitalization, or adjunctive basis to generate additional revenues. Both can create opportunities for psychologists in helping the hospital develop innovative and efficient patient treatment (Sciara, 1988).

The emphasis on fiscal accountability and cost containment also resulted in the growing trend toward the corporate ownership of hospitals for better financial control and management. Many small psychiatric hospitals have been bought by large corporate chains (Sciara, 1989). Corporate ownership has major implications for hospitals and practitioners. Health care professionals are now accountable not only for quality of care but also for the financial management of the patient's and hospital's resources. Corporate ownership entails the reporting of financial performance to stockholders and corporate financial officers. Fiscal considerations are now as important as clinical care. Control over hospital policies is rapidly being taken away from clinicians and given to fiscal intermediaries. Psychology is particularly vulnerable because corporate America does not understand psychological services. The challenge to psychology is to provide high-quality, cost-effective care to demonstrate the value of psychological services to the hospital and the parent corporation (Sciara, 1988).

Hospitals in the Marketing Era

Corporate ownership has created a different image of the hospital in the community. Corporate financial managers have begun marketing hospital services through television, billboards, radio, and newspapers. Hospital advertising has increased tenfold since the mid-1980s. Corporate managers of hospitals are appealing directly to the public to use their services and are maintaining control of this market through aftercare services, including nursing homes and home health care units. In addition to expanded inpatient services and eating-disorder programs, hospitals have added outpatient services, thereby competing directly with physicians, psychologists, and other health care practitioners. Many hospitals have established outpatient surgical services to compete with freestanding surgicenters. *Medical Economics* ("Where the Next Wave," 1989) reported that the number of hospitals offering home health care, wellness services, and occupational health programs has more than doubled. With even more hospitals planning to introduce or expand these services, profit-conscious corporate executives are beginning to look at other potential revenue enhancers, notably pain management and treatment of eating disorders. These new ventures are largely the result of the business approach to the economic survival of the hospital rather than to the clinical needs of practitioners. While designed to meet the needs of the patient-consumer, they are motivated more by economic conditions created by lowered bed occupancy.

As marketing of programs has increased, hospitals have realized a need for a behavioral management component in their health care services. Accordingly, they have organized employee assistance programs (EAPs) for small businesses, have offered preemployment screening with physical and psychological assessment, and have combined in marketing groups to offer prepaid health care services. Paradoxically, the expansion of hospital programs into behavioral health care has simultaneously created new markets and increased competition for psychology. Hospitals are now hiring social workers and "behaviorists" in order to avoid paying for more costly psychological expertise. The focus is not on quality of care and innovation, but on defining an economically viable service area. Consequently, psychologists must demonstrate that the quality of their services is worth the price.

Marketing to Hospitals

Marketing strategies by psychologists to hospitals must be individualized. Hospitals differ in a number of respects: Some specialize in psychiatric, chemical dependency, or rehabilitation services; others are general medical and surgical facilities with or without these specialty units. Some are independent or privately owned; others are part of a corporate chain or HMO (health maintenance organization) or are funded by the county or state. Urban and rural hospitals serve different populations. Military and Veterans Administration hospitals have their own regulations and priorities of care. Other inpatient sites include "stepdown" facilities for extended or skilled nursing care, rehabilitation, and adolescent treatment.

Within this range of facilities, the psychologist can function in the role of employee, as an independent contractor, or as the attending or consulting doctor with staff privileges. Staff privileges usually offer the greatest degree of

influence within the hospital, and independent contractors generally have more autonomy than hospital employees. Hospitals may be targeted as open, receptive, resistant, or closed to granting privileges to psychologists based on a variety of factors, including corporate policy and hospital politics. Although hospitals have been opening staff membership to psychologists, they have often had the status of "professional affiliate" rather than being awarded active, consulting, or courtesy staff privileges as specified by the JCAHO. The professional affiliate status does not define duties or clinical privileges. Consequently, the introduction of psychological expertise into hospitals has been very slow. Because hospitals are not aware of what psychologists can do and psychologists have limited opportunities to demonstrate their expertise, a new approach is required to appeal to both.

When marketing services to a hospital, psychologists must be cognizant of the formal and informal organization of the facility. Psychologists who want to practice in hospitals are frequently confronted with a broad array of interfacing political alliances, which are minimally confusing at best and economically frustrating at worst. Each hospital is accountable to its board of trustees. The administrative staff and the medical executive committee of the medical staff are accountable to the hospital administrator, who reports to the board of trustees. One of the main functions of the board of trustees is to resolve the economic and political tensions that often arise between the medical staff and the hospital administration, so that hospital services are available to the public and are affordable. Any service that psychologists wish to offer must satisfy the requirements of both the medical staff and the hospital administration.

It is politically important to know the individual members of the board of trustees, who are elected for specific terms; the members of the medical executive committee, who are elected on an annual or biennial basis; and the medical director. It is also essential to become acquainted with the hospital administrator and administrative staff, particularly those in key

133

positions where the psychologist is interested in providing services.

Strategies for Marketing

To market psychological services to hospitals, a comprehensive strategy is required, including:

1. Deciding the specific services to be offered.
2. Identifying the key persons and target groups to be contacted.
3. Developing a means of contacting and educating physicians, administrators, and other health professionals on the need for and value of the services to be offered.
4. Deciding what approaches will have the greatest appeal to these gatekeepers, integrating networking and accountability as part of the hospital team.
5. Developing a means of continually evaluating the clinical effectiveness, and cost efficiency of the services offered.
6. Maintaining good social and political relations within the hospital while evaluating evolving trends in hospital health care.

Such "target marketing," thoughtfully conceived and researched, is continually modified to meet the needs of the hospital and the community, as well as the patient-consumer.

Often, it is easier to market one's services as a member of the hospital staff rather than as an outside entrepreneur. By joining the staff, psychologists have more opportunities to learn of hospital plans that may be relevant to their own interests. In addition, psychologists with hospital privileges tend to have larger practices and incomes than practitioners who are solely office-based.

Psychologists with hospital privileges can be a significant referral source for hospitals. Dafter and Freeland (1988) conducted a survey of the rate of patient hospitalization by psychologists and its implications for their potential power in hospital markets in Los Angeles County. In this area, there are 2,400 licensed psychologists and a substantial number of hospitals permitting psychologists to admit and discharge patients in accordance with state law. Psychologists were found to refer about 4.1 patients for hospitalization annually, for an average of 48 patient days per year per practitioner. At an estimated reimbursement of $573 per day by private insurance, the respondents contributed an average of $27,500 each to hospital revenues. The *Psychiatric Times* reported, in contrast, that each psychiatrist contributes approximately $500,000 to hospital revenues, and that hospitals are currently facing a shortage of psychiatrists ("Hospitals Face Psychiatrist Shortage," 1989). Dafter and Freeland (1988) concluded that psychologists' market power is often either hidden or unappreciated since many psychologists refer to psychiatrists, who are then perceived by hospital administrators as the referral source. The shortage of psychiatrists may facilitate opening hospital practice opportunities for psychologists in a number of states. According to CHAMPUS data for mental health services, psychiatrists provide care to over 70% of inpatients, whereas psychologists treat only 18% (Dörken, 1989).

Illustrating the entrepreneurial approach, J. D. Cole (personal communication, February 28, 1989) described the marketing of a coping skills program, a group therapy approach to chronic pain developed for private practice, pain management inpatient and outpatient clinics, and rehabilitation units. After the program had been used in a private practice setting for two years, it was marketed to a local hospital and was integrated into the pain management program. Cole reported that the program has been accepted both by the medical profession, which refers patients, and by insurance companies, which pay for it. The program began by treating workers' compensation patients and now includes 40% private patients. In a later communication (March 1, 1989), Cole reported that the program had opened the doors to working in an inpatient pain management program. He concluded, "By having an entree, we as psychologists can benefit others and create a need for our service which increases the opportunity to develop hospital privileges."

There are at least 12 areas where psychological expertise is needed: alcoholism and substance abuse, cardiac rehabilitation, emergency room consultation, general hospitals, nursing homes, pain management, preemployment screening, psychiatric facilities, rehabilitation services, sleep disorders, preparation for surgery, and weight loss programs. Because each service tends to have its own advocates and infrastructure, it is important to respect protocols and political sensitivities. A number of these are described in Part IV of this book. Here, we discuss psychiatric facilities, general hospitals, and nursing homes as specific examples.

Psychiatric Facilities

Psychiatric facilities, including some 750 psychiatric hospitals nationwide and additional units attached to general hospitals, provide an opportunity for the expansion of psychological services. At present, psychiatric services are a profitable sector of the health care industry. The *Psychiatric Times* ("Hospitals Face Psychiatrist Shortage," 1989) reported an increased demand for services based on greater acceptance among employers, third-party payers, and the general public that eating disorders, drug abuse, and other disorders of psychological origin are legitimate health problems that should be treated. Although varying by size and location of the hospital, 71% of state or Veterans Administration hospitals reported a need for additional psychiatrists, followed by 68% of nonprofit hospitals, and 54% of for-profit hospitals.

In psychiatric facilities, psychologists can develop programs that deal with specific behavior problems. These may include cognitive-behavioral approaches to the assessment and management of aggressive behavior and other severe disturbances, such as psychosis, behavior therapy for children, adolescents, and the elderly, or consultation to nursing staff. As team members, psychologists can propose psychosocial interventions for patients and offer adjunctive and postdischarge care such as marital therapy, parent support groups, and family therapy to deal with the problems of children, adolescents, and the elderly. As a group therapy leader (e.g., in assertiveness training), the psychologist can be instrumental in defining group goals, deciding which patients are appropriate for the group, and dealing with psychosocial issues in order to enhance group effectiveness. As the psychiatric facility expands its range of services, psychologists can develop and run day treatment programs, outpatient programs, and other follow-up care that provides vertical integration of services and maintains contact with the hospital's market.

Psychologists can provide intake assessment and prescreening to determine whether the patient should be admitted to the hospital or referred elsewhere. They can assist prospective patients and their families in dealing with the psychosocial aspects of hospitalization. As diagnosticians, psychologists can assess personality dynamics and cognitive, behavioral, and neuropsychological functioning, and can provide mental status evaluations. Psychologists may have specified or informal therapy privileges in psychiatric units or facilities, so that, even though the patient is admitted under a physician, the psychologist has the primary responsibility for treatment.

Psychologists can also provide training for psychology interns and in-service education for nurses. They can provide outreach training and consultation to school psychologists and counselors who deal with the problems of children and adolescents, as well as clinical case conferences for other psychologists and practitioners in the community. As issues emerge in clinical care or the operation of the hospital, psychologists can offer research skills to answer questions, as in patient satisfaction surveys.

General Hospitals

In some psychiatric hospitals, psychologists are viewed as economic competitors of psychiatrists, with privileges limited to psychological testing. Psychology then becomes the captive of psychiatry because the individual hospital can decide the level of privileges to award. However, within general acute care hospitals, psychiatry is usually in the department of medicine. The other members of the department are internists with various specialties such as cardiovascular disease, gastroenterology, neurology, oncology, pulmonary disease, and dermatology, and they are apt to be tolerant of other differences in specialization, such as psychology versus psychiatry. Many internists have had difficult experiences with psychiatry and are looking for a new approach to treatment. Psychologists offer a conception of patient problems that internists find novel and refreshing. Psychologists can assist in obtaining patient compliance with kidney dialysis management and diabetic medication regulation. They can also assist in presurgical preparation and postsurgical assessment through hypnosis, crisis intervention, and family support.

For the psychologist who wants to provide consultation, regular attendance at department meetings is important. Psychiatrists rarely attend such meetings because of their more primary involvement in psychiatric facilities. Being one of the first to enter enhances visibility. Being one of the last to leave provides time to renew friendships with medical colleagues. Attendance at general staff meetings, golf outings, clambakes, and so on is also helpful. Holiday parties facilitate family social relationships that can be developed as a means of social interaction in the community.

Visibility can provide opportunities, which must then be solidified by competence and efficiency to ensure that the referring professional is well served. Because of pressures to discharge the patient as soon as possible, timeliness in hospital consultation is essential. To operate in a professional manner, the psychologist should (1) contact the referring physician to clarify the goals of the consultation and to establish a time for the evaluation; (2) inform the unit secretary of the time so that a consultation room can be reserved; (3) contact the patient concerning the time set for consultation; (4) fill out the hospital consultation form in the patient's chart on completing the evaluation; (5) notify the referring physician of the findings and discuss any questions; and (6) provide a formal report promptly. Some hospitals specify

that the report must be completed within a certain time (e.g., 48 hours).

It is important to maintain good relationships with the nursing staff and to conduct oneself as a professional colleague within the formal and informal "rules and regulations" of the hospital. Copies of the rules and regulations, as well as the bylaws, can be obtained from the medical staff secretary and can serve, among other things, as a helpful guide to the standard procedures for consultation.

Nursing Homes

Nursing homes, which some hospitals are building as adjuncts, are another market for psychological services (Smyer, 1986). Residents must be evaluated before admission and once a year during their residency. Psychological assessments are needed to differentiate a variety of disorders, including depression, dementia, and pseudodementia. Assessments can be helpful in distinguishing recent disorders and those representing chronic adaptation patterns. Behavioral approaches have been found helpful in decreasing paranoid behavior and disorientation as well as dependency behavior. Because psychological interventions have been shown to decrease demands on nurses' and physicians' time, they have become attractive in cost-conscious settings. Psychologists should take the initiative to clarify the need for psychological intervention for nursing-home residents and their families.

Psychology must establish itself with the medical and administrative staff as a professional specialty with valuable expertise. Although hospitals are more interested in and aware of the need for behavioral science services, financial pressures often result in hiring technicians (e.g., for substance abuse, eating disorders, and wellness programs) who are only marginally monitored by the professional staff. Many hospitals with psychiatric units are now offering outpatient services staffed primarily by social workers and mental health technicians. For the hospital as well as the patient-consumer, the decision to purchase or contract for health care is made on the basis of

(1) cost, (2) convenience, and (3) perceived quality. To remain competitive in the evolving health care market, psychologists must offer services that meet all three criteria.

Psychology and the Changing Philosophy of Hospital Care

Hospitals are now recognizing that the community looks to them not only to treat the sick but also to promote health. To meet the need for community health education, hospitals have developed patient support groups for various illnesses (e.g., cancer), have organized specific programs (e.g., to stop smoking), and have added outpatient facilities. In addition to expanding the range of services, this development has caused hospitals to look to the field of behavioral science to meet their responsibilities to the community. Hospitals are using psychological expertise to meet patient needs, to develop programs, and to improve quality control. A changing philosophy of health care and a new role for hospitals are developing as a result. Hospitals are integrating outpatient, day treatment, and partial hospitalization services to offer comprehensive packages of health care to their consumers. If hospitals are to be responsive to the needs of the entire community, they can no longer remain the exclusive fiefdom of physicians, as in the past. They must make available the full spectrum of health care services, including advances in behavioral science and health psychology as well as medical technology.

We believe that, because of changing economics and a changing philosophy of treatment, hospitals need psychologists and psychologists need hospitals. Hospitals are a central coordinating element of the health care delivery system. To respond effectively to changes in health care delivery, psychologists must be involved in all aspects of the health care system. Without an entry into hospitals through privileges or contract services, psychologists will have little impact on hospital policies and will be less able to compete, both individually and collectively, in the health care marketplace. In addition, psychologists are the proponents

of more humanistic treatment values. They believe in educating the patient to achieve control over the disorder, as in pain management, rather than increasing dependence on the physician or the facility, a philosophy more appropriate to today's economic policies in health care. Finally, the shift to stress-related disorders as the primary focus of treatment in hospitals, as opposed to the infectious disorders of yesteryear, requires psychological intervention.

We believe that the interface between psychology and hospitals is still too limited for a clear appreciation, by either party, of the value of psychological services. Psychologists must also adopt new perspectives to deal effectively with hospitals. They must be aggressive in identifying a market and meeting the needs of that market through contract services to hospitals and other agreements (Sciara, 1989).

In marketing to hospitals it is important to be organized. If the task of promoting hospital practice is left to individual practitioners, progress is likely to be piecemeal rather than the result of a systematic effort serving the public interest. Psychologists must advocate to hospitals and health care corporations. They must demonstrate that it is economically advantageous to the facility for them to have hospital privileges. In negotiating with the hospital, it is important to be informed; when psychologists are unaware of the types of contractual and financial arrangements they can develop with an inpatient facility and are not organized, they are unable to exert collective pressure for better privileges, working conditions, and remuneration.

Summary

Psychologists must learn to market their services to hospitals or to market for hospitals as part of their professional role. If psychologists do not learn to compete successfully as a profession in marketing services to hospitals, they risk being overlooked in the health care marketplace increasingly controlled by the hospital as gatekeeper of referrals. To influence inpatient facilities successfully, psychologists must develop a new concept of their role and responsibilities. Psychologists must be responsive to the hospital's agendas and interests. The services offered must be congruent with both the treatment philosophy and fiscal goals of the facility. Psychologists must be able to identify interests that are mutually rewarding for and beneficial to the hospital, the patients, and themselves. They must also be accountable to the patients and the facility in meeting program goals, in complying with policies and procedures, and in demonstrating creativity in service development (Sciara, 1988). Psychologists must view hospitals from a systems perspective and must be aware of the overlapping hierarchies of personnel in inpatient facilities. It is important to maintain good relationships with the wide range of roles in the hospital, from the patient, the treatment team, the hospital administrator, and the staff, to the medical staff and the medical director.

Psychologists must understand the fiscal pressures on hospitals. Their mission is to develop innovative models based on short-term treatment with aftercare. This approach must integrate quality in clinical care with budgetary considerations. Psychologists must be fiscally responsive to the hospital and the corporation if they are to help the hospital enhance its economic position. They must help the hospital make a profit or reduce costs. To be key players in the decisions being made in health care, psychologists must demonstrate to hospitals as well as to business and government that they can contribute to profitability. Finally, psychologists must understand the parameters, the pressures, and the rewards of hospital practice. In marketing and developing their services, they should assert themselves creatively and cost-effectively in the health care delivery system for the benefit of the patient, the hospital, the corporation, and their profession.

References

Binner, P. R. (1986). DRGs and the administration of mental health services. *American Psychologist, 41,* 64–69.

CAPP v. Rank. (1988, December 1). *California Psychologist,* Special Edition.

Cummings, N. A. (1986). The dismantling of our health system: Strategies for the survival of psychological practice. *American Psychologist, 41,* 426–431.

Dafter, R., & Freeland, G. (1988, December). Psychologists' unrealized power in the mental health marketplace. *The California Psychologist,* p. 14.

DeLeon, P. H. (1986). Increasing the societal contribution of organized psychology. *American Psychologist, 41,* 466–474.

Dörken, H. (1989). Hospital practice as gatekeeper to continuity of care: CHAMPUS mental health services, 1980–1987. *Professional Psychology, 20*(6), 419–420.

Dörken, H. & Associates (Eds.). (1986). *Professional psychology in transition: Meeting today's challenges.* San Francisco: Jossey-Brass.

Enright, M. F., Resnick, R. J., Sciara, A. D., DeLeon, P. H., & Tanney, F. M. (in press). The practicing of psychology in a hospital setting. *American Psychologist.*

Hospitals face psychiatrist shortage. (1989, July). *Psychiatric Times,* p. 4.

Ludwigsen, K. R., & Enright, M. F. (1988). The health care revolution: Implications for psychology and hospital practice. *Psychotherapy, 25,* 424–428.

Overcast, T. D., Salés, B. D., & Pollard, M. R. (1982). Applying antitrust laws to the professions. *American Psychologist, 37,* 517–525.

Resnick, R. J. (1985). The case against the blues: The Virginia challenge. *American Psychologist, 40,* 975–983.

Rosenberg, C. E. (1988). *The care of strangers: The rise of America's hospital system.* New York: Basic Books.

Sciara, A. D. (1988). Marketing, management and new perspective and responsibilities for psychologists. *Psychotherapy in Private Practice, 6*(2), 13–20.

Sciara, A. D. (1989). Private psychiatric hospitals: The new frontier. *Psychotherapy in Private Practice, 6*(4), 1–16.

Smyer, M. A. (1986). Providing psychological services in nursing homes. *The Clinical Psychologist, 39,* 105–107.

Where the next wave of hospital competition will come from. (1989, May 1). *Medical Economics,* p. 140.

Zaro, J. S., Batchelor, W. F., Ginsberg, M. R., & Pallak, M. S. (1982). Psychology and the JCAH: Reflections on a decade of struggle. *American Psychologist, 37,* 1342–1349.

CHAPTER 10

Computers in Psychological Practice

Historical and Current Uses

David E. Hartman and Benjamin Kleinmuntz

Introduction

In the 1970s and early 1980s, the birth of the Apple Computer and the IBM PC signaled the beginnings of a radical popularization of computer technology. In a few short years, personal computers became business necessities and invaded over 15% of American households (Squires, 1984). Nevertheless, practicing clinical psychologists have appeared to be ambivalent about computing. Although some psychologists eagerly and immediately welcomed computational methods into psychology, others were skeptical and were angered by the incursion of computers into clinical territories.

A fair judgment of clinical computing requires some degree of computer literacy. Whether this knowledge is then used to expanding one's clinical horizons or to "know one's enemy," psychologists may best decide their position by becoming aware of the historical and scientific, as well as the practical, clinical and legal

David E. Hartman • Department of Psychiatry, Cook County Hospital, Chicago, Illinois 60612-9985. Benjamin Kleinmuntz • Department of Psychology, University of Illinois at Chicago, Chicago, Illinois 60680.

implications of computerizing clinical psychology.

This chapter is addressed to the clinician who would like a basic guide to the history, the current uses, the advantages, and the possible pitfalls of computerizing psychological practice. The first section provides the reader with a historical primer; the rest of the chapter concentrates on current psychological practice issues, including clinical utility, validity, ethics, and the appropriate legal context of computer use.

History

During psychology's infancy (about the time that the 17th-century philosophers Descartes, Leibnitz, Locke, and Spinoza laid the groundwork for today's psychology), the computer had already been conceptualized. Descartes proposed an "information-processing" device (Gardner, 1985, p. 51) designed to represent external events. Leibnitz referred to a 12th-century "logical machine" (cited in McCorduck, 1979, p. 9). The others, no doubt also preoccupied with finding theological and practical solutions to mind–body dualism, were aware of the possibility of thinking machines that could, in Leibnitz's words, "bring reason to bear on all subjects" (in McCorduck, 1979, p. 9).

It was not until around 1812 that plans for such a machine were proposed by the English mathematician and inventor Charles Babbage. He proposed an "analytical engine" to "aid [in] the most complicated and abstruse calculations, [which "obstructed"] the progress of physical science." His idea was to construct a machine that could solve algebraic problems in order to economize on the "exhausting intellectual and mental labor indispensable [to the] advancement" of the sciences (quoted in Pfeiffer, 1962, p. 22). Unfortunately, Babbage never constructed his machine, feasible as it seems by today's standards, because the British government, having granted him $80,000 for a less ambitious project he never completed, decided that his engine was impractical. This "difference machine," as it was also called, contained strikingly prescient ideas on computer operation. For example, it would use punched cards to control its mathematical operations (Gardner, 1985, pp. 142–143), a method that was to be with us until the 1960s. The mathematical operations themselves would have been produced by the 18th-century equivalent of integrated circuits: some 50,000 wheels and gears within other wheels and gears, plus quantities of cams, camshafts, clutches, escapements, tangs, axles, and cranks.

The need for fast and accurate computational devices intensified during the 20th century, and World War II, mathematicians were in short supply and the need for computing machines was overriding financial conservatism. These factors led to the concrete realization of Babbage's dream in the form of an electromechanical system that was to become the prototype of contemporary digital computers (see Bowden, 1953; Goldstine, 1972). Shortly after, the now familiar names of Shannon (1950), Turing (1950), and von Neumann (1951, 1958) appeared in the emerging computer science literature; they proposed, once again, the use of machines to aid, if not replace, human thinking. The idea of computers as information ma-

chines was perhaps best expressed by the physicist Ridenour (1952), who, in this context, commented that these machines were called *computers* because computation was the only significant job that had been assigned to them thus far.

Shortly thereafter, Newell (1955), then Newell and Simon (1961), and then Newell and Simon in collaboration with Shaw (Newell, Shaw, & Simon, 1958), conceptualized a complex information-processing approach that went beyond using the computer as a numerical data processor and that actually simulated human thinking. Specifically, they listed the characteristics of the computer that rendered it a natural choice to perform noncomputational chores. These features included the ability to *read* a symbol, to *translate* it meaningfully for internal storage, and to *move* symbols rapidly from one storage location to another. Moreover, they noted the computer's facility for *comparing* two symbols and for *executing* one program command if the symbols are identical, and another if they are not. And finally, they described its ability to *associate* two or more symbols, allowing access to a symbol (or set of symbols) when the other(s) is given. Thus, in these authors' opinion, computer analogues of perception, memory storage, rule-based decision-making, and association should enable the computer to carry on intelligent activities.

In other words, instructions can be given to the machine (i.e., programs can be written) that combine these processes in such a way that the machine executes each of these instructions in order to solve quite complex problems. For example, programs can be written that command the computer to carry out operations such as the following: "Determine from a search of your memory whether you have encountered similar problems. If yes, what are the similarities and differences between the past problems and the present one? How were the past problems solved? What were the hit-and-miss rates of the major decision rules? Apply the best possible decision rules to the present problem and print out the various solutions."

Using the computer in this way, Newell and Simon (1972) demonstrated the machine's complex and intelligent problem-solving capabilities. For example, their computer programs successfully tackled such problems as theorem proving in symbolic logic, master's and grandmaster's level tournament chess-playing, discovering scientific laws, and unraveling complex cryptoarithmetic problems. Although the bulk of the early research on computer problem-solving was carried out with naive subjects, more recent work by the Carnegie-Mellon group (see Simon, 1979, 1986) has made ample use of the expert's knowledge as a model to serve as a basis for computer programs, knowledge that is elicited from the expert by any one of several experimental procedures and is later formalized for computer-programming purposes.

Now, as information-processing psychology is entering its fourth decade, its most recent forays into the sphere of human thinking include solving the Tower of Hanoi puzzle (Egan & Greeno, 1974; Kotovsky, Hayes, & Simon, 1985; Nilsson, 1971), constructing a world champion backgammon player (Berliner, 1980), and constructing several lesser programs that tackle the games of bridge (Stanier, 1975) and Go (Reitman & Wilcox, 1979). Other artificial intelligence programs that trace their origins to the work of Newell and Simon include applications designed for mineral prospecting (Duda, Gaschnig, & Hart, 1979), analogical problem-solving (Eliot, 1986), and outer-space-station operations (Leinweber, 1987). The most recent work directly inspired by Newell and Simon (e.g., Laird, Newell, & Rosenbloom, 1988; Waldrop, 1988) includes a computer program called *SOAR*, which is a self-correcting system based on the principle of chunking, and which can learn from its past mistakes. SOAR can solve problems using general principles of intelligence, which it can apply to task environments that it has never encountered. This innovative noncomputational approach is an improvement on an earlier GPS, or general problem-solving program, in that it operates on the basis of combining its production-rule, means-ends-analysis, and hill-climbing capabilities with a self-correcting procedure that never commits the same rule-application error twice.

Computers in Clinical Psychology

The history of computerized methods can be seen to converge with the concerns of psychologists because of two factors: An explosion of new computerized applications for both mathematical and verbal information processing, and the shrinking of computers from building to notebook size. It is the latter development that has changed psychological computation from an academic luxury to a desktop necessity.

Personal Computer Hardware

Hardware selection can be reduced to a basic premise: Select the programs of interest; *then* buy the computer that will run them. For psychologists, this choice usually boils down to software produced for either Apple or IBM–IBM-compatible computers; the two systems are incompatible. Other computers and relevant software exist, although from much smaller companies like Commodore and Atari, or else are marketed for expenisve engineering "workstations," available or planned through Sun, NeXT, and other companies. These latter systems have had little impact on psychological computing so far.

Apple manufacturers two lines of computers; the early Apple II series and the more technologically sophisticated Macintosh series. The Apple IIC and the IIGS have seen widespread adoption in elementary-school systems and therefore may be of interest to school and child psychologists. There is also a significant quantity of cognitive rehabilitation software written for Apple II computers; rehabilitation psychologists and other professionals who work with the brain-injured have good reason to investigate these machines. For general use, however, these systems are technologically dated and are probably not as attractive to the developers of new psychological applications as the newer and more powerful Macintosh systems.

Apple Macintosh computers, are "icon"-based systems, in which commands are displayed as graphic images rather than words (e.g., an "erase" command may be pictured as a pencil eraser and a "delete" command as a trash can). Users point to the command on the screen by moving a "mouse," a small rectangular object attached to the computer that allows the operator to point to the areas of interest. The Macintosh has found widespread support among novice and expert computer users alike for its ease of operation and the facility with which graphics can be manipulated on the screen.

IBM and IBM-compatible machines, the most common computer systems in business environments, have historically emphasized text manipulation over graphics. There is much less uniformity (and, some would say, more creativity) among programs on IBM and IBM-compatible computers than on Apple Macintosh systems. However, IBM-planned graphic systems (i.e., OS/2 with Presentation Manager) are becoming so similar to Apple's that some have suggested the letters of the company now stand for "I Built a Macintosh." In the long run, the choice between computers may be entirely academic, as the "cross-assemblers" currently under development may be able to translate all programs from one machine to another. In the short run, however, psychologists with finite funds must choose between the two.

When equally acceptable programs are available for each machine, the choice is determined by other important factors, including the "standard" in use at the psychologist's worksite, the preference of nearby colleagues who can provide instruction or support, the level of service provided by local vendors, and the cost. Currently, so-called IBM-compatibles, or "clones," are the best buy for the computing dollar; they are produced by many companies, whose competition brings down prices. Users who are not familiar with personal computers should probably choose among the better known clones (e.g., IBM, Zenith, Compaq, Dell, and Tandy), as the component quality and the support vary widely in less well known companies. Alternatively, psychologists who are knowledgeable about PCs can save substantial amounts of money by careful mail-order computer purchasing. *PC Magazine* is a useful "consumer's guide" to IBM and IBM-compatible hardware, and some research in this publication before purchase can also save time and money.

Apple Computer's patents have thus far prevented other companies from manufacturing Macintosh-compatible computers, so users who want Macintosh will have to stay with Apple. Some universities, however, may be able to provide faculty discounts on Apple (or IBM) products for hospital psychologists with academic appointments.

Within either the Apple or the IBM–IBM-compatible family of computers, more money buys more memory, processing speed, and data storage space. The programs used and the amount of information stored will determine system requirements. Psychologists who wish to build a computer system are strongly urged to query fellow psychologist computer users or professional consultants before buying a machine. The best results occur when the user is able to anticipate his or her computer requirements for the next several years and to purchase a computer that can be "grown into." Although the rapid cycle of computer obsolescence precludes planning any further in the future, it is probably unwise to automatically purchase the fastest, most powerful computer. This is a chimerical pursuit at best, as this year's' "state-of-the-art" will almost always become next year's "trailing edge." In addicting oneself to the yoke of technology, it is also easy to forget that almost no programs for psychologists require such powerful machines. Finally, the fastest, newest machines are also always far more expensive than their small increase in power warrants. It may pay either to wait a year for prices to go down or to purchase last year's leading-edge machine, which may now be quite affordable.

Personal Computer Software

Scientific Psychology Applications

Any scientific field that deals with large amounts of data, to be analyzed, compiled, and computed at rapid rates, invites the application of high-speed data-processing techniques. It is not surprising, therefore, to find that clinical psychologists have used digital computer technology almost from its inception in the 1950s, as modern "Boulder-model" psychology explicitly mandates clinical research in the context of practice. The analysis of collected data, which once meant dreaded sessions with manual or electric calculators, has been not only computerized but also "downsized" for the personal computers. In particular, several excellent statistical programs have been developed for personal computers. These programs are sufficient for all but the very largest projects, being constrained not by their statistical power, but by the limitations of data storage on the personal computer. Three of these packages, SYSTAT, SPSS/PC+, and Statgraphics, have been recommended by *PC Magazine* (Raskin, 1989). Of these, SYSTAT, the "system for statistics," was termed powerful and "unbeatable," excelling with linear models. Other advantages of the package include extremely powerful and accurate algorithms; an add-on graphics package, SYGRAPH, which produces high-quality graphs; and the capability of running on both large and small personal computers. There is also a Macintosh version of SYSTAT.

Other companies, including BMDP and SAS, have also brought their mainframe statistical programs to the personal computer. Psychologists already familiar with university applications of these systems may transfer that knowledge to personal computer implementations with relative ease. In general, however, statistical software is best described as "user-cordial," rather than user-friendly, and a working knowledge of programming and statistics is a prerequisite.

PC statistical software frees the user to work at home, to enter data interactively, and to run programs repeatedly without vying with irate engineers for expensive and precarious mainframe time. Although statistical software is not inexpensive ($400–$800), there is no additional cost per run, as is the case with mainframes. A missed semicolon in a command statement, for example, causes loss of time only, not precious mainframe computer funds. The psychologist who wishes to combine research and clinical

activity in a hospital practice (a goal supported by the American Psychological Association's Boulder model) has the power and capabilities of yesterday's university mainframe computers on a desktop.

Office Applications

Office applications of personal computers can be useful in almost any psychological practice, but their utility increases in direct proportion to the complexity of the psychological setting. A psychologist in part-time private practice who is unfamiliar with computer methods may not realize an appreciable benefit in time or convenience from computerization simply because of the steep learning curve in mastering the relevant equipment and software; a small number of patients may not warrant the investment in time and money.

The situation would be quite different for a mental health center, or for a hospital administrator who oversees the activities, scheduling, and billing for 30 or more full-time practitioners, or for psychologists working in hospitals. Most hospital psychologists are required by virtue of complex hospital hierarchies to process patient information rapidly. Referral services may require immediate reports about a particular patient, administrators may request billing records, and quality assurance directors may be interested in outcome information concerning a particular diagnostic category. Producing, tabulating, and organizing the information required by these clinical and administrative hierarchies are greatly facilitated by office computerization.

Further, office applications escape the controversy of clinical software because they are primarily *business* products capable of enhancing the personal productivity of psychological practice. The use of these products can save the psychologist time, money, and personnel. The available office software ranges from simple phone-dialing programs and calculators to software capable of managing patients' accounts in a large hospital-based practice.

However, although such a choice is available, computers cannot, of themselves, organize a psychological practice. In fact, the use of computerized office methods may require the psychologist to make changes in office systems, just to make his or her practice amenable to computerization. A realistic needs assessment in advance of implementing an automated office practice system is essential; information provided by colleagues who have developed similar systems or by a professional consultant can be of help in such an assessment. Cost alone is not a reliable index of utility, for either hardware or software. Burke and Normand (1987) cited "numerous situations in which computer service providers have purchased costly, unnecessary hardware that has hindered their ability to operate profitably" (p. 46). Similarly, psychologists might heed the story of a physician who drove his staff to distraction with an unworkable $1,000 practice management software package, only to discover a simple and powerful solution in a $300 program borrowed from his children's home computer (Quigley, 1984).

Felts (1984) divided general practice software applications into accounts receivable, management reports, general ledger systems, and clerical applications. Psychologists may also want to add several of the miscellaneous applications listed below, depending on their needs. The number of staff and the type of procedures performed will necessarily influence both hardware and software selection.

It has been chastening to come upon the antiquated suggestions of previous reviewers who have trod the path of recommending specific software packages. Software and hardware improve so rapidly that "state-of-the-art" systems are often discontinued anachronisms by the time a recommendation is actually published. The alternative, selected here, is to review the *categories* of software available to enhance the productivity of clinical practice without being tied firmly to specific products. Where specific products are listed, they should be taken as good *examples* of the genre at the time this chapter was written, rather than blanket *recommendations*. These examples are taken from the IBM–IBM-compatible literature, as that is the system most familiar to the authors,

but similar software is usually available for Apple Macintosh systems. Below is a listing of the most common office applications available to the clinical psychologist.

Office applications
A. Word processor
B. Data base
C. Spreadsheet
D. Accounting applications
E. Miscellaneous applications
1. Communications software
2. Appointment organizers
3. Personal information managers (PIMs)
4. Desktop publishing

Word Processors

A word-processing program is an absolute must for psychologists in clinical settings whose work requires correspondence, reports, or any form of written information. Word processors transform the computer into an intelligent typewriter, which moves, stores, and manipulates text typed into the computer. Many include routines that check spelling, analyze grammar, outline ideas, and give the writer a view of the page output before the text is printed.

On IBM–IBM-compatibles, word processors can be either "page-oriented" systems (e.g., Multi-Mate), which store every page separately and are more suitable for business letters than for multipage professional documents, or "document-oriented" (e.g., Wordperfect, Xywrite) processors. The latter are more appropriate for the variety of reports·and correspondence that psychologists generate.

Word processors can be further subdivided into those that display text that uses the computer's built-in fonts (text-based) and those that produce a graphic image of the words on the page (graphics-based). Text-based word processors are fast but are usually unable to display on the screen exactly how a document will appear. Large type, underlines, and italics are not displayed on the screen as they will eventually appear in print. Graphics-oriented word processors (e.g., Microsoft Word) show the document exactly as it will appear on the page, but they are inevitably slower than text-based processors because the computer must keep track of all of the dots, or pixels, that make up a letter on the screen, rather than just a single, prestored standard letter representation.

Another difference among word-processing programs is how much they help the user with on-screen directions, or menus. Some, like Wordstar, put menus on the screen to help the writer remember commands. Others, for clarity, present a nearly blank screen, and the user must remember commands (e.g., Xywrite, Wordperfect). The cost of word-processing programs ranges from $15 to $500 and thus can be a significant consideration. *Popularity* is another concern when one is trading documents with colleagues or soliciting advice. *Speed* of processing is an important factor for psychologists who write books or edit large hospital documents. For such "power users," Xywrite-III+ and Nota Bene are currently among the fastest and most feature-laden programs on the market. Wordperfect is somewhat slower but has many features of desktop publishing and is the best-selling word processor at the time of this writing. Wordstar is the program that many users learned on their first computer and has lately become a very powerful product in its own right, and Microsoft Word is also a major contender because of its graphic interface. Surprisingly, the program recommended by the American Psychological Association (APA), Manuscript Manager, although it automatically formats a document for APA style, was not seen to be competitive in speed and features when compared with other word-processing programs (Mendelson, 1989). The *PC Magazine* review of Manuscript Manager instead recommended Nota Bene, a fast and full-featured word processor that can be configured for APA styling conventions and can even convert from one style book to another.

Word processors save time and effort in the production of clinical reports and correspondence. Most have a "mail-merge" feature, which facilitates the generation of address labels, practice announcements, and thank you let-

ters. Writing and revising research articles are made unbelievably easier with programs that allow an electronic "cut-and-paste" movement of words, sentences, and paragraphs within the document.

The choice of word processors is a very personal one, not unlike the choice of that almost obsolete alternative, the pen. Actually trying out several programs and discussing one's requirements with a knowledgeable person are strongly recommended.

Data Base Programs

Data base programs allow users to catalog and sort the written word and can even be considered the verbal counterparts of statistical-research data-analyzers. For example, data base software can organize patient data by name, address, diagnosis, or whatever variables are of interest to the psychologist. What is more valuable, however, is the data base's facility in *retrieving* that information instantly, collating whatever variables are useful. Thus, a psychologist interested in the relationship between eye color and psychopathology could query the data base and sort all prior records according to these two variables. Data base information is particularly useful to the psychologist working within a hospital hierarchy. Although most hospitals already have computerized files, these are not likely to be in a format that fits the needs of the psychologist health provider. For example, rehabilitation psychologists may require data from longitudinal neurological and neuropsychological records to be accessible for statistical analysis. Administrative psychologists may focus on outcome data for research or insurance purposes. All psychologists who bill for their time require instant access to and updates of patient files. Patient information coded into a data base becomes easily accessible for clinical, diagnostic, billing, and quality assurance. With the appropriate software, the psychologist can answer questions like "How many clients have you seen with a diagnosis of 'borderline personality,' for how many sessions were they seen on average, and what percentage had Blue Cross?"

Patients can be cross-indexed under referral sources, treatment sites, or any variables of interest to the psychologist. Some typical items of interest are listed in Table 1.

Clinicians may also wish to keep lists of referral sources on a data base, enabling targeting advertising, letter writing, moving advertisements, and so on to be generated for specific mailing lists. Some commonly used databases for the PC environment include dBase IV, Paradox, and RBase.

Spreadsheets

Of use to the clinical psychologist whose position requires financial planning, budget analysis, or grant management, or simply a record of patient billing accounts, a spreadsheet program functions as an automatic accounts worksheet. Totals from different categories of expense or service are automatically recalculated as new data are entered. Powerful and popular spreadsheets at the time of this writing include Lotus 123, Microsoft Excel, Supercalc V, and Lucid 3D. One of the authors

Table 1. Categories of Information in a Psychologist's Hospital Data Base

---

Patient accounts
Patient registration information
  Name
  ID
  Date of birth
  Address(es)
  Phone number(s)
  Insurance carrier(s)
  Responsible party
  Referral source
  Release of information
  Close relative
  How often to update information
Billing information (e.g., costs, sessions, dates)
Diagnoses
Services (e.g., psychotherapy, tests, consultation, forensic services, neuropsychology, hypnosis)
Assessment data (e.g., test scores)
Treatment information
Other

---

uses Lucid 3D to completely manage his private and hospital practice.

Accounting

Accounting software functions as a special-purpose combination of data base and spreadsheet and automates tedious accounts receivable billing and account aging. Many programs will generate quarterly financial management according to the services performed, expenses, and income and will correctly place items in relevant tax schedules. Check writing, deposit slips, bill paying, and payroll can all be automated with computerized accounting methods. *DAC Easy* is one example of the many affordable accounting systems available and adaptable for psychologist billing. A word of caution here is that most of these general business packages also assume some knowledge of accounting principles and vocabulary (e.g., *debits*, *credits*, and *ledgers*). Psychologists unfamiliar with accounting vocabulary may wish to examine the specialized billing programs often advertised in the *APA Monitor*. Although less flexible, these single-purpose accounting and billing programs also demand fewer accounting skills of their users and may be easier to run.

Miscellaneous Applications

Communications. Communications software requires a modem, which is a sort of electronic telephone dialer for computers, to call up other users, large data bases, or other computers and to allow quick information gathering for patient care. One of the authors, who practices hospital health psychology and neuropsychology, routinely uses a modem with communication software to access large clinical data bases. After seeing a patient with an unusual medical syndrome, he obtains information on the disorder and its relationship to mental status by calling up the data base (e.g., MEDLINE or PsycINFO) and searching for information about the disease. Using communication software allows psychologists to "download" bibliographical references or research summaries into the personal computer. Several wholly electronic "bulletin boards" are also available so that psychologists can leave messages or questions for colleagues. Either way, the result can be increased efficiency and clinical accuracy for the practitioner, who can use communication software to greatly expand access to patient care information.

Appointment Organizers. These are the computerized equivalent of a "day-timer" appointment book. For individuals, the noncomputerized version is probably unbeatable. For psychologists who work in large practices that schedule groups of patients, and wherever individual practitioners must be assigned cases according to availability, computer organizers may be quite beneficial. One recently published program of that type is called Working Hours, published by Channelmark Corporation. The program creates a personnel record that can be used to track the appointments of an entire psychology staff and is capable of logging each clinician's appointment schedule in increments as small as 15 minutes. The program is alert to possible schedule conflicts and prevents the scheduling of clinicians when they are unavailable.

Personal Information Managers (PIMs). These programs are the computerized equivalent of "post-it" notes, those little yellow pieces of paper that have infested so many business and service-oriented offices. Essentially free-form data bases, these programs allow the user to make notes about people or situations and provide an easy search routine to get the information out again. For example, a psychologist may receive a call from a physician in another part of the hospital and discusses the possibility of having a patient undergo neuropsychological testing. Because no actual appointment is made, the information cannot be placed in a calender, but it can be entered into a PIM, to be filed under the patient's name, the physician's name, or the test procedures involved. Some PIMs, like the Lotus Agenda, have enough built-in artificial intelligence to automatically file such a request all three ways. Thus, at the end of the day, the psychologist might look in

Agenda's file of "patients," "calls from physicians," or "test procedures." Other PIMs, like Broderbund Software's Memory-Mate simply allow the user to type in whatever comes to mind, storing it and allowing the user to retrieve it by easy word or date searching. Psychologists with a computer on their desktop might conceivably use Memory-Mate or a similar PIM as an on-line clinical note organizer, writing a paragraph or two on each patient as she or he is seen and letting the program store the files automatically by date of service and patient.

Desktop Publishing

Desktop-publishing (DTP) programs are cousins of conventional word processors and allow the psychologist to produce newsletters, flyers, or professional documents that rival typeset quality. Usually used in combination with a laser printer, DTP software can generate attractive and eye-catching brochures, advertisements, and notices. These can be particularly effective marketing tools for hospital psychologists who need to advertise their service within the hospital and the community, but they may require expensive investments in learning time and hardware cost (over $10,000) to achieve professional-quality documents.

The foregoing categories of business software by no means exhaust what is available to the psychologist. There are over 12,000 commercially available programs on the market for IBM-type systems alone, and manufacturers continue to show great creativity at generating new applications with labor-saving functions. Psychologists who seek to automate aspects of their practice will easily find powerful, time-saving alternatives to paper and pencil.

Clinical Applications

A vast range of clinical functions are amenable to computerization. Many of these could scarcely be performed without the formidable "number-crunching" power and speed of the computer. Alternatively, some have argued that there are many other functions that should not be performed with the computer. In general, the objections tend to be highly correlated with the degree to which the computer program simulates the behavior of the clinician. Thus, psychologists who see no objection to a scoring program for the Minnesota Multiphasic Personality Inventory 2 (MMPI-2) may take exception to a computer program actually "writing" a psychological report. This issue has many dimensions, including the *adequacy* and the *accuracy* of the simulation, the *moral aspect* of having machine decision-making become influential in mental health care, and the *legal questions* regarding software that "treats" patients. These issues are addressed later in this chapter and have been covered in some detail elsewhere (e.g., Hartman, 1986a,b; Matarazzo, 1983, 1986; Matarazzo & Matarazzo, 1985).

Some categories of clinical software currently available are listed in Table 2.

Clinical Research: Personality Test Development, Validation, and Scoring

The use of computerized methods to *construct* a clinical assessment device has a longer and less controversial history then the use of that same computer to *administer* or *interpret* the test. One of the earliest and most extensive attempts to use the computer as a clinical research development tool occurred at R. B. Cattell's Institute for Personality and Ability Testing (IPAT) at the University of Illinois. The computer's use for factor analysis was readily recognized by these psychologists, who discovered that the computer reduced months of tedious intercorrelation calculation to hours—or even minutes—of machine time.

The two main uses of computerized factor analysis in personality assessment research occur in test item development and personality test validation. In the latter instance, factor analysis is a technique used to examine the meaning of a test by studying its correlations with many observed or empirically derived variables. The factors extracted from these analyses are constructs (or variables presumed to exist), and reference to a table of factor load-

Table 2. Clinical Software Applications

I. Clinical research
  A. Personality test development (e.g., factor analysis)
  B. Statistical analysis
II. Clinical practice
  A. Interview and history
    1. Interviewing
    2. History taking
    3. Symptoms
    4. Medical, health, and medication history
    5. Occupational history
    6. Psychological history
    7. Family and social history
    8. Therapy and progress notes
  B. Test administration and scoring
    1. Objective
      a. Neuropsychological
      b. Personality (e.g., MMPI)
      c. Intellectual
    2. Projective
      a. Rorschach (e.g., RIAP)
  C. "Report writers"
  D. "Therapy" programs
  E. Specialized clinical applications
    1. Vocational evaluation (e.g., Strong-Campbell)
    2. Biofeedback
    3. Training and job simulation
    4. Cognitive rehabilitation
    5. Psychoeducational training and teaching

ings discloses the number of constructs a test measures.

Personality test construction may also proceed by the use of factor analysis. In this case, the psychologist starts with a large pool of items, selected perhaps from his or her own observations of people or from textbook descriptions of behavior, and then intercorrelates the responses to these items until all possible pairs of items have been treated. The psychologist then factor-analyzes the resulting correlation coefficients to determine which items tend to cluster sufficiently to constitute a factor. An examination of the apparent characteristics that are involved in the structure of these item clusters determines the name that is to be assigned to particular clusters.

Generally speaking, the role of computers in the factor analysis of personality dimensions is only an ancillary one, which perhaps explains its relative lack of controversy. For example, the computer does not dictate which items are to be intercorrelated, nor does it select the items for test construction. These remain judgmental aspects of the procedure. Furthermore, certain facets of axis rotation and factor naming are determined intuitively by the clinical researcher. Finally, no amount of computational ease or skill makes it possible for the computer to prescribe which constructs and variables can serve as criteria for correlation with factor-analytically derived inventories. Only the experience of the trained psychologist can dictate these choices. In such cases, the real advantage of the computer lies in its freeing the psychologist from doing laborious calculations that previously used up much valuable time. The computer's calculation speed also allows many more variables, and methods for combining these variables, to arrive at factorial solutions.

Configurational Scoring

Computers have scored psychological tests since the 1940s, when Elmer Hankes, a Minneapolis electrical engineer, used keypunched cards to program an analogue computer to score the MMPI and the Strong Vocational Interest Blank (Fowler, 1985; Mooreland, 1987). More complex uses of computerized test-scoring awaited Meehl's (1950) suggestion that scoring patterns or configurations of answers rather than individual items would enhance the subtlety of an objective inventory (cf. Wilkinson, Gimbel, & Koepke, 1982). For example, some scales of the MMPI—especially the Paranoia, or Pa, scale—are relatively weak in the sense that some psychiatric patients easily evade detection. When items are scored configurationally, a considerably larger burden is placed on the person intending to evade detection on this test. Now, instead of being confronted with the choice of earning a score in the direction of paranoia, let us say, on the basis of a "true" or "false" response to one item, the respondent is faced with the problem of answering certain combinations of items that are valid for the dimension in question. For exam-

ple, in the method of configurational scoring, this respondent may earn 1 point in the direction of paranoia if he or she answers a certain pair of items TT ("true," "true") or FF ("false," "false") and no points for the patterns TF or FT. Three-, four-, or five-item combinations can also be used to complicate the task of faking. The precise choice and combinations of items that are deemed significant when answered in certain ways are, of course, empirically established by the prior statistical analysis of these items among particular patient and normal populations.

The patterns of responses, and the interitem correlations required to establish the joint validity of specific patterns of items, constitute an enormous computational task. For the MMPI, which consists of 566 items, this task would involve the computation of several million correlations for even a small sample of subjects. Even a large computer would require many hours of running time for such computations. However, it is now possible to accomplish this formerly prohibitive data-processing task because of the development of a "high-pass" filter technique and the advances in computer technology. Williams and Kleinmuntz (1969) made one of the earliest attempts to computerize the selection of configurational scores by using such a high-pass filter to detect cross-correlations among 566 MMPI items taken by 200 examiners (about 160,000 correlations). The detection system had some success with this limited sample but was never followed up in the field, probably because its computational burden and hardware cost exceeded what was available to 1960s academic psychologists.

Computer-Generated Narrative Interpretation of Test Results

A logical extension of automating configurational score selection is the automated production of verbal descriptors that correspond to those configurational scores. This use of the computer to interpret test profiles has been followed up extensively, controversially, enthusiastically, and in the main rather poorly (e.g., Hartman, 1986b; Matarazzo, 1983, 1986; Meehl,

1986); the deficiencies of this method are discussed later.

To the best of our knowledge, computer-based test interpretation programs, or CBTIs, had their start with the development of two programs for personality test interpretation: the Mayo Clinic program (Rome, Swenson, Mataya, McCarthy, Pearson, Keating, & Hathaway, 1962), which was designed to provide physicians with an overall understanding of their patients' personalities, and the Carnegie-Mellon University program (Kleinmuntz, 1963a, b), at about the same time, which had as its objective the identification of emotional maladjustment among entering college students. Here, we describe the latter program in some detail because it comes closer to a purely "clinical" use of computers, at the same time using some of the machine's brute force computational prowess.

In the Carnegie-Mellon University studies (Kleinmuntz, 1963a,b, 1968, 1969), several experienced MMPI interpreters were instructed to Q-sort (Stephenson, 1953, 1980) 126 MMPI profile sheets along a 14-step forced normal distribution. The MMPI expert who achieved the highest valid positive (80%) and valid negative (67%) success rates in predicting the criteria of "most" and "least" adjusted was selected for intensive study. He was instructed to think aloud while performing his Q-sort task and was encouraged to elaborate his precise reasons for sorting each profile into one or the other category. His performance was tape-recorded. For illustrative purposes, a flowchart of the decision rules based on his protocol is presented in Figure 1.

The sequential rules in Figure 1 that render an adjusted, maladjusted, or unclassified decision were based on many hours of taped verbalizations. These rules were then coded into a set of programmed instructions for processing by the computer. The success rates of the programmed rules, although lower, were nonetheless surprisingly similar to those of the Q sorter. The valid positive and valid negative hit percentages of the computer program were 63 and 88, respectively, and these compare favorably with the expert's rates (80% and 67%). These programmed rules were then subjected to a trial-and-error process of statistical searching and shuttling back and forth between intuitive hunches about combinations of various scales and their possible effects on the hit percentages in the student MMPI sample. On the basis of these statistical operations (later called *bootstrapping*); (Dawes, 1971; Dawes & Corrigan, 1974), an elaborate set of new rules was developed.

In the derivation of the new rules, the computer's capabilities for storing, retrieving, and rapidly manipulating large quantities of information were challenged, and its facility for processing these data at high speeds was taxed to the utmost. For example, one technique consisted of letting the computer apply all the rules to a particular MMPI profile and instructing it to withhold its maladjusted or adjusted decision until it computed a vote of the number of rules that favored one or the other of two classifications (for example, adjusted versus maladjusted). Further, on the basis of an empirical determination (printed out by the machine) of the relative strength or weakness of a particular decision rule as measured by its correct or incorrect application, the computer was programmed to attend to specific patterns of rules (for example, Rules 2, 8, and 13 versus the combination 3 and 11) and was instructed to break the tie in favor of the more powerful rules rather than just to attend to the number of votes that each profile received. The pattern-analytic approach to the rules themselves clearly removes this form of data processing from the realm of human capabilities.

Finally, the completed set of MMPI pattern-analytic rules included the original expert interpreter's information, a number of intraprofile slope characteristics that the expert failed to observed, and the initial set of rules. The valid positive and valid negative percentages achieved by the revised rules were 90 and 84, and the new set of rules thus outperformed the original ones. However, in a subsequent set of studies (Kleinmuntz, 1969) using about 1,000 new cases, they were shown to perform better than the average test interpreter, but no better than the best from among 10 clinicians.

Inspired by the successes of these and similar cookbook approaches to personality test

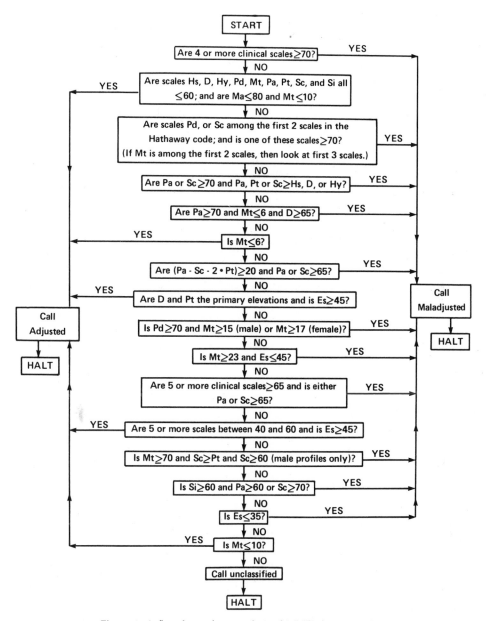

Figure 1. A flowchart of expert-derived MMPI decision rules.

interpretation (Marks & Seeman, 1963; Meehl & Dahlstrom, 1960), a multimillion-dollar cottage industry of interpretive personality-testing systems sprang up. Compared to those of Mayo and Carnegie, these programs aspire to offer physicians, psychiatrists, and clinical psychologists (and in some cases, anyone at all) a great deal more. They provide narrative interpretive statements such as the following, which is taken from the now defunct Roche Psychiatric Service Institute's MMPI report (Fowler, 1969): "The patient seems to be a person who had difficulty maintaining controls over her impulses. She behaves in a socially unacceptable manner, and is likely to experience guilt and distress" (p. 231).

Problems in Computer-Based
Test Interpretation

This and similar derivative narratives are not appreciably different from those that appear in the current MMPI report (Butcher, Keller, & Bacon, 1985; Fowler, 1985; Moreland, 1985) and are typical of the interpretations offered by many modern, commercially available computer-based test-interpretive systems. The good news about such systems is that the statements they print out are probably no worse than those of some psychologists, although we are put in mind of Fowler's statement that "Any psychologist who *can* be replaced by a computer *should* be." The bad news is that these systems are quite primitive, not only by human standards, but even in terms of current artificial intelligence (AI) technology.

One of the most glaring difficiencies of CBTIs, explicitly mentioned by several authors (e.g., Hartman, 1986b; Kleinmuntz, 1990; Matarazzo, 1983, 1986), has been an almost complete lack of empirical validation, or even concurrent validation against existing paper-and-pencil equivalents. For example, when Honaker (1988) reviewed the computerized MMPI literature, he suggested that current research has not demonstrated that computerized MMPIs produce absolute score values or variability similar to those of their paper-and-pencil equivalents. He also called attention to individual variation in subject responses to computerized testing. The failure of commercial psychological software companies to perform the needed validation studies appears to be an unfortunate outgrowth of the rush to commercialize a technology in advance of its scientific soundness.

A second obvious problem for CBTI recalls Newell's admonition that, no matter how erudite-appearing the output, expert systems are quite "shallow"; "they don't know what they know and why they know it" (cited in Wertheimer, 1985, p. 29). This deficiency is most obvious in the inability of many current programs to scan contextually for consistency in their narrative output. This shortcoming is less conspicuous with simpler depression or phobia inventories, where the univariate information

surveyed lessens the possibility of displaying contradictory combinations of output. However, when the software instrument is capable of surveying a wider range of psychopathology (e.g., the MMPI), dozens of rules and intrepetation statements relating to items and to indivual or configurational scores generate the final narrative profile.

Thus, although that final report is typically *prepared* as a grammatically parsed concatenation of these rule-based interpretations, the result is *read* as a *narrative document*; the more complex meanings and psycholinguistic implications are overlooked. Current interpretive programs do not have the artificial intelligence to resolve inconsistencies and maintain narrative flow in this collection of interpretations, a problem concisely noted by Golden (1985):

> In many of the programs that use actuarial approaches to interpret psychological tests, it is not unusual to see the computer contradict itself in the same report because it does not have the insight to see the meaning of putting two supposedly independent judgments together. (p. 44)

Further, the greater the number of independent interpretations capable of being generated, the greater the probability of generating reports with inconsistent, illogical combinations of these interpretations. In one computerized Clinical Analysis Questionnaire (CAQ) interpretive system, for example, low-, normal-, and high-score interpretive paragraphs are available for each scale, producing 4,096 possible combinations of paragraphs. When Tiffany (1988) ran this program over a client's scores, he noted the contradictory nature of the generated report:

> My patient's STABILITY paragraph indicated he "would be easily perturbed and prone to rumination or worry and avoidance of responsibilities and a tendency to give up easily" [while] the CONSCIENTIOUSNESS paragraph pointed out he would show a "highly conforming approach to life and be dominated by a sense of duty with a consistent and methodological approach to task." Such inconsistencies are not checked and resolved in an overall paragraph summing up this individual's behavior. (Tiffany, 1988, p. 2)

Another, more serious limitation of narrative test interpretation software is its inability to adjust a psychological interpretation for data gathered outside the test. An example can be found in the automated interpretations of cancer patients or other physically ill persons given the MMPI. Like the original paper-and-pencil test, the interpretive program would very likely have difficulty distinguishing true physical illness from hypochondria or somaticizing (see Reitan & Wolfson, 1985). Whereas the clinician could interpret an MMPI scale score in light of this information, a computer could not easily adjust its interpretations in this context. The problem of the computer's limited worldview is quite serious, as it can be argued that there are as many contexts as there are patients with MMPI profiles.

Matarazzo (1983, 1986) has forcefully pointed out this contextual nature of clinical evaluation, and Tallent (1987) emphasized the distinction between test taking and clinical evaluation, the latter requiring logical skills and information integration of a much higher order than the former.

More seriously, as suggested earlier, significant questions have been raised about the actual validity of current computerized psychological assessment and interpretation methods. Although computer assessment corporations have explicitly promoted their software as being capable of equalizing the skill of good and poor psychologists, there are scientific and ethical objections to this complete abrogation of psychological decision-making power to the computer. No less a luminary than Paul Meehl (1986), whose monograph on clinical versus statistical prediction was perhaps *the* seminal influence on computerized assessment, expressed regret that his 1954 classic work's "only sizable influence has been [on] proliferation of MMPI cookbook interpretations, *in excess of their demonstrated validity*" (p. 375; italics added).

Moreover, according to Geissman and Schultz (1988), it is far from clear that such software *can* be validated, as the ambiguous, complex, and multifactorial conditions examined by psychologists do not lend themselves well to expert system validation. They warned that "Expert systems (especially those that operate under uncertainty or with incomplete data) may have

so many possible states as to make exhaustive [validation] testing infeasible" (p. 30).

It is clear that psychological expert systems based on conventional testing procedures would incoporate this difficulty, since the theoretical models of pathology that they assess are often equally vague. Jackson (1987) reminded us that proponents of automated clinical methods have often

> overlooked the primitive or non-existent theoretical foundations and models underlying many of our most popular tests and psychological assessment methods. It is hardly surprising that it is difficult to specify standards for validating [computerized test inferences] when the inferences . . . have been developed in the absence of explicit rationales or models linking the aggregation of responses to scores, interpretations and behavioral predictions (pp. 71–72)

This problem is likely to become worse rather than better, especially if psychologists begin to construct expert systems based on a new technology called *neural networks*. Neural networks are software-simulated "neurons" linked together by a computer program. These neural networks usually incorporate both "digital" yes-no decisions that "fire" a neuron, as well as an "analogue" summation of input that causes "firing" after a certain "weight" of input has been achieved. Neural networks can *learn* correct decision-making by teaching the simulation how to categorize data. For example, one could instruct a neural network program to simulate human psychological test interpretation by feeding in a large number of MMPI profiles and telling the program whether each patient belongs to the normal, abnormal, schizophrenic, manic, or whatever other *a priori* categories are of interest. Based on experience with successive MMPIs, the neural network program would continuously attempt to adjust its internal connections and decision weights to identify correctly the maximum numbers of new and uncategorized MMPI profiles.

Both conventional artificial intelligence software and neural networks have as their goal the accurate simulation of human decision-making. Unlike in conventional artificial intelligence expert systems, however, no explicit set of rules is ever constructed to direct decision output. In fact, decision making in neural networks is essentially *inductive*; learning is through experience, and capabilities are stored in the form of weighted attributed or connections. For those reasons, even the programmers of the neural network *may have absolutely no idea* how the software makes its decisions. Unfortunately, this system is a bit like the human information processor, who often can answer a question without being able to recapture explicitly how a decision was made.

Although there do not appear to be any neural network programs performing psychological decision-making at this time, we believe that this prospect is inevitable—and quite troubling from an ethical and professional standpoint. For example, if Dr. Famous Psychologist instructs a neural network program to make decisions weighted just as she would weight them, could another user of that program claim to have consulted with Dr. Famous Psychologist? Could the second user claim that Dr. Famous Psychologist agrees with a particular interpretation of the MMPI. These ethical and professional dilemmas in the use of psychological software are discussed in greater detail later in this chapter.

Other Issues: Record Storage

The great advantages of compact, easily accessible microcomputer patient record storage are also its achilles heel (Bongar, 1988). First, stored records may be accessible by telephone to unauthorized personnel, from insurance companies to adolescent "hackers."

Questions about the longevity and stability of stored magnetic media have not been answered. In addition, the recent creation of so-called computer viruses poses a significant danger to computerized patient records. These subprograms, which are introduced into the system by being attached to other programs, are capable of irrevocably destroying the contents of a hard disk. They are the computerized equivalent of setting fire to a room full of filing cabinets.

Health professionals in general and psychol-

ogists in particular have been quite explicit about the need to preserve patient confidentiality (Keith-Spiegel & Koocher, 1985). Paradoxically, the very aspects of computerization that have allowed compact data storage and powerful methods of information access have endangered patient confidentiality.

Bongar (1988) suggested that, where data or computers are networked, confidentiality be ensured by "(a) coordinating access to all data, (b) training neophyte users in computer security (as well as giving them passwords) and (c) establishing the official policy, guidelines, and implementation procedures for network confidentiality" (p. 287).

Further Sources

Several sources are available to psychologists interested in business and clinically oriented microcomputer issues. For comprehensive reviews of computers and general-purpose software, *PC Magazine* is probably the most authoritative source for IBM–IBM-compatible equipment. *MacWorld* and other similar magazines fulfill the same function for Apple Macintosh computers.

To date, two "sourcebooks" have been published that list software for specific psychological purposes. The first is *Computer Use in Psychology: A Director of Software* (Stoloff & Couch, 1987), published by the American Psychological Association. The second, more extensive, book is Samuel E. Krug's *Psychware Sourcebook, 1978–1988* (1987), published by the Test Corporation of America. Both include descriptions of many types of psychological software, including academic, statistical, clinical, and business-related software. *Psychware Sourcebook* also includes sample listings of the report output of many clinical programs. It is likely that both works will continue to be updated as new software and technology become available.

Neither work attempts to analyze or critique the validity of the programs it reviews, leaving that task to the *Mental Measurement Yearbook* (BUROS) and similar sources. Another useful and readable source of psychological software reviews is the newsletter called *Computers in*

Psychiatry/Psychology, published by psychiatrist Marc Schwartz.

Several journals routinely publish psychological software reviews. *Computers in Human Behavior*, published by Pergamon Press, is a journal specialized in studies and reviews involving psychological software. Many other psychological journals also periodically publish studies of psychological software. A journal that publishes in the area of medical computing and that periodically publishes articles of interest to the psychological computer user is *M.D. Computing*, published by Springer-Verlag.

Noncomputational Issues: Computers in Clinical Psychology

Acceptance by Psychologists

Computerized simulations of clinical activity, including psychological testing, report generation, and psychotherapy simulations, have provoked reactions ranging from enthusiasm (Sagan, 1975) to outrage (Weizenbaum, 1976). This diversity of opinion reflects the capacity of computer technology to influence the practice of psychology for good or ill. Practice management software can be cost-effective and time-saving, much in the same way that computerized statistical programs are a vast improvement over the mechanical calculator.

The abuse of such powerful and useful tools is also possible. For example, computerized testing has led to high-profit mass-testing "mills," in which hundreds of patient's are "processed" at one time. "Overtesting" is a related danger, especially when tests are essentially self-administering. The psychologist may be tempted to inundate the patient with (and bill him or her for) reams of automated tests, producing more reams of unexamined output. Psychologists using the computer in these ways, make themselves into "peripherals" instead of service providers, giving up this crucial prerogative. The ethical and legal implications of abandoning the role of service provider are discussed below.

Several other undesirable side effects in psychological practice are possible, including making computerized patient data a procrustean bed. In such a case, a patient's diagnosis or treatment modality must fit the computer and not vice versa; the diagnosis, the service administered, and even whether the patient is to be treated at all depend on the pronouncement of the computer. A related risk is becoming addicted to technological arcana at the expense of patient care.

Yet another problem would be the delegation of psychological testing to paraprofessionals or nonpsychologists who have no familiarity with basic test principles or procedures. Test developers who market to untrained or undertrained service providers pose diagnostic risks to the patient and legal risks to themselves and to the end user of clinical software.

Acceptance by Other Hospital Personnel

Physicians and nurses in hospitals may be less enthusiastic about or may even obstruct the use of computers for clinical purposes. It is perhaps wise for psychologists intent on implementing such programs not to oversell the hardware, the programming elegance, or the technology of such a system. It may be preferable to emphasize the time-saving benefits of computers, which leave more time for prized activities such as individual psychotherapy. Other advantages that may be valued by hospital personnel include effort reduction, quality assurance facilitation, and increased clinical efficacy. Psychologists in hospitals should be ready to reassure staff members with unrealistic fears of job loss or loss of autonomy, as these issues have often been raised in conjunction with computerized methods.

Psychiatrists may have different reasons for avoiding computers. Schwartz (1984) cited a survey conducted by a psychiatrist who found that there is a role model clash between the psychiatrist's self-image as a professional and the assumed lower status of a computer user (Schwartz, 1984). Psychiatrists may also feel that their critically honed interpersonal skills are being exchanged for simple data collection or manipulation. Other reasons for psychiatric intransigence include an unwillingness to demonstrate lack of facility with this new technology and the feeling that computerized methods increase paperwork. Psychologists considering the introduction of computerized methods should be aware of these attitudes and ready to answer these potential objections before initiating computerized clinical methods in psychiatric hospitals.

Patient Acceptance

Patients' acceptance of computers as alternative input devices for their histories, problems, and so on is crucial to diagnostic accuracy and appropriate treatment. In fact, the majority of studies addressing the question have found that computer acceptability is high and increases as the areas to be investigated become more personal or sensitive (Burke & Normand, 1987). For example, one study compared subjects who were randomly assigned to receive the MMPI and an attitude survey in either a paper-and-pencil or a computerized format (Rozensky, Honor, Rasinski, Tovian, & Herz, 1986). Computerized MMPIs were seen by clinical subjects as more interesting and less anxiety-arousing than their paper-and-pencil equivalents. Even some chronic psychiatric patients appear to tolerate microcomputer applications well. One set of studies (Matthews, De Santi, Callahan, Koblenz-Sulcov, & Werden, 1987) noted that some very withdrawn and low-functioning schizophrenic inpatients appeared to enjoy using the computer for game playing. One patient with a history of elective muteness excelled in anagram solving and was taught to program in BASIC quite quickly. In another study in the same article, both clinicians and computers trained chronic schizophrenic inpatients in various educational tasks. Both groups improved over baseline, but the computer users' improvements were significantly greater than those of the human-trained patients.

The optimistic outlook of these studies is rather typical in the computer literature, which expresses curiously few worries about how fragile patients may react to running the gauntlet of automated evaluations and interviews.

The possibility that individuals without computer experience will experience anxiety about and a negative attitude toward computerized procedures is beginning to enter the literature. For example, Weinberg and Fuerst's survey of businesspeople and college students (1984) found that fully one quarter of their sample had at least mild "computerphobia." Five percent of that same sample would probably be classified as having computer-related anxiety disorders, with symptoms of nausea, dizziness, and high blood pressure. Similarly, Cruikshank (1982) compared standard doctor–patient interviews with interviews by a doctor who used a computer to code the patients' responses. The patients showed a 22% increase in self-reported stress in the computer-assisted interview, compared to a rise of only 8% when seeing the doctor without the computer. Further, although they are not necessarily "cyberphobic," sizable portions of the population may not understand or enjoy interacting with computers. Several studies have noted dislike of computers in older individuals, women (Cruikshank, 1982; Lucas, 1977), and those with higher education (Skinner & Allen, 1983).

Assessment of computer acceptability to patients is complex and undoubtedly depends on many variables, including an interaction between the personality of the subject and the stress or difficulty of the computer task. For example, when Heinssen, Glass, and Knight (1987) assigned university students to groups on the basis of a computer anxiety rating scale (CARS), the students with greater anxiety showed lower expectations, greater state anxiety, and poorer performance on an actual computer interaction that involved some stressful components, including a simulated "data error" message on-screen. Thus, the diagnostic label of the patient may determine whether he or she is capable of using computerized methods. Patients with organic brain syndrome, mania, or passive-aggressive traits, for example, cannot be counted on to produce adequate computer protocols. Then, too, one should question the use of computerized tools for clinical patients whose personality disorders may be due to inadequate, ineffective, or malevolent interpersonal contact. Spiesel (1984), for example, seriously doubts "the moral propriety of asking a frightened alienated patient to interact with a mechanical marvel of unfeeling, uncaring, stupid rigidity" (p. 242).

Ethical and Legal Issues: Secrecy versus Science

Computerized methods in psychology engender great controversy when they attempt to substitute for the clinical decision-making skills of the psychologist. Matarazzo (1983, 1985, 1986) and Hartman (1986a) have raised important ethical and legal questions surrounding the widespread use of such unvalidated interpretive services, indicating that such services are driven more by commercial motives than scientific and technical ones. In fact, scientific concerns about software validation may be antithetical to the proprietary nature of commercial software. Manufacturers have been unwilling to divulge logic and program codes that would enable psychologists to attempt validation. Indeed, several services have actively refused to cooperate in a validation study attempted by one of the authors (Kleinmuntz, 1986). Such secrecy can prevent the practicing clinical psychologist from discovering whether a particular interpretation is generated from actuarial data, clinical hunches, biorhythms, horoscopes, or any combination of actuarial, clinical, or pseudoscientific data. Without access to the logic of the software, there is no way for the practicing clinician to distinguish useful, valid software from the biorhythm or fantasy horoscopes available at the local supermarket.

Even if the clinician were to have such access, there are legal and ethical reasons why caution is warranted in accepting the interpretation of computer-based test interpretation or other types of "report-writing" software. Matarazzo and Matarazzo (1985) reminded us that the courts sanction the clinician's, but *not* the computer's, provision of expert opinions about mental health practice and theory. Thus, the psychologist who depends on the opinion of a CBTI is denying a responsibility that is prop-

erly his or hers alone. The medicolegal doctrine of a "reasonable standard of care" compares any decisions made by computer-using psychologists with the standard of care in the profession at large. Psychologists, as legally sanctioned independent health care providers, do not base important decisions on CBTI output. For example, the psychologist who orders the removal of a child from her home after her mother's MMPI interpretive program suggests that she may be an unfit parent is almost certainly both legally and ethically culpable. The psychologist sued for that decision cannot blame the computer for providing faulty data; he or she is ultimately accountable because the final legal decision for integrating and implementing decisions regarding patient welfare is the psychologist's alone.

Psychologists considering the implementation of computers in clinical practice should also be aware of the ethical principles and guides set down by the American Psychological Association's Ethical Principles of Psychologists (EPP) which apply to computer use. For example, Principle 2 requires psychologists to recognize the boundaries of their competence, and Principle 7a requires psychologists to understand and refer to other specialists as appropriate. These principles indicate that psychologists must already know how to score and interpret a particular test, like the Rorschach or the MMPI, before they can ethically substitute the computerized equivalent. Rather than use the test without such knowledge, the psychologist should refer the patient to a colleague for test administration and interpretation. Further, Principle 1a suggests that psychologists are responsible for any "professional action which may alter the lives" of individuals; this principle underlines the ultimate responsibility of psychologists for the impact of their assessments, whether carefully constructed or torn off the printer. Keith-Spiegel and Koocher (1985) reminded their readers that any psychologist who copies the output of a CBTI into a clinical report or signs that clinical report as his or her own is committing *plagiarism*.

Moreover, Principle 8e of the Ethical Principles of Psychologists, as well as the Guidelines for Computer-Based Tests and Interpretations (GCBTI; APA) reminds the user that computerized assessment is either a professional-to-professional consultation or one explicitly conducted in the context of the professional relationship. For this reason, psychologists might do well to exercise great caution when using interpretive software and basing their diagnostic or treatment decisions on its output. Although Peck (1983) suggested that psychologists "surrender the routine" to clinical computers, APA ethical principles and tort law require that the psychologist remain responsible for all decisions made in his or her name. Thus, the surrender of anything more serious than the most routine of duties may be both ethically improper and legally tenuous.

Advertising

The APA's Ethical Principle 4e explicitly prohibits psychologists from making public statements that claim unique, unusual, or unjustified abilities, and Principle 4g prohibits sensational, exaggerated, or superficial public statements. Though these principles appear to serve the public well, the Federal Trade Commission recently reviewed psychology's ethical principles and pronounced them in violation of laws pertaining to restraint of trade. Similar FTC mandates in support of free-market competition have liberalized advertising in other professions; they may now affect the APA's capacity to regulate the claims of software developers. Exactly how this ruling will affect computer advertising and use is unclear, but a recent U.S. Supreme Court decision giving attorneys the right to solicit clients by direct mail may be instructive. Attorneys in the state of Florida, which has such a law, have reportedly inundated "accident victims, bereaved spouses, parents and other victims of misfortune" with direct-mail solicitation of legal services (Tybor, 1988).

It could be argued that clinical psychology software requires *more*, not less, protection against irresponsible advertising, especially if a program like Neuralytic Systems' Dr. Shrink (recently renamed Mindviewer) used is an ex-

ample of this new breed of advertised software. Dr. Shrink is a program supposedly used by "trial lawyers, clinical psychologists, college professors, psychiatrists in private practice" and others who wish to use their "casual observations" to manipulate others. The program takes such observations and fashions a report detailing "how to gain their trust . . . persuade them . . . have power over them [or] sexually excite them." Dr. Shrink's "scientifically" constructed report should then, according to the advertisement's disclaimer, "either be shredded or placed under lock and key immediately" (presumably to limit liability to the software company). More responsible psychologists might prefer that the program—or better yet, the program's authors—be placed under lock and key immediately. As current licensing and regulatory statutes stand, programs like Dr. Shrink/Mindviewer are censurable if the developer is not only a psychologist but also a member of the American Psychological Association. Only time will tell whether the truth-in-advertising or other regulations will be applied to such software.

Conclusions

We hope that psychologists who review the foregoing complexities of computer use will find that there is no simple approach. The use of computational methods in psychology ranges across such a wide historical and conceptual span that simple conclusions are scarcely possible. We believe that psychologists can best use this developing technology by acquiring a rational understanding of its strengths and weaknesses. The clinician who yearns for the grail of true artificial intelligence software comparable to a trained psychologist is likely to be kept waiting for some time. It seems quite unlikely that human judgment and decision making will be easily programmable. It seems even more doubtful that the *artistic* components of clinical psychological judgment will prove easily amenable to computerization (see Kleinmuntz, 1990). Nonetheless, psychologists who are sensitive to both the professional benefits and the ethical, legal, and scientific pitfalls of computer use may improve their clinical practice with selective applications of present-day computers and software.

References

American Psychological Association. (1981). Ethical principles of psychologists. *American Psychologist, 36*, 633–638.

American Psychological Association. (1986). Guidelines for computer-based tests and intrepretation. Washington, DC: Author.

Berliner, H. J. (1980). Backgammon computer program beats world champion. *Artificial Intelligence, 14*, 205–220.

Bongar, B. (1988). Clinicians, microcomputers, and confidentiality. *Professional Psychology: Research and Practice, 19*, 286–289.

Bowden, B. V. (1953). *Faster than thought*. London: Isaac Pitman.

Burke, M. J., & Normand, J. (1987). Computrized psychological testing: Overview and critique. *Professional Psychology: Research and Practice, 18*, 42–51.

Butcher, J. N., Keller, L. S., & Bacon, S. F. (1985). Current development and future directions in computerized personality assessment. *Journal of Consulting and Clinical Psychology, 53*, 803–815.

Cruikshank, P. J. (1982). Patient stress and the computer in the consulting room. *Social Science Medicine, 16*, 1371–1376.

Dawes, R. M. (1971). A case study of graduate admissions: Application of three principles of human decision making. *American Psychologist, 26*, 180–188.

Dawes, R. M., & Corrigan, B. (1974). Linear models in decision making. *Psychological Bulletin, 81*, 95–106.

Egan, D. E., & Greeno, J. G. (1974). Theory of rule induction: Knowledge acquired in concept learning, serial pattern learning, and problem solving. In L. W. Gregg (Ed.), *Knowledge and cognition* (pp. 43–103). Hillsdale, NJ: Erlbaum.

Eliot, L. B. (1986). Analogical problem-solving and expert systems. *IEEE Expert, 1*, 17–30.

Felts, W. R. (1984). Choosing office practice systems for billing, accounting, and medical record keeping. *M.D. Computing, 1*(3), 11–17.

Fowler, R. D., Jr. (1969). Automated interpretation of personality test data. In J. N. Butcher (Ed.), *MMPI research developments and clinical applications* (pp. 105–126). New York: McGraw-Hill.

Fowler, R. D. (1985). Landmarks in computer-assisted psychological assessment. *Journal of Consulting and Clinical Psychology, 53*, 748–759.

Gardner, H. (1985). *The mind's new science: A history of the cognitive revolution*. New York: Basic Books.

Geissman, J. R., & Schultz, R. D. (1988) Verification and validation of expert systems. *AI Expert, 3*(2), 26–33.

Golden, C. (1985). Computer models and the brain. *Computers in Human Behavior, 1*, 35–48.

Goldstine, H. (1972). *The computer from Pascal to von Neumann*. Princeton, NJ: Princeton University Press.

Hartman, D. E. (1986a). Artificial intelligence of artificial psychologist? Conceptual issues in clinical microcomputer use. *Professional Psychology: Research and Practice, 17*, 528–534.

Hartman, D. E. (1986b). On the use of clinical psychology software: Practice legal, and ethical concerns. *Professional Psychology: Research and Practice, 17*, 462–465.

Hartman, D. E. (1990). The computerized clinician: Ethical, legal and professional issues in automating psychological services. In E. Margenau (Ed.), *The encyclopedic handbook of private practice*. New York: Gardner Press.

Heinssen, R. K., Glass, C. R., & Knight, L. A. (1987). Assessing computer anxiety: Development and validation of the computer anxiety rating scale. *Computers in human behavior, 3*, 49–59.

Honaker, L. M. (1988) The equivalency of computerized and conventional MMPI administration: A critical review. *Clinical Psychology Review, 8*, 561–577.

Jackson, D. N. (1987). Book review: *Computerized Psychological Assessment*, by J. N. Butcher (Ed.). *Computers in Human Behavior, 3*, 71–72.

Keith-Spiegel, P., & Koocher, G. P. (1985). *Ethics in psychology*. New York: Random House.

Kleinmuntz, B. (1963a). MMPI decision rules for the identification of college maladjustment: A digital computer approach. *Psychological Monographs, 77* (14, Whole No. 577).

Kleinmuntz, B. (1963b). Personality test interpretation by digital computer. *Science, 139*, 416–418.

Kleinmuntz, B. (1968). The processing of clinical information by man and machine. In B. Kleinmuntz (Ed.), *Formal representation of human judgment* (pp. 149–186). New York: Wiley.

Kleinmuntz, B. (1969). Personality test interpretation by computer and clinician. In J. N. Butcher (Ed.), *MMPI: Research developments and clinical applications* (pp. 97–104). New York: McGraw-Hill.

Kleinmuntz, B. (1986). *The automated psychologist: Computerizing psychological services: Historical foundations of computers in psychology*. Presented at the 1986 Annual meeting of the Illinois Psychological Association, Chicago.

Kleinmuntz, B. (1990). Can computers be clinicians? Theory and design of a diagnostic system. In M. Hersen, A. E. Kazdin, & A. S. Bellack (Eds.), *Clinical psychology handbook* (2nd ed.).

Kotovsky, K., Hayes, J. R., & Simon, H. A. (1985). Why are some problems hard? Evidence from Tower of Hanoi. *Cognitive Psychology, 17*, 248–249.

Krug, S. E. (1987). *Psychware Sourcebook 1987–1988*. Kansas City, MO: Test Corporation of America.

Laird, J. E., Newell, A., & Rosenbloom, P. S. (1987). Soar: An architecture for general intelligence. *Artificial Intelligence, 33*, 1–64.

Leinweber, D. (1987). Expert systems in space. *IEEE Expert, 2*, 26–38.

Lucas, R. W. (1977). A study of patients' attitudes to computer interrogation. *International Journal of Man–Machine Studies, 9,* 69–86.

Marks, P. A., & Seeman, W. (1963). *The actuarial description of abnormal personality.* Baltimore, MD: Williams & Wilkins.

Matarazzo, J. (1983, July 22). Computerized psychological testing. *Science, 221,* 263.

Matarazzo, J. (1986). Computerized psychological test interpretations: Unvalidated plus all mean and no sigma. *American Psychologist, 41*(1), 14–24.

Matarazzo, J., & Matarazzo, R. (1985). Clinical psychological test interpretations by computer: Hardware outpaces software. *Computers in Human Behavior, 1,* 235–253.

Matthews, T. J., De Santi, S. M., Callahan, D., Koblenz-Sulcov, & Werden, J. I. (1987). The microcomputer as an agent of intervention with psychiatric patients: Preliminary studies. *Computers in Human Behavior, 3,* 37–47.

McCorduck, P. (1979). *Machines who think: A personal inquiry into the history and prospects of artificial intelligence.* New York: W. H. Freeman.

Meehl, P. E. (1950). Configural scoring. *Journal of Consulting Psychology, 14,* 165–171.

Meehl, P. E. (1986). Causes and effects of my disturbing little book. *Journal of Personality Assessment, 50,* 370–375.

Meehl, P. E., & Dahlstrom, W. G. (1960). Objective configural rules for discriminating psychotic from neurotic MMPI profiles. *Journal of Consulting Psychology, 24,* 375–387.

Mendelson, E. (1989, February 14). Word processor specializes in APA style. *PC Magazine,* 46.

Moreland, K. L. (1985). Computer-assisted psychological assessment in 1986: A practical guide. *Computers in human behavior, 1,* 221–233.

Moreland, K. L. (1987). Computerized psychological assessment: What's available. In J. N. Butcher (Ed.), *A practitioner's guide to computerized assessment.* New York: Basic Books.

Newell, A. (1955). The chess machine: An example of dealing with a complex task by adaptation. *Proceedings of the Western Joint Computer Conference* (pp. 101–108).

Newell, A., & Simon, H. A. (1961). Computer simulation of human thinking. *Science, 134,* 2011–2017.

Newell, A., & Simon, H. A. (1972). *Human problem solving.* Englewood Cliffs, NJ: Prentice-Hill.

Newell, A., Shaw, J. C., & Simon, H. A. (1958). Elements of a theory of human problem solving. *Psychological Review, 65,* 151–166.

Nilsson, N. J. (1971). *Problem-solving methods in artificial intelligence.* New York: McGraw-Hill.

Peck, C. P. (1983). Surrendering the routine. *Journal of Clinical Psychology, 39,* 153–154.

Pfeiffer, J. (1962). *The thinking machine.* New York: J. B. Lippincott.

Quigley, E. J. (1984). Writing office practice systems without programming. *M. D. Computing, 1,* 18–23

Raskin, R. (Ed.). (1989, March 14). Statistical software for the PC: Testing for significance. *PC Magazine, 8*(5), 103–255

Reitan, R. M., & Wolfson, D. (1985). *The Halstead-Reitan Neuropsychological Test Battery.* Tucson: Neuropsychology Press.

Reitman, W., & Wilcox, B. (1979). The structure and performance of the interim 2 Go Program. In *Proceedings of the International Joint Conference of Artificial Intelligence,* Vol. 6, pp. 537–542.

Ridenour, L. N. (1952). Computers as information machines. *Scientific American, 187,* 116–118.

Rome, H. P., Swenson, W. M., Mataya, P., McCarthy, C. E., Pearson, J. S., Keating, F. R., & Hathaway, S. R. (1962). Symposium on automation techniques in personality assessment. *Proceedings of the Staff meetings of the Mayo Clinic. Mayo Clinic, 37,* 61–62.

Rozensky, R. H., Honor, L. F., Rasinski, K., Tovian, S. M., & Herz, G. I. (1986). Paper-and-pencil versus computer-administered MMPIs: A comparison of patients' attitudes. *Computers in Human Behavior, 2,* 111–116.

Sagan, C. (1975, January). In praise of robots. *Natural History, 89,* 8–20.

Schwartz, M. D. (1984). Why do psychiatrists avoid using the computer? In M. D. Schwartz (Ed.), *Using computers in clinical practice.* New York: Haworth Press.

Shannon, C. (1950). Programming a computer for playing chess. *Philosophical Magazine, 41,* 256–275.

Simon, H. A. (1979). Information processing models of cognition. *Annual Review of Psychology, 30,* 363–396.

Simon, H. A. (1986). Report of the research briefing panel on decision making and problems solving. *Research Briefings 1986, National Academy of Sciences.* Washington, DC: National Academy Press.

Skinner, H. A., & Allen, B. A. (1983). Does the computer make a difference? Computerized versus face-to-face versus self-report assessment of alcohol, drug, and tobacco use. *Journal of Consulting and Clinical Psychology, 51,* 267–275.

Spiesel, S. (1984). A skeptic's view of the computer as diagnostician. In M. D. Schwartz, (Ed.), *Using computers in clinical practice* (pp. 241–242). New York: Haworth Press.

Squires, D. R. (1984, December). Showdown at the PC Corral: What you need to know about the computer wars when shopping for a PC of your own. *Black Enterprise, 15,* 94–96.

Stanier, A. (1975). BRIBIP: A bridge bidding program. *Proceedings of the International Joint Conference on Artificial Intelligence, 4,* 215–235.

Stephenson, W. (1953). *The study of behavior: Q-technique and its methodology.* Chicago: University of Chicago Press.

Stephenson, W. (1980). Newton's fifth rule and Q methodology: An application to educational psychology. *American Psychologist, 35,* 882–889.

Stoloff, M. L., & Couch, J. V. (1987). *Computer use in psychology: A directory of software.* Washington, DC: American Psychological Association.

Tallent, N. (1987). Computer-generated psychological reports: A look at the modern psychometric machine. *Journal of Personality Assessment, 51,* 95–108.

Tiffany, D. W. (1988, May). Software review: 16PF Report. *Psychologists' Software Club Newsletter, 6*(5), 1–2.

Turing, A. (1950). Computing machinery and intellgience. In E. A. Feigenbaum & J. Feldman (Eds.), *Computers and thought* (pp. 11–35). New York: McGraw-Hill.

Tybor, J. R. (1988, June 14). Lawyers can target mail ads. *Chicago Tribune*, pp. 1, 10.

von Neumann, J. (1951). The general and logical theory of automata. In L. A. Jeffress (Ed.), *Cerebral mechanisms in behavior*. New York: Wiley.

von Neumann, J. (1958). *The computer and the brain*. New Haven, CT: Yale University Press.

Waldrop, M. M. (1988). Toward a unified theory of cognition. *Science, 241*, 27–29.

Weinberg, S. D., & Fuerst, M. (1984). *Computer phobia*. Effingham, IL: Banbury.

Weizenbaum, J. (1976). *Computer power and human reason*. San Francisco: W. H. Freeman.

Wertheimer, M. (1985). A gestalt perspective on computer simulations of cognitive processes. *Computers in Human Behavior, 1*, 19–33.

Wilkinson, L., Gimbel, B. R., & Koepke, D. (1982). Configural self-diagnosis. In N. Hirschberg & L. G. Humphreys (Eds.), *Multivariate applications in the social sciences* (pp. 103–113). Hillsdale, NJ: Erlbaum.

Williams, J. G., & Kleinmuntz, B. (1969). A process for detecting correlations between dichotomous variables. In B. Kleinmuntz (Ed.), *Clinical information processing by computer: An essay and selected readings* (pp. 100–128), New York: Holt, Rinehart & Winston.

PART IV

Practical Issues

General Clinical Issues

Before one can fully appreciate the direct impact of clinical psychology on specific medical disorders, an understanding of broader clinical issues raised by specific populations served in the medical setting is essential. The information provided by the following authors is important to an understanding of the scope and complexity of the factors facing program development. Heather C. Huszti and C. Eugene Walker (Chapter 11) detail critical concerns in consultation–liaison work in pediatrics, and Linas A. Bieliauskas (Chapter 12) focuses on similar issues for adult populations.

Women and geriatric populations are two major and separate groups served in medical settings and are of substantial interest to psychologists. Mary P. Koss and W. Joy Woodruff (Chapter 13) delineate emerging issues in women's health, including the sociological, epidemiological, psychological, and medical issues that pertain to women seeking treatment in the health care network. In addition, they focus on unique issues involving women employed as health care providers, Asenath La Rue and Charles McCreary (Chapter 14) delineate emerging issues in care for the elderly. Finally, because compliance to medical regimens is crucial to health care and involves many psychological variables, Dennis C. Turk and Donald Meichenbaum (Chapter 15) discuss the noncompliant patient. Despite the broad nature of these chapter topics, the authors offer specific guidelines for program development which may be included in establishing programs in any medical area.

CHAPTER 11

Critical Issues in Consultation and Liaison

Pediatrics

Heather C. Huszti and C. Eugene Walker

Introduction

An increasing number of psychologists are employed in medical settings. The most recent surveys suggest that 8%–10% of American Psychological Association (APA) members are currently working in medical settings (De-Leon, Pallak, & Hefferman, 1982; Dorken, Webb, & Zaro, 1982). This trend toward employment in medical settings appears to be continuing to increase (Stabler & Mesibov, 1984). A primary duty of many of the psychologists in medical settings is the provision of consultation services to various inpatient medical units and outpatient medical clinics.

The need for close collaboration between physicians and psychologists is very evident in pediatric settings. Studies conducted in pediatric outpatient clinics have suggested that a significant proportion of pediatric clinic patients have either a purely psychological prob-

lem or a mixture of physical and psychological problems (Duff, Rowe, & Anderson, 1973; Kempe, 1978; McClelland, Staples, Weisberg, & Begen, 1978; Wright, 1979). One frequently cited study suggested that only 12% of all patients had purely physical problems, whereas 36% had purely psychological problems, and 52% had a mixture of physical and psychological complaints (Duff *et al.*, 1973). Another study estimated that 37% of routine pediatrician visits involved some psychological issues and that an additional 19% of all visits involved academic problems (McClelland *et al.*, 1973). A survey reported by Kempe (1978) indicated that 50% of American families had sought some type of psychological help from their pediatrician. The results of these surveys suggest that a large percentage of the complaints that parents believe necessitate a visit to the pediatrician actually have some underlying psychological component. Given this interconnection between physical complaints and psychological issues, it seems logical to assume that psychologists can provide valuable consultation services to physicians involved in treating children.

Past research has also indicated that psychological intervention can reduce the utilization

Heather C. Huszti and C. Eugene Walker • Department of Psychiatry and Behavioral Sciences, University of Oklahoma Health Sciences Center, Oklahoma City, Oklahoma 73190.

of medical care (Follette & Cummings, 1967; Graves & Hastrup, 1981; Rosen & Wiens, 1979; Schlesinger, Mumford, & Glass, 1980). In a study that examined the effects of psychological consultations on medical utilization for low-income children and adolescents in which all medical care was completely subsidized, families that had received psychological consultations showed a significant reduction in the number of clinic visits over the year following the initial referral (Graves & Hastrup, 1981). This reduction was especially significant because the patients referred for psychological services had previously used the available medical services significantly more often than the nonreferred group.

Providing consultation services in a pediatric hospital presents new challenges for the psychologist. The medical system often operates by a different set of rules than those with which the psychologist is familiar. Medical systems use a different "language" than psychologists use among themselves. Medical systems also have a fairly rigid hierarchical power structure, which may be quite different from the somewhat more democratic power structures used in traditional mental health settings. In order to operate as an effective consultant, the psychologist needs to be aware of physicians' language and the rules by which the medical setting operates. On the other hand, the psychologist also needs to guard against the temptation to become overly immersed in the medical system (Elfant, 1985; Stabler & Mesibov, 1984). Psychological consultants need to be able to operate within the medical system while retaining those elements that make psychologists unique.

This chapter reviews the typical kinds of services that are provided by psychologists in pediatric hospitals. The components of receiving, accepting, and satisfactorally completing a consultation are discussed. Some of the common obstacles to successful consultations are also discussed. Finally, possible solutions to these obstacles are reviewed. Table 1 gives an outline of the steps used to perform a consultation. Each step is more fully explained in this chapter.

Table 1. Guidelines for Consultations

1. Consultant receives referral from physician or other professional.
2. Consultant completes referral card and confirms family has agreed to see psychological consultant.
3. Consultant contacts referral source for background information.
4. Consultant reviews medical chart.
5. Consultant interviews patient and/or relevant family members for history of referral problem.
6. Consultant writes initial chart note detailing impression of problem and preliminary treatment plan (see the appendix for an example).
7. Consultant directly communicates impressions and treatment plan to referral source.
8. Consultant communicates impressions and treatment plan to nurses involved in case. If nurses are involved in treatment plan, consultant leaves note in nurses' chart.
9. Consultant gives patient and family members feedback about impression and treatment plans.
10. Consultant talks with other professionals necessary to implement treatment plan.
11. Before patient is discharged, consultant writes final treatment summary in chart and arranges for follow-up care.

Referral Patterns

Several studies have examined the types of consultations psychologists received in pediatric hospitals. A study by Drotar (1977) examined the referrals to a psychological division of a children's teaching hospital over a three-year period. Approximately 46% of the total referrals were for the assessment of intellectual development. Approximately 20% were prompted by the patients' psychological adaptation to chronic illness, with problems such as depression, adjustment to home or school, or parental stress being most common. An almost equivalent percentage of consultations were to evaluate the role of psychological factors in the development of somatic symptoms. Approximately 8% were to evaluate behavioral problems in nonchronically ill children, and 8% of the referrals were for acute psychological crises, such as suicide attempts.

A more recent survey examined consultation requests to the division of pediatric psychology in a children's teaching hospital over a five-year period (Olson, Holden, Friedman, Faust, Kenning, & Mason, 1988). Psychological difficulties due to medical problems accounted for 46% of the referrals. Depression or suicide attempts accounted for 19% of the consultations. An additional 12% of referrals were to assess the patient's psychological adjustment to a chronic illness. The assessment and treatment of behavioral problems were the next most common referral request, accounting for 9% of the total referrals. Psychosomatic problems accounted for 8% of referrals. The need for psychological evaluation also accounted for 8% of the consultation requests.

The difference in the percentages of referrals for psychological evaluation in the Drotar (1977) and the Olson *et al.* (1988) studies is interesting. The study by Drotar covered the period during which consultation services were first available to the hospital. The study by Olson and colleagues was conducted approximately 14 years after psychological consultation services were first introduced into the hospital. It is interesting to speculate whether physicians' views of the function of psychologists have changed with the greater exposure to the work of psychologists. Certainly, the training of pediatricians has increasingly incorporated training in the psychological aspects that can affect medical treatment (Stabler, 1988). Therefore, pediatricians may be increasingly aware of how psychologists can be used in a medical setting. It would be interesting for Drotar to examine the differences between the types of referrals received currently and those received during the 1977 study by the pediatric psychology service in his hospital. Any differences might indicate how pediatricians' perceptions of psychologists change once a psychological consultation unit has been established in a hospital. Drotar's 1977 study did indicate that referrals to the consultation and liaison service had increased each year since its establishment three years previously, a finding suggesting that the physicians were satisfied with the services provided and increasingly used those services.

Understanding the reasons behind a physician's request for psychological consultation is important in the development of a successful consultation relationship with the physician. If psychologists understand what types of services physicians find most valuable, a context for the provision of initial consultation services can be developed. Initial consultations can be provided in this framework, which can then be used as a base from which the consultant can further educate the referring physicians about the additional services that psychologists can provide.

Certainly, physicians have misconceptions about psychologists, just as psychologists have misconceptions about physicians. The only way to correct the misconceptions of physicians is to demonstrate competence in a variety of areas and thus educate them about psychology. Appropriate referrals can be accomplished only when physicians understand what psychologists can and cannot accomplish. Because physicians do not necessarily receive psychological training, they cannot be expected to know either the variety or the limits of psychological interventions. A survey of pediatric and health psychologists indicated that, although over 60% of pediatric psychologists felt that physicians' lack of knowledge about psychological services was an obstacle to the provision of effective services (Stabler & Mesibov, 1984), less than 20% of these same psychologists felt that one goal of a consultant should be to help change physicians' attitudes toward psychologists. Unfortunately, until psychologists provide this type of education, physicians will remain unaware of the full range of the services that psychologists can provide.

Physicians' Perceptions of Psychological Consultations

Relatively few studies have examined physicians' perceptions of psychological consultation services. An early study of pediatricians found that the most valued psychological services were testing, consultation, and teaching (Stabler & Murray, 1973). Less value was placed on therapy, discharge planning, and in-service training. These perceptions suggest that pediatricians could use additional education about the types of services that psychologists provide. In a more recent survey, neurologists who used psychological consultation and liaison services said that psychologists were most useful in the evaluation of primarily psychological disorders, arranging psychological follow-up, performing mental status interviews, and providing emotional support for patients (Schenkenberg, Peterson, Wood, & DaBell, 1981).

It is encouraging that physicians who use consultation and liaison services appear to be satisfied with the services. The neurologists' chief complaint was that there were not enough psychological services available (Schenkenberg *et al.*, 1981). These neurologists wanted to see an expansion of services and an increase in the staff of the consultation and liaison service. Pediatricians in another study indicated that a discontinuation of the consultation and liaison services would have a severe negative impact on the patients' quality of care (Olson *et al.*, 1988).

Several surveys also evaluated what services physicians wanted offered. One survey found that pediatricians were most dissatisfied by what they perceived to be a lack of outpatient follow-up by the consultation and liaison service (Olson *et al.*, 1988). In this survey, the physicians estimated that 67% of all referrals needed some type of outpatient service. This perception could be due to a lack of clear communication between the physician and the consultant about the needs of the patient.

Physicians also indicated that they wanted psychologists to have some familiarity with the medical aspects of the illnesses that the psychological consultants would see most commonly (Meyer, Fink, & Carey, 1988; Schenkenberg *et al.*, 1981). In one survey, neurologists listed a moderate background in physiological body systems and medical illness as one of the most important professional qualifications for consultants (Schenkenberg *et al.*, 1981). A major concern expressed by family practitioners and internal medicine specialists about psychologists performing medical consultations was the perceived lack of training in the medical aspects of the cases seen (Meyer *et al.*, 1988).

These same physicians were also concerned that psychologists were not specifically trained to perform consultations and that this lack of training might interfere with their performance.

Consultation in Medical Settings

Fortunately, psychologists do not have to go to medical school in order to be effective consultants. A rudimentary understanding of the medical aspects of the diseases typically seen, of medical terminology, and of the medical system is sufficient. A great deal of this information can be acquired fairly rapidly.

In the traditional medical model, the attending physician assumes the responsibility for discovering the cause of the patient's illness and for treating it. The physician is the acknowledged expert on the patient's medical illness and serves as the manager of the case. In this capacity, the primary physician requests consultations with other specialists. The additional specialists assess the patient in terms of their particular area of expertise. The specialist attempts to answer the consultation question. For example, an oncologist who is following a child with leukemia may request a consultation with a cardiologist to assess the child's cardiac functioning if the treatment involves a chemotherapy drug that can cause cardiac damage. The cardiologist evaluates the child's cardiac functioning and reports the results of the tests. The communication between the specialists is typically accomplished through notes in the hospital chart, which follow a particular pattern, showing the person who referred the case, the reason for the referral, the consultant's observation of the patient, the tests performed, the results of the tests, the conclusions reached, and the recommendations for the patient's care.

Many of these traditional medical models of consultation are at odds with common psychological practices. Psychologists see many fewer patients than physicians because the time needed for each patient is much greater. The potential difficulties in establishing rapport with the patient are not always well understood by physicians, who can often gain the information they need within minutes. In completing a psychological assessment of a patient, psychologists are likely to examine a wide range of variables, as opposed to focusing on discrete elements. Additionally, psychologists are used to reporting the complexities and subtleties of clients' psychological functioning as opposed to using the format of a brief chart note. In the development of psychological treatment plans, direct communication between the psychologist and the attending physician is generally required.

Although it can be difficult to perform psychological consultations with pediatric patients within the traditional medical model, such consultations can be done successfully. Armed with an understanding of the traditional medical consultation model, psychological consultants can work within this framework to further educate physicians about how psychologists work with patients. Psychological consultants can also further educate physicians about the training and skills that psychologists have and how these skills may be used to the greatest benefit in health care settings. Through the ongoing education of referral sources, the services of psychologists can be used more effectively.

Concurrently, as psychologists spend time in medical settings, they can learn from physicians. Physicians and other medical personnel can help to educate the psychologist about the physical sequalae of the patient's illness and the subsequent treatment. Through further exposure to medical settings, psychologists can learn what types of information physicians need and how to effectively communicate that information. Additionally, by becoming familiar with the medical system, the consultant can help patients to better understand the unique aspects of the system, and to better negotiate their way through it.

There are basic elements common to most medical settings. However, each setting has its own particular organizational structure. It is well worth the time and effort to determine the unique interactions, or "politics," that occur within the setting in which one is working. By learning how members of the system interact with each other, the consultant can learn who holds the actual power in the setting, be it the hospital administrator or the inpatient ward nurse, and which members need to be informed about the consultant's activities. Additionally, it may be important to learn the organization's past history with mental health professionals. Different members of the medical setting may have preconceived ideas, both positive and negative, about the functions and abilities of mental health professionals. On occasion, these ideas may interfere with the consultant's activities. A useful book by Selvini Palazzoli, Anolli, DiBlasio, Giossi, Pisano, Ricci, Sacchi, and Ugazio (1986) provides explanations of how psychologists can effectively join different organizations as consultants.

Models of Consultation

The most typical kind of consultation is referred to as the *patient-centered model* (Faust, 1983), the *independent functions consultation* (Roberts & Wright, 1982), or the *resource consultation* (Stabler, 1979). In this model, the consultant is asked, via a consultation request by the pediatrician, to assess some aspect of the patient's functioning. Both the referring physician and the consultant interact with the patient and communicate test results and impressions through the hospital chart. This model is particularly appropriate for requesting diagnoses, for ruling out possible contributory factors, or for suggesting treatment options. Typically, a medical consultation is performed in one session.

At times, this model is entirely appropriate. However, there are disadvantages in performing psychological consultations based exclusively on this model. If the answer to the referral question is particularly complicated, as is often the case, additional contacts with the patient may be necessary. In addition, it may be difficult to communicate all of the nuances of the case through brief chart notes. When cases are particularly complicated, the opportunities for misunderstandings about the etiology and the treatment of the patient's problems are only multiplied by relying on a brief written summary of the consultant's findings. In order to ensure that the psychological complexities of the case will be fully communicated to the physician, direct communication is necessary. This direct communication can be difficult to arrange, given physicians' busy schedules and their lack of understanding of why direct communication may be necessary. It is also difficult to educate physicians further about the abilities of psychologists without being able to communicate directly with the referring physician. Without continued education, physicians may easily go on seeing psychologists' main abilities as being diagnosis and psychometric testing (Stabler & Murray, 1973).

A second model of consultation is the *indirect psychological consultation model* (Roberts & Wright, 1982), or *process-educative model* (Stabler, 1979). In this type of consultation, the psychological consultant works with the pediatrician or the medical professional directly to provide specific information about a particular patient. Generally, the consultant does not work directly with the patient but provides information to the pediatrician about specific psychological aspects of the patient. For example, the medical professional may want to know if certain behaviors of a child warrant a psychological referral for an assessment of depression. The presentation of psychological information in grand rounds, teaching seminars, and so on also constitutes indirect psychological consultation.

In indirect consultation, the psychologist typically relies on the information presented by the pediatrician. This information may be incomplete or faulty and may therefore hamper proper decision making by the psychologist. The psychologist also needs to trust that the recommendations made will be carried out accurately by the consulting medical professional. Because of these types of concerns, this kind of consultation is most successful in cases where a relationship of mutual trust of professional skills has been developed between the psychological consultant and the medical professional.

A third model of consultation is the *consultee-oriented model* (Faust, 1983), *collaborative team model* (Roberts & Wright, 1982), or *process consultation* (Stabler, 1988). In this model, the referring physician and the consultant interact with each other as well as with the patient. Within this model, the consultant can intervene in a number of ways. The circumstances of the case may suggest that further clarification for the referring physician is appropriate. Additionally, the consultant may need additional education about the medical aspects of the patient's illness. Alternatively, interventions may be made with the medical staff or the patient's family. Sometimes, the physician and the psychologist see the patient together and develop a joint medical and psychological treatment plan.

Using the above model of consultation, the psychologist is freer to evaluate the needs of each particular situation. Instead of focusing only on the patient, the psychologist is free to evaluate the relative contributions of the various elements of the child's system, such as the medical personnel treating the child, the child's parents, the child's siblings, and the child's school environment. In the process consultation model, the psychologist can then intervene at the level of the system deemed to be most appropriate.

Requests for Consultation

Requests for consultations can be received in a variety of ways. A formal request may be made to the psychologist by the attending physician, by the medical resident or student, or by the nursing service. Once the consultant is known to the medical staff, a physician may informally express concerns about a particular patient to the psychologist in the hall of the hospital. Nurses on the medical wards may also express concerns about other patients when the psychologist is present on the ward doing another consultation.

Consultation requests take many forms. They may range from the vague (i.e., "Please see T. J. and evaluate him") to the highly specific (i.e., "Please evaluate S. W. to determine if she is functioning at an age-appropriate intellectual level"). Some requests take the form of loaded requests (i.e., "Please do something with B.C.'s mother; she's crazy"). Many consultations are initiated for reasons other than those explicitly stated. A psychologist may be asked to "evaluate a patient" when the physician is really concerned that the patient's mother is suicidal, or that the child is being sexually abused. Also, a consultation may be requested because the staff is having a difficult time dealing with the patient. An interesting recent study of consultation requests for adult medical inpatients in the Netherlands attempted to determine what factors of a consultation request may predict whether medical staff issues are a part of the reason for the consultation request (Hengeveld, Rooymans, & Hermans, 1987). Patient–staff or intrastaff problems were rated by a team of hospital consultants based on the facts of each case. This team of consultants determined that 33% of the consultation requests assessed involved patient–staff or intrastaff problems. The medical staff completed a brief questionnaire about the consultation request. The request variables that most often predicted a consultation where staff problems were evident were an emotional tone or wording in the request, abnormal timing of the request, and a lack of clarity in the request. These variables correctly predicted 78% of staff problems in referrals in a discriminant analysis.

Many problems can be avoided by clarifying the reasons for the consultation before seeing the patient and the family. Many consultation services have devised standard consultation request forms that are completed for each consultation request. These forms usually include the patient's name, date of birth, chart number, and location in the hospital; the date of admission; the reason for the consultation; the medical diagnosis; the person making the referral; the person to contact for additional information; and the anticipated date of discharge. This type of consultation card provides a good beginning for the process of clarifying the reason for the consultation. Before seeing the patient, the consultant should contact the referring individual to determine the reasons for initiating the referral. This contact should in-

clude discussing what the referring physician would like to know explicitly as a result of the consultation. At this time, if the expectations are inappropriate for a psychological consultation, the consultant can help to reshape the request into a more appropriate form.

During the initial conversation with the referring physician, it is helpful to determine if the patient or the family has either directly requested psychological services or agreed to see a psychologist. One study of adult consultations estimated that 41% of the patients either had not been informed about the consultation or had not been well informed about the intent of the psychological consultation (Hengeveld, et al., 1987). If the family has not been informed of the consultation request, the psychologist may encounter an anxious or angry family. Families are often made anxious by the suggestion that they may need to see a "shrink." Families may assume that the referring physician thinks that they are "crazy." A family that is resistant to the idea of a psychological consult can make the evaluation very difficult to complete. Additionally, it is not ethical for a psychologist to see a child without a parent's permission to do so. Working with physicians on how to explain a psychological consultation to the family is worthwhile. Simply stating, "A psychologist is going to come and see your child," can raise family members' anxiety levels about the purpose of the visit. Instead, explanations such as, "It can be very stressful to have someone in the family who doesn't feel well," or "Sometimes, stress can make medical conditions worse," seem to make families feel more comfortable with the idea of a psychological consultation and more receptive to the consultant during the initial interview.

Before the Patient Consultation

Once the physician's or health care worker's request is fully clarified, the consultant should review the patient's medical chart. The medical chart contains useful background information about the course of the patient's illness and treatment. Medical records for the current hospitalization are usually kept on the same ward where the patient is; records of previous hospitalizations and outpatient visits are often stored in the medical records department or in the current medical inpatient unit. Current medical records generally contain the admissions note, progress notes from all of the medical specialists involved in the patient's care, nurses' chart notes, and the results of all tests ordered. Medical progress notes typically follow the *SOAP* format, detailing subjective reports by the patient (*S*), objective results or observations (*O*), the results of any assessments (*A*), and a plan for treatment (*P*).

When confronting medical charts for the first time, the consultant may be struck by the amount of medical jargon. Medical charts can be extremely difficult to understand initially without some guide to the meaning of the abbreviations and medical terms. There are several good resources to use to decipher medical charts. Each hospital uses slightly different abbreviations. A list of the approved abbreviations for the hospital should be available from the medical records department or the administration office. A chapter by Roberts and Wright (1982) provides a partial listing of the commonly used medical phrases and abbreviations, with lay definitions. A good medical dictionary can also be helpful. The *Physician's Desk Reference* (PDR), which is published yearly, provides information about prescription drugs and their effects and side effects. A basic textbook of pediatrics (Barnett, 1977; Nelson, Vaughan, & McKay, 1975) can be useful for general information about the child's diagnosis. The Merck manual (Berkow, 1982) also provides basic information about the symptoms, diagnosis, treatment, and prognosis of a wide variety of diseases. Additionally, it can be extremely helpful to develop a good working relationship with a physician, a physician's assistant (PA), or a nurse who can provide the consultant with explanations of various diseases or medical procedures. Although it is important to be familiar with medical terminology and diseases, it is also important to avoid the common mistake of overidentifying with the medical hierarchy (Roberts & Wright, 1982; Stabler, 1988).

Psychological consultants are most effective if they can maintain some independence from the medical system and can assess and intervene in the interactions between all members of the system (Drotar, 1983; Stabler, 1979).

In reviewing the medical chart, one must keep in mind the referring question and any hypotheses about the etiology of the referring problem. Information in the medical chart can suggest additional hypotheses or family areas in need of further explanation. The admissions note provides information about the admitting diagnosis, family and social history, and the history of the presenting complaint. The consultant should note if any of the admitting information has changed over the course of the hospitalization, as these changes, or the reasons for the changes, may provide some important information about the etiology of the behavior that led to the consultation request. The extent of the medical work-up may indicate if the consult was requested because a wide range of laboratory tests were within normal limits. Additionally, a review of what tests the patient has undergone may indicate whether the patient has experienced a number of invasive, nonconclusive tests and may, as a result, be feeling discouraged or depressed. As Faust (1983) recommended, a review of the medications being given can suggest whether any of the observed symptoms are due to the effects of medications (for example, the use of pain medications can cause children to become withdrawn and to appear depressed).

In addition to reviewing the medical chart, it can be extremely helpful to talk to the nurses who have had contact with the child and his or her family. Nurses often have the closest patient contact of any medical personnel. If the nurses' view of the child and/or the family differs significantly from that of the referring physician, it can raise questions about the contribution of the referring party's personal perceptions of the problem. Various people's perception of the problem may also suggest that the child is exhibiting different behaviors with different individuals, so that unique situation or person characteristics may serve as cues for the child's behavior.

If part of the consultation question appears to be a conflict between the patient and the medical system or within the medical system, it is important to avoid any appearance of taking sides. The consultant needs to act as a facilitator to obtain additional information, or to help each side understand the other's position. The consultant needs to avoid the trap of declaring one side "correct" or "incorrect," which will only cause the "incorrect" side to be angry. A more lasting solution can be obtained by helping to build bridges between the conflicting sides or by finding a compromise that allows each party to save face.

For example, many consultations are requested because the behavior of the patient or the patient's family is frustrating or distressing to the medical staff. In many cases, a contributing factor to the patient's behavior is frustration, confusion, uncertainty, or anger with the often overwhelming medical system. In the family's attempts to deal with the stresses inherent in having an ill member, in addition to the stresses of having to depend on an alien medical system to "fix" the illness, many parents and/or patients adopt coping behaviors that do not mesh well with the medical system. Parents may direct their anger about their child's illness toward the medical staff. Parents who cope with problems by questioning medical procedures and gathering additional information from other medical professionals may be seen as "noncompliant." Often, the consultant needs both to help the medical staff understand the parents' coping mechanisms and to help the parents understand the reason for the seemingly unreasonable behavior of the medical staff. Sometimes compromises can be made. For example, one mother was upset by the number of medical staff who disturbed her child's rest each morning. She initially attempted to solve the problem by screaming at each medical staff person who entered the room and refusing to allow any of them to touch her child. A consultation was requested to "get this mother to comply with medical treatment." The consultant was able to sympathize with the mother's concerns about her child's receiving enough rest and also to explain that, in a teaching hospital, many professionals are involved in a patient's care and that there are benefits in having more than one physician. Through a meeting with the attending physician and the mother, the consultant was able to explain the mother's concerns to the physician and vice versa. A compromise was worked out, with the mother's participation, so that the most important medical procedures were identified and only these procedures were performed in the morning. All the other medical personnel were asked either not to see this patient or to see her in the afternoon. Because both sides participated in the solution, both were satisfied that they had accomplished their goals.

Logistics of Performing a Consultation

The next decision is when and where to see the child. This seemingly simple question can cause multiple frustrations for the consultant. In a traditional hospital setting, there are few places that offer the privacy and the comfort typically found in traditional psychological settings. Patients are often in rooms with other patients. Different family members or friends may be present in the room and may interfere with the interview. Additionally, if the interview takes place in the patient's room, it can be interrupted by nurses or other medical personnel coming in to take vital signs or to perform physical therapy. Personnel may arrive to take the patient for medical tests. Another distraction may be the housekeeping staff's starting to clean the patient's room during a consultation. The consultant needs to learn to be assertive and to ask that other professionals wait until the consultation is over unless the procedure is vital.

It can be easy for the consultant to allow all other professionals a higher priority. This flexibility can have negative consequences for the outcome of the consultation. The willingness to defer to other professionals for access to the patient can suggest to the patient and to other medical professionals that there is less value to the psychologist's role than to other medical professionals' roles. Physicians may consequently view psychological explanations and treatments as less valid than medical ones. Patients and family members may see the psychological consultant as being less important than the physician and may be less likely to comply with his or her recommendations (Faust, 1983). Although it can be difficult to be assertive about the importance of the psychologist's role, it is essential for the consultant to believe in the efficacy of his or her role and to communicate that importance to both the patient and the medical staff.

It is always preferable to try to find a private setting for the interview. If the patient's room needs to be used for the interview, a "Do Not Disturb" sign can be placed on the door. It can be helpful to the new consultant to go over the physical setting of the hospital in advance to determine where available family or consultation rooms are located. Nurses on the hospital ward can be helpful in locating empty patient rooms or in making conference rooms available. Finding and locating these rooms when the consultant is already busy with the patient can be difficult.

Arranging a time to see the patient can also be a difficult and frustrating experience. Patients are often asleep or are being seen by other medical professionals for medical tests, physical therapy, or nutritional consultations. It can be a difficult decision either to let the patient sleep and to come back later or to wake the patient up. Certainly, a psychological consultation, which depends to some extent on building rapport with the patient, is not helped by first waking up a patient. At times, however, the patient must be seen and must therefore be disturbed.

As discussed previously, the pace of a hospital is rapid. It is not at all unusual to receive a consultation request on a patient who is scheduled to be released later the same day or "as soon as you've evaluated him." Generally, consultations are done on the same day as the request. One survey that evaluated the response time for consultation requests showed that 76% of patients were seen on the same day as the consultation request (Olson *et al.*, 1988).

This percentage was even higher for consultations to evaluate depression or suicide risk: 87% were seen on the same day. A survey of physicians suggested that one of the most positive perceptions of the consultation service was its *prompt response* to the consultation request (Schenkenberg *et al.*, 1981).

However, not all consultations can be evaluated quickly. With every consultation request, a number of variables need to be considered. These variables include whether the consultant has the necessary time to adequately perform the consultation, whether the consult is truly an emergency or part of a chronic problem, and what would be the level of the patient's risk if she or he were to be discharged before the consultation was finished. It can be appropriate to request that the patient be kept an extra day so that a thorough assessment can be made if necessary. Leaving a preliminary note in the chart detailing the plans being made to evaluate the psychological status of the patient can help the medical staff be aware of the reasons for keeping the patient hospitalized. An outpatient appointment can also be made, but often the compliance with such appointments is low, and this factor must also be considered.

Interviewing the Patient

Once all of the preliminary logistics have been sufficiently addressed, the consultant interviews the patient. The consultant should remain aware that many of the rules of psychotherapy learned in training do not apply when performing consultations. It is rare to receive a consultation request for a patient who has specifically asked to see a psychologist. Patients are referred by a third person and therefore are often less motivated to pursue psychotherapy or other treatment recommendations. In some cases, the referred patient is not the actual client. The referring physician, or the school system, or another outside party may ultimately be the one with whom the consultant works. Additionally, the definition of confidentiality may be different in consultation than in psychotherapy. Because the consultant has ties to both the patient and the medical system, the consultant may need to relay some of the information gained in an interview or a therapy session to the medical team. Information that is written in the medical chart cannot be considered confidential because any professional at the hospital can review any patient's chart. These issues need to be considered by the consultant before disseminating information to the medical staff. Any limits to confidentiality need to be conveyed to the patient and his or her family before the start of the consultation. Because mental health codes vary from state to state, the consultant should be familiar with the requirements of confidentiality in his or her state.

In work with pediatric patients, a decision needs to be made about who will be interviewed because sometimes the patient is too young to be able to provide all of the necessary information. Some research has suggested that, although children can report some problems, such as depression, they may report fewer symptoms than do their parents or other adult informants (Kazdin, 1987; Kazdin, French, Unis, & Esveldt-Dawson, 1983). However, adult informants may not always present an accurate view of the child either. Previous research has suggested that parental reports of child behavior may be influenced by numerous factors other than the child's true symptoms, such as parental depression and marital problems (Forehand, Lautenschlager, Faust, & Graziano, 1986; Griest, Wells, & Forehand, 1979; Mash & Johnson, 1983). In order to gain a clear picture of the child's symptoms, it may be helpful to gain as many different perspectives as possible about the child's presenting problem.

Often, the referral problem is maintained by family interactional patterns and necessitates talking to other family members. The consultant must evaluate which immediate and extended family members are involved in the development and/or remediation of the referral question. If consultation with family members is necessary, the consultant also needs to consider where and how to talk to the family. The feelings and fears of the child may need to be addressed before talking to family members

separately. The child may wonder why the doctor, as the consultant is invariably identified by the child, is talking to his or her parents in private. Children often fear that the doctor is giving the parents bad news about the child's condition. An older child or adolescent may be concerned that the consultant will disclose sensitive information about the patient to the parents. A brief explanation of the purpose of the talk to the parents, and of the rules of confidentiality, may help to allay these fears.

Often, a family interview format can be used so that the child can participate in the interview. The participation of all concerned individuals in one session may facilitate greater comfort with the treatment decisions made. A useful way to start this type of interview is to ask each individual present what she or he understands about the reason for the psychological consult.

The results of a recent study may be related to the above issue (Lewis, Knopf, Chastain-Larser, Ablin, Zoger, Matthay, Glasser, & Pantell, 1988). This study compared parents' and children's perceptions of bedside versus standard, outside-of-the-room medical rounds. The results suggested that both the parents and the children preferred bedside rounds. The older the children, the more they disliked having rounds performed outside their rooms. Several subjects reported that they either felt excluded or feared that bad news was being discussed during standard rounds. The parents also indicated that they felt more comfortable with bedside rounds. Interestingly, the parents felt that bedside rounds upset their child more than standard rounds, although the child indicated a preference for bedside rounds. Although the communication of medical information can be different from the communication of psychological information, some of the same principles may apply. The consultant must continue to use his or her best judgment about whom to interview and when.

Occasionally, it is difficult to locate family members. Sometimes, parents do not visit their child at all during the hospitalization, or only at night after work. It may be necessary to try to find a noncustodial parent who has not yet come to visit the child. Various solutions have been tried at one children's hospital. Often, if parents can be reached, they are agreeable to making a specific appointment with the psychological consultant. Consultants often need to stay late to catch those parents who work. Sometimes, enlisting the aid of other hospital professionals is useful. Nurses are often willing to contact the consultant when the parent appears, if the consultant first takes the time to explain the reason for the request to the nurses. Social workers may have additional resources for helping to locate parents. In some cases, social workers help by encouraging the parent to keep an appointment with the consultant. Good working relationships with these professionals are essential to enlisting their support when necessary.

The form of the interview of the patient or the family members depends on the referring question. An assessment of the patient or the family may require the use of psychometric tests if, for example, the referral question is about a possible developmental delay. Medical, interpersonal, social, school, and/or developmental variables may all be relevant. A thorough family history may also be helpful in identifying what variables are causing or supporting the continuation of the referral problem. When planning an interview with a patient or family member, it is helpful to consider the stated referral problem and to generate possible hypotheses about the cause of the problem. These hypotheses should include a wide range of possibilities. By considering possible causes before the interview, the consultant can explore the possible contributions of these causes during the interview, while gaining a full picture of the variables influencing the patient's life.

Diagnosis

The referral question will determine, to some extent, what type of diagnostic interview will be used. However, if the consultant remains too narrowly focused on the referral question, important explanations of the child's symptomatic behavior may be missed. For example,

a referral was made to assess a 16-year-old cancer patient's depression. The referring physician was concerned about the adolescent's relationship with her mother because the mother seemed overly involved with the adolescent's treatment. During a brief examination of the other facets of the child's life, it was found that the child had had significant social problems at school before the cancer diagnosis. The adolescent reported that she had had no friends and had been excluded from many social activities. These social problems had contributed to her depression. Her mother was aware that the adolescent was depressed but assumed it was because of the cancer diagnosis and was consequently trying to shield her daughter from much of the pressures of the treatment. If the consultant had explored only family relationship issues, a major contributing factor to the depression might have been missed.

Madow (1988) proposed using a matrix system for the assessment of adolescents in general hospital psychaitric units. This proposed matrix is also useful for evaluating the consultations requested in medical hospital settings. The proposed system evaluates a range of factors that may affect the patient. Using this system, the consultant explores additional areas beyond the presenting complaint. Each area that may affect the patient is evaluated from four perspectives: behavioral (what the child actually does), cognitive (how the child processes the world), affective (the emotional life of the child), and interpersonal (how the child interacts with other people). Madow proposed evaluating seven areas: individual, medical, family, legal, educational, peer, and biological.

For example, an 8-year-old female with a presenting complaint of depression might be evaluated behaviorally in the medical area as showing vegetative signs of depression along with a medical diagnosis of acute lymphocytic leukemia (ALL). An examination of the child's cognitions about her ALL shows a poor understanding of the prognosis for ALL: the child expects to die. The affective component of the medical area is the child's expressed depression and fear of dying, and the interpersonal area shows that the child has withdrawn from her peers because the treatment has caused her hair to fall out.

Obviously, there may be overlap between the areas, and not all areas may need to be considered in formulating an impression of the patient's problems. However, it is important, when assessing a consultation, to consider the wide variety of areas that can affect a child's functioning. Often, a variety of stressors contribute to the child's symptoms.

Many consultations involve the determination of whether the presenting medical symptoms are psychosomatic. The consultant should be wary of immediately concluding that the problem is psychosomatic, even if the referring physician has stated that he or she suspects it is psychosomatic. The consultant might do well to keep in mind a review by Dubowitz and Herser (1976), which found that 33%–46% of patients referred for a psychological evaluation of hysteric conversion were found on follow-up to have organic causes of their symptoms. The question of an organic basis for any unusual symptoms, even with psychosocial stressors present, should remain open if there is no conclusive medical evidence to the contrary. A psychosomatic diagnosis should be made because of strong, positive evidence of psychological causes; it should never be made by default.

It is easy to assume that patients with unusual symptoms and readily apparent psychological problems have psychosomatic causes for their physical symptoms. The psychologist, as the expert on psychological processes, is the one professional who can continue to remind physicians that even people with psychological problems may have coexisting physical illnesses. Often, the psychological consultant's role is that of the patient's advocate, because patients may not have much power in the medical system. In the case where the consultant suspects that a presenting problem may have a medical rather than a psychological etiology, he or she can tactfully discuss with the physician what diagnoses have been ruled out and why. Additionally, the consultant may ask the physician if other possible disease processes may be considered. At this point, the consultant should share with the physician why he or she is uneasy about calling the disorder psychosomatic. The consultant may then ask about other specialists who might be called on for additional input in the case. This type of dialogue may suggest alternative medical hypotheses to the physician. If the physician maintains that the disorder is psychosomatic, and if the consultant still feels uneasy about this diagnosis, the consultant should document in the chart his or her lack of significant psychological findings to support a psychosomatic etiology.

Obviously, a case where the consultant's opinion differs radically from the physician's may cause the medical staff to become angry with the consultant. Generally, a tactful and supportive approach to the physician will prevent this anger. However, a consequence of working with professionals from other disciplines is that they will occasionally be angry with you.

Treatment Recommendations

After interviewing the patient and/or the family, the consultant makes his or her recommendations. The consultant should make an initial chart note detailing the background information, the diagnosis, and the treatment recommendations (see the appendix for an example).

In making recommendations, it is important to remember the role of the consultant. The consultant works with the medical staff and consequently does not necessarily have to assume the responsibility for the outpatient care of all patients evaluated (Faust, 1983; Roberts & Wright, 1982; Walker, 1979). It is also helpful to keep in mind who referred the patient and who assumes primary responsibility for the patient. The referring physician and the primary physician (who may or may not be the same person) require feedback about the outcome of the consultation. Without this type of feedback, the physician may feel alienated from the patient's care and may be less likely to encourage the patient's compliance with the consultant's recommendations. Referrals to community outpatient settings are perfectly acceptable. In fact, a referral may be necessary if the family lives a distance from the hospital. It is advisable for the consultant to keep handy a reference list of local community mental health agencies to which patients can be referred before discharge. Consequently, the focus of the consultant's interventions should be more time-limited and aimed more at primary symptom relief than if the same patient were self-referred to an outpatient psychological clinic.

Many excellent articles exist in the pediatric psychology or pediatric literature that demonstrate treatment programs for specific pediatric medical problems. The *Journal of Pediatric Psychology* contains many useful treatment articles, and the recently published *Handbook of Pediatric Psychology* (Routh, 1988) contains many reviews of the treatment literature for a wide variety of commonly seen disorders. Pain control protocols exist for a variety of painful medical procedures, such as burn tankings (Elliott & Olson, 1983; Kavanaugh, 1983; Miller, Elliott, Funk, & Pruitt, 1988; Varni, Bessman, Russo, & Cataldo, 1980; Wakeman & Kaplan, 1978) and bone marrow aspirations (Jay, 1988; Jay, Elliott, Katz, & Siegel, 1987; Kuttner, Bowman, & Teasdale, 1988; Zeltzer & LeBaron, 1982). Pain control programs have included the use of hypnosis, relaxation training, or the acquisition of cognitive coping skills. Guidelines for the design and evaluation of feeding programs for failure-to-thrive infants are also well represented in the literature (Berkowitz, 1985; Drotar, 1988; Ramsay & Zelazo, 1988; Roberts & Maddux, 1982). These programs are behavioral, structuring positive and negative consequences for the child's attempts or failures to eat. These techniques are then taught to the parents. Many studies have also evaluated the reduction of children's anxiety through the use of preparation programs for dental procedures (see Melamed & Williamson in this text) or surgery or hospitalization (Atkins, 1987; Melamed & Siegel, 1975; Peterson & Mori, 1988). Behavioral treatment programs have used films, modeling, and explicit instruction in cognitive coping techniques, with varying results. Be-

havioral protocols also exist for the treatment of enuresis and encopresis (Walker, 1978; Walker, Milling, & Bonner, 1988). These treatment protocols may need to be modified because of the specific circumstances of the child and each specific medical setting.

Some consultations are quickly resolved by disseminating general information to the patient or the parents (Roberts & Wright, 1982). Many observed behavioral problems are the result of inconsistent parenting. Parents may simply need to read a basic book on parenting skills, or specific parent training may be necessary. Parents may need information about how to conduct toilet training, how to cope with chronic pediatric illnesses or disabilities, or how to explain death to children. A list of good readings on a number of commonly seen problems can be a valuable resource for parents and children. Roberts and Wright (1982) provided a list of good books on a variety of pediatric areas, some hospitals have family libraries that contain good resource books, and many children's bookstores are also a good source.

Before the psychologist implements any treatment program, the medical staff needs to be informed about the proposed treatment program and the rationale for the program. One should not rely on the referring physician or the nurse to read any chart notes made by the consultant. Consultation notes are typically short, less than one page; longer notes may not be read in their entirety. Therefore, it is useful to communicate the psychological subtleties of the consultation to the referring physician *in person*.

Enlisting the support and cooperation of the nursing staff is particularly important to the successful intervention of behavioral therapy programs. The nursing staff has the most sustained contact with patients. Often, the nurses are asked to help in the charting of the program. It is easy for medical staff to inadvertently, or even deliberately, sabotage a behavioral program. If the medical staff do not fully understand the rationale for the program, they may decide that the program is too time-consuming, given their other duties; that it won't be helpful to the child; or that it will harm the child. Also, nurses are generally taught a "tender-loving-care" approach to patients that may conflict with behavioral prescriptions. For example, in one behaviorally based feeding program, a 20-month-old was turned to the wall for 15 seconds if he didn't take a bite of the offered food. He was rewarded for eating the food with the opportunity to play with a novel toy. One nurse felt that turning the child to the wall was too cruel and refused to help with the program. Once the rationale for the program was explained and the success of the positive and negative consequences was demonstrated, the nurse became more willing to participate in the program. Taking the short amount of time necessary to identify the primary nurses of the child, inform them about the rationale for the treatment program, and describe the program can lead to the success of the program. A detailed description of the program can be inserted into the nurses' book of patients, cardex, or nurses' patient information book in order to reach nurses who work on the night shift or on weekends.

Many treatment programs necessitate the collaboration of medical and psychological professionals. Children with chronic illness should be evaluated and followed by a physician familiar with the disease process. For example, a noncompliant diabetic 13-year-old male was repeatedly admitted to the hospital for ketoacidosis (high blood sugar). The patient needed to be carefully monitored by a physician to determine his medical condition and to adjust insulin levels as necessary. The psychological consultant met with the family and found that the boy was noncompliant with medication partly because of normal adolescent needs to be independent. Although some of the struggle for independence was manifested between the boy and his mother, the medical staff also contributed to the problem by colluding with the mother and treating the adolescent as if he were still 6 or 7 years old by not allowing him to take the responsibility for his treatment and by dealing only with the mother when they talked about his disease process. The consultant worked with the family members in family therapy sessions, using some individual ses-

sions for specific problems. The adolescent received some individual sessions to help him learn more adaptive ways to assert his independence. In addition to the family, the consultant also worked with the medical team members to remediate their contributions to the power struggle.

The above scenario, which describes a type of problem that is not uncommon, demonstrates a number of strengths that the psychologist can bring to the pediatric setting. The consultant can evaluate the contribution of developmental issues to the exhibited symptoms. The psychologist can also evaluate the symptom of the identified patient from more than an individual perspective. For example, the relative contribution of the various systems with which the patient is involved may be assessed. These systems include the medical team, the family, the social support system, and the school. Interventions can be made on a systemwide level as well as on an individual level (Libow, 1985).

In order to maintain a systemwide perspective, it is important for the consultant to avoid an overidentification with the medical system. There is a tremendous pull for the psychologist to become enmeshed in that system. Hospitals generally move at a fast pace, and there is a great pressure to take immediate action and to resolve the patient's symptoms immediately (Elfant, 1985). It is easy for a psychologist to get caught up in the pace and to feel a need to intervene and "fix" the patient (Faust, 1983). In addition to personal pressures to solve the patient's problems, the consultant may also experience implicit or explicit pressure from the family, the patient, and other medical staff involved in the patient's care. It may become difficult to take the time necessary to evaluate the patient fully. At times, the symptoms that the medical staff see as dysfunctional may be adaptive for the family. For example, parents who continually question the decisions of the physician may be viewed as being "uncooperative," yet the questioning may be quite adaptive for the parents because it gives them a sense of control. The consultant who is too involved with the medical system may also label the parent as being "uncooperative" and may attempt to change the adaptive behavior, rather than to help the medical staff to understand the parents' coping mechanisms. The consultant should recognize that every system exerts a pull toward membership and to evaluate continually the degree to which he or she is overinvolved with either the medical system or the patient and family system.

Developing a Consultation Service

There are a number of innovative ways for the psychological consultant to become part of the hospital system and to educate physicians subtly about the roles that psychologists can play in a medical setting. The best education for physicians about the services a psychological consultant can provide is through a demonstration of those services. In addition to performing physician-referred inpatient consultations, the psychologist can provide a number of other services. The psychological consultant can offer to be in attendance at outpatient clinics during specific times in order to provide immediate consultation for patients with psychosocial problems. For example, in one hospital, psychologists attend the Adolescent Medicine Clinic one morning a week. Patients with suspected psychosocial problems are scheduled during this time. During this clinic, residents read medical charts, write chart notes, and consult with the attending physician and the consultant in one central room. The arrangement provides an opportunity for the psychologist to raise questions about potential psychosocial issues with residents on cases that may not be directly identified as including psychological problems. The consultant can also offer ideas about how residents can deal effectively with difficult patients. Through these informal interactions, the consultant is able to provide additional education about the interaction of psychological and physical symptoms. In our experience, medical residents quickly grow to appreciate the new perspective that the psychological consultant provides, and they begin to use the consultant's services more frequently.

Similar services can be provided in almost any outpatient clinic.

The psychological consultant can also attend the focused medical rounds held for inpatient units or for different disease services, such as hematology-oncology. In these meetings, patients are discussed, and treatment plans are developed. Psychological consultants can provide information about the psychosocial aspects of the patients, which may have an effect on treatment plans. By attending regular rounds, psychologists can also identify patients who may benefit from psychological intervention before a major crisis is reached. The psychologist's presence at regularly scheduled meetings helps to formalize their function in the unit (Koocher, Sourkes, & Keane, 1979).

A more formalized way to provide training about the interaction between psychological and physical issues for medical residents is for the residents to have a rotation through the hospital's psychiatry or psychology department. This rotation can include a number of training experiences. Medical residents can participate in group sessions on the inpatient psychiatry unit. The can also see patients with inpatient psychiatrists. The residents can also attend training seminars. These seminars may include participation in a family therapy team. In one hospital, family medicine residents sit behind a one-way mirror and participate as members of a family therapy resource team for the primary therapist, who actually sees the family. In some cases in which there are complex medical and psychological issues, the resident may also serve as a cotherapist with a psychologist or a psychology intern. Residents can participate in a variety of groups, such as therapy groups for sexual offenders or medical staff support groups for members of particularly stressful specialties such as hematology–oncology units. In one institution, residents and psychology interns see the residents' medical patients together. The psychology interns learn how physicians normally function in medical contexts, and the medical residents learn how different psychological interventions can be helpful with a wide variety of patients and medical problems. For instance, a psychology intern was able to demonstrate simple relaxation and distraction techniques while the medical resident was putting stitches in a patient's finger. Such experiences expose the residents to a wide variety of psychological services. The residents are not expected to learn how to provide sophisticated psychological interventions, but rather to learn when it may be appropriate to refer patients for psychological services and how to handle simple patient issues themselves more effectively.

Psychologists can also provide training for medical personnel through presentations to medical school classes and during grand rounds. Grand rounds are open presentations that are attended by attending physicians, medical residents and students, and community physicians.

Relationships with Other Professionals

It is important for the psychological consultant to develop a network system with all of the many professionals who work within the hospital setting. Generally, the relationship between child psychiatrists and pediatric consultation and liaison psychologists is a supportive one. The more psychodynamic model of child psychiatry is not as amenable to the quick interventions necessary in hospital consultation work. The behavioral intervention model, used most commonly by psychologists, is often better suited to the types of problems seen in inpatient medical settings. Therefore, many child psychiatry departments have opted not to pursue inpatient medical consultations. Often, psychologists and psychiatrists work cooperatively in the hospital setting because each professional model is more appropriate to different types of cases. The psychological consultant uses the child psychiatrist's expertise in long-term psychotherapy and psychotropic medication for children with severe psychopathology. The child psychiatrist uses the psychologist's expertise in short-term behavioral programs for children with well-defined behavioral disorders. This type of division of services helps to maintain a supportive relationship between the two disciplines. Unfortunately, in settings where there have been turf issues between psychiatrists and psychologists, the consultation and liaison service generally suffers. In these turf battles, the psychiatrists do not want psychologists to practice in the hospital, but they also frequently do not provide their own consultation service. Therefore, the patients are the ones who suffer. Friendly discussions between psychologists and psychiatrists about a division of labor, along with sincere demonstrations of mutual respect and frequent referrals to each other, go a long way toward preventing meaningless turf battles.

Other professionals can also be extremely helpful to the psychological consultant. Hospital social workers can find the necessary resources (such as low-income housing and food supplements) for patients and their families. Working in tandem with social workers can also alleviate some of the stresses that both kinds of professionals experience in today's financially strapped environment. Both may have too many patients to see. By combining resources, social workers and psychologists can provide specialized services to patients without duplicating their efforts. If the hospital has a child life service, it can also be extremely helpful. Child life workers are often skilled in medical play with children and through this play can help children express anxieties, familiarize them with different medical procedures, and help them to gain a greater sense of control and mastery. Again, by coordinating services, the different professionals avoid overlapping their services and instead complement each other's efforts. Each professional is able to obtain different information from the patient, and by sharing the information gained, all professionals can construct a more complete picture of the patient.

Likewise, schoolteachers who work in the hospital can be a valuable resource. For example, the teachers can help provide special attention to a child who needs additional structure during the day while in the hospital. A patient advocate can also help to find resources within the hospital system to help address the needs of patients and their families. Patient support groups can also provide useful services to patients, such as information about their disease or the support of others with the disease. For example, when one 9-year-old boy was hospitalized for two months with a number of infectious disease processes and was growing increasingly depressed and irritable, the different services got together and made out a daily schedule for him. With the cooperation of a child life worker, the schoolteacher, the social worker, a nurse, the chaplain, a psychology postdoctoral fellow, and a psychologist, the child's daily schedule became full. The psychologist served as the coordinator of the program and kept in touch with the various other professionals. Because of the variety of individuals working with the child, more information was gained than if only the psychologist had seen the child. For example, the child life worker was able to identify some of the child's medical fears, such as the fact that he was afraid of getting a blood transfusion. The psychologist was able to encourage the physician to explain to the child why blood was being given to him, and the psychologist helped the child to deal with his fears.

Each hospital setting has its own resources. It can be difficult to locate all of the services available to the patients. It also takes time to determine which resources are the most beneficial and reliable. When entering a new hospital system, the consultant may want to contact the directors of the various services to exchange information on the services that the psychological consultant can provide and the services provided by the contacted service. Professionals who have worked in hospitals for a time can also help identify which services are most helpful.

Summary

Providing psychological services to a pediatric hospital presents a number of challenges. The psychologist needs to learn how to function within the medical system, which has a different hierarchical structure, rules, and language from traditional psychological systems.

The entering consultant needs to maintain a delicate balance between understanding and being able to work within the medical system while still maintaining enough distance to be able to make interventions in that system. Seeing patients in an inpatient medical system is also different from seeing patients in a traditional psychological setting. Psychological consultants need to learn how to work within this new setting, which often necessitates great flexibility and creativity.

Certainly, working in a hospital provides a number of rewards. The setting continually provides new challenges, and the consultant is given the opportunity to work on a wide variety of cases. Additionally, the psychological consultant comes into contact with a wide variety of other health professionals. These contacts provide the consultant an opportunity to learn to view the patient from different perspectives. The consultant can continue to learn about psychology through the teaching of psychological principles to other professionals.

Appendix

Sample Psychological Consultant's Chart Note

Susan, a 16-year-old white female, was referred by Dr. Smith for an assessment of the level of her depression and her risk for suicide. Patient was admitted to the hospital for ingestion of 30 Tylenol tablets. Patient was interviewed alone for 30 minutes, and mother was interviewed with child for 15 minutes. Patient was initially reticent to talk and fearful. By the end of interview, patient was more open to discussing feelings and agreed to have psychologist return to continue discussion. Mother was cooperative and appeared to be very worried about daughter's feelings of depression.

Past History

Patient is oldest of four children in the family. Parents are divorced, and she has not seen the biological father in 10 years and stepfather in past 6 months. Patient reports she has made one previous suicide gesture. She reports she took 10 aspirins after a fight with her boyfriend. After taking the pills, she called her boyfriend. She was not hospitalized after this attempt.

Patient reports she has had several arguments with her mother concerning boyfriend, staying out late, and poor grades. She is now a sophomore at Central High School. Her grades have dropped from a "B" to a "D" average in the past 9 weeks. She broke up with her boyfriend approximately 1 week ago. Patient has lost 5 pounds and reports problems falling asleep at night in the past few weeks.

Present Problems

Patients reports taking the 30 Tylenol tablets after seeing her boyfriend at a movie with his new girlfriend. She reports feeling very sad and angry with her boyfriend and feeling like she would not "ever have anyone as awesome as he is." Patient returned home and argued with her mother about not doing household chores. Patient went to the bathroom, took the pills, and told her mother. She was then transported to Children's Hospital. Patient stated she was upset but did not want to die; she took the pills because she was angry. Patient appears tearful and expresses feelings of helplessness and hopelessness in terms of relationships. She often blames others for her problems (peers, school, etc.). Patient does have several close girlfriends and is planning to start a "part-time" job in 1 week.

Evidence of Moderate Depression

1. Weight loss
2. Sleep disturbance
3. Feeling hopeless/helpless in relationship.

Suicide Gesture—Low Risk

1. Patient did not view 30 Tylenol as a lethal dose.
2. Patient denies wanting to die.
3. Patient was in a situation in which detection was almost certain.
4. Patient immediately informed mother of taking pills.

Do not feel patient is high risk for suicide in hospital—does not require suicidal precautions on ward.

Patient stated she would not try to kill herself again and was anxious to be discharged from hospital. She has agreed to a no-suicide contract and to talk with friends or mother if feeling sad again.

Recommendations

Will make referral to guidance center. Mother has agreed to counseling.

References

Atkins, D. M. (1987). Evaluation of pediatric preparation programs for short-stay surgical patients. *Journal of Pediatric Psychology, 12*, 285–290.

Barnett, H. L. (1977). *Pediatrics* (16th ed.). New York: Appleton-Century-Crofts.

Berkow, E. (Ed.). (1982). *The Merck manual of diagnosis and therapy.* Rathway, NJ: Merck.

Berkowitz, C. (1985). Comprehensive pediatric management of failure to thrive: An interdisciplinary approach. In D. Drotar (Ed.), *New directions in failure to thrive: Implications for research and practice.* New York: Plenum Press.

DeLeon, P. H., Pallak, M. S., & Hefferman, J. A. (1982). Hospital health care delivery. *American Psychologist, 37*, 1340–1341.

Dorken, H., Webb, J. T., & Zaro, J. S. (1982). Hospital practice of psychology resurveyed: 1980. *Professional Psychologist, 13*, 814–829.

Drotar, D. (1977). Clinical psychological practice in a pediatric hospital. *Proffessional Psychologist, 8*, 72–79.

Drotar, D. (1983). Transacting with physicians: Fact and fiction. *Journal of Pediatric Psychology, 8*, 117–127.

Drotar, D. (1988). Failure to thrive. In D. K. Routh (Ed.), *Handbook of pediatric psychology* (pp. 71–107). New York: Guilford Press.

Dubowtiz, V., & Herser, L. (1976). Management of children with nonorganic (hysterical) disorders of motor function. *Dev. Med. and Child Neurology, 18*, 358–368.

Duff, R. S., Rowe, D. S., & Anderson, F. P. (1973). Patient care and student learning in a pediatric clinic. *Pediatrics, 50*, 839–846.

Elfant, A. B. (1985). Psychotherapy and assessment in hospital settings: Ideological and professional conflicts. *Professional Psychology: Reesarch and Practice, 16*, 55–63.

Elliott, C. H., & Olson, R. A. (1983) The management of children's distress in response to painful medical treatment for burn injuries. *Behavior Research and Therapy, 21*, 675–683.

Faust, D. S. (1983). Principles of consultation-liaison psychology. In C. E. Walker (Ed.), *The handbook of clinical psychology: Theory, research and practice* (Vol. 2). Homewood, IL: Dow Jones-Irwin.

Follette, W., & Cummings, N. A. (1967). Psychiatric services and medical utilization in a prepaid health plan setting. *Medical Care, 5*, 25–35.

Forehand, K., Lautenschalager, G. J., Faust, J. L., & Graziano, W. G. (1986). Parent perceptions and parent-child interactions in clinic-referred children: A preliminary investigation of the effects of maternal depressive moods. *Behavior Research and Therapy, 24*, 73–75.

Graves, R. L., & Hastrup, J. L. (1981). Psychological intervention and medical utilization in children and adolescents of low-income families. *Professional Psychology, 12,* 426–443.

Griest, D., Wells, K. C., & Forehand, R. (1979). An examination of predictors of maternal perceptions of maladjustment in clinic-referred children. *Journal of Abnormal Psychology, 88,* 277–281.

Hengeveld, M. W., Rooymans, H. G. M., & Hermans, J. (1987). Assessment of patient-staff and intrastaff problems in psychiatric consultations. *General Hospital Psychiatry, 9,* 25–30.

Jay, S. M. (1988). Invasive medical procedures: Psychological intervention and assessment. In D. Routh (Ed.), *Handbook of pediatric psychology* (pp. 401–425). New York: Guilford Press.

Jay, S. M., Elliott, C. H., Katz, E., & Siegel, S. E. (1987). Cognitive-behavioral and pharmacologic interventions for children's distress during painful medical procedures. *Journal of Clinical Child Psychology, 55,* 860–865.

Kavanaugh, C. (1983). Psychological intervention with the severely burned child: Report of an experimental comparison of two approaches and their effects on psychological sequelae. *Journal of the American Academy of Child Psychiatry, 22,* 145–156.

Kazdin, A. E. (1987). Assessment of childhood depression: Current issues and strategies. *Behavioral Assessment, 9,* 291–319.

Kazdin, A. E., French, H. H., Unis, A. S., & Esveldt-Dawson, K. (1983). Assessment of childhood depression: Correspondence of child and parent ratings. *Journal of the American Academy of Child Psychiatry, 22,* 157–164.

Kempe, C. H. (1978). *Report of the American Academy of Pediatrics,* prepared by the Task Force on Pediatric Education. Evanston, IL.

Koocher, G. P., Sourkes, B. M., & Keane, W. M. (1979). Pediatric oncology consultations: A generalizable model for medical settings. *Professional Psychology, 10,* 467–474.

Kuttner, L., Bowman, M., & Teasdale, M. (1988). Psychological treatment of distress, pain, and anxiety for young children with cancer. *Developmental and Behavioral Pediatrics, 9,* 374–382.

Lewis, C., Knopf, D., Chastain-Larser, K., Ablin, A., Zoger, S., Matthay, K., Glasser, M., & Pantell, R. (1988). Patient, parent, and physician perspectives on pediatric oncology rounds. *Journal of Pediatrics, 112,* 378–384.

Libow, J. A. (1985). The care of critically and chronically ill adolescents in a medical setting. In M. P. Mukin & S. L. Koman (Eds.), *Handbook of adolescents and family therapy.* New York: Gardner Press.

Madow, M. R. (1988). Issues in the diagnosis and treatment of adolescents in a general hospital inpatient unit. *General Hospital Psychiatry, 10,* 122–128.

Mash, E. J., & Johnson, C. (1983). Parental perceptions of child behavior problems, parenting self-esteem, and mother's reported stress in younger and older hyperactive and normal children. *Journal of Clinical Child Psychology, 51,* 86–99.

McClelland, C. Q., Staples, W. P., Weisberg, I., & Begen, M. E. (1978). The practitioner's role in behavioral pediatrics. *Journal of Pediatrics, 82,* 325–331.

Melamed, B. G., & Siegel, L. J. (1975). Reduction of anxiety in children facing hospitalization and surgery by use of filmed modeling. *Journal of Clinical Child Psychology, 43,* 511–521.

Meyer, J. D., Fink, C. M., & Carey, P. F. (1988). Medical views of psychological consultation. *Professional Psychology: Research and Practice, 19,* 356–358.

Miller, M. D., Elliott, C. H., Funk, M., & Pruitt, S. D. (1988). Implications of children's burn injuries. In D. Routh (Ed.), *Handbook of pediatric psychology* (pp. 426–447). New York: Guilford Press.

Nelson, W. E., Vaughan, V. C., & McKay, N. J. (1975). *Textbook of pediatrics.* Philadelphia: Saunders.

Olson, R. A., Holden, E. W., Friedman, A., Faust, J. L., Kenning, M., & Mason, P. J. (1988). Psychology consultation in a children's hospital: An evaluation of services. *Journal of Pediatric Psychology, 13,* 479–492.

Peterson, L. J., & Mori, L. (1988). Preparation for hospitalization. In D. Routh (Ed.), *Handbook of pediatric psychology* (pp. 460–491). New York: Guilford Press.

Ramsay, M., & Zelazo, P. R. (1988). Food refusal in failure-to-thrive infants: Nosogastric feeding combined with interactive-behavioral treatment. *Journal of Pediatric Psychology, 13,* 329–347.

Roberts, M. C., & Maddux, J. E. (1982). A psychosocial conceptualization of nonorganic failure to thrive. *Journal of Clinical Child Psychology, 11,* 216–226.

Roberts, M. C., & Wright, L. (1982). The role of the pediatric psychologist as consultant to pediatricians. In J. M. Tuma (Ed.), *Handbook for the practice of pediatric psychology* (pp. 251–289). New York: Wiley.

Rosen, J. C., & Wiens, A. N. (1979). Changes in medical problems and use of medical services following psychological intervention. *American Psychologist, 34,* 420–431.

Routh, D. K. (Ed.). (1988). *Handbook of pediatric psychology.* New York: Guilford Press.

Schenkenberg, T., Peterson, L., Wood, D., & DaBell, R. (1981). Psychological consultation/liaison in a medical and neurological setting: Physicians' appraisal. *Professional Psychology, 12,* 309–317.

Schlesinger, H. J., Mumford, E., & Glass, G. V. (1980). Mental health services and medical utilization. In G. VandenBos (Ed.), *Psychotherapy: Practice, research and policy.* Beverly Hills, CA: Sage.

Selvini Palazzoli, M., Anolli, L., Di Blasio, P., Giossi, L., Pisano, I., Ricci, C., Sacchi, M., & Ugazio, V. (1986). *The hidden games of organizations.* New York: Pantheon.

Stabler, B. (1979). Emerging models of psychology—pediatrician liaison. *Journal of Pediatric Psychology, 4,* 307–313.

Stabler, B. (1988). Pediatric consultation-liaison. In D. K. Routh (Ed.), *Handbook of pediatric psychology* (pp. 538–566). New York: Guilford Press.

Stabler, B., & Mesibov, G. B. (1984). Role functions of pediatric and health psychologists in health care settings. *Professional Psychology, 15,* 142–151.

Stabler, B., & Murray, J. P. (1973). Pediatricians' perceptions of pediatric psychology. *Clinical Psychologist, 27,* 12–15.

Varni, J., Bessman, B. A., Russo, D. C., & Cataldo, M. F. (1980). Behavioral management of chronic pain in children: Case study. *Archives of Physical Medicine and Rehabilitation, 61,* 375–379.

Wakeman, R. J., & Kaplan, J. Z. (1978). An experimental study of hypnosis in painful burns. *American Journal of Clinical Hypnosis, 21,* 3–12.

Walker, C. E. (1978). Toilet training, enuresis and encopresis. In P. Magrab (Ed.), *Psychological management of pediatric problems* (Vol. 1). Baltimore: University Park Press.

Walker, C. E. (1979). Behavioral intervention in a pediatric setting. In J. R. MacNamara (Ed.), *Behavioral approaches to medicine: Application and analysis.* New York: Plenum Press.

Walker, C. E., Milling, L., & Bonner, B. L. (1988). Incontinence disorders: Enuresis and encopresis. In D. K. Routh (Ed.), *Handbook of pediatric psychology* (pp. 363–397). New York: Guilford Press.

Wright, L. (1979). A comprehensive program for mental health and behavioral medicine in a large children's hospital. *Professional Psychology, 10,* 458–466.

Zeltzer, L., & LeBaron, S. (1982). Hypnosis and nonhypnotic techniques for reduction of pain and anxiety during painful procedures in children and adolescents with cancer. *Journal of Pediatrics, 101,* 1032–1035.

CHAPTER 12

Critical Issues in Consultation and Liaison

Adults

Linas A. Bieliauskas

Introduction

Before about 15 years ago, psychological consultation in medical settings was primarily service-oriented under the aegis of departments of psychiatry. Thus, its main function was seen as providing psychometric services in support of psychiatric interests, both in terms of clinical services and in terms of research. An ancillary service was brief intervention consultation, usually behavioral in nature, designed to alter specific patient or staff behaviors to encourage compliance with varying treatment regimens. The psychologist in such a setting was, by necessity, relegated to such interests, and most effort was directed to issues of psychopathology and interventions with psychiatric patients.

Since then, the situation has gradually changed, with relatively independent psychological services established in Veterans Administration (VA) hospitals, the establishment of semiautonomous sections of psychology in departments of psychiatry, the establishment of appointments for psychologists in medical departments other than psychiatry (such as family practice, neurology, and obstetrics-gynecology), the provision of specialized services (such as neuropsychology and pain management), and even the establishment of independent departments of psychology in medical schools and on hospital staffs.

It is particularly appropriate to focus on critical issues in psychological consultation in these changed relationships in medical settings because it then becomes clear that the research component of psychology has played at least as important a role as the clinical service component in fostering the developments outlined above. Our medical colleagues have come to recognize the important role of psychological factors in illness for understanding etiology, symptom presentation, and effective treatment, through a demonstration that careful research design and the effective measurement of behavior provide evidence that was previously

Linas A. Bieliauskas • Veterans Administration Medical Center and University of Michigan, Ann Arbor, Michigan 48105.

187

unattainable by an exclusive reliance on theoretical notions that would occasionally seem fanciful or inextricably complex. It is my thesis that it is the development of psychology within the scientist-practitioner model that has fostered the growth of its involvement in medical settings.

Though the practice of traditional clinical psychology still makes a strong and valuable contribution in medical settings, psychology has evolved much more into a role better described as behavioral science, in the context of studying and providing solutions to specific problems defined by particular medical approaches, illnesses, and specialty practice. To identify general critical issues in psychological consultation, one should ask, "What is needed and desired by the medical practitioner requesting consultation and how can this best be provided?" An outline developed by Weinman (1981) provides a good framework of reference. He divided psychological involvement in medicine into five areas:

1. Changes in behavior associated with such factors as aging, psychiatric illness, and neurological impairment.
2. The role of psychological factors in the etiology of medical problems.
3. Doctor–patient relationships.
4. The patient's response to illness and treatment.
5. Psychological approaches to treatment.

Because these areas no longer involve only traditional issues for clinical psychologists, involvement frequently requires reeducation in the types of entities that are the subjects of inquiry, and often, there is a need to become familiar with methodological procedures that are not traditionally learned in psychological training. As this chapter is designed to address practical day-to-day issues in consultation, I will draw on examples from our own clinical and research practice to illustrate specific questions and the approaches used to solve them, rather than survey the voluminous literature available. I will touch briefly on each of the points of the outline, with attention to the clinical and scientific issues raised.

Changes in Behavior Associated with Factors Such as Aging, Psychiatric Illness, or Neurological Impairment

The contributions of clinical psychology to understanding psychiatric illness and its treatment are well known, and I will not dwell on them here. Psychological contributions to the understanding of the behavioral changes in aging have ranged from understanding how the prevalence and appearance of psychopathology change in the elderly to how age-related changes in the functioning of the brain affect various behaviors. Because many of the latter contributions interrelate with neurological illnesses, let me focus on the types of research about patients with neurological impairment by citing examples of some of our own research, which uses psychological knowledge to elucidate the course of neurological disease.

It is clear that patients with Parkinson's disease (PD), a not uncommon degenerative neurological disease among the aged, have a high prevalence of psychological depression, ranging from 39% to 90% in different studies (Bieliauskas, Klawans, & Glantz, 1987). Some researchers also feel that this depression is related to changes in cognitive function associated with dementia in these patients (Passafiume, Boller, & Keefe, 1986). If such a relationship exists, it has significant implications for psychological consultation; that is, identification of the onset of depression may also be an indication of a more precipitous decline in cognitive abilities with an untoward prognosis.

We developed a protocol for cognitive evaluation and evaluation for depression in order to investigate the link between the two. I cannot overstate the importance of traditional psychological knowledge in the identification of the appropriate tests of clinical depression as well as more contemporary knowledge and measurement of the cognitive deficits typical in patients with PD, or elderly patients in general, as well as the importance of incorporating a respectable research design to answer questions that may be of critical clinical importance. We measured depression in these patients using the Minnesota Multiphasic Personality Inventory (MMPI) D scale, one that we felt was far superior to obvious self-report measures of depression, and we measured a variety of cognitive functions. We found that approximately 75% of our patients were depressed (a finding confirming earlier reports), but that the depression did not relate to cognitive impairment (a finding disconfirming earlier speculation that depression and dementia in PD were significantly related) (Bieliauskas & Glantz, 1989). In the course of this investigation, we also found that the MMPI D scale could be accurately approximated by the use of a shortened 20-item depression scale (Bieliauskas & Glantz, 1987). Finally, it is also known that many patients with PD who are undergoing standard medication treatment incur hallucinations after about four years. Using relatively standard MMPI-like measures, we developed a protocol that was able to predict accurately those patients who were at risk for hallucinations in the short run (Glantz, Bieliauskas, & Paleologos, 1986; also see Figures 1 and 2).

On a referral basis, when we are asked for consultation by neurology for patients with PD, we can now render a rapid assessment of psychological depression and the risk of hallucinosis within the context of the measurement of cognitive abilities. The development of the ability to provide such information, based on evidence, required an education in principles of neuropsychology, a knowledge of the cognitive aspects of PD, and a familiarity with traditional psychological information.

Whereas this is an example of the systematic development of a consultative information base, let me give a case example in which a specialized cognitive assessment, along with more traditional clinical psychological measurement, provided information crucial to consultation for a patient with neurological disease. A neurologist had a patient with amyotrophic lateral sclerosis (ALS), which, in its later stages, frequently involves the central nervous system. This young woman came to him complaining of memory problems. Cognitive testing documented the memory deficit—in this case, lowered subtest scores on the Wechsler Memory Scale. A year later, the patient complained that her memory was getting worse. Cognitive reevaluation documented the same memory problems as before, but with no progression (see Table 1). MMPI data, however, in addition to

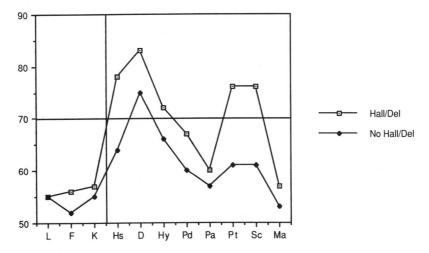

Figure 1. MMPI mini-mult profiles for patients with PD with (Hall/Del) and without (No Hall/Del) evidence of hallucinations/delusions.

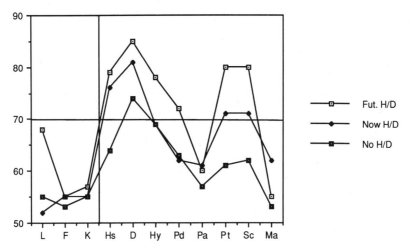

Figure 2. MMPI mini-mult profiles for patients with PD with subsequent (Fut. H/D), concurrent (Now H/D), and no (No H/D) hallucinations/delusions.

other tests such as the Rorschach, revealed the presence of significant distress and possible contributing interpersonal relationship difficulties (Figure 3). An interview then confirmed that there were significant marital relationship problems and that the patient's memory difficulties could be seen not as increasing but as reflecting her subjective impressions of interpersonal strife. A recommendation for referral for marital counseling was made. Consultation with the neurologist thus provided both reassurance about the possible progression of central nervous system involvement and a referral source for aiding the patient in dealing with her symptomatic complaints.

In terms of consultation regarding changes related to aging, psychiatric illness, or neurological impairment, the above examples emphasize the need for medical-specialty-based evaluations (in this case, neuropsychological assessment) that can document cognitive deficits and can compare levels of cognitive functioning over time, can evaluate the presence or development of psychopathology by means of traditional clinical psychology instruments, and can employ research protocols to develop evidence-based information that will enable the referring physician to provide optimal patient treatment.

The Role of Psychological Factors in the Etiology of Medical Problems

This is a controversial area of service and research. Psychological influences on traditional risk factors such as smoking and alcohol consumption are well accepted as contributory to diseases such as cancer or heart disease. However, direct psychological influence on disease etiology is not widely accepted by physicians and is often viewed with considerable skepticism. Nevertheless, the use of psychological approaches has begun to make meaningful contributions to questions of the etiology of physical changes.

Some time ago, we participated in a study that was designed to test the hypothesis that depression is a precursor of cancer. It was necessary to know that cancer frequently requires long incubation periods, that its occurrence can influence psychological responses, and that other risk factors may obscure any psychological roles. Which comes first, then, depression or cancer, and are they related? A review of the literature to date suggested that there was no good evidence that cancer patients were more depressed than anyone else (Bieliauskas & Garron, 1982), despite popular notions to the contrary. It also indicated that case control de-

Table 1. Cognitive Data for a 27-Year-Old Female with ALS[a]

		Initial testing	Testing one year later
PPVT IQ	(100 ± 15)	131	134
Wechsler Memory Scale Memory Quotient	(100 ± 15)	87	89
I. Information	(6)	6	6
II. Orientation	(5)	5	5
III. Mental Control	(9)	5	6
IV. Logical Memory	(13)	2.5	3.5
V. Digit Span	(10)	8	9
VI. Visual Memory	(13)	10	16
VII. Associate Language	(21)	16	16
A. Associate Recognition	(4)	4	4
B. Object Memory			
Naming	(10)	10	10
Recall	(10)	6	5
Recognition	(10)	10	10
C. Social Judgment	(7)	7	7
D. Sentences	(7)	7	7
E. Stereognosis			
Right	(3)	3	3
Left	(3)	3	3
F. Praxis			
Right	(3)	3	3
Left	(3)	3	3
G. Reading	(2)	2	2
Comprehension	(2)	2	2
Dictation	(2)	2	2
H. Figure Copying	(3)	3	3
I. Finger Tapping			
Right	(45)	57	51
Left	(40)	46	45
J. Right-Left	(12)	12	12

[a]PPVT and Wechsler Memory Scale tests indicate norms in parentheses; remaining tests indicate maximum expected performance in parentheses.

signs were inappropriate to answer the question of etiology. Instead, it was appropriate to use a prospective research design that, described remote premorbid data if that were possible. Luckily, we had such a sample available from a long-term study of heart disease, which included early psychological data and medical follow-up over a 17-year period. Using a high-point D profile on the MMPI as an indicator of depression, we found approximately a doubling of risk of death from cancer over the 17-year period associated with initially measured depression (Shekelle, Raynor, Ostfeld, Garron, Bieliauskas, Liu, Maliza, & Paul, 1981). Further analysis revealed that this doubling of risk did not change when the intervening years were divided into tertiles, a finding suggesting that the depression predated the cancer, rather than being a simultaneous state. Age, smoking, and alcohol consumption did not change this risk ratio. Subsequent analyses of the same data revealed that the doubling of risk applied to cancer morbidity as well as mortality (Persky, Kempthorne-Rowe, & Shekelle, 1987). It should be added that such a design required the use of odds ratios, as well as measures of prevalence that are not the same as the traditional statistics used in case control studies. In fact, our patients who died of cancer, as opposed to those who did not or who died of other causes, had a

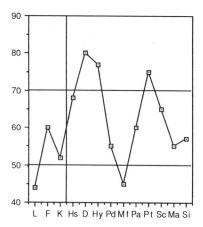

Figure 3. MMPI data for a 27-year-old female with ALS.

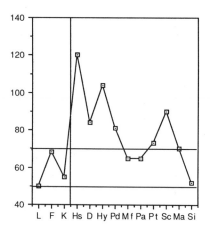

Figure 4. MMPI data for a 50-year-old male with multiple somatic complaints.

statistically higher score on the MMPI *D* scale at only the $p < .05$ level, a result that would hardly be meaningful in such a design. In fact, the depression measured seemed to apply to a state of chronic distress rather than clinically significant depression (Bieliauskas & Shekelle, 1983).

From a more traditional point of view, psychological conflict has been regarded as potentially etiological for various symptoms and patient complaints. Though perhaps not as prevalent as described during the zenith of psychoanalytic thought, such cases continue to occur. The example I will use here is a patient who presented with multiple physical complaints, including heart problems, breathing difficulties, skin irritation, and difficulties with memory and concentration. The patient was a strong, muscular man who felt that his symptoms were caused by exposure to a drain-cleaning chemical compound. The written list of symptoms he presented filled five pages of single-spaced text. A medical examination was inconclusive, and a psychological evaluation was requested. The neuropsychological protocols revealed no noticeable cognitive impairment. However, traditional personality test instruments such as the MMPI (see Figure 4) and the Rorschach revealed a hypochondriacal component in the patient's symptoms, which suggested a strong psychological etiology, well be-

yond that which might be expected from any physical cause. The patient was treated primarily on a psychological basis rather than subjected to an endless variety of medical tests searching for elusive physiological changes. It should be emphasized here that the absence of hard medical findings or the ambiguous medical findings that may be found in any segment of the general population is not sufficient to attribute causation to psychological-emotional factors alone. Psychometric assessment can provide the appropriate "rule-in" evidence that prevents mistakes made simply on the basis of an absence of medical findings, while simultaneously offering a rationale for psychological intervention.

It is thus emphasized that appropriate design and analysis methodologies are crucial in various aspects of studying medical disease and in arriving at clinically meaningful conclusions. It is also emphasized that the design approach was paired with traditional psychological assessment instruments in the examples I have given. The application of psychological techniques alone, without the pursuit of a meaningful question in an appropriate design, would not have provided the answers sought. In a clinical sense, it certainly made a difference in terms of how patients with cancer or ambiguous somatic symptoms were treated; instead of viewing depression as an inevitable consequence of disease, it can be viewed as a treatable entity. Somatic symptoms may also be viewed, in a more chronic sense, as a lifestyle component that may predispose one to health risk. In either case, the psychological contribution to physical change was clarified and provided useful tools that had not previously been available in consultation services.

Doctor–Patient Relationships

Great detail isn't necessary here, as many research strategies have been used to demonstrate that certain variables in the relationship of the physician to the patient are crucial to patient attitudes, to compliance with medical regimens, and even to patient desires. For example, in one of our studies, we found that, in patients with multiple sclerosis, attitudes such as personal life satisfaction, self-addressed coping effectiveness, and satisfaction with treatment and personnel facilities were related to life events and illness symptom stressors, but that this relationship varied by various demographic and psychological moderating factors, such as increased knowledge of the disease, decreased general anxiety, and aging (Counte, Bieliauskas, & Pavlou, 1983). The measures used were the Social Readjustment Rating Scale (Holmes & Rahe, 1967), the State-Trait Anxiety Scale (Spielberger, Gorsuch, & Lushene, 1970), and self-developed indexes. Beyond the use of traditional psychological tests and new ones, the number of variables was great enough and complex enough so that stepwise multiple-regression techniques needed to be used to define the relationships that were present. The techniques were selected because they offered both an estimate of the combined influence of all independent variables and the relative contribution of each independent variable, statistically controlling for the association of other variables. Though the study was descriptive, all the factors of design, a knowledge of the disease, the traditional and innovative use of psychological tests, and the use of appropriate statistical analysis were crucial in obtaining answers.

To discuss briefly several other general issues regarding physician–patient relationships, studies have found that, when physicians are surveyed about whether to tell a dying patient of the terminal nature of an illness, the majority of them feel that the patient should not be told. When patients are similarly surveyed, the overwhelming majority of them wish to be told. This is an example of research contributions that explore commonly held beliefs and that provide a clear illustration of how physicians relate to their patients. Similarly, studies of compliance with medical regimens have found that the application of operant conditioning paradigms significantly affects compliance with regimens. These kinds of research contributions augur well for the impact of psychological consultation on maximizing the benefits of patient care.

The Patient's Response to Illness and Treatment

The psychological effect of an illness on a patient and thus on the course of the illness is an important consideration in effective medical care. For example, Pancheri, Ballaterra, Matteoli, Cristofari, Polizzi, and Puletti (1978) reported that patients who present for treatment of myocardial infarction have significantly different (elevated) MMPI profiles if they are considered "well" 10 days later than if they are "not improved" 10 days later. Such approaches can yield significant data on which patients are at risk for medical complications on a psychological basis. On a case example level, let me illustrate how such an approach can be used. Figure 5 shows the MMPI profile of a 57-year-old male who was evaluated before a cardiac catheterization procedure. The MMPI suggested a considerable potential for distress and anxiety. When placed in the coronary intensive care unit after the procedure, the patient decompensated and caused considerable medical problems for himself and for the caring staff, such as tearing out intravenous lines and trying to jump out of his bed. Fortunately, the staff had been prewarned

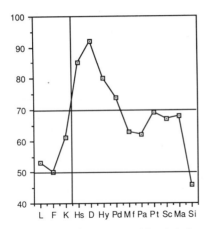

Figure 5. MMPI data for a 57-year-old male before cardiac catheterization.

and had the personnel and equipment available to handle the problem.

Let me also touch on the development of new instruments that measure patients' responses to illness and treatment. At our medical center, Leavitt, Garron, Whisler, and Sheinkop (1978) developed a questionnaire designed to measure lower back pain. This instrument consists of verbal descriptors of pain and is a much extended version of the well-known Melzack-Torgerson Pain Questionnaire (Melzack & Torgerson, 1971). As pain is a purely subjective phenomenon, such an instrument provides a reliable way to measure an entity that is quite ineffectively addressed by designations as gross as "mild," "moderate," or "severe." Patients with organic findings can be assessed pre- and postsurgery, for example, so that the success of the operation can be ascertained. Patients whom physicians elect to treat medically or palliatively can also be assessed at different points in time for the effectiveness of the intervention. The scale of Leavitt and Garron has been shown to differentiate patients with and without demonstrable organic disease (Leavitt, Garron, D'Angelo, & McNeill, 1979), to differentiate patients with genuine versus simulated pain (Leavitt, 1985), and to aid in assessing the contributions of stressing life events and psychological disturbance in patients with low back pain (Leavitt, Garron, & Bieliauskas, 1980). The clinical impact of developing such instruments is

obvious in determining and evaluating various treatment approaches and illustrates, again, how psychological measures can be used in treating medical or pseudomedical disease.

Approaches such as these are extremely important to effective medical care as they help to make subjective patient behaviors less ambiguous. Traditional instruments have their place, whereas in other cases, new, innovative measures need to be developed. Consulting psychologists, with their traditional and research skills, remain in the forefront of meeting these real needs in the medical setting.

Psychological Approaches to Treatment

This is perhaps the best currently known aspect of psychology's consultative role in medical care. The impact of biofeedback treatment for headache, back pain, neuromuscular re-education following stroke, and syndromes such as spastic dysphonia are well documented. Approaches to modifying behavioral risk patterns for heart disease, asthma, obesity, and possibly hypertension are strong areas of current research. Psychologists are also involved in providing special support services to patients with disabling diseases such as Parkinson's disease, multiple sclerosis, and cancer, and to families of patients with Alzheimer's disease. Patient protocols are generally part of any such treatment and are amenable to data analysis to determine the effectiveness or ineffectiveness of intervention methods.

Let me give just one example of how specific clinical methods can be used to ameliorate symptoms. A 45-year-old radiologist who had undergone a right temporal lobectomy to remove a tumor sought treatment for difficulties with attention and scanning, sustained concentration, leisure reading, driving, and self-monitoring in social situations. A neuropsychological evaluation conducted 2.5 years postsurgery revealed impaired visual-spatial memory, inefficiency in general visual scanning, a mild left-sided neglect, and diminished social perception. This pattern of deficits was seriously debilitating to an individual whose primary work depended on reading and interpreting brain scans. At this time, the radiologist also underwent an independent evaluation of his abilities at another medical center and was found to have approximately a 20% error rate in reading such scans. We initiated a four-month cognitive retraining program that was tailored to his specific neuropsychological deficits, using a pre- and post-test protocol, much as I described earlier, and tasks that we devised, based on the evaluation of his deficits. These included specific methods to improve attention and the substitution of verbal cues for visual ones in ensuring a complete scanning of radiological photographs. At the termination of treatment, improvements were observed in follow-up neuropsychological data, behavioral observations made by the patient's wife, and efficiency on work-related tasks. An independent follow-up evaluation of the patient's ability to read brain scans showed a 0% error rate (Rao & Bieliauskas, 1983; see Table 2).

This is only one example of detailed patient protocols, based on knowledge of cognitive aspects of the patient's dysfunction, and their effective use for the individual patient. I should add that the systematic collection of individual patient data not only provided the necessary information for treatment but gave evidence of the degree of change that had occurred. The variety of situations in which similar approaches can be used can be listed indefinitely, and similar approaches for a number of other specific medical conditions, such as diabetes, asthma, and sleep disorders, can be found in other chapters in this book.

Conclusions and Some Final Thoughts

I hope that the coverage of these five areas of interface between psychology and medicine in the consultation–liaison setting gives a flavor of the current picture. I have chosen to use examples in each area that we have found useful in our own work setting, but comprehensive descriptions of the many approaches used can be found in numerous books on this field, as well as within this book.

In conclusion, I would like to reemphasize

Table 2. Results of Pre- and Posttreatment Neuropsychological Evaluations

Variables	Pre	Post
1. Intelligence		
Wechsler Adult Intelligence Scale[a]		
Full Scale IQ	119	133
Verbal IQ	124	132
Information (age-corrected)	14	17
Comprehension	15	16
Arithmetic	14	15
Similarities	15	15
Digit Span	14	14
Vocabulary	12	13
Performance IQ	110	131
Digit Symbol	11	17
Picture Completion	15	19
Block Design	12	17
Picture Arrangement	11	11
Object Assembly	9	9
Progressive Matrices[b]		
Total score (centile)	40 (75)	53 (95)
2. Memory		
Wechsler Memory Scale[c,d]		
Memory Quotient (100 ± 15)	140	136
Information	6	6
Orientation	5	5
Mental Control	7	9
Logical Memory	17	13
Digit Span	13	13
Visual Reproduction	8	10
Associate Learning	19	18
Visual Retention Test[d,e]		
Number correct (8 expected)	5	7
Number of errors (2 expected)	7	4
Memory for Faces Test[f]		
Number correct (perfect score = 12)	9	10
3. Visual Scanning and Sustained Attention		
Cancellation Task[g]		
Total number of omissions	17	0
Left side omissions	10	0
Right side omissions	7	0
4. Visual-Spatial Functions		
Facial Recognition Test[d,h]		
Score (centile)	39 (8)	40 (11)
Bender-Gestalt Figure Drawings[i]		
Number correct (perfect score = 3)	3	3
5. Motor Speed		
Index Finger Tapping (IFT)		
Number of taps/10 sec. (centile)		
Right hand	58 (90)	NA[j]
Left hand	42 (22)	NA[j]

[a]Wechsler (1955). [b]Raven (1960). [c]Wechsler (1945). [d]Alternate forms used for each testing session. [e]Benton (1974). [f]Milner (1968). [g]Diller *et al.* (1974). [h]Benton and Van Allen (1973). [i]Bender (1938). [j]NA = not administered.

basic themes that extend throughout the effective application of psychology in medical settings as it is practiced today. These themes are heavily research-based. They include a solid grounding in traditional psychological theoretical content and methodology, a necessary knowledge of the specific medical entity with which one is dealing, the use and development of appropriate instruments to measure behavior, the consistent use of reliable patient protocols, and familiarity with the design and analysis techniques appropriate to the problem, for research and/or for clinical intervention. These purposes tend not to be mutually exclusive in medical settings.

As a final note, the clinician involved in medical consultation and liaison may be faced with numerous abbreviations or acronyms in reading patient charts or medical literature. A list of commonly used medical abbreviations is provided in Table 3 to aid in the consulting psychologist's understanding and integration of medical and psychological information.

I look forward to the continued and exciting growth of the role of psychology in providing optimal care to medical patients.

Table 3. Medical Abbreviations[a]

This is a list of commonly used medical abbreviations. Although abbreviations tend to be standard across settings, when comparing approved lists from different hospitals we have sometimes noted different usages. It is imperative that the practitioner obtain the accepted abbreviations before writing in any hospital's medical records. It is also important to known the conditions under which these abbreviations are used (e.g., usually not in discharge summaries).

A—assessment	bs—bowel sounds
a—before	BSO—bilateral salpingo-oophorectomy
AB—abortion	BUN—blood urea nitrogen
AD—right ear	Bx—biopsy
ADL—activities of daily living	c—with
ad lib—at pleasure	CA—carcinoma
AF—atrial fibrillation	Ca—calcium
AK—above knee	CAT—computerized axial tomogram
AL—left ear	CBC—complete blood count
AMA—against medical advice	CC—chief complaint
ANA—antinuclear factor	CCU—coronary care unit
ANS—autonomic nervous system	CHF—congestive heart failure
A&P—auscultation and percussion	CNS—central nervous system
ASA—aspirin	C/O—complaints of
ASCVD—arteriosclerotic heart disease	COPD—chronic obstructive pulmonary disease
AU—both ears	CP—cerebral palsy
A&W—alive and well	CPR—cardiopulmonary resuscitation
BAE, BE—barium enema	CrN—cranial nerve
B/C—birth control	CS, C/S—cesarean section
BCP—birth control pills	CSF—cerebrospinal fluid
bid—twice a day	CVA—cerebrovascular accident
BF—black female	CVD—cardiovascular disease
BK—below knee	Cx—cervix
BM—bowel movement	CXR—chest X ray
BMR—basal metabolic rate	d—diastolic
BO—bowel obstruction	D—dorsal spine
BOM—bilateral otitis media	D&C—dilation and curettage
BP—blood pressure	D/C'D—discontinued
BPH—benign prostatic hypertrophy	ID—intradermal
BS—breath sounds	I&D—incision and drainage
DIFF—differential blood count	IH—infectious hepatitis
DM—diabetes mellitus	IM—intramuscular
DOA—dead on arrival	IMP—impression

Continued

Table 3. (*Continued*)

DOB—date of birth
DOE—dyspnea on exertion
DTRs—deep tendon reflexes
Dx—diagnosis
EA—emergency area
ECG—electrocardiogram
EEG—electroencephalogram
EENT—eyes, ears, nose, throat
EMG—electromyogram
ENT—ears, nose, throat
EOM—extraocular movements
ESR—erythrocyte sedimentation rate
EUA—examination under anesthesia
FB—foreign body
FBS—fasting blood sugar
FH—family history
F/U—follow-up
FUO—fever of unknown origin
FVC—forced vital capacity
Fx—fracture
G, Gr—gravida
GB—gallbladder
GC—gonococcus
GE—gastroenterology
GG—gamma globulin
GI—gastrointestinal
gr—grain
GSW—gunshot wound
gt—drop
gtt—drops
GU—genitourinary
HA—headache
HBP—high blood pressure
HEENT—head, ears, eyes, nose, throat
H&L—heart and lungs
NHP—herniated nucleus pulposus
H&P—history and physical
HPI—history of present illness
htn—hypertension
hs—at bedtime
Hx—history
ICU—intensive care unit
NSSP—normal size, shape, position
N&V—nausea and vomiting
O—objective
OB—obstetrics
OBS—organic brain syndrome
Od—overdose
od—right eye
OM—otitis media
OPC—outpatient clinic
OS—left eye
ou—both eyes
p—after
p—pulse
p—plan

imp—improved
In situ—in normal position
IOP—intermittent positive-pressure breathing
IUP—intrauterine pregnancy
IV—intravenous
IVP—intravenous pyelogram
JRA—juvenile rheumatoid arthritis
KJ—knee jerk
KUB—kidney, ureter, bladder
L&A—light and accommodation
LAB—laboratory results
LAP—laparotomy
LLE—left lower extremity
LLL—left lower lobe
LLQ—left lower quadrant
LMD—local medical doctor
LMP—last menstrual period
LOC—level of consciousness
LP—lumbar puncture
LS—lumbosacral
LSK—liver, spleen, kidney
MH—marital history
MI—myocardial infarction
MM—malignant melanoma
MMR—measles, mumps, and rubella immunization
MOD—medical officer of the day
NA—not applicable
NAA—no apparent abnormalities
NB—newborn
NC—no change .
N/C—no complaints
NK—not known
NL—normal
NPO—nothing by mouth
NR—nonreactive
NSD—no significant difference
NSR—normal sinus rhythm
RTC—return to clinic
RTW—return to work
Rx—prescription, treatment
s—without
S—subjective
SB—stillbirth
SCC—squamous cell carcinoma
S&O—salpingo-oophorectomy
SOB—shortness of breath
SPP—suprapubic prostatectomy
SP—spinal
S/P—status post
SUBQ—subcutaneous
SRG—surgery
SX—symptoms
Sx—signs
T—temperature
T&A—tonsillectomy and adenoidectomy
TAB—therapeutic abortion

Continued

Table 3. (*Continued*)

PARA—number of pregnancies	TAH—total abdominal hysterectomy
PE—physical examination	TBLC—term birth, living child
PERLA—pupils equal, react to light and accommodation	TC—throat culture
PH—past history	TIA—transient ischemic attack
PI—present illness	tid—three times a day
PID—pelvic inflammatory disease	TL—tubal ligation
PM—postmortem	TPR—temperature, pulse, respiration
PMH—past medical history	TURP—transurethral resection of prostate
PMT—premenstrual tension	TVH—total vaginal hysterectomy
PNS—peripheral nervous system	UA—urinalysis
po—by mouth	U&C—usual and customary
prn—as needed	UCHD—usual childhood diseases
PS—prescription	UK—unknown
PTA—prior to admission	URI—upper respiratory infection
PVC—premature ventricular contraction	UTI—urinary tract infection
Px—physical examination	V—vein
Q—every	VDRL—venereal disease research laboratory (syphilis)
qd—every day	VS—vital signs
qh—every hour	VSS—vital signs stable
qid—four times a day	WM—white male
qm—every morning	WNL—within normal limits
qn—every night	XM—crossmatch
QNS—quantity not sufficient	YO—year old
qod—every other day	?—question of
qs—enough	
R—respiration	
RBC—red blood count	
REM—rapid eye movement	
R/O—rule out	
ROM—range of motion	
ROS—review of systems	
RR—recovery room	
W—widowed	
WBC—white blood cells	
WDWN—well developed, well nourished	
WF—white female	

[a]From C. Belar, W. Deardorff, and K. Kelly (1987). *The practice of clinical health psychology* (pp. 151–154). New York: Pergamon Press. Reprinted by permission.

References

Belar, C., Deardorff, W., & Kelly, K. (1987). *The practice of clinical health psychology.* New York: Pergamon Press.

Bender, L. A. (1938). A visual motor Gestalt test and its clinical use. *American Orthopsychiatry Association, Research Monograph*, No. 3.

Benton, A. L. (1974). *The Revised Visual Retention Test* (4th Ed.). New York: Psychological Corporation.

Benton, A. L., & Van Allen, M. W. (1973). *Test of Facial Recognition, manual.* Neurosensory Center Publication No. 287. Iowa City: University of Iowa.

Bieliauskas, L. A., & Garron, D. C. (1982). Psychological depression and cancer. *General Hospital Psychiatry, 4,* 187–195.

Bieliauskas, L. A., & Glantz, R. H. (1987). Use of the Mini-Mult D Scale in patients with Parkinson disease. *Journal of Consulting and Clinical Psychology, 55,* 437–438.

Bieliauskas, L. A., & Glantz, R. H. (1989). Depression type in Parkinson Disease. *Journal of Clinical and Experimental Neuropsychology, 5,* 597–604.

Bieliauskas, L. A., & Shekelle, R. B. (1983). Stable behavior associated with hi-point D MMPI profiles in a non-psychiatric population. *Journal of Clinical Psychology, 39,* 422–426.

Bieliauskas, L. A., Klawans, H. L., & Glantz, R. H. (1987). Depression and cognitive changes in Parkinson Disease: A review. In M. D. Yahr & K. J. Bergmann (Eds.), *Advances in neurology: Parkinson disease.* New York: Raven Press, *45,* 437–438.

Counte, M. A., Bieliauskas, L. A., & Pavlou, M. (1983). Stress and personal attitudes in chronic illness. *Archives of Physical Medicine and Rehabilitation, 64*, 272–275.

Diller, L., Ben-Yishay, U., Gerstmann, L. J., Goodkin, R., Gordon, W., & Weinberg, J. (1974). *Studies in cognition and rehabilitation*. Rehabilitation Monograph #50. New York: Institute of Rehabilitation Medicine, New York University Medical Center.

Holmes, T. H., & Rahe, R. H. (1967). The Social Readjustment Rating Scale. *Journal of Psychosomatic Research, 11*, 213–218.

Leavitt, F. (1985). Pain and deception: Use of verbal pain measurement as a diagnostic aid in differentiating between clinical and simulated low-back pain. *Journal of Psychosomatic Research, 29*, 494–505.

Leavitt, F. (1987). Detection of simulation among persons instructed to exaggerate symptoms of low back pain. *Journal of Occupational Medicine, 29*, 229–233.

Leavitt, F., Garron, D. C., Whisler, W. W., & Sheinkop, M. B. (1978). Affective and sensory dimensions of back pain. *Pain, 4*, 273–281.

Leavitt, F., Garron, D. C., D'Angelo, C. M., & McNeill, T. W. (1979). Low back pain in patients with and without demonstrable organic disease. *Pain, 6*, 191–200.

Leavitt, F., Garron, D. C, & Bielauskas, L. A. (1980). Psychological disturbance and life event differences among patients with low back pain. *Journal of Consulting and Clinical Psychology, 48*, 115–116.

Melzack, R., & Torgerson, W. S. (1971). On the language of pain. *Anesthesiology, 34*, 50–59.

Milner, B. (1968). Visual recognition and recall after right temporal lobe excision in man. *Neuropsychologia, 6*, 191–209.

Pancheri, P., Bellaterra, M., Matteoli, S., Cristofari, M., Polizzi, C., & Puletti, M. (1978). Infarct as a stress agent: Life history and personality characteristics in improved versus not-improved patients after severe heart attach. *Journal of Human Stress, 4*, 16–22.

Passafiume, D., Boller, F., & Keefe, N. C. (1986). Neuropsychological impairment in patients with Parkinson's Disease. In I. Grant & K. M. Adams (Eds.), *Neuropsychological assessment of neuropsychiatric disorders* (pp. 374–383). New York: Oxford University Press.

Persky, V. W., Kempthorne-Rowe, J., & Shekelle, R. B. (1987). Personality and risk of cancer: 20-year follow-up of the Western Electric study. *Psychosomatic Medicine, 49*, 435–449.

Rao, S. M., & Bieliauskas, L. A. (1983). Cognitive rehabilitation 2.5 years post right temporal lobectomy. *Journal of Clinical Neuropsychology, 4*, 313–320.

Raven, J. C. (1960). *Guide to the standard progressive matrices*. London: H. K. Lewis.

Shekelle, R. B., Raynor, W. J., Ostfeld, A. M., Garron, D. C., Bieliauskas, L. A., Liu, S. C., Maliza, C., & Paul, O. (1981). Psychological depression and 17-year risk of death from cancer. *Psychosomatic Medicine, 43*, 117–125.

Spielberger, C. D., Gorsuch, R. L., & Lushene, R. D. (1970). *STAI Manual, for the State-Trait Anxiety Inventory (Self-Evaluation Questionnaire)*. Palo Alto, CA: Consulting Psychologists Press.

Wechsler, D. (1945). A standardized memory scale for clinical use. *Journal of Psychology, 19*, 87–95.

Wechsler, D. (1955). *Wechsler Adult Intelligence Scale, manual*. New York: Psychological Corporation.

Weinman, J. (1981). *An outline of psychology as applied to medicine*. Briston, England: John Wright.

CHAPTER 13

Emerging Issues in Women's Health

Mary P. Koss and W. Joy Woodruff

Introduction

Women represent 51.3% of the total U.S. population: 123.8 million people (Bureau of the Census, 1987). They are the predominant consumers of health care in the United States, but they are a minority of health care providers (Bartuska, 1988; Davis, 1988; Weisman & Teitelbaum, 1989). This chapter explores the anatomy of women's health and health care. By considering data on life expectancy, mortality, morbidity, and lifestyle among different groups of American women, we gain information on the nature of women's health care needs. By examining the statistics on poverty and violence against women, we identify social conditions that engender poor health and create obstacles to effective health care. By documenting the status of women doctoral-level providers in medical schools, we raise concerns about the ability of these institutions to provide gender-sensitive health care and education.

Several limitations governed the preparation of this chapter. First, it addresses women's physical health and excludes issues directly related to mental health, which is an artificial distinction in practice. Readers interested in women's mental health should consult the report of the American Psychological Association's Task Force on Depression (1989), the Women's Mental Health Agenda (National Institute of Mental Health, 1987; Russo, 1985) and recent reviews (Klerman & Weissman, 1989; Nolen-Hoeksema, 1987). Second, the material is focused on adult women because issues related to adolescent and geriatric populations are considered elsewhere in this handbook. Although we have tried to highlight the most important health differences between minority and nonminority women and among the ethnic groups, a comprehensive review of ethnicity and health was beyond the scope of the chapter. In choosing material to include, we have used the definition of a women's health issue developed by the Public Health Service Task Force on Women's Health (Department of Health and Human Services [DHHS], 1985). Their criteria for defining health problems, conditions, or diseases as "women's issues" include the following:

1. diseases or conditions *unique* to women or some subgroups of women;
2. diseases or conditions *more prevalent* in women or some subgroups of women;
3. diseases or conditions *more serious* among women or some subgroups of women;

Mary P. Koss and W. Joy Woodruff • Department of Psychiatry, University of Arizona, Tucson, Arizona 85724.

4. diseases or conditions for which the *risk factors* are different for women or some subgroups of women; and

5. diseases or conditions for which the *interventions* are different for women or some subgroups of women. (p. 3)

An Epidemiologic Profile of Women's Health

Mortality

Women live longer than men (Cassell & Neugarten, 1988; DHHS, 1985; Strickland, 1988; Verbrugge, 1985). White female infants born in 1986 can be expected to outlive their same-aged male cohorts by seven years (Department of Commerce [DOC], 1989). The ratio of men to women is 68:100 for persons over 65 years (Cassell & Neugarten, 1988). At every point of development, girls and women live longer and maintain a biological advantage over males (Strickland, 1988). But life expectancy is shorter for minority women than for white women (DHHS, 1985). A black female infant born in 1986 can be expected to die five years earlier than a white infant born the same year (DOC, 1989). The divergence in life expectancies has been viewed as a critical social change that affects the health of women (DHHS, 1985).

Major Causes of Mortality

The top 10 leading causes of death for all women and for the general population are listed in Table 1. These rankings are based on age-adjusted mortality rates, which are the numbers of deaths from each condition per each 100,000 persons aged 17 to 65 (CDC, 1989g; Verbrugge, 1985). It can be seen from this table that the leading causes of women's deaths are almost identical to those for the population as a whole, although the rank ordering varies slightly. The major differences are childbirth-related fatalities, which are obviously limited to women, and AIDS, which has not yet entered the top 10 causes of death among women. The major causes of death remain the same in non-age-adjusted data, but the ordering changes somewhat. Among women, diabetes and suicide or homicide exchange places with pneumonia and COPD, declining to the seventh and eighth causes of death, respectively (DOC, 1989).

The number of deaths in 1986 for each cause of mortality among men and women is illustrated in Figure 1 (data from DOC, 1989, Table 117, p. 92). Examination of this figure immediately reveals that the deaths among women are lower for all conditions except cerebrovascular diseases, pneumonia or influenza, and

Table 1. Leading Causes of Death Based on Age-Adjusted Data for Women and for the General Population[a]

Women	Total population
1. Coronary disease	1. Coronary disease
2. Cancer	2. Cancer
3. Cerebrovascular diseases	3. Cerebrovascular diseases
4. Accidents/adverse effects	4. Accidents/adverse effects
5. Suicide/homicide	5. Chronic obstructive pulmonary disease (COPD)
6. Diabetes mellitus	6. Pneumonia/influenza
7. Pneumonia/influenza	7. Suicide/homicide
8. Chronic obstructive pulmonary disease	8. Diabetes mellitus
9. Reproductive events	9. Chronic liver disease/cirrhosis
10. Chronic liver disease/cirrhosis	10. Acquired immune deficiency syndrome (AIDS)

[a]Sources: CDC (1989g); Verbrugge (1985).

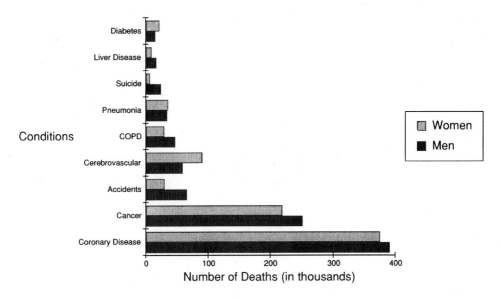

Figure 1. Leading causes of mortality by sex.

diabetes mellitus. In fact, women's death rates for COPD, accidents, suicides, homicides, and chronic liver disease are about half of men's.

Approximately one third of all deaths in women are from heart disease, but the mortality rate for this condition in women has declined at least 20% over the past 30 years (CDC, 1989b; DHHS, 1988; Leaf, 1988; Strickland, 1988; Wenger, 1985). Nevertheless, it is the leading cause of death for all women over 40 and the leading cause of death for black women under 40. Symptom onset occurs 10–15 years later among women than among men. Both the risk and the incidence of coronary disease rise among women after menopause (Collins, 1988; DHHS, 1988).

Almost half of the women's cancer deaths are accounted for by two types of malignant neoplasms: lung and breast (DOC, 1989). Lung cancer has now edged out breast cancer as the leading cause of cancer death in women (DHHS, 1988). It is dubbed an "equal opportunity tragedy," because women's smoking patterns, and hence their cancer rates, are now virtually identical to men's. Mortality from lung cancer among women has increased 600% over the past 30 years. In contrast, the number of breast cancer

deaths among women has remained relatively constant for the past 15 years (DHHS, 1988). About 10% of the U.S. female population is stricken with breast cancer during their lifetime (Carter, Jones, Schatzkin, & Brinton, 1989), with approximately 100,000 cases diagnosed each year (Grady, 1988). Another smoking-related cause of death is chronic obstructive pulmonary disease (COPD), which ranks as the eighth cause of death for women. Like lung cancer, the prevalence of COPD is increasing dramatically in women. Mortality rates for COPD in women rose one third between 1979 and 1986 (CDC, 1989b).

Several conditions kill more women than men. They include cerebrovascular diseases (the third leading killers of U.S. women), pneumonia or influenza (the seventh cause of death), and diabetes mellitus (the sixth cause of death). Over the past decade, fatalities from cerebrovascular diseases, especially stroke, have diminished by one-half among female patients, primarily because of decreased mortality among black women (Bankhead, 1989; CDC, 1989b; Garraway & Whisnant, 1987; Klag, Whelton, & Seidler, 1989). Pneumonia deaths occur most frequently among immunocompromised patients and the elderly; approximately 80%–90% of the victims are over 65 (Lui & Kendal, 1987). Diabetes is a condition that significantly affects life expectancy (Hartman-Stein & Reuter, 1988). In addition to the deaths directly attributed to diabetes, 100,000 lives are claimed annually by complications of diabetes (DHHS, 1985). Although death rates from diabetes have decreased slightly over the past decade, the lethality of the disease is twice as high for women as it is for men (CDC, 1989b; DHHS, 1985; Leaf, 1988; Nelson, Everhart, Knowler, & Bennett, 1988).

In contrast, women's mortality rates are approximately half of men's for three additional causes of mortality: accidents and adverse effects, suicide and homicide, and chronic liver disease and cirrhosis. Accidents include unintentional injuries such as motor vehicle accidents, falls, drownings, fires and burns, poisonings, chokings, firearm mishaps, and the adverse effects of medical and surgical procedures (American Medical Association Council on Scientific Affairs [AMA], 1989; CDC, 1989g). Auto accidents are the most frequent cause of accidental death. Although accidents are the fourth leading cause of death among women overall, they are the major cause of death for persons under age 45 (CDC, 1989g; Lichtenstein, Bolton, & Wade, 1989). Suicide, when combined with homicide, ranks fifth as a cause of death in women. Women make more suicide attempts than men, but the number of completed suicides is almost four times higher in men (CDC, 1986a). During the past 15 years, suicide rates among all young adults aged 20–24 have increased by one third (CDC, 1988d). However, the developmental apex for female suicides has remained at midlife (ages 35–54) (CDC, 1986a; Smith, Mercy, & Conn, 1988). Homicide deaths among women also are half of men's, although in some areas of the country firearms are the leading cause of death of young black females (aged 18–24) (AMA, 1989). Spouse homicides have constituted, on the average, 9% of all homicides since the mid-1970s (Mercy & Saltzman, 1989). In roughly 60% of case fatalities, the victim is the wife (Mercy & Saltzman, 1989). Since the mid-1970s, deaths from spouse homicide have decreased 17% for women; this

trend is attributable to a 46% reduction in the deaths of black wives (Mercy & Saltzman, 1989). Finally, mortality rates for chronic liver disease and cirrhosis among women are half of men's. However, the disease progression is faster and life expectancy is shorter among women with alcoholic liver disease than among men; approximately 90% of cirrhosis patients aged 40–59 are women (Sherlock, 1988). Since the late 1970s, case fatalities attributable to cirrhosis have declined (CDC, 1989b).

Conditions originating in the perinatal period occur frequently enough to constitute the ninth leading cause in women overall; 18,400 such deaths occurred in 1986 (DOC, 1989). The leading cause of maternal death for all women during the first trimester is ectopic pregnancy (CDC, 1988c; DHHS, 1985). Other leading causes of maternal death are embolism, hypertension, hemorrhage, cerebrovascular accidents and injuries, and anesthesia complications (CDC, 1988c; Koonin, Ellerbrock, Atrash, Rogers, Smith, Hogue, Harris, Chavkin, Parker, & Halpin, 1989).

Ethnic Trends in Mortality

In any discussion of women's health, it is important to recognize that they are a heterogeneous group. An epidemiological profile of "women" actually conceals a number of differences in mortality among the major ethnic groups in the United States. The outstanding instances are summarized in Table 2. It can be seen from this table that minority group status is associated with higher death rates for many conditions, including coronary disease, AIDS, endometrial and cervical cancer, and cirrhosis. The only conditions with lower death rates among minority groups are COPD and lung cancer, which claim fewer lives because minority women do not smoke as much as white women.

Gender Differences in Health-Related Behaviors

Examination of the leading causes of death reveals that many of them are related to individual choices in behavior or lifestyle, such as

Table 2. Ethnic Differences in Women's Mortality and Morbidity[a]

Disease	Risk compared to white women
1. AIDS	52% of women AIDS patients are black, 20% are Hispanic; deaths 13.6 times higher in black women, 10.2 times higher in Hispanic women
2. Cervical/endometrial cancer	Deaths 2 times higher in minority women
3. Childbirth/maternal mortality	Deaths 4 times higher in blacks; 3 times higher in Hispanics
4. Cirrhosis	Deaths almost 2 times higher in nonwhites
5. COPD	Less prevalent in nonwhites because of lower smoking
6. Diabetes mellitus	Prevalence is 25.4% of black women aged 55–66; 14.6% among whites; death rates are elevated among Native Americans and Hispanics
7. Digestive tract cancer	More prevalent in Asian women
8. Gallbladder cancer	More prevalent in Native American and Mexican-American women
9. Gallstones	Prevalence twice as high in Latinas
10. Homicide	Elevated rates for both black and Native American women
11. Hypertension	Prevalence is 2 times higher in black women
12. Lung cancer	Less prevalent in nonwhites because of lower smoking rates
13. Pneumonia	Excess mortality in Native Americans
14. Rheumatic fever	Excess mortality in Native Americans
15. Suicide	Incidence has risen 11.9% in whites and decreased 23.9% among nonwhites; nonwhite incidence is half the white rate

[a]Sources: Amaro (1988); CDC (1986a, 1989d,e,f); Cochran and Mays (1989); Cornoni-Huntley, La Croix, and Havlik (1989); Davis (1988); Diehl and Stern (1989); DHHS (1985, 1988); Doebbert, Riedmiller, and Kizer (1988); Haffner, Diehl, Stern, and Hazuda (1989); Mahoney, Michalek, Cummings, Nasca, and Emrich (1989); Makuc, Freid, and Kleinman (1989); Nelson, Becker, Wiggins, Key, and Samet (1989).

smoking, alcohol use, high-fat diets, and exposure to occupational stress (DHHS, 1988; Matarazzo, 1984; Monsen, 1989; Strickland, 1988). We now examine health-related behavior among women.

Smoking. Although more men than women smoke (34% versus 28%), the rate of decline has been more rapid for men than for women (Berman, 1989; Fiore, Novotny, Pierce, Hatrziandreu, Patel, & Davis, 1989). The group with the highest rate of smoking (40%) is young adult white women, whereas the lowest rate is among Hispanic women (DHHS, 1988). A prominent contributor to the volume of smoking appears to be the degree of acculturation among ethnic minorities; more assimilation leads to more consumption for both men and women (Marin, Perez-Stable, & Marin, 1989). Smoking cigarettes as a dieting strategy is still very much in vogue among women (Klesges & Klesges, 1988; Ockene, Sorensen, Kabat-Zinn, Ockene, & Donnelly, 1988; Rigotti, 1989). Some people also smoke cigarettes to buffer the CNS depressant effects of alcohol (Michel & Battig, 1989). Women who smoke have two to four times the risk of

heart attack of nonsmoking women (DHHS, 1988). Further, the risk factors for coronary disease (lipid disorders, hypertension, diabetes, and smoking) tend to cluster in the same patients (Sebastian, McKinney, & Young, 1989). Even passive smoke exposure negatively affects health by placing women at increased risk of both coronary disease and cervical cancer (Byrd, Shapiro, & Schiedermayer, 1989; Sandler, Comstock, Helsing, & Shore, 1989; Slattery, Robison, Schuman, French, Abbott, Overall, & Gardner, 1989; Tell, Howard, McKinney, & Toole, 1989).

Eating. Obesity in women is generally defined as body fat content greater than 30% (Ganley, 1989; Gray, 1989). Approximately 12% of men and 14% of women are obese (DOC, 1989). The prevalence of obesity is increasing in the United States (Kissebah, Freedman, & Peiris, 1989). Obesity is a health hazard because it is linked to an increased likelihood of hypertension, atherosclerosis, and diabetes, which are major risk factors for coronary disease (DHHS, 1988). In addition, the risk of both breast and colon cancer are increased with food intakes low in fiber and high in fat (DHHS, 1988; Monsen, 1989).

Also very evident in American culture are women who are preoccupied with weight control (Brehm, Kassin, & Gibbons, 1981; Schwartz, Thompson, & Johnson, 1982; van Strien & Bergers, 1988). Exercise and dieting are the common means of weight reduction (Hill, 1987; Polivy & Herman, 1987; Segal & Pi-Sunyer, 1989). But calorie-restricted diets that don't account for a woman's nutritional needs, coupled with excessive exercise, pose serious health risks to women (Dazzi & Dwyer, 1984; DHHS, 1988; Higgins, 1988; Mitchell, Pomeroy, & Huber, 1988; Owen, Kavle, Owen, Polansky, Caprio, Mozzoli, Kendrick, Bushman, & Boden, 1986; Philipp, Pirke, Siedl, Tuschi, Fichter, Eckert, & Wolfram, 1988; Schocken, Holloway, & Powers, 1989). The mortality rate among those women who adhere to extreme calorie restriction (anorexia nervosa) is 2%–8% (Herzog, Keller, & Lavori, 1988).

Alcohol and Drug Abuse. Alcohol is the Number 1 drug abused in the United States (Cook, Garvey, & Shukla, 1987; Moskowitz, 1989). The proportion of people who report that they had five or more drinks in one day during the previous year is 49% among men compared with 23% among women (DOC, 1989). This difference explains the overall lower rates of alcoholic liver disease in women. However, women appear to have an increased risk of multiple addictions, in which alcohol is combined with prescription drugs (DHHS, 1988; Matteo, 1988; Portans, White, & Staiger, 1989; Unger, 1988). Among those women who drink during pregnancy, fetal alcohol syndrome is a significant concern (CDC, 1988b; Keith, MacGregor, Friedell, Rosner, Chasnoff, & Sciarra, 1989; Zuckerman, Frank, Hingson, Amaro, Levenson, Kayne, Parker, Vinci, Aboagye, Fried, Cabral, Timperi, & Bauchner, 1989).

Occupational Stress. Women, including mothers with small children, have entered the labor force in unprecented numbers in recent years (Ford Foundation, 1989). In 1987, 72% of women between the ages of 25 and 54 were gainfully employed (Courtless & Lowe, 1989). Often, the

working woman performs multiple roles in life, including wife, mother, worker, maintainer of extended-family ties, and caretaker of elderly relatives (DHHS, 1988; Doty, 1986; Jacobs & McDermott, 1989; NIMH, 1986; Reisine & Fifield, 1988). This trend has raised questions of whether women's rates of heart disease would begin to rise as a response to their exposure to occupational stress and multiple-role tensions (Verbrugge & Madans, 1985). However, data indicate that women with multiple roles have fewer illness risks; working women have a slight health advantage over housewives (Kotler & Wingard, 1989).

Contrary to the notion that executive jobs produce high stress, heart disease is more common among women in clerical or low-status jobs. Clerical workers who reported having an unsupportive boss were at increased risk of developing coronary heart disease in an eight-year prospective study (Eaker, 1989). Finally, meta-analyses of studies of job involvement among women have indicated that it is unrelated to the Type A personality (a constellation of personality traits that has been linked to heart disease in men) (Booth-Kewley & Friedman, 1987).

Morbidity

Unfortunately, increased life expectancy has resulted in women's having more time than men to get sick because the onset age of physical disability is approximately the same for both sexes (Cassel & Neugarten, 1988).

Chronic Conditions

Chronic conditions include heart disease, arthritis, diabetes, and sensory impairment. These conditions can lead to functional limitations in activities and can create a need for long-term health care. The prevalence of chronic conditions in men and women is summarized in Figure 2 (data from DOC, 1989, Table 183, p. 114). Whereas it was shown earlier that mortality is generally higher among men, chronic conditions are more prevalent in women. Among the debilitating chronic conditions, only hearing and visual impairment (excluding cata-

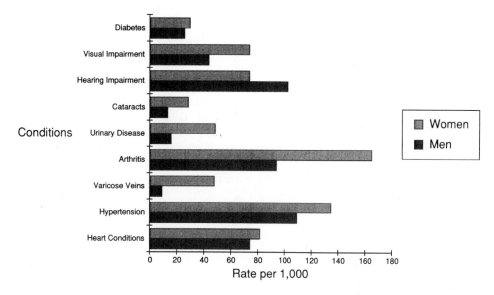

Figure 2. Prevalence of chronic conditions by sex.

racts) are found more frequently among men. Not shown in the figure are several less prevalent conditions that are far more likely to affect women than men. These include rheumatoid arthritis (3 times more common in women), systemic lupus (10 times more common in women), autoimmune liver disease (8 times more common in women), and chronic fatigue syndrome (Anderson, Bradley, Young, McDaniel, & Wise, 1985; Kaufman, Gomez-Reino, Heinicke, & Gorevic, 1989; Koo, 1989; Lerman, 1987; Schousboe, Koch, & Chang, 1988; Sherlock, 1988).

Acute Conditions

In addition to chronic conditions, people also experience acute illness, including infectious diseases, upper respiratory disease, digestive disease, and injuries. The rates at which acute conditions are reported are summarized in Figure 3 (data from DOC, 1989, Table 182, p. 114). This figure illustrates that all of the acute conditions except injuries, occur at higher rates among women. The possibility exists that women experience more respiratory and digestive conditions because of their close contact with children and their roles as health care workers and caretakers (DHHS, 1985, 1988).

Men have higher injury rates partly because of occupational and recreational exposure to dangerous situations. However, the injury rate for men and women is approximately equal after age 45 (DOC, 1989).

Conditions Unique to Women

Not included in the acute and chronic conditions reviewed above were the gynecological conditions, which are unique to women. In this section, we briefly review the status in several areas of gynecological health.

From 10% to 20% of women describe a debilitating level of symptoms premenstrually (DHHS, 1988; Wickes, 1988; York, Freeman, Lowery, & Strauss, 1989). The prevalence of endometriosis in the general population of premenopausal women is from 10% to 15% (Metzger & Haney, 1989). However, in gynecological surgery patients, 26.1% of women manifested signs of endometriosis (Guzick, 1989). Fatality rates for menstrual-related toxic shock syndrome are approximately 3% of reported cases (Markowitz, Hightower, Broome, & Reingold, 1987).

Likewise, between 10% and 15% of women suffer disabling symptoms at menopause; how-

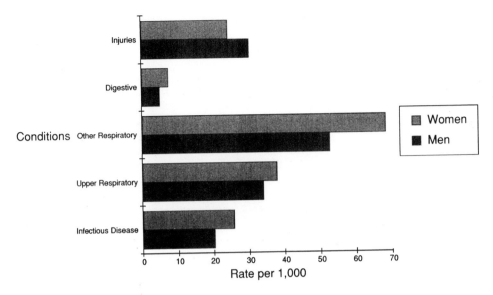

Figure 3. Prevalence of acute conditions by sex.

ever, it is unknown whether this group substantially overlaps with the group who experienced symptoms premenstrually (Collins, 1988). Though the onset of the menses in puberty has occurred earlier and earlier over the decades, the average age of menopause—51.4 years—has remained relatively stable over time (Morokoff, 1988). Most women spend one third of their lives in a postmenopausal state (Morokoff, 1988; Walsh & Schiff, 1989). Decreased estrogens affect the regularity of the menstrual cycle and the retention of calcium in bone tissue (DHHS, 1988). In addition to natural or surgical menopause, excessive exercise, smoking, and eating disorders (such as anorexia nervosa and bulimia nervosa) are common causes of lowered levels of estrogen production in women (DHHS, 1988; Hall, Hoffman, Beresford, Wooley, Hall, & Kubasek, 1989). Osteoporosis is expected to affect almost half of postmenopausal women (Cassel & Neugarten, 1988; Collins, 1988; DHHS, 1988; Lobo & Whitehead, 1989; Resnick & Greenspan, 1989; Walsch & Schiff, 1989). Long-term maintenance of physiological levels of estrogen remains the only proven effective way of reducing bone loss and the risk of fracture due to osteoporosis associated with menopause (Ettinger, 1988). However, the risk of endometrial cancer is two to four times greater in postmenopausal women receiving estrogen replacement therapy (Walsh & Schiff, 1989). This elevated risk may be reduced when estrogen is used with progestin, although the treatment remains controversial (DHHS, 1988).

Incidence rates for sexually transmitted diseases (excluding AIDS) in heterosexual women are skyrocketing. These diseases include syphilis, drug-resistant strains of gonorrhea, chancroid bacterial disease, chlamydia, genital warts, and genital herpes (Egerter, 1988; Kemeny, Cohen, Zegans, & Conant, 1989; Vanderplate, Aral, & Magder, 1988). Thus, it is not surprising that vaginitis is the presenting complaint for 10% of all the physician visits made by women in the United States (McCue, 1989). Women constitute 8% of the total AIDS population (Amaro, 1988; CDC, 1989f; Cochran & Mays, 1989). Heterosexual transmission of AIDS to the victim by an infected intravenous drug user or through intravenous drug use by the infected patient accounts for most cases in women (CDC, 1988a, 1989a; Glaser, Strange, & Rosati, 1989; Pollner, 1988).

Medical Utilization

Given the data on chronic and acute conditions, it is not surprising that women reported slightly more days than men during which their activities were restricted (mean of 17 versus 13 per year) or during which they stayed in bed (mean of 8 versus 5 per year) (DOC, 1989).

Physician, Hospital, and Nursing-Home Utilization

Women between the ages of 17 and 44 have twice as many physician visits as men (Verbrugge, 1985). Even after accounting for visits attributed to childbirth services, women's mean number of visits is 30% higher than men's; after age 45, women's outpatient physician visits still exceed men's by 10%–20% (Verbrugge, 1985). Because women live longer than men, their greater physician utilization could be interpreted as evidence of adaptive, self-protective health behavior.

Hospital utilization rates do not show the patterns seen in physician visits. Overall, women received more days of hospital care in 1986 than men did (849 days per 1,000 men compared with 1,003 days per 1,000 women) (DOC, 1989). However, the highest hospitalization rates for women occurred between the ages of 15 and 44 (the childbearing years). Both before age 15 and after age 44, men record higher rates of hospital care. In addition, the average length of stay in the hospital is shorter for women than for men at virtually every age.

So, although women have more chronic conditions and visit physicians more frequently, there is no suggestion that their rate of hospitalization is higher than that of men. Still, the absolute amount of hospital care used by women is greater after age 65 because there are 100 women alive at this point compared to 68 men (Cassel & Neugarten, 1988). Likewise, women constitute 75% of the persons who have resided in a nursing home longer than one month (DOC, 1989). On the basis of these data on medical utilization, women can be considered the predominant consumers of health care (Davis, 1988). We now turn to the treatments women receive from the medical care system.

Gender Differences in Treatments Rendered and Received

Physicians distinguish the characteristics of "difficult patients" along gender lines (Schwenk, Marquez, Lefever, & Cohen, 1989). Physicians' perceptions of difficult patients involve two factors: medical uncertainty (undifferentiated medical problems with vague, difficult-to-describe symptoms) and interpersonal difficulty (abrasive patient style) (Schwenk *et al.*, 1989). Difficult patient–physician relationships result in poor health care delivery as well as in patient dissatisfaction (Schwenk *et al.*, 1989). Difficult male patients are characterized as seeking disability or undeserved compensation, or as pursuing quack remedies. Difficult female patients are described as having a religious or personal belief system that interferes with treatment, as being too emotional, and as presenting a frequently changing symptom picture.

In some important respects, women receive different care from physicians than do men. Past data suggest that women's medical complaints have more often been attributed to psychogenic etiology and that male physicians have typically ordered less rigorous workups in response to common complaints than for men (Greer, Dickerson, Scheiderman, Atkins, & Bass, 1986). However, current data suggest that these practices are disappearing among younger physicians. Women at all ages obtain more prescription medications per year than do men (Verbrugge, 1985), especially between the ages of 19 and 34. Two thirds of the prescriptions for psychotropic drugs such as Valium and Xanax are for women. Some diagnostic and surgical interventions are performed exclusively for women: 30% of women over the age of 40 report having had a hysterectomy, for example (CDC, 1986b; Makuc, Freid, & Kleinman, 1989). In fact, of the 25 most common operative procedures, 11 are done only on women (Friedman, 1988).

Certain diagnostic and surgical interventions may be less effective with women than with men (Hertzer, Young, Beven, O'Hara, Graor, Ruschhaupt, & Maljovec, 1986; Pratt, Francis, Divine, & Young, 1989). For example, some angiography studies have shown that fewer than half of women presenting with typical anginal pain have significant occlusion of their coronary arteries, a finding that suggests that this test may be inappropriately prescribed for women (Eaker, 1989). Likewise, the proportion of negative abdominal explorations for females under age 40 years is consistently about twice the proportion for males (Luckmann, 1989).

Sometimes, these differences in response to medical treatments can be traced to initial empirical work that excluded women (Leaf, 1988). Causes of mortality that are more common in men tend to be rarely studied in women. Thus, most of the recommendations given to women to reduce the risk of coronary artery disease have been based on studies that involved only male subjects (Denmark, Russo, Frieze, & Sechzer, 1988). Likewise, techniques for measuring psychological hardiness as a barometer of health status and as a buffer against illness were developed without including female subjects (Allred & Smith, 1989; Funk & Houston, 1987; Hull, Van Treuren, & Virnelli, 1987). Women have also been conspicuously absent from or underrepresented in drug trial studies (Hamilton, Lloyd, Alagna, Phillips, & Pinkel, 1984; Kinney, Trautmann, Gold, Vesell, & Zelis, 1981). Even at the present time, there are no specific guidelines for drug testing on women (Miller & Young, 1989). Only since 1987 have principal investigators been required to justify any decision to exclude women from a proposed research study submitted for funding to the National Institutes of Health.

Women's Satisfaction with Medical Care

The information and instructions provided by physicians have been shown to affect patient satisfaction directly and to effect compliance with medical recommendations indirectly (Stoffelmayr, Hoppe, & Weber, 1989). Nevertheless, this is the area in which the most patient dissatisfaction is expressed (Adamson, Tschann, Gullion, & Oppenberg, 1989). Behav-ioral observation of physician–patient contacts has revealed that explanations consume an average of 1 minute during the typical 20-minute consultation (Stoffelmayr et al., 1989).

Older, better-educated female patients generally are more satisfied with their medical care than are younger, less-educated male patients (Adamson et al., 1989). Female medical patients who saw female internists were more satisfied with their care than any other gender mix of physicians and patients (Comstock, Hooper, Goodwin, & Goodwin, 1982). Women patients, if asked, express a preference for women physicians. Evidence has emerged from the National Ambulatory Care Survey data that documents greater sensitivity of women physicians as a group than of male physicians in the delivery of health services. Across all medical specialties, women physicians were found to spend more time with their patients (17 minutes per visit versus 13 minutes), to have a greater proportion of visits that lasted 16 minutes or more (55% versus 38% for internists), and to be more likely to engage in therapeutic listening and counseling (Arnold, Martin, & Parker, 1988).

Two areas of medicine where improvements have been suggested to address the needs of American women are the accessibility of health information and the sensitivity with which medical care is provided (DHHS, 1985). While the technical skills of physicians do not seem to differ by gender, women physicians seem better able to communicate sensitivity and caring to patients, and to listen without interrupting (Arnold et al., 1988). Some writers have concluded that the challenge confronting medicine in the coming decades is to learn from women physicians how to integrate better the art and science of medicine to improve patient satisfaction and care (Arnold et al., 1988).

Social Context of Women's Health

We now proceed to the social factors that may engender poor health and influence women's access to health care: poverty and violence. Although neither of these factors affects only women, they both affect women disproportionately.

The Feminization of Poverty

Women and children are the majority (78%) of the poor in the United States (Braveman, Oliva, Miller, Schaff, & Reiter, 1988; DHHS, 1985). Two trends account for the economic disadvantage of women: the increasing number of female-headed households and pay inequity. During the past 20 years, the number of single-parent families has doubled, and most of them (84%) are headed by a woman. Female-headed families represent 1 in every 5 families with children overall, 1 in 3 black families, and 1 in 4 Hispanic families (Braveman *et al.*, 1988; Courtless & Lowe, 1989).

Pay Inequity

The primary source of income for female-headed families is wages (78% of income); however, in 1987, the year-round, full-time median earnings of female workers were $16,909, compared with $26,008 for male workers (Courtless & Lowe, 1989). The before-tax income of female heads of households was less than one half that of men who were single parents: $16,620 versus $34,700. Further, minority individuals working full time to support a family earn one half the wages of their white counterparts (Courtless & Lowe, 1989). The financial stability of single-parent households has declined over the past 20 years because of the combination of eroding income and the rising costs of such necessities as food, housing, clothing, transportation, child care, and health care. Cases of expenditures exceeding income rose in the 1980s, particularly for those in poor health (Freeman, Blendon, Aiken, Sudman, Mullunix, & Corey, 1987).

There are a number of avenues by which poverty is related to poor health. Low socioeconomic status is usually associated with less healthy physical and social environments, altered self-perceptions and group identity, and altered capacity to adapt psychologically and behaviorally (Rodin & Salovey, 1989). The poor individual may lack the funds or knowledge to obtain preventive care and may face cultural barriers to understanding the importance of screening, diagnostic services, and follow-through (Rodin & Salovey, 1989). Three times as

many low-income adults (17–64 years of age) report unfavorable health status when compared to higher-income persons (Freeman *et al.*, 1987).

Employment-Linked Insurance System

Because of women's greater longevity, pensions and Social Security benefits are important sources of income for longer periods of time than they are for men. Yet women are often at a disadvantage within the benefit system, which is linked to employment and is thus biased against persons who are not in the paid labor force or who have had intermittent employment (DHHS, 1985). The problem of unpaid work illustrates the difficulties women face in gaining access to insurance benefits. The dominant mode of caregiving for the functionally disabled in this country occurs informally through family and friends (DHHS, 1988; Doty, 1986; Jacobs & McDermott, 1989; Reisine & Fifield, 1988). These caretakers are virtually all women, and they may work full time yet live in poverty (Ford Foundation, 1989). As a result, three times as many men receive income from private pensions and annuities as do women, and the pension revenues received by men are twice those of women. Through the year 2055, women's Social Security benefits will be two thirds those of men. Finally, the amount of worker's compensation received, which is based on wage rates, is influenced by entrenched pay inequity between women and men, minorities and nonminorities (DHHS, 1985).

The United States also has embraced a health insurance system that is linked to employment; health insurance is expensive and difficult to obtain unless it is received as a benefit of employment. Furthermore, the quality of health benefits is directly linked to the income level of the insured. Thus, occupational pay inequity translates into less comprehensive insurance benefits in female-headed households. In addition, many women lose their insurance coverage when they divorce.

It is often assumed that a "social safety net" exists to provide medical care for the poor. This is only true for the poorest of the poor. Left uncovered are 11 million people (50% of the povery population) whose incomes are below the federal poverty level but are too high to qualify them for Medicaid (Tallon, 1989; Thorpe, Siegel, & Dailey, 1989; Women's Research and Education Institute [WREI], 1987). Between 31 and 37 million Americans, almost all under 65 years old, have no health coverage at all (Ford Foundation, 1989; Thorpe *et al.*, 1989). Seventeen percent of the female population is uninsured (Braveman *et al.*, 1988). According to one conservative estimate, 13.5 million Americans (6%) reported forgoing medical treatment because of cost; care was denied to another 1 million because of their inability to pay (Freeman *et al.*, 1987; Tallon, 1989; WREI, 1987).

Women's health is directly related to their access to sound information and high-quality medical care (DHHS, 1985). One measure of unmet health care needs compares the number of physician visits made by insured persons with the number made by the uninsured. The gap between the two groups grew during the 1980s (Freeman *et al.*, 1987). One out of six Americans with a chronic disease (diabetes, cancer, heart disease, cerebrovascular disease) does not see a physician at all in a year's time. Of them, 18% are low-income, 25% are black, 22% are Hispanic, and 20% are uninsured (Freeman *et al.*, 1987). Although hospitalization rates have declined in the general population, the poor are more likely than the nonpoor to be hospitalized; this circumstance suggests a lack of access to appropriate preventive medicine.

Violence against Women

We have seen how poverty, which disproportionately impacts on women and children, both creates greater health care needs and at the same time limits access to quality health care. We now examine the impact of violence on women's health.

The Scope of Violence

The scope of violence against women is of stunning magnitude when crimes by both intimate and stranger perpetrators are considered: 1 in 2 women recalls sexual abuse before age 18,

1 in 5 women has been the victim of completed rape, and 1 in 4 women has been physically battered, according to the results of recent community-based studies (Kilpatrick, Saunders, Veronen, Best, & Von, 1987; Koss, Woodruff, & Koss, in press; Russell, 1982, 1983; Sorenson, Stein, Siegel, Golding, & Burnam, 1988; Strauss, Gelles, & Steinmetz, 1980; Wyatt, 1985). Many people, citizens and public officials alike, are unable to accept the fact that this amount of violence can exist without coming to public attention. Cover-up may be facilitated by the forced secrecy that is almost uniformly demanded by perpetrators of abuse. That the victims were silent is attested to by the small proportion of victims who informed authorities. In fact, only 2% of intrafamilial child sexual abuse, 6% of extrafamilial child sexual abuse, and 12%–25% of adult sexual assault cases were reported to police according to recent studies (Kilpatrick *et al.*, 1987; Koss *et al.*, in press; Russell, 1982, 1983).

More disturbing than the overall prevalence rates is the proportion of incidents that were perpetrated by close friends or family members. Acquaintances of the victim were implicated in up to 49% of child sexual assaults; romantic partners were implicated in 50%–57% of sexual assaults reported by adult women (Kilpatrick *et al.*, 1987; Koss *et al.*, in press; Russell, 1982). Family members perpetrated 90% of the physical and sexual assaults recalled by psychiatric patients (Carmen, Rieker, & Mills, 1984). Furthermore, the burden of violence falls disproportionately on women. For example, 78% of the substantiated cases of child sexual abuse involved girls, and more than 90% of the rape victims uncovered in the National Crime Survey were women (Flanagan & McGarrell, 1986; Wyatt & Powell, 1988).

The Impact of Violence on Health

Victimization by violence is a diagnosis that physicians are increasingly expected to make: both the Surgeon General's Workshop on Violence and Public Health and the Attorney General's Task Force on Family Violence recommended that the medical school curricula include education about domestic violence (DHHS, 1986; Department of Justice, 1984). The American College of Obstetricians and Gynecologists (ACOG) recently mailed information about battered women to its 28,000 members (ACOG, 1989). Victimization by violence has been shown to affect women's perceptions of their health as well as their utilization of medical services. Among women with a history of severe victimization, current medical utilization (in index year) was 33%–100% higher than nonvictims (Koss, Koss, & Woodruff, in press). In the year following rape, victims' physician visits increased 56% over preassault baseline (Koss *et al.*, in press).

Unfortunately, the likelihood is low that physicians currently realize their potential to initiate service provision for traumatized victims. The medical literature is mute on the long-term health impact of victimization. Instead, emergency and forensic treatment needs are emphasized (Hicks, 1988; Hochbaum, 1987; Martin, Warfield, & Braen, 1983). Further, only 53% of medical school curricula include teaching about domestic violence; an average of 1½ hours is devoted to the topic (CDC, 1989c). Nor do accepted clincial practices reflect a recognition of the significance of criminal violence in women's lives. Specifically, exposure to sexual violence fails to be included in the standard sexual history in clinical medicine (Ende, Rockwell, & Glasgow, 1984; Lewis & Freeman, 1987; Smith, 1989). Using current diagnostic techniques, personnel in a large metropolitan hospital correctly identified fewer than 5% of the episodes of domestic violence involving adult female patients (Stark, Flitcraft, Zuckerman, Grey, Robison, & Frazier, 1981). Standard psychiatric assessments also routinely fail to detect histories of sexual assault among inpatients (Jacobson & Richardson, 1987; Jacobson, Koehler, & Jones-Brown, 1987).

The failure to screen for victimization by violence communicates a lack of permission to discuss these issues in the medical setting. When psychosocial variables are ignored in diagnosis, somatic complaints, which are the ticket of admission to the medical system, may be inaccurately or inappropriately diagnosed

and treated as organic in etiology; this treatment, in turn, places an undue burden on the health services delivery system (Katon, Ries, & Kleinman, 1984). As well, opportunities are missed to facilitate simple confiding in a caring person, which has demonstrated therapeutic effects on the immune response (Pennebaker, Kiecolt-Glaser, & Glaser, 1988).

Providing Health Care for Women: Institutional Barriers

We have made considerable progress in drawing a picture of women's health status and their ability to gain access to needed health care. We now turn to the provision of gender-sensitive health care within medical school settings. Gender-sensitive health care and education include the following components: (1) didactic content about women's health receives proper emphasis in medical school curricula; (2) the strengths that women bring to health care provision are valued and rewarded; (3) women's unique experiences and needs are considered in the planning of treatment; and (4) women's health problems are considered as viable research specializations. We strongly believe that gender sensitivity ensues only when there is adequate representation of and institutional support for women faculty; these are uncharacteristic of predominantly male institutions. By considering the experiences of women professionals within medical schools in hiring, promotion, tenure, and pay equity, we estimate women providers' power within these institutions and the extent to which their contributions have been valued and rewarded.

Supply of Doctoral-Level Women

The American Psychological Association (1988) census found that 53% of doctoral-level psychologists are women (including 7% who are Asian, black, or Hispanic); this percentage reflects nearly a sixfold increase in female Ph.D.'s since 1950. In 1987, 54% of all Ph.D.'s awarded in clinical psychology went to female candidates. Roughly 5% of persons with doctorates in psychology work in academic medicine (Pion, Bramblett, & Wicherski, 1987). Nearly half of new doctoral recipients accept jobs in health care settings of some kind (Howard, Pion, Gottfredson, Flattau, Oskamp, Pfafflin, Bray, & Burstein, 1986).

The trend toward increasing the representation of women is more recent in medicine than in psychology. First-year female medical students in 1950 constituted 5% of the class; female enrollment today has increased to 37% (Bickel, 1988; Friedman, 1988; Nadelson, 1989). Residency programs are now 28% female; the proportion of female residents is highest in pediatrics (52%), obstetrics/gynecology (45%), psychiatry (41%), dermatology (40%), and pathology (37%) (Bickel, 1988; Martin, Parker, & Arnold, 1988). By contrast, less than 5% of surgical residents are women. Among the medical residents on duty on September 1, 1987, 4.8% were black and 5.1% were Hispanic (Crowley & Etzel, 1988).

Only 15% of active practicing physicians are women, compared with half of the practicing psychologists (Friedman, 1988; Weisman & Teitelbaum, 1989). In addition, the paucity of flexible work settings has resulted in the majority of female physicians being concentrated in part-time employment in institutional settings (Nadelson, 1989).

Hiring, Tenure, and Promotion

Total female medical school faculty membership grew from 13% in 1967 to 19% 1987 (Bartuska, 1988; Bickel, 1988). Among medical school faculty holding Ph.D.'s (and other health degrees), 21% are women. Among faculty holding M.D.'s, 15% are women, a figure which was 9% in 1967 (Bickel, 1988; Nadelson, 1989). The number of women with assistant professorships increased from 41% in 1978 to 49% in 1988, but the proportion of female full professors has remained relatively constant at 8%–9% (Bickel, 1988). These data suggest that availability and affirmative action hiring are not the major culprits in the underrepresentation of women in the upper ranks; instead, the blame lies in the retention and promotion of women once hired.

The average number of years required to ascend in rank is consistently higher for female faculty than for male faculty (Dickstein & Stephenson, 1987; Martin et al., 1988; Wallis, Gilder, & Thaler, 1981). Studies of career progression among medical school faculty hired in 1976 revealed that, by 1986, 25% of men compared with 19% of women were tenured or on a tenured-track appointment; 12% of men versus 3% of women had achieved the rank of full professor (Bickel, 1988). Fewer than one third of female academic physicians aged 47 years or older have attained the rank of full professor (Nadelson, 1989). Nadelson (1989) postulated that female physicians typically do not attain high academic status because they frequently prefer combined teaching, clinical, and administrative positions. Humanistic contributions to the academic medical environment do not result in a direct payoff for promotion (Bickel, 1988; Racy, Smith, & Ferry, 1989).

In the administration of academic medicine, the paucity of women is even more notable than among the upper ranks of faculty. In total, women represent 3% of the medical school academic administrative structure (Bickel, 1988). There are only 2 female deans in the 127 medical schools in the United States. Only one chair of a psychiatry department is filled by a woman, and psychiatry is a field with a relatively large pool of available women candidates (Bartuska, 1988; Nadelson, 1989). The grave underrepresentation of women in senior ranks and in administrative positions is difficult to explain when one recognizes that, since 1960, women have been more likely than men to enter academic medicine (Martin et al., 1988).

Pay Equity

Academic women, like other working women, receive less compensation and fewer other benefits than their male counterparts, even when they are in comparable positions (Nadelson, 1989). Specifically, the most recent biennial survey of doctorates in the United States, which is conducted by the National Research Council, reported that women scientists continue to be paid less than men in the same field and at the same experience level (Vetter, 1989). Women are offered lower starting salaries, and the gap widens with time. Over a 30-year professional lifetime, the differential totals $227,500 in 1987 dollars (Vetter, 1989). The field with the highest level of parity is psychology (women make 86% of men's salaries) and the field with the lowest level of parity is medical science (women make less than 75% of men's salaries) (Howard *et al.*, 1986; Vetter, 1989). Women also report more career obstacles, including less institutional or federal support for their education, more difficulty finding work, and less likelihood of finding full-time work. In the early 1970s, these trends were attributed to the recency of women's entry into the work force. But today, 40% of the women in the professional work force have more than 10 years of experience (Vetter, 1989).

Conclusions

The goal of this chapter was to review the anatomy of women's health. To recap the major points, it was concluded that women are quite healthy compared with men. They live longer and have lower mortality rates for many conditions. These differences are accounted for by sex-linked biological advantages favoring women, as well as lower levels of injurious health behaviors including smoking and drinking. In spite of greater longevity, women, not men, have a higher prevalence of both acute and chronic causes of morbidity, and health profiles on all dimensions are less favorable for minority women. Two important trends affecting women's health and the availablility of health care are the increasing rates of poverty, especially in female-headed households, and the staggering amount of violence perpetrated on women, often by offenders who are intimate acquaintances. Violence and poverty both increase women's needs for health services and at

the same time create obstacles to obtaining care. Women emerge as major consumers of health services primarily because of their greater longevity; their rates of hospitalization exceed those of men only during the childbearing years. If given the opportunity, women patients often express a preference for female health practitioners, who have been widely praised for their empathy and their ability to meld the art and science of healing.

In closing, we would like to identify some specific areas for action that offer the potential to improve women's health and their access to health care.

1. *Pay inequity.* Women are underpaid relative to men and increasingly must provide care to dependent children who are abandoned by their fathers. Other countries and various localities within the United States have adopted proposals to create pay equity. The absence of pay equity as a national policy promotes poor health among women.

2. *Employment-linked benefits and insurance system.* Both pension benefits and insurance coverage are linked to employment in the United States. This public policy affects women's access to adequate health care because it allows pay inequity, unpaid dependent care, and intermittent employment history to affect retirement and disability income and the quality of insurance coverage that is available. These inequities are especially troublesome for women, who must plan for a longer life span than men. Other countries have found methods by which to fund universal health care and pension benefits that are not linked to paid employment. Continuation of the status quo in the United States places women at a disadvantage in obtaining health care.

3. *Interpersonal violence.* Insidious effects ensue from the violence perpetrated on women within the American family. Strong cultural forces favor secrecy and foster illusions about the nature of the family. They have resulted in a long-term cover-up of intimate violence. Women do not need to have the aftereffects of violence turned into diseases; they need help to address the violence.

4. *Avoidance of women in research.* It has been demonstrated that women's physiology is not the same as men's. Thus, the present practices by which drugs reach the clinical trial stage before ever being tested on women is highly inappropriate. Yet, the restrictive testing guidelines currently in effect have not been revised since 1977. It is no longer allowable, according to federal policy, to exclude women from funded research studies in health (excluding drug studies) without explanation. But informed oversight is needed to ensure that this policy will become more than lip service; funding agencies must be expected to present research portfolios that address women's health issues.

5. *Obstacles for women providers.* The woman practitioner who chooses to work within a medical school faces several serious challenges. Women practitioners, regardless of specialization, face a longer, harder road to tenure and promotions, with far less likelihood of ever reaching full professorship or the administrative ranks. Without urgent attention to the mentoring and retention of women candidates, medical schools will find their faculty increasingly at odds with both health care consumers and the medical student body, which is fast approaching gender parity. The major consolation for the women currently on medical school faculties is that they may outlast the obstacles and live long enough to see significant changes.

References

Adamson, T. E., Tschann, J. M., Gullion, D. S., & Oppenberg, A. A. (1989). Physician communication skills and malpractice claims: A complex relationship. *The Western Journal of Medicine, 150*, 356–360.

Allred, K. D., & Smith, T. W. (1989). The hardy personality: Cognitive and physiological responses to evaluate threat. *Journal of Personality and Social Psychology, 56*, 257–266.

Amaro, H. (1988). Considerations for prevention of HIV infection among Hispanic women. *Psychology of Women Quarterly, 12*, 429–443.

American College of Obstetricians and Gynecologists. (1989). *The battered woman.* Washington, DC: American College of Obstetricians and Gynecologists. (ACOG technical bulletin No. 124).

American Medical Association Council on Scientific Affairs. (1989). Firearms injuries and deaths: A critical public health issue. *Public Health Reports, 104*, 111–120.

American Psychological Association, Task Force on Depression. (1989). *Research on women and depression.* Final report. Washington, DC: Author.

American Psychological Association, Women's Program Office Public Interest Directorate. (1988). *Women in the American Psychological Association 1988.* Washington, DC: Author.

Anderson, K. O., Bradley, L. A., Young, L. D., McDaniel, L. K., & Wise, C. M. (1985). Rheumatoid arthritis: Review of psychological factors related to etiology, effects, and treatment. *Psychological Bulletin, 98*, 358–387.

Arnold, R. M., Martin, S. C., & Parker, R. M. (1988). Taking care of patients—Does it matter whether the physician is a woman? *The Western Journal of Medicine, 149*, 729–733.

Bankhead, C. D. (1989, March 13). Americans are having as many strokes, but fewer prove fatal. *Medical World News*, p. 15.

Bartuska, D. G. (1988). Women in academic medicine: Equalizing the opportunities. *The Western Journal of Medicine, 149*, 779–780.

Berman, A. M. (1989). Facts on women and smoking. *Journal of the American Medical Women's Association, 44*, 55.

Bickel, J. (1988). Women in medical education: A status report. *New England Journal of Medicine, 319*, 1579–1584.

Booth-Kewley, S., & Friedman, H. S. (1987). Psychological predictors of heart disease: A quantitative review. *Psychological Bulletin, 101*, 343–362.

Braveman, P., Oliva, G., Miller, M. G., Schaff, V. M., & Reiter, R. (1988). Women without health insurance: Links between access, poverty, ethnicity, and health. *The Western Journal of Medicine, 149*, 708–711.

Brehm, S. S., Kassin, S. M., & Gibbons, F. X. (Eds.). (1981). *Developmental social psychology.* New York: Oxford University Press.

Bureau of the Census (1987). Money, income, and poverty status of families and persons in the United States 1986. *Current Population Reports* (Series P-60, No. 157). Washington, DC: U.S. Government Printing Office.

Byrd, J. C., Shapiro, R. S., & Schiedermayer, D. L. (1989). Passive smoking: A review of medical and legal issues. *American Journal of Public Health, 79*, 209–215.

Carmen, E. H., Rieker, P. P., & Mills, T. (1984). Victims of violence and psychiatric illness. *American Journal of Psychiatry, 141*, 378–383.

Carter, C. L., Jones, D. Y., Schatzkin, A., & Brinton, L. A. (1989). A prospective study of reproductive, familial, and socioeconomic risk factors for breast cancer using NHANES I data. *Public Health Reports, 104*, 45–50.

Cassel, C. K., & Neugarten, B. L. (1988). A forecast of women's health and longevity—Implications for an aging America. *The Western Journal of Medicine, 149*, 712–717.

Centers for Disease Control (CDC). (1986a). Annual summary 1984: Reported morbidity and mortality in the United States, *Morbidity and Mortality Weekly Report, 32*, 114–117.

217

Centers for Disease Control (CDC). (1986b). Hysterectomy among women of reproductive age, United States, update for 1981–1982. *CDC Surveillance Summaries: Morbidity and Mortality Weekly Report, 35,* 1–6.

Centers for Disease Control (CDC). (1988a). Distribution of AIDS cases, by racial/ethnic group and exposure category, United States, June 1, 1981–July 4, 1988. *Morbidity and Mortality Weekly Report, 37,* 1–10.

Centers for Disease Control (CDC). (1988b). Leading major congenital malformations among minority groups in the United States, 1981–1986. *Morbidity and Mortality Weekly Report, 37,* 17–24.

Centers for Disease Control (CDC). (1988c). Maternal mortality surveillance, United States, 1980–1985. *CDC Surveillance Summaries: Morbidity and Mortality Weekly Report, 37,* 19.

Centers for Disease Control (CDC). (1988d). Suicides among persons 15–24 years of age, 1970–1984. *CDC Surveillance Summaries: Morbidity and Mortality Weekly Report, 37,* 61–68.

Centers for Disease Control (CDC). (1989a). Acquired immunodeficiency syndrome associated with intravenous-drug use—United States, 1988. *Morbidity and Mortality Weekly Report, 38,* 165–170.

Centers for Disease Control (CDC). (1989b). Chronic disease reports: Mortality trends—United States, 1979–1986. *Morbidity and Mortality Weekly Report, 38,* 189–192.

Centers for Disease Control (CDC). (1989c). Education about adult domestic violence in U.S. and Canadian medical schools, 1987–88. *Morbidity and Mortality Weekly Report, 38,* 17–19.

Centers for Disease Control (CDC). (1989d). Impact of homicide on years of potential life lost in Michigan's black population. *Morbidity and Mortality Weekly Report, 38,* 4–6, 11.

Centers for Disease Control (CDC). (1989e). Unintentional poisoning mortality—United States, 1980–1986. *Morbidity and Mortality Weekly Report, 38,* 153–157.

Centers for Disease Control (CDC). (1989f). Update: Acquired immunodeficiency syndrome—United States, 1981–1988. *Morbidity and Mortality Weekly Report, 38,* 229–236.

Centers for Disease Control (CDC). (1989g). Years of potential life lost before age 65—United States, 1987. *Morbidity and Mortality Weekly Report, 38,* 27–29.

Cochran, S. D. & Mays, V. M. (1989). Women and AIDS-related concerns: Roles for psychologists in helping the worried well. *American Psychologist, 44,* 529–535.

Collins, J. B. (1988). Menopause. *Primary Care, 15,* 593–606.

Comstock, L. M., Hooper, E. M., Goodwin, J. M., & Goodwin, J. S. (1982). Physician behaviors that correlate with patient satisfaction. *Journal of Medical Education, 57,* 105–112.

Cook, B. L., Garvey, M. J., & Shukla, S. (1987). Alcoholism. *Primary Care, 14,* 685–697.

Cornoni-Huntley, J., La Croix, A. Z., & Havlik, R. J. (1989). Race and sex differentials in the impact of hypertension in the United States: The national health and nutrition examination survey I epidemiological follow-up study. *Archives of Internal Medicine, 149,* 780–788.

Courtless, J. C., & Lowe, S. (Eds.). (1989, February). *Family Economics Review, 2* (1).

Crowley, A. E., & Etzel, S. I. (1988). Graduate medical evaluation in the United States. *Journal of the American Medical Association, 260,* 1093–1101.

Davis, K. (1988). Women and health care. In S. E. Rix (Ed.), *The American woman 1988–89: A status report* (pp. 162–204). New York: Norton.

Dazzi, A., & Dwyer, J. (1984). Nutritional analyses of popular weight-reduction diets in books and magazines. *International Journal of Eating Disorders, 3,* 61–79.

Denmark, F., Russo, N. F., Frieze, I. H., & Sechzer, J. A. (1988). Guidelines for avoiding sexism in psychological research. *American Psychologist, 43,* 582–585.

Department of Commerce. (1989). *Statistical abstract of the United States 1989.* Washington, DC: U.S. Government Printing Office.

Department of Health and Human Services, U.S. Department of Justice. (1986). Surgeon General's Workshop on Violence and Public Health: Report. Washington, DC: Author.

Department of Health and Human Services, U.S. Public Health Service. (1985). *Report of the U.S. Public Health Service Task Force on Women's Health Issues.* (DHHS Publication No. PHS 85-50206). Washington, DC: U.S. Government Printing Office.

Department of Health and Human Services, U.S. Public Health Service. (1988). Women's health [supplement]. *Public Health Reports, 103,* 22–33, 38–50, 62–67, 74–99, 106–139, 141–144, 147–151, 366–375, 387–393, 575–577, 592–597, 628–637.

Department of Justice. (1984). Attorney General's Task Force on Family Violence: Final report. Washington, DC: Author.

Dickstein, L. J., & Stephenson, J. J. (1987). A national survey of women physicians in administrative roles. *Journal of the American Women's Association, 42,* 108–111.

Diehl, A. K., & Stern, M. P. (1989). Special health problems of Mexican-Americans: Obesity, gallbladder disease, diabetes mellitus, and cardiovascular disease. *Advances in Internal Medicine, 34,* 73–96.

Doebbert, G., Riedmiller, K. R., & Kizer, K. W. (1988). Occupational mortality of California women, 1979–1981. *The Western Journal of Medicine, 149,* 734–740.

Doty, P. (1986). Family care of the elderly: The role of public policy. *The Milbank Quarterly, 64,* 34–75.

Eaker, E. D. (1989). Psychosocial factors in the epidemiology of coronary heart disease in women. *Psychiatric Clinics of North America, 12,* 167–173.

Egerter, D. (1988, August 22). Boom in STD's setting off alarms. *Medical World News,* pp. 69–70.

Ende, J., Rockwell, S., & Glasgow, M. (1984). The sexual history in general medical practice. *Archives of Internal Medicine, 144,* 558–561.

Ettinger, B. (1988). A practical guide to preventing osteoporosis. *The Western Journal of Medicine, 149,* 691–695.

Fiore, M. C., Novotny, T. E., Pierce, J. P., Hatziandreu, E. J., Patel, K. M., & Davis, R. M. (1989). Trends in cigarette smoking in the United States. *Journal of the American Medical Association, 261,* 49–55.

Flanagan, T. J., & McGarrell, E. F. (1986). Sourcebook of criminal justice statistics—1985. (NCJ-100899). Washington, DC: U.S. Department of Justice, Bureau of Justice Statistics.

Ford Foundation. (1989). The common good. *Policy Recommendations of the Executive Panel of the Ford Foundation Project on Social Welfare and the American Future.* New York: Ford Foundation.

Freeman, H. E., Blendon, R. J., Aiken, L. H., Sudman, S., Mullunix, C. F., & Corey, C. R. (1987). Americans report on their access to health care. *Health Affairs, 6,* 6–18.

Friedman, E. (1988, April 25). Changing the ranks of medicine: Women MD's. *Medical World News,* pp. 57–68.

Funk, S. C., & Houston, B. K. (1987). A critical analysis of the Hardiness Scale's validity and utility. *Journal of Personality and Social Psychology, 53,* 572–578.

Ganley, R. M. (1989). Emotion and eating in obesity: A review of the literature. *International Journal of Eating Disorders, 8,* 343–361.

Garraway, W. M., & Whisnant, J. P. (1987). The changing pattern of hypertension and the declining incidence of stroke. *Journal of the American Medical Association, 258,* 214–217.

Glaser, J. B., Strange, T. J., & Rosati, D. (1989). Heterosexual human immunodeficiency virus transmission among the middle class. *Archives of Internal Medicine, 149,* 645–649.

Grady, K. E. (1988). Older women and the practice of breast self-examination. *Psychology of Women Quarterly, 12,* 473–487.

Gray, D. S. (1989). Diagnosis and prevalence of obesity. *Medical Clinics of North America, 73,* 1–13.

Greer, S., Dickerson, V., Schneiderman, L. J., Atkins, C., & Bass, R. (1986). Responses of male and female physicians to medical complaints in male and female patients. *The Journal of Family Practice, 23,* 49–53.

Guzick, D. S. (1989). Clinical epidemiology of endometriosis and infertility. *Obstetrics and Gynecology Clinics of North America, 16,* 43–59.

Haffner, S. M., Diehl, A. K., Stern, M. P., & Hazuda, H. P. (1989). Central adiposity and gallbladder disease in Mexican Americans. *American Journal of Epidemiology, 129,* 587–595.

Hall, R. C. W., Hoffman, R. S., Beresford, T. P., Wooley, B., Hall, A. K., & Kubasek, L. (1989). Physical illness encountered in patients with eating disorders. *Psychosomatics, 30,* 174–191.

Hamilton, J. A., Lloyd, C., Alagna, S. W., Phillips, K., & Pinkel, S. (1984). Gender, depressive subtypes, and gender-age effects on antidepressant response: Hormonal hypotheses. *Psychopharmacology Bulletin, 20,* 475–480.

Hartman-Stein, P., & Reuter, J. M. (1988). Developmental issues in the treatment of diabetic women. *Psychology of Women Quarterly, 12,* 417–428.

Hertzer, N. R., Young, J. R., Beven, E. G., O'Hara, P. J., Graor, R. A., Ruschhaupt, W. F., & Maljovec, L. C. (1986). Late results of coronary bypass in patients with peripheral vascular disease: 2. Five-year survival according to sex, hypertension, and diabetes. *Cleveland Clinic Journal of Medicine, 54,* 15–23.

Herzog, D. B., Keller, M. B., & Lavori, P. W. (1988). Outcome in anorexia nervosa and bulimia nervosa: A review of the literature. *The Journal of Nervous and Mental Disease, 176,* 131–143.

Hicks, D. J. (1988, November 30). The patient who's been raped. *Emergency Medicine,* 106–122.

Higgins, L. C. (1988, November 14). Cyclical dieting poses dangers: Yo-yo dieting appears to place patients at increased risk of cardiovascular disease and cancer. *Medical World News,* 25–26.

Hill, J. W. (1987). Exercise prescription. *Primary Care, 14,* 817–825.

Hochbaum, S. R. (1987). The evaluation and treatment of the sexually assaulted patient. *OB/Gyn Emergencies, 5,* 601–621.

Howard, A., Pion, G. M., Gottfredson, G. D., Flattau, P. E., Oskamp, S., Pfafflin, S. M., Bray, D. W., & Burstein, A. G. (1986). The changing face of American psychology. *American Psychologist, 41,* 1311–1327.

Hull, J. G., Van Treuren, R., & Virnelli, S. (1987). Hardiness and health: A critique and alternative approach. *Journal of Personality and Social Psychology, 53,* 518–530.

Jacobs, P., & McDermott, S. (1989). Family caregiver costs of chronically ill and handicapped children: Method and literature review. *Public Health Reports, 104,* 158–163.

Jacobson, A., & Richardson, B. (1987). Assault experiences of 100 psychiatric inpatients: Evidence of the need for routine inquiry. *American Journal of Psychiatry, 144,* 908–913.

Jacobson, A., Koehler, J. E., & Jones-Brown, C. (1987). The failure of routine assessment to detect histories of assault experienced by psychiatric patients. *Hospital and Community Psychiatry, 38,* 386–389.

Katon, W., Ries, R. K., & Kleinman, A. (1984). The prevalence of somatization in primary care. *Comparative Psychiatry, 25,* 208–215.

Kaufman, L. D., Gomez-Reino, J. J., Heinicke, M. H., & Gorevic, P. D. (1989). Male lupus: Retrospective analysis of the clinical and laboratory features of 52 patients, with a review of the literature. *Seminars in Arthritis and Rheumatism, 18,* 189–197.

Keith, L. G., MacGregor, S., Friedell, S., Rosner, M., Chasnoff, I. J., & Sciarra, J. J. (1989). Substance abuse in pregnant women: Recent experience at the perinatal center for chemical dependence of Northwestern Memorial Hospital. *Obstetrics and Gynecology, 73,* 715–720.

Kemeny, M. E., Cohen, F., Zegans, L. S., & Conant, M. A. (1989). Psychological and immunological predictors of genital herpes recurrence. *Psychosomatic Medicine, 51,* 195–208.

Kilpatrick, D. G., Saunders, B. E., Veronen, L. J., Best, C. L., & Von, J. M. (1987). Criminal victimization: Lifetime prevalence, reporting to police, and psychological impact. *Crime and Delinquency, 33,* 479–489.

Kinney, E. L., Trautmann, J., Gold, J. A., Vesell, E. S., & Zelis, R. (1981). Underrepresentation of women in new drug trials. *Annals of Internal Medicine, 95,* 495–499.

Kissebah, A. H., Freedman, D. S., & Peiris, A. N. (1989). Health risks of obesity. *Medical Clinics of North America, 73,* 111–138.

Klag, M. J., Whelton, P. K., & Seidler, A. J. (1989). Decline in U.S. stroke mortality: Demographic trends and antihypertensive treatment. *Stroke, 20,* 14–21.

Klerman, G. L., & Weissman, M. M. (1989). Increasing rates of depression. *Journal of the American Medical Association, 261,* 2229–2235.

Klesges, R. C., & Klesges, L. M. (1988). Cigarette smoking as a dieting strategy in a university population. *International Journal of Eating Disorders, 7,* 413–419.

Koo, D. (1989). Chronic fatigue syndrome: A critical appraisal of the role of Epstein-Barr virus. *The Western Journal of Medicine, 150,* 590–596.

Koonin, L. M., Ellerbrock, T. V., Atrash, H. K., Rogers, M. F., Smith, J. C., Hogue, C. J. R., Harris, M. A., Chavkin, W., Parker, A. L., & Halpin, G. J. (1989). Pregnancy-associated deaths due to AIDS in the United States. *Journal of the American Medical Association, 261,* 1306–1309.

Koss, M. P., Koss, P. G., & Woodruff, W. J. (in press). Deleterious effects of criminal victimization on women's health and medical utilization. *Archives of Internal Medicine.*

Koss, M. P., Woodruff, W. J., & Koss, P. G. (in press). Criminal victimization among primary care medical patients: Prevalence, incidence, and physician usage. *Behavioral Sciences and the Law.*

Kotler, P., & Wingard, D. L. (1989). The effect of occupational, marital and parental roles on mortality: The Alameda County study. *American Journal of Public Health, 79,* 607–612.

Leaf, D. A. (1988). A woman's heart: An update of coronary artery disease risk in women. *The Western Journal of Medicine, 149,* 751–757.

Lerman, C. E. (1987). Rheumatoid arthritis: Psychological factors in the etiology, course, and treatment. *Clinical Psychology Review, 7,* 413–425.

Lewis, C. E., & Freeman, H. E. (1987). The sexual history-taking and counseling practices of primary care physicians. *Western Journal of Medicine, 147,* 165–167.

Lichtenstein, M. J., Bolton, A., & Wade, G. (1989). Derivation and validation of a decision rule for predicting seatbelt utilization. *The Journal of Family Practice, 28,* 289–292.

Lobo, R. A., & Whitehead, M. (1989). Too much of a good thing? Use of progestogens in the menopause: An international consensus statement. *Fertility and Sterility, 51,* 229–231.

Luckmann, R. (1989). Incidence and case fatality rates for acute appendicitis in California: A population-based study of the effects of age. *American Journal of Epidemiology, 129,* 905–918.

Lui, K., & Kendal, A. P. (1987). Impact of influenza epidemics on mortality in the United States from October 1972 to May 1985. *American Journal of Public Health, 77,* 712–716.

Mahoney, M. C., Michalek, A. M., Cummings, K. M., Nasca, P. C., & Emrich, L. J. (1989). Mortality is a northeastern Native American cohort. *American Journal of Epidemiology, 129,* 816–826.

Makuc, D. M., Freid, V. W., & Kleinman, J. C. (1989). National trends in the use of preventive health care by women. *Journal of Public Health, 79,* 21–26.

Marin, G., Perez-Stable, E. J., & Marin, B. V. (1989). Cigarette smoking among San Francisco Hispanics: The role of acculturation and gender. *American Journal of Public Health, 79,* 196–198.

Markowitz, L. E., Hightower, A. W., Broome, C. V., & Reingold, A. L. (1987). Toxic shock syndrome: Evaluation of national surveillance data using a hospital discharge summary. *Journal of the American Medical Association, 258,* 75–78.

Martin, C. A., Warfield, M. C., & Braen, R. (1983). Physician's management of the psychological aspects of rape. *Journal of the American Medical Association, 249,* 501–503.

Martin, S. C., Parker, R. M., & Arnold, R. M. (1988). Careers of women physicians: Choices and constraints. *The Western Journal of Medicine, 149,* 758–760.

Matarazzo, J. D. (1984). Behavioral immunogens and pathogens in health and illness. In B. L. Hammonds & C. J. Scheirer (Eds.), *Psychology and Health* (pp. 5–44). Washington, D.C.: American Psychological Association.

Matteo, S. (1988). The risk of multiple addictions: Guidelines for assessing a woman's alcohol and drug use. *The Western Journal of Medicine, 149,* 741–750.

McCue, J. D. (1989). Evaluation and management of vaginitis: An update for primary care practitioners. *Archives of Internal Medicine, 149,* 565–568.

Mercy, J. A., & Saltzman, L. E. (1989). Fatal violence among spouses in the United States, 1976–85. *American Journal of Public Health, 79,* 595–599.

Metzger, D. A. & Haney, A. F. (1989). Etiology of endometriosis. *Obstetrics and Gynecology Clinics of North America, 16,* 1–14.

Michel, C., & Battig, K. (1989). Separate and combined psychophysiological effects of cigarette smoking and alcohol consumption. *Psychopharmacology, 97,* 65–73.

Miller, H. I., & Young, F. E. (1989). The drug approval process at the Food and Drug Administration: New biotechnology as a paradigm of a science-based activist approach. *Archives of Internal Medicine, 149,* 655–658.

Mitchell, J. E., Pomeroy, C., & Huber, M. (1988). A clinician's guide to the eating disorders medicine cabinet. *International Journal of Eating Disorders, 7,* 211–223.

Monsen, E. R. (Ed.). (1989). Costs and benefits of nutrition services: A literature review [Supplement]. *Journal of the American Dietetic Association, 89,* 22–29, 39–40.

Morokoff, P. J. (1988). Sexuality and perimenopausal and postmenopausal women. *Psychology of Women Quarterly, 12,* 489–511.

Moskowitz, J. M. (1989). The primary prevention of alcohol

problems: A critical review of the research literature. *Journal of Studies on Alcohol, 50,* 54–88.

Nadelson, C. C. (1989). Professional issues for women. *Psychiatric Clinics of North America, 12,* 25–33.

National Institute of Mental Health. (1986). *Family's impact on health: A critical review and annotated bibliography.* (DHHS Publication No. ADM 87-1461). Washington, DC: U.S. Government Printing Office.

National Institute of Mental Health. (1987). *Women's mental health: Agenda for research.* (DHHS Publication No. ADM 87-1542). Washington, DC: U.S. Government Printing Office.

Nelson, R. G., Becker, T. M., Wiggins, C. L., Key, C. R., & Samet, J. M. (1989). Ethnic differences in mortality from acute rheumatic fever and chronic rheumatic heart disease in New Mexico, 1958–1982. *The Western Journal of Medicine, 150,* 46–50.

Nelson, R. G., Everhart, J. E., Knowler, W. C., & Bennett, P. H. (1988). Incidence, prevalence, and risk factors for non-insulin-dependent diabetes mellitus. *Primary Care, 15,* 227–250.

Nolen-Hoeksema, S. (1987). Sex differences in unipolar depression: Evidence and theory. *Psychological Bulletin, 101,* 259–282.

Ockene, J. K., Sorenson, G., Kabat-Zinn, J., Ockene, I. S., & Donnelly, G. (1988). Benefits and costs of lifestyle change to reduce risk of chronic disease. *Preventive Medicine, 17,* 224–234.

Owen, O. E., Kavle, E., Owen, R. S., Polansky, M., Caprio, S., Mozzoli, M. A., Kendrick, Z. V., Bushman, M. C., & Boden, G. (1986). A reappraisal of caloric requirements in healthy women. *The American Journal of Clinical Nutrition, 44,* 1–19.

Pennebaker, J. W., Kiecolt-Glaser, J. K., & Glaser, R. (1988). Disclosure of traumas and immune function: Health implications for psychotherapy. *Journal of Consulting and Clinical Psychology, 56,* 239–245.

Philipp, E., Pirke, K., Siedl, M., Tuschi, R. J., Fichter, M. M., Eckert, M., & Wolfram, G. (1988). Vitamin status in patients with anorexia nervosa and bulimia nervosa. *International Journal of Eating Disorders, 8,* 209–218.

Pion, G. M., Bramblett, J. P., & Wicherski, M. (1987). *1985 Doctorate employment survey* (Preliminary Report). Washington, DC: American Psychological Association.

Polivy, J., & Herman, C. P. (1987). Diagnosis and treatment of normal eating. *Journal of Consulting and Clinical Psychology, 55,* 635–644.

Pollner, F. (1988, July 11). Heterosexual spread of AIDS poses growing threat. *Medical World News,* p. 14.

Portans, I., White, J. M., & Staiger, P. K. (1989). Acute tolerance to alcohol: Changes in subjective effects among social drinkers. *Psychopharmacology, 97,* 365–369.

Pratt, C. M., Francis, M. J., Divine, G. W., & Young, J. B. (1989). Exercise testing in women with chest pain: Are there additional exercise characteristics that predict true positive test results? *Chest, 95,* 139–144.

Racy, J. C., Smith, J. W., & Ferry, P. C. (1989). Tenure for a new age: Ideas for the turn of the century. *Archives of Internal Medicine, 149,* 1001.

Reisine, S. T., & Fifield, J. (1988). Defining disability for women and the problem of unpaid work. *Psychology of Women Quarterly, 12,* 401–415.

Resnick, N. M., & Greenspan, S. L. (1989). "Senile" osteoporosis reconsidered. *Journal of the American Medical Association, 261,* 1025–1029.

Rigotti, N. A. (1989). Cigarette smoking and body weight. *The New England Journal of Medicine, 320,* 931–933.

Rodin, J., & Salovey, P. (1989). Health psychology. *Annual Review of Psychology, 40,* 533–579.

Russell, D. E. H. (1982). The prevalence and incidence of forcible rape and attempted rape of females. *Victimology, 7,* 81–93.

Russell, D. E. H. (1983). The incidence and prevalence of intrafamilial and extrafamilial sexual abuse of female children. *Child Abuse & Neglect, 7,* 133–146.

Russo, N. F. (1985, September). *A Women's Mental Health Agenda.* Washington, DC: American Psychological Association.

Sandler, D. P., Comstock, G. W., Helsing, K. J., & Shore, D. L. (1989). Deaths from all causes in non-smokers who lived with smokers. *American Journal of Public Health, 79,* 163–167.

Schocken, D. D., Holloway, J. D., & Powers, P. S. (1989). Weight loss and the heart: Effects of anorexia nervosa and starvation. *Archives of Internal Medicine, 149,* 877–881.

Schousboe, J. T., Koch, A. E., & Chang, R. W. (1988). Chronic lupus peritonitis with ascites: Review of the literature with case report. *Seminars in Arthritis and Rheumatism, 18,* 121–126.

Schwartz, D. M., Thompson, M. G., & Johnson, C. L. (1982). Anorexia nervosa and bulimia: The socio-cultural context. *International Journal of Eating Disorders, 1,* 20–36.

Schwenk, T. L., Marquez, J. T., Lefever, R. D., & Cohen, M. (1989). Physician and patient determinants of difficult physician-patient relationships. *The Journal of Family Practice, 28,* 59–63.

Sebastian, J. L., McKinney, W. P., & Young, M. J. (1989). Epidemiology and interaction of risk factors in cardiovascular disease. *Primary Care, 16,* 31–47.

Segal, K. R., & Pi-Sunyer, F. X. (1989). Exercise and obesity. *Medical Clinics of North America, 73,* 217–236.

Sherlock, S. (1988). Liver disease in women: Alcohol, autoimmunity, and gallstones. *The Western Journal of Medicine, 149,* 683–686.

Slattery, M. L., Robison, L. M., Schuman, K. L., French, T. K., Abbott, T. M., Overall, J. C., & Gardner, J. W. (1989). Cigarette smoking and exposure to passive smoke are risk factors for cervical cancer. *Journal of the American Medical Association, 261,* 1593–1598.

Smith, J. C., Mercy, J. A., & Conn, J. M. (1988). Marital status and risk of suicide. *American Journal of Public Health, 78,* 78–80.

Smith, R. P. (1989). Sexual counseling made simple. *Resident and Staff Physician, 35,* 85–87.

Sorenson, S. B., Stein, J. A., Siegel, J. M., Golding, J. M., & Burnam, M. A. (1988). Prevalence of adult sexual assault: The Los Angeles Epidemiologic Catchment Area Study. *American Journal of Epidemiology, 126,* 1154–1164.

Stark, E., Flitcraft, A., Zuckerman, D., Grey, A., Robison, J., & Frazier, W. (1981). *Wife abuse in the medical setting: An introduction for health personnel.* (Domestic violence monograph series no. 7). Rockville, MD: National Clearinghouse on Domestic Violence.

Stoffelmayr, B., Hoppe, R. B., & Weber, N. (1989). Facilitating patient participation: The doctor-patient encounter. *Primary Care, 16,* 265–278.

Strauss, M. A., Gelles, R. S., & Steinmetz, J. K. (1980). *Behind closed doors: Violence in the American family.* Garden City, NJ: Anchor/Doubleday, 1980.

Strickland, B. R. (1988). Sex-related differences in health and illness. *Psychology of Women Quarterly, 12,* 381–399.

Tallon, J. R. (1989). A health policy agenda proposal for including the poor. *Journal of the American Medical Association, 261,* 1044.

Tell, G. S., Howard, G., McKinney, W. M., & Toole, J. F. (1989). Cigarette smoking cessation and extracranial carotid atherosclerosis. *Journal of the American Medical Association, 261,* 1178–1180.

Thorpe, K. E., Siegel, J. E., & Dailey, T. (1989). Including the poor: The fiscal impacts of medicaid expansion. *Journal of the American Medical Association, 261,* 1003–1007.

Unger, K. B. (1988). Chemical dependency in women: Meeting the challenges of accurate diagnosis and effective treatment. *The Western Journal of Medicine, 149,* 746–750.

VanderPlate, C., Aral, S. O., & Magder, L. (1988). The relationship among genital herpes simplex virus, stress, and social support. *Health Psychology, 7,* 159–168.

van Strien, T., & Bergers, G. P. A. (1988). Overeating and sex-role orientation in women. *International Journal of Eating Disorders, 7,* 89–99.

Verbrugge, L. M. (1985). Gender and health: An update on hypotheses and evidence. *Journal of Health and Social Behavior, 26,* 156–182.

Verbrugge, L. M., & Madans, J. H. (1985). Social roles and health trends of American women. *The Milbank Quarterly, 63,* 691–735.

Vetter, B. M. (1989, May 5). Bad news for women scientists—and the country. *The AAAS Observer,* 10.

Wallis, L. A., Gilder, H., & Thaler, H. (1981). Advancement of men and women in medical academia. *Journal of the American Medical Association, 246,* 2350–2353.

Walsh, B. W., & Schiff, I. (1989). Symptoms and treatment of the menopause. *Resident and Staff Physician, 35,* 33–40.

Weisman, C. S., & Teitelbaum, M. A. (1989). Women and health care communication. *Patient Education and Counseling, 13,* 183–199.

Wenger, N. K. (1985). Coronary disease in women. *Annual Review of Medicine, 36,* 285–294.

Wickes, S. L. (1988). Premenstrual syndrome. *Primary Care, 15,* 473–487.

Women's Research and Education Institute. (1987). Who cares? The health care gap and how to bridge it. *Proceedings of the April 30, 1986 Conference in Washington, D.C. sponsored by the Congressional Black Caucus Foundation, Congressional Hispanic Caucus Institute, and the Women's Research and Education Institute.* Washington, DC: Author.

Wyatt, G. E. (1985). The sexual abuse of Afro-American and White-American women in childhood. *Child Abuse and Neglect, 9,* 507–519.

Wyatt, G. E., & Powell, G. J. (1988). *Lasting effects of child sexual abuse.* Newbury Park, CA: Sage.

York, R., Freeman, E., Lowery, B., & Strauss, J. F. (1989). Characteristics of premenstrual syndrome. *Obstetrics and Gynecology, 73,* 601–605.

Zuckerman, B., Frank, D. A., Hingson, R., Amaro, H., Levenson, S. M., Kayne, H., Parker, S., Vinci, R., Aboagye, K., Fried, L. E., Cabral, H., Timperi, R., & Bauchner, H. (1989). Effects of maternal marijuana and cocaine use on fetal growth. *The New England Journal of Medicine, 320,* 762–768.

CHAPTER 14

Emerging Issues in the Care of the Elderly

Asenath La Rue and Charles McCreary

Introduction

Older adults comprise the largest subgroup of patients seen in general medical settings. Yet, most health care professionals, including psychologists, have had little or no training in normal aging or in diseases that affect the aged. According to an American Psychological Association survey conducted in 1982, less than 5% of psychological services were delivered to elderly clients, and only a fraction of a percentage of psychologists in health settings reported a primary focus on aging (VandenBos & Stapp, 1983). This pattern is changing to a degree, through improvements in graduate curricula and in clinical experience provided by internship and postdoctoral programs. However, there continues to be a need to integrate expertise in rapidly growing subspecialties such as neuropsychology, health psychology, and behavioral medicine, with an understanding of the psychology of aging.

This chapter is a combined effort of a geropsychologist and a health psychologist. Most of our work has been in a university-based

neuropsychiatric hospital with inpatient, outpatient, and day-treatment programs for older adults. Collectively, however, we have also had experience in behavioral medicine programs in medical hospitals and in consultation to geriatric long-term-care facilities. Our first aim in the chapter is to highlight some of the basic issues that are influencing health care for older patients, including demographic patterns, health care needs and utilization, and attitudinal barriers to quality care. The second aim is to identify the most common reasons that older patients might be referred to a psychologist in a medical setting, and to outline basic strategies that we have found useful in responding to these referrals.

Older Adults and Health Care: Some Widely Recognized Trends

Demography and Medical Morbidity

Older adults, especially the very old, are rapidly increasing in number. People 65 years or older now comprise 12% of the U.S. population, and those 55 and above account for 21% (U.S. Bureau of the Census, 1987). By 2030, one in five Americans will be 65 or older, and one in three will be at least 55 (Spencer, 1984). The

Asenath La Rue and Charles McCreary • Department of Psychiatry and Biobehavioral Sciences, University of California at Los Angeles, Los Angeles, California 90024-1759.

ratio of old to young people has shifted significantly since 1900, and by the year 2030, the proportion of children and elderly adults is expected to be nearly equal (Spencer, 1984). Very old people (more than 85 years) are one of the fastest growing subgroups within our population; the life expectancy for this group has increased 24% since 1960, and people who are 85 and older are expected to increase proportionally from about 1% of the population to more than 5% between 1986 and 2050 (Spencer, 1984).

The likelihood of suffering from a chronic or disabling medical condition increases rapidly with age. More than four out of five people over the age of 65 have at least one chronic medical illness, and many have multiple conditions. According to data provided by the National Center for Health Statistics (1987a), the five most prevalent categories of illness, and their rates per 1,000 for people aged 65 and older versus age 45–64, are arthritis (480 vs. 285); hypertension (394 vs. 251); hearing impairment (296 vs. 136); heart conditions (277 vs. 123); and orthopedic conditions (277 vs. 123). The difference in prevalence rates is striking for these conditions; for example, the odds of suffering from arthritis are 68% higher and of suffering from hypertension are 57% higher in the older group.

Older people use medical personnel and facilities more often than younger adults. People aged 65 and over are hospitalized twice as often as the younger population, stay 50% longer, and use twice as many prescription drugs (National Center for Health Statistics, 1987b). Elderly patients also exceed younger adults in outpatient physician visits, by a ratio of about three to two (U.S. Senate Special Committee on Aging, 1987–1988).

In 1985, persons over the age of 65 accounted for 30% of all hospital discharges and 41% of all short-stay hospital days of care. People aged 75 and above, who constitute only 5% of the current population, accounted for 16% of hospital discharges and 22% of short-stay hospital days (National Center for Health Statistics, 1987b). Most older people come to hospitals because of an acute episode of a chronic condition. The most common categories of discharge diagnoses for elderly patients are diseases of the circulatory system (31%); heart disease (20%); digestive diseases (12%), respiratory disease (11%); neoplasms (19%) and pneumonia (4%). On average, 4 diagnoses are assigned to elderly patients, as opposed to only 2.4 diagnoses per younger patient (National Center for Health Statistics, 1987c).

Although older people make frequent use of medical facilities and personnel, many of their day-to-day needs for assistance are met by relatives and friends. For disabled older people living in the community, relatives provide 84% of the care; about 80% of these caregivers provide assistance on a daily basis, and 64% do so for at least a year (U.S. Senate Special Committee on Aging, 1987–1988).

Those with chronic needs that cannot be met at home generally receive care within nursing homes. At any given time, only about 5% of the elderly population are residing in nursing homes, but the lifetime risk for this type of institutionalization is much higher (estimated at 52% for women and 30% for men who have reached the age of 65; U.S. Senate Special Committee on Aging, 1987–1988). Residents of nursing homes are most often very old (22% ≥85 years vs. 1% aged 65–74 years). There are more women than men in these settings (75% vs. 35%), and more people without children (37% vs. 19% for older people in the community; U.S. Senate Special Committee on Aging, 1987–1988).

People over the age of 65 account for one third of the country's total personal health care expenditures. Per capita health care spending in 1984 was $4,200 per old adult, of which 45% was spent for hospitalization, nearly 21% for physician services, and an additional 21% for nursing-home care (U.S. Senate Special Committee on Aging, 1987–1988). Medicare provides the largest single source of funding for the health care costs of older people, and the greatest proportion (69%) of the Medicare funds is spent on short-stay hospital care. For older people in nursing homes, one of the primary funding sources is Medicaid, which covered 42% of the costs of long-term care in 1985. It is also important to note that the elderly paid one third of their own medical costs in 1984 through direct payments or through insurance premiums, and that more than one half (52%) of nursing-home costs in 1985 had to be covered by direct patient payments (U.S. Senate Special Committee on Aging, 1987–1988).

Aging and Mental Disorders

Data on mental disorders in older adults run counter to the trends for physical disease; that is, most forms of mental illness are no more prevalent in the aged than in younger adults, and for a variety of reasons, the old are receiving fewer mental health services than younger age groups.

Information provided by the National Institute of Mental Health (NIMH) Epidemiologic Catchment Area Survey suggests that community-resident older adults may have slightly lower rates of diagnosable mental disorder than younger people. For example, the one-month prevalence rate for all forms of mental disorder was reported to be 12.3% for people aged 65 years and older, compared to 16.9%, 17.3%, and 13.3% for people of the ages 18–24, 25–44, and 45–64, respectively (Regier, Boyd, Burke, Rae, Myers, Kramer, Robins, George, Karno, & Locke, 1988). Only one category of illness (i.e., severe cognitive impairment) was observed significantly more often in the old than in the young (4.9% for the ≥65 group vs. 0.6% to 1.2% for younger groups).

Although one in every eight older people suffers from mental disorder, only a small proportion of older adults receive any community-based mental health services. For example, elderly people have been estimated to account for only 2% of the patients seen by private therapists, and for only 6% of those served by community mental health centers (Roybal, 1988). The gap between mental health needs and service utilization has been attributed to several factors. The most tangible, and probably the most important, are limited reimbursement, limited access, and staffing patterns (Gatz & Pearson, 1988; Knight, 1986). Attitudes may also play a role, among both professionals and older patients. Although there is little evidence that mental health service providers, or Americans in general, are strongly or per-

vasively ageist (see review by Gatz & Pearson, 1988), more specific types of biases (e.g., over-diagnosis of organic conditions or less frequent referrals for psychotherapy) are sometimes present. There is also growing concern that ageism may be assuming new forms. Braithwaite (1986) described an "antidiscrimination response" among health care professionals, in which a specific older person's problems are ignored or minimized because professionals have lowered expectations of elderly people in general; in effect, by "going out of their way not to denigrate the aged" (Gatz & Pearson, 1988), professionals may be failing to recognize legitimate psychological problems with potential for remediation. A second area of concern is that both the general public and health care professionals may be forging too strong a mental association between old age and possible Alzheimer's disease. Gatz and Pearson (1988) cited evidence that older adults, college students, and caregivers of patients with dementia all greatly overestimated the prevalence of Alzheimer's in the aged.

Older patients themselves contribute to underutilization of mental health services. It is clear that contemporary old people are more likely to consult medical specialists, such as their family doctor, a general practitioner, or an internist, than they are to seek out the services of a mental health professional (Gatz, Popkin, Pino, & VandenBos, 1985). Many older people focus heavily on the reporting of physical phenomena, as opposed to psychological events, and if psychological complaints are raised, these, too, are most often communicated to a primary-care physician (e.g., Shapiro, Skinner, Kessler, Von Korff, German, Tischler, Leaf, Benham, Cottler, & Regier, 1984).

These observations suggest that psychologists who work in medical settings are more likely to encounter older patients and their mental health problems than are those who work in private practice or in community facilities. The exact prevalence of mental illness in older medical patients is difficult to estimate because of the lack of large-scale investigations with adequate methodology. Studies often focus on a single category of illness (e.g., delirium), provide little information about methods for case identification, or report prevalence only within series of patients who have already been referred for psychiatric evaluation (e.g., Small & Fawzy, 1988). However, a recent study by Rapp and colleagues (Rapp, Parisi, & Walsh, 1988; Rapp, Parisi, Walsh, & Wallace, 1988) with a medically-ill aged sample can be used to illustrate how often mental problems are likely to be encountered. The subjects were a random sample of more than 300 inpatients over the age of 65 who had been admitted to the medical-surgical services of a Veterans Administration hospital. Nearly a third of these patients had signs of cognitive impairment, and more than a quarter without such deficits were found to have other psychiatric disorders when the research diagnostic criteria (RDC) were used (Spitzer, Endicott, & Robbins, 1978). These estimates are in line with those provided by the literature on psychiatric consultations in medical settings, where 37%–54% of aged patients are diagnosed as having cognitive disorder and 23%–27% as having affective disorder (see Small & Fawzy, 1988, for a review).

When input from mental health professionals is not available, the literature suggests that mental disorders are typically overlooked in medical settings. German, Shapiro, Skinner, Von Korff, Klein, Turner, Teitelbaum, Burke, and Burns (1987), who studied an ambulatory care population of more than 1,200 patients of varying ages, found that physicians assigned psychiatric diagnoses or made note of emotional distress for only 41% of the elderly patients who had scored in the critical range on a mental health screening questionnaire; attempts at treatment (counseling, contact with human services agencies, referral to a mental health specialist, or the prescription of psychoactive medications) were observed for only 32% of the emotionally distressed older patients. These investigators and many others (e.g., Lipowski, 1987; Small & Fawzy, 1988) have emphasized that underrecognition of mental illness is a particularly serious problem for older medical patients. There are many reasons why this may be so: compared to younger patients, the elderly tend to rely more heavily on primary-care physicians; their multiple medical illnesses may divert physicians' attention away from

psychiatric signs and symptoms; depression and anxiety may be viewed as unremarkable in people with serious medical illness; and those without specific geriatrics training may find it hard to distinguish normal aging changes from those that result from illness.

Common Referral Questions for Older Patients in Medical Settings

Psychologists most often consult on medical cases when staff suspect that major mental illness is present, or when there are specific behavioral management problems (Small & Fawzy, 1988). Because cognitive problems and depression predominate as reasons for referral, much of our discussion pertains to procedures for improving the identification and treatment of these conditions. Pain management and sleep disturbance are also discussed because they are particularly promising areas for psychological consultation and intervention with older patients.

Cognitive Impairment: Identification and Treatment

When an older medical patient is evaluated for confusion or memory disturbance, the desired outcome is likely to be a specific diagnostic impression (e.g., Alzheimer-type dementia or multi-infarct dementia). However, in acute care settings, the ability to make specific diagnostic statements is limited by many practical considerations. Patients generally stay in the hospital for only a few days and may be too fatigued or in too much pain to participate in a thorough assessment. Often, an adequate history is lacking, and without such information, it may be impossible to reasonably evaluate the relative contributions of active systemic illness or medical treatment to current cognitive problems. Ambulatory care provides a better setting for neuropsychological testing, but here, too, the likelihood of completing thorough evaluations is restricted for geriatric patients. Many older patients are discharged to nursing homes, where further evaluation of mental status is unlikely, and even those who return to their own or their families' homes may have financial or transportation limitations that will interfere with outpatient evaluations.

Medical psychologists' most positive contributions to cognitively impaired elderly patients are likely to come in three areas: (1) providing reliable and valid indications of the extent of global cognitive dysfunction, in particular, indicating whether current problems exceed those expected with normal aging; (2) providing input on the temporal and topical pattern of cognitive dysfunction that may contribute to the detection of delirium; and (3) educating staff and family members about environmental and behavioral management techniques that can minimize the excess disabilities that so often accompany cognitive impairment. To fill these roles, we need to be informed about the broad range of disorders that may present with cognitive impairment, to have an excellent working knowledge of the distinctions between dementia and delirium, and to be realistically aware of the prognosis for cognitive-impairment syndromes.

Common Forms of Cognitive Impairment

Next to normal aging, which often results in the mild compromise of fluid intellectual abilities and the learning and recall of new information, dementia is by far the most common form of old-age cognitive impairment. Of people over the age of 65, 10%–15% have at least mild dementia, and 4%–6% have moderate to severe impairment (Terry & Davies, 1980). The primary risk factor for dementia is advancing age, and in people who are 80 years old or older, prevalence estimates generally exceed 20%. In hospitals and in long-term care, delirium is also a common disorder in elderly patients. Liston (1982) cited several studies with elderly patients on medical-surgical wards in which the rates of delirium were reported to range from 14% to 30%. Lipowski (1987) described two additional investigations in which 30%– 50% of medically hospitalized patients aged 70 years or older displayed symptoms of delirium either initially or at some point in the hospital stay. Delirium is often underdiagnosed, and problems in its recognition and management

are more common for elderly than for younger patients (Lipowski, 1987).

Diagnostic Criteria for Dementia. The term *dementia* denotes a state of generalized and persistent cognitive decline, with prominent memory impairment, that is severe enough to interfere with important everyday activities. Such cognitive incapacity was once called *senility* and was attributed to old age or to attendant conditions such as "hardening of the arteries." Currently, dementia is regarded as an illness, distinct from normal aging, that may have a variety of specific causes, as discussed below.

It is important to recognize that dementia is a *behavioral* syndrome. When correctly applied, a diagnosis of dementia indicates that certain types of cognitive disturbance are present. In the revised third edition of the American Psychiatric Association's *Diagnostic and Statistical Manual of Mental Disorders* (DSM-III-R; APA, 1987), the positive criteria for dementia are (1) demonstrable evidence of impairment in short- and long-term memory; (2) at least one of the following: impairment of abstract thinking, impaired judgment, personality change, or other disturbance of higher cortical function, such as aphasia, apraxia, agnosia, and constructional difficulty; and (3) significant interference of the disturbance with work or usual social activities or relationships with others.

Certain exclusionary criteria must also be met before the label of dementia can be applied. In the DSM-III-R, the primary basis for exclusion is that dementia cannot be diagnosed when cognitive disturbance occurs exclusively during the course of delirium. One of two conditions pertaining to etiology must also obtain: (1) there is evidence from history, physical examination, or laboratory tests of a specific organic factor judged to be etiologically related to the disturbance; or (2) in the absence of such evidence, an etiological organic factor can be presumed, if the disturbance cannot be accounted for by any nonorganic mental disorder (e.g., major depression).

Although the DSM-III-R criteria for dementia leave much to be desired, they do address some issues of importance in evaluations taking place in medical settings. These criteria place restraints on the type and extent of deficits that can be considered sufficient to raise the question of dementia and to prompt the clinician to consider certain explanations. Explicit recognition is given to the fact that "dementia is not synonymous with aging" (APA, 1987, p. 106) and that a diagnosis of dementia is warranted only when the cognitive deficits are quite severe.

Delirium. Like dementia, delirium results in a general compromise of intellectual functions, but the core deficit is in attention, and the symptoms have a more abrupt onset and fluctuating course. In the DSM-III-R, the behavioral features necessary for a diagnosis of delirium consist of the following: (1) reduced ability to maintain attention to external stimuli and to shift attention appropriately; (2) disorganized thinking, demonstrated by rambling, irrelevant, or incoherent speech; and (3) two or more of the following problems: reduced level of consciousness, perceptional disturbances, increased or decreased psychomotor activity, disorientation, and memory impairment, either recent or remote.

An additional criterion is that the clinical features develop over a short period of time and tend to fluctuate. Also, there must be evidence of a specific organic factor judged to be the cause of the mental disturbance, or all reasonable nonorganic causes (e.g., mania) must have been ruled out.

An important aspect of these criteria is that they preclude the diagnosis of dementia in the presence of significant delirium, except in the case where there is a clear history of preexisting dementia. When in doubt as to the differentiation of dementia and delirium, the recommendation is to assign a provisional diagnosis of delirium, which "should lead to a more active therapeutic approach" (APA, 1987, p. 102).

"Reversible" and "Currently Irreversible" Causes

At one time, the term *dementia* was used to connote an irreversible organic mental disorder, whereas *delirium* (or *acute organic brain syndrome*) was used to designate reversible organic conditions. As indicated above, the current diagnostic criteria make no assumptions regarding the reversibility of cognitive impairment symptoms, partly because clinical research has found that either dementia or delirium can sometimes result from specific, potentially treatable conditions. Table 1 lists some of the conditions that have been associated in the literature with dementia, delirium, or both (National Insitute on Aging Task Force, 1980). In many instances, the evidence that a condition is a causative factor is largely anecdotal, or is based on a few case examples; also, both *dementia* and *delirium* are terms that are used quite loosely, the former referring to relatively pervasive cognitive problems that have a gradual onset, and the latter to more transient, rapidly developing states in which confusion is a prominent feature. The original aim of this listing was simply to raise practitioners' consciousness to the fact that generalized cognitive impairments *may* have a specific cause and, at times, a specific treatment.

It is noteworthy that therapeutic drug intoxication is at the top of the list. The list of nonpsychiatric medications that can produce psychiatric symptoms, either alone or in interaction, is daunting (see The Medical Letter on Drugs and Therapeutics, 1984, 1986, for some useful listings), and psychoactive medications are among the most common precipitants of adverse reactions. Drug effects are of particular concern in the elderly because older people consume more physician-prescribed medications than younger adults and many take multiple medications (Ouslander, 1981). The increased potential for drug toxicity in older individuals is well documented, as is the heightened possibility of drug–drug interactions (Hicks, Dysken, Davis, Lesser, Ripeckyj, & Lazarus, 1981). Acute toxicity is the most common cause of delirium, as discussed below, but adverse medication reactions can also present as dementia. This dementia may result when an individual has been taking a medication at a fixed dose for many years but is no longer metabolizing it with the same degree of efficiency, or when a new drug is added without adequate consid-

Table 1. Reversible Causes of Mental Impairment[a,b]

Therapeutic drug intoxication[d,D]
Depression[d]
Metabolic

a. Azotemia or renal failure[d,D]
b. Hyponatremia[d,D]
c. Hypernatremia[D]
d. Volume depletion[d,D]
e. Acid-base disturbance[D]
f. Hypoglycemia[d,D]
g. Hyperglycemia[D]

h. Hepatic failure[d,D]
i. Hypothyroidism[d,D]
j. Hyperthyroidism[d,D]
k. Hypercalcemia[d,D]
l. Cushing's syndrome[d]
m. Hypopituitarism[d,D]

Infection, fever, or both

a. Viral[d,D]
b. Bacterial
   Pneumonia[D]
   Pyelonephritis[D]
   Cholecystitis[D]

   Diverticulitis[D]
   Tuberculosis[d,D]
   Endocarditis[d,D]

Brain disorders

a. Vascular insufficiency
   Transient ischemia[D]
   Stroke[d,D]
b. Trauma
   Subdural hematoma[d,D]
   Concussion/contusion[D]
   Intracerebral hemorrhage[D]
   Epidural hematoma[D]

c. Infection
   Acute meningitis[D]
   Chronic meningitis[d,D]
   Neurosyphilis[d,D]
   Subdural empyema[d,D]
   Brain abscess[d,D]
d. Tumors
   Metastatic to brain[d,D]
   Primary in brain[d,D]
e. Normal pressure hydrocephalus[d]

Cardiovascular

a. Acute myocardial infarct[D]
b. Congestive heart failure[d,D]
c. Arrhythmia[d,D]

d. Vascular occlusion[d,D]
e. Pulmonary embolus[d,D]

Pain

a. Fecal impaction[d,D]
b. Urinary retention[D]

c. Fracture[D]
d. Surgical abdomen[D]

Sensory deprivation states such as blindness or deafness[d,D]
Hospitalization

a. Anesthesia or surgery[d,D]

b. Environmental change and isolation[d,D]

Alcohol toxic reactions

a. Lifelong alcoholism[d]
b. Alcoholism new in old age[d,D]

c. Decreased tolerance with age, producing increasing intoxication[d,D]
d. Acute hallucinosis[D]
e. Delirium tremens[D]

Anemia[d,D]
Tumor—Systemic effects of nonmetastic malignant neoplasm[d,D]
Chronic lung disease with hypoxia or hypercapnia[d,D]
Deficiencies of nutrients such as vitamin B, folic acid, or niacin[d]
Accidental hypothermia[D]
Chemical intoxications

a. Heavy metals such as arsenic, lead, or mercury[d,D]

b. Consciousness-altering agents[d,D]
c. Carbon monoxide[d,D]

[a]Adapted from "Senility Reconsidered" by the National Institute of Aging Task Force, 1980, *Journal of the American Medical Association, 244,* pp. 261–262.
[b]d = dementia; D = delirium; d,D = either or both.

eration of the set of medications that the individual is already taking.

Depression is placed second on the list. The fact that a substantial minority of older depressed people experience severe cognitive deficits is well known among geriatric professionals, but the extent of the cognitive problems that can result from depression may be hard for clinicians to appreciate if they are not experienced in treating depression in older patients.

The list also includes sensory deprivation, pain, and hospitalization as possible primary causes of either dementia or delirium. The fact that these conditions can result in global (and sometimes persistent) cognitive problems may be an example of what Roth (1971) and Rothchild (1942) have referred to as altered compensation or "threshold" effects, in which an older person's brain function is viewed as less adaptive than that of a younger person, so that the old are more vulnerable to precipitous losses in function when faced with abrupt environmental changes. This is a common experience for older patients with a prior history of brain impairment, but it can also occur in individuals without signs of cognitive decline before hospitalization. It is important to plan for this possibility when discussing elective hospitalization with an older person and his or her family, and to minimize the amount of relocation whenever possible.

How often can dementia be attributed to specific medical, psychiatric, or environmental conditions? A commonly cited figure, based on autopsy and clinical studies, is that 10%–20% of older demented patients have potentially reversible disorders (e.g., National Institute on Aging Task Force, 1980). These estimates are being reexamined in prospective studies, such as the one conducted at the University of Washington, Seattle (Larson, Reifler, Sumi, Canfield, & Chinn, 1985). The subjects were 200 individuals aged 60 or older who represented consecutive referrals to a geropsychiatry outpatient program specializing in the diagnosis and treatment of cognitive impairment. Clincial diagnoses were assigned by a multidisciplinary team, supplemented by neuropsychological evaluation and a medical diagnostic

work-up. In this carefully evaluated series, 9% had no dementia or had only age-associated forgetfulness, and 17% had specific medical causes other than Alzheimer-type or multi-infarct dementia. Relatively common specific causes were dementia due to drugs (5%), alcohol-related dementias (4%), hypothyroidism (1.5%), and other metabolic disorders (2.5%). Of the individuals judged to be without dementia, two thirds (5% of the total sample) were diagnosed as having depression.

As these statistics illustrate, even when complete diagnostic evaluations are performed, the specific causes of dementia are likely to remain unidentified for a majority of patients. Under these circumstances, the diagnosis of exclusion—Alzheimer-type dementia—becomes appropriate. This disorder is the most common of the "currently irreversible" dementias, although careful attention to medical and psychosocial problems can attenuate some of the symptoms (see below). In medical settings, emphasis is understandably placed on the evaluation of the presenting systemic problem, and a thorough diagnostic examination for specific causes of dementia may not be performed. In addition, many patients also have acute mental status changes that complicate the evaluation of baseline cognitive ability. For these reasons, great caution is recommended in the use of the label of Alzheimer-type dementia in medically ill individuals.

Structured Assessment Techniques

Clinical interview supplemented by a structured mental status exam is the most useful procedure for assessing cognitive deficit in medically ill aged patients. The results of these procedures can provide a valuable baseline for assessing treatment-related cognitive changes and can also be used to determine if a referral for neuropsychological assessment is desirable or feasible (see Chapter 18 by Sweet, in this book). Many cognitive mental status examinations are available, and several, including the Information-Memory-Concentration Test (Blessed, Tomlinson, & Roth, 1968), the Mini-Mental State Examination (Folstein, Folstein, & McHugh, 1975), the Mattis Dementia Rating Scale (Mattis, 1976), the Kahn-Goldfarb Mental Status Questionnaire (Kahn, Goldfarb, Pollack, & Peck, 1960), and the Short Portable Mental Status Questionnaire (Pfeiffer, 1975), have been widely used with geriatric patients (see Kane & Kane, 1981; Poon, 1986; Roca, 1987 for additional examples and critiques). Two such instruments, the Mini-Mental State and a newer scale, the Neurobehavioral Cognitive Status Examination (Kiernan, Mueller, Langston, & Van Dyke, 1987), are discussed in detail below.

The Mini-Mental State Examination (MMSE). The MMSE was developed as "a practical guide for grading the cognitive state of patients" (Folstein *et al.*, 1975). It is a 30-point scale that includes orientation items and brief tests of memory, concentration, language, and motor skills.

The original work on the MMSE (Folstein *et al.*, 1975) was based on a study of 206 hospitalized patients diagnosed as having dementia syndromes, affective disorder, affective disorder with cognitive impairment ("pseudodementia"), mania, schizophrenia, and personality disorders, as well as 63 normal older subjects. The mean score for elderly normal subjects was 27.6 ($SD = 1.7$, range $= 24–30$), whereas the mean score for patients with dementia was only 9.6 ($SD = 5.8$, range $= 0–22$). Age-matched groups with affective disorder scored in between these two extremes. Those with uncomplicated depressive disorder had a mean score of 26.1 ($SD = 4.4$, range $= 17–30$), and those judged to have depression with cognitive impairment averaged 18.4 correct ($SD = 5.7$, range $= 9–27$). These preliminary data led to the recommendation that scores of 0–23 be used to raise the question of cognitive impairment (i.e., none of the normal older adults had scored less than 24).

The MMSE has generally been found to have adequate reliability, and some concurrent validation studies have been reported (see Roca, 1987, for a review). However, several studies have underscored the point that "the MMSE does not make a diagnosis. Rather, a low MMSE score indicates a need for further evaluation" (Folstein, Anthony, Parhad, Duffy, & Gruenberg, 1985, p. 232). For example, Anthony and

others (Anthony, Le Resche, Niaz, Von Korff, & Folstein, 1982) administered the MMSE to 99 patients consecutively admitted to a general medical ward at a university hospital; each patient was independently evaluated by a psychiatrist who judged the presence or absence of mental disorder. Although the MMSE was found to have an overall sensitivity of 87% and a specificity of 82% relative to clinical diagnosis, the specificity was found to be particularly low for patients with an eighth-grade education or less (63% vs. 100% for those with more education) and for individuals aged 60 years and above (65% vs. 92% for younger adult subjects). As a result, these investigators concluded, "If the MMSE is to be used as a screen for delirium and dementia in the general population . . . a second stage procedure may be necessary for a definitive diagnosis . . ., particularly in persons who have less than an eighth-grade education, or who are elderly" (p. 406). Unacceptably high rates of false-negative errors have also been reported for the MMSE, generally when it has been used with younger and less globally impaired neurological and neurosurgical patients (Schwamm, Van Dyke, Kiernan, Merrin, & Mueller, 1987).

These findings suggest both substantial strengths and significant weaknesses in the use of the MMSE as a screening test for older adults. The primary advantages are that it probes a broad range of cognitive functions in a brief and reliable manner, and that it is sensitive to the gross impairments in cognitive state that are observed in older patients with dementia, delirium, and depression. In addition, because the MMSE is now commonly used in clinical settings where geriatric patients are treated, its findings are interpretable by a relatively wide audience of health care professionals. However, the absence of age- and education-adjusted cutting scores is an important limitation, as is the insensitivity of the scale to focal brain impairment.

The Neurobehavioral Cognitive Status Examination. Because brief tests such as the MMSE have sometimes yielded high false-negative rates, especially for individuals with mild or focal brain impairment, several new tests have been developed that attempt to provide a differentiated index of various cognitive domains while still retaining brevity. The most recent of these measures is the Neurobehavioral Cognitive Screening Examination (NCSE) developed by Kiernan *et al.* (1987), based on their experience with patients in a general medical hospital.

An examiner first rates a patient's consciousness, attention, and orientation. Then, graded series of questions are asked in five areas: language (assessing fluency, comprehension, repetition, and naming); constructional ability; memory (recall of four words at 10 minutes with category prompts, followed by multiple choices); calculation; and verbal reasoning (similarities and comprehension items). Each of the sections begins with a difficult item (called the *screen*); if the subject passes this item, no other questions of this type are given. If the screen is failed (as is reported to be the case with 20% of normal subjects), the remaining items in that section (the *metric*) are administered. Because normal subjects often need to be given only the screens for most sections, the test is said to take less than 5 minutes to administer to these individuals. For patients with impairment, administration time ranges from 10 to 20 minutes. The results are summarized in profile format, with separate points plotted for alertness, orientation, and attention, and for language, memory, calculation, visuoconstruction, and reasoning performance. Preliminary standardization data have been reported for several groups, including older adult volunteers (aged 70–92). Lower mean scores were observed in the older adult group in three areas (constructions, memory, and similarities), and cutoffs for clinical interpretation are adjusted downward on these scales for elderly patients.

The NCSE shows some promise as a screening instrument for medically ill older patients. The profile of outcomes is potentially useful, as a single summary score may mask important individual areas of weakness. No independent validation studies have been published as yet, and there are no data on reliability. It will be important for performance on this test to be separately examined in elderly patients with psychiatric disorder or medical illness before

abnormal scores are considered suggestive of neurological impairment.

In a preliminary examination of the NCSE in an inpatient geropsychiatric sample (Osato, Yang, & La Rue, 1988), we found that the NCSE was effective in distinguishing individuals with organic brain disorders from those with affective disorder, based on the number of clinical subscales yielding scores in the impaired range. However, even in depressed patients with no history or indication of neurological disorder, only a small proportion (4 of 14) scored within the limits recommended for normal aged subjects on all of the clinical scales. At present, a reasonable procedure would be to use the NCSE in conjunction with some more established mental status measure, or as an antecedent to more complete neuropsychological testing.

A Delirium Rating Scale. Although specific structured mental status exams may help to elicit cognitive and attentional deficits, scores on these exams do not distinguish delirium and dementia (e.g., Folstein *et al.*, 1985; Lipowski, 1987). Lipowski's recommended diagnostic procedure (1987) is an informal bedside mental status evaluation designed to elicit key symptoms of delirium, such as inconsistencies, disorganization, and distractibility. He emphasized that "acute onset of cognitive and attentional deficits and abnormalities, whose severity fluctuates during the day and tends to worsen at night, is practically diagnostic" (p. 1791).

For an experienced clinician familiar with elderly patients, these procedures may well prove adequate. However, many physicians and psychologists working in medical settings lack specialized experience with the aged and may work under severe time constraints that limit their ability to elicit and identify the inconsistencies and performance failures to which Lipowski (1987) referred. More objective screening measures would be of great assistance in ensuring that delirium will not be overlooked.

A step toward the development of such a measure has recently been published by Trzepacz, Baker, and Greenhouse (1988). These investigators reported data for 20 medical-surgical patients with delirium and 9 control subjects in each of three groups (dementia, schizophrenia, and affective or personality disorder) on a 10-item Delirium Rating Scale (DRS). The items pertain to the temporal onset of the symptoms, perceptual disturbance, the hallucination type, delusions, psychomotor behavior, cognitive status during formal testing, the physical disorder, sleep–wake cycle disturbance, lability of mood, and the variability of symptoms. Each is introduced by a brief statement of rationale for its selection and some rough guidelines for the differentiation of the symptoms produced by different types of disorders (e.g., affective disorders or schizophrenia vs. delirium). Mean DRS scores were found to be markedly elevated for the delirium group relative to any of the comparison groups, with no overlap in scores. Scores on the DRS correlated significantly with MMSE and Trails B scores in the delirium group, and there were no significant correlations between DRS scores and age in any of the patient groups. The interrater reliability between two independent raters familiar with the scale's use was high ($r = 0.97$).

Independent replication studies and an examination of the scale's psychometric properties are needed before its use can be recommended, but the DRS at least provides a list of potentially relevant clinical features to evaluate when delirium is suspected. Other diagnostic observations were reported by Weicker (1987), who compared the performance of elderly patients diagnosed with delirium, dementia, or depression on an objective mental status examination. Mental control items (e.g., counting backward from 20 or reciting the months or the alphabet) and clock-drawing tasks were found to be particularly useful in distinguishing dementia and delirium, with delirious patients more often losing track or perseverating on these simple tasks.

Treating Cognitive Impairment

Dementia. How often does detection of the specific medical and environmental causes of dementia lead to a reversal of or an improve-

ment in the symptoms? In the prospective study by Larson and colleagues (1985), the subjects were followed clinically over a one-year period. Of the 200 patients, 55 (27.5%) improved in cognitive function for a month or more, as determined by clinical assessment and/or improvement in psychometric and functional ratings. Most of the positive changes were attributed to the effective treatment of coexisting illnesses, such as depression or congestive heart failure. In 28 patients (14%), the improvement was sustained over the year, but in 27 (13.5%), it was transient. In only 2 cases was a completely reversible dementia observed (i.e., where the patient improved from a demented state to normal mental function).

Patients with dementia and their family members are often sustained by the small positive changes that can result from the effective treatment of coexisting conditions; as in other chronic or progressive illnesses, even a little improvement can make the difference between hope and despair. By attending to the total picture of a patient's problems, medical as well as behavioral, psychologists can strengthen their credibility and effectiveness in therapeutic work with caregivers. Contributing to support groups for family caregivers and starting such a group for staff who deal with demented patients on a daily basis are roles that psychologists are well prepared to take. Training family caregivers in behavioral management techniques can also be beneficial (Zarit, Anthony, & Boutselis, 1987). If time constraints make such involvement impossible, psychologists can at least be prepared to provide information about relevant resources in the community, such as the Alzheimer's Disease and Related Disorders Association.

Delirium. The importance of prompt treatment for delirium is underscored by the high mortality rates associated with this condition. For elderly patients, death has been reported to occur within the first month of illness for 17%–25% of patients (Liston, 1982). Younger patients appear to have a better chance of recovery from delirium than the elderly, and a more positive prognosis is observed if the episode is brief (Liston, 1982).

Lipowski (1983, 1987) emphasized the frequency with which delirious states in the elderly are induced by medications and recommended that current medications be stopped, or dosages reduced, to determine if these are the cause of the delirium. Identifying and treating systemic illnesses that may be etiologically related is also obviously an immediate priority. The categories of the physical illnesses that are most often associated with delirium in the elderly include cardiovascular disorders (e.g., congestive heart failure, myocardial infarction, and cardiac arrhythmias), infections (e.g., pneumonia and urinary tract infection), metabolic encephalopathies (e.g., electrolyte and fluid imbalance, diabetes, and hepatic, renal, and pulmonary failure), cerebrovascular disorders, and cancer (Lipowski, 1983; also see Table 1).

There are important environmental and psychological components of treatment that can affect the course of recovery. Having a familiar person such as a relative stay with the patient, providing a modest and consistent amount of stimulation, and providing frequent reorientation have all been recommended as useful (Liston, 1982). Skilled nursing care or supervision by family caregivers is crucial for preventing falls and injuries that may result from disorganized and agitated behavior. Low doses of major tranquilizers such as haloperidol may be useful in controlling extreme agitation (Lipowski, 1987), but it is often crucial to avoid psychoactive medications, particularly if the original delirium was drug-induced. It can be helpful to inform family members that the symptoms of delirium may resolve slowly, the acute phase being followed by a transitional phase in which there are residual problems with cognition, affect, or behavior, even when attention appears to have returned to normal.

Depression in Elderly Medical Patients

The shades of gray that differentiate normal and abnormal mood in a young adult persist, and may even be amplified, in old age. Particularly among the medically ill elderly, it is easy to come to expect some degree of depression and, perhaps, to underestimate the worth of

formal evaluation. Without such evaluation, however, it appears likely that serious depression will be overlooked. Research indicates that many middle-aged patients seen in general medical settings have significant symptoms of depression that are not detected or treated by primary-care physicians (e.g., Katon, Berg, Robins, & Risse, 1986), and current cohorts of older people are believed to be at an even greater risk of misdiagnosis, as they may be especially predisposed to view affective illness in somatic terms (e.g., Katon, Kleinman, & Rosen, 1982).

Rapp et al.'s recent study (1988) with aged, medically ill veterans gives some indication of the prevalence of depression in geriatric medical patients and of the underrecognition of this disorder on general medical services. Several self-report scales of depression were administered to 150 elderly patients (>65 years) who were randomly selected from consecutive admissions to medical-surgical units. After discharge, the medical charts were reviewed for indications of depression as recognized by the treating physicians (e.g., diagnosis of depression, use of the word depressed or related words such as despondent, or orders for psychiatric treatment or consultation). Based on RDC diagnoses, 27% of the patients received diagnoses of current mental disorder, and 15% were found to have depression. None of these patients was assigned a psychiatric diagnosis by house staff; only 2 of 23 were described in terms suggestive of depression; and 6 false positives were observed, most of whom had some notation about depression in their chart from previous admissions. Consistent with the underrecognition of depression, none of the patients was treated with antidepressant medications during the hospital stay.

Age Differences in Depressive Symptoms

The DSM-III-R does not provide separate criteria for diagnosing depression in older adults. However, under the heading of age-specific features of mood disorders, the manual notes, "In elderly adults some of the symptoms of depression, e.g., disorientation, memory loss, and distractibility, may suggest Dementia" (APA, 1987, p. 220). As many as one in every five older depressed patients has significant cognitive problems, but as discussed in greater detail below, these problems often improve substantially with treatment or time.

There is also a substantial literature suggesting that somatic complaints are more prominent in older depressed patients than in younger ones. In part, this finding results from the objectively higher rates of illness in the aged. On occasion, however, somatic complaints may serve as depressive equivalents, taking the place of mood disturbance, or these complaints may be amplified out of proportion to the underlying physical state. In both of these circumstances, a diagnosis of depression may be appropriate even if an actual medical illness complicates the picture, particularly when the depressive symptoms exceed age-adjusted cutoffs on depression rating scales (see below).

Physical Illness as a Risk Factor for Depression in Old Age

Many studies suggest that physical disease may affect the development and persistence of depressive disorders in older people. Blazer and Williams (1980) reported that almost 15% of a large, stratified random sample of older adults in the community had "substantial depressive symptomatology," and in nearly one half of this subgroup, the depressive symptoms were judged to be related to medical illnesses. When community-resident elderly are followed longitudinally, those who are in poorer initial health or who become ill in the interim are more likely to develop depression than those whose health is better (Phifer & Murrell, 1986). In a one-year naturalistic follow-up of older adults who had been hospitalized for depression, Murphy (1983) observed that patients with major chronic physical health problems were significantly more likely to relapse or to remain continuously ill than elderly depressives with minimal or less severe health problems.

Physical illness can be considered a risk factor for depression at any age, but there are

suggestions in the literature that the link between medical illness and depression may be stronger in the old than in the young. In a longitudinal investigation, Turner and Noh (1988) compared predictors of depression among three adult age groups (18–44 years, 45–64 years, and 65 years and above). Physical disability measures (pain and functional limitation) were significantly related to depression in the oldest age group, but not in the two younger groups. By contrast, stressful life events were more closely linked to depression in the young than in the old, and social support and a sense of mastery were significant contributors at all ages.

Depression Rating Scales

Although several studies have shown that elderly subjects' total scores on depression rating scales are inflated by their responses to somatic items, recent investigations indicate that depression rating scales can be effectively used with the medically ill elderly (e.g., Norris, Gallagher, Wilson, & Winograd, 1987; Rapp et al., 1988a).

The Beck Depression Inventory (BDI). The BDI (Beck, Ward, Mendelson, Mock, & Erbaugh, 1961) is a 21-item, multiple-choice survey that includes items evaluating mood, sense of pessimism and guilt, social withdrawal, sleep disturbance, loss of energy, and weight and appetite. Like most standardized assessment instruments, it was initially developed for use with younger adults, and good normative data for elderly populations are lacking. However, the BDI has been used in several studies with elderly depressives, where it appears to meet the minimum standards for reliability and validity. For example, Gallagher, Nies, and Thompson (1982) administered the BDI to both normal and depressed older adults and obtained values of .90, .84, and .91, respectively, for test–retest, split-half, and alpha reliabilities. Satisfactory validity relative to RDC diagnoses were observed in a second study (Gallagher, Breckenridge, Steinmetz, & Thompson, 1983).

A 13-item short form of the BDI is available

that may be particularly useful in medical settings (Beck & Beck, 1972). Scogin, Beutler, Corbishley, and Hamblin (1987) obtained Spearman-Brown split-half and alpha coefficient reliabilities of .84 and .90, respectively, for the short form in a sample of depressed and normal older adults. Using a cutoff score of 5, these authors found the BDI short form to correctly identify 97% of the depressed sample and 77% of the nonpatient sample.

The Geriatric Depression Scale (GDS). Developed by Yesavage, Brink, Rose, Lum, Huang, Adey, and Leirer (1983), the GDS consists of 30 items selected for their "appropriateness and performance with the aged population" (p. 154). A yes-no answer format is used to simplify the self-rating procedure. Mood is most extensively surveyed, and there are items assessing cognitive complaints and social behavior. The GDS has eliminated most somatic items ("Do you often get restless and fidgety?" and "Do you feel full of energy?" are the closest approximations).

Norris *et al.* (1987) recently validated the BDI and the GDS against RDC and clinical diagnoses of depressive disorders in a geriatric medical outpatient population. When the BDI was used, the proportion of patients correctly diagnosed was .84 for both a conservative and a more lenient BDI cutoff score (17 and 10, respectively). When the GDS was used, the accuracy of diagnosis was .77 and .84 for lenient and more conservative cutoff scores (10 and 14, respectively).

Further support for the use of these scales in general medical settings was provided in Rapp and colleagues' research (1988) with patients on medical-surgical units. Both the BDI and the GDS were found to have moderately high sensitivity (83% and 70%) and specificity (65% and 89%) relative to RDC diagnoses using cutoffs of 10 on each scale; near-perfect sensitivity to major depression, as opposed to minor or intermittent depression, was observed for both scales. To increase specificity, Rapp *et al.* recommended raising the cutoff on the GDS to 13 or using the BDI Psychological subscale, with a cutoff of 5. These investigators underscored the

fact that such screening instruments were *much* more sensitive to depression than the procedures being used by medical house staff (which resulted, as discussed above, in the overlooking of nearly all cases of depression).

The Hamilton Rating Scale for Depression (HAMD). Self-rating scales may not be appropriate for some elderly patients, particularly those with visual deficits, limited education, and poor English-language proficiency. For example, Toner, Gurland, and Teresi (1988) used the Zung Self-Rating Depression Scale (Zung, 1965) with older patients at a general internal medical group practice and found that 35% were unable to complete the scale because of visual problems, illiteracy, lack of motivation, or other reasons. Self-rating scales may also be insensitive to "masked depressions," where subjective dysphoria is minimal and somatic symptoms predominate. or where an individual wishes deliberately to minimize the reporting of psychological distress.

An observer-rated depression scale, such as the HAMD (Hamilton, 1960), is likely to be most useful in these situations. This 23-item instrument was originally designed as an outcome measure for drug studies of depression in the general adult population. As such, it is weighted for the type of variables that medications are able to alter (i.e., sleep, weight change, psychomotor speed, and other biological concerns), although it also taps depressive mood, anxiety, guilt, loss of libido, paranoia, obsessional symptoms, and suicidal ideation.

Kochansky (1979) noted that normative studies of the HAMD with elderly populations are lacking and indicated that this and other depression rating scales may fail to distinguish between dementia and depression as a basis for certain symptoms (e.g., changes in work and everyday activities and difficulties sleeping). However, the HAMD has been used extensively in clinical research investigations with older adults, where it has been shown to be sensitive to changes in depressive symptoms (e.g., Jarvik, Mintz, Steuer, & Gerner, 1982), and some rough normative reference points are provided in studies such as that of Yesavage and colleagues (1983). Concerns have also been raised about the need to train raters in order to ensure adequate reliability and validity. Recently, a structured interview guide has been developed for the HAMD (Williams, 1988), which may be of some help in increasing reliability in busy clinical settings.

Distinguishing Depression and Dementia

Problems with concentration, memory, and psychomotor slowing are common in major depression (Weingartner & Silberman, 1982), but these deficits are usually mild in severity and less striking clinically than disturbances in mood or vegetative symptoms such as poor appetite or sleep (Friedman, 1964). However, in elderly depressed patients, cognitive deficits induced by depression are superimposed on age-related changes, increasing the odds that intellectual function will be significantly compromised (Siegfried, 1985).

Between 10% and 20% of depressed older adults have cognitive problems severe enough to rival the deficits produced by organic dementia (McAllister, 1983; Rabins, 1983). This combination of cognitive and affective symptoms has been referred to as *depressive pseudodementia* (Caine, 1981; Wells, 1979) or, more accurately, as *dementia syndrome of depression* (Folstein & McHugh, 1979). The correct diagnosis of the dementia syndrome of depression is complicated by the fact that many patients with Alzheimer-type or multi-infarct dementia develop depressive symptoms (e.g., Liston, 1977; Reifler, Larson, & Hanley, 1982). Thus, combinations of cognitive and depressive symptoms can have different underlying causes.

The dementia syndrome of depression can be clearly identified only on a *post hoc* basis, that is, when cognitive performance returns to normal as depression lifts. However, a monitoring of certain clinical features may facilitate the prospective identification of patients with this condition. According to Wells (1979), depressive pseudodementia is likely to be characterized by some or all of the following features: a history of prior depressive episodes, mood disturbance antedating cognitive problems in the current episode, abrupt onset and rapid progression of cognitive deficits, circumscribed as opposed to global impairment on mental status examinations, frequent "don't know" answers in cognitive testing, and incongruity between performance on testing and everyday cognitive function.

These guidelines for detecting depressive pseudodementia have not been validated in any large-scale longitudinal investigations. However, Reynolds, Hoch, Kupfer, Buysse, Houck, Stack, and Campbell (1988) recently described the pretreatment clinical presentations of 14 patients diagnosed with depressive pseudodementia and of 28 diagnosed with primary degenerative dementia (PDD). The pseudodemented subjects were selected on the basis of several criteria, including a positive response to treatment with antidepressant medication, with parallel cognitive benefits. Compared to the patients with PDD, those with pseudodementia had initially presented with a milder level of global cognitive impairment (a mean MMSE score of 23 as opposed to 17), more severe depressive symptomatology (a mean HAMD score of 20 as opposed to 16), and fewer problems with everyday activities such as finding one's way around familiar streets. On the MMSE, the pseudodemented patients had as many problems as the PDD patients on delayed recall, the repetition of a phrase, and following the three-stage command, but they outperformed the demented group on orientation, registration, calculations, and the language and visuographic items. The pattern of cognitive test findings lends some support to Caine's suggestion (1981) that depressive pseudodementia may impair "subcortical" functions to a greater extent than it impairs "cortical" abilities. On the HAMD, the pseudodemented patients were rated as having more anxiety, insomnia, and loss of libido than the PDD patients.

This study and many others show that some patients with mixed symptoms of depression and dementia improve substantially with treatment and maintain these gains for at least a year or two (e.g., La Rue, Spar, & Hill, 1986; Post, 1966; Rabins, Merchant, & Nestadt, 1984).

However, it is unlikely that brief screening instruments will prove sufficient to distinguish patients with the dementia syndrome of depression from those who have depressive symptoms in addition to mild dementia of the Alzheimer or multi-infarct type. More extensive neuropsychological evaluation can be helpful in this differentiation in some cases. Often, however, the best course is to treat the depressive symptoms and to carefully monitor the treatment response. In our own research in this area, we found that cognitively impaired depressed patients benefited as much from treatment as those who were cognitively intact, but that a longer and more aggressive course of treatment was required to achieve a positive outcome in the patients with cognitive problems (La Rue et al., 1986).

Treating Geriatric Depression

Psychologists working in medical settings often find themselves in a position of needing to advocate treatment for depression in an elderly patient. Older patients may be viewed as poor psychotherapy candidates, and antidepressant medications may be withheld because of concern about possible side effects. The assumption may also be made that if the medical illness is eliminated, or its effects ameliorated, the depressive symptoms will disappear without specific intervention.

Psychotherapy. Recent studies of psychotherapy outcome in geriatric depression have produced encouraging results. Thompson, Gallagher, and Breckenridge (1987) studied the effects of individual behavioral, brief psychodynamic, and cognitive therapies for older adults diagnosed as having major depressive disorder according to the RDC. At a six-week interim evaluation using an extensive battery of symptom rating scales, including the BDI, the GDS, and the HAMD, the treated patients had improved significantly relative to wait-list controls; there was no indication of spontaneous improvement in the untreated control condition. At the end of three to four months of active therapy, 70% of the treated subjects had improved substantially, and 52% were judged to be in remission relative to the RDC. No significant differences were obtained between the therapy modalities. The investigators concluded that professionals should encourage older patients to seek psychotherapy, even if they are initially reluctant, as there was little evidence of spontaneous improvement in untreated older patients.

Thompson and colleagues did not provide information about the medical status of the participants in their study, and it is reasonable to question whether this high degree of efficacy would be obtained in physically ill depressed individuals. Jarvik et al. (1982) studied the effectiveness of two forms of group psychotherapy—cognitive-behavioral and psychodynamic—for elderly depressed patients who had medical contraindications for antidepressant medication. At 26 weeks of treatment, an average improvement in HAMD scores of 30% was observed, compared with a 19% decrement on this symptom index in a placebo control condition. Twelve percent of these patients were in remission from depression at this point in treatment.

The differences in psychotherapy efficacy in these two studies may have resulted from procedural differences or from sampling differences unrelated to physical health. Although Jarvik and colleagues observed less dramatic results, both studies found that psychotherapy was successful in reducing depressive symptoms compared to an absence of specific intervention.

Drug Treatment. Antidepressant medications have also been found to be effective in treating geriatric depression. For example, in a companion investigation of the psychotherapy study noted above, Jarvik and colleagues (1982) studied the efficacy of two tricyclic antidepressants—imipramine and doxepin—in a double-blind placebo-controlled study of elderly outpatients. A 50% improvement in HAMD scores was observed at 26 weeks for the drug-treated patients, and 45% of the subjects were in remission at this treatment point. A bimodal distribution was observed, with subjects either responding well to the drug early in treatment or failing to show an appreciable response at any point in the investigation.

Much attention has been paid to the negative effects of tricyclic antidepressants, including possible orthostatic hypotension, anticholinergic effects, and cardiac conduction disturbances. However, if careful monitoring is available, evidence is accumulating that this class of drugs can be successfully used to treat depression in older patients, even if chronic heart disease is present (e.g., Veith, Raskind, Caldwell, Barnes, Gumbrecht, & Ritchie, 1982). Concern has also been expressed about the long-term use of antidepressants by older patients. Georgotas, McCue, Cooper, Nagachandran, and Chang (1988) recently examined the effectiveness and safety of antidepressant medications as maintenance therapy for elderly depressed patients who had initially shown a positive response to these drugs. Over 70% of the patients followed over four to eight months remained well; 18% had relapses, and 5% discontinued therapy because of side effects. For some older patients, therefore, even long-term drug therapy may prove to be an effective treatment option.

Pain Problems in the Elderly

Even though there is a lack of representative and well-documented studies, there is evidence to suggest a high prevalence of pain problems in older people. In one study of a psychiatric sample, the patients reported that they had spent about one quarter of their average waking time during the past week in pain (Hyer, Gouveia, Harrison, Warsaw, & Coutsouridis, 1987), rating their usual pain intensity levels at around 3.2 on a 10-point scale. The authors found that the pain measures were correlated with levels of psychosocial distress, so that higher pain intensity was related to more somatic concern, depression, and anxiety.

There are more degenerative changes and pain-related illnesses in higher age groups (disc disease, arthritis, diabetic peripheral neuropathy, cancer, and neurological disorders), but the incidence of older patients requesting treat-

ment in interdisciplinary pain clinics is unexpectedly low: fewer than 10% of the patients seen in pain clinics are 65 or older (Kwentus, Harkins, Lignon, & Silverman, 1985). There is no clear explanation for older patients' not using pain treatment facilities as readily as younger patients even though they suffer from more pain-producing illnesses; however, use may be affected by some of the same factors (e.g., limited access, limited referrals, and negative attitudes) that result in the underutilization of mental health services.

There is some controversy about whether there are decreases in sensitivity to painful stimuli as one ages. The commonly accepted stereotype that pain is a relatively natural consequence of aging may lead older patients to endure more pain before reporting it to health professionals for treatment (Butler & Gastel, 1980). This stereotype may also lead physicians and other health practitioners to neglect the emotional and physical discomforts of older patients in comparison to younger age groups. The lack of attention to pain symptoms in the aged has lead to conditions such as appendicitis not being readily diagnosed in the older population (Kwentus et al., 1985).

Types of Pain Problems

Arthritis and Other Degenerative Conditions. Chronic joint disease is common in the elderly, and many people experience some type of joint pain soon after middle age. Two of the more frequently occurring pain disorders are osteoarthritis and rheumatoid arthritis. Osteoarthritis is a wear-and-tear disease characterized by a loss of joint cartilage and hypertrophy of bone. Although the majority of patients over the age of 50 show radiological evidence of osteoarthritis, particularly in weight-bearing areas of the knees and ankles, many persons with such evidence do not report pain (Sturgis, Dolce, & Dickerson, 1987). Rheumatoid arthritis is often more widespread and disabling than osteoarthritis and typically has a more significant life impact on the older person.

Another degenerative condition causing chronic pain in the older person is osteoporosis. This entails a reduction in the density and weight of bone throughout the body. Eventually, bones can no longer tolerate mechanical forces, and the result is fractures in areas such as the lumbar spine, the hip, the wrists, and the long bones. Women seem to be at a specially great risk for osteoporosis; 50% of women over the age of 65 show evidence of bone mass loss. It should also be noted that many patients show multiple diseases and painful combinations of degenerative conditions (Sturgis *et al.*, 1987).

Cancer. Cancer occurs more frequently in people 65 and older than in the younger adult populations. For example, the risk of dying from breast cancer dramatically increases with age (Hassell, Guiliano, Thompson, & Zarem, 1985). Pain is a very frequent complaint found in the older cancer patient. It may be associated with the disease itself, as when tumors press on nerves. In addition, the pain may result from treatment for cancer. Certain chemotherapies lead to intense discomfort (Redd, 1982), and there is a high incidence of postsurgical pain in cancer patients. Finally, the progression of metastatic disease accounts for a high incidence of pain complaints in patients with cancer (McGivney & Crooks, 1984). Chapter 20 by Tovian in this book provides additional information on cancer-related problems and treatment.

Headaches. Headaches are another common pain problem in the elderly. Tension headaches may be the most frequent type of head pain in the elderly patient, the significant contributing factors including cervical osteoarthritis, psychological stress, and depression (Sturgis *et al.*, 1987). Although migraine headaches are less common in the elderly, other vascular headaches do exist, and these disorders are often related to conditions such as congestive heart failure, transient ischemia, and various metabolic disturbances (Sturgis *et al.*, 1987). The accurate diagnosis of headache in the elderly person is very important because head pain complaints may be related to a complex interaction of psychological and physical findings.

Treatment Considerations

In general, many of the same treatments that work for the younger patient with pain problems also work for the older patient (see Chapter 23 by Chapman in this book); however, certain changes may be necessary that take into account the special needs of the older person. Some of the reviews of headache problems have suggested that there is a negative relationship between increases in age and the response to treatments such as relaxation training and biofeedback (Holroyd & Penzien, 1986). However, there have been very few prospective studies on the relaxation training of geriatric headache sufferers. The one study that could be identified (Arena, Hightower, & Chong, 1988) showed that relaxation therapy for tension headache in the elderly was effective. The subjects were men and women between the ages of 62 and 80 who had had tension headaches for at least 10 years. The treatment consisted of seven sessions of a modified progressive relaxation therapy. The patients were asked to keep a diary of headache activity and to note the number of headache-free days. Retrospective studies may not have found that relaxation training and other similar treatments were as effective in the elderly because the treatments were not specifically designed to make sure that the older person could understand the treatment instructions and comply with the prescribed exercises.

The treatment of pain problems in the elderly is often complicated by a number of special problems. Some patients have failing memory, which may interfere with their ability to give an accurate history of the onset and progression of their pain problem or to comply with complicated pain treatments. Another problem that affects the treatment of pain in the elderly is the high incidence of depression (Fordyce, 1978; Romano & Turner, 1985). Depression increases the intensity of the pain and may lead to decreased socialization and activity, which, in turn, can influence coping with chronic pain problems. Older persons who are neglected by family and friends may obtain attention only when they report pain problems or other medical difficulties.

Another problem is that pain medications affect older people differently than younger patients. The elderly are significantly more sensitive to the pain-relieving effects of narcotics because of alterations of receptors, changes in plasma protein binding, and the longer time it takes the system to clear these medications (Kaiko, Wallenstein, Rogers, Brabinski, & Houda, 1982). Also, older patients may experience more side effects from medication, and close monitoring of drug levels is important in order to ensure proper therapeutic concentrations and to prevent toxic side effects (Hall & Beresford, 1984). Aspirin, the most commonly used drug for pain in the treatment of arthritis, is responsible for a large number of adverse drug reactions that can result in hospitalization (Pfeiffer, 1982). Older patients may take large amounts of aspirin and may experience side effects such as gastric bleeding, bronchospasm, and coagulation difficulties (Sturgis *et al.*, 1987). Finally, the narcotics, as well as some nonnarcotic pain medications such as the benzodiazepines, may lead to problems with depression and drug dependence, all of which can have a negative impact on the proper management of a chronic pain problem (Sturgis *et al.*, 1987).

Sleep Problems

Many older people report a variety of difficulties in sleeping, including problems in falling asleep, frequent awakenings during the night, excessive daytime tiredness, and the overall subjective experience of poor sleep. Insomnia, or the group of disorders of the initiation and maintenance of sleep, is the most common sleep disorder and is especially prevalent in the elderly and in women (Nino-Murcia & Keenan, 1988). The prevalence of serious insomnia has been found to increase from about 14% in 18- to 34-year-olds to about 25% for 65- to 79-year-olds (Bootzin & Engle-Friedman, 1987). Almost one half of persons aged 65–79 were reported to have some difficulty with insomnia in the previous 12 months. Survey results also show a disproportionately high use of sleep medications in older adults. Sixty-nine percent of those taking prescription medications for sleep were persons between the ages

of 50 and 79 (Mellinger, Balter, & Uhlenhuth, 1985).

The use of hypnotic or sedative medications is even greater among the institutionalized elderly. In a careful study of 180 nursing-home residents, almost 40% received hypnotics daily (Cohen, Eisdorfer, Prinz, Breen, Davis, & Gadsby, 1983). There was some evidence that these older persons were consuming disproportionate amounts of sleep medications, even though they may not have had specific sleep disturbances.

Despite the heavy use of such sleep medications, there is evidence that their use not only is ineffective but also may be potentially dangerous for the person with a chronic sleep difficulty (Bootzin & Engle-Friedman, 1987). Tolerance of hypnotics develops quickly, so that larger and larger doses are needed to induce sleep, and continuous use eventually results in less deep sleep, more fragmented sleep, and reduced rapid-eye-movement (REM) sleep (Bootzin & Engle-Friedman, 1987). Furthermore, hypnotics may lead to serious problems of dependency (Kales, Scharf, & Kales, 1978). Because hypnotics are central nervous system depressants, they can have other deleterious side effects; they may impair respiration and motor and intellectual functioning, and they may lead to depression. Daytime sleepiness is also reported as a side effect of the inappropriate use of hypnotics (Bootzin & Engle-Friedman, 1987). The daytime side effects of hangover, sleepiness, impaired motor and intellectual functioning, and dysphoric mood seem to be related to the continued pharmacological actions of these medications, working as central nervous system depressants. The relatively long-lasting effects of these drugs may be observed for several days after the older person has stopped taking them (Bootzin & Engle-Friedman, 1987). The elderly seem to be particularly susceptible to adverse side effects because they may have more serious medical disorders, such as liver disease, kidney disease, and cardiac or respiratory diseases. The elderly person also seems to be more likely to have problems with toxic drug interactions following multiple drug uses, as when hypnotics are combined with other prescribed drugs (Miles & Dement, 1980).

Cause of Sleep Problems in the Elderly

There are a variety of causes of sleep problems in the elderly. Many problems are related to the above-mentioned difficulties with the overuse of sleep medications. Many of the elderly use over-the-counter medications that have an antihistamine as an active ingredient. A frequent side effect of antihistamines is drowsiness. However, carefully controlled studies of these medications have not found them to be more effective than placebos (Kales, Tan, Swearingen, & Kales, 1971). There are also potential side effects with these over-the-counter medications, such as memory disturbances and confusion, and if they are combined with alcohol or prescribed medications, they may create even more problems (Bootzin & Engle-Friedman, 1987).

Other chemical substances that can create problems with sleep are alcohol, caffeine, and nicotine. Alcohol is a depressant and it can decrease REM sleep. Heavy drinkers have been found to have fragmented sleep with frequent awakenings. In addition, withdrawal from alcohol often produces REM rebound, nightmares, and poor-quality sleep (Bootzin & Engle-Friedman, 1987). Although there is no evidence that the moderate use of alcohol presents a problem, heavy drinking before sleep does lead to sleep disturbances rather than to improvement in the quality of sleep. Another problem with alcohol is that it may potentiate the effects of hypnotics or other sleep medications. Both caffeine and nicotine are central nervous system stimulants that produce lighter and more fragmented sleep. Complaints of sleep difficulties may be related to an excessive use of these substances. Because caffeine has a blood half-life of approximately six hours, the older person may experience sleep difficulties throughout the night after eating or drinking foods or beverages that contain caffeine. Reducing or eliminating the intake of caffeine and stopping smoking may lead to major improvements in sleep (Bootzin & Engle-Friedman, 1987).

Sleep-related respiratory difficulties can also result in insomnia. The most critical of these disorders is sleep apnea, which affects approximately 40% of the older patients seen in sleep disorder clinics (Coleman, Miles, Guilleminault, Zarcone, van den Hoed, & Dement, 1981). Apnea is considered a problem when there are five or more episodes per hour of reduction of airflow for 10 seconds or longer that disrupt sleep. Sleep apnea difficulties increase with age and are associated with obesity (Jamieson, 1988).

Poor sleep habits can also contribute to sleep difficulties. A major problem in the older person is a generally inactive lifestyle. This seems to be especially true of older persons in institutions. In addition to inactivity, the older person may frequently take naps during the day, and because afternoon and evening naps may contain more deep sleep and less REM sleep, the sleep during the night may consist of more light sleep and more frequent awakenings. Persons who fall into the habit of sleeping late in the morning or taking naps whenever fatigue overwhelms them are also likely to develop circadian rhythm disturbances. This happens when the sleep cycle is desynchronized and when there is difficulty in falling asleep (Bootzin & Engle-Friedman, 1987). There may be a tendency for older adults to return to the more polyphasic alteration of sleep and wakefulness that is characteristic of early life.

Sleep difficulties can reflect emotional concerns accompanied by anxiety and worry. Older persons may worry about losses, health, and finances, and these worries may precipitate a temporary sleep problem; the situation may be further compounded by a worry about getting to sleep in itself. The bedroom then becomes a cue for the anxiety and depression associated with frustrated attempts to fall asleep. Sometimes, this reaction is noted in persons who say that they can fall asleep in places other than their own beds. These sleeping difficulties are a special problem because they can continue long after the initial temporary stresses have been alleviated. Finally, more serious psychopathology, such as major depression, is often related to a significant sleep disturbance.

Treatment of Sleep Problems in the Elderly

Some studies have evaluated psychological interventions that seem to be quite helpful in treating sleep problems in the elderly (Bootzin, Engle-Friedman, & Hazelwood, 1983; see also Chapter 25 by Cartwright in this book). These studies suggest that one important aspect of effective treatment is providing good information and reassurance. It is especially important to try to help people avoid excessive worry about why they cannot sleep. Specific sleep information can also be helpful. This information includes the possible disruptive effects of nonprescription medications, alcohol, caffeine, nicotine, stress, inactivity, daytime naps, and other environmental factors (Bootzin & Engle-Friedman, 1987). Patients are told that, in general, people get the amount of sleep that they need, that there are great individual differences in sleep needs, that a few nights of sleep deprivation do not produce serious impairments, and that some people who have lived long, productive, satisfying lives report less than two hours of sleep a night (Meddis, 1977).

Sometimes, specific training in relaxation skills helps an older person achieve better sleep. One method is progressive muscle relaxation, which involves teaching people how to tense and then relax various muscle groups in order to learn to induce a state of comfort and calm. A potential difficulty in using progressive muscle relaxation with an older person is that those with arthritis may experience increased pain as a result of tensing and releasing various muscle groups (Bootzin *et al.*, 1983), and specific instruction should be given not to tense muscle groups that produce discomfort. In general, relaxation strategies, with some of the other procedures, have been found to benefit older patients significantly in achieving a better quality of sleep.

Bootzin and Engle-Friedman (1987) proposed some stimulus control instructions to help patients develop a consistent sleep rhythm, to strengthen the idea of the bedroom as a place for sleep, and to eliminate unhealthy sleep habits:

1. Lie down intending to go to sleep only when you are sleepy.
2. Do not use your bed for anything except sleep; that is, do not read, watch TV, eat, or worry in bed. Sexual activity is the only exception to this rule. On such occasions, the instructions are to be followed afterward, when you intend to go to sleep.
3. If you find yourself unable to fall asleep, get up and go into another room. Stay up as long as you wish, and then return to the bedroom to sleep. Although we do not want you to watch the clock, we want you to get out of bed if you do not fall asleep immediately. Remember, the goal is to associate your bed with falling asleep quickly. If you are in bed more than about 10 minutes without falling asleep and have not got up, you are not following this instruction.
4. If you still cannot fall asleep, repeat Step 3. Do this as often as is necessary throughout the night.
5. Set your alarm and get up at the same time each morning irrespective of how much sleep you got during the night. Following this rule will help your body acquire a consistent sleep rhythm.
6. Do not nap during the day.

Finally, some older persons with serious emotional and interpersonal difficulties also have interrelated sleep problems. Psychotherapy for these difficulties is often necessary in addition to the specific treatment of the sleep problem.

Some Issues of Special Concern for Medical Psychologists

Older adults' multiple medical illnesses and their predilection for medical services make it difficult to provide effective mental health care. For psychologists who work in medical settings, these drawbacks present challenges as well as opportunities. Older patients are available in abundance, and they are likely to have a variety of problems in which psychological expertise may be helpful. One challenge is to adapt what we know to the preferences and abilities of older patients. Another is to arrange to be paid for our services.

There is growing evidence that contemporary older adults cope with illness in different ways than younger patients and have different expectations of health care professionals. A number of studies have suggested that today's elderly patients desire less control over their health care than younger individuals, and that they frequently prefer that professionals make decisions for them (e.g., Smith, Woodward, Wallston, Wallston, Rye, & Zylstra, 1988; Woodward & Wallston, 1987). Age differences in styles of coping with chronic illness have also been documented: older adults, particularly those with serious illness, make greater use of minimization and less use of information seeking and emotional expression in coming to grips with sickness (Felton & Revenson, 1987). Whether these differences are due to cohort influences or aging effects is not clear, as the pertinent studies have all been cross-sectional. However, the finding is a robust one, having been observed across samples differing in education, health status, and involvement with the medical establishment (Smith *et al.*, 1988).

Woodward and Wallston (1987) found that lowered desire for control in health care situations is mediated by older adults' diminished sense of self-efficacy, whereas their desire for control in other situations is not so clearly linked to self-perceptions. These investigators emphasized the complexity of the medical system to the layperson and the potential costliness of making a mistake in explaining why older people may be particularly lacking in confidence in health-related situations. The possibility that comprehension and decision making may be compromised by age-related cognitive changes was also noted as a potentially important factor.

There are many implications of these findings for psychologists in medical settings. Because older patients may not initiate requests for information about different treatment options, psychologists need to take the responsibility for educating older patients and their families about behaviorally oriented treatments. Also, because many of the most effective behavioral interventions involve increasing the patient's sense of control over physical functions, these techniques may need to be adapted for people whose preferred coping style is to

vest control in others. What this amounts to is harder work, and more clinical research is needed to help to define how adaptations in treatment may best be accomplished.

Psychologists also need to be more active politically if they hope to be adequately reimbursed for their work with older patients (Carr, 1987). Medicare's coverage for mental health services has been poor from its inception, and improvements have been negligible (Roybal, 1988). Therefore, for psychologists who see a wide range of patients, and whose salaries derive in whole or in part from fee collection (e.g., in pain clinics or on consultation–liaison services), there is little incentive to see older patients whose only health coverage is Medicare. Obviously, without improvement in funding patterns, the aged will continue to receive less mental health care than they need, and few will have recourse to treatments other than medications or somatic therapies.

Summary and Conclusions

The upward shift in this country's age distribution, the correlation between aging and chronic illness, and the escalating costs of medical care are prompting close scrutiny of the types of health programs made available to older people and of the funding for these programs. Some benefit may eventually accrue to clinical psychologists from these revisions, as increasing emphasis is placed on cost reduction and on preventive interventions. At present, however, comparatively few psychologists are able to specialize in work with the aged in medical settings. Priority must therefore be placed on increasing the level of knowledge of old-age disorders within the larger group of medical psychologists who come in contact with older patients, and on identifying roles that can be uniquely complemented by psychological training.

Psychologists can contribute significantly to the assessment and treatment of cognitive impairment and depression, the two most prevalent forms of mental disorder in older people. Cognitive disorders and depression are both often overlooked in the medically ill aged, and particularly for delirium and depression, detection is a crucial first step in ensuring that treatment will be given. Psychotherapy is clearly effective in alleviating depression for many older people, and in delirium, management of the environment can play a key role in prompt recovery. Even when dementia is present, psychologists in medical settings need to be aware that some cases of apparent dementia can be completely reversed with appropriate treatment, and that currently irreversible dementing illnesses can be compounded by adverse coexisting conditions, either medical, psychiatric, or environmental.

Pain and sleep problems are very common in older adults, and in both of these areas, effective behavioral interventions have been developed. Elderly patients commonly take medications for pain and insomnia, but drugs may have limited long-term effectiveness and often lead to adverse side effects. The literature on treating older patients with behavioral and psychotherapeutic interventions is still quite limited, and we do not know how best to adapt these techniques to older patients, whose style of coping with health problems may be more passive and other-directed than that of younger patients. Procedures also need to be varied depending on memory difficulties, physical limitations, lower energy level, and social difficulties, as well as transportation and financial limitations. Yet, the overall goal of psychological approaches to the management of pain, sleep difficulties, and other psychological problems is similar to the goals for younger persons, that is, to increase independence, to maximize patients' ability to control their problems, and to increase their quality of life.

References

American Psychiatric Association (1987). *Diagnostic and statistical manual of mental disorders* (3rd ed. rev.). Washington, DC: Author.

Anthony, J. C., Le Resche, L., Niaz, U., Von Korff, M., & Folstein, M. (1982). Limits of the "Mini-Mental State" as a screening test for dementia and delirium among hospital patients. *Psychological Medicine, 12*, 397–408.

Arena, J. G., Hightower, N. E., & Chong, G. C. (1988). Relaxation therapy for tension headache in the elderly: A prospective study. *Psychology and Aging, 3*, 96–99.

Beck, A. T., & Beck, R. W. (1972). Screening depressed patients in family practice: A rapid technique. *Postgraduate Medicine, 52*, 81–85.

Beck, A. T., Ward, C. H., Mendelson, M., Mock, J., & Erbaugh, J. (1961). An inventory for measuring depression. *Archives of General Psychiatry, 4*, 561–573.

Blazer, D. G., & Williams, C. D. (1980). Epidemiology of dysphoria and depression in an elderly population. *American Journal of Psychiatry, 137*, 439–444.

Blessed, G., Tomlinson, B. E., & Roth, M. (1968). The association between quantitative measures of dementia and of senile change in the cerebral grey matter of elderly subjects. *British Journal of Psychiatry, 114*, 797–811.

Bootzin, R. R., & Engle-Friedman, M. (1987). Sleep disturbances. In L. L. Carstensen & B. A. Ededstein (Eds.), *Handbook of clinical gerontology* (pp. 238–251). New York: Pergamon Press.

Bootzin, R. R., Engle-Friedman, M., & Hazelwood, L. (1983). Insomnia. In P. J. Lewinsohn & L. Teri (Eds.), *Clinical geropsychology: New directions in assessment and treatment*. Elmsford, NY: Pergamon Press.

Braithwaite, V. A. (1986). Old age stereotypes: Reconciling contradictions. *Journal of Gerontology, 41*, 353–360.

Butler, R. N., & Gastel, B. (1980). Care of the aged. In L. Ng & J. J. Bonica (Eds.), *Pain, discomfort, and humanitarian care* (pp. 297–311). New York: Elsevier.

Caine, E. (1981). Pseudodementia: Current concepts and future directions. *Archives of General Psychiatry, 38*, 1359–1364.

Carr, J. E. (1987). Federal impact on psychology in medical schools. *American Psychologist, 9*, 869–872.

Cohen, D., Eisdorfer, C., Prinz, P., Breen, A., Davis, M., & Gadsby, A. (1983). Sleep disturbances in the institutionalized aged. *Journal of the American Geriatrics Society, 31*, 79–82.

Coleman, R. M., Miles, L. E., Guilleminault, C. C., Zarcone, V. P., van den Hoed, J., & Dement, W. C. (1981). Sleep wake disorders in the elderly: A polysomnographic analysis. *Journal of the American Geriatrics Society, 29*, 289–296.

Felton, B. J., & Revenson, T. A. (1987). Age differences in coping with chronic illness. *Psychology and Aging, 2*, 164–170.

Folstein, M. F., & McHugh, P. R. (1979). Dementia syndrome of depression. In R. Katzman, R. D. Terry, & K. L. Bick (Eds.), *Alzheimer's disease: Senile dementia and related disorders* (pp. 87–96). New York: Raven Press.

Folstein, M. F., Folstein, S., & McHugh, P. R. (1975). Mini-Mental State: A practical method of grading the cognitive state of patients for the clinician. *Journal of Psychiatric Research, 12*, 189–198.

Folstein, M. F., Anthony, J. C., Parhad, I., Duffy, B., & Gruenberg, E. M. (1985). The meaning of cognitive impairment in the elderly. *Journal of the American Geriatrics Society, 33*, 228–235.

Fordyce, W. E. (1978). Evaluation and managing chronic pain. *Geriatrics, 33,* 59–62.

Friedman, A. S. (1964). Minimal effects of severe depression on cognitive functioning. *Journal of Abnormal and Social Psychology, 69,* 237–243.

Gallagher, D., Nies, G., & Thompson, L. W. (1982). Reliability of the Beck Depression Inventory with older adults. *Journal of Consulting and Clinical Psychology, 50,* 152–153.

Gallagher, D., Breckenridge, J., Steinmetz, J., & Thompson, L. (1983). The Beck Depression Inventory and Research Diagnostic Criteria: Congruence in an older population. *Journal of Consulting and Clinical Psychology, 51,* 945–946.

Gatz, M., & Pearson, C. G. (1988). Ageism revised and the provision of psychological services. *American Psychologist, 43,* 184–188.

Gatz, M., Popkin, S. J., Pino, C. D., & VandenBos, G. R. (1985). Psychological interventions with older adults. In J. E. Birren & K. W. Schaie (Eds.), *Handbook of the psychology of aging* (2nd ed.) (pp. 755–785). New York: Van Nostrand Reinhold.

Georgotas, A., McCue, R. E., Cooper, T. B., Nagachandran, N., & Chang, I. (1988). How effective and safe is continuation therapy in elderly depressed patients? *Archives of General Psychiatry, 45,* 929–932.

German, P. S., Shapiro, S., Skinner, E. A., VonKorff, E., Klein, L. E., Turner, R. W., Teitelbaum, M. L., Burke, J., & Burns, B. (1987). Detection and management of mental health problems of older patients by primary care providers. *Journal of the American Medical Association, 257,* 489–493.

Hall, R. C., & Beresford, T. P. (1984). Tricyclic antidepressants. *Geriatrics, 39,* 81–93.

Hamilton, M. (1960). A rating scale for depression. *Journal of Neurology, Neurosurgery, and Psychiatry, 23,* 56–62.

Hassell, C. M., Guiliano A. E., Thompson, R. W., & Zarem, H. A. (1985). Breast cancer. In C. Hassell (Ed.), *Cancer treatment* (pp. 137–180). Philadelphia: W. B. Saunders.

Hicks, R., Dysken, M. W., Davis, J. M., Lesser, J., Ripeckyj, A., & Lazarus, L. (1981). The pharmacokinetics of psychotropic medication in the elderly: A review. *Journal of Clinical Psychiatry, 42,* 374–385.

Holroyd, K. A., & Penzien, D. B. (1986). Client variables and the behavioral treatment of chronic headache: A meta-analytic review. *Journal of Behavioral Medicine, 9,* 515–536.

Hyer, L., Gouveia, I., Harrison, W., Warsaw, J., & Coutsouridis, D. (1987). Depression, anxiety, paranoid reactions, hypochrondriasis, and cognitive decline of later-life inpatients. *Journal of Gerontology, 42,* 92–94.

Jamieson, A. O. (1988). Obesity and sleep disordered breathing. *Annals of Behavioral Medicine, 10,* 107–112.

Jarvik, L. F., Mintz, J., Steuer, J., & Gerner, R. (1982). Treating geriatric depression: A 26-week interim analysis. *Journal of the American Geriatrics Society, 30,* 713–717.

Kahn, R. L., Goldfarb, A. I., Pollack, M., & Peck, A. (1960). Brief objective measures for the determination of mental status in the aged. *American Journal of Psychiatry, 117,* 326–328.

Kaiko, R. F., Wallenstein, S. L., Rogers, A. G., Brabinski, P. Y., & Houda, R. W. (1982). Narcotics in the elderly. *Medical Clinics of North America, 66,* 1079–1089.

Kales, J., Tan, T., Swearingen, C., & Kales, A. (1971). Are over-the-counter sleep medications effective? All-night EEG studies. *Current Therapeutic Research, 13,* 143–151.

Kales, A., Scharf, M. B., & Kales, J. D. (1978). Rebound insomnia: A new clinical syndrome. *Science, 201,* 1039–1040.

Kane, R. A., & Kane, R. L. (1981). *Assessing the elderly: A practical guide to measurement.* Lexington, MA: Lexington Books.

Katon, W., Kleinman, A., & Rosen, G. (1982). Depression and somatization: A review, Part 2. *The American Journal of Medicine, 72,* 241–247.

Katon, W., Berg, A. O., Robins, A. J., & Risse, S. (1986). Depression—Medical utilization and somatization. *The Western Journal of Medicine, 144,* 564–568.

Kiernan, R. J., Mueller, J., Langston, J. W., & Van Dyke, C. (1987). The Neurobehavioral Cognitive Screening Examination: A brief but quantitative approach to cognitive assessment. *Annals of Internal Medicine, 107,* 481–485.

Knight, B. (1986). Management variables as predictors of service utilization by the elderly in mental health. *International Journal of Aging and Human Development, 23,* 141–147.

Kochansky, G. E. (1979). Psychiatric rating scales for assessing psychopathology in the elderly: A critical review. In A. Raskin & L. Jarvik (Eds.), *Psychiatric symptoms and cognitive loss in the elderly.* Washington, DC: Hemisphere.

Kwentus, J. A., Harkins, S. W., Lignon, N., & Silverman, J. J. (1985). Current concepts of geriatric pain and its treatment. *Geriatrics, 40,* 48–57.

Larson, E. B., Reifler, B. V., Sumi, S. M., Canfield, C. G., & Chinn, N. M. (1985). Diagnostic evaluation of 200 elderly outpatients with suspected dementia. *Journal of Gerontology, 40,* 536–543.

La Rue, A., Spar, J., & Hill, C. (1986). Cognitive impairment in late-life depression. *Journal of Affective Disorders, 11,* 179–184.

Lipowski, Z. J. (1983). Transient cognitive disorders (delirium, acute confusional states) in the elderly. *American Journal of Psychiatry, 140,* 1426–1436.

Lipowski, Z. J. (1987). Delirium (acute confusional states). *Journal of the American Medical Association, 258,* 1789–1792.

Liston, E. (1977). Occult presenile dementia. *Journal of Nervous and Mental Disease, 164,* 263–267.

Liston, E. H. (1982). Delirium in the aged. *Psychiatric Clinics of North America, 5,* 49–66.

Mattis, S. (1976). Mental status examination for organic mental syndrome in the elderly patient. In L. Bellak & T. B. Karasu (Eds.), *Geriatric psychiatry* (pp. 79–121). New York: Grune & Stratton.

McAllister, T. W. (1983). Overview: Pseudodementia. *American Journal of Psychiatry, 140,* 528–533.

McGivney, W. T., & Crooks, G. M. (1984). The care of patients with severe chronic pain in terminal illness. *Journal of the American Medical Association, 251*, 1182–1188.

Meddis, R. (1977). *The sleep instinct*. London: Routledge & Kegan Paul.

Medical Letter on Drugs and Therapeutics. (1984, February). *Drug interactions update*. New Rochelle, NY: Medical Letter.

Medical Letter on Drugs and Therapeutics (1986, August). *Drugs that cause psychiatric symptoms*. New Rochelle, NY: Medical Letter.

Mellinger, G. D., Balter, M. B., & Uhlenhuth, E. H. (1985). Insomnia and its treatment. *Archives of General Psychiatry, 42*, 225–232.

Miles, L. E., & Dement, W. C. (1980). Sleep and aging. *Sleep, 3*, 119–220.

Murphy, E. (1983). The prognosis of depression in old age. *British Journal of Psychiatry, 142*, 111–119.

National Center for Health Statistics. (1987a). Current estimates from the National Health Interview Survey, United States, 1986. *Vital and Health Statistics* Series 10, No. 164.

National Center for Health Statistics. (1987b). Family use of health care, United States, 1980. *National Medical Care Utilization and Expenditure Survey* Series B, Descriptive Report #10, DHHS Pub. No. 87-20210.

National Center for Health Statistics. (1987c). Utilization of short-stay hospitals, United States, 1985, Annual Summary. *Vital and Health Statistics* Series 13, No. 91.

National Institute on Aging Task Force. (1980). Senility reconsidered. *Journal of the American Medical Association, 244*, 259–263.

Nino-Murcia, G., & Keenan, S. (1988). A multicomponent approach to the management of insomnia. *Annals of Behavioral Medicine, 10*, 101–106.

Norris, J. T., Gallagher, D., Wilson, A., & Winograd, C. H. (1987). Assessment of depression in geriatric medical outpatients: The validity of two screening measures. *Journal of the American Geriatrics Society, 35*, 989–995.

Osato, S., Yang, J., & La Rue, A. (1988). *Diagnostic utility of the NCSE in a geropsychiatric sample*. Unpublished manuscript.

Ouslander, J. G. (1981). Drug therapy in the elderly. *Annals of Internal Medicine, 95*, 711–722.

Pfeiffer, E. (1975). A short portable mental status questionnaire for the assessment of organic brain deficit in elderly patients. *Journal of the American Geriatrics Society, 23*, 433–441.

Pfeiffer, R. F. (1982). Drugs for pain in the elderly. *Geriatrics, 37*, 67–76.

Phifer, J. F., & Murrell, S. A. (1986). Etiologic factors in the onset of depressive symptoms in older adults. *Journal of Abnormal Psychology, 95*, 282–291.

Poon, L. (Ed.). (1986). *Handbook for clinical memory assessment of older adults*. Washington, DC: American Psychological Association.

Post, F. (1966). Somatic and psychic factors in the treatment of elderly psychiatric patients. *Journal of Psychosomatic Research, 10*, 13–19.

Rabins, P. V. (1983). Reversible dementia and the misdiagnosis of dementia: A review. *Hospital and Community Psychiatry, 34*, 830–835.

Rabins, P. V., Merchant, A., & Nestadt, G. (1984). Criteria for diagnosing reversible dementia caused by depression: Validation by 2-year follow-up. *British Journal of Psychiatry, 144*, 488–492.

Rapp, S. R., Parisi, S. A., & Walsh, D. A. (1988). Psychological dysfunction and physical health among elderly medical inpatients. *Journal of Consulting and Clinical Psychology, 56*, 851–855.

Rapp, S. R., Parisi, S. A., Walsh, D. A., & Wallace, C. E. (1988). Detecting depression in elderly medical inpatients. *Journal of Consulting and Clinical Psychology, 56*, 509–513.

Redd, W. H. (1982). Behavioral analysis and control of psychosomatic symptoms of patients receiving intensive cancer treatment. *British Journal of Clinical Psychology, 21*, 351–358.

Regier, D. A., Boyd, J. H., Burke, J. D., Jr., Rae, D. S., Myers, J. K., Kramer, M., Robins, L. N., George, L. K., Karno, M., & Locke, B. Z. (1988). One-month prevalence of mental disorders in the United States. *Archives of General Psychiatry, 45*, 977–986.

Reifler, B. V., Larson, E., & Hanley, R. (1982). Coexistence of cognitive impairment and depression in geriatric outpatients. *American Journal of Psychiatry, 139*, 623–626.

Reynolds, C. F., Hoch, C. C., Kupfer, D. J., Buysse, D. J., Houck, P. R., Stack, J. A., & Campbell, D. W. (1988). Bedside differentiation of depressive pseudodementia from dementia. *American Journal of Psychiatry, 145*, 1099–1103.

Roca, R. P. (1987). Bedside cognitive examination. *Psychosomatics, 28*, 71–76.

Romano, J. M., & Turner, J. A. (1985). Chronic pain and depression: Does the evidence support a relationship? *Psychological Bulletin, 97*, 18–34.

Roth, M. (1971). Classification and aetiology in mental disorders of old age: Some recent developments. In D. W. K. Kay & A. Walk (Eds.), *Recent developments in psychogeriatrics: A symposium* (pp. 1–18). Ashford, England: Headley Brothers.

Rothchild, D. (1942). Neuropathologic changes in arteriosclerotic psychoses and their psychiatric significance. *Archives of Neurology and Psychiatry, 48*, 417–436.

Roybal, E. R. (1988). Mental health and aging: The need for an expanded Federal response. *American Psychologist, 43*, 189–194.

Schwamm, L. H., Van Dyke, C., Kiernan, R. J., Merrin, E. L., & Mueller, J. (1987). The Neurobehavioral Cognitive Status Examination: Comparison with the Cognitive Capacity Screening Examination and the Mini-Mental State Examination in a neurosurgical population. *Annals of Internal Medicine, 107*, 486–491.

Scogin, F., Beutler, L., Corbishley, A., & Hamblin, D. (1987, August). *Reliability and validity of the Short-Form Beck Depression Inventory with older adults*. Paper presented at the Annual Convention of the Gerontological Society of America, Washington, DC.

Shapiro, S., Skinner, E. A., Kessler, L. G., Von Korff, M., German, P. S., Tischler, G. L., Leaf, P. J., Benham, L., Cottler, L., & Regier, D. A. (1984). Utilization of health and mental health services. *Archives of General Psychiatry, 41*, 971–978.

Siegfried, K. (1985). Cognitive symptoms in late-life depression and their treatment. *Journal of Affective Disorders*, Supplement 1, S33–40.

Small, G. W., & Fawzy, F. I. (1988). Psychiatric consultation for the medically ill elderly in the general hospital: Need for a collaborative model of care. *Psychosomatics, 29*, 94–103.

Smith, R. A. P., Woodward, N. J., Wallston, B. S., Wallston, K. A., Rye, P., & Zylstra, M. (1988). Health care implications of desire and expectancy for control in elderly adults. *Journal of Gerontology, 43*, P1–P7.

Spencer, G. (1984). Projections of the population of the United States by age, sex, and race: 1983 to 2080. *Current Population Reports* Series P25, No. 952, U.S. Bureau of the Census.

Spitzer, R. L., Endicott, J., & Robbins, E. (1978). Research Diagnostic Criteria: Rationale and reliability. *Archives of General Psychiatry, 35*, 773–782.

Sturgis, E. T., Dolce, J. J., & Dickerson, P. C. (1987). Pain management in the elderly. In L. L. Carstensen & B. A. Ededstein (Eds.), *Handbook of clinical gerontology* (pp. 190–203). New York: Pergamon Press.

Terry, R. D., & Davies, P. (1980). Dementia of the Alzheimer type. *Annual Review of Neuroscience, 3*, 77–95.

Thompson, L. W., Gallagher, D., & Breckenridge, J. S. (1987). Comparative effectiveness of psychotherapies for depressed elders. *Journal of Consulting and Clinical Psychology, 55*, 385–390.

Toner, J., Gurland, B., & Teresi, J. (1988). Comparison of self-administered and rater-administered methods of assessing levels of severity of depression in elderly patients. *Journal of Gerontology: Psychological Sciences, 43*, P136–140.

Trzepacz, P. T., Baker, R. W., & Greenhouse, J. (1988). A symptom rating scale for delirium. *Psychiatry Research, 23*, 89–97.

Turner, R. J., & Noh, S. (1988). Physical disability and depression: A longitudinal analysis. *Journal of Health and Social Behavior, 29*, 23–37.

U.S. Bureau of the Census. (1987). Estimates of the population of the United States, by age, sex and race: 1980–1986. *Current Population Reports* Series P-25, No. 1000.

U.S. Senate Special Committee on Aging. (1987–1988). *Aging America: Trends and projections.* Washington, DC: U.S. Department of Health and Human Services.

VandenBos, G. R., & Stapp, J. (1983). Service providers in psychology: Results of the 1982 APA Human Resources Survey. *American Psychologist, 38*, 1330–1352.

Veith, R. C., Raskind, M. A., Caldwell, J. H., Barnes, R. F., Gumbrecht, G., & Ritchie, J. L. (1982). Cardiovascular effects of tricyclic antidepressants in depressed patients with chronic heart disease. *The New England Journal of Medicine, 306*, 954–959.

Weicker, W., (1987, October). *Diagnosis of delirium in older adults.* Paper presented at the meeting of the Canadian Gerontological Society, Calgary, Alberta.

Weingartner, H., & Silberman, E. (1982). Models of cognitive impairment: Cognitive changes in depression. *Psychopharmacology Bulletin, 18*, 27–42.

Wells, C. F. (1979). Pseudodementia. *American Journal of Psychiatry, 136*, 895–900.

Williams, J. B. W. (1988). A structured interview guide for the Hamilton Depression Rating Scale. *Archives of General Psychiatry, 45*, 742–747.

Woodward, N. J., & Wallston, B. S. (1987). Age and health care beliefs: Self-efficacy as a mediator of low desire for control. *Psychology and Aging, 2*, 3–8.

Yesavage, J. A., Brink, T. L., Rose, T. L., Lum, O., Huang, V., Adey, M., & Leirer, O. (1983). Development and validation of a geriatric depression screening scale: A preliminary report. *Journal of Psychiatric Research, 17*, 37–49.

Zarit, S. H., Anthony, C. R., & Boutselis, M. (1987). Interventions with caregivers of dementia patients: Comparison of two approaches. *Psychology and Aging, 2*, 225–232.

Zung, W. W. K. (1965). A self-rating depression scale. *Archives of General Psychiatry, 12*, 63–70.

CHAPTER 15

Adherence to Self-Care Regimens

The Patient's Perspective

Dennis C. Turk and Donald Meichenbaum

Introduction

Early in the history of medicine, Hippocrates (*On Decorum*, 1923 translation) cautioned against relying on patient's self-reports of their compliance with medical regimens because he believed that patients typically do not tell the truth. He suggested that patients try to avoid recriminations, embarrassment, or rejection by the practitioner by responding in a socially desirable manner.

Despite the admonition of Hippocrates, patients' acceptance of the health care provider's recommendations and their performance of recommended sets of prescribed behaviors did not received much serious empirical attention until the 1970s. Before this time, the failure of patients to behave in accordance with the health care provider's advice was noted, and bemoaned, but not systemically evaluated, nor were there many attempts to improve patients' adherence to recommendations. Before the ad-

vent of "modern" medicine and the availability of active medications for infectious diseases, adherence was perhaps less of a concern. Failure to ingest preparations with few active, or even sometimes pernicious, ingredients (e.g., oil of ants and ground sheep gall stones) would not lead to harmful consequences and was perhaps even a wise course. With the availability of expanded scientific knowledge and the advancement of pharmacological knowledge, adherence became, and continues to be, a major concern in the provision of health care.

To illustrate the magnitude of the problem, we can note some rather sobering statistics regarding adherence rates. For example, it has been estimated that 50%–60% of patients do not adhere to appointments for preventive programs (DiMatteo & DiNicola, 1982). It has been noted that 20%–80% of participants drop out of various lifestyle change programs designed to treat obesity, smoking, or stress (Dunbar & Agras, 1980). Buckalew and Sallis (1986) estimated that, out of the 750 million new prescriptions written each year in the United States and England, there are over 250 million cases of partial or total nonadherence. In one study, Boyd, Covington, Stanaszek, and Coussons (1974) found that 18% of 134 patients in a hospital outpatient clinic never filled their prescrip-

Dennis C. Turk • Pain Evaluation and Treatment Institute, University of Pittsburgh School of Medicine, Baum Boulevard at Craig Street, Pittsburgh, Pennsylvania 15213. **Donald Meichenbaum** • Department of Psychology, University of Waterloo, Waterloo, Ontario N2L 3G1, Canada.

tions and that 56% of those who filled the prescription used improper dosage intervals. In another study Mattar, Markello, and Yaffe (1975) reported that only 5 of 100 parents of children with otitis media fully adhered to the prescribed 10-day medication regimen. Among hypertensive patients, up to 50% of patients failed to follow referral advice; over 50% dropped out of care within one year. These statistics, which indicate that the incidence of nonadherence is high, may actually be an underestimate because studies of adherence are often based on those who volunteer to participate. This group of volunteers is probably more likely to adhere to medical regimens than those who decide not to participate in the studies.

Beyond illustrating the magnitude of the problem, the statistics above note the range of areas in which adherence is a concern, that is, from recommendations that promote or maintain health, to following treatment plans for acute conditions, to the performance of diverse self-care behaviors in chronic illnesses. But the concerns about patient adherence are not limited to the medical domain. Phillips (1988) noted that adherence problems affect clients attending mental health clinics as well. From about 40% to 55% of those clients undergoing intake in a mental health clinic do not return for the first (agreed-upon) interview. In outpatient psychotherapy, 70% to 80% of the cases drop out by the fifth session. Put in other words, about 10% of the remaining number of persons after each session, from Session 1 onward, leave the system on their own. The flow through the delivery system of both medical and mental health can be considered one of the best predictors of the level of patient adherence.

The skeptic might retort that nonadherence is the patient's problem, not that of the health care provider, and that, if patients choose not to follow the "sage advice" of the health care provider, the consequences will be on their head alone. From this perspective, the responsibility for compliance rests solely with the individual patient who has sought treatment. Yet, the consequences of nonadherence with prescribed medical treatment may be grave, including the

exacerbation and progression of disability, the development of secondary complications, more frequent medical emergencies, the unnecessary prescription of more potent and/or toxic drugs, and the failure of the treatment. Taking drugs in an inappropriate manner may lead to medical problems, toxic or otherwise. For example, consider some of the consequences attributed to nonadherence. Anderson (1974) noted that 7% of hospital admissions are due to drug reactions. One seventh of all hospital days are reportedly devoted to the care of drug toxicity at an estimated cost of $3 billion annually (Kayne & Cheung, 1973). Hurwitz (1969) found adverse drug reaction to be 10.2%. A high percentage of these illnesses arise from improper use and not from inherent drug toxicity (Steward, Cluff, & Leighton, 1972). These adverse reactions are largely due to patients' deciding to take medications other than those prescribed, or to take them in ways other than prescribed.

The assumption seems to be that the failure of a prescribed treatment regimen is solely the patient's decision—this being a version of health care providers' "blaming the victim." Although it is true that the effectiveness of treatment depends in large part on the efficacy of the treatment and the extent of patient adherence, even full compliance does not always guarantee symptom relief or illness recovery. Because many factors other than medication influence the clinical course, treatment adherence or nonadherence does not always correlate with clinical outcome. A few examples will highlight the challenge both for health care providers and especially for patients, who may come to believe that nonadherence will not lead to negative consequences, and that compliance does not invariably lead to favorable consequences. For chronic illnesses such as diabetes, with complex management requirements, the relationship between full adherence and disease control is modest at best (e.g., Becker, Radius, Rosenstock, Drachman, Shuberth, & Teets, 1978). Maintaining good diabetic control, for example, *may* help prevent or forestall serious disease complications such as retinopathy, renal disease, or vascular difficulties, although good control is not a guarantee of future health (Cahill, Etzwiler, & Freinkel, 1976; Drash & Becker, 1978). Similarly, Glasgow, McCaul, and Schafer (1987) found no clear relationship between adherence and glycemic control in insulin-dependent diabetics.

Thus, even sticking tightly to the health care providers' recommendations is no guarantee of therapeutic success. And as we shall see, many factors affect an individual's adherence to recommendations that are not a direct result of the patients' capricious disregard of expert advice. In fact, Haynes (1979) implicated some 250 factors in patient nonadherence.

Initial considerations of nonadherent behavior were focused on the so-called defective or deviant behavior of the patient or the individual seeking information about health care, disease, or illness. A host of individual demographic and individual difference measures were examined in order to identify those patient factors that could possibly be used to predict noncompliance. Many health care providers believed that they could determine who was likely to be noncompliant; however, these efforts proved not to be accurate. Neither age, nor socioeconomic status, education level, marital status, religion, and so on proved to be good predictors (Dunbar & Stunkard, 1979). Similarly, no personality variables have been identified that can accurately predict lack of motivation and consequently noncompliance behaviors (Meichenbaum & Turk, 1987). Roth (1987) noted that health care providers are quite poor at estimating the level of their patients' adherence, usually overestimating the level of compliance.

When patient factors could not be identified as predictors of nonadherence, the search turned to other factors, including those that characterize the patient's disease or disorder, the nature and demands of the treatment regimen, the characteristics of health care providers, the nature of the relationship between providers and patients, and the features of the health care system. Table 1 summarizes the many factors that have been implicated in patient noncompliance.

Before we consider the influence of these factors on treatment nonadherence, especially from a patient perspective, we need to consider what is meant by the terms *compliance* and *adherence*, and how they are measured.

What Is Adherence?

Up to this point, we have used the terms *adherence* and *compliance* interchangeably. These two terms, however, have different connotations and implications. *Compliance* usually refers to the extent to which patients are obedient and faithfully follow health care providers' instructions, proscriptions, and prescriptions. A number of authors (Eisenthal, Emery, Lazare, & Udin, 1979; Kristeller & Rodin, 1984) have argued that the term *compliance* connotes a passive patient role. The term *noncompliance* contains an evaluative component that implies a negative or prejudicial attitude toward the patient and often presumes that failure to comply is the patient's fault.

In contrast, the term *adherence* implies a more active, voluntary collaborative involvement of the patient in a *mutually acceptable* course of behavior that produces a desired preventive or therapeutic result (Meichenbaum & Turk, 1987). The term *adherence* implies choice and mutuality in treatment planning and implementation. Patients who are adherent are viewed as acting on a consensually agreed-upon plan that they have had a part in designing, or at least as accepting the importance of performing the specific recommended treatment actions (DiMatteo & DiNicola, 1982). Because our perspective emphasizes active patient participation that enhances and facilitates the performance of appropriate and recommended behavior, we will use the term *adherence* throughout the remainder of this chapter.

In considering the concept of adherence, it is important not to impose a dichotomy of "adherers" versus "nonadherers." Such a dichotomy is an oversimplification because patients may exhibit different adherence rates for different aspects of treatment, or they manifest variable adherence at different times. As Blackwell (1979) observed, patients may exhibit different adherence rates for different reasons for each

Table 1. Factors Related to Treatment Nonadherence[a]

Patient variables
  Characteristics of the individual
  Type and severity of psychiatric diagnosis (in particular, diagnosis of schizophrenia, bipolar affective disorder, paranoia, or personality disorder)
  Sensory disabilities
  Lack of understanding
  Inappropirate or conflicting health beliefs (e.g., misconceptions about the disorder, lack of understanding of prophylaxis, and beliefs that medicine is necessary only when symptoms are present)
  Inappropriate expectations about treatment
  Competing sociocultural and ethnic folk concepts of disease and treatment
  Implicit models of illness
  Apathy and pessimism
  Failure to recognize that one is ill or in need of medication
  Previous or present history of nonadherence to other regimens
  Dissatisfaction with practitioner or treatment
  Characteristics of individual's social situation
  Lack of social supports
  Family instability or disharmony
  Environment that supports nonadherent behavior (e.g., residential instability)
  Competing or conflicting demands (e.g., poverty or unemployment)
  Lack of resources (e.g., transportation, money, or child care)
Disease or disorder variables
  Chronic condition with lack of overt symptomatology
  Stability of symptoms
  Disorder-related characteristics (e.g., confusion, visual distortion, or psychological reactions)
Treatment variables
  Complexity of treatment regimen (e.g., multiple medications)
  Long duration of treatment
  Degree of behavioral change required (e.g., interference with typical behavior patterns or major alterations in lifestyle)
  Side effects of medication or medication that alters behavior (e.g., sedation or extrapyramidal involvement)
  Expense
  Preparation, form, and mode of administration (e.g., capsule, tablet, or injection)
  Color and size of pill
  Inadequate labels
  Awkward container design
Setting variables
  Absence of continuity of care
  Long waiting time
  Long elapsed time between referral and actual appointment
  Absence of individual appointment times (i.e., mass scheduling)
  Lack of cohesiveness of treatment delivery system
  Inconvenience associated with operation of clinic (e.g., hours, location relative to public transportation, or parking availability)
  Poor reputation of treatment facility
  Inadequate supervision by professionals
  Expense
Relationship variables: Patient–health-care-provider interactions
  Inadequate or inappropriate communication
  Attitudinal and behavioral (verbal and nonverbal) faults on the part of either provider or patient
  Failure of provider to elicit patient concerns and negative feelings
  Patient dissatisfaction
  Inadequate supervision

[a]Adapted from D. Meichenbaum and D. Turk. (1987). *Facilitating Treatment Adherence: A Practitioner's Guidebook.* New York: Plenum Press.

component of a complex regimen. For example, consider insulin-dependent diabetic patients, who not only must take medication in the prescribed manner (i.e., the regular injection of insulin) but must also carefully control their diet, modify their exercise patterns, regularly check their levels of sugar by urine or blood testing, be vigilant for signs of hyperglycemia and hypoglycemia and know what actions to take for either, monitor their feet for signs of infection, and so forth (Turk & Speers, 1983). Moreover, the diabetic is responsible for self-care 24 hours a day, 365 days a year for the remainder of his or her life.

There are few strong relationships between the diabetic patient's adherence to one aspect of the self-care regimen and the extent to which they adhere to other regimen tasks (Glasgow *et al.*, 1987). Moreover, there is no one-to-one relationship between glycemic control and patient adherence. In fact, regimen adherence may be only one of several factors influencing glycemic control (e.g., the diabetic patient's stress level, his or her individual metabolic rate, and the appropriateness of the treatment regimen). Such complexities are poorly summarized by dichotomous judgments of "adherers" versus "nonadherers," or by arithmetic means of level of adherence, or by arbitrarily weighted averages of adherence assigned to various treatment components. These complexities underscore the challenging task not only of how to conceptualize adherence, but also of how to assess adherence.

Assessment of Adherence

Following from the preceding paragraphs, when studies report on the prevalence of non-adherence, it is important to ask what criteria are being used to make this determination. As Epstein and Cluss (1982) noted, it is not known what constitutes an adequate level of adherence for most medical problems. For example, is 100% of the prescribed medication for a particular illness necessary for recovery, or will 80% or 50% be sufficient for a positive therapeutic outcome?

In some studies, investigators have noted that less than 100% adherence was adequate to bring about the desired health effects. For example, Luscher, Vetter, Siegenthaler, and Vetter (1985) reported that 80% adherence to a medication regimen for hypertensives resulted in the normalization of blood pressure, whereas 50% or less of adherence proved ineffective. Olson, Zimmerman, and Reyes de la Rocha (1985) reported that children diagnosed as having streptococcal pharyngitis required an adherence rate of 80% to achieved therapeutic results, whereas children taking oral penicillin as a prophylactic for rheumatic fever required only 33% of the medication to reduce the rate of streptococcal infection. Thus, it is important to determine the minimum standards necessary to achieve the desired health benefits. Given the lack of precision in much of health care delivery, investigators have recommended the use of multiple indicators of adherence (e.g., patient self-report, pill counts, medication monitoring, appointment keeping, and clinical outcome). Each of these measures has its own unique set of advantages and disadvantages. Each of the existing methods suffers from various levels of imprecision, sensitization, and oversimplification. The imprecision reflects primarily the use of static averages rather than dynamic measurement. We review some of the merits and problems inherent in each of these approaches below.

Patient Self-Report

The most easily obtained and most frequently used method for assessing adherence is asking patients directly whether they have taken their medication or asking them to record the frequency, duration, and number of specific behaviors performed (e.g., the practice of relaxation exercises). Patients' subjective self-reports have been challenged because they are often inaccurate and are likely to be biased in a socially desirable direction. Moreover, the simple act of self-monitoring may serve as a cue and may thus alter the behavior.

Evidence of the validity of patient self-report of adherence is quite mixed. Roth (1987) summarized several studies with such diverse groups as ulcer patients, psychiatric patients, and mothers who administered penicillin to their children, and in each case, self-reports of adherence were not corroborated by other objective indicators of adherence. Based on these and other related studies, Roth concluded:

> Thus, when a patient states that the medication is being taken regularly, it often is not. When a patient states that occasional doses are being missed, that is usually an understatement of the extent of deviation from the regimen. However, when a patient states that the drug is not being taken, this is usually corroborated. (p. 110)

Although there are clear limitations to patient self-reports, a number of authors have noted that they can often predict actual adherence behavior with a fair degree of accuracy (e.g., Dunbar & Agras, 1980; Morisky, Green, & Levine, 1986). For example, Tebbi, Cummings, Zevon, Smith, Richards, and Mallon (1986) reported that levels of serum corticosteroids measured by bioassay "corroborated in every case" self-reports of medication adherence of pediatric and adolescent cancer patients.

Sackett (1976) used the following question when working with hypertensive patients: "Most people have trouble remembering to take their medication. Do you have trouble remembering to take yours?" The individuals who admitted to low adherence demonstrated greater response to adherence-improving strategies (in terms of achieving treatment goals). Sackett concluded that the busy health care provider can identify at least 50% of the patients who will benefit most from adherence enhancement strategies and can target their efforts appropriately without having to resort to pill counts, chemical markers, or other time-consuming techniques.

In sum, despite limitations, self-report has the important advantage of being easy to implement, and it may actually enhance adherence by encouraging a discussion of adherence difficulties between the health care provider and the patient. Thus, patient self-reports of adherence may be of particular use in clinical practice. In research, it is most appropriate to combine self-report with the other more objective assessment procedures discussed below.

Pill Counts

The behavioral method most commonly used to measure adherence to taking medication is the pill or bottle count. In this case, the physician may provide for an oversupply of medication, and patients are asked to return the unused portion at specific time periods. The physician subtracts the quantity returned from the prescribed dosage to determine the degree of adherence. However, the number of pills present does not guarantee that those missing were actually ingested. Thus, like self-report, pill counts are subject to distortion.

There are other problems with pill counts as well. For example, a pill count revealing that 75% of the prescribed doses are missing at the end of an observation period can be interpreted in various ways: (1) 75% of the doses each day were taken as prescribed; (2) all of the doses on 75% of the days were taken as prescribed; (3) some intermediate combination of 1 and 2; or (4) frank patient duplicity in removing the doses before the appointment. Quite often, the 75% represents only an average, obscuring potentially important patterns. Roth (1987) recommended that, in order to obtain more precise information, physicians ask their patients to maintain a record of the amounts of medication taken and the times when it was taken.

Physiological and Biochemical Monitoring

Biochemical markers are useful in measuring adherence because they are less subject to bias than self-reports and pill counts. Chemical tracers incorporated into medication are not discernible to patients but are readily detected by chemical assay of blood or urine. Another biochemical procedure involves the bioassay of medication or its metabolites in urine or blood.

Although this procedure is more objective and thus less subject to bias than other procedures, a variety of factors unrelated to treatment adherence may be significant determinants of assay values, including total daily dose, dosage schedule, and age. For example, urine tests are sensitive only to pill taking two to three days before the test and are affected by individual variations in the metabolism of the drug. Moreover, even biological assays or tracer compounds may measure only the last, or the last several, doses, which may be atypical of the entire treatment interval since the last visit.

These measurement problems in biological assays are compounded by their high cost, their limited practicality in clinical practice, their limited availability, and their vulnerability to variations in drug half-life and metabolic conditions. Estimates of the extent of adherence based on the results of biological assays may produce misleading results because of such factors as individual variations in serum levels of absorbed drugs and difficulty in specifying the optimum serum levels (Soutter & Kennedy, 1974). Additionally, if the marker, metabolite, or drug is rapidly excreted, biochemical assessment will provide information only about recent doses. It does not measure adherence over time. Problems are also posed by food and medication combinations that affect nutrient absorption, drug absorption or excretion, and electrolyte balance (Hartshorne, 1977). Thus, the validity of the results regarding the patient's adherence is open to question.

Clinical Outcome

On the surface, it may seem that the best way to assess adherence is to consider the treatment outcome. Treatment outcome may be used as a criterion measure, but as we noted earlier, and as has been noted by others, "there is often not a straightforward link between health outcome and compliance" (Eraker, Kirscht, & Becker, 1984, p. 259).

Because of the many factors other than the medication that may be operative, it is difficult to draw conclusions about adherence from clinical course. For example, Inui, Carter, and Pecoraro (1981) reported that 40% of well-controlled hypertensives were nonadherent by pill count. Clinical course can be affected by a change in behaviors other than the use of prescribed medications. Moreover, treatment protocols are largely based on patients' typical response to a regimen and do not take into account individual variability (Dunbar, 1983). When patients in a treatment program are getting better, the tendency is to assume high rates of adherence; when they are not doing well, adherence problems may or may not be invoked as explanations (Dunbar & Stunkard, 1979). But the fact is that patients may get better despite low rates of adherence and may not improve despite high rates of adherence. The Sackett, Haynes, Gibson, Hackett, Taylor, Roberts, and Johnson (1975) study of 134 steelworkers treated for hypertension found that only 54% of the subjects fell within the cells of high-adherence good blood pressure control and low-adherence low blood pressure control, whereas over 33% of those with low rates of adherence achieved control equivalent to those who had high levels of adherence.

In a related study, Fontana, Kerns, Rosenberg, Marcus, and Colonese (1986) attempted to explain the relationships between adherence and fitness; they reported that "neither the extent of participation in exercise (beyond 6 weeks of the 12 week program), nor the continuation of exercise activity was related significantly to improvement in fitness" (p. 13). DeBusk, Haskell, Miller, Berra, and Taylor (1985) demonstrated a spontaneous increase in exercise tolerance after discharge from the hospital in low-risk patients with myocardial infarcts (MI), a finding suggesting that the degree of improvement ascribed to supervised rehabilitation may be exaggerated. The real benefit of exercise rehabilitation may in fact be related less to changes in exercise tolerance than to improvements in psychological well-being and quality of life (Fontana et al., 1986; Oldridge, 1986).

The occurrence or nonoccurrence of physiological changes may be due to a number of factors other than a prescribed exercise program (Haynes, 1984). For example, Kavanaugh, Shephard, Pandit, and Doney (1970) found a change in physical fitness in groups who participated in exercise treatment equal to the changes in groups that practiced meditation control. In addition, outcome research involving other medical treatments has sometimes shown that patients who do not comply with either treatment or placebo show improvement, whereas a substantial number of those who do comply with treatment fail to show improvement (Sackett, Haynes, Gibson, & Johnson, 1976). Finally, the physiological effects of training may not be linearly related to exercise participation (moderate levels of exercise may produce changes no greater than low levels). Thus, the degree of adherence may not be related to the degree of change in fitness.

Kavanaugh *et al.* (1970) randomized post-MI subjects into groups taking one of two treatments: aerobic exercise or meditation. Although very different changes in fitness would be expected, the results showed equivalent effects for each procedure on fitness and cardiovascular end points. It is plausible that components common to both exercise and meditation—namely, the credibility of the procedure, experimenter attention, and social support—had more significant effects on the end points than the procedure itself. Rechnitizer, Cunningham, Andrew, Buck, Jones, *et al.* (1983) showed no difference in reinfarction rates between patients taking high-intensity and low-intensity exercise, despite significant improvements in fitness for only the high-intensity group. Finally, Epstein, Wing, Koeske, and Valoski (1985) showed that aerobic and calisthenics groups matched for therapist attention did not differ in weight loss over a two-year period even though there were significant differences in energy expenditure in the two groups.

Even if there were a known isomorphic relationship between medication or a set of prescribed behaviors and positive health outcome, a number of problems would have to be acknowledged before accepting the results as indicating treatment adherence. Consider the recommendation to lose weight. A patient may lose weight while gorging one day and starving for the rest of the week. Although the patient may weigh less at the end of the week, this health outcome will not accurately reflect the extent to which he or she has adopted the

dietitian's recommendations. On the other hand, hypertensive patients may adhere strictly to a low-sodium diet, but if they fail to take their medicine, modify their exercise pattern, and reduce their smoking, the clinical outcome of their dietary adherence may be difficult to detect.

It is important to separate the degree of health change due to increased exercise *per se* from that due to concurrent changes in other health behaviors that may influence the target health outcome (Blair, Jacobs, & Powell, 1985; Sallis, Haskell, Wood, Fortmann, & Vranizan, 1986). For example, adherence interventions with cardiac patients may be successful in increasing exercise behavior, but they may also lead to decreased smoking, changes in diet, or increased adherence to the taking of previously prescribed medications. Therefore, the observed change in health cannot be attributed solely to the increased exercise behavior, and one would have to monitor changes in these other health behaviors in order to separate their effects.

The search for a "gold standard" method of assessing adherence is still in progress. As we have previously noted (Meichenbaum & Turk, 1987):

> The absence of reliable, valid, clinically sensitive indices of adherence is an important problem because it can compromise clinical trials, lead to ordering of unnecessary diagnostic test or use of alternative medications, inhibit the identification of reliable determinants, and consequently, hinder attempts to establish appropriate treatment regimens. (p. 38)

Individual Patient's Perspective

It is quite evident that medical and health care advice is not simply accepted and carried out as ordered. In one study, Davis (1968) asked her patients their intention of following their physicians advice; 15% indicated that they did not plan to, and 45% who said they intended to follow their physician's advice did not. We have suggested that demographic and patient background variables do not appear to predict adherence to a range of recommended behaviors, some with such serious consequences that nonperformance seems irrational. For example, Vincent (1971) reported that only 42% of patients being treated for glaucoma, who had been informed that, if they did not administer eye drops three times a day, they "would go blind," actually adhered to the recommended regimen. Moreover, even when these patients were at the point of becoming legally blind in one eye, their rates of adherence improved to only 58%. In another study, Strelzer and Hassell (1988) noted that, even when dialysis patients knew they would ultimately suffer congestive heart failure as a result of fluid overloading, only 25%–50% continued the pattern of adherence. Faberow (1986) suggested that, in some instances, such nonadherence should be viewed as indirect self-destructive behavior. Although self-destruction may account for nonadherence on some occasions, it hardly seems likely that this explanation is sufficient to explain the majority of the instances of reported nonadherence.

How can a person who has paid the money and taken the time to consult a health care professional not comply with the treatment recommendations? This behavior appears to be irrational—or is it? Health care providers need to consider the self-care regimen and the disease itself from the patient's perspective. Patients do not define themselves exclusively as patients when they are not in direct contact with the health care system. People with diabetes or hypertension do not identify themselves by the presence of the disease. If they did, we might be concerned about their preoccupation with their bodies or the disease.

Individuals who happen to have a disease define themselves by many different roles and responsibilities. The coauthors of this chapter both wear glasses, being myopic, but they do not define themselves as patients; rather, they view myopia and the wearing of corrective lens, although inconvenient, as only a part of their lives. Their ophthalmologists view them primarily from the perspective of myopia. Similarly, a diabetologist sees a person with diabetes as a "diabetic," and an internist views a person with hypertension as a "hypertensive,"

but they lose sight of the patients in the context of their concerns as parents, employed persons, marital partners, homeowners, and so on. Each role makes demands, some of which may compete with the self-care demands of being a "diabetic," a "hypertensive," and the like.

Health care advice is received in a specific psychosocial context, evaluated according to personal estimates of appropriateness and potential efficacy, and implemented in varying levels and degrees within a particular sociocultural setting. Many external factors impinge on the original prescriptions and recommendations until they are adjusted to each individual's needs and situation. Clinicians apply a "structural" definition founded on physical function and damage, basing their therapy recommendations on these findings, whereas patients use a "functional" measure of health based on school attendance, athletic participation, activity level, and the presence or absence of symptoms to decide whether they will adhere to the prescribed regimen.

Patients often have their own ideas about taking medication, which comes only in part from physicians; these ideas affect their level of adherence. Patients evaluate both the physician's actions and the prescribed drugs in comparison to what they themselves know about illness and medication. For example, in a study of arthritis patients, Arluke (1980) found that they evaluated the therapeutic efficacy of drugs against the achievement of specific outcomes. Medicines were judged ineffective when a salient outcome was not achieved, usually in terms of the patient's expected time frames. The very nature of arthritis, however, can militate against adherence. For instance, in rheumatoid arthritis, the natural course of the disease in most patients is characterized by exacerbations and remissions. These may occur regardless of the level of adherence to the treatment regimen. Therefore, if a patient suffers an exacerbation of symptoms despite adherence to a prescribed regimen or, conversely, improves when not adhering, subsequent adherence may be adversely affected.

Consider the self-care regimens recommended for people with various chronic disorders, such as arthritis. Arthritis treatment regimens demand that patients perform multiple, complex procedures of unknown efficacy that must be carried out over an extended period. Often, there is little immediate effect on the patient's symptoms, and some medications produce unpleasant or even life-threatening side effects. Similarly, think of the patient with diabetes. The self-care regimen makes major demands on the individual's life, and there is no guarantee that careful adherence to all aspects of the proposed regimen will have any long-term beneficial effect. Moreover, 90% of diabetic patients, while in the hospital, have unexplained blood glucose fluctuations. Thus, even the tight adherence program maintained in the hospital cannot control fluctuations; the resulting feelings of frustration and discouragement undermine adherence. Families in which there is a member with cystic fibrosis are in an even more serious situation. Strict adherence to the prescribed regimen can take many hours of time and greatly impact family functioning, yet adherence only prolongs the inevitable death.

Thus, in the case of most chronic illness, patients can often expect only an amelioration of the symptoms, or a delay of the inevitable fatal outcome, rather than cure and a return to healthy functioning. In these cases, the small benefits of adherence may not justify the large costs of inconvenience and discomfort. Because even total adherence cannot guarantee symptom relief or illness recovery, many patients must decide how to balance their health needs delicately with their needs to achieve near-normal behavioral and psychological functioning for themselves and their families.

Nonadherence may very well be the most logical, rational response to professional instruction in such cases (Bellasari, 1988), and it may reflect what Deaton (1985) labeled "adaptive noncompliance." Weintraub (1976) distinguished between *capricious adherence*, which involves "irregular therapeutic behavior based on false theory and misinformation" (p. 129), from *intelligent nonadherence*, in which patients have valid, although not necessarily the medically most appropriate, reasons for not adhering, such as possible unpleasant side effects, confusion about dosage, or concerns about ingesting too much prescribed medicine. The "intelligent nonadherer" may reject medications because of concerns about a deterioration in the quality of his or her life. In the patient's frame of reference, treatment nonadherence is a decision based on judgment. For example, Cooper, Love, and Raffoul (1982) indicated that 73% of nonadherence was intentional. The most common reason given for intentional nonadherence was that the patient did not believe that the drug was needed in the dosage prescribed by the physician. Only 15% indicated that they took less of the drug because of side effects or other negative consequences. To the patient, when a medication is no longer seen as efficacious, it is likely to be stopped. The logic is that, if there are no symptoms and the medication makes no difference, there is no reason to continue to take it.

Given the disheartening failures of the treatment to control negative effects, is it surprising that patients with diverse diseases conclude, "What's the use?" Add to this the sense of helplessness and the demands and disruptions created by adhering to the treatment regimen, and we can more fully appreciate the implicit calculation by patients of the ratio of the perceived benefits of adherence to its perceived inconvenience (i.e., disruption of lifestyle).

On a somewhat more positive note, the patient's decision to stop taking medications may sometimes be a rational-empirical method of determining the efficacy of the prescribed treatment. Patients with chronic disorders tend to be nonadherent in the taking of medications, especially when they are asymptomatic most of the time, and when the consequences of noncompliance are delayed in appearance. From the patient's perspective, the issue is more one of self-control than of adherence. From a medical perspective, the issue is one of finding the "right" treatment for the "right" patient. In short, the patient understands the regimen in terms of the way it will affect his or her life. The health professional understands the regimen in terms of the way it will affect the patient's health.

Let us consider some of the research that has

examined the rationales that patients offer in making decisions not to adhere to recommendations. Medication nonadherence is a major concern in psychiatry, as well as in all other areas of medicine. Jamison, Gerner, and Goodwin (1979) asked 42 manic-depressive patients to rank in importance the factors in potentially stopping treatment. The three highest rankings were given to discomfort in having mood controlled by medication, distress about having a chronic illness, and side effects. These concerns inhibited adherence to the prescribed psychotropic medication.

In a study of nonadherence among epileptic patients, Conrad (1985) examined the rationales used by patients in making the decision not to adhere. Physicians typically alter doses of anticonvulsive medication in times of increased seizure activity or troublesome drug side effects. It is difficult to strike the optimum level of anticonvulsive medications. To people with epilepsy, it seems as if physicians engage in a certain amount of trial-and-error behavior. Medications are viewed, both by doctors and by patients, as an indicator of the degree of disorder. If seizure activity is not controlled or if it increases, patients see doctors respond by raising dosage or changing medications. The more medicine prescribed, the "worse" the epilepsy, and conversely, reduction in the dosage is taken as a sign of "improvement" in the epileptic condition.

If the epilepsy is getting worse, patients may take more medication; if it is getting better, they may take less medication. This is the pattern of physician behavior that the epileptics have observed. Thus, if the physician reduces or raises the dose or the strength of the medication, patients may take the liberty of experimenting with the dosage themselves. Epileptics may reduce their dosage in order to determine if they are able to get along on a lowered amount of medication, or to determine whether they have improved and no longer need the drug.

For most people with epilepsy, there is not a one-to-one correspondence between taking or missing medication and seizure activity. People who take medication regularly may have seizures, and some people who discontinue their medication may be seizure-free for months

or longer. Medical experts say that a patient may well miss a whole day's medication, yet still have enough of the drug in the bloodstream to prevent a seizure during this period (Conrad, 1985).

How can one know whether a period without seizures is the result of medication or of a spontaneous remission of the disorder? How can one know if epilepsy is "getting better," while still taking medication? In order to test to see if epilepsy is "still there," people may take themselves off medication as an experiment to see "if anything will happen." Thus, regulating medication may represent an attempt to assert some degree of control over a condition that appears at times to be completely beyond control. What appears to be nonadherence from a medical perspective may actually be a form of the patient's asserting control over his or her disorder. The issue is more clearly one of responding to the meaning of medications in everyday life than of adherence to physicians' orders and recommended medical regimens.

In a study of pediatric asthma patients, it was found that patient and parent nonadherence to medication schedules may actually be a positive, adaptive response (Deaton, 1985). Based on parents' knowledge of their child's disease severity, the seasonal variability in asthma symptoms, concerns about limited effectiveness and potentially severe side effects of prescribed medication, and consideration of the therapeutic costs in the family's quality of life, nonadherence decisions were part of the patient's strategy for coping with chronic illness. In this particular study, adaptive nonadherence was found to be significantly correlated with better asthma control, whereas stricter adherence to the medical regimen was not: "A conscious decision which is based on adequate information about the regimen and intimate knowledge of the child may in fact be superior to passive compliance with the regimen" (p. 12).

Health care providers need to ask themselves several questions in thinking about the likelihood of a patient's adhering to a particular regimen. Does the health care provider believe that the patient will be susceptible to continuing problems if the recommended behavior is not adopted? Does the patient perceive the benefits of adopting the recommended behavior to be greater than the perceived risks, costs, side effects, barriers, and hassles? These questions form the basis of the influential health belief model as applied to adherence behavior (Becker & Maiman, 1975).

Where incongruity or dissonance arises between different aspects of a disease—for example, between the elimination of concrete signs of disease (symptoms) and the necessity for continuing distressing treatment (Nerenz, Leventhal, Love, & Ringler, 1984), or between the diagnostic symptoms experienced and the side effects of the treatment (Ringler, 1981)—individuals attempt to resolve the conflict. How patients resolve such incongruities influences treatment adherence. For instance, the match between the etiological explanations offered by health workers and the extent to which such explanations agree or conflict with patients' causal attributions may modify patient attributions, may generate or resolve dissonance, and may thereby influence adherence behaviors.

Patients with a chronic illness such as rheumatoid arthritis, diabetes, or epilepsy frequently get discouraged by extended treatment that may produce limited therapeutic results, and some consequently become less adherent over time.

In sum, if health care providers hope to facilitate adherence rather than simply to bemoan its occurrence, they need to consider the patient's perspective and expectations, such as the patient's (1) expectations about the clinical encounter; (2) beliefs and misconceptions about the cause, severity, or symptoms of the illness and susceptibility to complications or exacerbations; (3) perceptions of the goals of the treatment; (4) perception of the cost and risks versus the benefits of the treatment; (5) existing health-related knowledge, skills, and practices; (6) degree of adaptation to the disease; (7) sense of hopelessness or lack of self-efficacy; (8) learning limitations; and (9) family involvement.

Health Care Provider Perspective

A survey of physicians conducted by Stone (1979) revealed that only 25% acknowledged that they might in any way contribute to their patients' treatment nonadherence. The small proportion of physicians who acknowledged their important role runs counter to the available data. In a study conducted by Cochran (1984), patients with bipolar affective disorders expressed several common concerns in the course of intervention, including discomfort with the psychosocial changes produced by lithium treatment, worries about the future course of the illness, distress over being a chronic mental patient, and concern about the safety of long-term lithium treatment. In this study, the patients' normative beliefs about their psychiatrists' expectations predicted adherence to lithium treatment. The most appropriate targets for such interventions may be a patient's normative beliefs—that is, a belief that relevant individuals (family and friends), particularly the psychiatrist, want the patient to adhere to the lithium regimen—and the patient's desire to do as the physician wishes (see Cochran & Getlin, 1988).

A number of studies have underscored the discrepancy between physicians' and patients' understanding and appraisal of physician–patient interactions. In one study, Taylor, Burdette, and Edwards (1980) asked patients and physicians, after an interview, to indicate the main purpose of the visit. Even though the categories were broad (e.g., the continuation of health maintenance and the care of a physical problem), there was disagreement in one third of the encounters. Waitzkin and Stoeckle (1976) recorded the interactions of more than 300 patients and their physicians in both offices and hospitals. They found that, during a visit averaging 20 minutes, little more than 1 minute was actually spent by the physician in giving information. But when asked to estimate the time they spent giving information, the physicians estimated nearly one quarter to one half of the visit. Later, when asked to estimate their patients' desire for information, 65% of them underestimated how much information the patients wanted. In addition, Merkel, Rudisill, and Nierenberg (1983) found a correlation of .1 between patients' satisfaction with a consultation and physicians' perceptions of their satisfaction.

The discrepancy between physicians and patients when it comes to reasons for nonadherence is illustrated in another set of studies. House, Pendelton, and Parker (1986) found a major discrepancy between patients' and physicians' views of the reasons for non-insulin-dependent diabetic patients' dietary adherence difficulties. The physicians overwhelmingly (80%) perceived motivational problems as the source of dietary nonadherence. The patients, on the other hand, mentioned environmental causes (e.g., life circumstances such as family, job, or economic conditions, 38%) slightly more frequently than they did motivational causes (lack of incentive or desire, 34%) and also placed considerable emphasis (26%) on somatic and physiological factors (physical limitations such as visual or ambulatory restrictions that interfered with food preparation). These results strongly indicate that the physicians perceived dietary nonadherence as being largely under the patient's control, whereas the patients maintained that their nonadherence was out of their control to a large extent. For physicians to have maximum impact in increasing patient adherence, they must be aware that patients have a strong tendency to disclaim personal responsibility for the failure to adhere.

Svarstad (1976) found that 50% of patients could not correctly report how long they were supposed to continue taking their medication. Two high-blood-pressure surveys (Bohnstedt, Leonard, Trudeau, & Bal, 1985; Levine, Morisky, Bone, Lewis, Ward, & Green, 1982) indicate that the major reason that patients stopped taking their antihypertensive medications was that "the doctor told me to stop." Further probing in to what the doctor had actually said suggested problems of misunderstanding and confusion, such as "The doctor said my blood pressure was normal," or "The doctor said my blood pressure was under control." Often, what the patient hears is not what the health care provider says or intends to say.

Svarstad (1986) proposed a human communication model of treatment adherence. This model assumes that adherence requires comprehension and recall of the regimen, as well as motivation to follow it. The model suggests that patients experience a variety of problems and concerns that undermine their willingness and ability to adhere, and health care providers need to ask patients specifically about the problems they are having. Attention to patients' understanding should help to identify likely problem areas, fears, doubts, and hesitations about the diagnosis and treatment and should enable the practitioner to individualize the regimen and resolve whatever problems or concerns are undermining patients' desire to adhere.

Rarely do health care professionals provide patients with specific information or practical advice regarding such factors as what the patient should do if side effects occur, if doses are forgotten, or if symptoms disappear. Health care providers often assume that patients understand what might seem to the provider to be simple instructions. However, Massullo, Lasagna, and Grinar (1974) asked patients to interpret 10 prescription labels and found that many patients made errors when interpreting common instructions. What is the physician's intent when the prescription indicates that the medication should be taken every six hours? Should the patient set an alarm to wake him or her up to take the medication? Should the patient take a double dose if he or she misses one? There are many other examples of confusion in apparently simple instructions as well as cryptic and insufficient information regarding proscribed behaviors.

Adherence of Health Care Providers

Despite the knowledge that sound investigations have produced during the last two decades concerning the enhancement of adherence, many practitioners remain unaware of basic adherence management principles (Logan & Haynes, 1986). For example, in a survey of general practitioners, Logan (1978) found that 59% *never* called patients who missed scheduled appointments, and that only 20% claimed that they always did.

Although the literature has focused on the adherence of patients, attention also needs to be directed to adherence by health care providers. For example, in a study conducted by the American Society of Internal Medicine, Hare and Barnoon (1973) found little association between the established criteria for the delivery of treatment and physicians' performance.

It has been noted that, even in facilities where the supervision of medication taking is left to professionals, the issue of adherence to medication regimens is a problem. A study conducted by the University of Southern California School of Pharmacy (Kayne & Cheung, 1973) evaluated the drug preparation and administration system in three extended-care facilities that provide skilled-nursing care after discharge from an acute-care hospital. In 2,001 doses of medications that were administered to 275 patients, 397 medication errors were observed (an error rate of 20%). Missed doses accounted for 53% of all medication errors. Of the 397 errors, 47 were the wrong drug, 47 were the wrong dose of the drug, 20 were the wrong route of administration, and 69 were inappropriate dosing intervals.

Meichenbaum and Turk (1987) reviewed the literature on the adherence behaviors of health care providers, and although somewhat pessimistic in their conclusion, they provided an analysis of the likely causes of health care provider nonadherence and some suggestions to reduce these.

What Can Be Done to Facilitate Adherence?

Health care providers who adopt an authoritarian approach generally assume that patients' underlying fears, anxieties, and hesitations about clinical treatment are misdirected or trivial, and that patients should either accept the recommended treatment regimen or find another practitioner. Thus, they respond by ignoring or dismissing many of the patient's complaints, demanding adherence, and becoming angry or hostile (Danziger, 1981; Davis, 1966; Svarstad, 1976). Ignoring patients' fears and concerns about the regimen does not make them go away, and coercive tactics eventually lead to patient concealment and dissatisfaction, errors in clinical judgment, and higher dropout rates (Svarstad, 1976).

Clinicians who adopt the collaborative patient participatory approach take a broader view of patient feedback and its role in the treatment process. They assume not only that patients have the right to question and complain, but that negative feedback is a critical part of the communication and treatment processes because it allows the practitioner to individualize or adjust the regimen and to resolve whatever physical, psychological, or social problems are interfering with the patient's adherence.

The practitioner may ask whether the amount of attention needed to facilitate adherence is cost-effective. Weinstein and Stason (1976) developed the first quantitative evidence of the importance of adherence to medical regimens (for hypertensives). They showed quite convincingly that funds expended on improving adherence to antihypertensive regimens produce a greater impact on disability and death than does an equal amount spent on the detection of new cases and the initiation of treatment. Thus, efforts to enhance treatment adherence were economical, as well as beneficial to health. Several recent books and papers have described adherence enhancement strategies in some detail. Because we have recently reviewed and critiqued these strategies in some detail elsewhere (see Meichenbaum & Turk, 1987), we have merely listed them in Table 2 and will highlight some of the general adherence enhancement intervention guidelines.

Common to many of the successful adherence enhancement intervention programs is the observation that patients should be encouraged to participate actively in their own health care. They should be involved in treatment planning and in decisions through the use of mutual goal-setting techniques to ensure that their priorities, lifestyle, resources, and possible adherence barriers will be considered. In addition to these general guidelines, Haynes, Wang, and Da Mota Gomes (1987) provided a short list of adherence-improving actions based on their review of the adherence literature. This straightforward list provides a checklist that health care providers to can use to evaluate their clinical practice:

Table 2. Summary of Adherence Enhancement Interventions[a]

Keeping appointments
Give specific appointment time with name of specific
health care professional to be seen
Use reminders (mail, telephone)
Use efficient clinic scheduling to minimize waiting time
Use short referral time
Ensure continuity of care
Keep an active follow-up appointment file
Discuss reasons for previously missed appointments
Following acute medical regimens
Improve patient–professional communication by involv-
ing patient in planning and implementation of treat-
ment program
Customize treatment plan
Simplify treatment regimens
Use patient education and evaluate patient understand-
ing
Anticipate management of side effects
Use special reminders and special medication packaging
Use adjunctive services (e.g., health educator, pharma-
cist counseling)
Be aware of medication engineering and packaging (e.g.,
color or shape of a pill can transmit a specific message
both in general and to different ethnic groups)
Following chronic medical regimens
Improve patient–provider communication, and pick up
on any negative feedback
Use behavior modification procedures (e.g., self-mon-
itoring, goal setting, behavioral contracting, commit-
ment and reinforcement procedures)
Teach self-management skills (e.g., problem solving, decisional balance sheet construction, relapse prevention, attribution training)
Use graduated regimen implementation
Involve significant others (e.g., spouse, children, rela-
tives, neighbors)
Use supervision (e.g., home visits, brief hospitalization)
Use adjunctive services (e.g., health educators, psychotherapeutic services)
Integrate treatment regimen into normal life
Use role playing, and paradoxical techniques when ap-
propriate

[a]Adapted from D. Meichenbaum and D. Turk. (1987). *Facilitating Treatment Adherence: A Practitioner's Guidebook*. New York: Plenum Press. Table organization based on Haynes (1979).

1. Listen to the patient.
2. Ask the patient to repeat what has to be done.
3. Keep the prescription as simple as possible.
4. Give clear instructions on the exact treatment regimen, preferably in writing.
5. Make use of special reminder pill containers and calendars.
6. Call if an appointment is missed.
7. Prescribe a self-care regimen that takes into account the patient's daily schedule.
8. Emphasize the importance of adherence at each visit.
9. Titrate the frequency of visits to the patient's adherence needs.
10. Acknowledge the patient's efforts to adhere at each visit.
11. Involve the patient's spouse or other partner.

Adherence to long-term treatments is more difficult to achieve: no single intervention has been shown to be useful on its own to date. Studies of self-monitoring (Johnson, Taylor, Sackett, Dunnett, & Shimizu, 1977; Shepard, Foster, Stason, Solomon, McArdle, & Gallagher, 1979), home visits (Johnson, 1977), tangible rewards (Shepard *et al.*, 1979), peer group discussion (Shepard *et al.*, 1979), counseling by a health educator (Levine, 1982) or a nurse (Shepard *et al.*, 1979), and special-unit dose-reminder pill-packaging (Becker, Glanz, Sobel, Mossey, Zinn, & Knott, 1986) have shown no benefit when these maneuvers have been applied in isolation. Rather, enhancing long-term adherence require combinations of the following: clear instructions; recalling nonattenders; patient self-monitoring of adherence and/or treatment outcomes; enhancement of social support; contingency contracting and rewards or reinforcement for high adherence; and group discussion and supervised self-management. None of these interventions is self-sustaining: they must continue to be applied as long as adherence is required, especially in the instance of permanent lifestyle changes (see Cameron & Best, 1987).

A Final Note

Changing the long-term behavior of health care providers to manage adherence successfully cannot be done by simply informing or instructing practitioners about efficacious interventions (Evans, Block, Steinberg, & Penrose, 1986). As Haynes *et al.* (1987) noted, to overcome these barriers will require the application of proven techniques of continuing education, including an audit of performance, feedback on deviations from expected standards of care, apprenticeship programs for practitioners, and targeted training by practitioners who are regarded by their peers as the source of innovations. In the future, computerization of health records may become widespread enough to automate some aspects of adherence monitoring and reminder systems, thus reducing the burden on practitioners. This chapter has argued that what is also required is that health care providers take the patient's perspective as well as develop more effective interventions.

ACKNOWLEDGMENTS

Completion of this chapter was supported in part by grants DE 07514 from the National Institute of Dental Research and ARNS 38698 from the National Institute of Arthritis, Musculoskeletal and Skin Diseases awarded to Dennis C. Turk.

References

Anderson, W. F. (1974). Administration, labelling and general principles of drug prescription in the elderly. *Gerontology Clinics, 16,* 4–9.

Arluke, A. (1980). Judging drugs: Patients' conceptions of therapeutic efficacy in the treatment of arthritis. *Human Organism, 39,* 84–88.

Becker, M. H., & Maiman, L. A. (1975). Sociobehavioral determinants of compliance with health and medical care recommendations. *Medical Care, 13,* 10–25.

Becker, M. H., Radius, S. M., Rosenstock, I. M., Drachman, R. H., Shuberth, L. C., & Teets, K. C. (1978). Compliance with medical regimens for asthma: A test of the Health Belief Model. *Public Health Reports, 93,* 268–277.

Becker, M. H., Glanz, K., Sobel, E., Mossey, J., Zinn, S. L., & Knott, K. A. (1986). A randomized trial of special packaging of antihypertensive medications. *Journal of Family Practice, 22,* 357–361.

Bellasari, A. (1988). Owning CF—Adaptive noncompliance with chest physiotherapy in cystic fibrosis. In H. A. Baer (Ed.), *Encounters with biomedicine: Case studies in medical anthropology.* New York: Breach Science.

Blackwell, B. (1979). Treatment adherence: A contemporary viewpoint. *Psychosomatics, 20,* 27–35.

Blair, S. N., Jacobs, D. R., & Powell, K. E. (1985). Relationship between exercise or physical activity and other health behaviors. *Public Health Reports, 100,* 172–180.

Bohnstedt, M., Leonard, A. R., Trudeau, M. J., & Bal, D. G. (1985, April). *Patient noncompliance in hypertension control.* Paper presented at the National Conference on High Blood Pressure Control, Chicago, IL.

Boyd, J. R., Covington, J. R., Stanaszek, W. F., & Coussons, R. T. (1974). Drug defaulting: 2. Analysis of noncompliance patterns. *American Journal of Hospital Pharmacy, 31,* 485–491.

Buckalew, L. W., & Sallis, R. E. (1986). Patient compliance and medication perception. *Journal of Clinical Psychology, 42,* 49–53.

Cahill, G. F., Jr., Etzwiler, D. D., & Freinkel, N. (1976). "Control" and diabetes. *New England Journal of Medicine, 294,* 1004–1005.

Cameron, R., & Best, J. A. (1987). Promoting adherence to health behavior change interventions: Recent findings from behavioral research. *Patient Education and Counseling, 10,* 139–154.

Cochran, S. D. (1984). Preventing medical noncompliance in the outpatient treatment of bipolar affective disorders. *Journal of Consulting and Clinical Psychology, 52,* 873–876.

Cochran, S. D., & Getlin, M. J. (1988). Attitudinal correlates of lithium compliance in bipolar affective disorders. *Journal of Nervous and Mental Disease, 176,* 457–464.

Conrad, P. (1985). The meaning of medication: Another look at compliance. *Social Science and Medicine, 20,* 29–37.

Cooper, J. K., Love, D. W., & Raffoul, P. R. (1982). Intentional prescription nonadherence (noncompliance) by the elderly. *Journal of the American Geriatric Society, 30,* 329–333.

Danziger, S. K. (1981). The uses of expertise in doctor-patient encounters during pregnancy. In P. Conrad & R. Kern (Eds.), *The sociology of health and illness.* New York: St. Martin's Press.

Davis, M. S. (1966). Variations in patients' compliance with doctors' advice: Analysis of congruence between survey responses and results of empirical observations. *Journal of Medical Education, 41,* 1037–1048.

Davis, M. S. (1968). Variations in patients' compliance with doctors' advice: An empirical analysis of patterns of communication. *American Journal of Public Health, 58,* 274–288.

Deaton, A. V. (1985). Adaptive noncompliance in pediatric asthma: The parent as expert. *Journal of Pediatric Psychology, 10,* 1–14.

DeBusk, R. F., Haskell, W. L., Miller, N. H., Berra, K., & Taylor, C. B. (1985). Medically directed at-home rehabilitation soon after clinically uncomplicated acute myocardial infarction: A new model of patient care. *American Journal of Cardiology, 55,* 251–257.

DiMatteo, M. R., & DiNicola, D. D. (1982). *Achieving patient compliance: The psychology of the medical practitioner's role.* New York: Pergamon Press.

Drash, A. L., & Becker, D. (1978). Diabetes mellitus in the child: Course, special problems and related disorders. In H. W. Kazen & R. J. Mahler (Eds.), *Diabetes, obesity and vascular disease: Metabolic and molecular interrelationships.* Indianapolis: Lilly.

Dunbar, J. M. (1983). Compliance in pediatric populations: A review. In P. J. McGrath & P. Firestone (Eds.), *Pediatric and adolescent behavioral medicine: Issues in treatment.* New York: Springer.

Dunbar, J. M., & Agras, W. S. (1980). Compliance with medical instructions. In J. M. Ferguson & C. B. Taylor (Eds.), *Comprehensive handbook of behavioral medicine* (Vol. 3). New York: Spectrum.

Dunbar, J. M., & Stunkard, A. J. (1979). Adherence to diet and drug regimen. In R. Levy, B. Rifkind, B. Dennis, & N. Ernst (Eds.), *Nutrition, lipids, coronary heart disease.* New York: Raven Press.

Eisenthal, S., Emergy, R., Lazare, A., & Udin, H. (1979). Adherence and the negotiated approach to patienthood. *Archives of General Psychiatry, 36,* 393–398.

Epstein, L. H., & Cluss, P. A. (1982). A behavioral medicine perspective on adherence to long-term medical regimens. *Journal of Consulting and Clinical Psychology, 50,* 960–971.

Epstein, L. H., Wing, R., Koeske, R., & Valoski, A. (1985). A comparison of lifestyle exercise, aerobic exercise, and calisthenics on weight loss in obese children. *Behavior Therapy, 16,* 345–356.

Eraker, S. A., Kirscht, J. P., & Becker, M. H. (1984). Understanding and improving patient compliance. *Annals of Internal Medicine, 100,* 258–268.

Evans, D. A., Block, M. R., Steinberg, E. R., & Penrose, A. M. (1986). Frames and heuristics in doctor-patient discourse. *Social Science and Medicine, 22,* 1027–1034.

Faberow, N. L. (1986). Noncompliance as indirect self-destructive behavior. In K. E. Gerber & A. M. Nehemkis (Eds.), *Compliance: The dilemma of the chronically ill.* New York: Springer.

Fontana, A. F., Kerns, R. D., Rosenberg, R. L., Marcus, J. L., & Colonese, K. L. (1986). Exercise training for cardiac patients: Adherence, fitness, and benefits. *Journal of Cardiopulmonary Rehabilitation, 6,* 4–15.

Glasgow, R. E., McCaul, K. D., & Schafer, L. C. (1987). Self-care and glycemic control in Type I diabetes. *Journal of Chronic Diseases, 40,* 399–412.

Hare, R. L., & Barnoon, S. (1973). *Medical care appraisal and quality assurance in the office practice of internal medicine.* Paper presented at the meeting of the American Society of Internal Medicine, San Francisco.

Hartshorne, E. A. (1977). Food and drug interactions. *Journal of the American Dietetic Association, 70,* 15–19.

Haynes, R. B. (1979). Strategies to improve compliance with referrals, appointments and prescribed medical regimens. In R. B. Haynes, D. W. Taylor, & D. L. Sackett (Eds.), *Compliance in health care.* Baltimore: Johns Hopkins University Press.

Haynes, R. B. (1984). Compliance with health advice: An overview with special reference to exercise programs. *Journal of Cardiopulmonary Rehabilitation, 4,* 120–123.

Haynes, R. B., Wang, E., & Da Mota Gomes, M. (1987). A critical review of interventions to improve compliance with prescribed medications. *Patient Education and Counseling, 10,* 155–166.

Hippocrates. (430 B.C./1923). *On decorum and the physician* (Vol. 2). W. H. S. Jones, trans. London: William Heinemann.

House, W. C., Pendelton, L., & Parker, L. (1986). Patients' versus physicians' attributions of reasons for diabetic patients' noncompliance with diet. *Diabetes Care, 9,* 434.

Hurwitz, N. (1969). Predisposing factors in adverse reactions to drugs. *British Medical Journal, 1,* 536–539.

Inui, T. S., Carter, W. B., & Pecoraro, R. E. (1981). Screening for noncompliance among patients with hypertension: Is self-report the best available measure. *Medical Care, 19,* 1061–1064.

Jamison, K. R., Gerner, R. H., & Goodwin, F. K. (1979). Patient and physician attitudes toward lithium: Relationship to compliance. *Archives of General Psychiatry, 36,* 866–869.

Johnson, A. L., Taylor, D. W., Sackett, D. L., Dunnett, C. W., & Shimizu, A. G. (1977). Self-blood pressure recording—An aid to blood pressure control? *Annals of the Royal College of Physicians and Surgeons of Canada, 10,* 32–36.

Kavanaugh, T., Shephard, R., Pandit, V., & Doney, H. (1970). Exercise and hypnotherapy in the rehabilitation of the coronary patient. *Archives of Physical Medicine and Rehabilitation, 51,* 578–587.

Kayne, R. C., & Cheung, A. (1973). An application of clinical pharmacy in extended care facilities. In R. H. Davis & W. K. Smith (Eds.), *Drugs and the elderly.* Los Angeles: University of Southern California Press.

Kristeller, J., & Rodin, J. (1984). The function of attention in cognitive models of behavior change and maintenance. In A. Baum, S. E. Taylor, & J. E. Singer (Eds.), *Handbook of psychology and health: Vol. 4. Social psychological aspects of health.* Hillsdale, NJ: Erlbaum.

Levine, D. M., Morisky, D. W., Bone, L. R., Lewis, C. E., Ward, W. B., & Green, L. W. (1982). Data-based planning for educational interventions through hypertension control programs for urban and rural populations in Maryland. *Public Health Reports, 97,* 102–112.

Logan, A. G. (1978). *Investigation of Toronto general practitioners' treatment of patients with hypertension.* Toronto: Canadian Facts.

Logan, A. G., & Haynes, R. B. (1986). Determinants of physicians' competence in the management of hypertension. *Journal of Hypertension, 4* (Suppl. 5), S367–S369.

Luscher, T. F., Vetter, H., Siegenthaler, W., & Vetter, W. (1985). Compliance in hypertension: Facts and concepts. *Journal of Hypertension (Supplement), 3,* 3–10.

Mattar, M. E., Markello, J., & Yaffe, S. J. (1975). Inadequacies in the pharmacologic management of ambulatory children. *Journal of Pediatrics, 87,* 137–141.

Mazzullo, J. V., Lasagna, L., & Grinar, P. F. (1974). Variations in interpretation of prescription instructions: The need for improve prescribing habits. *Journal of the American Medical Association, 227,* 929–931.

Meichenbaum, D., & Turk, D. C. (1987). *Facilitating treatment adherence: A practitioner's guidebook.* New York: Plenum Press.

Merkel, W. T., Rudisill, J. R., & Nierenberg, B. P. (1983). Preparing patients to see the doctor: Effects on patients and physicians in a family practice center. *Family Practice Research Journal, 2,* 147–163.

Morisky, D. E., Green, L. W., & Levine, D. M. (1986). Concurrent and predictive validity of a self-reported measure of medication adherence. *Medical Care, 24,* 67–74.

Nerenz, D. R., Leventhal, H., Love, R. H., & Ringler, K. E. (1984). Psychological aspects of cancer chemotherapy. *International Review of Applied Psychology, 33,* 521–530.

Oldridge, N. B. (1986). Cardiac rehabilitation, self-responsibility, and quality of life. *Journal of Cardiopulmonary Rehabilitation, 6,* 153–156.

Olson, R. A., Zimmerman, J., & Reyes de la Rocha, S. (1985). Medical adherence in pediatric populations. In A. R. Zeiner, D. Bendell, & C. E. Walker (Eds.), *Health psychology: Treatment and research issues.* New York: Plenum Press.

Phillips, E. L. (1988). *Patient compliance.* Lewiston, NY: Hans Huber.

Rechnitzer, P. A., Cunningham, D. A., Andrew, G. M., Buck, C. W., Jones, N. L., *et al.* (1983). Relation of exercise to recurrence rate of myocardial infarction in men: Ontario Exercise-Heart Collaborative Study. *American Journal of Cardiology, 51* 65–69.

Ringler, K. E. (1981). *Process of coping with cancer chemotherapy.* Doctoral dissertation, University of Wisconsin, Madison, cited in D. R. Nerenz, & H. Leventhal. (1983). Self-regulation theory in chronic illness. T. G. Burish & L. A. Bradley (Eds.), *Coping with chronic illness.* New York: Academic Press.

Roth, H. P. (1987). Measurement of compliance. *Patient Education and Counseling, 10,* 107–116.

Sackett, D. L. (1976). Priorities and method for future research. In D. L. Sackett & R. B. Haynes (Eds.), *Compliance with therapeutic regimens.* Baltimore: Johns Hopkins University Press.

Sackett, D. L., Haynes, R. B., Gibson, S., Hackett, D., Taylor, D. W., Roberts, R. S., & Johnson, R. (1975). Randomized clinical trials of strategies for improving medication compliance in primary hypertension. *Lancet, 1,* 1205–1207.

Sackett, D. L., Haynes, R. B., Gibson, E., & Johnson, A. (1976). The problem of compliance with hypertensive therapy. *Practical Cardiology, 2,* 35–39.

Sallis, J. F., Haskell, W. L., Wood, P. D., Fortmann, S. P., & Vranizan, K. M. (1986). Vigorous physical activity and cardiovascular factors in young adults. *Journal of Chronic Diseases, 39,* 115–120.

Shepard, D. S., Foster, S. B., Stason, W. B., Solomon, H. S., McArdle, P. J., & Gallagher, S. S. (1979). Cost-effectiveness of intervention to improve compliance with antihypertensive therapy. *Prevention Medicine, 8,* 229.

Soutter, R. B., & Kennedy, M. C. (1974). Patient compliance in drug trials: Dosage and methods. *Australian and New Zealand Journal of Medicine, 4,* 360–364.

Stewart, R. B., Cluff, L. E., & Leighton, E. D. (1972). Commentary: A review of medication errrors and compliance in ambulatory patients. *Clinical Pharmacology Therapy, 13,* 463–468.

Stone, G. C. (1979). Patient compliance and the role of the expert. *Journal of Social Issues, 35,* 34–59.

Streltzer, J., & Hassell, L. H. (1988). Noncompliant hemodialysis patients: A biopsychosocial approach. *General Hospital Psychiatry, 10,* 255–259.

Svarstad, B. L. (1976). Physician-patient communication and patient conformity with medical advice. In D. Mechanic (Ed.), *The growth of bureaucratic medicine: An inquiry into the dynamics of patient behavior and the organization of medical care.* New York: Wiley.

Svarstad, B. L. (1986). Patient-practitioner relationships and compliance with prescribed medical regimens. In L. H. Aiken & D. Mechanic (Eds.), *Applications of social science to clinical medicine and health policy.* New Brunswick, NJ: Rutgers University Press.

Taylor, R. B., Burdette, J. A., & Edwards, J. (1980). Purpose of the medical encounter: Identification and influences on process and outcome in 200 encounters in a model family practice center. *Journal of Family Practice, 10,* 495–500.

Tebbi, C. K., Cummings, K. M., Zevon, M. A., Smith, L., Richards, M., & Mallon, J. (1986). Compliance of pediatric and adolescent cancer patients. *Cancer, 58,* 1179–1184.

Turk, D. C., & Speers, M. A. (1983). Diabetes mellitus: A cognitive functional analysis of stress and adherence. In T. Burish & L. A. Bradley (Eds.), *Coping with chronic disease.* New York: Academic Press.

Vincent, P. (1971). Factors influencing patient noncompliance: A theoretical approach. *Nursing Research, 20,* 509–516.

Waitzkin, H., & Stoeckle, J. D. (1976). Information control and the micropolitics of health care. *Journal of Social Issues, 10,* 263–276.

Weinstein, M. C., & Stason, W. B. (1976). *Hypertension: A policy perspective.* Cambridge: Harvard University Press.

Weintraub, M. (1976). Intelligent noncompliance and capricious compliance. In L. Lasagna (Ed.), *Patient compliance.* Mt. Kisco, NY: Futura.

PART V

Practical Issues

Research

Psychologists are trained as scientists, as well as clinicians. Not only can they conduct meaningful research, but clinical psychologists tend to have a critical, empirical perspective of their clinical work. As one of many professions providing service in the medical setting, only clinical psychology makes research training a core feature of its discipline. James Malec (Chapter 16) outlines the steps that can be taken to develop and implement psychological research in medical settings, including writing proposals, funding projects, conducting research, and publishing the results. Malec also discusses the important "products" of clinical psychology research in medical settings.

Research in the Medical Setting

Implementing the Scientist-Practitioner Model

James Malec

Introduction

Clinical psychology has been unique among the health care professions in adopting a model of training that requires competency in both research methodology and clinical practice (Strickland, 1988). Clinical psychologists are most often trained in university-based graduate schools and receive a Ph.D. to signify the attainment of journeyman skills in the profession. The Ph.D. requires the completion of a doctoral dissertation. The satisfactory completion of dissertation research requires knowledge in the fundamentals of research: experimental methodology, design, and statistical analysis. A Ph.D. in clinical psychology also requires competency in the basic skills of the profession and additional development of these skills during a one year full-time clinical internship. This model of training, which combines research and clinical training, has come to be known as the *scientist-practitioner model*.

The scientist-practitioner model first received

James Malec • Department of Psychiatry and Psychology, Mayo Clinic and Foundation, Rochester, Minnesota 55905.

official sanction from the academic community in clinical psychology at a meeting in Boulder, Colorado, in 1949 sponsored by the National Institute of Mental Health and the American Psychological Association (Raimy, 1950). Barlow, Hayes, and Nelson (1984, pp. 7–8) specified five reasons identified in the Boulder conference that support this model of training: (1) to develop graduate trainees' interest in both research and practice; (2) to encourage research that will expand the limited knowledge base of clinical psychology; (3) because the number of applicants exceeds the number of positions in clinical psychology graduate training programs, to give preference to those candidates who show potential for both research and clinical practice; (4) to alert researchers, through their involvement in clinical practice, to important clinical issues for investigation; and (5) to provide financial support for research through clinical activities.

In the aftermath of the Boulder Conference, the scientist-practitioner model has experienced considerable criticism and controversy (Frank, 1984; Goldfried, 1984). Most Ph.D.'s in clinical psychology make a clear decision to engage almost exclusively in clinical practice

(Norcross & Prochaska, 1982), and research productivity among clinical psychology practitioners is low (Conway, 1988). Such evidence suggests that research training is unduly emphasized in clinical psychology training programs, given the career choices of most graduates. Such observations led to the development of professional schools for training clinical psychologists. Professional schools emphasize clinical as opposed to research training for the completion of a doctor of psychology degree (Psy.D.). It is interesting, however, that most professional schools in clinical psychology continued to require some demonstration of research competencies of their graduates, and some even require a formal dissertation (Jones, 1987).

In response to the threat posed by professional schools, traditional clinical psychology graduate programs increased preinternship clinical training and have shown some movement toward accepting nontraditional dissertation projects, that is, projects that consist of a series of single-case designs or that focus on program evaluation. This has been a healthy movement in traditional graduate programs. As Barlow *et al.* (1984) pointed out, the most critical threat to the scientist-practitioner model may be the lack of relevance of the results of standard experimental group designs to the clinical practice of psychology. These authors supported the use of single-case designs in clinical research, citing F. C. Thorne's ideal description of each patient's clinical care as a "single and well-controlled experiment" (Barlow *et al.*, 1984, p. 5).

Thus, with professional schools in psychology requiring research training and graduate schools emphasizing applied research and increased training relevant to clinical practice, the scientist-practitioner model survives and continues to be the most accepted model for training in clinical psychology. In surveying groups of psychologists who were primarily scientists, primarily practitioners, or scientist-practitioners, Conway (1988) found that the majority of each group basically supported the scientist-practitioner model.

Controversy regarding the scientist-practitioner model may also find a basis in the overlapping, but distinct, value systems held by professionals whose careers are devoted mainly to scientific research as compared to those devoted mainly to clinical service (Conway, 1988; Frank, 1984). Few would argue that the primary goal of research is the expansion of scientific knowledge. For most scientific researchers, this primary goal also serves as a fundamental value. The pursuit of scientific knowledge is a very long-term goal and, consequently, a highly abstract value. Realistically, the expansion of scientific knowledge is a goal that must extend across generations. Historically, the most important scientific discoveries have resulted in entire new agendas for scientific investigation. The most significant answers in the history of science have created large numbers of new questions. In integrating extant knowledge in the physical sciences, Einstein created work for generations of new scientists in subatomic physics and astrophysics. Crick and Watson answered a fundamental question about the nature of life in describing DNA structure, only to find a list of additional questions currently being pursued in the fields of molecular biology, immunology, and genetics.

Short of this primary long-term goal of expanding knowledge, good scientific research is also distinguished by the creation of more tangible and immediate products and by-products. Einstein's theories have allowed for the harnessing of atomic power. Crick and Watson's discovery has led to tests for birth defects and the development of new drugs whose effectiveness is based on knowledge of the genetic structure of the pathogenic organism.

The creation of valuable products is likewise a characteristic of good applied behavioral research in the medical setting. The introduction of behavioral research into medical settings can result in many products, only one of which is the expansion of scientific knowledge. For many medical practitioners, these products have more salience in terms of their value system than the pursuit of scientific knowledge. Historically, medical practice has been devoted to arriving at beneficial short-term outcomes for patients. Medical knowledge has been acquired, in part, through scientific discipline, but just as substantially through systematic practice and case analysis.

The immediacy of the positive outcome is paramount in the value system of the medical practitioner. Whether or not the partial removal of a tumor affects five-year survival rates, such surgery has been highly valued in American medical practice if the short-term positive consequences of decreased pain and suffering for the patient result from the surgery. In contrast, the impact on five-year survival rates of a surgical procedure is of primary importance to the medical researcher. A moment's reflection reveals that the *basic* value system of the medical practitioner is not scientific, but humanistic.

These two value systems, that of the scientist and that of the humanistic practitioner, are not mutually exclusive. In fact, the scientist-practitioner subscribes to the values of both by identifying mechanisms for obtaining good outcomes for patients in the context of a larger effort to expand scientific knowledge.

Although the scientist-practitioner model is not the primary model for medical school training, a contingent of physicians with a particular interest in medical research implicitly, and sometimes explicitly, espouse the scientist-practitioner model. In his presidential address to the American Society of Clinical Investigation, William N. Kelley (1984) advised young physician researchers to actively protect their time allocated to both research and clinical service. Kelley stressed that involvement in patient care is essential for physician researchers to be able to "think creatively about clinical problems" (p. 1122).

The production of some positive short-term outcomes is also critical for the survival of a training model. As a model of training, the scientist-practitioner model ineffably melds into a model of professional life. However, the ideal of the scientist-practitioner will not be maintained following training if this model does not result in more immediate and concrete products and rewards for those who embrace it, in addition to advancing scientific knowledge. The next section of this chapter explores the unique fruits of professional practice as a behavioral scientist-practitioner in medical set-

tings, giving examples of each. At least four major categories of products of research related to the practice of clinical psychology in medical settings can be identified:

1. Demonstration of the effects of treatment.
2. Demonstration of individual treatment effects in collaborating with patients for behavior change.
3. Definition of need for clinical services.
4. Acquisition of knowledge not otherwise obtainable.

Products of the Scientist-Practitioner Approach

Demonstration of the Effects of Treatment

Gaining acceptance of behavioral interventions for medical patients through the scientific demonstration of treatment effectiveness has been an important role for the behavioral scientist-practitioner. Keefe's review (1984) of research in behavioral medicine communicates a sense of the broad scope of treatment studies in behavioral health psychology. These types of studies also provide a clear example of the blend of values characterized by the scientist-practitioner model. Hence, treatment outcome studies have focused not only on the short-term outcome valued by the practitioner, but on long-term outcomes that demonstrate the specificity of a treatment effect. Positive short-term outcomes may result for nonspecific reasons, including the placebo phenomenon, the focused motivation of the patient, patient expectations, and social influences. Positive outcomes that are maintained for a substantial period of time are typically associated with interventions that have specific value in changing a pathological process.

Recall that the primary value of the practitioner is a short-term positive outcome for the patient. Consequently, gaining acceptance for a given treatment will depend primarily on its short-term outcome. The scientist, conversely, has most confidence in treatments that offer positive outcomes that are maintained in the long run. The scientist side of the scientist-practitioner may experience considerable frustration when the treatments that have the best track record on the long-term side fail to gain acceptance because their short-term outcomes are not distinguished or are perhaps even negative.

This type of scenario has been evident in the history of research in smoking cessation. After six months to one year, most smoking-cessation interventions show only a 15%–20% success rate of individuals who continue to abstain from smoking. Although no intervention has been extremely effective in helping people stop smoking, interventions that have included an aversive behavioral technique, called *rapid smoking*, have typically achieved a success rate that is at least twice the norm, that is 30%–40% (Hall & Hall, 1987; Lichtenstein, 1982).

In rapid smoking, the smoker is asked to puff cigarettes at a regular rate that is much more frequent than is normal in his or her smoking pattern. A typical program requires the smoker to puff a cigarette every 5 seconds for 10 minutes. With short breaks in between, three 10-minute rapid-smoking periods would be included in a treatment session.

Despite strong research evidence of the superior long-term effectiveness of rapid smoking, this technique has not gained broad acceptance. From the practitioner's point of view, it has a number of negative short-term consequences. First, it is aversive. Second, it is counterintuitive: people are encouraged to smoke more, in order to end up smoking less. Third, it runs counter to common sense in another way; that is, for a period of time, people who should quit smoking for health reasons are asked to take in more smoke than they normally would. In fact, this last concern has been addressed by additional research, which indicates that rapid smoking can be safely used with healthy individuals and even with patients with a history of myocardial infarction (Hall & Hall, 1987). Instead of rapid smoking, practitioners continue to direct their patients to self-help groups or educational programs, to use other behavioral techniques, or simply to go "cold turkey."

The lack of acceptance of a technique of proven effectiveness may appall the scientist side of the scientist-practitioner. Nonetheless there is an inherent wisdom in the humanism of the practitioner side that should not be overlooked in the focus of scientific research. Clinical work on smoking cessation has suggested to many that there are subgroups of smokers who respond differently to different intervention strategies. Schachter (1982) suggested that the subgroup of smokers who have shown the poorest long-term outcome (and the best results for rapid smoking) appears to be a group that has a chronic history of failure in attempts to quit smoking. It is the smoker who has been most frustrated in quitting who is most likely to volunteer for studies of new stop-smoking techniques, particularly aversive ones like rapid smoking. There appear to be other smokers who benefit from much less intense or aversive interventions. In fact, the American Cancer Society (1986) reported that 90% of smokers who quit do so unaided.

The saga of smoking cessation thus offers an example of the benefits of the balance of the scientific-practitioner model in identifying effective treatments for clinical problems. The scientist-practitioner learns from the results of experimental research as well from experiential knowledge gained from humanistic clinical practice. This combination of knowledge bases maximizes the probability of the development and validation of effective treatments.

Demonstration of Individual Treatment Effects

As mentioned before, the scientific perspective is a long-term perspective. Establishing confidence in a behavioral treatment technique through scientific research requires multiple studies conducted in different settings with a range of appropriate subjects. If the treatment technique is robust enough to prove effective despite variations in settings and subjects, confidence is established. Finding that a treatment will survive the gamut of repeated investigation may take a lifetime. Even then, the question remains whether the treatment will be as effective in a particular case as it is in the majority of cases.

In day-to-day practice, the scientist-practitioner has the opportunity to evaluate the effectiveness of a treatment in the individual case. One example of this type of treatment monitoring occurs in the context of serial cognitive or behavioral assessments of patients involved in medical treatments:

M.G. is a young woman who sustained a severe traumatic brain injury. To monitor her recovery of attentional and memory abilities, I ordered the administration of a brief battery of neuropsychological tests for M.G. every two to four weeks. As part of the rehabilitative treatment, M.G. was given baclofen, a drug that can help to reduce spasticity. The introduction of baclofen appeared to affect M.G.'s performance on a simple test of focused attention, a digit span task that requires the repetition of progressively longer series of numbers in a forward sequence and then in a backward sequence. Over a period of 45 days, M.G.'s performance on the digit span task increased to normal limits. Baclofen, introduced at Day 43, did not have an immediate effect. However, with the continued administration of baclofen, M.G.'s digit span performance decreased, as assessed at Day 57 and Day 67.

Independent of the neuropsychological data, physical medicine therapists had begun to report increased inattentiveness in M.G. following the introduction of baclofen. Therapists also reported only minimal effects of the drug in terms of reduced spasticity and muscle tone. Neuropsychological data provided more objective evidence of the reduced capacity to focus attention reported by the therapists. Baclofen was discontinued and subsequent neuropsychological testing showed improvement in M.G.'s digit span performance, offering support for the hypothesis that baclofen had indeed resulted in decreased attention in this particular case.

The scientist-practitioner model presents itself in an even clearer light in clinical settings in which a scientific approach is used to monitor the effects of a behavioral treatment. In the tradition described by F. C. Thorne, the scientist-practitioner considers each clinical case a "single and well-controlled experiment" (Barlow *et al.*, 1984, p. 5). Data are collected that show the diagnostic problem. Additional data are collected to show that the treatment has been effective in diminishing the problem. Ideally, further evidence is acquired to show that the treatment effects generalize beyond the treatment setting, that is, that the problem remains under control after the patient leaves treatment and returns to his or her daily activities.

Biofeedback treatment offers a good example of this "*n* of 1" experimental approach to clinical treatment:

H.L. is a 44-year-old male who had been referred to the Mayo Clinic Behavioral Medicine Service with a diagnosis of tension headaches exacerbated by stress. He had no history of other medical or psychological problems. His history was also negative for substance abuse. He was on no medications.

H.L. was first oriented to behavioral treatment and to the biofeedback procedure. Then baseline biofeedback trials were conducted. EMG surface electrodes that monitor electrical activity in muscle were placed on the skin over the frontalis (forehead) muscles and in an occipital placement on the back of the head. The skin temperature in the fingertip (an indication of more generalized arousal and activation) was also monitored. During baseline, H.L. was given no feedback or information about levels of muscle tension. At the start of baseline recording, H.L. reported headache pain of 3 to 4 on a scale of 5 of increasing intensity.

Figure 1 shows surface EMG levels in microvolts as

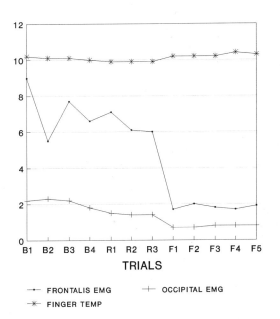

Figure 1. Physiological data for H.L. during initial biofeedback session.

recorded from the frontalis and occipital electrode placements. The recorded levels during baseline (indicated by the trials labeled "B" in Figure 1) show abnormally high muscle activity, compared to that of other individuals. For the next series of trials (labeled "R" in Figure 1), H.L. was instructed to relax with his eyes closed. As Figure 1 illustrates, relaxation instructions without feedback had no effect on the levels of muscle tension.

In the last phase (trials labeled "F"), H.L. was given auditory feedback about his levels of head muscle tension. With the addition of feedback, H.L. was rapidly able to obtain control of these muscles and to reduce his levels of tension as Figure 1 shows. Figure 1 also shows little change in finger temperature over the course of the baseline and treatment trials. (To allow finger temperature to be placed on the same graph as EMG levels, degrees Fahrenheit minus a constant of 69 are reported in Figure 1.) At the end of the session, H.L. reported a subjective level of headache pain of 1 to 2 on a scale of 5.

Thus, a reduction in head muscle tension coincided with a decrease in head pain in this case, even though a relationship between frontalis EMG levels and headache pain has not been reliably observed in group studies (Blanchard, 1979). Furthermore, in the case of H.L., a reduction in head pain appeared to be specifically associated with a reduction in head muscle tension and not to be dependent on a more generalized relaxation response, as would have been indicated by a simultaneous increase in skin temperature.

Data indicating that he was able to control head muscle tension and that the reduction of head muscle tension reduced headache pain was shared with H.L. This evidence increased H.L.'s confidence in biofeedback training as an effective method for him to use in reducing headache pain. Treatment continued with H.L. until he was proficient in regulating his head muscle tension (and associated headaches) without the assistance of biofeedback monitoring.

This case illustrates how the clinical psychologist who is both scientist and practitioner can use the scientific method to monitor the effectiveness of her or his clinical practice in the individual case. In the process, the patient also becomes a scientist and practitioner, learning to use technical data (EMG biofeedback data and self-report data) to appreciate when his or her efforts at self-control are effective.

This approach to treatment stands in contrast to traditional medical practice, in which

the patient puts complete confidence in the doctor to offer effective remedies that may be beyond the patient's understanding. In the scientist-practitioner model of patient care, the patient is asked only to have enough confidence in the care provider to participate in treatment according to the provider's recommendations. After that, the patient is given clear evidence of treatment effectiveness (or lack thereof). There is no need for the patient to take the provider's word for it. The provider–patient relationship becomes increasingly collegial as the two scientist-practitioners (provider and patient) investigate the patient's problem and discover effective solutions. Ultimately, the junior colleague (the patient) gains sufficient experience and confidence to continue the management of the problem independently.

Compliance with medical treatments is critical for the effectiveness of many treatments. A doctor can prescribe a drug regimen, but it will not be effective if the patient does not take the drug as prescribed. The preceding example of biofeedback treatment illustrates how the scientist-practitioner approach to clinical treatment elicits patient compliance by engaging the patient as a collaborator and colleague in investigating the problem and its solution.

Although the creation of a collaborative treatment atmosphere enhances treatment compliance in many cases, the effects of this approach are not universally positive. The collegial, collaborative atmosphere created by the scientist-practitioner model of assessment and treatment is most effective in eliciting compliance from individuals who highly value their own ability to manage themselves and who tend to resist advice from authority figures, particularly when it is given with little rationale.

On the other hand, behavioral clinicians practicing in this model have all been frustrated by more dependent patients who resist taking personal responsibility in their own care and who may indeed be more comfortable in treatments where they can passively accept the prescribed intervention, be it medication or surgery, recommended by a respected authority. When behavioral interventions are the only appropriate form of treatment for these latter types of personalities, these interventions may be delivered in a more directive fashion. Self-monitoring techniques and behavioral practice are prescribed authoritatively, and collaborative sharing of data with the patient is deemphasized. Williams, Thompson, Haber, and Raczynski (1986) described in greater detail how behavioral interventions for headache can be adapted to the specific needs dictated by characteristics of the patient's personality and emotional status.

In practice, modifying the basic scientist-practitioner approach to be more authoritative must occur only in a minority of cases, very likely because of the basic consistency of this approach with fundamental and traditional cultural values in the United States of personal independence, self-reliance, and self-determination.

Definition of Need for Clinical Services

A scientific approach to clinical practice offers a methodology for documenting needs for clinical services. I am now involved with an institutional committee in developing a research protocol to evaluate the need for more extensive follow-up of patients with very mild head injury. The committee includes a physiatrist, a psychiatrist, an anaesthesiologist, a behavioral neurologist, a neurosurgeon, an emergency medicine specialist, a nursing supervisor, and a neuropsychologist. The patient group in question are patients who have had a concussion with brief loss of consciousness or amnesia, but no structural evidence of brain injury (e.g., no evidence of intracranial bleeding or swelling, normal CT scan).

Although the literature has raised concern regarding possible long-term cognitive and/or behavioral problems resulting from this type of injury, this possibility remains controversial (Colohan, Dacey, Alves, Rimel, & Jane, 1986; Levin, Mattis, Ruff, Eisenberg, Marshall, Tabaddor, High, & Frankowski, 1987). Some clinicians feel that behavioral sequelae of this mild type of head trauma occur very infrequently, and then only in cases where emotional problems have preceded the injury or where the

possibility of an advantageous legal settlement is present after the injury (McMordie, 1988). In our setting, where most patients are from rural or small-town communities, the possibility of an additional factor is raised. Most of the studies documenting problems after mild head injury have been done in urban settings, where life stress may be higher and family and community support less reliable. It may be that our patients show sequelae of mild head injury even less frequently because of the excellent support that they typically receive from family, friends, and employers in their home communities.

In this atmosphere of controversy, it is difficult to justify implementing a program of consistent follow-up for patients with mild head injury, many of whom may not need these services or who cannot afford the additional expense of these services. After months of committee meetings that focused on diverse indiviudal clinical experiences relevant to the need for follow-up services for this patient group, the committee ultimately arrived at a rational solution: to conduct a research study. The proposed study would seek to determine the frequency of sequelae of mild head injury and also to document any premorbid psychiatric or substance abuse problems that might be associated with problems in the aftermath of mild head trauma. Whether or not the results of this project will be of interest to the scientific community on a national basis, the implementation of this project will allow practitioners in our setting to plan clinical service for a given patient population on a rational basis derived from a carefully planned research study.

Another example of needs assessment through scientific investigation has been our work at the University of Wisconsin–Madison (Malec, Romsaas, & Trump, 1985; Wolberg, Tanner, Romsaas, Trump, & Malec, 1989) examining mood and psychopathology associated with curable cancers (i.e., testis cancer and breast cancer). In the planning of this research some years ago, controversy characterized estimates of the frequency of psychological problems associated with testis or breast cancer. Opinions based on individual clinical experiences ranged from feelings that nearly every patient with testis or breast cancer experiences severe psychological distress to beliefs that most of these patients cope satisfactorily. Our studies of these patient groups indicated that, of both breast cancer patients and testis cancer patients, most experienced very minimal or transient distress. Twenty to twenty-five percent of both groups experienced mild distress of a type that would be expected to respond to one of a number of interventions, such as brief individual or group counseling, relaxation or stress management training, or participation in an organized support group. The frequency of this type of mild distress appeared to be greater for patients while they were involved in chemotherapeutic treatment. Only a small percentage (5%–10%) demonstrated severe psychiatric disorders.

These studies have offered data that should be useful in rational planning to meet the emotional needs of patients with curable cancer. Additional research will of course be needed to verify the effectiveness of the recommended referral and treatment approaches to patients showing specified degrees of emotional problems associated with curable cancer.

Acquisition of Knowledge Not Otherwise Obtainable

Introducing scientific methodology into clinical practice offers an opportunity to learn about human capacities and functions. In some instances, this knowledge cannot be acquired in any other fashion. The history of neuropsychology—the study of relationships between brain function and behavior—is the history of a discipline that has relied heavily on the study of clinical cases (Grant & Adams, 1986; Walsh, 1978). These cases (studied individually or in series) have provided experiments of a nature that cannot be replicated in a laboratory setting.

Although animal studies have provided some hypotheses about relationships between the function of specific brain regions and behavior, it is inconceivable to conduct similar experiments with humans. Naturally occurring brain trauma and disease, however, have created opportunities to examine the effects of specific brain lesions on human behavior. The case of Phineas Gage, for instance, provided a classic illustration of the frontal lobe syndrome, and that of H.M. provided a classic illustration of the dissociation between motor skill learning and memory in an amnestic patient. (The interested reader will find more detailed descriptions of the cases of Gage and H.M. in Walsh, 1978.) These experiments of nature form the backbone of the science and practice of neuropsychology.

A particularly current and exciting area in which scientific investigation focuses on naturally occurring disease states and their association with psychological states is that of psychoimmunology. As described by Jemmott and Locke (1984, p. 78), the field of psychoimmunology subscribes to a "complex and holistic model" that examines interrelationships among psychological, social, and biological factors in disease and the treatment of disease.

Like neuropsychology, psychoimmunology has a basis in animal models. Evaluation of these animal models in the human species depends on the study of psychological and social factors in naturally occurring disease states. Some experiments have been conducted in which relatively benign disease factors (e.g., rhinovirus, the "cold" virus) have been introduced into human volunteers who have been randomly assigned to conditions of high or low psychological stress (Jemmott & Locke, 1984). However, the evaluation of psychological factors in more virulent diseases requires a careful investigation of the natural course of these diseases. Both experimental and naturalistic observation have provided substantial evidence of psychological mediation in immune system competency.

The importance of naturalistic investigation in psychoimmunology is exemplified by the psychoimmunological study of AIDS. Preliminary evidence suggests that psychological distress and loss of social support result in a diminished immune respone and are associated with the more rapid progression of AIDS (Kiecolt-Glaser & Glaser, 1988). Other studies reviewed by Kiecolt-Glaser and Glaser (1988) suggest that psychological interventions, particularly conditioning paradigms, may be useful in enhancing the immune response to the AIDS virus. If the immune function can be bolstered through classical conditioning, this offers a means for prolonging life after HIV infection. Psychological treatments may offer an adjunct to pharmacological treatments that have significant side effects and may allow chemotherapeutic treatments to be delivered over longer periods of times at lower doses.

Relationships between behavior, illness, and immune function have also become clear in studies of cancer patients (Levy, 1985). Levy, Herberman, Lippman, and d'Angelo (1987), for example, showed that general psychological adjustment, lack of social support, and symptoms of fatigue and depression are related to the function of one aspect of the immune system: natural killer cell activity. These same authors also reported that natural killer cell activity, in turn, is related to tumor activity, as evidenced by the presence of positive lymph nodes.

An Additional Product

One product of the scientist-practitioner approach in a medical setting that has not been mentioned is perhaps the most concrete: jobs. As the practice of medicine becomes more "complex and holistic," the job market will increase for scientist-practitioners who are able to simultaneously offer methods of diagnosis and treatment of the psychological components of health and illness as well as to evaluate the validity of these methods. Jobs for psychologists working in the fields of health psychology and behavioral medicine, neuropsychology and neuroscience, rehabilitation psychology, and pediatric health psychology have been on a growth gradient since the late 1970s, and this trend is expected to continue (Altman & Cahn, 1987). Since 1976, employment for psychologists in university settings has declined, whereas employment in hospitals has increased (Stapp & Fulcher, 1983). A survey of the Health Psychology division of the American Psychological Association indicated that psychologists practicing in health care settings typically function as scientist-practitioners, sharing time among clinical and research activities as well as administrative and teaching duties (Morrow, Carpenter, & Clayman, 1983). In a paraphrase of William N. Kelley's advice (1984) to young physicians considering a career in clinical investigation, employment as a scientist-practitioner in a health care environment can be recommended to young clinical psychologists for (1) the challenge of reducing human suffering directly through clinical practice and indirectly through research; (2) the opportunity of working with bright, creative people at the frontiers of the science and the profession; and (3) the prospect of a reasonable financial livelihood for oneself and one's family.

Developing and Implementing Behavioral Research in Medical Settings

The Proposal

The written research proposal is a giant step toward making your project a reality. The act of translating your research idea into written form requires you to begin to address the concrete operations of the project. Methodology must be defined. Measurement techniques, data collection procedures, and proposed statistical analyses must be outlined. The realities of the costs of the project are recognized as a budget is developed. Subject availability and recruitment must be considered. The proposal involves you in the process of finding solutions to the problems posed by the selected methodology and the projected costs. Finally, the proposal opens your project to peer review and critique.

The precise form of the research proposal will vary from institution to institution. Save yourself time by determining beforehand what the standard form for a proposal is in your institution and by writing the proposal in this format. A different format is likely to be required if the proposal is to be submitted to an external funding agency for review. It is particularly important to follow the format required by external funding agencies carefully to be competitive for funding.

Most formats for proposals emphasize an elaboration of the significance of the study, that is, the unique contribution that the study will make to the advance of scientific knowledge and/or clinical practice. Careful elaboration of methodology, data collection, and statistical analyses is also typically requested. The literature review is usually of less importance and need not be exhaustive. Nonetheless, a strong proposal presents enough of a literature review to clarify the contribution of the proposed study to the field. A review of your own studies and/or case analyses done preliminary to the proposed research results in a particularly strong presentation.

Writing the proposal also offers an opportunity to clarify items like authorship. Studies done in medical settings are notorious for having numbers of authors that rival the numbers of subjects. Individuals are sometimes included in the list of authors simply as a "thank you" for referring patients to the study. In recent years, many institutions, particularly those with university affiliations, have adopted more rigorous rules of authorship, which discourage the use of authorship as a *quid pro quo* for patient referral. An example of formal guidelines for authorship can be found in the most recent edition of the *Publication Manual of the American Psychological Association* (American Psychological Association, 1983, p. 20).

It seems reasonable and ethical to expect that an author on a paper will participate in a study at least to the degree of reviewing and commenting on the proposal and of reviewing and commenting on the final report of the completed project. Such minimal expectations can be set at the time of the written proposal. Some of the coinvestigators will, of course, make more substantial contributions. This is also a good time to negotiate the order of authorship for the final paper, which then will be the same as for the proposal.

Even though some expectations for participation need to be set for all authors of a proposal for professional and ethical reasons, do not expect large amounts of reasonableness, fairness, and collegiality in accomplishing applied research. If you are sincere about pursuing a line of research, be prepared to do the lion's share of the work in writing the proposal, realizing the project, and writing the final paper. You may find one or two collaborators who are of much help, but it is naive and self-defeating to think that a viable research project can be effected by committee with everyone doing their fair share. So be prepared to do the work and share the credit. If over time you are typically the one who is sharing the credit, rest assured that this will be at least implicitly recognized by your colleagues.

Funding the Project

Very respectable research can be conducted on a shoestring budget by careful data collection in the course of clinical practice. The "*n* of 1" model for scientific clinical practice discussed previously can result in significant, publishable reports of single cases or series of cases. Indeed, some would argue that documenting effective treatments through replicated positive results in a series of cases is a more viable approach to clinical research than experimental designs with randomized assignment to treatment groups, which may result in statistically, but not clinically, significant findings (Barlow & Hersen, 1984). Retrospective or prospective studies of patient groups with distinguishing characteristics can also be accomplished with meager budgets through a careful collection of relevant data in the course of clinical practice.

Nonetheless, a funded research project has a special worth. Obtaining funding for a project means that the project has entered a competition with other studies and won. A funded study is one in which you have the resources to collect the data carefully and analyze them completely. A funded study requires no explanations to administrators (who may believe that your research interests are somehow interfering with your commitment to generating revenue, even though the time you spend in analyzing data and writing is coming out of your social life, not patient care hours). When your project is funded, your time is paid for. No apologies are necessary.

Funding, particularly federal (i.e., National Institutes of Health) funding, is becoming increasingly scarce. More and more federal money appears to be diverted into ostensibly saving and (with enthusiastic paranoia) monitoring money. Despite these unfortunate, but oh-so-predictable, cultural changes as the pendulum continues to swing from a liberal to a conservative mentality, federal money is still accessible if you have a great project and your interests coincide with hot areas of investigation (e.g., AIDS).

Often overlooked are other sources of potential funding. There are a number of private funding agencies for projects exploring the interface between medicine and behavior. The Robert Wood Johnson Foundation is one of the better known of such organizations. Get your name on the mailing list of research bulletins, such as the *Science Agenda* published by the American Psychological Association Science Directorate, which will alert you to new funding sources. If you have a computer and a modem, you can access the Science Directorate's bulletin board of RFPs (requests for proposals).

Look close to home as well. Your own institution may have funds available to support research projects of merit. If you are a recent graduate, explore the possibility of special small grants for new researchers offered through private and public agencies. If you are an experienced clinician who has established a reputation in a particular area of treatment, do not overlook the possibility of establishing a research fund from the donations of grateful patients.

Individuals who have benefited from your services may sincerely, but indirectly, search for some way to repay you. To respond that no repayment (other than the standard fee) is necessary may actually disappoint some patients who feel that your standard fee does not begin to compensate you for the dramatic changes that you have helped them make in their lives. Establishing a research fund where donations

from grateful patients can be directed meets a dual purpose of supporting your research and allowing your patients to express their sincere appreciation in a culturally acceptable manner. It should go without saying that ethical standards need to be maintained scrupulously in working with patients who may be potential contributors to your research fund. Contributions cannot be overtly encouraged. Any suggestion that the quality of treatment is contingent on contributions would be grossly unethical. In some cases, you may choose to discourage contributions by intepreting these in terms of therapeutic transference. Ethical dilemmas are minimized if there is an administrative section in your institution that can manage the research fund for you and to whom you can direct interested patrons for information.

Committee Review

Your institution will require a committee review of experimental and prospective research. At a minimum, an institutional committee (typically called the institutional review board or the human subjects committee) with the legal mandate to protect the rights of human subjects will review the proposal. Often, a departmental research committee will additionally review the quality of the proposal. If you are seeking funding from the institution, another committee, which establishes priorities for funding, may review the project.

Each of these committees reviews the proposal with a specific focus. The human subjects committee is primarily concerned with ensuring that the research will not violate the rights of human volunteers. Rights of human subjects and ethical concerns in the conduct of research with human subjects are elucidated in the Ethical Principles of Psychologists (American Psychological Association, 1981). This committee will make sure that appropriate consent is obtained from research subjects, that the subjects are appropriately apprised of any risks involved in their participation in the experiments and that the subjects experience no degree of coercion to participate. The overall merit of the study is of less concern to this committee than simply whether the experimental protocol includes appropriate protections for human volunteers. In many institutions, a streamlined procedure for review by the human subjects committee is available to minimal risk protocols. Minimal risk protocols are those in which the experimental interventions are not physically invasive and pose no risk of physical harm or severe psychological harm to the subject. Many behavioral studies fall into the minimal risk category. Institutions vary regarding the necessity of a review by the human subjects committee of retrospective or history review studies, and most institutions do not require such a review. Whether human subjects review is required or not, the fundamental ethical and legal concern in retrospective studies is that the confidentiality and the anonymity of subjects are maintained.

A departmental research committee that reviews all research conducted in your department may review your protocol. Such committees may make some suggestions about human subjects considerations but can be expected to focus on the scientific merit and the methodological integrity of the proposal. Psychologically, this is probably the most anxiety-producing review, as the spotlight really is on the quality of the proposal. Criticisms of the proposal may, of course, be arbitrary and may be based on a lack of knowledge of the field by the reviewers. On the other hand, many times good suggestions for improving the methodology will come from the departmental research committee. Once the departmental review is complete, the primary investigator has the job of nondefensively considering the criticisms and suggestions offered and revising the protocol to capitalize on the good suggestions.

If a funding committee also reviews the proposal, this committee will also be concerned with the quality of the project. However, it is important to keep in mind that this committee will be concerned only with the scientific merit of the project *vis-à-vis* other projects that it has before it. The funding committee has a finite budget with which to sponsor projects. If your project does not get funding, the reason may be that it is not as good as it could have been.

Alternatively, nonfunding may mean that other projects were given priority because they were of better quality or were directed at research questions perceived to be of greater value in terms of current institutional objectives or societal needs.

Conducting the Research

With your approved proposal as a blueprint, actually conducting the research should be straightforward. Take a little time to set up a system for making sure the research protocol is followed precisely. A few well-constructed forms can be very helpful in making sure that exclusion criteria are examined and that data are collected systematically. Your preliminary investigation at the time of the proposal to ensure the availability of appropriate subjects should obviate problems in subject recruitment.

Publication

A primary goal of scientific investigation is sharing the results with colleagues in the scientific community. Of course, findings need to be presented in an objective fashion, and the style of the writing should adhere to the standards for scientific publication. Scientific or not, writing is always directed at an identifiable audience. Professional journals tend to have a specific readership. The readership or audience of each journal varies in terms of its background, interest, and sophistication in your field of inquiry. No matter how significant your research findings, these findings will be rejected if they are not addressed to an audience that has the background, interest, and sophistication to appreciate your results.

Do some investigation of journals in which your results might be published. Ask colleagues for suggestions about where to find a receptive audience for your results. Examine the journals themselves. Has the journal published other articles similar to yours? Is the methodology of the studies published in the journal similar to yours? Look at the reference list that you collected for your literature review. Where have the studies been published that preceded yours?

No audience will have as great a familiarity with the focus of your research as you do. To find acceptance for your research, you will need to communicate your findings in a way that assists your audience's understanding of and enthusiasm for your efforts. Tie your findings into prior research that is familiar to your audience. Build some suspense in your introduction by presenting the critical questions that your project addresses. Pique interest with figures and tables that clearly present your most significant findings. Carefully explain methodology or analyses that may be unfamiliar to your audience. Allow for a convincing climax by answering your research questions with straightforward conclusions. *Offer clear conclusions*—even though, in scientific writing, the statement of your conclusions will almost always need to be followed by a series of caveats indicating the limitations of your study and possible alternative interpretations of your data. Use this opportunity to build suspense regarding your next project that will address the controversies left unanswered by the study you are reporting.

Give yourself some time to think about your next project. This should come naturally because, as a rule, good research presents more questions than it answers. And above all, enjoy the process. It is not that science *can* be fun. Science *is* fun. It's basically the same process—taken to a higher degree of complexity and sophistication—that led us at the age of 3 to poke at a puddle with a stick to see the reaction of the water. Gil French, the retired chairperson of psychology at the University of South Dakota, would say that, in order to be creative, scientists need to be allowed to play. In allocating resources for science, society perennially faces the problem of determining which purported scientists play productively and which tend to play in a gratuitous and nonproductive manner. Assuming that you are, or are in the process of becoming, a conscientious, productive scientist-practitioner: enjoy. Have some fun in your work. Allow yourself some awe, wonder, and surprise. And in this way, keep your conscientiousness from restricting your creativity.

ACKNOWLEDGMENT

I am grateful to Donald E. Williams, Ph.D. of the Department of Psychiatry and Psychology at the Mayo Clinic for sharing data and information relevant to the case of H.L.

References

Altman, D. G., & Cahn, J. (1987). Employment options for health psychologists. In G. C. Stone, S. M. Weiss, J. D. Matarazzo, N. E. Miller, J. Rodin, C. D. Belar, M. J. Follick, & J. E. Singer (Eds.), *Health psychology: A discipline and a profession* (pp. 231–244) Chicago: University of Chicago Press.

American Cancer Society. (1986). *Cancer facts and figures*. New York: Author.

American Psychological Association. (1981). Ethical principles of psychologists (revised). *American Psychologist, 36*, 633–638.

American Psychological Association. (1983). *Publication manual of the American Psychological Association* (3rd ed.). Washington, DC: Author.

Barlow, D. H., & Hersen, M. (1984). *Single case experimental designs: Strategies for studying behavior change* (2nd ed.). New York: Pergamon Press.

Barlow, D. H., Hayes, S. C., & Nelson, R. O. (1984). *The scientist practitioner: Research and accountability in clinical and educational settings*. New York: Pergamon Press.

Blanchard, E. B. (1979). Biofeedback: A selective review of clinical applications in behavioral medicine. In J. R. McNamara (Ed.), *Behavioral approaches to medicine: Application and analysis* (pp. 131–156). New York: Plenum Press.

Colohan, A. R. T., Dacey, R. G., Alves, W. M., Rimel, R. W., & Jane, J. A. (1986). Neurologic and neurosurgical implications of mild head injury. *Journal of Head Trauma Rehabilitation, 1*, 13–21.

Conway, J. B. (1988). Differences among clinical psychologists: Scientists, practitioners, and scientist-practitioners. *Professional Psychology: Research and Practice, 19* 642–655.

Frank, G. (1984). The Boulder model: History, rationale, and critique. *Professional Psychology: Research and Practice, 15*, 417–435.

Goldfried, M. R. (1984). Training the clinician as scientist-professional. *Professional Psychology: Research and Practice, 15*, 477–481.

Grant, I., & Adams, K. M. (Eds.), (1986). *Neuropsychological assessment of neuropsychiatric disorders*. New York: Oxford University Press.

Hall, S. M. & Hall, R. G. (1987). Treatment of cigarette smoking. In J. A. Blumenthal & D. C. McKee (Eds.), *Applications in behavioral medicine and health psychology: A clinician's source book* (pp. 301–323). Sarasota, FL: Professional Resource Exchange.

Jemmott, J. B., & Locke, S. E. (1984). Psychosocial factors, immunologic mediation, and human susceptibility to infectious diseases: How much do we know? *Psychological Bulletin, 95,* 78–108.

Jones, N. F. (1987). Schools of professional psychology as sites for training health psychologists. In G. C. Stone, S. M. Weiss, J. D. Matarazzo, N. E. Miller, J. Rodin, C. D. Belar, M. J. Follick, & J. E. Singer (Eds.), *Health Psychology: A discipline and a profession* (pp. 501–511). Chicago: University of Chicago Press.

Keefe, F. J. (1984). Research methods in behavioral medicine. In A. S. Bellack & M. Hersen (Eds.), *Research Methods in Clinical Psychology* (pp. 283–323). New York: Pergamon Press.

Kelley, W. N. (1984). Clinical investigation and the clinical investigator: The past, present, and future. *Journal of Clinical Investigation, 74,* 1117–1122.

Kiecolt-Glaser, J. K., & Glaser, R. (1988). Psychological influences on immunity: Implications for AIDS. *American Psychologist, 43,* 892–898.

Levin, H. S., Mattis, S., Ruff, R. M., Eisenberg, H. M., Marshall, L. F., Tabaddor, K., High, W. M., & Frankowski, R. F. (1987). Neurobehavioral outcome following minor head injury: A three-center study. *Journal of Neurosurgery, 66,* 234–243.

Levy, S. M. (1985). *Behavior and cancer.* San Francisco: Jossey-Bass.

Levy, S., Herberman, R. Lippman, M., & d'Angelo, T. (1987). Correlation of stress factors with sustained depression of natural killer cell activity and predicted prognosis in patients with breast cancer. *Journal of Clinical Oncology, 5,* 348–353.

Lichtenstein, E. (1982). The smoking problem: A behavioral perspective. *Journal of Clinical Psychology, 50,* 804–819.

Malec, J. F., Romsaas, E., & Trump, D. (1985). Psychological and personality disturbance among testis cancer patients. *Journal of Psychosocial Oncology, 3,* 55–64.

McMordie, W. R. (1988). Twenty-year follow-up of the prevailing opinion of the posttraumatic or postconcussional syndrome. *The Clinical Neuropsychologist, 2,* 198–212.

Morrow, G. R., Carpenter, P. J., & Clayman, D. A. (1983). A national survey of health psychologists: Characteristics, training, and priorities. *The Health Psychologist Newsletter, 5,* 6–7.

Norcross, J. C., & Prochaska, J. O. (1982). A national survey of clinical psychologists: Characteristics and activities. *The Clinical Psychologist, 35,* 5–8.

Raimy, V. C. (1950). *Training in clinical psychology.* New York: Prentice-Hall.

Schachter, S. (1982). Recidivism and self-cure of smoking and obesity. *American Psychologist, 37,* 436–444.

Stapp, J., & Fulcher R. (1983). The employment of APA members: 1982. *American Psychologist, 35,* 861–866.

Strickland, B. R. (1988). Clinical psychology comes of age. *American Psychologist, 43,* 104–107.

Walsh, K. W. (1978). *Neuropsychology: A clinical approach.* New York: Churchill Livingstone.

Williams, D. E., Thompson, J. K., Haber, J. D., & Raczynski, J. M. (1986). MMPI and headache: A special focus on differential diagnosis, prediction of treatment outcome, and patient-treatment matching. *Pain, 24,* 143–158.

Wolberg, W. H., Tanner, M. A., Romsaas, E. P., Trump, D. L., & Malec, J. F. (1989). Psychosexual adaptation to breast cancer surgery. *Cancer, 63,* 1645–1655.

PART VI

Templates for Program Development

Thus far, the reader has been provided with the foundations of clinical psychology in medical settings with chapters on education, politics, professionalism, practice management, and general clinical issues. The reader is now presented with the clinical core of the handbook: specific programmatic development as applied to specific medical disorders. Medical areas were chosen on the basis of extant work done by psychologists in those areas. This section has many chapters reflecting the breadth and scope of psychologists' involvement in medical settings. Each clinical chapter is divided into reviews of the literature on psychological problems secondary to the specific medical disorder. Assessment and intervention procedures follow in an effort to provide the clinician with guidelines for patient involvement. Finally, programmatic guidelines are offered by each of the authors that integrate psychological services into the existing medical system. These guidelines can serve as templates for future program development and can provide suggestions for the modification and expansion of existing programs.

The editors (Chapter 17) begin by integrating the science and the service of psychology in medical settings and by demonstrating that psychologically based programs have developed to meet a multiplicity of patient care and institutional needs. Diagnostic evaluations and psychometric testing have long been a basic clinical function of the psychologist. Jerry J. Sweet (Chapter 18) discusses the development of a psychological evaluation and testing service in a medical setting. The subsequent chapters relate to specific medical problems. David L. Tobin, Craig Johnson, and Kevin Franke (Chapter 19) present their ideas for program development for the complex issue of eating disorders. Steven M. Tovian (Chapter 20) integrates clinical psychology into both adult and pediatric oncology programs. Stephen M. Weiss, Roger T. Anderson, and Sharlene M. Weiss (Chapter 21) develop programs for cardiovascular disorders and hypertension. The development of rehabilitation programs for patients with brain and spinal cord injuries is discussed by Joseph Bleiberg, Robert Ciulla, and Bonnie L. Katz (Chapter 22). Pain and headache programs are the focus of Stanley L. Chapman (Chapter 23). Jean C. Elbert and Diane J. Willis (Chapter 24) address specific pediatric populations as they discuss programs for behavioral and learning disorders. Rosalind D. Cartwright (Chapter 25) presents program developments involving sleep disorders. Daniel J. Cox, Linda Gonder-Frederick, and J. Terry Saunders (Chapter 26) integrate clinical psychology programs into the treatment of diabetes. Thomas L. Creer, Russ V. Reynolds, and Harry Kotses (Chapter 27) apply cognitive-behavioral approaches to the problems of adult and pediatric asthma. Harry S. Shabsin and William E. Whitehead (Chapter 28) discuss clinical psychology programs for gastrointestinal disorders. Barbara G. Melamed and David J. Williamson (Chapter 29) discuss psychological programs for the treatment of dental disorders. Daniel S. Kirschenbaum (Chapter 30) integrates clinical psychology into hemodialysis programs. Finally, Kathleen Sheridan (Chapter 31) focuses on psychological services for people with AIDS.

CHAPTER 17

Toward Program Development

An Integration of Science and Service in Medical Settings

Ronald H. Rozensky, Jerry J. Sweet, and Steven M. Tovian

Great Hospitals, with their Schools, are something more than blocks of buildings where patients are doctored, and students and nurses are taught. I do believe in the spirit of a place. To me, the *genius loci* is really there: and the *Religion Discipuli*, the student's obedience to the spirit of Hospital life, is a very important part of his education. . . . Or will anybody say that the *genius loci* is all nonsense, and that a great Hospital is only a big machine? My answer is, that I know what I am talking about. Sickness, as Lucretius says of impending death, shows us things as they are: the mask is torn off, the facts remain. That is the spiritual method of the Hospital: it makes use of sickness, to show us things as they are . . . this plunge into the actual flood of lives is a fine experience.

—*Confessio Medici*, 1908, pp. 19–20

Introduction: The *Genius Loci* of the Hospital and Clinical Psychology

Hospitals and medical settings, like any organization or institution, are made of innumerable spokes that emanate from the hub and support the structure of the wheel. The hub is the philosophy or mission, the spokes the pro-

fessional people within the hospital. As the wheel moves forward, the spokes blend in the eye of the observer into a solid that carries the momentum of the hub forward to its goal, in this case, quality patient care, the *genius loci*. Modern clinical psychology in the medical setting has become a strong spoke in that organizational wheel. The clinical psychologist and psychological programs can play an integral part in carrying out the mission of each hospital or medical setting.

As we study the role of clinical psychologists

Ronald H. Rozensky, Jerry J. Sweet, and Steven M. Tovian • The Evanston Hospital and Northwestern University and Medical School, Evanston, Illinois 60201.

and how they bear their share of the hospital's mission and momentum, we can see from the first 16 chapters of this text that the profession has its own history and philosophy as well as its own educational, professional, research, and general clinical activities within the health care system. The economic pressures, quality of care, and political issues presented in those chapters also help define the complexity of the medical environment in which so many clinical psychologists now work. As we shall see from the second half of this text, psychologically based programs have developed to meet a multiplicity of patient care and institutional needs.

The Structure of Knowledge and Psychology

Our identity as clinical psychologists in medical settings and the relationships and communications that we have with others (physicians, other psychologists, nurses, social workers, ancillary service personnel, administrators, and our patients and their families) must be founded on some definitions from our field's knowledge base as they are applied to both science and practice. The historical, political, economic, and emotional arguments regarding the scientist-practitioner or professional models of training and practice (Boulder or Vail) have been addressed elsewhere (Frank, 1984; Stern, 1984; Strickland, 1985; Tyler & Spiesman, 1967). Conway (1988) detailed the personal characteristics, personality and cognitive variables, and critical incidents that have influenced the development of psychologists oriented to the scientist, practitioner and scientist-practitioner models. Finally, Mittelstaedt and Tasca (1988) offered a constructive, collaborative model for the blend of scientist and practitioner.

The present discussion focuses on a philosophy concerned with the presentation of professional communications. The structure of knowledge on which that communication is based has a practical utility for the clinical psychologist in the medical setting. Psychology, according to Piaget and Kamii (1978), is a science that presupposes the rules of the other natural sciences and, like them, continues to look for explanations. The *clinical* psychologist, through both training and experience, combines the nomothetic and idiographic approaches within the questioning attitude of the scientist, tempered by the sensitivity of the humanist (Shakow, 1976). Matarazzo's definition (1987) of health psychology helps in summarizing the present description of clinical psychology in the medical setting as "the application of the common body of knowledge of psychology . . . to any number of venues in the arena of health" (p. 899). Such (clinical) applications of the science of psychology should be based on explanations and formulations that are "clear, coherent, and parsimonious" (Fiegl, 1959, p. 122). Additionally, it is the clinical psychologist's comfort with the scientific model and hypothesis testing in general that allows for the accurate presentation of both research data and clinical hypotheses (either diagnoses or treatment plans). The clinical psychologist's comfort with the probabilistic nature of the relationship between the true state and the judged states of nature is most likely maximized by those individuals with the "mathematical wherewithal to reason quantitatively" (Kleinmuntz, 1984, p. 122).

The Application of Scientific Method to Clinical Data

The direct logic of the scientific method can be translated *clinically* in the medical setting by the use of the presentation of information via the problem-oriented medical record (POMR) concept and methodology (Sandlow, Hammet, & Bashook, 1974; Weed, 1970). Platt's rules (1964) of strong inference in the scientific method can be seen in Figure 1 (adapted from Sandlow *et al.*, 1974), illustrating the importance of hypothesis testing. Further, that same figure presents the translation of scientific methodology to the structured logic of the hypothesis testing carried out by the clinician who uses the problem-oriented method.

With the POMR, the clinician is asked to view his or her lists of patient problems or

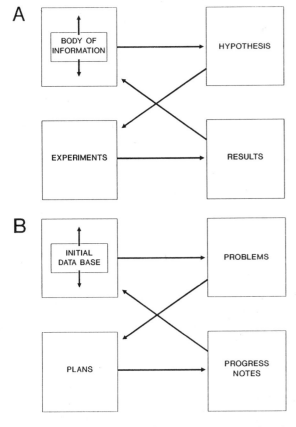

Figure 1. The application of scientific method to clinical data: A. the scientific method; B. the problem-oriented record. From L. Sandlow, W. Hammet, & P. Bashook (1974), *Problem Oriented Medical Records: Guidelines for Format and Forms* (p. 2). Chicago: Michael Reese Medical Center. Reprinted by permission.

diagnoses as "hypotheses" whose purpose is to guide the formulation of "experiments," or clinical treatments. Those treatments then bring about changes or "results." Results are then charted as part of the clinical record as "progress notes" and thus become part of the "data base" on which future hypotheses are formulated.

It is the discipline of the scientific method, the "better monitoring by a system that mobilizes criticism" (Weed, 1970, p. 4), and the preparation of the research-based thesis that prepare the scientist in developing a respect for order, logic, and consistency (Weed, 1970). In the clinical environment, there is, in effect, not a single thesis, according to Weed, but thousands of hypotheses with "variables that are exceptionally difficult to enumerate and control" (p. 5). It is the disciplined thinking, trained by the scientific method, and the inherent comfort with being a hypothesis tester that prepares the clinician to organize and challenge his or her own clinical thinking.

Clinical Data and the Medical Setting

In a survey of 80 physicians (family practice, internal medicine, pediatrics, and psychiatry), Liese (1986) found that physicians were as likely to seek consultation from psychologists as they were from psychiatrists and social workers. Meyer, Fink, and Carey (1988) reported that the 139 of 500 physicians returning their survey viewed psychological consultations as helpful and generally available, although they did express concern about the adequacy of psychologists' training to provide such consultations. They also noted that physicians frequently commented on not receiving feedback from psychologists regarding patients referred for consultation.

Given the usefulness and availability of psychological consultation and treatment, Wright (1982) offered some suggestions to the psychologist working within the medical environment:

> Nonpsychiatric physicians respect critical thinking more than exotic intuition. They find it difficult to carry on discussions around third-level constructs such as "object cathexis" or "'infantile sexuality." . . . psychologists are seldom found making statements about which there is only say a 55% level of confidence as though they were 95%–99% confident. All of this serves to make for a more natural union of psychology with nonpsychiatric medicine. (p. 3)

As a result, can we make a case for the natural interplay between clinical medicine and clinical psychology in the medical setting?

> The perceptive student will notice certain similarities between the clinical method and the scientific method. Each begins with observational data which suggest a series of hypotheses. These latter are tested in the light of further observations, some

clinical, other contrived laboratory procedures. Finally, a conclusion is reached, which in science is called a *theory* and in medicine a *working diagnosis*. The modus operandi of the clinical method, like that of the scientific method, cannot be reduced to a single principle or a type of inductive or deductive reasoning. It involves both analysis and synthesis, the essential parts of Cartesian logic. The physician does not start with an open mind any more than does the scientist, but with one prejudiced from knowledge of recent cases; and the patient's first statement directs his thinking in certain channels. He must struggle constantly to avoid the bias occasioned by his own attitude, mood, irritability, and interest. (Wintrobe, Thorn, Adams, Bennett, Braunwald, Isselbacher, & Peterdordf, 1970, p. 5)

In discussing the mathematical concepts of Bayesian conditional probabilities and clinical decision-making, Wulff (1981) reminded the reader that "diagnosis is not an end in itself; it is only a mental resting-place for prognostic considerations and therapeutic decisions" (p. 80).

Scientifically trained clinical psychologists might remember that

> frequently the hypothesis that is tested is stated in such a way that, when data tend to contradict it, the experimenter is actually demonstrating what it is that he is trying to establish. In such cases the experimenter is interested in being able to reject or nullify the hypothesis being tested. . . . Rejection of the hypothesis being tested is equivalent to supporting one of the possible alternative hypotheses which are not contradicted. (Winer, 1971, p. 11)

The *working diagnosis* described by Wintrobe *et al.* (1970) above is often stated in the medical record as a "rule out" of some stated diagnosis. The scientifically minded clinician, with the training and comfort with the null hypothesis, H_o, can function more scientifically as a diagnostician within the medical environment. After a series of studies, Lehman, Lempert, and Nisbett (1988) concluded that the statistical and methodological training received by graduate-level psychologists increases their ability to reason about problems of a conditional or biconditional nature and that statistical training affects thinking about everyday-life events. Medical students studied by Lehman *et al.* were taught that "the confounded variable principle underlies much medical 'sleuthing'" (p. 440) and were found to have an improved ability to solve conditional and biconditional problems as well.

Wickramasekera (1989) discussed the role of the psychologist functioning as a psychophysiological therapist in concert (both philosophically and geographically) with physicians: "the clinical health psychologist can function as a peer and expert consultant to the MD on issues of mind-body interaction. The psychologist's role is that of an expert who does not duplicate the physician's training but has a different and useful set of psychosocial investigative skills" (p. 109).

The Scientifically Minded Clinician

The clear presentation of organized clinical data, diagnostic impressions, treatment plans, and ultimately, positive treatment outcomes is the product offered by the clinical psychologist within the medical milieu. It is training as a hypothesis tester, or what Sheridan and Choca refer to in this book as a scientifically minded clinician, with an orientation to accountability, comfort with and training in measuring the effectiveness of interventions, and an objectivity in presenting data, that forms the foundation of the field of clinical psychology in medical settings.

Summary

Data to be shared and used clinically must be gathered via a clinical relationship with the patient, the patient's family, and other care providers. The professionalism of the clinical psychologist plus the genuineness, warmth, empathy, and acceptance of the patient by the clinician, combined with his or her scientific-mindedness, forms the art of the practice of clinical psychology.

While discussing the growth role of psychology within the medical environment, Sterling (1982) cautioned that we not "lose sight of our own area [of competence as psychologists]" (p. 793). Further, he warned that "if we lose our uniqueness, the patients suffer for lack of a champion who is aware of their (psychological) needs, and we are in danger of losing a sense of who we are" (p. 793).

The following chapters, each detailing an area of practice of clinical psychologists in medical settings, illustrate this unique relationship between science and service. As the early Gestalt psychologists would remind us, the whole is different (and in this case greater) than the sum of the parts.

References

Confessio Medici. (1908). New York: Macmillan.

Conway, J. B. (1988). Differences among clinical psychologists: Scientists, practitioners, and scientist-practitioners. *Professional Psychology: Research and Practice, 19*, 642–655.

Fiegl, H. (1959). Philosophical embarrassment of psychology. *American Psychologist, 14*, 115–128.

Frank, G. (1984). The Boulder Model: History, rationale, and critique. *Professional Psychology: Research and Practice, 15*, 417–435.

Kleinmuntz, B. (1984). The scientific study of clinical judgment in psychology and medicine. *Clinical Psychology Review, 4*, 111–126.

Lehman, D. R., Lempert, R. O., Nisbett, R. E. (1988). The effects of graduate training on reasoning: Formal discipline and thanking about every-day life. *American Psychologist, 43*, 431–442.

Liese, B. S. (1986). Physicians' perceptions of the role of psychology in medicine. *Professional Psychology: Research and Practice, 17*, 276–277.

Matarazzo, J. D. (1987). There is only one psychology, no specialties, but many applications. *The American Psychologist, 42*, 893–903.

Meyer, J. D., Fink, C. M., & Carey, P. F. (1988). Medical views of psychological consultation. *Professional Psychology: Research and Practice, 19*, 356–358.

Mittelstaedt, W., & Tosca, G. (1988). Contradictions in clinical psychology training: A trainee's perspective of the Boulder Model. *Professional Psychology: Research and Practice, 19*, 353–355.

Piaget, J., & Kamii, C. (1978). What is psychology? *American Psychologist, 7*, 648–652.

Platt, J. R. (1964). Strong inference. *Science, 146*, 347–353.

Sandlow, L. J., Hammet, W. H., & Bashook, P. G. (1974). *Problem oriented medical records: Guidelines for format and forms.* Chicago: Michael Reese Medical Center.

Shakow, D. (1976). What is clinical psychology? *American Psychologist, 30*, 553–560.

Sterling, M. C. (1982). Must psychology lose its soul? *Professional Psychology: Research and Practice, 13*, 789–796.

Stern, S. (1984). Professional training and professional competence: A critique of current thinking. *Professional Psychology: Research and Practice, 2*, 230–243.

Strickland, B. R. (1985). Over the Boulder(s) and through the Vail. *The Clinical Psychologist, 38*, 52–56.

Tyler, F. B., & Spiesman, J. C. (1967). An emerging scientist-professional role in psychology. *American Psychologist, 22*, 839–847.

Weed, L. L. (1970). *Medical records, medical education, and patient care.* Cleveland: Press of Case Western Reserve University.

Wickramasekera, I. (1989). Somatizers, the health care system, and collapsing the psychological distance that the somatizer has to travel for help. *Professional Psychology: Research and Practice, 20*, 105–111.

Winer, B. J. (1971). *Statistical principles in experimental design.* New York: McGraw-Hill.

Wintrobe, M. M., Thorn, G. W., Adams, R. D., Bennett, I. L., Braunwald, E., Isselbacher, K. J., Peterdordf, R. G. (Eds.). (1970). *Harrison's principles of internal medicine.* New York: McGraw-Hill.

Wright, L. (1982). Incorporating health care psychology into independent practice. *The Independent Practitioner, 2*, 1–4.

Wulff, H. K. (1981). *Rational diagnosis and treatment.* Oxford, England: Blackwell Scientific Publications.

CHAPTER 18

Psychological Evaluation and Testing Services in Medical Settings

Jerry J. Sweet

Introduction

Psychological testing has been an integral part of the history of clinical psychology and plays an increasingly important role in the lives of individuals in our society (Dahlstrom, 1985; Klopfer, 1983). Within medical settings, psychological testing has grown rapidly in recent years and remains one of the most common practices of psychologists. The growth of such assessment-oriented specialties as clinical neuropsychology has even outdistanced the enormous growth in the treatment-oriented field of behavioral medicine (Wedding & Williams, 1983). Although more traditional personality assessment (especially projective testing) experienced some disfavor among clinicians and subsequent decreased use in the 1960s (Jackson & Wohl, 1966; Shemberg & Keeley, 1970), such testing again achieved widespread acceptability and emphasis in clinical training in the

1970s and 1980s (Ritzler & Alter, 1986). In fact, based on a survey of practicing clinical psychologists, (Wade & Baker, 1977; Wade, Baker, Morton, & Baker, 1978), it has been reported that the Rorschach ranks first among the tests most frequently recommended for clinical psychology students to learn.

This chapter focuses on the historical and current importance, as well as the practical aspects, of clinical psychologists' performing formal testing with medical and psychiatric patients. It is not possible in this single chapter to detail or critically evaluate the specific methods, interpretations, and decisions involved in individual evaluations of patients. Rather, guidelines and recommendations for establishing and maintaining a psychological evaluation and testing service within a medical setting are provided. This discussion includes definitions of assessment, a historical overview, the importance of clinical interviews, descriptions of relevant training, important characteristics of clinicians, typical referral questions, the common tests in use today, and specific considerations in establishing and maintaining an evaluation service in a medical setting.

Jerry J. Sweet • The Evanston Hospital and Northwestern University and Medical School, Evanston, Illinois 60201.

What Is Psychological Assessment?

A number of definitions exist concerning psychological tests. Consider the following:

> As used in psychology, the term *test* denotes a set of stimulus materials, together with both explicit procedures about the circumstances, manner, and sequence in which they are to be presented to a test subject, and detailed instructions that the subject is to be given about what he or she is to do with these materials, in order to draw from the subject a series of actions or reactions by means of which he or she may be typified or characterized in regard to the attribute(s) under consideration. (Dahlstrom, 1985, p. 63)

> A test is a systematic procedure for observing behavior and describing it with the aid of numerical scales or fixed categories, while a *standardized test* is one in which the tester's words and acts, the apparatus, and the scoring have been fixed so that the scores collected at different times and places are fully comparable. (Cronbach, 1984, pp. 26–27)

> A psychological test is essentially an objective and standardized measure of a sample of behavior. (Anastasi, 1988, p. 23)

Although varying in amount of detail, the above definitions by well-known experts in the field are consistent in describing the nature of psychological testing. As is readily evident from these definitions, the intent is to measure behavior objectively and reliably by using procedures and materials that are standardized. In this respect, psychological measurement follows the assumptions and goals of measurement in the physical sciences, with the same attendant concerns regarding reliability and validity. Cronbach (1984) noted that the universal assumption of all areas of measurement is captured in the old pronouncement, "If a thing exists, it exists in some amount. If it exists in some amount, it can be measured" (p. 41).

Many diagnostic medical tests are highly reliable. Interestingly, many have simply assumed psychological measurement to be inherently fraught with much greater difficulty than diagnostic medical studies (e.g., radiology). However, one need only observe a group of physicians trying to determine a consensual diagnosis from a set of spinal studies or from the results of a physical exam to realize that this is not necessarily the case. Indeed, because

psychologists have perceived a need to prove their assessment field sound by being more rigorous than other health professionals, most commonly used psychological test procedures are at least as clinically reliable and objective as many medical measures, especially when one also recognizes that *interpretation* of both sets of data must take place before the data that have been collected are of any value. One reason is that psychologists have always assumed that different observers may not view the same information in the same way; hence, there has been an emphasis on ensuring interrater reliability, which is not evident in most areas of medical practice.

Psychological assessment is more than just psychological testing. As described by Goldstein and Hersen (1984), other approaches to evaluating individual differences include interviewing, behavioral observations, and physiological recordings. Further, the clinical psychologist engaging in psychological assessment is not narrowly focused on test scores. Instead, he or she typically attempts to understand the individual within the broad context of his or her social and cultural environment, family system, psychological and medical history, and educational and vocational attainment.

Historical Overview of Psychological Assessment

Several detailed histories of psychological testing are available in the literature (cf. Anastasi, 1988; Dahlstrom, 1985; Dubois, 1970). A brief summary here may provide a useful context within which to consider the other topics covered in this chapter.

The available historical evidence suggests that, for thousands of years, civilized societies have had a deep interest in and have made repeated attempts toward the formal evaluation of individual differences in abilities and personal characteristics. Efforts to identify, through rigorous mental and physical examinations, individuals with the best vocational qualifications for government positions within the Chinese empire date back to 2200 B.C. However, more influential in the early development of evaluation and selection methods in Western cultures were the beliefs of the early greek writers, such as Aristotle, Hippocrates, Pythagoras, and Galen, who attempted to understand individual differences through "various anatomic, physiognomic, or physiologic indicators" (Dahlstrom, 1985). In response to increased knowledge about human bodily functions and an increased interest in understanding mental functions (particularly, mental capability and mental retardation as they related to the results of educational efforts), anthropometric testing and sensorimotor testing by the English biologist Francis Galton and the American psychologist James Mckeen Cattell set the stage for modern-day psychological testing.

At the International Exposition of 1884, Galton set an interesting precedent in charging people threepence to have their reaction times and their physical and sensorimotor functions measured, believing that sensory discriminative capacity would be higher among the most intellectually capable and impaired among the intellectually deficient. In 1890, Cattell first coined the term *mental test* in the psychological literature. In 1896, Lightner Witmer, the father of clinical psychology, established the first psychological clinic (essentially a clinic for children with learning problems). In this clinic, Witmer established a major precedent in the use of, but not complete reliance upon, formal assessment instruments in the practice of *clinical* psychology (McReynolds, 1987). Further advances occurred at the turn of the century, when experimental psychologists, such as Cattell, Thorndike, and Spearman, began to develop psychometric theory and technique, and others, such as Ebbinghaus, Binet, Simon, and Terman, began constructing intellectual measures for use with school-aged children. In 1916, Terman's revision of Binet's work led to the first major "Americanized" individual test of intelligence, the Stanford-Binet. World War I soon created a need for group testing instruments, to which the American Psychological Association (APA) responded by assigning members, under the direction of APA president Robert Yerkes, to construct such instru-

ments. The tests eventually developed by army psychologists were known as the Army Alpha and Beta tests. After the war, the army tests, in modified forms, became available for public use and led to an era of IQ consciousness in school systems and colleges.

Experience with the processing of recruits during World War I also brought about the beginnings of objective personality assessment in the form of Woodworth's Personal Data Sheet, which was used to identify seriously neurotic men. In 1921, apparently realizing the enormous growth and influence of psychological testing, Cattell and 200 other psychologists founded the Psychological Corporation. This corporation had a strong influence in the then young field, which at that time had little in the way of established test development, distribution, and use guidelines. David Wechsler, a student of both Cattell and Thorndike, found the intellectual tests of the day to be lacking and, beginning in 1939, published his important series of individual intellectual tests with this company.

Additional important historical events include the development, and eventual release in 1940, of the Minnesota Multiphasic Personality Inventory by Starke and Hathaway and, in 1935, the beginning of neuropsychological assessment in the United States, beginning in Halstead's University of Chicago laboratory. Both of these events continue to have a profound impact on current assessment practices. More recently, the application of computer technology to the administration, scoring, and interpretation of psychological and neuropsychological instruments has become a major influence on clinical practice (see Chapter 10 in this book, by Hartman and Kleinmuntz).

For the first 30 or so years of the current century, clinical psychologists were almost exclusively engaged in performing psychological assessments, with virtually no involvement in treatment. However, by the beginning of World War II, one third of clinical psychologists listed psychotherapy among their clinical activities (Kirsch & Winter, 1983). Immediately after the war, government funding of clinical psychology training programs through the Veterans

Administration greatly increased the numbers of clinicians engaging in psychotherapy. Even though treatment is currently a major focus of both training and practice, psychological testing also remains an important activity of clinical psychologists, and one that continues to have an increasing impact on society.

Although the emphasis thus far in this chapter has been on formal testing, in practice the role of the individual clinician in gathering information has always been essential. Observations and information obtained by the clinician provide an important part of the context within which test data can be understood.

Clinical Interviewing

Keeping in mind that psychological assessment consists of more than the use of psychological tests, the importance of clinical interviewing needs to be underscored. In contrast to the *psychometric* approach, described earlier, the emphasis on interviewing as a means of gathering important patient information largely comes from the *impressionistic* approach to mental health evaluation. The impressionistic approach is based on the belief that an understanding of the patient can be achieved by a sensitive observer who looks for significant cues and integrates them into a total impression that attempts to appreciate the often presumed relationship of the individual's history to the current presentation (Cronbach, 1984). An analog to these very different approaches to assessment can be found in comparing placing the back of the hand to the forehead versus using a thermometer to determine an individual's body temperature. Although one can sometimes obtain a general subjective impression of whether body temperature is elevated on the basis of touch alone, this impression will not be precise.

With regard to evaluations of cognitive status these two approaches can be illustrated in the use of a mental status examination versus the use of neuropsychological measures. However, this comparison should not be interpreted as suggesting that psychologists do not use or rely on clinical interviews. Indeed, the use of psychological tests in a "blind" manner, without either interview or other historical information, is an undertaking fraught with peril. This type of practice is perhaps acceptable only to clinicians with either an excessively high index of trust in test scores or a high threshold for the acceptance of mistakes both in clinical decision making and in the subsequent negative professional consequences that stem from incorrect diagnoses and recommendations. It is difficult to imagine a situation in which a psychologist, using the approach that a radiologist may use in examining an X ray of the leg for evidence of fracture, can expect routinely to draw conclusions from test scores alone. Even in neuropsychological assessment, which is often *presumed* to be a more objective, straightforward type of assessment and in which a seemingly higher reliance on technicians may reduce face-to-face contact with the patient in question, it is *not* commonplace for the test scores to speak so loudly and clearly that the need for further information to provide a context for interpretation is obviated (cf. Adams & Jenkins, 1981). In fact, when technicians are used to gather data, the interview becomes even more important because it may be the only contact that the clinician has with the patient. In teaching assessment to our graduate students, we explicitly attempt to provide a model of practice that *does not* elevate test scores to a level of importance where they supercede understanding the larger picture of what the patient is all about. Because we are sampling behavior, the more information available to shed light on the meaning of the behavior the better, including observations of the patient during the gathering of the test data and while interacting with significant others.

Clinical interviewing is beginning to attain a more esteemed status throughout the mental health field, in part because there has been a deliberate attempt to change the basic approach to gathering interview information in order to make the results less subjective and more reliable (Goldstein & Hersen, 1984). Particularly for clinicians who do not use formal testing, whether psychologists, psychiatrists, social workers, nurses, or others, a number of structured interview formats for particular settings or patient populations now exist. For example, the Schedule for Affective Disorders and Schizophrenia (SADS) (Spitzer & Endicott, 1977) and the SCID-II, which is designed to facilitate diagnosis of Axis II disorders (see Frances, 1987), and others have become relatively popular for gathering information relevant to diagnosis in psychiatric settings.

the major companies publishing psychological test materials.

Physiological measures are not discussed in this chapter, as traditionally they have not been considered "psychological tests" and are considered elsewhere in discussions of clinical biofeedback in this book (see Part VI). Aptitude tests, which can be described as measures of specialized vocational or academic skills or abilities, are not included here because of their low frequency of use by clinical psychologists.

Types of Tests in Use

Both the type of tests used and the frequency of testing vary with the type of practice setting (Lubin, Larsen, Matarazzo, & Seever, 1986). In considering the various types of tests currently in use by clinical psychologists, a discussion of each may be easier if they are categorized by purpose. Accordingly, we can consider formal: intellectual, academic (achievement and learning-disability), vocational interest, personality, and neuropsychological measures. Each of these categories of tests is important to psychologists performing assessments of patients in medical settings. A brief description of the nature of these tests, examples of instruments within each of these categories, and estimates of frequency of usage by clinicians, where available, are all that current space limitations will allow. Table 1 contains an address listing of most of

Intellectual

As can be inferred from the brief historical comments earlier in this chapter, the assessment of intellectual abilities has been a major activity of clinical psychologists since the field began. Although many instruments exist that attempt to estimate levels of intellectual functioning, the *sine qua non* of this area of assessment is the Wechsler Adult Intelligence Scale—Revised (WAIS-R) (Wechsler, 1981) and the Wechsler Intelligence Scale for Children—Revised (WISC-R) (Wechsler, 1974). Before the publication of the Wechsler series, the standard against which all other intelligence measures were evaluated was the Stanford-Binet. Although it is still in existence in the form of a fourth edition (Thorndike, Hagen, & Sattler, 1986), relatively few clinical internships, even among those specializing in child clinical train-

---

Table 1. A Partial Listing of Major Psychological Test Publishers and Distributors

American Guidance Service Publishers' Building P.O. Box 99 Circle Pines, MN 55014–1796	National Computer Systems P.O. Box 1416 Minneapolis, MN 55440	Reitan Neuropsychology Laboratory 1338 E. Edison St. Tucson, AZ 85719
DLM Teaching Resources P.O. Box 4000 One DLM Park Allen, TX 75002–1302	PRO-ED 5341 Industrial Oaks Blvd. Austin, TX 78735	Riverside Publishing Company 8420 Bryn Mawr Ave. Chicago, IL 60631
Jastak Assessment Systems P.O. Box 4460 Wilmington, DE 19807	Psychological Assessment Resources P.O. Box 998 Odessa, FL 33556	Stoelting Company 1350 S. Kostner Ave. Chicago, IL 60623
	The Psychological Corporation P.O. Box 9954 San Antonio, TX 78204–0954	Western Psychological Services 12031 Wilshire Blvd. Los Angeles, CA 90025

ing, offer substantial exposure to a version of this instrument (Elbert & Holden, 1987). An examination of survey data on psychological test usage since 1935 (Lubin, Larsen, & Matarazzo, 1984) indicates that the Stanford-Binet was ranked Number 1 in 1935 and in 1946, following which it has steadily declined to the point of relative obscurity today. In contrast, the predecessor of the adult version of the Wechsler (the Wechsler-Bellevue) was ranked Number 2 in 1935 and subsequently, in the form of the WAIS beginning in 1955 and in revised form as the WAIS-R beginning in 1981, had risen to Number 1 in 1969 and 1982 surveys.

Basically, the WAIS-R elicits measures of intellectual abilities through an evaluation of specific skills primarily related to (1) verbal comprehension and reasoning and (2) visual-spatial reasoning and organization. A concise discussion of the WAIS-R literature can be found in Matarazzo and Matarazzo (1984). Although also relevant to other areas of psychological assessment, nowhere in the field do concerns regarding minority bias in testing get as much deserved attention as in the WAIS-R literature. Because tests are designed to determine what a person can or cannot do, but do not necessarily help us understand why, factors such as cultural background, motivation, and emotional state need to be considered carefully (Anastasi, 1988).

Academic

Testing specific to learning experiences in the classroom typically involves achievement testing or, in some instances, learning-disability testing. In the overall evaluation of school and academic performance, test procedures from many other areas (e.g., intellectual, personality, and neuropsychological) are also used frequently. In contrast to the group-administered achievement tests which at a given academic level have been popular in school systems over the years (e.g., California Achievement Tests, Iowa Tests of Basic Skills), the tests used for individual evaluations by clinical psychologists have tended to be brief and less comprehensive. Such tests typically provide standard scores for basic academic abilities such as reading recognition, reading comprehension, arithmetic, and spelling, rather than specific information learned in science or other curriculum areas. Intended primarily for screening purposes, these measures include the Wide Range Achievement Test—Revised (Jastak & Jastak, 1984) and the Peabody Individual Achievement—Revised (Markwardt, 1989). More specific and detailed diagnostic tests are typically used when one is attempting to identify or rule out specific learning disabilities. Numerous tests with narrow scope, such as the Woodcock Reading Mastery Test (Woodcock, 1973) and the Key Math Test (Connolly, Nachtman, & Pritchett, 1976), as well as a few broader tests such as the Woodcock Johnson Psychoeducational Battery (Woodcock & Johnson, 1977), have been developed for this purpose and are now well known to school psychologists and learning-disability specialists. A few of these tests with sound psychometric properties seem to be growing in popularity among clinical psychologists as well. As with intellectual measures, it is only with a careful consideration of multiple factors (e.g., background, motivation, and emotional state) that the reasons for performance at a given academic level can be understood. It should also be noted that some neuropsychologists support the use of traditional neuropsychological measures, rather than "psychoeducational" measures for an understanding of learning disabilities (e.g., Rourke, Fisk, & Strang, 1986).

Vocational Interest

Clinical psychologists may be asked to provide information relevant to career choice and the pursuit of vocational alternatives. Although such tests are used less frequently than other types of testing within medical settings, this need for vocational interest assessment may be identified in the course of counseling an injured worker, a college student, or a disenchanted employee who lacks direction; vocational interest may also be a presenting referral question. As might be expected, the tests used to assess vocational interests have been self-report measures. A few tests have dominated this area, so that, at present, this area is perhaps best represented by the Strong-Campbell Interest Inventory (SCII) (Campbell & Hansen, 1981) and the very similar Career Assessment Inventory (CAI) (Johannson, 1982). Both of these measures provide (via computer scoring, which is a must because of the amount of "number crunching") descriptions of relevant job characteristics and broad categories of interest, as well as extensive comparisons of the individual's degree of interest in specific occupations with that of actual individuals successfully employed in those specific occupations. Although not a guarantee that an individual will find happiness in or be capable of performing a specific job, the type of assessment of vocational interest represented by the SCII appears promising and psychometrically sound (Anastasi, 1988).

Personality

Traditional Testing Originally Intended for Psychiatric Patients

Among the tests used in clinical evaluation, the prototypical examples of *objective* and *projective* personality tests are easy to identify. For decades, the single most frequently used objective personality measure has been the Minnesota Multiphasic Personality Inventory (MMPI). The MMPI consists of 566 questions in a true-false format. The responses produce numerous scales, subscales, and critical item lists that assess mood, psychotic experience, perception of health problems, interpersonal difficulties, and other clinical domains. The sheer number of the scales that can be derived has caused many users to rely on computers to score, and to even interpret, the results of the MMPI. The test was originally devised by Hathaway and McKinley in 1940 and first distributed for use in 1947, and by 1975, approximately 6,000 reference citations had appeared in the literature (Greene, 1980). Given the growth of the profession since that time, the number of published research studies today could easily be double the 1975 figure. Dahlstrom (1985)

compared the universal acceptance and worldwide use of the MMPI to the spread of the Binet scales as measures of intelligence in the early 20th century. Although currently under revision, the MMPI has to date been used in the same basic form since being released. The establishment in 1984 of a new normative data base from which t scores (standardized scores with a mean of 50 and a standard deviation of 10) could be derived is the only major alteration of the psychometric properties of the scale that has taken place. However, dramatic changes have taken place through the years with regard to interpretation of MMPI responses. For example, Scale 5 (Masculinity-Femininity) is no longer used to identify individuals engaging in homosexual behavior or having homosexual concerns (Greene, 1980), and Scales 1 (Hypochondriasis) and 3 (Hysteria) are no longer used to discriminate between so-called organic and functional pain (Strassberg, Reimherr, Ward, Russell, & Cole, 1981; Sweet, 1981).

The extensive research data base of the MMPI has been one of the reasons it has remained in favor for so long. Effective validity scales that can detect patient response styles, such as defensiveness and exaggeration of symptoms, have no doubt also played a major role in the continued use of the MMPI in psychiatric, medical, and legal cases. Frequent current uses include the evaluation of personality characteristics, mood, and adjustment to medical conditions.

The Rorschach has been synonymous with projective testing for decades, although other measures, such as the Thematic Apperception Test (TAT), are also used frequently. The elicitation of responses in an unstructured manner to ambiguous stimuli (inkblots in the case of the Rorschach) is the basis of all true projective assessment. As mentioned earlier, the Rorschach was viewed negatively in the 1960s and the early 1970s but has currently been returned to good favor within the field. At present, 88% of APA clinical psychology training programs have at least one course emphasizing the Rorschach, and the majority of course instructors teach the Exner Comprehensive System. In fact, the resurgence of the Rorschach has been largely due to the works of John Exner and his colleagues (Exner, 1978, 1986; Exner & Weiner, 1982), who established an empirically based scoring and interpretive system. This system has essentially replaced the various diverse scoring systems, which previously had not been accepted uniformly. As a result, the research interest in and the clinical use of the Rorschach in the United States have been quite high in recent years (Lubin *et al.*, 1984; Sweeney, Clarkin, & Fitzgibbon, 1987). Exner's scoring system is extensive and complex, but allows experienced users to make sound inferences with regard to the presence of psychotic process, personality dynamics, interpersonal style, mood, and coping ability.

Of course, as is true of the MMPI, few diagnostic conclusions can be derived solely on the basis of the personality or psychological test data alone. Rather, a specific diagnosis such as that found within the revised third edition of the *Diagnostic and Statistical Manual of Mental Disorders* (DSM-III-R) of the American Psychiatric Association (1987) typically requires careful interviewing, observation, and review of the patient's history.

Specialized Testing Intended for Use with Medical Patients

Although tests such as the MMPI, originally intended for use with psychiatric patients, have also been used with medical patients with some degree of success, some tests have been specifically designed for use with medical patients. These typically have the advantage of greater brevity and a more carefully selected normative data base. Conversely, their disadvantages include an inability to answer broad questions involving psychiatric disorder and potential inappropriateness for other than the narrow band of medical patients for which the specific instrument was designed.

Along these lines, Gatchel and Baum (1983) concisely summarized a few of the measures relevant to clinical health psychology in general. These measures consist primarily of self-report rating scales, including:

1. The SCL-90, a 90-item self-report rating scale designed to measure psychopathology in medical outpatients (Derogatis, Lipman, & Covi, 1973). The scale produces nine dimensions of primary psychopathology and three global indices of pathology.
2. The Schedule of Recent Experiences by Holmes and Rahe (1967), a self-report checklist of recent stressors.
3. The Life Experiences Survey (Sarason, Johnson, & Siegel, 1978), a 57-item self-report rating of significant recent life events.
4. The Behavior Pattern Interview (Friedman & Rosenman, 1974), a structured interview intended to identify personality characteristics associated with heart disease.
5. The Jenkins Activity Survey (Jenkins, Rosenman, & Friedman, 1967), a self-report inventory designed to provide a measure of behavioral characteristics associated with Type A behavior.

Although measures such as the above can be useful in the efficient gathering of helpful clinical information, such instruments typically have very little to offer in the way of psychometric properties. One of the few psychological measures that was specifically designed for medical patients and that also strives to fulfill basic psychometric requirements is the Millon Behavioral Health Inventory (MBHI—Millon, Green, & Meagher, 1969). The MBHI is a 150-item self-report inventory fashioned along the lines of the MMPI. The intent of the authors was to provide a description of personality style, psychogenic attitudes, psychosomatic correlates, and prognostic indices specifically normed on and intended for use with various medical populations, such as chronic pain patients. Despite the attractive premise in the stated purpose of the MBHI, there has been relatively little validation research to date in which the MBHI has been used with specific clinical populations. Reviews by Allen (1985) and Lanyon (1985) have been critical of the initial validation research and the psychometric properties of the MBHI.

In the area of chronic pain (see also the discussion in Chapter 23 by Chapman, in this book), an initial validation study of the MBHI in an outpatient multidisciplinary pain center by Sweet, Breuer, Hazlewood, Toye, and Pawl (1985) suggested the following areas of difficulty: (1) a lack of specificity of the MBHI scales, in that most intercorrelations were extremely high; (2) the failure of the Pain Treatment Responsivity scale of the MBHI to predict the outcome of pain treatment better than the MMPI D Scale or the MBHI Allergic Inclination Scale; (3) the high degree of correlations between MBHI scales and either admission or denial of psychopathology on the MMPI; (4) the absence of strong relationships between MBHI scales and the hypochondriasis and hysteria scales of the MMPI; and (5) the relative absence of sensitive indicators of validity with regard to patient response style in completing the MBHI (i.e., because only three items are considered for a "reliability check," a number of response biases and random profiles are considered "valid"). A study by Wilcoxson, Zook, and Zarski (1988) reported results very similar to the Sweet *et al.* (1985) findings with regard to the prediction of pain treatment outcome and statistical overlap between MBHI scales. Lee-Riordan and Sweet (1990) have also found problems in the statistical relationships between the MMPI and the MBHI, as well as high MBHI intracorrelations.

A variety of questionnaires and rating scales currently exist for use with specific clinical populations within medical settings. These measures are too numerous to mention here and, again, often are without psychometric foundation, even though they can provide an efficient means of gathering very relevant information. The interested reader will find mention of some specific informative nonpsychometric methods of obtaining information within the clinical program chapters in this book. Readers interested in a further discussion of psychometric issues as applied to medical settings might consult the special supplemental issue of *Cancer* (Vol. 53, No. 10, May 15, 1984).

Neuropsychological

In recent years, interest in and demand for clinical neuropsychological assessment have grown rapidly (D'Amato, Dean, & Holloway, 1987; Wedding & Williams, 1983), especially in medical settings. Although some measures commonly used in neuropsychological evaluations are actually tests of intellectual ability, academic achievement, and personality such as the WAIS-R, the WRAT-R, and the MMPI, respectively (Seretny, Dean, Gray, & Hartlage, 1986), the number of specific neuropsychological measures available to the practicing clinician has grown enormously as the field has expanded. Because of limited space, only the major batteries are discussed here (see Lezak, 1983, for a compendium of tests and techniques).

The Halstead-Reitan Neuropsychological Battery dates back to the first systematic neuropsychological research in the United States, initiated by Halstead in 1935 and culminating in a relatively intact battery first described in the literature in 1947 (Parsons, 1986). The Halstead-Reitan remains a popular assessment tool today (Seretny *et al.*, 1986). The Luria-Nebraska Neuropsychological Battery was developed more recently (Golden, Hammeke, & Purisch, 1978) and has been available for clinical use since about 1979. Since that time, the Luria-Nebraska has become the second most frequently used neuropsychological battery. Differences between these two batteries in terms of content, length of administration, and relevance to a theory of brain function led to a number of comparative clinical and statistical studies beginning in 1981 (Golden, Kane, Sweet, Moses, Cardellino, Templeton, Vicente, Kennelly, & Graber, 1981; Kane, Sweet, Golden, Parsons, & Moses, 1981). All *empirical* comparative studies to date have reported comparable diagnostic accuracy and strong statistical relationships between these batteries.

In order to be considered truly comprehensive, a neuropsychological battery must include an evaluation of such important neuropsychological functions as: verbal and nonverbal reasoning, expressive and receptive language, cognitive efficiency, sustained attention, learning and memory, visual-spatial abilities, sensory processes, and motor abilities. Although meant to be comprehensive, *both* the Halstead-Reitan and the Luria-Nebraska batteries have weaknesses that typically require the use of complementary measures (e.g., of learning and memory, complex reasoning, complex motor, and sustained attention with the Luria-Nebraska; of learning and memory, language, and academic with the Halstead-Reitan). Excellent descriptions of both of these extensive batteries, as well as reviews of published clinical and research comparisons, can be found in Incagnoli, Goldstein, and Golden (1986). Clinicians experienced with both batteries can begin to identify case situations in which a comprehensive evaluation is indicated that favors one or the other battery (e.g., the Halstead-Reitan with alcoholics, who may have significant subclinical deficits, and the Luria-Nebraska with elderly, potentially demented patients).

It is noteworthy that Seretny, Dean, and Seretny (1985) reported that the percentage of surveyed neuropsychologists using portions of the Halstead-Reitan was higher (59%) than the number using the entire battery (37%). In keeping with this finding, Sweet, Tal, and Shain (1985) found that, among a sample of 52 neuropsychologists, 33% used standardized batteries routinely, 43% endorsed the use of *flexible batteries* (i.e., variable but routine groupings of tests for different *types* of patients: head-injured, alcoholic, elderly, etc.), and 23% endorsed a completely individualized, flexible test selection without any uniformity of data collection across patients. A larger survey of 184 American Board of Professional Psychology (ABPP) Diplomates in clinical neuropsychology and non-ABPP clinical neuropsychologists found that, given the same three alternatives, approximately 18% endorsed the routine use of standardized batteries, 54% endorsed *flexible batteries*, and 29% endorsed an individualized flexible approach (Sweet & Moberg, 1990). In a hospital-based general testing service in which a broad range of patients is seen for an equally broad range of reasons, and in which the referral sources have very different needs, we have found the flexible battery position to be the most satisfactory.

A time-worn expression suggests that, when one has only a hammer to work with, the world looks like a nail. In the not too distant past, *the* tools of psychological assessment were the "big three" (WAIS, Rorschach, and Bender-Gestalt). Nearly all referral questions (psychological *and* neuropsychological) were answered on the basis of these few measures. Certainly, with the various excellent "tools" available today, and with ever more diverse clinical opportunities opening up, neuropsychologists have been able to make more and more sophisticated contributions in their practices.

Although there are also clinical advantages to the completely individualized, flexible approach, with the trend in the health market toward accountability and the increasing demand within the profession and in society in general for scientific bases for practice methods, it is difficult to feel comfortable with this approach. Unfortunately, there is insufficient space here to discuss further the practical implications of different philosophical approaches to neuropsychological testing.

A note of explicit caution with regard to the interpretation of neuropsychological tests is in order. Despite the enormous information value of these measures, poor scores are not synonymous with brain dysfunction. Numerous alternative hypotheses need to be considered and discarded before brain dysfunction is diagnosed (cf. Newman & Sweet, 1986; Puente & McCaffrey, 1990; Sweet, 1983). In this regard, the concept of *comorbidity* (Feinstein, 1985) is extremely relevant to the work of psychological diagnosticians.

Essentials in Establishing and Maintaining a Hospital-Based Testing Service

Schenkenberg, Peterson, Wood, and DaBell (1981) determined the positive characteristics looked for by 112 physicians who used psychologists for consultation-liaison services. Among the *personal* characteristics that these physicians deemed important for psychologists to have were a pleasant demeanor, compassion, interest, availability, effective communication, cooperation, intelligence, openness, and common sense. Among the *professional* qualifications of a psychologist rated highly by the physicians were knowledge, background in medical illness, verbal and written communication skills, knowledge of local psychological resources, and diagnostic acumen. Clearly, a number of these characteristics would also describe the successful psychologist providing formal evaluation and testing services.

Belar, Deardorff, and Kelly (1987) suggested a number of excellent points that may aid psychologists working in medical settings: do not overidentify with the traditional medical model; provide prompt follow-through in providing service to the patient and communication back to the referral source; and understand local referral customs (e.g., who is allowed to refer and when patients are to be given feedback by the consultant). These and other salient points are elaborated below.

Establishing Professional Relationships

As the profession and the number of clinical psychology practitioners have grown, and as the health care environment has changed, the importance of professional relationships in one's practice has increased. As described by Sweet and Rozensky (1991), the changes in the health care marketplace appear to be increasing competitiveness both within and between disciplines, an increased monitoring of both practice and economics by colleagues and third-party payers, and the provision of incentives for group and institutional practice. One net effect of these changes is the need for clinicians to be ever mindful of the importance of interpersonal relationships in their practice. A hospital-based testing service is truly a person-oriented endeavor in which professional relationships with consumers (i.e., referral sources, patients and their families, and the general public), clinical and research colleagues, employees, and third-party payers are among our most important commodities.

In practical terms, the establishment and maintenance of professional relationships require both clinical skill and good common sense. In medical settings that do not have established clinical psychological services, the first step is to discern the needs of the potential referral base. Typically, this is best done in the course of making initial contacts (i.e., at time of personal introductions, usually with the help of someone you know on the staff, state your intention of establishing an evaluation service). At this point, one's skills as a good listener and a sensitive interviewer are extremely important. Typical referral questions from different medical specialists are listed in Table 2. Having found out the needs of at least the primary potential referral sources, one can provide realistic expectations, educate potential referral sources, describe the planned service, and seek feedback regarding the plans from those who would use the service.

In the event that the medical setting already has established psychological services, find out directly whether the services you plan are already provided and in what ways the planned services may affect the established services. In most cases, the existence of established services in some other practice area does not preclude the initiation of testing services. However, the presence of preexisting testing services within the medical setting is quite another matter and raises several serious questions: To the extent that such services will compete with one another, will the referral base be able *and willing* to support both? Can the planned service coexist in a collaborative effort with the established service? And is there some salient reason that this particular setting, rather than another with no established service, should be pursued? Again, as psychology practitioners increase in numbers, the potential for competition within a given setting also increases. However great the competition, we should nevertheless strive to establish and maintain good professional relationships (Sweet & Rozensky, 1991).

An important dimension of relating effectively to medical professionals is being able to "speak the language." In psychology graduate programs, the professional language is often

Table 2. Frequent Referral Questions by Medical Specialty

1. Internal medicine and family practice
   Does patient with chronic illness need treatment for depression?
   Where there are no positive medical findings, is there psychological disturbance that explains physical complaints?
   Is elderly patient demented?
2. Neurology and neurosurgery
   Is there a psychological component to chronic pain or headache problem?
   Establish a baseline of functioning prior to brain surgery. Does seizure patient have associated psychological disturbance or neuropsychological deficit?
3. Pediatrics
   Are psychological factors affecting child's physical condition?
   Do psychological factors, brain dysfunction, learning disability, or attention deficit disorder provide explanation of poor learning in school?
   Does child have conduct disorder?
   Has medical illness produced a treatable psychological disorder?
   Is family stress causing deleterious emotional consequences?
   What are psychological effects of this case of child abuse?
   What are treatment recommendations for this case of school phobia?
4. Psychiatry
   How much of elderly patient's presentation is depression rather than dementia?
   What are psychodynamics of patient in treatment?
   What type of therapeutic strategies might be most effective with this patient?
   Is there risk of suicide?
   Does patient have a thought disorder?
   Is patient capable of benefiting from insight-oriented therapy?
5. Physiatry (rehabilitation)
   Will patient have difficulty coping with family or daily activities following discharge?
   Has known stroke or head injury caused significant cognitive impairment?
   How does patient's disability affect her or his ability to work?
   Is brain-damaged patient a candidate for cognitive retraining, and if so, what strategies and modalities are indicated?

couched in terms applicable only to the social sciences and the study of overt human behavior and internal emotional and cognitive processes. However, in medical settings, the language is often anatomically oriented and replete with medical abbreviations, terms, and phrases that require some familiarity and understanding of human anatomy and physiology, as well as of the nature and purpose of medical diagnostic and treatment procedures. The more one specializes in working with a particular medical population (e.g., those with brain injuries, chronic back pain, or headaches), the more precise the psychologist's medical knowledge needs to be. Of course, the reason for acquiring this knowledge is not to engage in medical practice, which would be illegal and unethical, but to be able to communicate effectively with the physician and the patient. Also, any psychologist working within a medical setting, regardless of the type of program or

department, needs to have a good working knowledge of the standard psychiatric nomenclature, currently represented by the DSM-III-R (American Psychiatric Association, 1987). However, the psychologist should not assume that nonpsychiatry physicians are familiar with current psychiatric nomenclature.

Educating Referral Sources

When interacting with mental health professionals, one can safely assume that they have had a reasonably high degree of exposure to the training of psychologists and are familiar with at least the general areas of practice in which psychologists are typically engaged. However, one should not assume that even our mental health colleagues have a prior understanding of or exposure to such recent specialty practice areas as health psychology or neuropsychology, or of such specifics as what constitutes an appropriate referral question to a psychological evaluation and testing service. With non-mental-health professionals, there may be a substantial need for basic educative information pertaining to the capabilities and practice areas of psychologists. Such information can usually be conveyed in informal one-to-one meetings, by in-services or more formal lectures to staff, or by the distribution of salient articles on the topic of interest. In particular, the legitimate uses of psychological and neuropsychological testing, as well as the limits of these procedures, may need to be made explicit.

Within medical settings, two of the referral questions that most often lead to inappropriate expectations of a testing service are (1) Is this a "functional" or "organic" patient? and (2) What is the etiology of the patient's dementia? Logically, the former question cannot be answered solely on the basis of psychological testing, and the latter question can be answered only partially with respect to the similarity or lack of similarity of the data with test results known to be associated with certain known causes of dementia. Once the limitations of testing are clear to the referral source, a more effective use can be made of the psychologist's talents, as well as of the psychologist's and the patient's time.

Providing Efficient and Responsive Services

In an era of utilization review of inpatient stays, capitated health plans (e.g., many Health Maintenance Organizations and Preferred Provider Organizations), and attempts by government to cut back health care costs, the length of hospital stays has been decreasing. Thus, any services that are offered to inpatients must be efficient and responsive with regard to the attending physician's need to understand the patient and to plan for follow-up care before discharge. More than one competent psychologist has failed in efforts to establish a practice within a medical setting simply because reports took too long to reach the referral source.

Providing efficient services involves both timeliness and cost. Responsive services begin with a clear-cut, easy system of making referrals to the service. Flexibility in how referrals can be made is likely to be important to busy physicians. Physicians seeing patients on inpatient units can easily make use of a standard referral sheet, which has a well-defined route to the evaluation service (i.e., perhaps hand-carried by the unit clerk to the service mailbox). An example of the referral sheet used in our inpatient psychiatry units can be found in Figure 1. An outpatient referral process handled in this manner would not be efficient. Instead, telephone referral to a secretary or a staff member who fills in the written form by obtaining information over the telephone is a good alternative. Some referral sources prefer to have the patient make contact to set up the appointment, and then to send referral and medical information once it seems that the patient will actually follow through with the recommended evaluation.

If a referral question can be answered in a two- to three-hour evaluation, rather than a four- to five-hour evaluation, then it can be considered efficient; the results are obtained faster and at less cost. More is not necessarily better. Lengthy, comprehensive evaluations are "better" only if the added information is necessary and can justify the added time and expense. Thus, to perform *comprehensive* testing of each patient in a reflexive manner would seem to be a disservice to both the patient and the referral source, as well as a poor use of program resources. The bottom line is whether the evaluation and the feedback are responsive to the referral question: Was the necessary information gathered to enable an intelligent response to the question, and was the feedback given in a timely, comprehensible manner?

An example of the appropriate application of a screening battery can be found in the neuropsychological evaluation of psychiatric and elderly patients. Clinical research has shown that screening batteries perform at a level comparable to comprehensive batteries in differentiating psychiatric from brain-damaged patients (e.g., Goulet Fisher, Sweet, & Pfaelzer-Smith, 1986; Wysocki & Sweet, 1985), as well as with demented and nondemented elderly (cf. Poon, 1986). In both clinical situations, there is a real

The Evanston Hospital

DEPARTMENT OF PSYCHIATRY

PSYCHOLOGICAL EVALUATION AND TESTING SERVICE (P.E.T.S.)

PSYCHOLOGICAL ASSESSMENT REFERRAL FORM

PATIENT NAME: REFERRED BY:

AGE: ROOM:

WORKING DIAGNOSIS:

RELEVANT HISTORY:

REFERRAL QUESTION:

TESTS FOUND HELPFUL IN PAST
WITH SIMILAR PATIENTS:

COMMENTS TO FACILITATE
WORKING WITH PATIENT:

THANK YOU   _____   _____
                           DATE   AUTHORIZED SIGNATURE

PLEASE RETURN TO P.E.T.S. ROOM 5208

_____

QUALITY ASSURANCE DATES (OFFICE USE)

Rec'd _____   Dictated _____

Initiated _____   Report Sent_____

Complete _____

Procedures: Time:

Figure 1. Example of inpatient referral sheet.

need to keep procedures as brief as possible because of the limited tolerance and endurance for formal testing of abilities. A thorough rationale for and review of neuropsychological screening batteries are contained in Berg, Franzen, and Wedding (1987).

Report Writing

Although it may be accurate to state that writing psychological and neuropsychological reports is an art, for graduate psychology students and beginning clinicians who do not consider themselves artistic writers it will hopefully be anxiety-reducing to know that it is also a concrete skill that can be learned. Much has been written on this essential task of clinical psychologists (see Kellerman & Burry, 1981; Klopfer, 1983; Tallent, 1976; vanReken, 1981). In many cases, report writing is situation-specific (i.e., the type of format and even the report content are influenced by the circumstances surrounding the need for the report). To meet the diverse informational and time demands of different referral sources, flexibility in providing feedback is essential. Prompt feedback in the form of verbal reports or preliminary handwritten reports in the chart need to be considered, as well as brief formal reports. A "one size fits all" lengthy, detailed report is neither responsive nor timely for most medical referrals. Because there is no single correct format, but a number of formats that will accomplish the purpose of facilitating communication, it is most important that the writer choose a format that will be conducive to organizing and articulating his or her thoughts about a case. Examples of the formats used most often in our service are provided in Table 3.

In those instances when neuropsychological evaluations do not require that comprehensive measures be administered, there is no need for subheadings within the test findings section. When traditional psychological evaluations incorporate different types of tests, the test findings section can be divided by appropriate subheadings (e.g., "Intellectual and Academic," "Personality and Emotional," and "Vocational"). If one does a relatively large number of evalua-

Table 3. Examples of Test Report Formats

Example 1. Traditional psychological testing referral

Psychological Evaluation Testing Service
Report of psychological evaluation
    Name: Date:
    Date of birth: Status:
    Referred by:
    Evaluated by:
    Referral information:
    Procedures:
    Behavior and observations:
    Relevant interview information:
    Test findings:
    Summary and conclusions:

Example 2. Neuropsychological testing referral requiring comprehensive evaluation

Report of neuropsychological evaluation (all the same identifying information at the top of the page, as well as the same major headings; test findings section has the following subheadings)
Test findings
    Intellectual, cognitive, and language functioning
    Memory and learning
    Visual-spatial functioning
    Sensorimotor functioning
    Personality and emotional functioning

tions on a regular basis, flexibility in report formats and content allows the clinician to do his or her job with greater ease.

Formal reports of psychological and neuropsychological evaluations may be the most important product of an evaluation service. Bruner's description (1942) of research writing errors and inaccurate referencing as "an annoyance to future investigators and a monument to the writer's carelessness" (p. 68) applies directly to clinical report writing as well. The consumers of the reports that we write today may include the health care professionals, rehabilitation counselors, attorneys, or teachers who will come in contact with the patient at any time in the future. Table 4 compares the general characteristics of what can be considered "good," "poor," and "unacceptable" reports. Given that writing styles vary a great deal, and that judgments of what is acceptable in professional writing may vary even more, there are

Table 4. Characteristics of Good, Poor, and Unacceptable Reports

Good report
- Identifies and describes patient accurately.
- Writer's degree, title, and speciality indicated.
- Well-organized report.
- Clear, variable sentence structure.
- Well typed, or legible if hand-written.
- Consistent, clear relationship between data and interpretations and conclusions; all able to be discerned by reader.
- Recommendations for treatment, if appropriate, are specific and related directly to the data and conclusions in the report.

Poor report
- Insufficient or incorrect identifying information on patient (e.g., age and education level).
- Writer not identified well (e.g., degree and speciality).
- Poor organization of report.
- Poor grammar and unusual phrasing.
- Poorly typed, or not legible if hand-written.
- Relationship between data, interpretations, and conclusions difficult to discern.
- Treatment recommendations, if appropriate, not specific enough and do not seem to follow from the rest of the report.

Unacceptable report
- Inappropriate or inconsistent test interpretations. To the reader, results seem at odds with conclusions.
- Incorrect use of diagnostic terminology and/or statistical information.
- Idiosyncratic and arbitrary interpretive statements (e.g., a digit span score of 2 means right occipital brain damage caused by toxins).
- Inappropriate treatment recommendations, or not treatment recommended despite obvious need.

relatively few general guidelines that can be expected to hold:

1. Most of the report can be in the past tense (e.g., "Ms. Smith's injury was received while she was working as a pipe fitter"). Obvious exceptions are current, relatively stable facts about the patient, such as stature (e.g., "Mr. Jones is a tall, thin Caucasian male"). Although tenses may change with content through the report, mixing tenses within a single sentence can and should be avoided.

2. Repetitive use of phrasing and similar sentence structure should be avoided. It is not uncommon in reports to see nearly every sentence within the relevant interview information beginning with "The patient reported. . . ." Because the section heading we use already indicates that the source is the patient, this redundant phrasing is not necessary and can be reserved for content that the examiner wishes to underscore as being the perception of the patient (e.g., the patient's description of how an accident happened, or the patient's characterization of other people, such as a spouse) and not objective fact. In addition, altering the grammatical constructions of sentences creates a much more readable report.

3. Quotes from the patient should be used to describe key symptoms or important historical details. A quote is somewhat like a picture in the sense that both can be "worth a thousand words" if used judiciously. If psychologically important information would be lost by restating what the patient said, a quote should be used instead (e.g., "This crushing, nagging pain feels like the weight of the world on my shoulders").

4. Information contained in different sections should not be redundant. It is commonplace for students to repeat information unnecessarily in the report. This repetition may be due to the inexperienced clinician not knowing what to say, or to an inadequate conceptualization of the report sections. For example, the referral information section can simply state the basic reason and context of the evaluation; it need not contain the historical information obtained from the patient, which, in our opinion, belongs in the relevant interview information section. Another common practice is to include interview information in the behavior and observations section and then to report it again in the interview section. When deciding whether to elaborate historical points in great detail, a basic consideration is whether the referral source already knows these facts or needs to know them.

5. The patient's privacy should be respected. Highly sensitive information should not be put in a report simply because it is available. Because, ultimately, we cannot predict or control who will obtain and read a given report once it has left the office, care should be taken in deciding which personal historical facts will be included in a report. In many cases, a general statement regarding sensitive information may be all that is needed (e.g., that an adult surgery patient had an emotionally difficult and chaotic childhood, without saying that she was a victim of incest), with more specific information relayed verbally as needed.

6. Jargon should be avoided in a report. As with the preceding item, it is not possible to determine in whose hands a report will end up. Thus, it is best to be as descriptive as possible, without using theory-bound concepts and jargon.

7. The data should not be overinterpreted. As Klopfer (1983) stated, "The fact that false feedback is readily accepted along with true, makes it necessary for the psychologist to be especially vigilant about the nature of his or her remarks, both in written and in oral form" (p. 501). Although a useful teaching exercise in graduate assessment classes, seeing pathology everywhere (e.g., overlocalizing in neuropsychological reports) in a real clinical situation will have significant negative consequences.

8. It is not necessary to give a diagnosis if the data do not warrant doing so. Rather, it is quite acceptable to indicate to the referral source that, in a particular case, the data are indeterminate and do not allow a diagnostic conclusion. As a matter of keeping this issue in perspective, all medical diagnostic specialty areas share this reality, even neuroradiology, with all of its sophisticated technology.

9. When working with patients who are involved in worker's compensation, civil, or criminal proceedings, it is best to anticipate that the procedures used and each statement in the report may have to be justified *in court*. This is *not* the time to experiment. Legal cases can be very strenuous and require careful and thoughtful attention to detail. The question of malingering is often raised by the defendant in civil litigation and may require special consideration in the selection of test procedures. Unfortunately, good clinical research on malingering is very sparse. Thus, even the most experienced clinicians have difficulty with this thorny issue and must often rely solely on their own clinical judgment and prior experiences. The interested reader is referred to Melton, Petrila, Poythress, and Slobogin (1987) and Weiner and Hess (1987) for in-depth material pertaining to psychological evaluations in legal cases.

Building the Service to Match the Setting

The environment in which the service is housed will, to a large extent, determine the makeup of the staff, the most common referral sources, the most common referral questions, and the most frequently used evaluation procedures. In a medical setting, a psychological evaluation and testing service is typically housed within a particular department, such as neurology, rehabilitation, or psychiatry. Some programs are intended primarily to serve only their own department's needs, and others may provide evaluation services for the hospital at large. The need for specialty procedures or for working with specialty populations varies from setting to setting. For example, a program that accepts referrals only from an adult rehabilitation setting has little or no need for learning-disability or projective testing, whereas a program working exclusively in a pediatrics department will have no need for adult testing capabilities and instead will need to be capable of providing specialized assessment procedures exclusive to young children. However, the general testing service is likely to be called on at one time or another to perform assessments of all types for the entire age range.

Some information pertaining to the Psychological Evaluation and Testing Service (known within the hospital as PETS) of the Evanston Hospital may help to illustrate the characteristics of a general hospital-based evaluation service that is intended to meet the needs of its environment. Our service is housed within the Department of Psychiatry. In any given month, approximately 60% of our referrals come from within the department and 40% from outside the department, with a roughly even split between inpatients and outpatients. The PETS staff provides both neuropsychological assessment and traditional psychological testing, again in roughly equal numbers.

The high-volume demand for both projective testing and neuropsychological assessment has made it essential that we have staff who are highly specialized in and devoted full time to both of these areas. We have also found that it is helpful to have both male and female staff, as well as staff who are particularly experienced with older children and adolescents. Although we provide personality and basic intellectual and academic testing to young children, we have chosen not to perform learning-disability and neuropsychological testing on young children (8 years and younger), as there are other psychologists in the hospital who specialize in these referrals.

Within the sphere of neuropsychological assessment, the most common referrals concern differentiating depression and dementia in the elderly and the diagnosis, delineation of deficits, and rehabilitation in cases of closed head injury. Within the more traditional psychological testing sphere, common referrals involve adolescents who are performing poorly in school and are abusing drugs or are suicidal, as well as depressed or psychotic adults whose diagnoses have been unclear or whose treatment has not been effective. With adolescents, a fair amount of learning-disability and academic testing is required.

Because treatment issues have been a major part of most referral questions we have found it important for the PETS staff to be well versed in diverse types of treatments and the treatments offered by local referral sources. It would be most difficult to make appropriate treatment recommendations and to explain what can be expected of a treatment without experience in following patients from the initial assessment phase through the treatment phase.

Maintaining Quality Assurance

Quality with regard to psychological evaluation services involves a number of dimensions (e.g., reliability, content, and efficiency). To some clinicians, the data collection process may seem an unlikely place to begin thinking about quality assurance. Once having learned how to collect data, many assume that, like "riding a bicycle," they will never forget how to do it. In fact, much clinical assessment research and practice presumes that the individual gathering the data did so in the correct, standardized manner, without checking whether this actually occurred. However, any experienced clinician who has been involved in closely supervising staff or in teaching the trainees over the course of a year has undoubtedly seen the frequent problem of examiners' drifting away from the standardized administration and scoring procedures over time. In behavioral assessment research, this phenomenon is well known as "observer drift" (Johnson & Bolstad, 1973; Kazdin, 1977). If self-observant, clinicians and students can even notice such examiner drift in themselves. Thus, there is an ongoing need for clinicians, even those who use their skills daily, to continually refresh their memory of accepted, standardized procedure. In an organized evaluation service, staff should be encouraged to periodically observe one another as patients are seen, or, alternatively, demonstrate and discuss proper assessment procedures together in a group. Such periodic checks on the reliability of data-gathering procedures and scoring procedures are particularly important for students and clinicians who use their newly learned assessment skills only intermittently and thus have not had a chance to overlearn them (see Barton & Ascione, 1984).

Also, there are variables related to carrying out the evaluations efficiently. How long after the referral was made was the patient seen? After the patient was seen, how long did it take to complete and send the report? Beyond speed, what type of feedback was received about the results of the evaluation from the referral source (e.g., criticisms, praise, or requests for more specific information)? What feedback do patients give to referral sources about their evaluations (e.g., regarding expense, convenience in scheduling, and the sensitivity of the examiner)?

Ultimately, quality assurance data may be reflected in the number of referrals and the number of referral sources across time. Although the process is time-consuming, data on the above dimensions are worth keeping in order to establish a reasonable degree of quality. Requesting routine feedback from referral sources on the usefulness and the accuracy of data not only helps with quality assurance, but also simultaneously builds rapport with the referral sources.

Equally important is the content of the report. Are the specifics mentioned in the report accurate? Does the report convey the information in an organized, thoughtful, and readable manner that answers the referral questions? In a large service with a diversity of senior and junior staff, assurance of *report* quality may require that all reports be reviewed by the most experienced person in that area (e.g., projectives, neuropsychological, or learning disability). A fairly consistent general report format that can be altered as needed for different kinds of evaluations will allow regular consumers of the service to find specific information more easily in one or another part of the report. For some evaluations, such as those of hospitalized school-aged children who need alternative academic placement, there may be a need to include a fair number of specific test scores so that the treatment staff can work with placement settings. Such special needs can be dealt with easily by attaching an addendum sheet of relevant scores to the report.

Maintaining and Storing Patient Records

According to Hall (1988), the length of time that psychologists must keep records on their patients is determined by state, federal, and professional guidelines. The federal and American Psychological Association guidelines are in agreement in stating that full records must be maintained for 3 years after the last contact with the patient, and that at least a summary record must be maintained for an additional 12 years. Hall also noted that, when old records are sent out in response to a request, the psychologist sending the records should indicate whether the data in the records have become obsolete and thus do not apply to the patient at present. Because there is no way to control the distribution of reports once they are sent from the office, the patient consent form should state this clearly. An example of a release of information form is presented in Figure 2.

Summary

Psychological testing and formal evaluations have played an important part in the history and development of clinical psychology as a profession. Much of this activity has occurred within medical settings, where conducting such evaluations continues to be a central role for many psychologists. Economic changes and increased competitiveness within the health care marketplace greatly affect, but do not threaten, this type of practice. In fact, in this age of accountability, the desire to document and objectify problems may even lead to greater reliance on formal test procedures.

Within medical settings, there is typically a need for intellectual, personality, neuropsychological, academic or learning-disability, and vocational interest testing. In establishing a psychological evaluation and testing service within a medical setting, numerous factors deserve attention, including establishing professional relationships, educating referral sources, providing efficient and responsive services, writing reports, building a service that matches the environment, maintaining quality assurance (particularly with regard to data gathering and report writing), and maintaining records.

The Evanston Hospital

PSYCHOLOGICAL EVALUATION AND TESTING SERVICE (P.E.T.S.)

DEPARTMENT OF PSYCHIATRY

<u>RELEASE OF INFORMATION</u>

Date Prepared_____

I authorize_____ to release the following
              (Facility/Therapist)

information_____
                (Nature of information to be disclosed)

concerning_____, to

_____
        (Name of facility/therapist)   (Address)

for the specific purpose of _____

_____

I understand that I may revoke this authorization at any time
except to the extent that action has been taken on this
authorization. I further understand that this authorization
shall expire without my express revocation on:

_____, 19_____.

I further understand that the agency or individual which receives
this information, in accordance with State/Federal laws will not
disclose this information without further consent.

The Psychological Evaluation and Testing Service cannot guarantee
that agencies or individuals receiving this information will act
in compliance with these laws.

_____   _____
        (Signature)   (Relationship to patient)

_____   _____
        (Date)   (Witness)

Figure 2. Example of release of information form.

ACKNOWLEDGMENTS

The author gratefully acknowledges Drs. Erin Bigler and Robert Heaton for their helpful critiques of the final manuscript for this chapter, as well as Dr. Robert Prescott for his critique of an earlier version of the manuscript. Thanks are also extended to Paul Moberg, research fellow supported by the Evanston Hospital, for his assistance in the preparation of the manuscript.

References

Adams, R., & Jenkins, R. (1981). Basic principles of the neuropsychological examination. In C. E. Walker (Ed.), *Clinical practice of psychology: A guide for mental health professionals*. New York: Pergamon Press.

Allen, M. (1985). Review of Millon Behavioral Health Inventory. In J. V. Mitchell, Jr. (Ed.), *The ninth mental measurements yearbook* (Vol. 1). Lincoln, NE: University of Nebraska Press.

American Psychiatric Association. (1987). *Diagnostic and statistical manual* (3rd ed. rev.). Washington, DC: Author.

Anastasi, A. (1988). *Psychological testing* (6th ed.). New York: Macmillan.

Barton, E., & Ascione, F. (1984). Direct observation. In T. Ollendick & M. Hersen (Eds.), *Child behavioral assessment: Principals and procedures*. New York: Pergamon Press.

Belar, C., Deardorff, W., & Kelly, K. (1987). *The practice of clinical health psychology*. New York: Pergamon Press.

Berg, R., Franzen, M., & Wedding, D. (1987). *Screening for brain impairment: A manual for mental health practice*. New York: Springer.

Bruner, K. (1942). Of psychological writing: Being some valedictory remarks on style. *Journal of Abnormal and Social Psychology, 37*, 52–70.

Campbell, D. & Hansen, J. (1981). *Manual for the SVIB-SCII Strong-Campbell Interest Inventory* (3rd ed.). Stanford, CA: Stanford University Press.

Connolly, A. J., Nachtman, W., & Pritchett, E. M. (1976). *KeyMath Diagnostic Arithmetic Test: Manual*. Circle Pines, MN: American Guidance Service.

Cronbach, L. (1984). *Essentials of psychological testing* (4th ed.). New York: Harper & Row.

Dahlstrom, W. (1985). The development of psychological testing. In G. Kimble & K. Schlesinger (Eds.), *Topics in the history of psychology*. Hillsdale, NJ: Erlbaum.

D'Amato, R., Dean, R., & Holloway, A. (1987). A decade of employment trends in neuropsychology. *Professional Psychology: Research and Practice, 18*, 653–655.

Derogatis, L., Lipman, R., & Covi, L. (1973). The SCL-90: An outpatient psychiatric rating scale. *Psychopharmacology Bulletin, 9*, 13–28.

Dubois, P. (1970). *A history of psychological testings*. Barton: Allyn and Bacon.

Elbert, J., & Holden, E. (1987). Child diagnostic assessment: Current training practices in clinical psychology internships. *Professional Psychology: Research and Practice, 18*, 587–596.

Exner, J. (1978). *The Rorschach: A comprehensive system* (Vol. 2). New York: Wiley.

Exner, J. (1986). *The Rorschach: A comprehensive system* (Vol. 1) 2nd ed.). New York: Wiley.

Exner, J., & Weiner, I. (1982) *The Rorschach: A comprehensive system* (Vol. 3). New York: Wiley.

Feinstein, A. (1985). *Clinical epidemiology: The architecture of clinical research*. Philadelphia: Saunders.

Frances, A. (1987). *DSM-III personality disorders: Diagnosis and treatment*. New York: Guilford Press.

Friedman, M., a& Rosenman, R. (1974). *Type A behavior and your heart*. New York: Knopf.

Gatchel, R., & Baum, A. (1983). *An introduction to health psychology*. Reading, MA: Addison-Wesley.

Golden, C., Hammeke, T., & Purisch, A. (1978). Diagnostic validity of a standardized neuropsychological battery derived from Luria's neuropsychological tests. *Journal of Consulting and Clinical Psychology, 46*, 1258–1265.

Golden, C., Kane, R., Sweet, J., Moses, J., Cardellino, J., Templeton, R., Vicente, P., Kennelly, D., & Graber, B. (1981). The relationship of the Halstead-Reitan Neuropsychological Battery to the Luria-Nebraska Neuropsychological Battery. *Journal of Consulting Psychology, 49*, 410–417.

Goldstein, G., & Hersen, M. (1984). Historical perspectives. In G. Goldstein & M. Hersen (Eds.) *Handbook of psychological assessment*. New York: Pergamon Press.

Goulet Fisher, D., Sweet, J., & Pfaelzer-Smith, E. (1986). Influence of depression on repeated neuropsychological testing. *International Journal of Clinical Neuropsychology, 8*, 14–18.

Greene, R. (1980). *The MMPI: An interpretive manual*. Orlando, FL: Grune & Stratton.

Hall, J. (1988). Records for psychologists. *Register Report: Newsletter for Psychologist Health Service Providers, 14*, 3–4.

Holmes, T., & Rahe, R. (1967). The social readjustment rating scale. *Journal of Psychosomatic Research, 11*, 213–218.

Incagnoli, T., Goldstein, C., & Golden, C. (1986). *Clinical application of neuropsychological test batteries*. New York: Plenum Press.

Jackson, C., & Wohl, J. (1966). A survey of Rorschach teaching in the university. *Journal of Projective Techniques and Personality Assessment, 30*, 115–134.

Jastak, S., & Jastak, G. (1984). *The Wide Range Achievement Test—Revised: Administration Manual*. Wilmington, DE: Jastak.

Jenkins, C., Roseman, R., & Friedman, M. (1967). Development of an objective psychological test for the determination of the coronary-prone behavior pattern in employed men. *Journal of Chronic Diseases, 20*, 371–379.

Johannson, C. (1982). *Manual for Career Assessment Inventory* (2nd ed.). Minneapolis: National Computer Systems.

Johnson, S., & Bolstad, O. (1973). Methodological issues in naturalistic observations: Some problems and solutions for field research. In L. Hammerlynck, L. Handy, & E. Mash (Eds.), *Behavior change: Methodology, concepts, and practice.* Champaign, IL: Research Press.

Kane, R., Sweet, J., Golden, C., Parsons, O., & Moses, J. (1981). Comparative clinical diagnostic accuracy of the Halstead-Reitan and Standardized Luria-Nebraska Neuropsychological Batteries. *Journal of Consulting and Clinical Psychology, 49,* 484–485.

Kazdin, A. (1977). Artifact, bias, and complexity of assessment: The ABCs of reliability. *Journal of Applied Behavior Analysis, 4,* 7–14.

Kellerman, H., & Burry, A. (1981). *Handbook of psychodiagnostic testing: Personality analysis and report writing.* New York: Grune & Stratton.

Kirsch, I., & Winter, C. (1983). A history of clinical psychology. In C. E. Walker (Ed.), *The Handbook of Clinical Psychology* (Vol. 1). Homewood, IL: Dorsey.

Klopfer, W. (1983). Writing psychological reports. In C. Walker (Ed.), *The handbook of clinical psychology* (Vol. 1). Homewood, IL: Dorsey.

Lanyon, R. (1985). Review of Millon Behavioral Health Inventory. In J. V. Mitchell, Jr. (Ed.), *The ninth mental measurements yearbook* (Vol. 1). Lincoln, NE: University of Nebraska Press.

Lee-Riordan, D., & Sweet, J. (1990). *Evaluation of the Millon Behavioral Health Inventory and the MMPI in the assessment of back pain patients.* Unpublished manuscript.

Lezak, M. (1983). *Neuropsychological assessment* (2nd ed.). New York: Oxford University Press.

Lubin, B., Larsen, R., & Matarazzo, J. (1984). Patterns of psychological test usage in the United States: 1935–1982. *American Psychologist, 39,* 451–454.

Lubin, B., Larsen, R., Matarazzo, J., & Seever, M. (1986). Selected characteristics of psychologists and psychological assessment in five settings. *Professional Psychology: Research and Practice, 17,* 155–157.

Markwardt, F. (1989). *Peabody Individual Achievement Test—Revised: Manual.* Circle Pines, MN: American Guidance Service.

Matarazzo, R., &b Matarazzo, J. (1984). Assessment of adult intelligence in clinical practice. In P. McReynolds & G. Chelune (Eds.), *Advance in psychological assessment* (Vol. 6). San Francisco: Jossey-Bass.

McReynolds, P. (1987). Lightner Witmer: Little-known founder of clinical psychology. *American Psychologist, 42,* 849–858.

Melton, G., Petrila, J., Poythress, N., & Slobogin, C. (1987). *Psychological evaluations for the courts: A handbook for mental health professionals and lawyers.* New York: Guilford Press.

Millon, T., Green, C., & Meagler, R. (1979). The MBHI: A new inventory for the psychodiagnostician in medical settings. *Professional Psychology, 10,* 529–539.

Newman, P., & Sweet, J. (1986). The effects of clinical depression on the Luria-Nebraska Neuropsychological Battery. *International Journal of Clinical Neuropsychology, 8,* 109–114.

Parsons, O. (1986). Overview of the Halstead-Reitan Battery. In T. Incagnoli, G. Goldstein, & C. Golden (Eds.), *Clinical applications of neuropsychological test batteries.* New York: Plenum Press.

Poon, L. (Ed.). (1986). *Handbook for clinical memory assessment of older adults.* Washington, DC: American Psychological Association.

Puente, A., & McCaffrey, R. (in press). *Handbook of neuropsychological assessment: A biopsychosocial perspective.* New York: Plenum Press.

Ritzler, B., & Alter, B. (1986). Rorschach teaching in APA-approved clinical graduate programs: Ten years later. *Journal of Personality Assessment, 50,* 44–49.

Rourke, B., Fisk, J., & Strang, J. (1986). *Neuropsychological assessment of children.* New York: Guilford Press.

Sarason, I., Johnson, J., & Siegel, J. (1978). Assessing the impact of life changes: Development of the Life Experiences Survey. *Journal of Consulting and Clinical Psychology, 467,* 932–946.

Schenkenberg, T., Peterson, D., Wood, D., & DaBell, R. (1981). Psychological consultation/liaison in a medical and neurological setting: Physician's appraisal. *Professional Psychology, 12,* 309–317.

Seretny, M., Dean, R., & Seretny, S. (1985, October). Second survey of the National Academy of Neuropsychology. Paper presented at the annual meeting of the National Academy of Neuropsychology, Philadelphia, PA.

Seretny, M., Dean, R., Gray, J., & Hartlage, L. (1986). The practice of clinical neuropsychology in the United States. *Archives of Clinical Neuropsychology, 1,* 5–12.

Shemberg, K., & Keeley, S. (1970). Psychodiagnostic training in the academic setting: Past and present. *Journal of Consulting and Clinical Psychology, 34,* 205–211.

Spitzer, R., & Endicott J. (1977). *Schedule for affective disorders and schizophrenia.* Technical Report. New York: New York State Psychiatric Institute.

Strassberg, D. S., Reimherr, F., Ward, M., Russell, S., & Cole, A. (1981). The MMPI and chronic pain. *Journal of Consulting and Clinical Psychology, 49,* 220–226.

Sweeney, J., Clarkin, J., & Fitzgibbon, M. (1987). Current practice of psychological assessment. *Professional Psychology: Research and Practice, 18,* 377–380.

Sweet, J. (1981). The MMPI in evaluation of response to treatment of chronic pain. *American Journal of Clinical Biofeedback, 4,* 121–130.

Sweet, J. (1983). Confounding effects of depression on neuropsychological testing: Five illustrative cases. *Clinical Neuropsychology, 5,* 103–109.

Sweet, J., & Moberg, P. (1990). A survey of practices and beliefs among ABPP and non-ABPP clinical neuropsychologists. *The Clinical Neuropsychologist, 4,* 101–120.

Sweet, J., & Rozensky, R. (1991). Professional relations. In M. Hersen, A. Kazdin, & A. Bellack (Eds.), *The clinical psychology handbook* (2nd ed.). New York: Pergamon Press.

Sweet, J., Breuer, S., Hazlewood, L., Toye, R., & Pawl, R. (1985). The Millon Behavioral Health Inventory: Concurrent and predictive validity in a pain treatment center. *Journal of Behavioral Medicine, 8,* 215–226.

Sweet, J., Tal, C., & Shain, M. (1985). *A survey of practices and beliefs in clinical neuropsychology.* Unpublished manuscript.

Tallent, N. (1976). *Psychological report writing.* Englewood Cliffs, NJ: Prentice-Hall.

Thorndike, R., Hagen, E., & Sattler, J. (1986). *Stanford-Binet Intelligence Scale (Fourth Ed.).* Chicago: Riverside.

vanReken, M. (1981). Psychological assessment and report writing. In C. Walker (Ed.), *Clinical practice of psychology.* New York: Pergamon Press.

Wade, T., & Baker, T. (1977). Opinions and use of psychological tests: A survey of clinical psychologists. *American Psychologist, 32,* 874–882.

Wade, T., Baker, T., Morton, T., & Baker, L. (1978). The status of psychological testing in clinical psychology: Relationships between test use and professional activities and orientations. *Journal of Personality Assessment, 42,* 3–10.

Wechsler, D. (1955). *Manual for the Wechsler Adult Intelligence Scale.* New York: Psychological Corp.

Wechsler, D. (1974). *Manual for the Wechsler Intelligence Scale for Children—Revised.* New York: Psychological Corp.

Wechsler, D. (1981). *Manual for the Wechsler Adult Intelligence Scale—Revised.* New York: Psychological Corp.

Wedding, D., & Williams, M. (1983). Training options in behavioral medicine and clinical neuropsychology. *Clinical Neuropsychology, 5,* 100–102.

Weiner, I., & Hess, A. (1987). *Handbook of forensic psychology.* New York: Wiley.

Wilcoxson, M., Zook, A., & Zarski, J. (1988). Predicting behavioral outcomes with two psychological assessment methods in an outpatient pain management program. *Psychology and Health, 2,* 319–333.

Woodcock, R. (1973). *Manual for the Woodcock Reading Mastery Tests.* Circle Pines, MN: American Guidance Service.

Woodcock, R., & Johnson, B. (1977). *Woodcock-Johnson Psycho-Educational Battery: Examiner's Manual.* Allen, TX: DLM Teaching Resources.

Wysocki, J., & Sweet, J. (1985). Identification of schizophrenic, brain-damaged, and normal medical patients using a brief neuropsychological screening battery. *International Journal of Clinical Neuropsychology, 7,* 40–44.

CHAPTER 19

Development of an Eating-Disorder Program

David L. Tobin, Craig Johnson, and Kevin Franke

Introduction

Paralleling the rise in incidence of anorexia nervosa and bulimia, there has been a proliferation of programs whose purpose is to provide treatment to patients with eating disorders. These programs present a wide range of treatment philosophies that reflect the diversity of presentation within this patient population. The circumscribed symptom patterns that identify bulimia and anorexia nervosa are often accompanied by a wide spectrum of concomitant symptoms that complicate the clinical picture. Thus, patients must undergo a thorough medical and psychological evaluation, and treatment programs must be prepared to meet a variety of patient needs. The purpose of this chapter is to provide a comprehensive model for treatment programs attempting to accomplish this goal.

Clinical Features

The clinical features of eating-disorder patients as described in the revised third edition of the American Psychological Association's *Diagnostic and Statistical Manual of Mental Disorders* (DSM-III-R; APA, 1987) reflect the following:

1. Anorexia nervosa is characterized by attempts to maintain body weight 15% below what is normally expected for a given age and height. Despite successful efforts to remain thin, patients maintain an intense fear of gaining weight or becoming fat, so that even the accumulation of a few pounds provokes tremendous anxiety. Perhaps most remarkable is the disturbance in body image, in which patients believe that they are fat even when they are emaciated. In addition, female patients must be amenorrheic for at least three months.

2. Bulimia is characterized by recurrent episodes of binge eating (at least twice weekly) followed by purging behavior that may include self-induced vomiting; the use of laxatives, diet pills, or diuretics; or rigorous dieting. During the eating binges, there is a feeling of lack of control over

David L. Tobin • Department of Psychiatry, University of Chicago, Chicago, Illinois 60637. **Craig Johnson** • Laureate Psychiatric Clinic and Hospital, Tulsa, Oklahoma 74147-0207, and Department of Psychiatry, Northwestern University Medical School, Chicago, Illinois 60611. **Kevin Franke** • Department of Psychiatry, Northwestern University Medical School, Chicago, Illinois 60611.

eating and a persistent overconcern with body shape and weight.

These symptoms are easily identified, and diagnosis of the two disorders is relatively straight-forward. In 180 consecutive intakes at our eating-disorders clinic, 139 patients met the DSM-III-R criteria for bulimia, and 29 patients met the DSM-III criteria for bulimia or the DSM-III-R criteria for atypical eating disorder. This latter group tended to demonstrate most but not all of the DSM-III-R criteria for bulimia. Thirteen patients met the DSM-III-R criteria for anorexia nervosa, and 1 of these patients also met the criteria for bulimia. Many bulimic patients who do not fully meet the criteria for anorexia also engage in considerable restrictive dieting. Only 7% of our patient sample had a primary diagnosis of anorexia nervosa, and 3% of the sample were male.

In addition to bulimic and anorexic symptoms, the clinical picture is also complicated by concomitant psychiatric symptoms that include but are not limited to affective disorders and personality disorders. Approximately half our patients present with concomitant depression and anxiety. Substance abuse is not uncommon, though it is probably underrepresented in our sample of patients because other programs in our geographic area do a better job with chemically dependent patients, and these patients tend to be screened out in the initial phone contact. Approximately 50% of our patients demonstrated evidence of a concomitant Axis II diagnosis, usually from personality disorder cluster II (Gwirtsman, Roy-Byrne, Yager, & Gerner, 1983; Johnson, Tobin, & Enright, 1989; Levin & Hyler, 1986). Patients occasionally present with symptoms of schizophrenia.

What is most interesting about these concomitant diagnoses is that there is absolutely no relationship between the severity of the eating symptoms and the level of depression or the presence of a personality disorder. Nonetheless, it is these concomitant psychiatric diagnoses that often determine the course of treatment and thus play a primary role in determining treatment decisions. For example, most

patients in our clinic who are free of personality disturbance or severe depression achieve significant reductions in eating symptoms within a year of entering treatment, whereas fewer than half of our patients with personality disorders or severe depression achieve such reductions. In addition, it has generally been our experience that patients who present with substance abuse and eating-disorder symptoms need to have their substance difficulties treated before attempting to manage their eating disorder. Thus, both clinical and programmatic success depends on successfully diagnosing and managing these concomitant symptoms.

Etiology and Epidemiology

Bulimia

Etiological models of bulimia have tended to reflect unidimensional hypotheses about disease onset. Hypotheses commanding significant attention in the literature are that bulimia is (1) a variant of affective disorder (e.g., Herzog, 1982; Hudson, Pope, Jonas, & Yurgelun-Todd, 1983; Lee, Rush, & Mitchell, 1985; Pope & Hudson, 1984); (2) a variant of obsessive-compulsive disorder (e.g., Rosen & Leitenberg, 1982; Williamson, Kelly, Davis, Ruggerio, & Veitia, 1985); (3) a variant of anorexia nervosa (e.g., Casper, Eckert, Halmi, Goldberg, & Davis, 1980; Garfinkel, Moldofsky, & Garner, 1980; Russel, 1979); or, finally, (4) a distinct diagnostic entity (e.g., Mitchell & Pyle, 1982). The empirical support for any one of these models has been mixed, so that investigators have speculated on less parsimonious explanations of disease onset.

Alternatively, investigators have suggested that bulimia is a heterogeneously determined disorder with psychological (e.g., Bruch, 1973; Goodsitt, 1984; Johnson & Connors, 1987; Sours, 1980), biological (e.g., Brotman, Herzog, & Woods, 1984; Hudson *et al.*, 1983; Hudson, Laffer, & Pope, 1982; Strober & Goldenburg, 1981; Walsh, Roose, & Glassman, 1983), familial (e.g., Humphrey & Stern, 1988; Johnson & Flach, 1985; Strober, Salkin, Burroughs, & Morrel, 1982), and sociocultural factors (e.g., Crandall, 1988; Garner, Garfinkel, & Olmstead, 1983; Garner, Garfinkel, Schwartz, & Thompson, 1980; Striegel-Moore, Silberstein, & Rodin, 1986) that interact (cf. Johnson & Connors, 1987).

These factors are combined in a biopsychosocial model that hypothesizes the following pathway to the development of bulimic symptoms (Johnson & Connors, 1987; Johnson & Maddi, 1986): (1) sociocultural values on thinness encourage restrictive dieting in young women (e.g., Garner *et al.*, 1983); (2) continued dieting leads to weight loss and also to a psychobiological impasse of excessive hunger (Johnson & Maddi, 1986; Williamson *et al.*, 1985); (3) dietary restraint breaks down, and binge eating occurs (Ruderman, 1985); (4) loss of control leads to anxiety about weight gain and purging behavior (Rosen & Leitenberg, 1982); (5) bingeing and purging lead to changes in self-efficacy, self-esteem, and levels of depression (Johnson & Maddi, 1986), which lead to (6) reinstatement of the dietary restraint.

As the sociocultural values that reinforce this sequence of behaviors affect most, if not all, young women, it is not surprising that most young women have tried to diet at some point in their lives. It is somewhat more striking that bingeing behaviors have been reported in as high as 90% and vomiting in as high as 16% of high school or college females (Halmi, Falk, & Schwartz, 1981; Johnson, Lewis, Love, Lewis, & Stuckey, 1984). A significant percentage of these females never develop bulimia; the estimates of prevalence suggest that approximately 5% binge-eat on a weekly or more frequent basis, and that 1% both binge-eat and purge (Johnson *et al.*, 1984; Pyle, Mitchell, Eckert, & Halverson, 1983). Females who develop bulimia are likely to demonstrate vulnerability in one or more of the above biological, psychological, systemic, or sociocultural risk factors. Thus, although a significant percentage of females who develop the disorder are also at risk for the development of affective illness or personality disorder, a sizable percentage of patients are relatively free of these complications. Only about 1% of bulimics are male, so that this subgroup is relatively unstudied at the present time.

Anorexia Nervosa

Etiological hypotheses about anorexia have also reflected both unidimensional and multifactoral models (for a review, see Garfinkel & Garner, 1982; Johnson, Thompson, & Schwartz, 1984). There is considerable literature concerning psychological and family influences (e.g., Bruch, 1973; Kog & Vandereycken, 1981; Palazzoli, 1974; Sours, 1980), as well as biological findings that include twin studies showing a higher rate of concordance among monozygotic than among dizygotic twins, a family incidence of psychiatric illness, and disturbances in hypothalamic-pituitary function (for a review, see Garfinkel & Garner, 1982). A cultural preoccupation with thinness obviously serves to increase the risk of developing this disorder, but anorexia has not increased in prevalence at the same rate as bulimia since the mid-1960s. Current estimates of prevalence are about 1% of young middle- to upper-class women (for a review, see Szmukler, 1985), with a much decreased incidence in lower socioeconomic classes. Only 5% of anorexics are male.

Perhaps because of the psychological characteristics needed to maintain dietary restraint below 15% of normal body weight or because of the psychophysiological effects of starvation (cf. Keys, Brozek, Henschel, Mickelson, & Taylor, 1950), anorexics have more consistency in their clinical presentation than do bulimics. In addition to an unrelenting drive to achieve thinness, Bruch (1973) described the typical anorexic as possessing (1) "a disturbance of delusional proportions in the body image"; (2) "a disturbance in the accuracy of the perception or cognitive interpretation of stimuli arising in the body"; and (3) "a paralyzing sense of ineffectiveness." These features are especially characteristic of anorexics who present shortly after the onset of puberty. Both clinical (e.g., Bruch, 1973) and empirical (Humphrey, 1986, 1987) findings suggest that a disturbance in the family system plays a crucial role in the etiology of this subgroup of anorexic patients. Anorexic patients who maintain the disorder to adulthood or whose symptoms do not appear until adulthood often present with rigid personality traits, such as obsessive-compulsive personality disorder, paranoid personality disorder, and borderline personality disorder (Johnson & Connors, 1987; Sours, 1980), suggesting a link between the presence of personality disturbance and the onset and maintenance of anorexic symptoms.

Even when patients do not fully meet DSM-III-R criteria for both bulimia and anorexia, it is quite common for patients to present with concomitant symptoms. It is extremely common, for example, to find bulimic patients who engage in extensive periods of restrictive dieting. In fact, excessive dieting is one of the forms of purging that can be used in making a diagnosis of bulimia. In summarizing the theoretical speculation about the etiology of these two disorders, we would suggest that they represent a final common pathway for biological, family, psychological, and sociocultural risk factors resulting in a wide spectrum of clinical presentations. Consequently, eating-disorder programs need to do a comprehensive intake evaluation that considers the heterogeneity of these patient groups.

Standardized Assessment

Although it is beyond the scope of this chapter to present the intake assessment in great detail, it is extremely important that programs pay particular care to this aspect of their clinical services (see Johnson & Connors, 1987, Chapter 8, for a more detailed description of this process). Based on the mistaken assumption that eating-disorder patients are homogeneous in their symptom presentation, some programs offer little more than an administrative screening for entry into standardized treatment protocols. This screening offers patients a pretense of expertise that may result in treatment failure and the mistaken assumption by patients that they are untreatable.

Given the relatively recent and limited empirical data base for eating-disorder patients, it is important to include an array of standardized measures in the intake evaluation. These measures provide a generalization of clinical experiences from other patient populations and promote the identification of patient characteristics that will affect the course of treatment. Our standardized assessment packet is divided into the following categories:

Eating Symptoms

The Diagnostic Survey for Eating Disorders— Revised (DSED-R)

The DSED-R (Johnson, 1984) is a standardized intake survey that focuses on various aspects of anorexia nervosa and bulimia. The questionnaire is divided into 12 sections, which provide information on demographic factors, weight history, body image, dieting behavior, binge-eating behavior, purging behavior, exercise, related behaviors, sexual functioning, menstruation, medical and psychiatric history, life adjustment, and family history. The survey can be used as a self-report instrument or as a semistructured interview guide.

Eating Disorders Inventory

The Eating Disorders Inventory (EDI; Garner, Olmstead, & Polivy, 1984) is a multiscale measure that assesses traits common in anorexia nervosa and bulimia, such as a drive for thinness, body dissatisfaction, and perfectionism.

Other Symptoms

SCL-90

The SCL-90 (Derogatis, Rickles, & Rock, 1976) is a widely used multidimensional symptom inventory that measures the following dimensions of psychopathology: somatization, anxiety, obsessive-compulsiveness, phobia, depression, hostility, paranoia, psychoticism, interpersonal sensitivity, and several indices of global distress.

Beck Depression Inventory

The BDI (Beck, Ward, Mendelson, Mock, & Erbaugh, 1961) is an easily administered and widely used measure of depression.

Personality Disorders

Borderline Syndrome Index

The BSI (Conte, Plutchik, Karasu, & Jerrett, 1980) measures characteristics thought to be central in borderline psychopathology, including poor impulse control, the absence of a consistent self-identity, and impaired object relations.

Structured Clinical Interview for DSM-III-R

The SCID-II Personality Questionnaire is a self-report instrument designed to assess DSM-III-R personality disorders (Spitzer & Williams, 1985).

Social Environment

Social Adjustment Scale

The SAS (Weissman, 1975) measures performance over the last two weeks in the areas of work, social activities, relationship with extended family, and roles as spouse, parent, and family member.

Family Environment Scale

The FES (Moos, 1974) measures the social environment characteristics of families in terms of relationship dimensions (e.g., cohesion and conflict), personal growth dimensions (e.g., independence and achievement orientation), and system maintenance dimensions (e.g., organization and control).

The Intake Interview

The intake interview is designed to be comprehensive in covering factors that have influenced the course of the disorder and that will affect the course of treatment. During the first five minutes, we attempt to establish a sense of collaboration with the patient and to explore the patient's expectations and concerns regarding the intake evaluation. Because the advantages of conducting the interview in a structured format appear to outweigh the disadvantages, we explain to patients our agenda: to assess their eating difficulties as thoroughly as possible in a relatively limited amount of time (1 to 1½ hours). The interview is roughly divided into the following sections.

Areas of Focus

Weight History

The structured interview begins with questions regarding the patient's current, highest, lowest, and desired weight. The answers often provide a historical record of how much weight preoccupations and fluctuations have affected the patient's self-esteem and life adjustment.

Body Image

Body image perception ranges from mild distortion to severely delusional and can reflect the patient's overall adjustment.

Dieting Behavior

It is important to know at what age a patient began dieting, the frequency of dieting attempts, the degree of restriction, the use of fad dieting techniques, and the general pattern of dieting behavior.

Binge Eating

The assessment of binge-eating behavior involves the macroassessment of the major life circumstances surrounding the onset of the behavior as well as the microassessment of the details of the patient's daily routine and the specific pattern of the binge-eating episodes. We are interested in knowing the onset, precipitants, duration, and frequency of the binges and the types of foods on which the patient binges.

Purging Behavior

In addition to the macro- and microassessment that we do for binge eating, we are also interested in the method of purging. The commonly used methods involve vomiting; the use of laxatives, diuretics, and diet pills; restrictive dieting; and excessive exercise. Though all of these behaviors are attempts to purge unwanted calories, different behaviors can serve somewhat different adaptive functions. For example, restrictive dieting may give a patient a more anorectic presentation, whereas the use of laxatives frequently adds a very self-punitive function to the purging behavior.

Medical Issues

In addition to assessing the adaptive function of the binge-purge cycle, it is important to evaluate the patient's physical condition for the medical complications of bulimia. All patients are expected to receive a complete physical on entering our treatment program. We facilitate the medical evaluation by having a nurse do a preliminary medical screening at the time of the psychological assessment and by referring patients to an internist who works closely with our program when patients do not have their own physician. Excessive dieting, binge eating, and purging influence endocrine function, disturb blood chemistry, destroy tooth enamel, irritate the esophagus, and disturb the gastrointestinal system. Severe medical complications make hospitalization the only treatment option.

Personality Disorder

It is important to evaluate the level of personality disturbance during the initial assessment, as the presence of a personality disorder forecasts a slow, difficult treatment (Johnson *et al.*, 1989). Some of this information can be gathered by history. Evidence of stormy, chaotic interpersonal relationships suggests that it may be difficult for a patient to engage in a helping relationship. The patient's approach to the interview is also an important source of information in assessing personality. For example, the level of the patient's mistrust during the intake interview may suggest the presence of personality disturbance.

To a certain degree, assessment in this area is guided by a theoretical orientation. For our more dynamically oriented clinic, we are most concerned about whether patients are capable of forming a stable, lasting relationship and can adequately test reality in their object relationships. Patients' potential for self-destructive behavior is also of great concern. Patients who have difficulties in these areas (and who are not psychotic) are likely to be diagnosed as having a borderline personality disorder (Adler, 1985; Johnson & Connors, 1987; Kernberg, 1988; Summers, 1988). Other patients may have stable relationships but have difficulty in identifying and working towards the treatment goals. These patients may look relatively undisturbed, as they can readily adapt to their external environment to meet the needs of others. They have difficulty, however, in meeting their own needs and have been described as having a "false self" or a narcissistic personality disorder (Goodsitt, 1984; Johnson & Connors, 1987; Kohut, 1971). A formal assessment battery composed of projective instruments (e.g., the Rorschach and the TAT) and objective instruments (e.g., the MMPI and the MCMI) can be very helpful in specifying the personality profile (see Sweet, Chapter 18), but some preliminary decisions about the level of personality disturbance must be made on the basis of the limited testing of our initial battery and the interviewer's impressions.

As the formation of a stable therapeutic alliance and goal setting is important to behaviorally oriented clinicians, they are also interested in diagnosing personality disorders. Behavioral psychologists tend to rely on the descriptive taxonomy of the third edition of the American Psychiatric Association's *Diagnostic and Statistical Manual* (DSM-III; APA, 1980) and DSM-III-R, which provide concrete guidelines for identifying personality disorders. The DSM-III-R offers a potentially more observable set of criteria for assessing the level of personality disturbance than previous diagnostic systems. There are a number of structured interviews and self-report instruments that are designed to diagnose DSM-III personality disorders. However, there is little empirical evidence to support their use (for review see Widiger & Frances, 1987).

The identification of personality disorders and character pathology is crucial to the disposition process. In a sample of 55 of our clinic patients with bulimia, 21 were identified as having borderline personality disorder, and 19 were diagnosed as being free of personality disorder. Although only 21% of the non-personality-disordered patients remained symptomatic at the end of one year, 62% of the borderline patients continued to meet the DSM-III-R criteria for bulimia. Thus, both the clinic and the patient must be prepared to make a long-term commitment to the treatment process. Non-personality-disordered patients may benefit from brief therapy; personality-disordered patients will probably not benefit from brief therapy.

Family Characteristics

Family history, family dynamics, and the patient's current level of family involvement can also play a crucial role in the onset and maintenance of eating-disorder symptoms. It is important to rule out the extent to which symptomatic behavior is promoted by the family system as well as to elicit information about communication patterns within the family that may influence treatment (Minuchin, Rosman, & Baker, 1978). We try to assess the family's cohesiveness, communication style, and method of conflict resolution and behavior control, as well as the role of the patient's symptoms within the family system.

As in the assessment of personality disturbance, the interview provides two important sources of information: the family history and the interview process itself. If the patient is an adolescent or a young adult living at home, we insist that the parents come to the evaluation. When family members refuse to attend the evaluation, we can infer a limited amount of family support for the patient's attempts to change symptomatic behavior.

General Level of Adaptive Functioning

In addition to obtaining information that is relatively specific to the onset and maintenance of eating-disorder symptoms, we are interested in assessing the patient's general level of functioning. This assessment overlaps with the level of personality disturbance, as has previously been discussed. Even when there is no sign of personality disturbance, we are interested in the extent to which patients are able to work, go to school, or engage in interpersonal relationships. We wish to examine their capacity to be alone and the extent to which time alone is overwhelming and provokes their disregulated eating patterns. Are the patients willing and motivated to change? What intellectual, emotional, and financial resources can they bring to treatment?

Disposition

The above is a description of our standard intake interview. The process takes approximately 1½–2 hours. At the end of the interview, the evaluator offers the patient a formulation and offers a recommendation for treatment. Sometimes, the patient needs to gather additional information (e.g., to explore family support for treatment), or the evaluator feels the need to seek consultation with the clinic staff. In proposing an initial treatment plan, the evaluator must weigh all the available information and attempt to match patient need with program activities. Sometimes, the fit between a given patient (e.g., a patient with both eating and substance abuse difficulties, who may benefit from a 12-step program) and our program is not good, and a referral to a program that more adequately meets the patient's needs must be made. As can be seen in the next section, the treatment interventions that are necessary for this patient population are quite varied, and it is difficult for a program to excel in every area.

Clinical Intervention

Following a thorough assessment of all the systems (i.e., physiological, intrapsychic, familial, marital, interpersonal, and sociocultural) that have contributed to the onset or maintenance of a person's eating-disorder symptoms, the treatment team should assess the adaptive function of the symptoms. Restrictive dieting, low weight, and bingeing and purging may be used by the patient in many different ways: (1) to bolster self-esteem; (2) to regulate dysphoric affects; or (3) to resolve intrapsychic and interpersonal conflicts. The long-term impact of these symptoms is quite debilitating and can be fatal, and the symptoms themselves must be treated as aggressively as possible. Approaches to helping patients manage their symptoms include cognitive-behavioral techniques (e.g., Fairburn, 1984; Garner & Bemis, 1984), psychoeducational techniques (e.g., Connors, Johnson, & Stuckey, 1984), self-management techniques (e.g., Kanfer & Scheft, 1986), and relapse prevention techniques (e.g., Marlatt & Gordon, 1978).

Although some patients can make use of these direct attempts to manage symptoms, others cannot. We believe, therefore, that it is important to understand the origins and the nature of the psychological problems that predisposed the patient to the development of the symptomatic behavior and to be able to convey this understanding to the patient in the context of relationship-oriented therapy (e.g., Bruch, 1973; Goodsitt, 1984; Johnson & Connors, 1987; Sours, 1980). The successful program must be able to integrate psychodynamic and cognitive-behavioral approaches to treatment (Johnson, Connors, & Tobin, 1987).

Understanding the adaptive function(s) of the symptoms also aids the treatment team in selecting the type and the intensity of the interventions that will be necessary. What follows is a description of the kinds of interventions that have been demonstrated to be effective with eating-disordered patients.

Ambulatory Care

Psychoeducational Group Therapy

Groups should be composed of from five to nine members and a therapist or a cotherapy team. These groups are most useful for bulimics who are 18–30 years old, who have high motivation for change, who are at or near normal weight, and who do not have other significant diagnoses, such as major depression or personality disorder (Johnson & Connors, 1987). Groups for more severely disturbed patients can also be successful when used in conjunc-

tion with other treatments (e.g., individual therapy and hospitalization). Psychoeducational groups help patients to understand the antecedents of their symptoms (e.g., what triggers a binge), to develop normal eating patterns, and to develop alternative coping strategies (Garner, Rockert, Olmstead, Johnson, & Coscina, 1984; Johnson, Connors, & Stuckey, 1983). The intervention strategies include self-monitoring, goal setting, education about nutrition and the consequences of restrictive dieting, and challenging irrational beliefs about thinness. Most groups are time-limited (12–14 weeks) and are conducted in 90-minute sessions. The groups help patients overcome the shame that often accompanies eating-disorder symptoms and help them to make use of the social supports. Patients are able to mobilize more quickly when they learn that they are not alone and that they are not "bad" or "weak" because they have these eating-disorder symptoms.

Individual Psychotherapy

Individual psychotherapy is used to treat both eating symptoms and the underlying psychological and emotional problems. Patients who do not have significant character pathology and who are capable of rapidly forming a working alliance with the therapist can make use of educational, behavioral, and cognitive techniques such as those mentioned above to reduce and eliminate bingeing and purging rather quickly (in one to four months) (Fairburn, 1984). Eliminating restrictive eating in low-weight bulimics and anorexics usually requires a longer, more dynamically oriented course of individual therapy in which maladaptive patterns of attachment are addressed. These low-weight bulimic and anorexic patients usually present with more rigid defenses against experiencing the dysphoric affect that arises whenever they begin to develop intimacy with another person. For example, extreme restriction of food intake and related problems of preoccupation with food, weight, and social isolation in the anorexic may represent the patient's effort to use the multiple defenses of regression, isolation of affect, and withdrawal to avoid the inevitable fear and pain they would experience in bringing their emotional needs into a relationship with another person (Sugarman, Quinlan, & Devenis, 1981; Sugarman & Kurash, 1982). For the bulimic, bingeing and purging may be a concrete representation of a central conflict concerning how to be true to one's own feelings and thoughts and to be related to a needed other at the same time. The bulimic patient may be convinced that, in order to preserve the approval of others, she or he must not feel or express certain needs or feelings. In this context, bingeing and purging become a way of regulating what the bulimic believes are unacceptable impulses or moods (e.g., Goodsitt, 1984; Swift & Letvin, 1984). Without intensive dynamic therapy, these issues are likely to remain unresolved.

Nutritional Counseling

Anorexic and bulimic patients have often based their eating and dieting habits on inaccurate information or incorrect assumptions about food, digestion, and weight gain or loss. For example, many patients believe that muscle tissue turns into fat in the absence of a vigorous exercise regimen or that any fat that is consumed becomes a permanent part of the body. These false assumptions maintain and strengthen the unhealthy eating patterns. Counseling by a nutritionist, or by a well-informed nurse or therapist, is an important component in the overall confrontation of the patient's illogical assumptions about food and weight. For example, patients can be given specific meal plans to follow. They can be guided in how often they weigh themselves as a protection against the day-to-day fluctuations of body weight and can be provided a reasonable expectation of what their normal body weight should be.

Medical Monitoring

The role that medical monitoring plays in the recovery process depends on the type and severity of the symptoms. Any patient who has not had a thorough medical examination within the past year should be advised to do so.

Patients with very low weight, obese patients, patients who binge and purge multiple times a day, patients whose laxative abuse is severe or prolonged, and patients who have a medical illness such as diabetes that may interact with the eating-disorder symptoms to create potentially life-threatening emergencies should be seen regularly by an internist familiar with eating disorders.

It is most helpful if the treatment program cultivates a close working relationship with one physician or a small group of physicians who have time in their practice and an interest in eating disorders. Ideally, this physician(s) should be on-site during the initial evaluation and available for consultation. For example, when deciding if a patient with severe symptomatology needs hospitalization, it is particularly helpful to be able to assess medical risk. For a review of the medical complications of eating disorders, the reader is urged to see Zucker (1989).

Psychopharmacological Treatment

Some anorexic and bulimic patients have major depressive episodes or panic attacks and chronic debilitating anxiety. Medication can sometimes be useful in helping these patients manage their dysphoric moods so they can participate more effectively in other components of their treatment plan. Although the link between bulimia and major affective disorder remains unclear, there is some evidence that a significant percentage of bulimic patients develop bulimia secondary to the occurrence of unipolar or bipolar affective disorder (Glassman & Walsh, 1983; Hudson et al., 1982). The biological symptoms of major depression suggest the strong possibility that pharmacological treatment may be an important component of treatment.

Family and Marital Therapy

Under certain conditions it may be crucial to involve the identified patient's family in the treatment (Humphrey & Stern, 1988). The family may undermine the patient's progress if the patient's symptoms serve an adaptive function for the family and if the family system lacks the resources and support it needs to allow the patient to give up her or his symptoms (Roberto, 1986; Strober & Humphrey, 1987). The need for ongoing family sessions is suggested (1) when the patient lives with the family of origin; (2) when the patient demonstrates little evidence of true emotional separation from the family; (3) when there is another serious problem in the family (i.e. alcoholism, or physical or sexual abuse); and (4) if the patient requires hospitalization. Treatment goals include interventions (1) to reorganize dysfunctional relationships and structures; (2) to better tolerate important affects; (3) to encourage individuation; and (4) to resolve conflicts between family members.

Similar to work in the family of origin, a patient's spouse can facilitate both symptom management and emotional healing. Working with the marital couple can help each partner identify the intrapsychic and interpersonal experiences that trigger eating-disorder symptoms. Once these "triggers" are identified, alternative coping strategies can be developed with the couple working in partnership.

Self-Help and Other Support Group Meetings

Learning new attitudes about self and others and learning new behaviors that obviate the need for eating-disorder symptoms often require more support than is currently present in the patient's family system or social milieu. Support from the patient's therapist, spouse, or family members notwithstanding, support from others struggling with the same symptoms and underlying emotional issues can often facilitate recovery. Group support helps patients process day-to-day stressors and increases their understanding of the links between their feelings, their attitudes, and their food-related behaviors. Not all patients can benefit from these groups, however. Patients with severe ego deficits may not find enough structure in peer-led self-help groups and may disturb the support function of such groups for other patients.

The National Association of Anorexia Nervosa and Associated Disorders (ANAD), a non-profit corporation based in Highland Park, Illinois, has organized a self-help group program called Applying New Attitudes and Directions (ANAD). The organization has published guidelines for setting up and running ANAD self-help groups. There are many such groups in many parts of the country and anyone interested in recovering from an eating disorder is free to participate. Information on where to find groups can be obtained from local hospitals or by contacting the ANAD's national headquarters, 312-931-3438 (Meehan, Wilkes, & Howard, 1984).

Hospital Care

Under certain conditions, ambulatory care may be insufficient to get a patient started on the road to recovery. These conditions may include (1) additional Axis I or Axis II diagnoses that preclude the patient from using the behavioral, cognitive, or relationship interventions offered in outpatient treatment; (2) medical complications from severe or prolonged eating-disorder symptomatology that endanger the patient's life; or (3) evidence of self-injurious or suicidal behavior (Johnson & Connors, 1987, pp. 195–225). When one or more of these conditions exist, more intensive interventions such as hospitalization or partial hospitalization (i.e., day or evening treatment programs) can be used. The program must be able to provide continuity in the treatment in order to provide quality care. When patients are accepted into our inpatient unit, it is usually with the understanding that they will be followed in our outpatient program. Without continuity, the hospitalization is undermined.

Inpatient Milieu

The primary goals of inpatient treatment should be (1) to normalize the patients' eating habits and weight and (2) to lay the groundwork for ongoing recovery following discharge to outpatient treatment. The first goal is accomplished by having each patient move through a structured, well-monitored eating protocol, in which patients assume increasing responsibility for their nutritional needs. The second goal is accomplished by helping patients to identify self-destructive behaviors and by using the holding capacity of the milieu to reinforce norms of self-care and to work through intense negative transferences. The milieu quickly provides patients with the environment they need to learn the difference between acting out and working through their feelings. The holding capacity of the milieu disrupts the self-defeating cycle of acting out, followed by negative feedback (i.e., you are "bad" or "sick"), followed by more shame and self-hate, followed by more acting out, and so on.

Partial Hospitalization

As very little has been written about day or evening hospital programs, the following comments reflect the clinical experience of the authors. There is a subgroup of patients for whom inpatient hospitalization is too restrictive and may interfere with the maintenance or development of ego strengths and coping behavior, but for whom psychotherapy once or twice a week is insufficient. Placement in a day or evening hospital program can sometimes be used as an alternative to inpatient hospitalization when the patient needs or wishes to continue working, going to school, or living at home during treatment. A highly structured evening or day hospital program can fill the need for an intensive treatment without disrupting the social, occupational, or educational functioning that occurs with inpatient admission. However, the powerful affects that are evoked in attempts to refeed some anorexic patients or in attempts to help a bulimic eliminate the use of laxatives) may be too anxiety-provoking to manage in even a partial hospital program.

Placement in a partial hospital program can also be used as part of a discharge plan for hospitalized patients, as they can be most vulnerable to symptom relapse when they leave the structure of the inpatient milieu. The partial program becomes a transitional space in which the former inpatient can continue working through the interpersonal and intrapsychic dynamics that were brought into focus in the hospital milieu.

Developing and Maintaining Programs

Choosing and Training Staff

As can be seen from the previous sections, eating-disorder programs must have a multidisciplinary health care team in order to address the needs of this patient population. There needs to be expertise in cognitive-behavioral therapy, psychodynamic psychotherapy, family therapy, nutrition, pharmacotherapy, inpatient milieu, and medical care. Therapists must be flexible in their orientation and must be able to combine intervention strategies (e.g., combined cognitive-behavioral–psychodynamic psychotherapy and family systems work). The team must know enough about the other specialties in the program to communicate with other team members and with the medical community at large. Acquiring such a staff often requires that experts in one area be hired and then cross-trained in other areas (e.g., a cognitive-behavioral therapist trained in nutrition and psychodynamic psychotherapy). Because of the diverse needs of these patients, the clinic director needs to have specialized training in eating disorders.

Sources of Referral

The foundation of our program and the most important marketing tool for patient referrals consists of (1) the program's setting in a medical school; (2) the broad clinical and academic training of the program staff, much of which has taken place in the medical school; and (3) clinical and academic community involvement, on a local, national, and international level. The setting in a medical school is advantageous not only because of the recognition the program receives in the community, but also because we are able to offer the wide range of adjunct services that eating-disorder patients often require. Thus, we are considered a comprehensive tertiary-care facility. Because of the specialized expertise of the clinic staff, interactions with potential sources of referral are likely to be positive and to promote the likelihood of an ongoing relationship and continued referrals. Part of this expertise involves knowing when patients would be most appropriately referred to another program or encouraged to remain in treatment with their current therapist. Finally, there is a firm commitment in our clinic to sharing our experience with other clinicians and with the academic community. Thus, we are very involved in teaching and research, both in and outside the program, and the resulting recognition and relationships promote patient referrals. Although we advertise in the local media from time to time, the bulk of our referrals come from the professional community.

Programs that are in the early stages of formation do not have the marketing advantages of a mature program, but they can nevertheless endeavor to emulate the above-mentioned program characteristics. Program affiliation with either a medical school or some other strong community medical setting promotes a referral base and facilitates the provision of comprehensive services. The hiring of staff that has trained in a medical setting facilitates integration into a medical setting. And finally, if program staff are not yet able to participate nationally as leaders of the scientific and clinical advances in the treatment of eating disorders, they can participate as consumers on a national level and can provide expertise to the local community. Giving lectures and doing workshops in community organizations and schools alert potential clients and their families to their need for treatment as well as to the availability of program services.

Educating Physicians

The role of the specialized program in the larger medical community often involves the education of practitioners in other specialties. This takes time and courtesy but is necessary both to promote potential referral sources and to guide other medical practitioners in the spe-

cial needs of eating-disorder patients. For example, the adequate monitoring of endocrine functions requires knowledge of the impact that bingeing and purging can have on hormone levels and may determine the physician's choice of whether or not to implement pharmacological therapy. These decisions are often very complex, and the best possible scenario involves establishing ongoing relationships with other medical specialties, perhaps by bringing a practitioner into the program on a formal basis or by arranging a more informal relationship that is, nevertheless, consistent.

Many medical specialties are not equipped to handle the behavioral difficulties of patients with bulimia or anorexia. Although the typical medical patient presents with physical symptoms that can be treated with surgical or pharmacological interventions, our patients present with many behaviors that sabotage the physician's attempts to cure. These behaviors may present themselves as noncompliance with the medical regimen (e.g., being unable to stick to the prescribed diet) or as the undoing of medical care (e.g., vomiting up medications). Eating-disorder patients may demand inordinate amounts of time and energy and may be immensely frustrating to practitioners who do not specialize in their treatment. They may also demand inordinate amounts of time from practitioners who do specialize in their treatment, but at least these practitioners expect frustration and know the guidelines for structuring the treatment to maximize patient compliance. By educating practitioners, the eating-disorder specialist can be enormously helpful as consultant to the general medical community.

Program Evaluation and Research

Because the high morbidity of bulimia is a relatively recent phenomenon, many early programs grew out of an interest in research questions centered on the epidemic rise of this symptom presentation. Thus, in many mature programs, there is strong commitment to integrating research on and clinical involvement with patients. Developing programs can also benefit from this strategy.

Research questions can be used to make

both programmatic and individual treatment decisions. Data that are clinically relevant can be collected in the normal course of an intake evaluation. This information can then be used both as a tool for making treatment disposition and as a starting point for an ongoing evaluation of the patient's progress in the program. Annual evaluations of eating symptoms and of psychological and social adjustment can be conducted to monitor treatment progress and document the clinical course of eating-disorder patients.

For example, it is clear that some bulimic patients benefit from brief psychoeducational group therapy, whereas other patients are unable to achieve symptomatic improvement for many years despite intensive individual psychotherapy. Thus, it would be valuable to identify patient characteristics predicting treatment outcome. Asking this and other important questions in a formal, empirical fashion encourages program involvement in the academic and clinical community and strengthens the clinical foundation of the program. For examples of clinical and academic programs that have found ways to integrate research and practice the reader is invited to see the following journals: the *International Journal of Eating Disorders*, the *Journal of Consulting and Clinical Psychology*, the *Journal of Abnormal Psychology*, the *Journal of Clinical Psychiatry*, and the *American Journal of Psychiatry*.

Patient Flow through the Clinic: Case Example

Susan was a 24-year-old, single white woman who was urged by her brother to contact our clinic for treatment of her bulimic symptoms. Susan was bingeing and purging twice a day and taking laxatives on a daily basis. She was 5′ 5″ tall, weighed 106 pounds, and maintained this low weight by her persistent vomiting.

Susan came to the initial evaluation with her mother and her brother. She reported being depressed and having difficulty sleeping and concentrating on her job as a secretary. She was not suicidal at the time of her phone call. She lived at the home of her parents, who did not seem to notice Susan's difficulties.

Susan had been anorexic when she was 14 years old for about nine months but had recovered without treatment. At the age of 16, Susan had a difficult time with a boyfriend who said something about her weight, provoking Susan into a pattern of restrictive dieting that she had maintained. She began bingeing and purging when she was 17 after attending a cheerleading camp in the summer. Her bingeing and purging was exacerbated when she left for college, and she was unable to finish her freshman year of college and returned to her parents' home.

Though the above description is a fictionalized account, it is a typical presentation and represents many of the parameters already discussed in the chapter. By *typical*, we mean that many of the questions that are salient to any intake evaluation are represented in the case. The severity of the patient's bingeing and purging behavior, her low weight, the absence of any food retention, the laxative abuse, and the history of depression would make hospitalization the likely treatment of choice. Although the appearance of any single set of symptoms may not indicate hospitalization (e.g., bingeing and vomiting twice a day), the combination of symptoms would make outpatient treatment a difficult endeavor. Medical complications might not appear in this patient, despite the extensive purging, laxative abuse, and low weight. Nevertheless, the patient's extensive reliance on restrictive dieting and purging behavior to organize her day would make response to outpatient interventions questionable. The rebound edema from discontinuing laxatives is likely to provoke extreme anxiety, which can extinguish attempts to manage symptoms. Even if the eating symptoms are successfully targeted, it is quite possible that removal of the patient's eating symptoms would unmask a severely depressed mood and would increase the risk of suicidal behavior.

In responding to such a patient's request to be involved with the clinic, we would collaborate with her to examine her motivation, her resources, and her commitment to treatment at this time. We would also assess her willingness to come into the hospital and her financial resources. Attempts would be made to involve this patient's family in treatment, beginning with the initial evaluation. We would want to know if her family actively supported treatment, would stand passively in the wings, or would attempt to sabotage the patient's efforts.

In this hypothetical account, Susan's brother and mother seemed concerned and motivated to do anything they could, including family therapy. Susan was very reluctant to consider coming into the hospital but was very interested in pursuing outpatient treatment. The evaluator was able to clarify for the family that this was a crisis in the patient's life and for the family, and that only under a very clear contract would the clinic consider treatment on an outpatient basis. The patient would have to followed medically by the clinic internist, and to come in twice a week for individual psychotherapy and once a week for family therapy. A specific contract would be arranged regarding progress with the patient's symptoms, and hospitalization would be mandated if this progress was not seen.

As can be seen in the above description, a tremendous amount of work takes place during the initial phone call and intake evaluation. Though not every diagnostic question can be answered in the initial evaluation, we usually feel confident in initiating a treatment plan (e.g., hospitalization versus some form of outpatient care). When a patient is brought into the hospital, formal testing is ordered to clarify the patient's dynamics in order to rapidly mobilize the inpatient milieu. Outpatient therapists typically have more time to construct a dynamic formulation and so may or may not request a formal assessment.

The emergence of the patient's family as a source of support was unanticipated during the phone screening and proved to be the glue that held this outpatient intervention together. Resistance by the family would certainly have mandated that the evaluator insist on hospitalization, and the stage was set to provide hospitalization should the outpatient trial prove unsuccessful. We feel that specialized programs need to be able to provide this level of sophisticated assessment.

Conclusions

The rapid increase in the incidence of anorexia and bulimia make these patients an attractive target for program development. The seemingly homogenous presentation of bulimic and anorexic symptoms has facilitated their identification in both the clinical literature and the popular culture. Consequently, it is easy to market services to these patients. Unfortunately, their seeming diagnostic homogeneity may make treatment strategies also appear homogeneous. It is our experience that such mistaken assumptions lead to treatment failures and staff frustration and affect the viability of the program.

As can be seen from this chapter, we believe that developing an eating-disorder program is very complicated. The overriding complication results from the diversity of the concomitant psychopathology in these patient groups. Programs must provide expertise in a wide range of services to meet a wide range of patient needs. Inpatient and ambulatory care, the evaluation and treatment of medical difficulties, nutritional education, symptom management, relationship-oriented treatment, and pharmacological therapy are the diverse components of the comprehensive program. Though specialization in the treatment of eating disorders is an essential component of a successful program, the program staff must be broadly trained and must have experience with a wide range of psychiatric difficulties in order to meet the heterogeneous needs of these patients.

References

Adler, G. A. (1985). Borderline psychopathology and its treatment. Northvale, NJ: Jason Aronson.

American Psychiatric Association. (1980). *Diagnostic and statistical manual of mental disorders* (3rd ed.). Washington, DC: Author.

American Psychiatric Association. (1987). *Diagnostic and statistical manual of mental disorders* (3rd ed. rev.). Washington, DC: Author.

Beck, A. T., Ward, C. H., Mendelson, M., Mock, J., & Erbaugh, J. (1961). An inventory for measuring depression. *Archives of General Psychiatry, 5,* 561–571.

Brotman, W. W., Herzog, D. G., & Woods, S. W. (1984). Antidepressant treatment of bulimia: The relationship between bingeing and depressive symptomology. *Journal of Clinical Psychiatry, 45,* 7–9.

Bruch, H. (1973). *Eating disorders: Obesity, anorexia nervosa, and the person within.* New York: Basic Books.

Casper, R. C., Eckert, E. D., Halmi, K. A., Goldberg, S. C., & Davis, J. M. (1980). Bulimia: Its incidence and clinical importance in patients with anorexia nervosa. *Archives of General Psychiatry, 37,* 1030–1035.

Connors, M. E., Johnson, C. L., & Stuckey, M. K. (1984). Treatment of bulimia with brief psychoeducational group therapy. *American Journal of Psychiatry, 141,* 1512–1516.

Conte, H. R., Plutchik, R., Karasu, T. B., & Jerrett, I. (1980). A self-report borderline scale: Discriminative validity and preliminary norms. *Journal of Nervosa and Mental Disease, 168,* 428–435.

Crandall, C. S. (1988). The social contagion of binge eating. *Journal of Personality and Social Psychology, 55,* 589–599.

Derogatis, L. R., Rickles, K., & Rock, A. F. (1976). The SCL-90 and the MMPI: A step in the validation of a new self-report scale. *British Journal of Psychiatry, 128,* 280–289.

Fairburn, C. G. (1984). Cognitive behavioral treatment for bulimia. In D. M. Garner & P. E. Garfinkel (Eds.), *Handbook of psychotherapy for anorexia nervosa and bulimia* (pp. 160–192). New York: Guilford Press.

Garfinkel, P. E., & Garner, D. M. (1982). Anorexia nervosa: A multidimensional perspective. New York: Brunner/Mazel.

Garfinkel, P. E., Moldofsky, H., & Garner, D. M. (1980). The heterogeneity of anorexia nervosa: Bulimia as a distinct subgroup. *Archives of General Psychiatry, 37,* 1036–1040.

Garner, D. M., & Bemis, K. M. (1984). Cognitive therapy for anorexia nervosa. In D. M. Garner & P. E. Garfinkel (Eds.), *Handbook of psychotherapy for anorexia nervosa and bulimia* (pp. 107–146). New York: Guilford Press.

Garner, D. M., Olmstead, M. P., & Polivy, J. (1984). *Eating Disorder Inventory Manual.* Psychological Assessment Resources.

Garner, D., Garfinkel, P., Schwartz, D., & Thompson, M. (1980). Cultural expectations of thinness in women. *Psychological Reports, 47,* 483–491.

Garner, D. M., Garfinkel, P. E., & Olmstead, M. (1983). An overview of sociocultural factors in the development of anorexia nervosa. In P. L. Darby, P. E. Garfinkel, D. M. Garner, & D. V. Coscina (Eds.), *Anorexia nervosa: Recent developments in research* (pp. 65–82). New York: Alan R. Liss.

Garner, D. M., Rockert, W., Olmstead, M. P., Johnson, C., & Coscina, D. V. (1984). Psychoeducational principles in the treatment of bulimia and anorexia nervosa. In D. M. Garner & P. E. Garfinkel (Eds.), *Handbook of psychotherapy for anorexia nervosa and bulimia* (pp. 513–572). New York: Guilford Press.

Glassman, A. H., & Walsh, B. T. (1983). Link between bulimia and depression unclear. *Journal of Clinical Psychopharmacology, 3,* 203.

Goodsitt, A. (1984). Self-psychology and the treatment of anorexia nervosa. In D. M. Garner & P. E. Garfinkel (Eds.), *Handbook of psychotherapy for anorexia nervosa and bulimia* (pp. 55–82). New York: Guilford Press.

Gwirtsman, H. E., Roy-Byrne, P., Yager, J., & Gerner, R. H. (1983). Neuroendocrine abnormalities in bulimia. *American Journal of Psychiatry, 140,* 559–563.

Halmi, K. A., Falk, J. R., & Schwartz, E. (1981). Binge-eating, and vomiting: A survey of a college population. *Journal of Psychological Medicine, 11,* 697–706.

Herzog, D. B. (1982). Bulimia: The secretive syndrome. *Psychosomatics, 23,* 481–487.

Hudson, J. I., Laffer, P. S., & Pope, H. G., Jr. (1982). Bulimia related to affective disorder by family history and response to the dexamethasone suppression test. *American Journal of Psychiatry, 137,* 695–698.

Hudson, J., Pope, H., Jr., Jonas, J., & Yurgelun-Todd, D. (1983). Family history study of anorexia nervosa and bulimia. *British Journal of Psychiatry, 142,* 133–138.

Humphrey, L. L. (1986). Structural analysis of parent-child relationships in eating disorders. *Journal of Abnormal Psychology, 95,* 395–402. Humphrey, L. L. (1987). Comparison of bulimic-anorexic and nondistressed families using structural analysis of social behavior. *Journal of the Academy of Child and Adolescent Psychiatry, 26,* 248–255.

Humphrey, L. L., & Stern, S. (1988). Object relations and the family system in bulimia: A theoretical integration. *Journal of Marital and Family Therapy, 14,* 337–350.

Johnson, C. (1984). The initial consultation for patients with bulimia and anorexia nervosa. In D. M. Garner & P. E. Garfinkel (Eds.), *Handbook of psychotherapy for anorexia nervosa and bulimia* (pp. 19–51). New York: Guilford Press.

Johnson, C., & Connors, M. (1987). *Bulimia nervosa: A biopsychosocial perspective.* New York: Basic Books.

Johnson, C., & Flach, A. (1985). Family characteristics of 105 patients with bulimia. *American Journal of Psychiatry, 142,* 1321–1324.

Johnson, C. L., & Maddi, K. L. (1986). The etiology of bulimia: A bio-psycho-social perspective. *Annals of Adolescent Psychiatry, 13,* 253–273.

Johnson, C., Connors, M., & Stuckey, M. (1983). Short-term group treatment of bulimia. *International Journal of Eating Disorders, 2,* 199–208.

Johnson, C., Lewis, C., Love, S., Lewis, L., & Stuckey, M. (1984). Incidence and correlates of bulimic behavior in a female high school population. *Journal of Youth and Adolescence, 13,* 15–26.

Johnson, C., Thompson, M., & Schwartz, D. (1984). Anorexia nervosa and bulimia: An overview. In W. J. Burns & J. V. Lavigne (Eds.), *Progress in pediatric psychology.* New York: Grune & Stratton.

Johnson, C., Connors, M., & Tobin, D. L. (1987). Symptom management of bulimia. *Journal of Consulting and Clinical Psychology, 56,* 668–676.

Johnson, C., Tobin, D. L., & Enright, A. (1989). Prevalence and clinical characteristics of borderline patients in an eating disordered population. *Journal of Clinical Psychiatry, 50,* 9–15.

Kanfer, F. H., & Scheft, B. K. (1986). Self-management

theory and clinical practice. In N. S. Jacobsen (Ed.), *Cognitive and behavior therapists in clinical practice*. New York: Guilford Press.

Kernberg, O. F. (1988). Object relations theory in clinical practice. *Psychoanalytic Quarterly, 57*, 481–505.

Keys, A., Brozek, J., Henschel, A., Mickelson, O., & Taylor, H. L. (1950). *The biology of human starvation* (Vol. 1). Minneapolis: University of Minnesota Press.

Kog, E., & Vandereycken, W. (1981). Family characteristics of anorexia nervosa and bulimia: A review of the research literature. *Clinical Psychology Review, 5*, 159–180.

Kohut, H. (1971). *The analysis of the self*. New York: International Universities Press.

Lee, N. F., Rush, A. J., & Mitchell, J. E. (1985). Bulimia and depression. *Journal of Affective Disorders, 9*, 231–238.

Levin, A. P., & Hyler, S. E. (1986). DSM III personality diagnosis in bulimia. *Comparative Psychiatry, 27*, 47–53.

Marlatt, G. A., & Gordon, J. R. (1978). Determinants of relapse: Implications for maintenance of behavior change. In P. Davidson (Ed.), *Behavioral medicine: Changing health lifestyles* (pp. 410–452). New York: Brunner/Mazel.

Meehan, V., Wilkes, N. J., & Howard, H. L. (1984). *Applying new attitudes and directions*. National Association of Anorexia Nervosa and Associated Disorders, Inc., Box 271, Highland Park, IL.

Minuchin, S., Rosman, B. L., & Baker, L. (1978). *Psychosomatic families: Anorexia nervosa in context*. Cambridge: Harvard University Press.

Mitchell, J. E., & Pyle, R. L. (1982). The bulimic syndrome in normal weight individuals: A review. *International Journal of Eating Disorders, 1*, 61–73.

Moos, R. (1974). *Family Environment Scale Manual*. Palo Alto, CA: Consulting Psychologists Press.

Palazzoli, M. S. (1974). *Self-starvation*. London: Chaucer.

Pope, H. G., Jr., & Hudson, J. I. (1984). *New hope for binge eaters*. New York: Harper & Row.

Pyle, R. L., Mitchell, J. E., Eckert, E. D., & Halverson, P. A. (1983). The incidence of bulimia in freshman college students. *International Journal of Eating Disorders, 2*, 75–85.

Roberto, L. G. (1986). Bulimia: The transgenerational view. *Journal of Marital and Family Therapy, 12*, 231–240.

Rosen, J., & Leitenberg, H. (1982). Bulimia nervosa: Treatment with exposure plus response prevention. *Behavior Therapy, 13*, 117–124.

Ruderman, A. (1985). Restraint, obesity, and bulimia. *Behavior Research and Therapy, 23*, 151–156.

Russel, G. (1979). Bulimia nervosa: An ominous variant of anorexia nervosa. *Psychological Medicine, 9*, 429–448.

Sours, J. A. (1980). *Starving to death in a sea of objects*. New York: Jason Aronson.

Spitzer, R., & Williams, J. (1985). *Structured interview of DSM-III-R personality disorders*. New York: Biometrics Research Department, New York State Psychiatric Institute.

Striegel-Moore, R. H., Silberstein, L. R., & Rodin, J. (1986). Towards an understanding of risk factors for bulimia. *American Psychologist, 41*, 246–263.

Strober, M., & Goldenburg, I. (1981). Ego boundary disturbance in juvenile anorexia nervosa. *Journal of Clinical Psychology, 37*, 433–438.

Strober, M., & Humphrey, L. L. (1987). Familial contributions to the etiology and course of anorexia nervosa and bulimia. *Journal of Consulting and Clinical Psychology, 55*, 654–659.

Strober, M., Salkin, B., Burroughs, J., & Morrel, W. (1982). Validity of the bulimia-restrictor distinction in anorexia nervosa. *Journal of Nervous and Mental Disease, 170*, 345–351.

Sugarman, A., & Kurash, C. (1982). The body as a transitional object in bulimia. *International Journal of Eating Disorders, 1*, 57–67.

Sugarman, A., Quinlan, D., & Devenis, L. (1981). Anorexia nervosa as a defense against anaclitic depression. *International Journal of Eating Disorders, 1*, 44–61.

Summers, F. (1988). Psychoanalytic therapy of the borderline patient: Treating the fusion-separation contradiction. *Psychoanalytic Psychology, 5*, 339–355.

Swift, W. J., & Letvin, R. (1984). Bulimia and the basic fault: A psychoanalytic interpretation of the bingeing-vomiting syndrome. *Journal of the American Academy of Child Psychiatry, 23*, 489–497.

Szmukler, G. I. (1985). Review: The epidemiology of anorexia and bulimia. *Journal of Psychiatric Research, 19*, 113–120.

Walsh, B. T., Roose, S. P., & Glassman, A. H. (1983). *Depression and eating disorders*. Paper presented at the annual meeting of the American Psychiatric Association, Los Angeles.

Weissman, M. M. (1975). The assessment of social adjustment. *Archives of General Psychiatry, 32*, 357–364.

Widiger, T. A., & Frances, A. (1987). Interviews and inventories for the measurement of personality disorders. *Clinical Psychology Review, 7*, 49–75.

Williamson, D. A., Kelly, M. L., Davis, C. J., Ruggerio, L., & Veitia, M. C. (1985). The psychophysiology of bulimia. *Advances in Behavior Research and Therapy, 7*, 163–172.

Zucker, P. (1989). Medical complications of bulimia. *Journal of College Student Psychotherapy, 3*, 27–40.

CHAPTER 20

Integration of Clinical Psychology into Adult and Pediatric Oncology Programs

Steven M. Tovian

Introduction

This chapter is about the integration of clinical psychology into adult and pediatric oncology programs. Oncology programs in medical settings are important areas of practice for clinical psychologists because of the numerous psychosocial problems secondary to cancer. This chapter will identify these psychosocial problems, offer guidelines for assessment and intervention, and discuss the development and maintenance of the clinical psychology program within the cancer treatment center.

Cancer is understood to be a disease of cells proliferating with disregard to the body's regulatory signals (DeVita, Hellman, & Rosenberg, 1985). This proliferation causes tumors, or neoplasms, which are either benign or malignant (cancerous). The malignant tumors invade surrounding tissue, organs, and organ systems and/or give rise to metastases, or secondary growths, to other parts of the body.

Cancer, the second most common cause of

death in the United States, kills about 494,000 Americans annually. For children aged 3 to 14 and women aged 30 to 54, cancer is the leading cause of death. Overall, cancer death rates have declined slightly in the past 25 years, except for lung cancer, which is rising in incidence and mortality in both men and women. Recent estimates indicate that 985,000 Americans develop cancer annually (Silverberg & Lubera, 1988).

Cancer affects every age group, from newborns to the elderly, although more than one half of all cancer deaths occur in persons over 65. From age 20 through 40 cancer is more common in women than in men, but between ages 60 and 80 more cancers occur in men. Overall, more men than women die of cancer. Despite these statistics, one of three persons diagnosed is cured (DeVita *et al.*, 1985).

It is estimated that, in 1985, cancer claimed about 1,700 children's lives from 6,000 diagnosed cases (Silverberg & Lubera, 1988). There are important differences between childhood and adult cancer. The most common cancers of adults, including lung, breast, colorectal, and skin cancer, are uncommon in children. Children develop malignancies of rapidly growing body systems such as the central nervous sys-

Steven M. Tovian • The Evanston Hospital and Northwestern University and Medical School, Evanston, Illinois 60201.

tem and connective tissue. Another difference is that childhood cancer is usually a highly malignant disease that has a significant impact on a growing and developing individual with immature organ systems and emotional development (DeVita *et al.*, 1985).

Psychosocial Issues Associated with Cancer

Becoming a "cancer patient" results in facing the possibility of a disabling illness, pain, altered goals, roles, plans, body disfigurement, separation from family, alienation of friends and peers, loss of or change in physical functioning, and death. Cancer and its medical treatment are stressful life events (Burish & Lyles, 1983). Inherent in the illness and the treatment are changes in appearance and physical function, reduction in the sense of personal control, and a threat to one's life. Each forms a serious challenge to an individual's self-concept (Kaufman, 1989; Wellisch & Cohen, 1985). Holland and Rowland (1989) provide an in-depth discussion of the psychological care of the patient with cancer.

Quality of Life

The changes in physical functioning and comfort, psychological and social functioning, and a person's satisfaction with himself or herself are multifactorial constructs that help to define quality of life (Mulhern, Horowitz, Ochs, Friedman, Armstrong, Copeland, & Kuhn, 1989). The disease itself can threaten the quantity of life, and the changes and distress brought on by the diagnosis and treatment of cancer can threaten the quality of life (Rozensky, 1983). Psychologists, by virtue of their professional expertise and their role on the oncology team, are involved in assessment and interventions along quality-of-life dimensions.

Cancer involves many distinct malignant diseases, each affecting a different organ or organ system. Depending on the type, location (site), and extent (stage) of the individual's cancer, the therapy may include surgery, chemotherapy, radiation therapy, hyperthermia treatment, or a combination of these modalities, concurrently or sequentially. All adult and pediatric patients experience a disruption of normal activities during the course of diagnosis and treatment. Treatment effects such as alopecia (hair loss), nausea, vomiting, mouth sores, anorexia, pain, and malaise may be associated with either chemotherapy or radiation therapy; pain may be associated with hyperthermia treatments; and altered body image and function may be associated with surgery. The time of onset and the duration of such effects are determined by the specific therapeutic modality used. Other psychological problems are determined by unique interactions of the cancer, its treatment, and premorbid demographic and psychosocial features such as age, gender, and family constellation and functioning, as well as the psychological functioning and personal coping resources manifested by the patient before the diagnosis of cancer (Meyerowitz, Heinrich, & Schag, 1983; Mulhern *et al.*, 1989).

Therefore, cancer can pose a wide variety of psychosocial problems, acute and chronic, for both the adult patient (e.g., Andrykowski & Redd, 1987; Friedenbergs, Gordon, Hibbard, Levine, Wolf, & Diller, 1981–1982; Gordon, Friedenbergs, Diller, Hibbard, Wolf, Levine, Lipkins, Ezrachi, & Lucido, 1980; Holleb, 1988; Lehman, DeLisa, Warren, deLateur, Bryant, & Nicholson, 1978; Wellisch, Landsverk, Guidera, Pasnau, & Fawzy, 1983) and the pediatric patient (e.g., Carpenter & Onufrak, 1984; Koocher, 1986; Koocher & O'Malley, 1981; S. B. Lansky, List, & Ritter-Sterr, 1986; Van Dongen-Melman, Pruyn, Van Zanen, & Saunders-Woudstra, 1986; Varni & Katz, 1987). For example, in a study of 570 adult cancer patients, Wellisch *et al.* (1983) found the following prevalence rates: somatic side effects (83%); mood disturbance (44%); medical equipment problems (34%); impaired family relationships (32%); cognitive impairment (27%); treatment compliance problems (26%); financial difficulties or worries (20%); role difficulties in the family (17%); and problems with body image (8%).

Koocher (1985) identified points of emotional vulnerability in childhood cancer at which strategic interventions are appropriate: receiving the diagnosis; the onset of treatment; aversive side effects from treatment; reaching the end of a prescribed treatment phase; reentry into school, family, work, and social life; recurrence or relapse; the terminal phase; and anniversary phenomena or life marker events for the survivor. The association between possible psychosocial issues and the phase of cancer is presented in Table 1.

Developmental Perspective

The age of the cancer patient has an important effect on the type of assessment and intervention used by the psychologist.

Childhood

The majority of the literature on childhood cancer is family-focused; a minority is focused on the child alone. Specifically, the emphasis is on the family's meeting the child's needs. The psychologist working in pediatric oncology needs to consider several factors. The presenting psychosocial problem will be primarily reactive to the cancer and its treatment. In addition, the psychologist must consider the child's developmental level in understanding the concepts of separation, illness, death, and mourning, as well as the child's defense mechanisms, coping styles, and fantasies, both positive and negative (e.g., illness as a punishment for bad behavior). There is the need for the child to have predictable daily routines for the family to maintain a cohesive unit to reduce the child's anxiety and depression. The use of play therapy should be used as an individual means of expressing anger, hostility, and so on. During remission, it may be difficult for parents to treat their child normally, and management problems, possibly involving overprotectiveness, overpermissiveness, and overindulgence, may result (Goldberg & Tull, 1983; Wellisch, 1981). In addition, the psychologist may need to maintain close liaison with school personnel on such issues as cognitive changes secondary to the disease process and the treatment effects that may affect learning and neuropsychological functioning (Copeland, Fletcher, Pfefferbaum-Levine, Jaffe, Reid, & Maor, 1985). In

Table 1. Phase of Cancer and Possible Psychosocial Issues[a]

Prediagnostic	Constant overconcern with the possibility of having cancer
	Denial of the disease's presence and delay in seeking treatment
Diagnostic	Shock, disbelief
	Initial and partial denial
	Anxiety
	Anger, hostility
	Depression
Initial treatment	
Surgery	Grief reaction to changes in body image
	Postponement of surgery (avoidance)
	Search for nonsurgical alternatives
Radiation therapy	Fear of x-ray machines and side effects
	Fear of abandonment during treatment
Chemotherapy	Fear of side effects
	Conditioned emotional response
	Changes in body image
	Altruistic feelings
	Anxiety, isolation
Follow-up	Return to normal coping patterns
	Fear of recurrence
Recurrence and retreatment	Shock, disbelief, denial
	Anxiety
	Anger
	Depression
Disease progression	Frenzied search for new information, consultants, alternative and unproven cures
Terminal palliation	Fear of abandonment
	Loss and anger
	Anticipatory mourning
	Acceptance by patient and survivors

[a]Adapted from L.C. Rainey, D. K. Wellisch, F. I. Fauzy, D. Wolcott, and R. O. Pasnau (1983).

such cases, the role of neuropsychological evaluation and referral of the patient (child or adult, when relevant) is important. Specific psychosocial problems related to childhood and adolescence are presented in Table 2.

Adolescence

In work with the adolescent, the intervention issues may be similar to the child's. However, there are unique developmental features inher-

**Table 2. Psychosocial Problems
in Childhood and Adolescence[a]**

Infant and toddler
  Damage to the developing CNS
  Lack of opportunity to practice motor and interpersonal
    skills
  Separation-related anxieties
Preschooler
  Damage to the developing CNS
  Lack of interaction with other children
  Lack of cognitive stimulation
  Feelings of guilt
  Lack of understanding of the disease
School-aged children
  Learning disabilities
  Absence from school
  Diminished social interaction
  Parental overprotectiveness and lack of discipline
  Sibling rivalry
Adolescent
  Perceived alienation of peers
  Compromised intellectual and physical well-being
  Absence from school
  Anxiety and uncertainty about the future
  Parental overprotection and lack of expectations
  Negative risk-taking behavior

[a]Adapted from S. K. Maul-Mellott & J. N. Adams (1987). *Childhood Cancer: A Nursing Overview.* Boston: Jones & Bartlett.

ent in adolescence, involving the establishment of autonomy (emotional and economic) from the parents; psychosexual development, including the acceptance of new sexual roles, new peer relations, and new social responsibilities; and the development of a future orientation, including preparation for an occupation and for marriage. Besides these unique developmental features, the psychologist may need to anticipate power struggles between the adolescent and the medical staff. The resolution of such power struggles may require including the adolescent in as many treatment decisions as possible, so that the adolescent is an active participant in his or her own care. The psychologist may also need increased sensitivity to the assault on psychosexual identity by treatment effects (e.g., loss of hair and possible sterility). This assault may require more honest and open communications between treatment providers and the adolescent (Wellisch, 1981).

Adulthood

There are significant developmental landmarks, such as marriage, the presence of children, and retirement, that characterize adulthood. Mages and Mendelsohn (1980) discussed the effects of cancer in adulthood based on a life-stage developmental approach. The authors divided adulthood into three stages: young adult (ages 18–35); midlife (ages 36–55); and older adult (ages 56–75). Different adaptive tasks are inherent in each developmental stage, which partly determine the modes of experiencing cancer and the personal resources available for coping. For example, the effects of sterility in testicular cancer are different for a 25-year-old male who is engaged to be married and was planning to have children and for a 48-year-old male with two grown children. This life-stage developmental approach provides a framework for patients to define reactions to cancer in terms of who they are, what they have experienced, and what they hope for the future.

Biopsychosocial Model

An important model proposed by several authors (e.g., Andrykowski & Redd, 1987; Barofsky, 1981; Belar, Deardorff, & Kelly, 1987) for understanding the complex psychosocial problems experienced by the cancer patient involves a biopsychosocial orientation. According to this model, the patient's behavior and physical symptoms are related to (1) the disease process itself; (2) the treatment modalities; (3) the patient's interpersonal milieu; and (4) the environment in which the medical treatment is received. Such a model lends itself well to a precise format that can be used in any evaluation. This format is summarized in Table 3.

Psychological and physical symptoms may occur as a result of the cancer itself or as side effects of treatments, mostly chemotherapy and radiation. For example, Table 4 lists psychological conditions associated with chemotherapy. Medical management of these symptoms is a primary concern of the oncologist. The psychologist should be in close consulta-

Table 3. Problem Format from Biopsychosocial Model

Behavior
  Isolation and withdrawal
  Loss of control
  Self-management
  Noncompliance
Affect
  Depression
  Anxiety
  Anger
  Malaise
Somatic conditions
  Nausea and vomiting
  Pain
  Sexual dysfunction
  Nutritional problems
  Alopecia (hair loss)
  Mouth sores
  Obesity
  Anorexia
  Cachexia
Interpersonal conditions
  Communication with family
  Communication with medical staff
Cognitive
  Cognitive impairment
  Lack of knowledge about the disease
  Health beliefs
  Perceived meaning of the illness
  Attitudes and expectations about the illness, the health
    care system

tion with the oncologist in assessing whether these symptoms are directly affected by psychosocial factors and can be targets of psychosocial interventions (Andrykowski & Redd, 1987).

In Table 4, it can be seen that depression is a common iatrogenic effect of chemotherapy. Depressive symptoms can also be a result of disease-related factors such as hypercalcemia, liver failure, endocrine imbalances, tumor-produced psychoactive hormonal substances, decreased blood-oxygen levels, and central nervous system metastases (Andrykowski & Redd, 1987; Goldberg & Tull, 1983). Medical factors should be considered first when treating psychological symptoms such as depression or anxiety. Depression that stems from psychosocial origins is characterized by more overt cognitive symptoms (e.g., expressions of guilt, helplessness, and hopelessness), intermittent symptomology (e.g., periodic expressions of normal mood), an absence of neurovegetative symptoms (e.g., lethargy, insomnia, and anorexia), and no previous or family history of a major affective disorder (Andrykowski & Redd, 1987).

Clinical Assessment

Goals

The goals of psychological assessment in oncology are to provide the psychologist with an understanding of the patient in his or her physical and social environment; the patient's relevant psychological strengths and weaknesses; the evidence of psychopathology; the response to the disease and the treatment regimen; and an identification of the coping skills being used (Belar *et al.*, 1987). The psychologist would do well to follow the adaptive tasks that must be accomplished by any medical patient (Moos, 1977). The psychologist should be aware of how the patient is coping with pain, incapacitation, and other symptoms. The patient's coping style in the hospital and in response to special treatment procedures, as well as how the patient is developing and maintaining adequate relations with the health care staff, is important. Whether the patient is maintaining a reasonable emotional balance, preserving a satisfactory self-image, and maintaining a sense of competence and mastery should be assessed. Also, whether the patient is preserving relationships with family and friends and how the patient is preparing for an uncertain future need to be determined.

Assessment of the adult and pediatric oncology patient is often an ongoing process that depends on the referral question posed to the psychologist. The extent of the referral question depends on the role of the psychologist in the oncology program. From this author's experience at a cancer care center in a hospital, a majority of the referrals involve psychological and psychophysiological reactions to cancer

Table 4. Possible Psychological Reactions Associated with Chemotherapy Agents[a]

Agent	Nature of reaction
Steroids	
Prednisone	Depression, irritability, euphoria, psychosis
Dexamethasone (Decadron)	Depression, irritability, euphoria, psychosis
Methylprednisone (Medrol)	Depression, irritability, euphoria, psychosis
Alkylating agents	
Dacarbazine	Depression
Hexamethylmelamine	Depression
Mechlorethamine (Mustargen)	Delirium, toxic encephalopathy
Cyclophosphamide (Cytoxan)	Transient delirium
Antimetabolites	
Methotrexate	Dementia, depression
Fluorouracil	Delirium, ataxia
Enzymes	
L-Asparaginase	Confusion, depression, paranoia, psychosis
Antibiotics	
Mithramycin	Agitation, delirium
Vinca alkaloids	
Vinblastine	Agitated depression
Vincristine (Oncovin)	Alterations in sensations, muscle weakness, depression
Vindesine	Muscle weakness
Miscellaneous agents	
Tamoxifen citrate	Weight gain, depression
Procarbazine hydrochloride	Depression, confusion
Mitoxantrone	Fatigue
Interferon	Anorexia, weight loss
Procarbazine	Delirium
Mitotane	Delirium, severe depression
Cyclophosphamide + Methotrexate + 5-Fluorouracil	Agitated depression
5-Azacytidine	Depression

[a]Adapted from R. Goldberg and R. M. Tull (1983); D. K. Wellisch and R. S. Cohen (1985).

and cancer treatment. Secondary types of referrals involve the somatic effects of psychological distress and the psychological complications of cancer. The parameters for possible referrals for psychological intervention are summarized in Table 5. These parameters can be used in suggesting to medical and nursing staff when to make referrals (Rainey, Wellisch, Fawzy, Wolcott, & Pasnau, 1983).

Methods of Assessment

Clinicians should not be wedded to one method or process of assessment, although a good clinical interview is the core method. Multimeasure methods, including formal observations and patient diaries, with convergent hypothesis testing are often most appropriate

(Belar *et al.*, 1987). Traditional, broad-based psychological assessment instruments used in personality assessment (e.g., the MMPI, the Rorschach, and the TAT) are used only when referral questions concerning psychosomatic factors are involved and a past history of psychiatric dysfunction is suspected (Barofsky, 1981).

Interview

The interview can elicit current and historical data across the biopsychosocial areas. The interview can also be used to develop a supportive working relationship with the patient, the family, and significant others. It is important for the psychologist to understand his or her own stimulus value with cancer patients when

Table 5. Guidelines for Possible Referrals for Psychological Intervention[a]

1. The patient's emotional response interferes with his or her ability to seek appropriate treatment or to cooperate with the necessary procedures.
2. The patient's emotional response causes greater distress than the disease itself or increases disease-related impairment.
3. The patient's emotional response interferes with his or her activities of daily living at home, work, or school.
4. The patient's emotional response results in a curtailing of his or her usual sources of gratification.
5. The patient's emotional response results in personality or behavioral disorganization and is so severe and grossly inappropriate that it results in the misinterpretation and distortion of environmental events.
6. The patient has a history of psychiatric disorders, substance abuse, or suicide attempts.
7. The patient has limited social support.

[a]Adapted from D. K. Wellisch and R. S. Cohen (1985).

interpreting interview data (Belar *et al.*, 1987). Depending on the referral question, the preference of the psychologist, the time constraints, and the mental status of the patient, interviews may be structured, semistructured, or unstructured.

Goldberg and Tull (1983) provided an excellent example of a data base and semistructured interview for cancer patients and significant others. The interview questions should include information about how the patient (or a significant other) became aware of the cancer and what were the factors or people that influenced the decision to seek a medical diagnosis. The patient's ideas about the cause of the cancer are crucial. Information about the patient's (or the family's) reactions to the diagnosis is also very important. How the diagnosis was explained and what the patient's (or the family's) thoughts and feelings were in response to the diagnosis are questions necessary to ask in the interview. Further information about who in the family was most affected and about whether or not everyone in the patient's family is aware of the diagnosis is important. Questions about how the patient (or a significant other) feels he or she will do with cancer, as well as about the patient's view of the future, should be included,

along with whether the illness has been discussed with the patient's employer or teacher and what the reactions have been at work or at school or with friends. Questions identifying how relationships with friends and family have changed and whether the patient knows others with cancer and how they managed should be included. Identifying what has been the biggest problem(s) since diagnosis is crucial. For example, have financial concerns been important? Questions about the importance of religion in the patient's life and whether religious or philosophical beliefs have changed as a result of cancer need to be included. Identifying major crises or loses in the patient's life before the cancer diagnosis and noting how the patient managed or coped with these can offer predictions about how the patient may respond in the future. Inquiries about the personal and family history of substance abuse and/or psychological disturbance requiring treatment are recommended. Finally, questions should be asked about how treatment has been at this particular center and about whether the patient (or the family) has thoughts about seeking other treatments.

In assessing family functioning, Welch-McCaffrey (1988) recommended that interviews assess family structure (e.g., how the family is organized and what patterns characterize their interactions), family boundaries (e.g., who is allowed to participate in the family system and whether they will permit outside intervention), and family function (e.g., what the process is by which the family operates). The latter involves assessing family communication patterns, the environment where communication takes place, the health status of the other members (including their resources and the proximity of family members), whether unresolved problems existed before the cancer diagnosis, and whether the roles within the family are assigned or assumed. Wellisch and Cohen (1985) identified variables that characterize high-risk families and that often require referral for therapeutic intervention: minimal contacts with organizations or groups in the community; inflexible role relationships; an imbalance of power in making decisions; and low autonomy within family relationships resulting in inhibited personal growth for the members.

Questionnaires

There have been a number of attempts to develop narrowly focused questionnaires normed on adult cancer populations. However, given the heterogeneity of pediatric cancer, it is difficult to obtain, even at larger institutions, sufficiently large disease- or treatment-specific samples to permit controlled studies of the factors affecting both the quality of life and psychosocial reactions to cancer in children (Mulhern *et al.*, 1989). The Play Performance Scale for Children (PPSC: L. L. Lansky, List, Lansky, Cohen, & Sinks, 1985) has been used with pediatric oncology patients but requires further validational studies (Mulhern *et al.*, 1989). The Quality of Life Index (Spitzer, Dobson, Hall, Chesterman, Levi, Sheperd, Battista, & Catchlove, 1981) and the Functional Living Index for Cancer (Schipper & Levitt, 1985) are two quality-of-life measures normed on adult cancer populations; they measure physical well-being, physical abilities, emotional states, social support, and future outlook.

The Cancer Inventory of Problem Situations (CPIS; Ganz, Rofessart, Polinsky, Schlag, & Heinrich, 1986; Schag, Heinrich, & Ganz, 1983) is a questionnaire containing 144 statements describing problems that adult cancer patients may face daily. The problems are divided into subscales that include functional health, difficulty communicating with medical staff, chemotherapy-related problems, communications at work, cognitive problems, body image, weight maintenance, communication with family and friends, compliance, anxiety in medical situations, and sexual dysfunction. The Psychological Adjustment to Illness Scale—Self Report (PAIS-SR; Derogatis, 1986) has been recently normed on lung cancer and mixed-diagnosed adult cancer patients (Derogatis, 1989). The PAIS-SR is a 46-item questionnaire developed to reflect adjustment in seven psychosocial domains; health care orientation, work, family, sexual relationships, extended family relationships, social and leisure areas, and psychological distress. The Profile of Mood States (POMS) has also been recently normed on adult cancer patients on entering treatment protocols for various cancer diagnoses (Cella, Tross, Orav, Holland, Silberfarb, & Rafla, 1989). The POMS is a 65-item adjective checklist developed to measure the following moods: tension, depression, anger, vigor, fatigue, and confusion. Morrow (1986) developed a questionnaire, the Morrow Assessment of Nausea and Emesis, that assesses nausea and emesis in adult patients undergoing chemotherapy. The questionnaire can serve as a pre- and postmeasure of change when specific behavioral interventions are used for this problem. (These interventions are discussed in greater detail later in this chapter.) According to the biopsychosocial model discussed previously, specific variables such as depression, anxiety, and pain, for example, can be assessed by existing and validated questionnaires reviewed by Belar *et al.* (1987).

The purpose of a clinical assessment of the oncology patient is to understand the patient and his or her problem so as to develop a treatment strategy or a management decision. The psychologist need not be wedded to a particular assessment strategy. With the interview as the core technique, supported by any of the previously mentioned questionnaires, flexibility in approaches may be an asset in view of the complexity of the psychosocial variables involved.

Clinical Interventions

Goals

A goal of any clinical intervention is to enable the patient to function physically, socially, and emotionally at the highest level possible within the constraints of the disease and the treatment. Interventions can also enable the patient to make adaptive decisions regarding the treatment, to mobilize and potentiate her or his support system, to increase her or his locus of personal control, and to mourn losses (e.g., dreams, plans, goals, relationships, physical functioning, and body parts). Finally, clinical interventions can help the patient to make realistic plans for the future and can assist the patient in issues around terminality and mortality, if appropriate (Wellisch & Cohen, 1985).

Koocher (1985) identified additional goals for intervention with children and adolescents, such as assisting the child and the family to anticipate events that may or may not happen. This anticipation may involve preparing for hospitalization or stressful medical procedures (e.g., rehearsal through play, desensitization, and stress inoculation). Unfamiliarity can elicit anxiety in children.

Providing educational material about cancer geared to the child's own language, developmental stage, and unique viewpoint is important for both children and their families. Koocher recommended actively soliciting the child's feedback about what he or she understands about the illness and the treatment. Finally, interventions should emphasize open communication and should increase, sustain, and enhance the child's or adolescent's internal locus of control. Increasing locus of control may include allowing the child or the adolescent an increased role in scheduling activities or treatments in the inpatient milieu as well as in discharge planning.

Special Considerations

Clinical interventions with cancer patients require understanding of the patient's age and type of problem. Interventions also require flexibility compared to traditional interventions in clinical psychology. Patients and their families may need to be seen at regular intervals incorporating planned, and at times unplanned, interruptions, depending on the stage of illness, the treatment regimen, and the patient's needs. The psychologist often needs to be prepared for relatively rapid changes in the physical appearance of patients (e.g., hair loss and weight loss) after the resumption of treatment. The psychologist can also expect outpatient cancellations due to low stamina, low blood counts, and periodic days of poor health. The duration of many sessions in the course of an intervention may have to be reduced for

these same reasons. Also, the scheduling of outpatient appointments are best made well after chemotherapy sessions.

The psychologist can also expect shorter duration in inpatient sessions due to patient fatigue, the schedule of numerous medical tests, and visitors. Scheduling inpatients can be difficult because patients are often off the unit for radiation treatments or diagnostic tests. It is best to schedule inpatient sessions with the primary nurse or the unit secretary and then with the patient. The psychologist should also be prepared to see inpatients, at time, partially clothed. Touching patients, including gently holding a patient's hand for reassurance or helping a patient in and out of bed, differs dramatically from established norms in traditional clinical psychology interventions (Belar et al., 1987).

Finally, cancer can be a terminal illness. The reader is advised to review the psychological literature on terminal illness (e.g., Rando, 1984; Sobel, 1981; Worden, 1982). The dynamics of death involve not only the patient, but the patient's family and friends and the medical staff as well. Grief and mourning processes go well beyond the point of death, and the psychologist can be helpful to family, friends, and staff during this period. The psychologist should recognize how his or her own unresolved feelings regarding death may influence countertransference and the delivery of services to the dying cancer patient.

Treatment

Individual

Wellisch (1981) noted that the overall goals of individual psychotherapy with the cancer patient are to strengthen defenses, enhance coping skills, and reduce the fear of isolation. The author cautioned against taking too passive a stance in treatment and interpreting or reducing patient defenses, because they protect the patient from deep, primitive fears associated with mortality and death. Wellisch and Cohen (1985) recommended that the therapist assume a more active, more supportive, and more self-revealing stance with the cancer patient than in more traditional individual psychotherapy. Weisman (1979) reinforced the active use of empathy, the clarification of patient responses, open-ended questioning, and respect for the patient's resistance. Weisman, Worden, and Sobel (1980) took an active, direct individual psychotherapeutic approach that includes the identification of primary affect, training in problem solving, relaxation training, and inquiry into existential issues.

Gordon *et al.* (1980) compared the effects of psychosocial counseling with groups of patients with three kinds of cancer (breast cancer, lung cancer, and melanoma) after participation in a systematic program of psychosocial rehabilitation versus a no-treatment control group. The results of the study suggest that those patients in counseling showed a more rapid decline in negative affect scores (e.g., anxiety and hostility) and a more realistic outlook on life than the control subjects. The treatment group also showed a greater (but statistically nonsignificant) number of subjects who returned to work. Compared with controls, the treatment group also showed a pattern of more active use of time. The authors also found that lung cancer patients needed the most psychosocial care, and the melanoma patients the least. The authors suggested that cancer is not a unitary psychological problem, and that the stage of illness should be assessed for its unique effect on each patient.

Feinstein (1983) thoroughly reviewed the literature on individual psychotherapy with cancer patients and identified eight variables in cases in which improvement in disease status has been associated with individual psychotherapeutic intervention. The variables include altering stress conditions and their management; working through unresolved grief; stimulating the will to live; promoting realistic, positive expectations; mobilizing psychological capacities for psychophysiological control; constructively handling denial; increasing appropriate emotional expression; and strengthening specific personal traits of patients who go into remission. Feinstein concluded that more research is needed to match specific in-

terventions with each variable and to investigate whether changes in a variable influence the course of cancer.

The effect of emotions on the etiology and progression of cancer is an important issue that emerges frequently in working with the cancer patient. This is most likely a result of the popular support of such books as *Getting Well Again: A Step-by-Step Guide to Overcoming Cancer for Patients and Their Families* (Simonton, Mathews-Simonton, & Creighton, 1978) and *Love, Medicine and Miracles* (Siegel, 1986). The clinical psychologist is well advised to maintain a consistent scientist-practitioner approach to this issue and to keep well versed in the area of psychoneuroimmunology (Kiecolt-Glaser & Glaser, 1987; Levy, 1986; Levy, Herberman, Lippman, & d'Angelo, 1987). The role of psychosocial factors in the development and progression of cancer reflects, in part, our culture's attitudes toward health and illness. These attitudes include a belief in the power of the individual over adverse forces or circumstances, the need to understand and control the unknown, the way in which the public reacts to research data, and the way in which our society regards illness, particularly cancer (Dreher, 1987).

The role of behavior in the etiology of cancer is clear. Lifestyle factors are known to increase the risk of getting many cancers. Tobacco use, dietary factors, alcohol use, overexposure of skin to sunlight, and sexual practices contribute to the development of approximately 75% of cancers (Levy, 1986). However, there is still not enough evidence from reliable prospective studies to show that personality factors, emotional states, or life events are contributing causes of cancer. In terms of behavior and illness progression, variables such as noncompliance or delays in screening or detection activities can result in cancer progression. As far as results from controlled studies are concerned, there is a difference between the use of relaxation to control certain side effects of chemotherapy and the use of relaxation for directing the immune cells to destroy cancer cells. The emerging field of psychoneuroimmunology will hopefully begin to resolve questions about emotions, cancer, and immune response.

Group

Group therapy with cancer patients has been used in various circumstances, settings, and styles. Group therapy can include patients only or patients plus family members (Wellisch & Cohen, 1985). Various forms of group therapy include patient education, coping-skill training, and support groups (Lieberman, 1987; Monaco, 1987). Groups should not be started or run casually. They should be thought through and custom-designed in a fashion that takes into account the training and supervision of the psychologist and the nature and problems of the population served (Goldberg & Tull, 1983; Wellisch, 1981). Groups can serve more patients than individual therapy, thus making better use of the psychologist's time. The patient benefits include mutuality, feeling less alone with problems, and increasing the bases of social support. Goldberg and Tull (1983) offered the following guidelines in establishing group therapy within oncology programs: (1) identify the goals of the group (e.g., education, support, and problem solving); (2) decide whether the group will be opened or closed (e.g., ongoing or time-limited); (3) decide how the patients will be screened (e.g., number of group members, age, diagnosis, and stage); (4) determine how the group will be offered to potential members (e.g., peer support groups or multiple family groups); (5) decide what supervision is available; (6) decide what record keeping is involved; (7) decide what the nature of the group process is (e.g., information and education, or exploratory); and (8) decide how the group fits into the treatment process (e.g., the physician and nurse input).

Telch and Telch (1985) provided a thorough critical review of support group therapy, medical education groups, and problem-oriented approaches with coping-skill training. These authors compared support groups, no-treatment controls, and coping-skill training groups across measures of behavioral and affective dysfunction and found the coping-skill training groups superior. Telch and Telch used coping with stressful medical procedures, communicating with physicians, increasing the

activities of daily living, the personal management of time and physical appearance, and the management of distressing affect as problem areas in the coping-skill groups. Behavioral strategies such as homework assignments, self-monitoring, goal setting, behavioral rehearsal, and modeling were also used in the coping-skill training groups. Problem-oriented group (or individual) intervention approaches (e.g., Sobel & Worden, 1982) show great promise in integrating clinical psychology into oncology programs.

Family

Family therapy involves meetings with the whole or parts of the patient's family unit. Like individual functioning, family functioning may be conceptualized in terms of developmental issues (e.g., newlywed phase, parental adaptation phase, nurturing phase, and maturity phase), and these issues will determine the number of members involved in therapy (Welch-McCaffrey, 1988). Wellisch and Cohen (1985) outlined two key issues in family therapy with the cancer patient. First, family therapy can improve communication about the illness and its impact on family relations. Second, therapy can develop more favorable family-based support for both the patient and other family members. Wellisch (1981) identified four central tasks in family therapy with cancer patients: (1) teaching the family how to communicate with the patient; (2) giving attention to and managing affirmative intimacy and emotional boundaries (e.g., focusing on the "forced" togetherness that can develop between an adolescent and his or her father with cancer when separation would be foremost in the absence of the illness); (3) determining the effects of cancer on the dependence–independence axis of the family and/or marital relationship; and (4) managing family frustrations regarding the burdens and limits of cancer, which may be turned on the medical staff and on each family member. Wellisch also cautioned the psychologist who is doing family therapy to avoid overidentification with the family and their working through unresolved personal loss, pushing too much change too quickly, overprotecting the cancer patient, and alliance with the family against the oncologist in a bid to compete with the physician for "control" of the patient and the family.

Marital Therapy. Significant others can play a major role in helping the cancer patient to cope. Although much of the literature on marital therapy involves the effects of breast cancer on couples (e.g., Pederson & Valanis, 1988), many guidelines inherent in family therapy are appropriate to the marital dyad as well. The spouse or the significant other of the cancer patient faces many extreme stresses and is often overlooked in evaluation and intervention.

Sex Therapy. An important area of the marital relationship involves sexual function. Cancer patients and their partners often require education and support as physical problems or psychological problems caused by either the disease or its treatments emerge (Glasgow, Halfin, & Althausen, 1987; Schain, 1987). Sex therapy is generally short-term (6–12 sessions) and typically includes a combination of education, support, and behavioral techniques directed at resolving conflicts that inhibit sexual functioning and at enhancing sexual activity within the limits of capability for both partners (Derogatis & Kourlesis, 1981; Schover, 1988a,b; Wellisch & Cohen, 1985).

Cognitive-Behavioral Therapy

Andrykowski and Redd (1987) offered a thorough review of cognitive-behavioral interventions in cancer care. According to these authors, the primary goals and features of cognitive-behavioral interventions are (1) the remediation of concrete symptoms; (2) individualization in accordance with the patient's experience and preferences; (3) a time-limited course; (4) evaluation in terms of observable changes in patient behavior; and (5) change or maintenance, based on the patient's response. Cognitive-behavioral interventions tend to be less intrusive than psychopharmacological interventions, to increase the patient's locus of personal control, and to decrease the patient stigma of having

"psychiatric problems" and can serve as a means of developing the initial rapport needed to intervene with other problems. The authors also noted that cognitive-behavioral interventions are typically unlikely to challenge the competencies of other members of the health care team and thus deemphasize "turf" issues. Because of the training of clinical psychologists in behavioral techniques, such interventions appear to be a "natural" contribution of the clinical psychologist to the oncology team.

Aversion Reactions to Chemotherapy. No other single problem area in oncology has used more cognitive-behavioral interventions than aversion reactions to chemotherapy. In addition to serve nausea and emesis following chemotherapy treatments, at least 25% of chemotherapy patients develop anticipatory side effects (Andrykowski & Redd, 1987). Nausea or emesis may occur as patients approach the hospital for treatment, when they enter the oncology clinic, when they see the oncology nurses, or as the nurse prepares the infusion. Some patients report feeling nauseated when they talk about treatment (cognitive mediation); others report feeling nauseated an entire day before treatment. Anticipatory nausea or emesis is the result of respondent or classical conditioning. Through repeated association with chemotherapy and its aversive aftereffects, previously neutral stimuli (e.g., the sights, smells, and sounds of the treatment environment) acquire nausea- or emesis-eliciting properties (Andrykowski & Redd, 1987). In reviews of over 30 individual-case and controlled studies on the biobehavioral treatment of anticipatory nausea and vomiting (ANV), Morrow and Dobkin (1988) and Morrow (1989) concluded that progressive relaxation training appears to be effective in controlling posttreatment nausea and vomiting and that systematic desensitization appears to be effective in controlling both anticipatory and posttreatment nausea and vomiting. The authors found that studies on hypnosis, although methodologically weak, provide some support for its use with children experiencing ANV. Multiple-muscle-site EMG biofeedback has also been used successfully to augment the relaxation training in both progressive relaxation and systematic desensitization (Burish, Shartner, & Lyles, 1981).

In using this type of intervention with adult cancer patients experiencing ANV, it is important to discuss the notion that the patient is not responsible for the problem and is not being noncompliant with the medical regimen. Explaining the classical conditioning paradigm, as well as offering prevalence data to patients, helps in reducing the guilt and despondency connected with not being an "ideal" patient. This author has also found that using a flexible approach in developing a hierarchy for systematic desensitization, as well as hierarchy that includes imagery of times before the day of treatment, activities in preparation for the day of treatment, the car ride to the treatment center and that involves multiple senses (e.g., sight, sound, taste, and smell) in the waiting room and the treatment room, increases the chances of success. In designing the cancer center in which this author works, care was taken to design each treatment room with a different color and decor to address and minimize repeated exposure to the same environmental stimuli from session to session. Research has yet to be carried out to determine whether this design has reduced the conditioning effects.

Stressful Procedures. The use of cognitive-behavioral interventions with stressful medical procedures, a problem for both the patient and the staff administering the procedure, was reviewed by Gill (1984). A comprehensive intervention package by Jay and Elliot (1984) was found to be especially useful for children who feared painful bone-marrow aspirations and lumbar punctures. This approach included filmed modeling, reinforcement for the child, breathing exercises, distractive emotive imagery, and behavioral rehearsal with dolls so that the child identifies with the staff, becomes desensitized to the actual treatment, and practices novel coping techniques. This author has used breathing exercises and distractive emotive imagery in assisting adult cancer patients to tolerate increasing amounts of time in often painful hyperthermia treatment (Rozensky, Honor, Tovian, & Herz, 1985).

In summary, intervention strategies with cancer patients based on cognitive-behavioral and problem-oriented techniques have been effective in controlled studies. Such interventions offer the psychologist a solid basis for developing a professional presence in oncology.

Developing and Maintaining the Program

Purpose and Goals

The general purpose of integrating clinical psychology into adult and pediatric oncology programs is to enhance patient care. The parameters for integrating clinical psychology into oncology programs include who does what, to whom, when, and with what effects (Silberfarb & Bloom, 1982). The targets for intervention may include the patient, the family, both, and/or the professional staff. Interventions may include only high-risk patients, for example, but all interventions should be based on a sound assessment of the patients' problems, needs, deficits, and strengths. Interventions may be most effective when provided on demand, on early identification of problems, at common crisis points, or as primary prevention at the time of diagnosis or of treatment initiation. The clinical psychologist needs to consider the possible discrepancy between the demands of a particular oncology program and his or her ability to respond to those demands. Inpatient interventions, for example, often require more immediate solutions in the context of a time-limited hospitalization, whereas outpatient interventions may involve somewhat longer time periods.

Scope of Services

Oncology programs in the medical center offer the clinical psychologist opportunities to apply the full spectrum of clinical psychological, neuropsychological, and health psychology services. In addition to providing interventions for psychological reactions to cancer, the psychologist may need to provide services (e.g., the traditional mental health service) to those patients with psychological dysfunctions that predate the cancer diagnosis. The psychology program can offer neuropsychological expertise (e.g., neuropsychological assessment) when cancer, as well as its treatments, affect the central nervous system. Problems involving compliance with the medical regimen, the management of stressful medical procedures, pain problems, and the iatrogenic emotional effects of chemotherapy involve health psychology services. Finally, programs involving smoking and tobacco cessation, breast self-examination, and other lifestyle modification approaches also use health psychology services.

New Program Guidelines

In developing a model for inpatient pediatric oncology consultation, Koocher, Sourkes, and Keane (1979) found the following guidelines essential in the establishment of their program: (1) assess the staff's perceived needs by evaluating the types of psychological problems encountered and the types of interventions that are of greatest value; (2) offer a routine, predictable presence so that the staff can plan for the optimal use of the psychologist's time, and so that the permanent staff get a sense of "stability" in a place where frequent staff turnover is common; (3) maintain adequate role definition and separation via open communication to prevent the formation of manipulative relationships between patients and staff; (4) teach the staff to be more attuned to patient psychological issues; (5) use teaching conferences to involve the staff in psychological intervention; and (6) stress a collaborative team effort. Prompt, timely follow-up of referrals and concrete, succinct, practical communications via charting and reports are also encouraged in developing either inpatient or outpatient programs (Belar *et al.*, 1987; Meyer, Fink, & Carey, 1988).

Psychologists' Role

The goals of a clinical psychology program in oncology can be accomplished by the psychologist acting as a direct service provider, a consultant, a program administrator, a researcher,

and a teacher and trainer. These roles are not mutually exclusive and may depend on the number of professionals and the resources available. The role of the psychologist also depends on the needs and qualities of any specific medical center and on whether a psychology program has been established.

Stone (1983) outlined graduate training requirements in health psychology that are applicable to work in oncology. The psychologist choosing to work in oncology should strive to develop a high tolerance of frustration, an understanding of his or her own stimulus value (as rapport with patients must often be established quickly), an avoidance of professional fanaticism, an acceptance of some reliance on another field (oncology) for working with diverse sets of data, and expertise in working with death and dying (Belar *et al.*, 1987).

Direct Service Providers. All cancer patients do not require psychotherapeutic intervention (Weisman, 1979). When a need for intervention exists, a number of patients do not desire intervention (Liss-Levinson, 1982a,b; Worden & Weisman, 1980). For these reasons, as well as because of staffing constraints, psychologists may not see every patient in a major oncology program. When providing direct service to oncology patients, the psychologist often sees patients formally referred by the oncologist and/or by the nursing staff. The merits of having the psychologist see all the patients within a program versus seeing patients on referral only are discussed by Bleiberg, Ciulla, and Katz in Chapter 22 of this book.

Whether the psychologist sees all patients or only those referred, developing professional alliances with physicians and nursing staff is crucial. The psychologist needs to be aware of referral patterns and customs within the medical center. Identifying those physicians and nurses in oncology and related fields who are sensitive to psychosocial issues is important. Power struggles with the medical staff should be avoided. The objective of good patient care should be overriding. The psychologist can make positive use of being "outside" the medical hierarchy as a guideline for improving communication with physicians, as well as for serving as a patient advocate on psychosocial issues.

Participating in case conferences and providing in-service programs for oncology staff are excellent methods of establishing and reinforcing referral patterns. These activities enable the psychologist to understand the needs of potential referral sources, to sensitize the medical staff to the psychological factors in oncology, and to educate the staff in his or her areas of interest and expertise. Maintaining flexibility in intervention approaches (i.e., combining individual and family therapy when appropriate) appeals to many oncologists. The appeal of such flexibility tends to be in avoiding excessive referrals, possible delays and overlap in treatment, and confusion in implementing a treatment plan. Additional guidelines for participating in case conferences include having recent knowledge of the patient's chart and keeping abreast of the literature (including specialized journals such as the *Journal of Psychosocial Oncology, Cancer, CA—A Journal of the American Cancer Society*, and *Behavioral Medicine Abstracts*).

Indirect Consultation. The psychologist does not actually see the patient in the indirect consultation. Indirect consultation takes several forms. In one model, the psychologist presents information in seminars, conferences, oncology section meetings, and in-service training or continuing-education programs. This model was used successfully by Koocher *et al.* (1979) in establishing a program on an inpatient pediatric oncology unit. The model can be used to gain recognition by medical staff and can be used to prepare staff for making appropriate referrals for direct intervention. This model also involves consultation to medical-center-based hospice programs.

A second form of indirect consultation involves assisting another professional in the use of psychological interventions within traditional medical treatments. For example, the psychologist consults with an oncology nurse about implementing a variation of progressive relaxation exercise with a particular patient before chemotherapy. Another example involves

the psychologist consulting with a physical therapist about initiating a contingency management program to increase ambulation in a youngster with a below-the-knee amputation. The constraints on this model involve an absence of financial reimbursement for the psychologist's time, as well as possible limitations in the scope of the services that can be offered by the psychologist.

The psychological needs of both the patient and the staff inevitably become part of the treatment process in oncology. The psychologist can provide clarification of and insight into such issues in staffing conferences. At these conferences, the psychologist also validates the staff's concerns about particularly difficult patients or families. In addition, the psychologist can provide a context for talking about staff countertransference feelings and can provide mutual understanding and support. Within this model, the psychologist can serve specifically as a consultant to the staff and monitor their reactions to the stresses of being health care providers in oncology. The psychologist must be careful to administer services either to patients or to staff and not to both.

A final model of indirect consultation involves on-the-spot or hallway consultations. In this model, the medical staff establish brief contacts with the psychologist to discuss problem patients and/or families. The need for such consultation often stems from the oncologist's report that the patient refuses to meet with the psychologist for fear of the stigma attached to meeting with a mental health professional. The oncologist decides not to confront the referral issue with the patient and possibly to jeopardize the physician–patient relationship. This type of indirect consultation also stems from the oncologist's attempt to screen the appropriateness of a referral (e.g., what the psychology service can do for a retarded cancer patient who has attention deficit disorder). The disadvantages of this approach are that the psychologist never sees the patient, is unable to judge the reliability of the physician's psychological diagnosis, and makes accurate decisions difficult. On the positive side, such hallway consultations offer confirmation that the psychologist is a part of the oncology team. Roberts (1986) discussed indirect consultation models in further detail.

Research. Psychosocial research in oncology affords the psychologist scholarly recognition in the medical setting. Research can also provide initial funds for future program and staff expansion. Training in the scientist-practitioner model enables the psychologist to engage in research from a solid foundation. In doing research, psychologists must try to balance general priorities in oncology research and their own interests in psychosocial oncology and must be sensitive to the politics and priorities of research at their particular medical center. Working with an oncologist as a coinvestigator on a psychosocial research project, for example, is also an effective means of establishing and maintaining referral sources.

Training. Maintaining faculty rank in a medical center affords the psychologist opportunities for the teaching and supervision of clinical psychology externs and interns as well as trainees from other disciplines, such as psychiatry, medicine, and nursing. The general goals of psychology training include increasing awareness of the psychosocial problems of oncology patients, mastery of the psychological assessment and intervention approaches used with oncology patients and their families, and recognition of the trainees' own professional limitations in working with oncology patients. Such goals can be attained through the use of seminars, individual and group supervision, and observational consultation (Rainey *et al.*, 1983).

Program Administrator. Miller (1981) discussed methods appropriate to the evaluation of health care programs, a peer review model for the evaluation of health service professionals' competencies, and the procedures necessary for the implementation of an evaluation program in a medical center. Additional tasks for the psychologist as program administrator involve quality assurance, the maintenance and establishment of professional relationships, the education of referral sources, financial and budget issues, and the building of a program that will match the setting.

Quality assurance with regard to clinical psychology services involves a number of dimensions. The patient flow through psychological services involves intake, assessment, intervention, termination, and follow-up. Intake involves how referrals are received by the psychology service (e.g., by telephone, by referral form, or from the staff) and (may require) a standard demographic form (e.g., Goldberg & Tull, 1983). Assessment refers to those techniques used to identify problem areas (e.g., interview, instruments, scales, and inventories). The intervention is the techniques used to solve problems. Termination may involve readministration of instruments used at intake to monitor intervention progress. Follow-up involves checking patient progress in the problem area three to four weeks after termination and serves as another means of evaluating the effectiveness of the interventions used.

Quality assurance may be reflected in volume indicators across a time period (e.g., per month). Sample indicators used in the health psychology program and other programs in the department of psychiatry at the Evanston Hospital include the number of referral sources, hours of consultation with patients and staff, referrals sent to other programs and/or agencies, patients (inpatients and outpatients), patient hours, new patients, and treatment hours (individual, group, family or marital, and biofeedback). These parameters can be used to evaluate productivity, to build services to match the needs of the setting, to expand services and personnel, and to ascertain whether the staff is available to meet the clinical needs of the patients.

Data can also be reported on quality indicators, such as the number of unplanned terminations, the number of complaints received from staff and patients, the percentage of charts with no progress note per session, the percentage of inpatient consultations completed within 24 hours of receiving the referral, and the percentage of patients indicating less than the expected improvement on a symptom change scale. When quality thresholds are not met, a narrative analysis of the problem is made along with a plan for corrective action. Quality thresholds are subject to change depending on trend analysis and reasonable levels of patient care.

Establishing Professional Relationships

Establishing and maintaining professional relationships is crucial in integrating clinical psychology programs into the oncology programs. The following guidelines are important: knowing the needs of the potential referral base, communicating effectively with oncologists and oncology nurses by "speaking their language" (e.g., medical terminology and abbreviations) and educating referral sources about the services offered. The last includes the areas of specialty and the limits of practice, the objective being a more efficient use of the psychologist's talents and time and the flexibility to develop the service to match the needs of the oncology program.

Particular attention must be paid to developing good relations with the primary-care nurse. Psychologists and nurses have many similarities in their position as nonphysicians in the power hierarchy of the hospital system. The reputation of the psychologist with patients and physicians can be immeasurably enhanced by a good working relationship with the nurses. Such a relationship is crucial to receiving referrals and implementing inpatient treatment interventions according to plan.

Relations with Other Mental Health Professionals

The psychosocial problems of cancer patients can be approached through a variety of models, and therefore, many other professionals have become involved in the psychological aspects of oncology (Goldberg, 1988). The use of counseling and psychotherapeutic services is often determined not by sound reasoning, but by local conditions involving funding, the availability of various professionals, and the biases of staff and patients. Lack of clarity

about what a psychologist can contribute to cancer care adds to the ambiguity for both patients and staff, not to mention confusion for third-party payers. In situations where many professionals are involved in the psychosocial care of oncology patients, it may be helpful for all the professionals to meet in a psychosocial staff meeting to discuss cases and professional issues, to make appropriate referrals, and to prevent overlap. It is important for the psychologist, and not hospital administrators or other mental health professionals, to define the unique contributions that the psychology service can make to the oncology program. This definition requires "positioning" (Charns & Schaefer, 1983) of the psychology program in the medical center; establishing good professional relationships, in which physicians and nurses are educated about those situations best managed by psychologists; and collaborating with other mental health professionals. Ideally, referrals to other professionals should be based on the limits of the psychologist's expertise and professional boundaries, should be initiated as a result of the psychologist's clinical judgment, and should be completed by the psychologist through direct contact so that an overlap of services is avoided.

Psychiatrists. Referral to consultation–liaison psychiatrists is appropriate when the psychologist suspects that a patient will require psychopharmacological management for a major affective disorder, an active psychosis, and/or delirium. The psychiatrist can also serve as a diagnostic consultant for cases in which an underlying organic basis may explain rapid mood, behavior, or sensorium changes in the cancer patient (Goldberg, 1988).

Social Workers. Referral to medical social workers is appropriate when the patient will require the use of outside agencies and/or specialized medical equipment to aid in discharge planning. The medical social worker often has expertise in coordinating home health care services and a knowledge of the financial aspects of social health care delivery and of community resources outside the medical center.

Nurse–Clinical-Specialists. The nurse–clinical-specialist is typically a registered nurse who has obtained a master's degree in a specialized nursing area (e.g., psychiatric nursing). Nurse-clinicians are often employed in large medical centers and teaching hospitals to work directly with primary-care nurses on difficulties with problem patients and on nurses' response to the stresses of health care delivery in oncology. Depending on the referral customs and the political roles in the departments of nursing in a medical center, it is not uncommon for nurse-clinicians to engage in individual, group, and family psychotherapy with cancer patients (Goldberg & Tull, 1983). The psychologist is wise to develop a good professional relationship with the nurse-clinician as an extension of the crucial psychologist–primary-nurse relationship.

Case Studies

Case 1

An 11-year-old male, the oldest in a sibship of three, was diagnosed four months previously with a brain tumor and underwent surgery. Despite completed trials of radiation and a good prognosis, the pediatrician reported that the child was withdrawn and overly dependent and was doing poorly in school. The child also reported severe headaches, although medical tests revealed no apparent physical etiology. The pediatrician also reported that a younger sibling had been caught stealing candy at a local store, and he observed that the mother had gained considerable weight since the child's diagnosis. The child and the family were referred for psychological intervention.

The child was seen in 12 initial sessions of individual psychotherapy focusing on his fear of the tumor's recurrence and his fear of death. Biofeedback training was also used to help teach him pain management skills and to gain better control of his bodily response to stress. The child was also referred for neuropsychological evaluation to assess his learning skills and revealed a slight learning problem in his visual processing skills. The parents were seen in six sessions of marital therapy to discuss their reactions to their child's illness. The stress of their child's recent illness resulted in their own fears of tumor recurrence and a disruption of parenting patterns within the family and communication between themselves. Both parents acknowledged that their 11-year-old child had and was receiving extreme attention, much to the anger of the other children. Five sessions of family therapy were completed in an effort to establish the previous patterns of family relationships. On termination, the child reported significantly fewer headaches and improved in school performance. The family's awareness of how the child's illness had affected their functioning led to improved communications and less anxiety.

This case illustrates the effectiveness of a flexible treatment approach to the multiple problems affecting a child diagnosed with cancer and his family. In this particular case, the problems had emerged after diagnosis and medical treatment were completed, and they required several modes of psychological intervention.

Case 2

A 16-year-old male inpatient, receiving both radiation and chemotherapy for bone cancer, was referred for psychological intervention. The nursing staff reported that the patient was hostile, belligerent, and noncompliant in his refusal to go to treatments and to engage in self-care activities. The patient was interviewed and was observed on the inpatient unit.

A treatment plan was developed and discussed in great detail with the oncologist, the pediatrician, the nursing staff, and the patient. The recommendations focused on initiating a contingency management program based on the patient's desire for more privacy and for permission to wear street clothes rather than a hospital gown secondary to compliance with treatments and decreased belligerence toward the nursing staff. It was also recommended that the patient be allowed to shower at night, as he did at home, instead of in the early morning, as is typically dictated by staffing patterns on the inpatient unit. Finally, contingent on his appropriate behavior, the patient was allowed choices in setting the times for his chemotherapy treatments. The outcome of the plan was the completion of 90% of the treatment sessions, as compared to a base rate of less than 50%, and a significant decrease in hostile behavior reported by the nursing staff.

This case illustrates an inpatient intervention with an adolescent. The assessment of the patient involved both interviewing and observation in the inpatient milieu. The intervention was based on developmental concerns both inherent in adolescence (i.e., the need for autonomy and individuality) and adversely affected by hospitalization. The success of the intervention was enhanced by input from the patient and the staff, especially the nurses.

Case 3

A 49-year-old single female, never married and without children, diagnosed with breast cancer, was referred for psychological intervention. She was postmastectomy with two positive nodes and was receiving cyclophosphamide, methotrexate, and 5-fluorouracil. After two treatments, she complained of extreme anxiety, agitation, and nausea approximately two days before her treatments. Her symptoms were severe enough to inhibit her work as an office manager in a large law firm. During the assessment interview, she revealed no history of psychiatric difficulties or substance abuse. In addition, she revealed that a close relationship with a male had ended after her mastectomy several months before. She also reported that her reaction days before treatment made her feel as if she were "losing her mind," and that she was fearful that her employer would become impatient with her deteriorating job performance. It was evident that she gained feelings of satisfaction, mastery, and autonomy through employment, and that when her work was jeopardized, she became more anxious and despondent.

This patient was seen in 20 sessions of individual psychotherapy. The initial treatment sessions focused on systematic desensitization for her anticipatory anxiety and nausea. The later treatment sessions focused on changes in her self-image secondary to her mastectomy and cancer and used a problem-oriented approach. As an outcome of the treatment, minimal anxiety and nausea before the treatments were reported by the patient and her nurse. The patient reported no disruption in work before her treatments. Follow-up also revealed that the patient had started social dating after psychotherapy.

This case illustrates the potential benefits of combining several cancer-specific therapeutic approaches to the multiple psychosocial problems of the adult patient. The success of the systematic desensitization for ANV was most likely enhanced by this patient's desire for autonomy. The success of the systematic desensitization also enhanced her ability to use the problem-oriented approach and to cope effectively with the changes brought about by her mastectomy.

Summary

This chapter has described cancer as a very intrusive chronic illness that places enormous stress on patients and their families. Cancer is often associated with major quality-of-life changes and, for nearly half of cancer patients, eventual death. In addition to the psychosocial and existential problems associated with the disease, patients often experience significant aversive side effects from the treatment. Indeed, for some patients, "the treatment is worse than the disease itself" (Adrykowski & Redd, 1987, p. 315). For these reasons, the comprehensive care of the cancer patient requires the use of numerous medical and psychosocial resources. A variety of methods of psychological assessment and intervention have been described in this chapter. To date, major contributions from the field of psychology have been to study the impact of cancer and its treatments on the psychological function of the patient, the patient's family, and staff, as well as the role that psychological and behavioral variables may have in cancer risk and survival. This chapter also described the development and maintenance of the clinical psychology program in oncology. Those clinical psychologists involved and interested in psychosocial oncology are to be encouraged; our methods have a great deal to offer to patients, their families, and the hospital staff. Clinical psychology has emerged as an important field within psychosocial oncology for both applications and research.

References

Andrykowski, M. A., & Redd, W. H. (1987). Life threatening disease: Biopsychosocial dimensions of cancer care. In R. L. Morrison & A. S. Bellack (Eds.), *Medical factors and psychological disorders*. New York: Plenum Press.

Barofsky, I. (1981). Issues and approaches to the psychosocial assessment of the cancer patient. In C. K. Prokop & L. A. Bradley (Eds.), *Medical psychology: Contributions to behavioral medicine*. New York: Academic Press.

Belar, C. D., Deardorff, W. W., & Kelly, K. E. (1987). *The practice of clinical health psychology*. New York: Pergamon Press.

Burish, T. G., & Lyles, J. N. (1983). Coping with the adverse effects of cancer treatments. In T. G. Burish & L. A. Bradley (Eds.), *Coping with chronic disease*. New York: Academic Press.

Burish, T. G., Shartner, C. D., & Lyles, J. N. (1981). Effectiveness of multiple muscle-site EMG biofeedback and relaxation training in reducing the aversiveness of cancer chemotherapy. *Biofeedback and Self-Regulation, 6*, 523–535.

Carpenter, P. J., & Onufrak, B. (1984). Pediatric psychosocial oncology: A compendium of the current professional literature. *Journal of Psychosocial Oncology, 2*, 119–136.

Cella, D. F., Tross, S., Orav, E. J., Holland, J. C., Silberfarb, P. M., & Rafla, S. (1989). Mood states of patients after the diagnosis of cancer. *Journal of Psychosocial Oncology, 7*, 45–54.

Charns, M. P., & Schaefer, M. J. (1983). *Health care organizations: A model for management*. Englewood Cliffs, NJ: Prentice-Hall.

Copeland, D. R., Fletcher, J. M., Pfefferbaum-Levine, B., Jaffe, N., Reid, H., & Maor, M. (1985). Neuropsychological sequelae of childhood cancer in long-term survivors. *Pediatrics, 75* ;745–753.

Derogatis, L. R. (1986). The psychosocial adjustment to illness scale (PAIS). *Journal of Psychosomatic Research, 1*, 77–91.

Derogatis, L. R. (1989). *The Psychosocial Adjustment to Illness Scale-Self Report (PAIS-SR)*. Baltimore: Clinical Psychometric Research.

Derogatis, L. R., & Kourlesis, S. M. (1981). An approach to the evaluation of sexual problems in the cancer patient. *Ca—A Cancer Journal for Clinicians, 31*, 46–50.

DeVita, V. T., Hellman, S., & Rosenberg, S. A. (1985). *Cancer: Principles and practice of oncology*. Philadelphia: Lippincott.

Dreher, H. (1987). Cancer and the mind: Current concepts in psycho-oncology. *Advances, 4*, 27–43.

Feinstein, A. D. (1983). Psychological interventions in the treatment of cancer. *Clinical Psychology Review, 3*, 1–14.

Friedenbergs, I., Gordon, W. A., Hibbard, M., Levine, L., Wolf, C., & Diller, L. (1981–1982). Psychosocial aspects of living with cancer: A review of the literature. *International Journal of Psychiatry and Medicine, 11*, 303–329.

Ganz, P. A., Rofessart, J., Polinsky, M. L., Schag, C. C., & Heinrich, R. L. (1986). A comprehensive approach to the assessment of cancer patients' rehabilitation needs: The cancer inventory of problem situations and a companion interview. *Journal of Psychosocial Oncology, 4*, 27–42.

Gill, M. (1984). Coping effectively with invasive medical procedures: A descriptive model. *Clinical Psychology Review, 4*, 339–362.

Glasgow, M., Halfin, V., & Althausen, A. F. (1987). Sexual response and cancer. *Ca—A Cancer Journal for Clinicians, 37*, 322–333.

Goldberg, R. J. (1988). Psychiatric aspects of psychosocial distress in cancer patients. *Journal of Psychosocial Oncology, 6*, 139–163.

Goldberg, R. J., & Tull, R. M. (1983). *The psychosocial dimensions of cancer: A practical guide for health-care providers*. New York: Free Press.

Gordon, W. A., Friedenbergs, I., Diller, L., Hibbard, M., Wolf, C., Levine, L., Lipkins, R., Ezrachi, O., & Lucido, D. (1980). Efficacy of psychosocial interventions with cancer patients. *Journal of Clinical and Consulting Psychology, 48*, 743–759.

Holland, J. C., & Rowland, J. H. (Eds.). (1989). *Handbook of psychooncology: Psychological care of the patient with cancer*. New York: Oxford University Press.

Holleb, A. I. (1988). Psychosocial issues and cancer. *Ca—A Cancer Journal for Clinicians, 38*, 130–192.

Jay, S. M., & Elliot, C. H. (1984). Psychological intervention for pain in pediatric cancer patients. In G. B. Humphrey, G. B. Grindey, L. P. Denher, R. T. Acton, & T. J. Pysher (Eds.), *Adrenal and endocrine tumors in children*. Boston: Martinus Nijhoff.

Kaufman, M. (1989). Cancer: Fact vs. feelings. *Newsweek*, (April 26), p. 10.

Kiecolt-Glaser, J. K., & Glaser, R. (1987). Psychosocial moderators of immune function. *Annals of Behavioral Medicine, 9*, 16–20.

Koocher, G. P. (1985). Promoting coping with illness in childhood. In J. C. Rosen & L. J. Solomon (Eds.), *Prevention in health psychology*. Hanover, NH: University Press of New England.

Koocher, G. P. (1986). Psychosocial issues during the acute treatment of pediatric cancer. *Cancer, 58*, 468–472.

Koocher, G. P., & O'Malley, J. E. (1981). *The Damocles syndrome: Psychological consequences of surviving childhood cancer*. New York: McGraw-Hill.

Koocher, G. P., Sourkes, B. M., & Keane, M. (1979). Pediatric oncology consultations: A generalizable model for medical settings. *Professional Psychology*, 10, 467–474.

Lansky, L. L., List, M. A., Lansky, S. B., Cohen, M. E., & Sinks, L. F. (1985). Toward the development of a play performance scale for children (PPSC). *Cancer*, 56, 1837–1840.

Lansky, S. B., List, M. A., & Ritter-Sterr, C. (1986). Psychosocial consequences of cure. *Cancer*, 58, 529–533.

Lehman, J. F., DeLisa, J. H., Warren, C. G., deLateur, B. J., Bryant, P. L., & Nicholson, C. G. (1978). Cancer rehabilitation: Assessment of need, development, and evaluation of model of care. *Archives of Physical Medicine*, 59, 410–419.

Levy, S. M. (1986). *Cancer and behavior*. San Francisco: Jossey-Bass.

Levy, S. M., Herberman, R., Lippman, M., & d'Angelo, T. (1987). Correlation of stress factors with sustained depression of natural killer cell activity and predicted prognosis in patients with breast cancer. *Journal of Clinical Oncology*, 5 348–353.

Lieberman, M. A. (1987). The role of self-help groups in helping patients and families cope with cancer. *Ca—A Cancer Journal for Clinicians*, 38, 162–168.

Liss-Levinson, W. S. (1982a). Clinical observations on the emotional responses of males to cancer. *Psychotherapy: Theory, Research, and Practice*, 19, 225–330.

Liss-Levinson, W. S. (1982b). Reality perspectives for psychological services in a hospice program. *American Psychologist*, 37, 1266–1270.

Mages, N. L., & Mendelsohn, G. A. (1980). Effects of cancer on patient's lives: A personalogical approach. In G. C. Stone, F. Cohen, & N. E. Adler (Eds.), *Health psychology: A handbook*. San Francisco: Jossey-Bass.

Meyer, J. D., Fink, C. M., & Carey, C. F. (1988). Medical views of psychological consultation. *Professional Psychology: Research and Practice*, 19, 356–358.

Meyerowitz, B. E., Heinrich, R. L., & Schag, C. C. (1983). A competency-based approach to coping with cancer. In T. G. Burish & L. A. Bradley (Eds.), *Coping with chronic diseases*. New York: Academic Press.

Miller, T. W. (1981). Professional service evaluation in a medical setting. In C. K. Prokop & L. A. Bradley (Eds.), *Medical psychology: Contributions to behavioral medicine*. New York: Academic Press.

Maul-Mellott, S. K., & Adams, J. N. (1987). *Childhood cancer: A nursing overview*. Boston: Jones & Bartlett.

Monaco, G. P. (1987). Parent self-help groups for the families of children with cancer. *Ca—A Cancer Journal for Clinicians*, 38, 169–175.

Moos, R. H. (1977). *Coping with physical illness*. New York: Plenum Press.

Morrow, G. R. (1986). Behavioral management of chemotherapy-induced nausea and vomiting in the cancer patient. *The Clinical Oncologist*, 113, 11–14.

Morrow, G. R. (1989). Chemotherapy-related nausea and vomiting: Etiology and management. *Ca—A Cancer Journal for Clinicians*, 39, 89–104.

Morrow, G. R., & Dobkin, P. L. (1988). Anticipatory nausea

and vomiting in cancer patients undergoing chemotherapy treatment: Prevalence, etiology and behavioral interventions. *Clinical Psychology Review*, 8, 517–556.

Mulhern, R. K., Horowitz, M. E., Ochs, J., Friedman, A. G., Armstrong, F. D., Copeland, D., & Kuhn, L. E. (1989). Assessment of quality of life among pediatric patients with cancer. *Psychological Assessment: A Journal of Consulting and Clinical Psychology*, 1, 130–138.

Pederson, L. M., & Valanis, B. G. (1988). The effects of breast cancer on the family: A review of the literature. *Journal of Psychosocial Oncology*, 6, 95–118.

Rainey, L. C., Wellisch, D. K., Fawzy, F., Wolcott, D., & Pasnau, R. O. (1983). Training health professionals in psychosocial aspects of cancer: A continuing education model. *Journal of Psychosocial Oncology*, 1, 41–59.

Rando, T. A. (1984). *Grief, dying, and death: Clinical interventions for caregivers*. Champaign: IL: Research Press.

Roberts, M. C. (1986). *Pediatric psychology: Psychological interventions and strategies for pediatric problems*. New York: Pergamon Press.

Rozensky, R. H. (1983). Psychotherapy, quality of life, and the cancer patient. *Cancer Focus*, 5, 41–46.

Rozensky, R. H., Honor, L. F., Tovian, S. M., & Herz, G. (1985). Pain reactions to hyperthermia cancer treatment. *Journal of Psychosocial Oncology*, 3, 75–83.

Schag, C. C., Heinrich, R. L., & Ganz, P. A. (1983). Cancer inventory for problem situations: An instrument for assessing cancer patients' rehabilitation needs. *Journal of Psychosocial Oncology*, 1, 11–24.

Schain, W. S. (1987). The sexual and intimate consequences of breast cancer. *Ca—A Cancer Journal for Clinicians*, 38, 154–161.

Schipper, H., & Levitt, M. (1985). Measuring quality of life: Risks and benefits. *Cancer Treatment Reports*, 69, 1115–1123.

Schover, L. R. (1988b). *Sexuality and cancer: For the woman who has cancer, and her partner*. New York: American Cancer Society.

Schover, L. R. (1988a). *Sexuality and cancer: For the man who has cancer, and his partner*. New York: American Cancer Society.

Siegel, B. S. (1986). *Love, medicine and miracles*. New York: Harper & Row.

Silberfarb, P., & Bloom, J. (1982). Research in adaptation to illness and psychosocial intervention. *Cancer*, 50, 1926–1927.

Silverberg, E., & Lubera, J. (1988). Cancer statistics, 1988. *Ca—A Cancer Journal for Clinicians*, 38, 5–22.

Simonton, O. C., Mathews-Simonton, S., & Creighton, J. (1978). *Getting well again: A step-by-step guide to overcoming cancer for patients and their families*. Los Angeles: Tarcher.

Sobel, H. J. (1981). *Behavior therapy in terminal care: A humanistic approach*. Cambridge, MA: Ballinger.

Sobel, H. J., & Worden, J. W. (1982). *Helping cancer patients cope: A problem-solving intervention for health-care professionals* New York: Guilford Press.

Spitzer, W. O., Dobson, A. J., Hall, J., Chesterman, E., Levi, J., Sheperd, R., Battista, R. N., & Catchlove, B. R. (1981).

Measuring the quality of life of cancer patients: A concise QL-index for use by physicians. *Journal of Chronic Diseases, 34*, 585–597.

Stone, G. C. (Ed.). (1983). National working conference on education and training in health psychology. *Health Psychology, 2*, 1–153.

Telch, C. R., & Telch, M. J. (1985). Psychological approaches for enhancing coping among cancer patients: A review. *Clinical Psychology Review, 5*, 325–344.

Van Dongen-Melman, J. E. W. M., Pruyn, J. F. A., Van Zanen, G. E., & Sanders-Woudstra, J. A. R. (1986). Coping with childhood cancer: A conceptual view. *Journal of Psychosocial Oncology, 4*, 147–157.

Varni, J. W., & Katz, E. R. (1987). Psychological aspects of childhood cancer: A review of research. *Journal of Psychosocial Oncology, 5*, 93–119.

Weisman, A. D. (1979). *Coping with cancer*. New York: McGraw-Hill.

Weisman, A. D., Worden, J. W., & Sobel, H. J. (1980). *Psychosocial screening and intervention with cancer patients: Research report*. Cambridge, MA: Shea.

Welch-McCaffrey, D. (1988). Family issues in cancer care: Current dilemmas and future directions. *Journal of Psychosocial Oncology, 6*, 199–211.

Wellisch, D. K. (1981). Intervention with the cancer patient. In C. K. Prokop & L. A. Bradley (Eds.), *Medical psychology: Contributions to behavioral medicine*. New York: Academic Press.

Wellisch, D. K., & Cohen, R. S. (1985). Psychosocial aspects of cancer. In C. M. Haskell (Ed.), *Cancer treatment*. Philadelphia: Saunders.

Wellisch, D. K., Landsverk, J., Guidera, K., Pasnau, R. O., & Fawzy, F. (1983). Evaluation of the psychosocial problems of the homebound cancer patient: 1. Methodology and problem frequencies. *Psychosomatic Medicine, 45*, 11–21.

Worden, J. W. (1982). *Grief counseling and grief therapy: A handbook for the mental health practitioner*. New York: Springer.

Worden, J. W., & Weisman, A. D. (1980). Do cancer patients really want counselling. *General Hospital Psychiatry, 2*, 100–103.

CHAPTER 21

Cardiovascular Disorders

Hypertension and Coronary Heart Disease

Stephen M. Weiss, Roger T. Anderson, and Sharlene M. Weiss

Introduction

The cardiovascular disorders, particularly hypertension and coronary heart disease, have received increased attention from the psychological community since the early 1970s (Shepherd & Weiss, 1987). Dissatisfaction with the psychoanalytically oriented "psychosomatic" approach to mind–body relationships prompted a "dissident" group of behavioral and biomedical scientists and clinicians to develop the concept of *behavioral medicine* to portray more accurately the interrelationship and interdependence of these two broad areas of inquiry in their efforts to understand, to treat, and ultimately to prevent the major chronic diseases of our time, for example, heart disease, cancer, and stroke (Schwartz & Weiss, 1978).

Cardiovascular disease, particularly coronary heart disease, is the leading cause of death and disability in this country. This fact in itself would be sufficient reason for the scientific and clinical communities to give cardiovascular disease a high priority. In this context, there is good news and bad news. The good news concerns the consistent decline in cardiovascular mortality (over a 35% decrease since 1968). The bad news concerns our inability to reach a consensus on what is the reason for this decline. Part of the explanation lies in the acknowledged multifactorial nature of the disease process itself, which involves genetic, physiological, biochemical, and environmental factors in some still-to-be apprehended configuration. A related explanation has to do with the difficulties in crossing disciplinary boundaries to study the organism as an integrated "whole" simultaneously, at all levels at which the disease process presents itself. Thus, we think of a "mosaic," whose pattern can be appreciated only when all of the constituent parts (colors) are represented in the appropriate con-

Stephen M. Weiss and Roger T. Anderson • Behavioral Medicine Branch, National Heart, Lung, and Blood Institute, Bethesda, Maryland 20892.   Sharlene M. Weiss • Health Promotion and Disease Prevention Branch, National Center for Nursing Research, Bethesda, Maryland 20892.

figuration, the whole being greater than as well as contextually different from the sum of its parts.

As the behavioral and biomedical scientific communities have become increasingly comfortable in addressing these issues jointly, new insights into the facilitating, inhibiting, synergistic, and catalytic nature of the interaction of biological and environmental variables have led to the development of *biobehavioral* approaches to disease treatment and prevention. For example, although literally hundreds of studies have compared the effects of pharmacological and nonpharmacological treatments on blood pressure reduction, only recently have a few investigations begun to consider how combining these approaches might overcome their respective shortcomings. It has also become obvious that individual differences may account for at least some of the variability associated with various treatments and that diagnostic assessments may be broadened (e.g., ambulatory monitoring) so that the constellation of factors governing blood pressure regulation may be better understood. These and related issues will be more fully addressed in the section of this chapter devoted to the treatment of hypertension.

As our biotechnology and treatment strategies have achieved greater prolongation of life, choices among therapies have required increasing consideration of the impact on the patient's *quality of life*. Cardiovascular diseases are a leading cause of functional limitations (Collins, 1986), including deficits in physical, psychological, and social functioning. The determinants of these deficits are complex and involve a variety of factors, including cardiovascular functional capacity, the side effects of treatment, and the psychological response to illness and functional loss.

Cardiac transplantation, artificial hearts and related organs, coronary artery bypass surgery, and various drug regimens for hypertension, coronary heart disease (CHD), and other cardiovascular disorders are among the treatment options that may, to a greater or lesser degree, affect the various dimensions of life functioning. Salient quality-of-life issues in the treatment of cardiovascular disease are discussed in a separate section in this chapter.

As the biobehavioral treatment of cardiovascular disease has been the subject of many volumes, this chapter must be limited to selected issues in the prevention and control of hypertension and coronary heart disease. The interested reader can obtain additional information on these topics from Schwartz, Shapiro, Redmond, Ferguson, Ragland, and Weiss (1979), Surwit, Williams, and Shapiro (1982), Gentry (1984), Matarazzo, Weiss, Herd, Miller, and Weiss (1984), Peterson (1983), Friedman and Ulmer (1984), Chesney and Rosenman (1985), Matthews, Weiss, Detre, Dembroski, Falkner, Manuck, and Williams (1986), Craig and Weiss (1989), Houston and Snyder (1988), Wenger, Mattson, Furberg, and Elinson (1984a), and Spilker (1990).

Hypertension

Hypertension is now recognized as perhaps the most important public health problem in the United States today. Approximately 60 million adult Americans are estimated to be at increased health risk because of elevated blood pressure; approximately two thirds of those with diastolic hypertension have "mild" hypertension (90–105 mm Hg). Well-controlled clinical trials in several countries have established the morbidity and mortality benefits associated with lowering blood pressure (e.g., Australian Therapeutic Trial in Mild Hypertension, 1980; Hypertension Detection and Follow-up Program, 1979). All studies to date have focused primarily on reducing blood pressure pharmacologically; thus, the recommended treatment regimens today tend to be pharmacological. It has become obvious, however, that such therapies are not without risk. For example, in the Multiple Risk Factor Intervention Trial (MRFIT, 1982), higher mortality rates were noted in the treated group of hypertensive patients; subsequent analysis of the data; these rates appeared to be related to resting EKG abnormalities in the "special intervention" group. Diuretic medications have also come under scrutiny with respect to their potential role in stimulating lipid mobilization (Lasser, Grandits, & Caggiula, 1984).

Multifactorial Origin

It is generally accepted that blood pressure is multifactorially determined and that age, sex, diet, family history, and social and physical environment may all play some role in determining the blood pressure in a given individual. Recently, attention has been focused on the role of physical and psychological stressors and their role in the regulation (or disregulation) of blood pressure (e.g., Matthews & Fredrikson, 1990; Manuck, Morrison, Bellack, & Polefrone, 1985). Because 90% of the patients with elevated blood pressure are characterized as having "essential" hypertension (i.e., hypertension of unknown etiology), it is obvious that the unique constellation of factors producing elevated blood pressure in a given individual will, to some extent, determine her or his response to any given therapy. Thus, individual differences, such as constitutional, personality, and environmental characteristics, must be considered singly and in combination in identifying the most effective and appropriate treatment strategy for a given individual.

In the light of our willingness to regard hypertension as multifactorial in origin, it is indeed curious that our standard measurement and treatment procedures have generally been lacking in breadth and variety. This lack would, of course, be understandable if we were considering a specific bacteria, vector, or virus as the source of disease, as there undoubtedly would be a one best approach to measurement, treatment, and prevention. The multifactorial perspective, however, assumes that different *patterns* of variables operating in each individual express themselves through the common pathway of blood pressure values (Schwartz *et al.*, 1979). For example, the once-hallowed "resting, clinic measurements" are now being challenged by ambulatory monitoring strategies and measurements of the *range* of blood pressure excursions under standardized conditions of physical and psychological challenge (De-

vereux, Pickering, Harshfield, Kleinert, Denby, Clark, Pregibon, Jason, Kleiner, Borer, & Laraugh, 1983; Perloff, Sokolow, & Cowan, 1983). As Pickering, Harshfield, Kleinert, Blank, and Laraugh (1982) demonstrated, significant differences between home, work, and clinic blood pressures suggest that the specific pattern of blood pressure values under different environmental conditions must be better understood if we are to develop treatment strategies relevant to the needs of the individual patient.

In casting a broader net, measurement and treatment technologies must account for the unique patterns of causation and response. Such approaches must also be sensitive to the synergistic potential of combinations of contributing factors that may be overlooked when these contributors are examined sequentially, rather than simultaneously. Our new measurement technologies, including ambulatory and telemetric monitoring and noninvasive visualization, combined with sophisticated computer-based statistical packages, permit the precise measurement and the detailed analysis of individual response patterns (Herd, Gotto, Kaufmann, & Weiss, 1984). A better understanding of these patterns will permit us to devise more effective, targeted treatment strategies that, one would hope, will use pharmacological agents as one approach of many, rather than the primary approach, to hypertension control.

Major unresolved controversies exist about the relative contribution of neurogenic factors to the development and maintenance of high blood pressure. These issues were extensively reviewed by Julius, Weder, and Egan (1983) and need not be repeated here. As behavioral treatment strategies ostensibly reduce sympathetic tone to environmental and psychological stressors, one might hypothesize that such therapies would be most effective with patients identified as having neurohormonal hyperresponsivity. Unfortunately, substantive evidence to support such hypotheses is lacking. Experimental studies by Benson, Dryer, and Hartley (1978) and DeQuattro, Shkhvatsabaya, Yurenev, Salenko, Khramelashvili, Foti, and Allen (1984) are suggestive, but larger scale prospective studies are required.

Behavioral Treatment Strategies

In considering behavioral treatments, one must differentiate among the various forms of biofeedback and relaxation therapies inasmuch as they may involve different physiological processes and sites of action (much as one would find with pharmacological agents). Autogenic relaxation procedures, for example, rely on more central, cognitive processes to achieve their effects, whereas "progressive muscle relaxation" strategies focus on peripheral, skeletal-muscle innervation. "Direct" biofeedback procedures involve moment-to-moment monitoring of blood pressure (or some direct derivative, such as pulse wave velocity), whereas "indirect" biofeedback methods rely on information from more peripheral sources such as thermal, sweat-gland, or muscle-tension measurements (for a more extensive discussion, see Fahrion, Norris, Green, & Snarr, 1988; Seer, 1979).

In an in-depth review of relaxation and biofeedback studies, the Health and Public Policy Committee of the American College of Physicians (1985) compared the results of 18 biofeedback studies (average decline: 7.8 mm Hg systolic, 5.6 mm Hg diastolic blood pressure) with 25 studies using a variety of relaxation therapies (average decline: 11 mm Hg systolic, 7.1 mm Hg diastolic blood pressure). Although the variation in procedures across studies suggests caution in drawing definitive conclusions, these results appear to be consistent with other reports (modest, yet significant decreases in blood pressure), relaxation procedures being somewhat more effective when one procedure is compared directly with the other. When the treatments were sequenced in a well-controlled counterbalanced design, however, the biofeedback–relaxation therapy sequence appeared to be more effective than either therapy alone or than the reverse sequence (Engel, Glasgow, & Gaarder, 1983).

Another treatment variation has been to combine the two procedures (e.g., Fahrion et al., 1988). In addition to lowering blood pressure, relaxation produces other measurable physiological changes, such as decreasing muscle tension and skin conductance. Using biofeedback measures related to these changes (e.g., electromyographic activity, and finger temperature), one can enhance the acquisition of relaxation skills through the reinforcement of appropriate responses.

Recent studies have also noted particular success with a biofeedback–assisted relaxation procedure using *thermal* rather than direct blood pressure feedback as the most effective measure in obtaining and maintaining lower blood pressure (Blanchard, McCoy, Musso, Gerardi, Pallmeyer, Gerardi, Cotch, Siracusa, & Andrasik, 1986; Fahrion et al., 1988). Relaxation training also appears amenable to group therapy settings as demonstrated in successful studies by Patel, Marmot, Terry, Carruthers, Hunt, and Patel (1985) and Fahrion et al. (1988). Both Green, Green, and Norris (1980) and Patel (1975) successfully used combinations of pharmacological and behavioral therapies for lowering blood pressure and at the same time reducing or eliminating dependence on medication in a significant proportion of their patients. Unfortunately, there has not yet been a clinical trial that compares behavioral, pharmacological and pharmacological–nonpharmacologic "combination" therapies on which risk–benefit and cost–benefit analyses can be based or by which the relative efficacy of these treatments can be determined in lowering blood pressure and maintaining that effect over time.

Finally, a variety of treatment-condition and "therapist" variables must be considered in evaluating the treatment efficacy of home practice; the use of aids such as cassette tapes, self-instructional materials, and self-monitoring equipment; group versus individual treatment (e.g., Patel, 1975; Patel, 1984; Patel et al., 1985); and time-limited versus goal-limited treatment. For example, behavior therapy research typically standardizes the number of treatment sessions for all patients. Although this standardization is generally recognized as a sound research methodology, it ignores a critical cognitive variable: people acquire skills at different rates. As relaxation training has been likened to learning to ride a bicycle or to play tennis, one would expect that the rate of skill acquisition plays an important role in achieving (and maintaining) goal blood pressure. Thus, if one were to account for different rates of learning by substituting goal-limited for time-limited protocols, one would, in all likelihood, improve the therapeutic efficacy of such treatment (e.g., Engel *et al.*, 1983).

Preliminary evidence suggests that "therapist" characteristics, which are generally ignored in pharmacological trials, may play a significant, perhaps even a decisive, role in the outcome of behavioral therapies, which tend to involve more intensive patient–provider interaction than do pharmacological therapies. The "self-management" emphasis on the individual rather than the pill as the active agent in the treatment regimen requires the patient (and the therapist) to take a more active role in achieving blood-pressure-reduction goals. Thus, we hear of the "Patel effect" (referring to the extraordinary clinical skills of Dr. Chandra Patel) in explanations of her consistently superior results as compared to those of her similarly trained colleagues. Such variables should be measurable, quantifiable, and hopefully, trainable for most aspiring therapists. They must also be recognized as significant, perhaps critical, factors in assessing the efficacy of treatment through comparative effectiveness studies under carefully controlled conditions (rather than the usual "nonspecific" or "placebo" effect appellation of the traditional drug trial, which seeks to eliminate this source of variance rather than capitalizing on it).

As noted in reviews by Seer (1979), Johnston (1982, 1985), Patel (1975), and McCaffrey and Blanchard (1985), the antihypertensive effects of biofeedback and relaxation therapies are highly dependent on the variables discussed above. In general, the more effective treatment studies have demonstrated statistically significant blood pressure reductions of 10–26 mm Hg systolic and 5–15 mm Hg diastolic, the magnitude of the effect depending, in part, on initial blood pressure values (Irvine, Johnston, Jenner, & Marie, 1986; Patel, Marmot, & Terry, 1981; Peters, Benson, & Peters, 1977).

Adherence and Compliance

Adherence to the therapeutic regimen is a major concern in all forms of antihypertensive therapy. Particular attention must be given to enhancing the motivation to maintain effective levels of therapy. Pharmacological measurements have traditionally suffered low rates of adherence, largely because of the aversive side effects of medication (Sackett & Snow, 1979). Nonpharmacological therapies have had similar compliance rates because of the time and effort required for daily (or periodic) practice. Two innovative approaches to achieving and maintaining adherence to therapy using "state-of-science" learning principles have been successful in generating the requisite motivation to maintain adherence to therapy.

In the first approach, the motivation generated by the desire to reduce or eliminate antihypertensive medication is used to encourage patients currently on medication to maintain adherence to both a pharmacological *and* a nonpharmacological regimen, with "step-down" drug withdrawal procedures being contingent on adequate blood pressure control by the nonpharmacological therapy. Thus, the patient is "rewarded" for diligent practice of the nonpharmacological procedures by medication reduction, the ultimate goal being the reduction and/or elimination of all medications when control continues to be successful with nonpharmacological methods (Green *et al.*, 1980; Patel, 1975). Such programs tend to enhance both pharmacological *and* nonpharmacological adherence through patient participation in a goal-directed program of adequate blood pressure control with minimal (or no) medication. The reinforcement of the step-down procedure appears to be a powerful motivating factor in continued adherence to the program, even when medications are not completely eliminated (Weiss, 1983).

The second approach focuses on patient "contracting" as a means of establishing a "therapeutic alliance" between patient and health care provider in achieving and maintaining the goal blood pressure (Steckel, 1982). Rather than focusing on any one treatment strategy, the patient is advised of the various pharmacological and nonpharmacological means available for lowering blood pressure, and specific blood pressure reduction goals are established for each session. Desired rewards selected by the patient are made contingent on the successful accomplishment of the goal. The initial setting of very modest goals allows a high rate of success, establishing a "pattern" of success; the subsequent blood pressure goals become increasingly difficult to achieve and maintain. Contingencies are then particularly sensitive to individual differences in terms of rewards and learning rates; experimental programs have demonstrated remarkably high rates of success compared with various traditional treatment modes. Of particular interest has been the finding that the successful achievement and maintenance of goal blood pressure have depended on a willingness to participate actively in the program rather than on specific adherence to pharmacological therapy. This finding supports the potential efficacy of "combination" therapy approaches that emphasize the systematic enhancement of patient motivation to maintain the activities necessary to control blood pressure.

Individual Differences

The study of individual differences as determinants of responses to treatment has also been underrepresented in hypertension research. Characteristics of the *individual*, such as cognitive, personality, genetic, constitutional (e.g., obesity), and lifestyle (e.g., smoking, diet, and activity level); *interpersonal* factors, such as the quality and quantity of family, work, and social relationships; and *environmental* circumstances, such as urban or rural residences, crowding, noise, and work setting—all are potential contributors to blood pressure variation. Pharmacological and nonpharmacological interventions can be most effectively used when the configuration of the contributors is well understood for a given individual. Thus, we must compare those who have been successful on a given treatment with those who

have not in our efforts to understand the interaction of the demographic, psychosocial, behavioral, and environmental factors that may account for differences in treatment effectiveness.

A comparison of treatment strategies illustrates how individual differences in the cognitive dimension may affect treatment outcome. For example, the pharmacological treatment of blood pressure typically follows a general protocol (e.g., stepped care), but the specific dosage and combination of drugs are determined by the individual's response to the treatment program. Curiously, when using nonpharmacological therapies, the typical protocol attempts to "standardize" the treatment regimen by specifying a set number of sessions for all patients. Although this is generally recognized as sound research methodology, it ignores a critical cognitive variable: We tend to acquire skills at different rates. As relaxation training has been likened to learning to ride a bicycle or to play tennis, one would expect that the rate of skill acquisition would play an important role in achieving and maintaining goal blood pressure. Thus, if one accounted for individual differences in skill acquisition by using *goal*-limited rather than *time*-limited protocols, one would undoubtedly improve therapeutic efficacy (e.g., Green *et al.*, 1980).

Placebo Effect

Expectancies of treatment and nonspecific effects such as the placebo pill and the "white-coat" syndrome are actually different forms of conditioned responses to stimuli that have become associated with therapeutic benefits (White, Tursky, & Schwartz, 1985). Unfortunately, previous trials have sought to minimize or eliminate this source of variance. We submit that one should study the parameters of such variables with the objective of maximizing these effects to enhance treatment regimens (e.g., to achieve therapeutic goals with minimal medication). Once understood, these nonspecific effects can be maximized in all studies of therapeutic efficacy (as well as in clinical practice), thus eliminating a source of uncontrolled variance from the trials, as well as encouraging the best clinical practice behavior on the part of health care providers.

In summary, research on stress management treatment techniques has demonstrated modest but clinically significant blood pressure reductions in patients at four-year follow-up after treatment. The range of expected blood pressure reductions, as noted in the more successful studies, is 10–26 mm Hg systolic and 5–15 mm Hg diastolic blood pressure; the magnitude of the effect varies in proportion to the initial blood pressure values. Increasingly effective behavioral treatment techniques in the hands of competent, well-trained clinicians have produced results that encourage the continued use of such techniques alone or in association either with other nonpharmacological approaches (as "step 0" therapy) or with pharmacological regimens so as to maximize blood pressure reduction and maintenance with a minimal adverse impact on the quality of life.

Coronary Heart Disease

There are several major threads in the biobehavioral literature pertaining to coronary heart disease. Lifestyle, personality, and social factors (e.g., Type A behavior) and environmental stressors have been identified as major potential contributors to CHD and therefore have been targeted for intervention strategies. Unfortunately, most efforts to understand the relationships between biobehavioral factors and CHD have been correlational in nature. Only recently have investigators begun systematically investigating the underlying pathways connecting such factors directly to the development and progression of CHD.

Basic Research

Recent studies attempting to understand the *mechanism of action* underlying the behavior–CHD relationships have focused on the responsivity of the cardiovascular system to behavioral–environmental challenges (Matthews & Fredrikson, 1990). Objective assessments of the impact of physical and psychological stressors, alone and in interaction with other environmental variables, on the cardiovasculature in both animal and human studies are confirming the relevance of these variables to the development and progression of CHD. For example, hyperreactivity to environmental stressors (e.g., with respect to blood pressure, heart rate, glucose or lipoprotein metabolism, or fibrinogen) may be potentiated by high-fat diets, smoking, salt or caffeine, or a sedentary lifestyle. These potential interactions may help us to better comprehend the complex nature of the behavior–CHD relationship and what options may be successful in CHD prevention and control. Here again, the issue of individual differences may be illuminated by clinical assessment techniques that identify those individuals who may be hypersensitive to various substances—and how environmental stressors may potentiate these sensitivities as assessed by cardiovascular reactivity. Although this area of research is still in its infancy, we believe it important to stimulate the thinking of our readers about the critical links in the hopes that they will broaden their clinical approaches to health behavior in terms of both their underlying assumptions and the strategies they use.

Lifestyle

As the economic toll of the treatment of cardiovascular disease continues its breathtaking pace, efforts to prevent as well as to cure coronary heart disease have become major elements in the national health strategy. Lifestyle, which for our purposes translates into health-related behaviors, is related to the concept of *risk factors* (i.e., probabilistic statements based on epidemiological observational data that provide "relative risk" estimates of morbidity and mortality, comparing populations with and without the independent variable under consideration). Smoking, a sedentary lifestyle, obesity, Type A behavior, hypertension, diabetes, elevated serum cholesterol, caffeine, and "stress," among other factors, have been regarded as enhancing the risk of CHD. Only three of these factors—smoking, hypertension, and elevated serum cholesterol—are generally accepted as *consistently* related to CHD risk (Gordon, Castelli, Hjortland, Kannel, & Dawber, 1977). As these three account for only approximately 50% of the variance associated with CHD, however, investigators are attempting to cast a wider net in efforts to achieve a better understanding of and control over the disease process itself.

The so-called secondary risk factors mentioned above are still in question because of the inconsistent findings of studies conducted thus far. Such inconsistencies can be interpreted in two ways: (1) as spurious findings (Type I or II errors) or (2) as being caused by uncontrolled variances in the form of other unidentified variables (perhaps other known primary or secondary risk factors) episodically interacting with the independent variables under consideration to produce inconsistent results on the dependent variables (i.e., cardiovascular end points related to CHD). It is beyond the scope of this chapter to attempt to deal with the specific treatment programs for the issues identified above; the reader is directed to excellent sources of information on treatment programs (e.g., Blumenthal & McKee, 1987; Brownell, 1986; Cataldo & Coates, 1986; Matarazzo *et al.*, 1984). However, even a cursory review of the efficacy of health behavior intervention programs reveals that we are much more successful in changing behavior than in *maintaining* such changes. Although there may be many reasons for this frustrating circumstance, one of the major ones may be that the behavioral scientist–clinician traditionally approaches behavior change from a *topographic* perspective, that is, a focus on external manipulations to effect and maintain change. Little effort is devoted to attempting to understand how such manipulations affect the *internal* (i.e., physiological and biochemical) environment and homeostasis and whether other factors (e.g., diet) may enhance or inhibit the likelihood of success. To better understand the nature of these relationships to health-related behaviors, we must become more aware of the underlying mechanisms in the interplay of internal and external influences on health and health behav-

ior if our interventions are to match the complexity and sophistication of the systems we are dealing with. For example, the successful maintenance of smoking cessation depends in part on the level of biological sensitivity to nicotine and other tobacco-related substances as well as on effects on caloric metabolism, in addition to effective behavior modification strategies. Understanding this mix of interacting variables provides therapists a broader array of combination options for intervention.

Lifestyle Assessment: Health Risk Appraisal

To aid in counseling individuals in lifestyle modification, an innovative technique called the *health risk appraisal* (HRA) has been developed. Data from epidemiological studies used to calculate the probabilistic factors for disease and mortality are translated into an *individual* risk profile, comparing the individual's family history, health habits, and demographic data with these population findings. Originally developed by Robbins and Hall (1970), HRAs are now available in over 70 published versions.[1]

The data obtained from the HRA are typically calculated in terms of "health age" to determine the relative health status of the individual with respect to his or her actual chronological age. A third determinant, "attainable health age," is also calculated in some versions of the HRA; it removes immutable characteristics from the "health age" computation (e.g., previous illnesses and family history). Thus, the difference of "attainable health age" from "actual chronological age" would be amenable to change, based on a modification of health-related lifestyle behaviors. These calculations are then reviewed with the individual to stimulate her or his interest and participation in activities (e.g., health promotion programs) aimed at changing her or his lifestyle in a health-enhancing direction.

These instruments can be divided into three

---

[1]A list of selected HRA instruments and a reading list can be obtained from the Office of Disease Prevention and Health Promotion National Health Information Center, P.O. Box 1133, Washington, D.C. 20013-1133.

general categories: self-scored, which are usually brief and can be scored by the individual; microcomputer usage, which can be analyzed on a personal computer; and computer-scored, which are mailed to a control facility for processing.

The benefits of the HRA for the individual are as follows:

1. It is a credible source of health information, based on scientific data and presented in a highly personalized form.
2. It assists the individual to understand the concept of personal health risk and the role of individual health practices in the etiology and prevention of disease.
3. It demonstrates the quantitative nature of risk-taking behavior and the synergistic potential of individual risks added together.
4. It quantifies the relative importance of various health practices, so that the individual can choose which ones to work on and where to begin.
5. It provides a periodic quantitative measure or progress (risk reduction) if health practices are changed in health-protective directions.

The benefits of the HRA on the organizational level are as follows:

1. It provides a structure for health personnel to use in focusing discussions of health and behavior.
2. It relies on questionnaires, physiological measurements, and computer-assisted calculations, so that its application to large groups is feasible, efficient, and relatively inexpensive.
3. It enhances the development of a data base for epidemiological research, health planning, and program evaluation.
4. The data-gathering devices, computer software, and other aspects of the program can be marketed as a package so that commercial firms are stimulated to become involved.

Many studies concerning the effectiveness and impact of the HRA in changing health-related behaviors have been undertaken, with equivocal results. The following conclusions are based on the evidence currently available:

1. The HRA is more suitable for use with middle-aged than with younger or older individuals. The HRA provides little statistical incentive to those under 40 to change poor health habits, and for persons over 65, the risk factors are not good predictors (Safer, 1982).
2. Individuals given the HRA as part of an educational process or in health-counseling sessions make more beneficial changes than individuals exposed only to the results of the questionnaire (i.e., with no counseling). Concerns have been raised about providing HRA findings without an opportunity for discussion and clarification with a health professional (Schoenbach, Wagner, & Berry, 1987).
3. The existing HRAs are not suitable for minority, ethnic, or blue-collar populations because of vocabulary and language limitations, as well the lack of epidemiological data on risk factors in minority populations (Schoenbach, Wagner, & Berry, 1987).

In summary, techniques for measuring personal health risk have emerged at a time of recognized need as a way of translating knowledge about the epidemiology of disease and mortality into activities that promote personal health. Their use has expanded rapidly, and they clearly have considerable potential for educating the public about the prospective implications of personal lifestyle risks to health and for structuring the content of risk factor educational messages (Defriese, 1987; Smith, McKinlay, & Thorington, 1987).

Personality

Since 1959, when Friedman and Rosenman first proposed what was later to be called the *Type A behavior pattern* (TABP), there has been extensive controversy over the role of behavioral and personality factors in the development and treatment of CHD.

The Western Collaborative Group Study (WCGS) prospectively demonstrated a doubling of the relative risk of CHD and related phenomena among middle-aged white males characterized as "hard driving, competitive, time urgent, aggressive, hostile, impatient" (i.e., Type A individuals) as compared to a similar cohort of men lacking such characteristics (i.e., Type B individuals) (Rosenman, Brand, Jenkins, Friedman, Straus, & Wurm, 1975).

Over the years, cross-sectional angiographic studies have come down on either side of this question in approximately equal numbers, whereas several subsequent studies (e.g., MRFIT, 1982; a subsequent 22-year follow-up of the WCGS data) have failed to confirm earlier findings (Ragland & Brand, 1988; Shekelle, Hulley, Neaton, Billings, Borhani, Gerace, Jacobs, Lasser, Mittlemark, & Stamler, 1985b).

Many questions have been posed concerning the equivalence of the assessments, the similarity of the populations, and other methodological considerations across studies in efforts to resolve the apparent contradictions in findings. Several issues have emerged from these efforts:

1. The division of individuals into dichotomous "at-risk" and "not-at-risk" groups lacks both sensitivity and specificity.
2. The measurement instruments are not well-correlated with one another.
3. Large constellations of diffuse personality and behavioral characteristics ostensibly are given equal weight in determining whether an individual is or is not characterized as Type A.
4. Whatever predictive validity Type A behavior possesses as a risk factor for CHD appears to be limited to males under the age of 50, when the more traditional risk factors have not yet achieved clinical expression. Beyond age 50, Type A behavior appears to exert a modestly *protective* influence (Ragland & Brand, 1988; Williams, Barefoot, Haney, Harrell, Blumenthal, Pryor, & Peterson, 1988).

The more promising approach of analyzing the various components of the TABP has yielded surprisingly strong correlations between hostility and unexpressed anger and CHD. These findings have been sustained in the reanalysis (Dembroski, MacDougall, Costa, & Granditis, 1989) of the major negative study for Type A (MRFIT, 1982) and has demonstrated a consistently significant relationship to the occurrences of CHD and total mortality in several other studies (e.g., Barefoot, Peterson, Harrell, Hlatky, Pryor, Haney, Blumenthal, Siegler, & Williams, 1989; Shekelle, Gale, & Norusis, 1985a) including the WCGS where it was the only component to survive the multivariate analysis of all Type A components (Chesney, Hecker, & Black, 1988).

Identifying hostility and unexpressed anger as "toxic" personality characteristics begs the questions: (1) Can one change such "traits"? and (2) Will change result in a lowered risk of CHD? The one substantive intervention study (Friedman, Thoresen, Gill, Powell, Ulmer, Thompson, Price, Rabin, Breall, Dixon, Levy, & Bourg, 1984, 1986) using post myocardial-infarction patients focused on changing Type A behavior to reduce the risk of reinfarction. One of the major assessments in the study concerned "reduction of hostility." Significant reductions in morbidity were noted in the experimental group; mortality differences were also observed in the experimental-group–control-group comparisons, particularly among those patients with less severe cardiac damage (low Peel Index scores). It should be noted for the record that the intervention consisted of both behavioral change and cardiac counseling; the control group received cardiological counseling only. It was underscored in the study that although it had been hypothesized that behavior modification would improve on the results obtained with standard cardiological counseling (which proved to be the case), it was in no way offered as a *replacement* for such counseling (cardiological counseling being necessary, although not in itself sufficient).

Again, to review the literature in this area more fully, we refer the reader to an excellent review published as a special issue of the *Annals of Behavioral Medicine* ("Area Review," 1988). Suffice it to say that recent events have shifted the focus of research attention from Type A behavior to hostility and unexpressed anger (Chesney, 1988). Conceptual and measurements problems must be resolved before further substantive progress is likely to occur, however. An observed relationship between hostility and cardiovascular reactivity in both human and animal studies is encouraging (Dembroski *et al.*, 1990; Manuck, Kaplan, & Matthews, 1986) but is too preliminary at this point to allow definitive conclusions. Further research will undoubtedly clarify the nature and strength of this relationship, including association, pathways, and potential interventions.

Public Health Approaches
to Lifestyle Modification

Although psychologists traditionally deal with clinical interventions in their efforts to treat and control disease, prevention efforts must be considered from a population perspective as well as from the usual individual or small-group approach. Worksites, schools, and communities are the most likely places for the development of such large-scale programs, in which the unit of investigation is the entire community, worksite, or school. Although one-to-one and small-group health counseling and interventions take place in these settings as well, public health strategies also rely heavily on environmental modifications as a means of changing behavior (risks) in entire populations. Media (print and electronic), legislative and policy changes, and the restructuring of the physical environment are the principal interventions used (e.g., DeLeon & Vandenbos, 1984; Maccoby & Alexander, 1980; Syme, 1987). Job restructuring may also be undertaken to provide more autonomy and control over work schedules, which have a demonstrated relationship to CHD prevalence (Karasek, Baker, Marxner, Ahlbom, & Theorell, 1981).

Such large-scale studies are also necessary to determine (1) whether the behavior of populations is amenable to change by behavioral, psychological, and communications strategies and (2) whether such changes will bring about significant reductions in event rates (i.e., mor-

bidity and mortality). At present, preliminary data from several ongoing large-scale community risk-reduction programs suggest that risk-factor-related behavior can be modified in the desired direction. Whether such differences will be sufficient to reduce population risk remains to be seen. It will be several years before all of the data on this question will be available. Nonetheless, community and worksite programs continue to proliferate throughout the country and will require well-trained health-behavior-change professionals to ensure that such programs will maximize their potential for disease prevention and health promotion.

Quality of Life and
Cardiovascular Disorders

The role of quality of life in the treatment of disease is twofold: (1) it characterizes the biobehavioral sequelae of disease and identifies specific areas of intervention in the disease process, and (2) it provides information on the functional outcomes of various therapies for particular disorders and for subsets of patients, so as to help patients and health care providers make the most appropriate choice of treatment or treatment combinations.

Generically, quality of life denotes an individual's ability to derive satisfaction from a variety of roles in physical, psychological, and social functioning (Wenger *et al.*, 1984a). Health-related quality of life refers specifically to functional outcomes that occur as a *consequence* of illness and its treatment.

Initially, quality of life research in cardiovascular disease (CVD) focused primarily on return to work (Cay, Vetters, Philips, & Dugard, 1973; LaMendola & Pellegrini, 1979), and on work activity mainly as an independent measure of the medical management of physical symptoms and myocardial function (Kellerman, 1975) and as ancillary data to determine the economic impact of an intervention. Because health denotes a state of physical, mental, and social well-being (World Health Organization, 1948), the concept of quality of life has been broadened to include the following components (adapted from Shumaker, in press): (1) physical functioning (e.g., self-care, ambulation, and somatic symptoms); (2) social functioning (e.g., at work, at home, and in the community); (3) cognitive functioning (e.g., memory, reaction time, and problem-solving ability); (4) emotional functioning; (5) personal productivity (e.g., employment status and the pursuit of rewarding activities such as home-making, gardening, and volunteer work); and (6) intimacy, which includes sexual functioning and interpersonal relations. The importance placed on assessing each of these quality-of-life outcomes in CVD is determined by whether the disease is symptomatic or asymptomatic, the severity of the disease, and the type and range of interventions (i.e., surgical, pharmacological, and behavioral), as well as characteristics of the patient's lifestyle. See the appendix for an overview of quality-of-life assessment methods and strategies.

Hypertension

Quality-of-life issues in hypertension typically pertain to the need for long-term adherence to therapy for an asymptomatic disorder. Because patients with mild hypertension have no overt manifestations of illness, therapies that have adverse effects on quality of life can jeopardize long-term adherence. For mild hypertension, where there is some controversy about the benefits of treatment, adverse quality-of-life changes may also raise ethical concerns.

Patients undergoing pharmacological treatment therapies for hypertension report more depressive symptoms (Huapaya & Ananth, 1980) and anxiety symptoms (Wheatley, Balter, Levine, Lipman, Bauer, & Bonato, 1975) than the general population. There is an appreciable body of evidence associating various antihypertensive drugs with changes in emotional, cognitive, and physical functioning, plausibly through interference with adrenergic nervous system functioning. Early evidence of adverse effects surfaced over three decades ago when hypertensive patients treated with the adrenergic inhibitor reserpine were found to be at increased risk of clinical depression (Fries,

1954; Wallace, 1955). More recently, beta-adrenergic blockers (e.g., propranolol) and centrally acting adrenergic inhibitors (e.g., clonidine and methyldopa) have been found to be associated with depression (Prichard & Owens, 1983); cognitive, psychomotor, and perceptual impairments (Medical Research Council, 1981; Soloman, Hotchkiss, Saraway, Bayer, Ramsey, & Blum, 1983); and behaviorally in terms of declines in energy, sexual activity, and increased absenteeism at work (Croog, Levine, Testa, Brown, Bulpitt, Jenkins, Klerman, & Williams, 1986; Jachuck, Brierley, Jachuck, & Wilcox, 1982). For example, Croog et al. (1986), who compared the effects of captopril, methyldopa, or propranolol in mild and moderate hypertension, found that the captopril patients fared better than the other two groups in terms of general well-being, physical symptoms, and sexual dysfunction. The addition of a diuretic worsened symptoms in all three regimens; however, captopril retained its superiority over the other two drugs.

Patel and Marmot (1987) reported positive effects of behavioral treatment of hypertension on quality-of-life variables, as part of a four-year follow-up study on the behavioral treatment of hypertensive patients. Of the nine areas evaluated, four showed significant differences, favoring the treatment group over the control group. Those four areas were relationships at work, general health, enjoyment of life, and personal and family relationships. The continued practice of relaxation skills appears to have progressive treatment effects beyond the initial treatment period. In the treatment group that continued regularly to practice their relaxation skills as opposed to those who did not, two additional areas of improvement were noted: concentration at work and mental well-being. However, for some patients, the time and effort needed to achieve the daily practice of relaxation may be burdensome and result in compliance difficulties.

This kind of clinical data is invaluable in efforts to reconcile the demands for quality-of-life maintenance and for therapeutic effectiveness. Wenger (1988) concluded that, based on the available evidence, hypertensive patients whose lifestyle requires high levels of activity or whose occupations require a high level of mental alertness may not be ideal candidates for beta-blocking drugs; patients with a history of depression may not be ideal candidates for treatment by beta-blockers, reserpine, or methyldopa; and patients with a history of sexual problems may not be ideal candidates for antihypertensive agents known to impair sexual functioning.

Coronary Heart Disease

Myocardial Infarction

Myocardial infarction (MI) is a life-threatening event, resulting in acute and sometimes severe limitations on physical and psychosocial functioning. Quality-of-life concerns become prominent once the patient has been stabilized and involve treatment strategies that relieve physical symptoms and emotional distress, and that improve cardiac functional aerobic capacity and efficiency.

Physical rehabilitation plays a major role in quality of life by facilitating return to daily activities, which, in turn, is related to emotional adjustment. However, it is clear that, by themselves, physical capacity and return to work are insensitive measures of quality of life. Psychological distress, which is common early post-MI, may persist for reasons independent of physical capacity and may prevent resumption of premorbid levels of functioning (Cay, Vetter, & Philips, 1972; Ibrahim, Feldman, Sultz, Staiman, Young, & Dean, 1974; Lloyd & Cawley, 1982; Mayou, Foster, & Williamson, 1978; Trelawny-Ross & Russell, 1987; Wiklund, Sanne, Vedin, & Wilhelmsson, 1984), particularly in the areas of work, family, and marital relations. Psychosocial interventions involving education and counseling of MI patients have achieved lasting beneficial effects on quality of life as an inpatient treatment (Oldenberg, Perkins, & Andrews, 1985), and as an adjunct to a rehabilitation program for patients experiencing physical and emotional complications from MI (Stern, Gorman, & Kaslow, 1983).

In the Stern et al. (1983) study, exercise and counseling were compared for their effects on patient rehabilitation. The results show that different interventions influenced unique aspects of quality of life. Although both interventions lessened emotional distress, the counseling group fared better than either the control or the exercise groups in social skills, interpersonal relationships, and degree of independence in daily functioning. Currently, there is little information to suggest what specific psychosocial processes or moderating factors (e.g., social support, coping skills, and personality variables) are important in predicting quality-of-life outcomes.

The assessment of quality of life of MI patients should address different issues along the distinct phases of hospital and follow-up care. During acute care, the symptoms associated with physical functioning include pain, fatigue, breathlessness, and activity limitations, and those associated with psychological functioning include emotional distress, sleep disturbance, and deficits in cognitive and intellectual functioning. In the intermediate and final hospital phases, quality-of-life assessments should focus on psychological functioning at the time of transfer from the coronary care unit (Cay, Vetters, Philips, & Dugard, 1972; Wenger et al., 1984b); during the remainder of the hospitalization, assessments should focus on the patient's physical and psychological readiness to resume normal living, including patient and family expectations regarding recovery in various roles and in cognitive and emotional functioning. At follow-up, quality-of-life assessments should include the extent of the resumption of roles and activities connected with work, family, and community, as well as life satisfaction in these areas.

Coronary Artery Bypass Graft Surgery

For patients for whom medical and surgical interventions will have comparable effects on morbidity and mortality, the choice of a particular intervention will be determined by its potential for improving the patient's quality of life. The relative benefits of surgical over medical therapies generally involve enhanced physical symptom relief and increased exercise tolerance. However, these results must be weighed against other outcomes that may affect quality of life: physical suffering and psychological disturbance due to surgery; the rate of return to normal activity; and how long various functional gains are maintained. In addition, knowledge of what subsets of patients are likely to obtain quality-of-life benefits from a particular procedure is critical.

The CVD surgical procedure most often discussed as a beneficial alternative to medical therapy is coronary artery bypass graft surgery (CABG). Since 1974, the annual number of patients who have undergone CABG surgery has more than quadrupled; over 200,000 patients per year undergo what is now the second most commonly performed major elective surgery in the United States. The vast majority of CABG patients have decreases in angina and increases in exercise tolerance, and many report improvements in daily activities (Jenkins, Stanton, Savageau, Delinger, & Klein, 1983; Kornfeld, Heller, Frank, Wilson, & Malm, 1982; Langeluddeke, Fulcher, Baird, Hughes, & Tennant, 1989). The long-term quality-of-life benefits of surgery have not been extensively studied. In the Coronary Artery Surgery Study (CASS, 1983), a large clinical study comparing the medical and surgical treatment of patients with coronary artery disease, the benefits from surgery were, for the most part, limited to patients who were symptomatic presurgery, that is, those most likely to have at least moderate activity limitations.

The adverse effects of CABG surgery are short-term and often include incisional pain and psychological disturbance. An estimated 15%–35% of patients show at least moderate neuropsychological dysfunction lasting up to eight weeks post-CABG surgery (Mayou, 1986; Raymond, Conklin, Schaeffer, Newstadt, Matloff, & Gray, 1984; Savageau, Stanton, Jenkins, & Klein, 1982) characterized by delirium, confusion, memory loss, or difficulty in thinking or perception. These deficits are not explained by anxiety and/or depression and may be due to anesthesia, medications, or cerebral hypoxia. Longer term deficits in subtle areas of neuropsychological functioning appear to occur in less than 5% of patients (Mayou, 1986).

Although increased exercise tolerance and relief of physical symptoms can be seen as major advantages of CABG surgery, in a significant subset of patients these benefits are not transferred to other functional domains. Despite good operative outcomes, an estimated 15%–65% of patients experience emotional distress six months to one year postsurgery (CASS, 1983; Gundle, Bozman, Tate, Raft, & McLaurin, 1980; Heller, Frank, & Kornfeld, 1974; Mayou *et al.*, 1978; Rabiner & Wilner, 1976). Fewer patients are working by one year postsurgery than before surgery, a worse outcome than after uncomplicated MI (Bruce, Bruce, Hossack, & Fusako, 1983; CASS, 1983; Mayou & Bryant, 1987), and medical therapies (CASS, 1983). Decreases in sexual activity have been reported in a significant subset of CABG patients (Frank *et al.*, 1972; Gundle *et al.*, 1980; Kornfeld *et al.*, 1983; Langeluddeke *et al.*, 1989).

A comprehensive rehabilitation program that addresses the psychosocial needs of the patient and the family (e.g., education and supportive counseling), in addition to traditional goals of restoring strength and improving functional capacity, may be valuable in building on or enhancing the benefits of CABG.

Cardiac Transplantation

Similar quality-of-life issues apply to the cardiac transplant. In those select medical centers involved in cardiac transplantation, psychologists participate in the selection and care of heart recipients (S. M. Tovian, personal communication, August 30, 1989). Heart transplantation consists of several stressful aspects for patients, including a work-up period to establish eligibility for a new heart; a waiting period for a donor if the patient is accepted into a transplant program; readmission to the hospital for transplantation; and postdischarge recovery. Psychological variables appear to play a role in mediating the outcome of heart transplantation, and psychologists may be involved in assessment and intervention with the transplant patient

(Hecker, Norvell, & Hills, 1989; Olbrisch, Levenson, & Hamer, 1989). A further discussion of the psychological variables involved in cardiac transplantation can be found in Beidel (1987).

In summary, coronary heart disease results in a number of disruptions in the normal functioning of an individual in the six dimensions of quality of life described above, which have increasingly become of clinical interest in terms of both CHD interventions and rehabilitation programs. Medical and nonmedical factors may moderate the net effect of disease on quality of life. Medical treatments act indirectly by altering underlying disease or disease processes and directly on emotional and cognitive functioning. A strong moderating role of nonmedical factors is also indicated by the data and may be due to influences on the motivation to resume functioning, the satisfaction of psychological needs, and health beliefs. These influences may belong to the social environment (e.g., social settings and economic structure); interpersonal processes (e.g., patient education, social support, stress, and economic strain); and intrapersonal processes (e.g., coping skills and stress management, personality variables, and physical health and capacity) (see Shumaker, 1990). As there are currently few data on the links between medical and nonmedical factors in quality of life, this subject continues to deserve a high research priority.

Summary

We have selectively addressed specific, representative research and practice issues of interest to health psychologists in the area of cardiovascular disease. We have limited this chapter to hypertension and coronary heart disease because of the extensive data base available that supports specific diagnostic and treatment options for these disorders. Because lifestyle behaviors cut across the cardiovascular spectrum, clinicians are cautioned not to be oversimplistic in their assessment of the behavior change issue, and to be sensitive to the interrelated nature of behavioral and biological variables. Generally, our information on effec-

tive treatment strategies is limited to hypertension, which, given its prevalence of 40–60 million persons in this country alone, could keep the majority of clinical health psychologists fully engaged for life if the treatment of choice were "biobehavioral."

We need to know more about patient and therapist variables as they relate to the effectiveness of treatment, so we can become more specific and selective in patient–treatment matching. Our knowledge of the relationships of behavior and CHD is also developing steadily. Although our intervention strategies are still rudimentary and await additional clarification from our scientific colleagues, several promising avenues for intervention appear to be worth exploiting, although with due regard and caution, because of the lack of definitive efficacy data.

Quality-of-life assessment has already become a major influence on the choice of a treatment strategy in instances where an adequate data base has been developed. As these instruments become increasingly sensitive to the multidimensional nature of "levels of functioning," it is expected that quality-of-life determinations will become integral to all treatment decision making.

We currently have in hand or are in the process of developing increasingly sophisticated bioinstrumentation that noninvasively (or minimally invasively) provides comprehensive biobehavior assessments, both in the laboratory and during ambulatory daily activities. Such information, combined with health risk appraisal data and other psychometric assessments (e.g., of Type A behavior, hostility, anger, and quality of life), can provide a complete psychosocial-demographic-psychophysiological profile for differential diagnosis, for selecting the most appropriate treatment of choice, and for providing an assessment of the progress and effectiveness of the treatment.

Although we are far from having all of the desired information on the behavior–CVD relationship, a sizable body of theoretical approaches, intervention strategies, and the related hardware and software is available to the well-trained clinician. We believe that, by adopting the "biobehavioral" perspective, such clinicians can substantially improve on the traditional treatment approaches in meeting the needs of their clients at *all* relevant levels of inquiry and remediation.

Appendix

Measurement of Quality of Life

Assessment of quality of life in clinical and epidemiological studies generally involves either an index or a battery. The merits of each approach, along with examples, are discussed below. Readers interested in a critique and more detailed discussion of QL instruments should consult Wenger, Mattson, Furburg, and Elinson (1984), Kaplan (1985), and Spilker (1990).

Indices. One approach is to use a single instrument to obtain a single number, or set of factor scores, to characterize QL. This can be accomplished either by simple or weighted cumulation of a set of scores assigned to responses in an inventory. Proponents of the single summary score approach hold that although QL is a multidimensional construct, it is actually experienced as a subjective evaluation of functioning, that is, as an "overall desirability of the aggregate" (Kaplan, 1988). This focus is especially well suited to health policy or population-based research used to derive estimates of a total illness burden. The Quality of Well-Being Scale (Kaplan & Bush, 1982) is the major example of decision-theory approach to QL assessment.

A second approach in QL indices is to score and interpret separately the various dimensions that make up quality of life. Proponents of this approach argue that dimensions of quality of life are distinct and independent, and cannot be meaningfully collapsed into a single estimate; positive and negative changes among quality-of-life components cannot be represented. Separate scores are desirable in clinical applications where the focus is on specific treatment effects, or in the selection of an optimal treatment program based on a variety of patient characteristics, e.g., lifestyle or preference, and in identifying specific health-care needs; and with regard to treatment programs where specific side effects must be monitored (Stewart, 1988). The Sickness Impact Profile (Bergner, Bobbitt, & Pollard, 1976), and the McMaster Health Index Questionnaire (MHIQ; Chambers, 1988) are widely used examples of this approach.

In brief, the appropriateness of either approach will depend on the research need. Estimating the gross impact of a disease or therapy on quality of life within a population; or relative to other diseases or therapies, i.e., in cost/utility studies, may be better suited to decision theory approach. Studies on specific health care needs of the patient or subset of patients will require separate component scores.

Battery. A second, *ad hoc* approach to measure quality of life is to assemble a battery of separate instruments for each of the dimensions of QL. The major advantage of this approach in clinical populations is that QL assessment can be tailored, using the best instruments available, to a particular research goal. Some disease or treatment groups may require a more sensitive measurement of various dimensions of quality of life may be needed than possible with general health instruments, or aspects which are not an included in these instruments. For example, in CABG, and for patients taking medication, assessment of intellectual and cognitive functioning may be desirable. In studies where affective involvement is a problem (e.g., MI patients), a researcher may choose from several standardized anxiety scales, such as the Taylor Manifest Anxiety Scale (Taylor, 1953), the Cattell and Scheier IPAT Anxiety Scale (Cattell & Scheier, 1961), and the Spielberger State Trait Anxiety Index (Spielberger, 1983); depression scales such as the Beck Depression Inventory (Beck, Ward, Mendelson, Mock, & Erbaugh, 1961) and the Center for Epidemiologic Studies-Depression scale (CES-D; Radloff, 1977); and assessment of mood states using the Profile of Mood States (McNair, Lorr, & Droppleman, 1971). These scales have been widely used and extensively validated against psychiatric norms. However, because there are limits on the amount of time, the number of resources, and the burden which may be placed upon the patient, it is important to construct a parsimonious battery, including only the dimensions of quality of life relevant to the medical condition at hand.

The Quality of Well-Being Scale (QWB)

The QWB is widely used and is a hallmark of the decision-theory approach to quality-of-life measurement. This instrument was designed to evaluate functional, symptomatic, and mental health outcomes of all types of acute and chronic diseases and injuries. It contains three dimensions of daily activity: mobility, physical activity, and social activity; and a separate dimension which measures medical symptoms and problems that might inhibit function. Responses are scored along a stepped gradient of dysfunction: no limitation–dependent for activity scales; and death–no symptoms for symptom/problem complexes scales. These in turn are then integrated as unique combinations of "well states," and are weighted in terms of their relative desirability to determine a respondent's QWB score. The QWB score can be further adjusted by formula to reflect prognoses for any disease or disability, yielding an estimate of "well life expectancy," which Kaplan (1988) defines as the current life expectancy adjusted for the diminished quality of life associated with dysfunctional states and the duration of stay in each state. By use of the QWB mortality, morbidity, and weighted behavioral function states can be simultaneously considered in evaluating the impact of disease or a treatment on quality of life.

Sickness Impact Profile (SIP)

The SIP is one of the most widely used separate components measures of quality of life. The SIP was designed to provide a descriptive profile of changes in behavioral function due to sickness, and is applicable to any disease or disability. The instrument contains 136 items, forming 12 daily activity categories, which cluster into three factors: independent, including sleep and rest, eating, work, home management, recreation, and pastimes; physical, including ambulation, mobility, and body care; and psychosocial, including social interaction, alertness, emotional behavior, and communication. A mean SIP score and the factor scores express impairment in function percentage; the SIP does not include mortality outcomes.

The McMaster Health Index Questionnaire (MHIQ)

The MHIQ (Chambers, 1988) is similar to the SIP in that it measures separately physical, social, and emotional functional components of quality of life. Items in the physical function dimension cover mobility, self-care, communication, and global physical functioning. The social dimension includes general well-being, work and social role functioning, personal and interpersonal relations, and global social function. The emotional dimension includes self-esteem, personal relationships, thoughts about future, critical life events, and global emotional function.

References

Area review: Coronary prone behavior: Continuing evolution of the concept. (1988). *Annals of Behavioral Medicine, 10*(2).

Australian therapeutic trial in mild hypertension. (1980). *Lancet, 1*, 1261.

Barefoot, J. C., Peterson, B. L., Harrell, F. E., Hlatky, M. A., Pryor, D. B., Haney, T. L., Blumenthal, J. A., Siegler, I. C., & Williams, R. B. (1989). Type A behavior and survival: A follow-up study of 1467 patients with coronary artery disease. *American Journal of Cardiology, 64*, 427–432.

Beck, A. T., Ward, C. H., Mendelson, M., Mock, J., & Erbaugh, J. (1961). An inventory for measuring depression. *Archives General Psychiatry, 4*, 561–571.

Beidel, D. C. (1987). Psychological factors in organ transplantation. *Clinical Psychology Review, 7*, 677–694.

Benson, H., Dryer, T., & Hartley, L. (1978). Decreased oxygen consumption during exercise with elicitation of the relaxation response. *Journal of Human Stress, 4*, 38–42.

Bergner, M., Bobbitt, R. A., Carter, W. B., & Gilson, C. (1981). The Sickness Impact Profile: Development and final revision of a health status measure. *Medical Care, 19*, 787–805.

Blanchard, E., McCoy, G., Musso, A., Gerardi, M., Pallmeyer, T., Gerardi, R., Cotch, P., Siracusa, K., & Andrasik, F. (1986). A controlled comparison of thermal biofeedback and relaxation training in the treatment of essential hypertension: 1. Short-term and long-term outcome. *Behavioral Therapy, 17,* 563.

Blumenthal, J. A., & McKee, D. C. (Eds.). (1987). *Applications in behavioral medicine and health psychology: A clinician's source book.* Sarasota, FL: Professional Resource Exchange.

Brownell, K. D. (1986). Public health approaches to obesity and its management. In L. Breslow, J. E. Fielding, & C. B. Lave (Eds.), *Annual Review of Public Health , 7,* 521–533.

Bruce, E. H., Bruce, R. A., Hossack, K. F., & Fusako, K. (1983). Psychosocial coping strategies and cardiac capacity before and after coronary artery bypass surgery. *International Journal of Psychiatry in Medicine, 13*(1), 69–83.

Cataldo, M. F., & Coates, T. J. (1986). *Health and industry: A behavioral medicine perspective.* New York: J. Wiley.

Cattell, R. B., & Scheier, I. H. (1961). *The meaning and measurement of neuroticism and anxiety.* New York: Ronald Press.

Cay, E. L., Vetter, N., Philips, P., & Dugard, P. (1972). Psychosocial status during recovery from a heart attack. *Journal of Psychosomatic Research, 16,* 425–435.

Cay, E. L., Vetter, N., Philips, P., & Dugard, P. (1973). Return to work after a heart attach. *Journal of Psychosomatic Research, 17,* 231–243.

Chambers, L. W. (1988). The McMaster Health Index Questionnaire: An update. In S. R. Walker & R. M. Rosser (Ed.), *Quality of life assessment and applications.* Boston: MTP Press.

Chesney, M. (1988). Evolution of coronary-prone behavior. *Annals of Behavioral Medicine, 10*(2), 43–45.

Chesney, M. A., & Rosenman, R. H. (1985). *Anger and hostility in cardiovascular and behavioral disorders.* Washington, DC: Hemisphere.

Chesney, M. A., Hecker, M. H., & Black, G. W. (1988). Coronary-prone components of Type A behavior in the WCGS: A new methodology. In M. B. K. Houston & C. R. Snyder (Eds.), *Type A behavior pattern: Research, theory and intervention* (pp. 168–188). New York: J. Wiley.

Collins, J. G. (1986). Prevalence of selected chronic conditions, United States, 1979–1981, in National Center for Health Statistics. DHHS (PHS) #86-1583, Government Printing Office.

Coronary Artery Surgery Studies, Principal Investigators and Their Associates. (1983). Coronary Artery Surgery Study (CASS): A randomized trial of coronary artery bypass survival data. *Circulation, 68,* 951–960.

Craig, K. D., & Weiss, S. M. (Eds.), 1990. *Behavioral medicine: Prevention and early intervention.* New York: Plenum Press.

Croog, S. H., Levine, S., Testa, M. A. Brown, B., Bulpitt, C. J., Jenkins, C. D., Klerman, G. L., & Williams, G. H. (1986). The effects of antihypertensive therapy on the quality of life. *New England Journal of Medicine, 314,* 1657–1664.

Defriese, G. (1987). A research agenda for personal health risk assessment. *Health Services Research, 22,* 4.

Deleon, P., & Vandenbos, G. (1984). Public health policy and behavioral health. In J. D. Matarazzo, S. M. Weiss, J. A. Herd, N. E. Miller, & S. M. Weiss (Eds.), *Behavioral health: A handbook of health enhancement and disease prevention* (pp. 150–163). New York: J. Wiley.

Dembroski, T. M., MacDougall, J. M., Costa, P. T., & Granditis, G. A. (1989). Components of hostility as predictors of sudden death and myocardial infarction in the multiple risk factor intervention trial. *Psychosomatic Medicine, 51,* 514–522.

DeQuattro, V., Shkhvatsabaya, I., Yurenev, A., Salenko, B., Khramelashvili, V., Foti, A., & Allen, J. (1984). Left ventricular hypertrophy and neural tone in hypertension: Divergent effects of diuretic and relaxation therapy. In *USA-USSR Joint Symposium. Hypertension: Psychophysiological, Biobehavioral, and Epidemiological Aspects* (pp. 57–68). NIH Publication No. 86-2704, Washington, DC: U.S. Government Printing Office.

Devereux, R., Pickering, T., Harshfield, G., Kleinert, H., Denby, L., Clark, L., Pregibon, D., Jason, M., Kleiner, B., Borer, J., & Laragh, J. (1983). Left ventricular hypertrophy in patients with hypertension: Importance of blood pressure response to regularly occurring stress. *Circulation, 68,* 470.

Engel, B. T., Glasgow, M. S., & Gaarder, K. (1983). Behavioral treatment of high blood pressure: 3. Follow-up results and treatment recommendation. *Psychosomatic Medicine, 45,* 23.

Fahrion, S., Norris, P., Green, A., & Snarr, C. (1986). Biobehavioral treatment of essential hypertension: A group outcome study. *Biofeedback and Self-Regulation, 11*(4), 257–277.

Frank, K. A., Heller, S. S., & Kornfeld, D. S. (1972). A survey of adjustment to cardiac surgery. *Archives of Internal Medicine, 130,* 735–738.

Friedman, M., & Ulmer, D. (1984). *Treating Type A behavior—and your heart.* New York: Knopf.

Friedman, M., Thoresen, C. E., Gill, J. J., Powell, L. H., Ulmer, D., Thompson, L., Price, V., Rabin, D., Breall, W. S., Dixon, T., Levy, R., & Bourg, E. (1984). Alteration of Type A behavior and reduction in cardiac recurrences in post-myocardial infarction patients. *American Heart Journal, 108,* 237–248.

Friedman, M., Thoresen, C. E., Gill, J. J., Powell, L. H., Ulmer, D., Thompson, L., Price, V., Rabin, D., Breall, W. S., Dixon, T., Levy, R., & Bourg, E. (1986). Alteration of Type A behavior and its effect on cardiac recurrences in postmyocardial infarction patients: Summary results of the Recurrent Prevention Project. *American Heart Journal, 112*(4), 653–665.

Fries, E. D. (1954). Mental depression in hypertensive patients treated for long periods with large doses of reserpine. *New England Journal of Medicine, 275,* 1006–1008.

Gentry, W. D. (Ed.), (1984). *Handbook of behavioral medicine.* New York: Guilford Press.

Gordon, T., Castelli, W. P., Hjortland, M. C., Kannel, W. B.,

& Dawber, T. R. (1977). Predicting coronary heart disease in middle-aged and older persons: The Framingham Study. *Journal of the American Medical Association, 238*, 497–499.

Green, E. E., Green, A. M., & Norris, P. A. (1980) Self regulation training for control of hypertension. *Primary Cardiology, 6*, 126.

Gundle, G. J., Bozman, B. R., Tate, S., Raft, D., & McLaurin, L. P. (1980). Psychosocial outcome after coronary artery surgery. *American Journal of Psychiatry, 137*, 1591–1594.

Health and Public Policy Committee, American College of Physicians. (1985). Biofeedback for hypertension. *Annals of Internal Medicine, 102*(5), 709–715.

Hecker, J. E., Norvell, N., & Hills, H. (1989). Psychologic assessment of candidates for heart transplantation: Toward a normative data base. *Journal of Heart Transplantation, 8*, 171–176.

Heller, S., Frank, K., & Kornfeld, D. (1974). Psychological outcome following open heart surgery. *Archives of Internal Medicine, 135*, 908–914.

Herd, J., Gotto, A., Kaufmann, P., & Weiss, S., (1984). *Cardiovascular Instrumentation*. NIH Publication No. 84-1654, Washington, DC: U.S. Department of Health and Human Services.

Houston, B. K., & Snyder, C. R. (Eds.) (1988). *Type A behavior pattern: Research, theory, and intervention*. New York: Wiley.

Huapaya, L., & Ananth, J. (1980). Depression associated with hypertension: A review. *Psychiatric Journal of the University of Ottawa, 5*, 58–62.

Hypertension Detection and Follow-Up Program Cooperative Group (1979). Five year findings of the hypertension detection and follow-up program. *Journal of the American Medical Association, 242*, 2562–2571.

Ibrahim, M. A., Feldman, J. G., Sultz, H. A., Staiman, M. G., Young, L. J., & Dean, D. (1974). Management after myocardial infarction: A controlled trial of the effect of group psychotherapy. *International Journal of Psychiatry in Medicine, 5*, 253–268.

Irvine, M. J., Johnston, D. W., Jenner, D. A., & Marie, G. V. (1986). Relaxation and stress management in the treatment of essential hypertension. *Journal of Psychosomatic Research, 30*, 437.

Jachuck, S. J., Brierley, H., Jachuck, S., & Wilcox, P. M. (1982). The effect of hypotensive drugs on the quality of life. *Journal of the Royal College of General Practitioners, 32*, 103–105.

Jenkins, C. D., Stanton, B. A., Savageau, J. A., Delinger, P., & Klein, M. D. (1983). Coronary artery bypass surgery: Physical, psychological, social and economic outcomes six months later. *Journal of the American Medical Association, 250*, 782–788.

Johnston, D. W. (1982). Behavioral treatment in the reduction of coronary risk factors: type A behavior and blood pressure. *British Journal of Clinical Psychology, 21*, 281–294.

Johnston, D. W. (1985). Psychological intervention in cardiovascular disease. *Journal of Psychosomatic Research, 29*, 447.

Julius, S., Weder, A. & Egan, B., (1983). Pathophysiology of early hypertension: Implication for epidemiologic research. In F. Gross & T. Strasser (Eds.), *Mild hypertension: Recent advances* (p. 219). New York: Raven Press.

Kaplan, R. M. (1985). Quality of life assessment. In P. Karoly (Ed.), *Measurement strategies in health psychology*. New York: Wiley.

Kaplan, R. M. (1988). Health-related quality of life in cardiovascular disease. *Journal of Consulting and Clinical Psychology, 56*, 382–392.

Kaplan, R. M., & Bush, J. W. (1982). Health-related quality of life measurement for evaluation research and policy analysis. *Health Psychology, 1*, 61–80.

Karasek, R., Baker, D., Marxner, F., Ahlbom, A., & Theorell, T. (1981). Job decision latitude, job demands, and cardiovascular disease: A prospective study of Swedish men. *American Journal of Public Health, 71*, 694–705.

Kellerman, J. (1975). Rehabilitation of patients with coronary heart disease. *Prognosis in Cardiovascular Disease, 17*, 303.

Kornfeld, D. S., Heller, S. S., Frank, K. A. Wilson, S. H., & Malm, J. R. (1982). Psychological and behavioral responses after coronary artery bypass surgery. *Circulation, 66*(Suppl. 3), 24–28.

LaMendola, W. F., & Pellegrini, R. V. (1979). Quality of life and coronary artery bypass surgery patients. *Social Science and Medicine, 13A*, 457–461.

Langeluddeke, P., Fulcher, G., Baird, D., Hughes, C., & Tennant, C. (1989). A prospective evaluation of the psychosocial effects of coronary artery bypass surgery. *Journal of Psychosomatic Medicine, 33*, 37–45.

Lasser, N. L., Grandits, G., & Caggiula, A. W. (1984). Effects of antihypertensive therapy on plasma lipids and lipoproteins in the Multiple Risk Factor Intervention Trial. *American Journal of Medicine, 76*, 52–66.

Lloyd, G. C., & Cawley, R. H. (1982). Psychiatric morbidity after myocardial infarction. *Quarterly Journal of Medicine, 201*, 33–42.

Maccoby, N., & Alexander, J. (1980). Use of media in lifestyle programs. In P. O. Davidson & S. M. Davidson (Eds.), *Behavioral medicine: Changing health lifestyles* (pp. 351–370). New York: Brunner/Mazel.

Manuck, S.B., Morrison, R. L., Bellack, A. S., & Polefrone, J. M. (1985). Behavioral factors in hypertension: Cardiovascular responsivity, anger and social competence. In M. A. Chesney & R. H. Rosenman (Eds.), *Anger and hostility in cardiovascular and behavioral disorders*. Washington, DC: Hemisphere.

Manuck, S. B., Kaplan, J. R., & Matthews, K. A. (1986). Behavioral antecedents of coronary heart disease and atherosclerosis. *Arteriosclerosis, 6*, 2–14.

Matarazzo, J. D., Weiss, S. M., Herd, J. A., Miller, N. E., & Weiss, S. M. (Eds.), (1984). *Behavioral health: A handbook of disease prevention and health enhancement*. New York: Wiley.

Matthews, K., & Fredrikson, M. (1990). CV responses to behavioral stress and hypertension: A meta-analytic review. *Annals of Behavioral Medicine, 12*(1), 30–39.

Matthews, K. A., Weiss, S. M., Detre, T., Dembroski, T. M.,

Falkner, B., Manuck, S. B., & Williams, R. W. (Eds.), (1986). *Handbook of stress, reactivity, and cardiovascular disease.* New York: J. Wiley.

Mayou, R. A. (1986). The psychiatric and social consequences of coronary artery surgery. *Journal of Psychosomatic Research, 30,* 255–271.

Mayou, R., & Bryant, B. (1987). Quality of life after coronary artery surgery. *Quarterly Journal of Medicine, 239,* 239–248.

Mayou, R., Foster, A., & Williamson, B. (1978). Psychosocial adjustment in patients one year after myocardial infarction. *Journal of Psychosomatic Research, 22,* 447–453.

McCaffrey, R. J., & Blanchard, E. B. (1985). Stress management approaches to the treatment of essential hypertension. *Annals of Behavioral Medicine, 7,* 5.

McNair, D. M., Lorr, M., & Droppleman, L. F. (1971). *EITS Manual for the Profile of Mood States.* San Diego, CA: Educational and Industrial Testing Service.

Medical Research Council Working Party on Mild to Moderate Hypertension. (1981). Adverse reactions to Benzofluorazide and propranolol for the treatment of mild hypertension. *Lancet, 2,* 539–543.

Multiple Risk Factor Intervention Trial Research Group. (1982). Multiple Risk Factor Intervention Trial: Risk factor changes and mortality results. *Journal of the American Medical Association, 248,* 1465–1477.

Olbrisch, M. E., Levenson, J. L., & Hamer, R. (1989). The PACT: A rating scale for the study of clinical decision-making in psychosocial screening of organ transplant candidates. *Clinical Transplantation, 3,* 1–6.

Oldenberg, B., Perkins, R. J., & Andrews, G. (1985). Controlled trial of psychological intervention in myocardial infarction. *Journal of Consulting and Clinical Psychology, 53,* 852–859.

Patel, C. H., (1975). Twelve month follow-up of yoga and biofeedback in the management of hypertension. *Lancet, 1,* 62–64.

Patel, C. H. (1984). A relaxation-centered behavioral package for reducing hypertension. In J. D. Matarazzo, S. M. Weiss, J. A. Herd, N. E. Miller, & S. M. Weiss (Eds.), *Behavioral health: A handbook of disease prevention and health enhancement* (pp. 846–861). New York: J. Wiley.

Patel, C., & Marmot, M. G. (1987). Stress management, blood pressure, and quality of life. *Journal of Hypertension, 5*(Suppl. 1), S21–S28.

Patel, C. H., Marmot, M. G., & Terry, D. J. (1981). Controlled trial of biofeedback-aided behavioral methods in reducing mild hypertension. *British Medical Journal, 282,* 2005–2008.

Patel, C. H., Marmot, M., Terry, D., Carruthers, M., Hunt, B., & Patel, M. (1985). Trial of relaxation in reducing coronary risk: Four year follow-up. *British Medical Journal, 290,* 1103–1106.

Perloff, D., Sokolow, M., & Cowan, R. (1983). The prognostic value of ambulatory blood pressures. *American Medical Association, 249,* 2792.

Peters, R. K., Benson, H., & Peters, J. M. (1977). Daily relaxation breaks in a working population: 2. Blood pressure. *American Journal of Public Health, 67,* 954–959.

Peterson, L. H. (Ed.), (1983). *Cardiovascular rehabilitation: A comprehensive approach.* New York: Macmillan.

Pickering, T., Harshfield, G., Kleinert, H., Blank, S., & Laragh, J. (1982). Comparisons of blood pressure during normal daily activities, sleep, and exercise. *Journal of the American Medical Association, 247,* 992–996.

Prichard, B. N. C., & Owens, C. W. I. (1983). Drug treatment of hypertension. In J. Genest, O. Kuchel, P. Hamet, & M. Cantin (Eds.), *Hypertension: Pathophysiology and treatment* (2nd ed.). New York: McGraw-Hill.

Rabiner, C., & Wilner, A. (1976). Psychopathology observed on follow-up after coronary bypass surgery. *Journal of Nervous and Mental Diseases, 163,* 295–310.

Radloff, L. S. (1977). The CES-D scale: A self-report depression scale for research in the general population. *Journal of Applied Psychological Measurement, 1,* 385–401.

Ragland, D. R., & Brand, R. J. (1988). Coronary heart disease mortality in the Western Collaborative Group Study: Follow-up experience of 22 years. *American Journal of Epidemiology, 127,* 462–475.

Raymond, M., Conklin, C., Schaeffer, J., Newstadt, G., Matloff, J., & Gray, R. (1984). Coping with transient intellectual dysfunction after coronary bypass surgery. *Heart and Lung: The Journal of Critical Care, 13,* 531–539.

Robbins, L. C., & Hall, J. H. (1970). *How to practice prospective medicine.* Indianapolis: Methodist Hospital of Indiana.

Rosenman, R. H., Brand, R. J., Jenkins, C. D., Friedman, M., Strauss, R., & Wurm, M. (1975). Coronary heart disease in the Western Collaborative Group Study: Final follow-up experience of 8½ years. *Journal of the American Medical Association, 233,* 872–877.

Sackett, D., & Snow, J. (1979). The magnitude of compliance and noncompliance. In R. Haynes, D. Taylor, & D. Sackett (Eds.), *Compliance in health care.* Baltimore: Johns Hopkins University Press.

Safer, M. A. (1982). An evaluation of the Health Hazard Appraisal based on primary data for a randomly selected population. *Public Health Reports, 97*(1), 31–37.

Savageau, J. A., Stanton, B., Jenkins, C. D., & Klein, M. D. (1982). Neuropsychological dysfunction following elective cardiac operation: 2. A six month reassessment. *Journal of Thoracic and Cardiovascular Surgery, 84,* 595–600.

Schoenbach, V. J., Wagner, E. H., & Berry, W. L. (1987). Health risk appraisal: Review of evidence for effectiveness. *Health Services Research, 22,* 4.

Schwartz, G. E., & Weiss, S. M. (1978). Behavioral medicine revisited: An amended definition. *Journal of Behavioral Medicine, 1*(3), 249–251.

Schwartz, G. E., Shapiro, A. P., Redmond, D. P., Ferguson, D. C. E., Ragland, D. R., & Weiss, S. M. (1979). Behavioral medicine approaches to hypertension: An integrative analysis of theory and research. *Journal of Behavioral Medicine, 2/4,* 311.

Seer, P. (1979). Psychological control of essential hypertension: Review of the literature and methodological critique. *Psychological Bulletin, 86,* 1015.

Shekelle, R. B., Gale, M., & Norusis, M. (1985a). Type A

score (Jenkins Activity Survey) and risk of recurrent coronary heart disease in the Aspirin Myocardial Infarction Study. *American Journal of Cardiology, 56,* 221–225.

Shekelle, R. B., Hulley, S., Neaton, J., Billings, J., Borhani, N., Gerace, T., Jacobs, D., Lasser, N., Mittlemark, M., & Stamler, A., & the MRFIT Research Group. (1985b). The MRFIT behavior pattern study: 2. Type A behavior pattern and incidence of coronary heart disease. *American Journal of Epidemiology, 122,* 559–570.

Shepherd, J. T., & Weiss, S. M. (1987). *Behavioral medicine and cardiovascular disease: Proceedings from the National Conference. Circulation, 76*(Suppl. 1).

Shumaker, S. A. (in press). Quality of life research in NHLBI-sponsored trials. In S. A. Shumaker & C. Furberg (Eds.), Research on quality of life and cardiovascular disease. Special Issue of the *American Journal of Preventive Medicine.*

Smith, K. W., McKinlay, S. M., & Thorington, D. (1987). Health risk appraisal instruments for assessing coronary heart disease risk. *American Journal of Public Health, 77*(4), 419–424.

Soloman, S., Hotchkiss, E., Saraway, S. M., Bayer, C., Ramsey, P., & Blum, R. S. (1983). Impairment of memory function by antihypertensive medication. *Archives of General Psychiatry, 40,* 1109–1112.

Spielberger, C. D. (1983). Manual for the State-Trait Anxiety Inventory (Form Y). Palo Alto, CA: Consulting Psychologists Press.

Spilker, B. (Ed.), (1990). *Quality of life assessments in clinical studies.* New York: Raven Press.

Steckel, S. B. (1982). *Patient contracting.* Norwalk, CT: Appleton-Century-Crofts.

Stern, M. J., Gorman, D. A., & Kaslow, L. (1983). The group counseling versus exercise therapy study. *Archives of Internal Medicine, 143,* 1719–1725.

Surwit, R. S., Williams, R., & Shapiro, D. (1982). *Behavioral approaches to cardiovascular disease.* New York: Academic Press.

Syme, S. L. (1987). Coronary artery disease: A sociocultural perspective. *Circulation, 72*(Suppl.), I-112–I-116.

Taylor, J. (1953). A personality scale of manifest anxiety. *Journal of Abnormal Psychology, 48,* 285–290.

Trelawny-Ross, C., & Russell, O. (1987). Social and psychological responses to myocardial infarction: Multiple determinants of outcome at six months. *Journal of Psychosomatic Research, 31,* 125–130.

Wallace, D. C. (1955). Treatment of hypertension: Hypotensive drugs and mental changes. *Lancet, 2,* 116.

Weiss, S. M. (1983). Health and illness—The behavioral medicine perspective. In L. S. Zegans, L. Temoshok, & C. VanDyke (Eds.), *Emotions in health and illness: Foundations of clinical practice.* New York: Academic Press.

Wenger, N. K. (1988). Quality of life issues in hypertension: Consequences of diagnosis and considerations in management. *American Heart Journal, 116,* 628–631.

Wenger, N. K., Mattson, M. E., Furberg, C. D., & Elinson, J. (Eds.), (1984a). *Assessment of quality of life in clinical trials of cardiovascular therapies.* New York: Le Jacq.

Wenger, N. K., Mattson, M. E., Furberg, C. D., & Elinson, J. (1984b). Assessment of quality of life in clinical trials of cardiovascular therapies. *American Journal of Cardiology, 54,* 908–913.

Wheatley, D., Balter, M., Levine, J., Lipman, R., Bauer, M. L., & Bonato, R. (1975). Psychiatric aspects of hypertension. *British Journal of Psychiatry, 117,* 327–336.

White, L., Tursky, B., & Schwartz, G. E. (Eds.), (1985). *Placebo: Theory, research, and mechanisms.* New York: Guilford Press.

Wiklund, I., Sanne, H., Vedin, A., & Wilhelmsson, C. (1984). Psychosocial outcome one year after a first myocardial infarction. *Journal of Psychosomatic Research, 28,* 309–321.

Williams, R. B., Barefoot, J. C. Haney, T. L., Harrell, F. E., Blumenthal, J. A., Pryor, D. B., & Peterson, B. (1988). Type A behavior and angiographically documented coronary atherosclerosis in a sample of 2,289 patients. *Psychosomatic Medicine, 50,* 139–152.

World Health Organization. (1948). *Constitution of the World Health Organization.* Geneva, Switzerland: WHO Basic Documents.

CHAPTER 22

Psychological Components of Rehabilitation Programs for Brain-Injured and Spinal-Cord-Injured Patients

Joseph Bleiberg, Robert Ciulla, and Bonnie L. Katz

Introduction

This chapter is about clinical psychology as practiced in the rehabilitation hospital. Rehabilitation settings are fertile areas of practice for psychologists, and a wide range of clinical and consultation services is applicable. To practice most effectively, however, psychologists should understand the organization and "culture" of the rehabilitation hospital and the special needs of its patients and staff. The chapter is broken into three sections. First, the clinical and administrative environment of the rehabilitation hospital is described, with emphasis on how it affects the role definition and mode of practice of psychologists. The second and third sections examine two specific populations, spinal-cord-injured and brain-injured patients, to identify and illustrate the issues specific to the psychological care of each group.

Joseph Bleiberg and Bonnie L. Katz • National Rehabilitation Hospital, 102 Irving Street N.W., Washington, DC 20010. Robert Ciulla • Islip Mental Health Center, 1747 Veterans Highway, Islandia, New York 11769.

The Rehabilitation Hospital

The environment of the rehabilitation hospital and the psychologist's role are shaped by many factors. At the broadest level are sociopolitical and medical-economic factors that determine who will receive insurance coverage for rehabilitation, for how long a period of treatment, for what types of treatment, and for treatment provided by which professions. At the level of a specific rehabilitation hospital, psychological practice is affected by the administrative structure and medical staff bylaws of the hospital, the "history" of psychology within that hospital (or the lack of it), and the way the different rehabilitation professions have defined their professional boundaries, or "turf." Our discussion begins with a review of the goals of rehabilitation hospitalization.

Goals of Rehabilitation

Patients in rehabilitation hospitals have experienced catastrophic and life-altering disease or injury. The diagnoses most frequently seen in the rehabilitation setting include traumatic brain injury, cerebrovascular accident, brain tumor, spinal cord injury, amputation secondary to trauma or to diseases such as diabetes or peripheral vascular disease, severe arthritis and other degenerative diseases of the joints or muscles, advanced multiple sclerosis and other progressive or degenerative neurological diseases, severe burns, and chronic pain.

By the time a patient is referred to the rehabilitation hospital, acute-care medical services to reverse or arrest the injury or disease have typically already been performed. Cure, therefore, usually is not a realistic goal of the hospitalization, and the more typical goal is to ameliorate the functional limitations on independent living resulting from the injury or disease. Moreover, the major "gatekeeper" for entry into the rehabilitation hospital—insurance reimbursement—almost always requires that the patient have functional goals related to enhanced independence in daily living activities. For example, the "Three-Hour Rule" of Medicaid specifies that patients must have functional goals, that achieving the goals must require at least three hours per day of combined physical and occupational therapy, and that patients must be discharged from the hospital when they stop making progress toward achieving the goals.

Thus, the rehabilitation patient is in the hospital primarily to reduce functional limitations in daily living. These limitations consist of three related, but in important ways independent, factors: "impairments," "disabilities," and "handicaps" (Diller & Ben-Yishay, 1987). Impairments consist of deficits in specific motor, sensory, or cognitive abilities, such as the inability to move one's legs or to remember new information. Disabilities, on the other hand, represent the effects of impairments on one's capacity to perform specific tasks, such as going to the grocery store, getting to the second floor in one's house, reading a newspaper, or remembering a phone number. Handicaps are the cumulative effect of disabilities on one's capacity to function effectively in various societal roles, such as employment, recreation, parenting, or housekeeping.

Although the rehabilitation patient almost invariably has some specific area(s) of impairment, impairment only partially determines whether there will be disability or handicap. For example, a paraplegic patient has impairment in the use of his or her legs, and this results in a disability in walking. However, there may or may not be disability in *mobility*, depending on whether the patient is skilled in the use of a wheelchair and is in a wheelchair-accessible environment, has access to and has been trained in the use of an automobile with hand controls, and has achieved a sufficient level of psychological adaptation to the alterations in capacities imposed by the paraplegia to be willing to use such devices and accessible environments. Even if the patient does have a disability with respect to mobility, there may not be a handicap in vocational functioning, depending on whether the patient's vocation can be pursued productively despite the presence of certain impairments and disabilities. Impairments, therefore, do not invariably result in disability or handicap, nor is it possible to generalize precisely about how a specific impairment will disable or handicap a given patient.

It also is important to note that the reduction of a specific impairment or disability does not invariably result in the reduction of a handicap. Let us assume that a paraplegic patient has an incomplete spinal cord injury, so that, with appropriate exercises and training he or she will be able to walk using braces and crutches. Such a mode of walking, however, may be so slow and effortful that it results in less mobility than proficiency in the use of a wheelchair. Let us assume further that the patient has great psychological investment in walking even though walking does not give as good mobility as a wheelchair, as in the case of a patient we saw who was a sales representative and who insisted that he greet his clients "standing tall and at eye level, not sitting down and having them bend down to shake my hand, and especially not having them look down at me while we are talking." For this patient, the psychological benefits of walking and standing, in terms of enhanced self-esteem, motivation, and self-confidence, may be sufficient to offset the negative effects of reduced mobility.

The selection of rehabilitation goals, therefore, often involves a complex set of compromises and trade-offs to identify a group of goals that, in aggregate, provides the greatest reduction in the patient's handicap. This selection requires the coordinated and integrated application of multiple and diverse approaches, ranging from exercises and other treatments to reduce specific impairments, to learning to use compensatory techniques and devices to reduce disabilities, to restructuring or developing new roles at work or in the family in order to create domains in which life can be pursued with the least possible degree of handicap. All of these levels of intervention have large psychological components.

The Rehabilitation Team

The rehabilitation team includes representatives from medicine (usually a physician from the specialty of "physical medicine and rehabilitation," termed a *physiatrist*, but sometimes from related specialties such as neurology or orthopedics), rehabilitation nursing, physical therapy, occupational therapy, speech and language pathology, vocational rehabilitation, biomedical engineering, social work, recreational therapy, respiratory therapy, nutrition, and prosthetics and orthotics. The efforts of this large team are organized on the basis of the patient's functional goals. Thus, if a severely brain-injured patient is incapable of eating independently, the speech pathologist may train the patient to swallow, the physical therapist may work on the patient's posture and seating position to facilitate eating, the biomedical engineer may modify the patient's wheelchair to better sustain the eating posture, the occupational therapist may work on developing the necessary hand function and on adapting eating utensils to accommodate impaired hand or finger function, the nutritionist may formulate diets of appropriate consistency given the patient's impaired swallowing or chewing ability, the respiratory therapist may manage secretions, and so on. Because most patients have many such functional goals, the rehabilitation team is characterized by the need for extensive

collaboration and a high degree of mutual interdependence. This extends to include the patient's family, who in many rehabilitation settings are considered to be members of the team. A case manager orchestrates the interdependent activities of the team, and often, but not always, this is the physician. It is common practice, as well as an established standard of accreditation bodies such as the Commission for Accreditation of Rehabilitation Facilities (CARF), for the team to meet at least biweekly to develop and then monitor and revise a comprehensive, multidisciplinary care plan that centers on explicit functional goals. Patients and families are encouraged to participate in such team conferences, as well as in family conferences, to ensure that the functional goals pursued in the rehabilitation hospital will address the practical needs that the patient and the family will face on the patient's discharge from the hospital.

The pursuit of functional goals is typically a highly practical affair: teaching the patient to tie one-handed knots or to walk with a prosthetic leg, modifying the kitchen and bathroom of a home to make them wheelchair-accessible, and so on. Similarly, the rehabilitation team has a practical and utilitarian focus, tending much more toward action than toward "feelings." As a result, the psychologist who is new to the rehabilitation setting may feel out of place. Throughout his or her training in mental health settings, the psychologist will have worked with teams who viewed their primary mission as addressing the emotional state of their patients. In the rehabilitation setting, however, the treatment team may view patients' emotional issues simply as impediments to the real purpose of the hospitalization ("He's so depressed that we can't get him motivated for physical therapy"), and patients may be puzzled about why they are seeing a psychologist ("I came here to learn how to walk!").

The rehabilitation team includes many professions to which the psychologist has had little or no exposure during traditional mental health training, and it pursues a different agenda from that of the mental health setting. Similarly, patients and families (and certainly their insurance companies) often do not view the purpose of the rehabilitation hospitalization as being the pursuit of mental health goals. However, although psychological goals are not the overt or explicit reason for the rehabilitation hospitalization, they can be central to its success. Moreover, psychological issues are unavoidable in the rehabilitation setting: the intense interdependence between the patient and the team, and among the individual members of the team, produces a level of intimacy that inevitably arouses strong needs, wishes, and conflicts on the part of all involved.

Just as relationships within psychotherapy dyads are bipersonally active (Langs, 1988), relationships between physical, occupational, and other therapists and their patients are influenced strongly by the intrapsychic responses of both parties (Gunther, 1977). The many binary relationships that the patient forms with individual therapists on the team are influenced by his or her current and historical relationships with others. Similarly, how individual therapists relate to patients is determined by factors such as the therapist's degree of psychological insight, the appropriate management of interpersonal boundaries, the capacity to establish health symbiotic relationships, and the ability to tolerate and contain, rather than destructively act on, patients' hurtful projective identifications.

As Gans (1983) noted, projection, identification, and denial may go back and forth between both parties, creating unrealistic views and distorted expectations. Some patients may view the team as benevolent and omnipotent, existing solely to promote the patient's well-being and recovery, and surely able to effect a cure if only the patient is sufficiently obedient to and compliant with the team's recommendations. Other patients may view the team as a group of adversaries, there only to thwart the patient's efforts at independence and autonomy. At different times, some patients hold both views of the team, or one view toward some team members and another view toward others. Team members may view patients' failure to make anticipated therapeutic gains as personal failures in their healing ability and may project hatred onto the patient as a result

(Gans, 1983). Or, team members may identify with patients (Judd & Burrows, 1986). Take, for example, a recently admitted 22-year-old female patient, quadriplegic from an auto accident during summer vacation after graduation from college, who was about to start a career and be married: it is not unlikely that the physical or occupational therapist treating this patient will share many of these demographics. The patient may become the recipient of an overzealous therapist who copes with the anxiety stimulated by identification with the patient by having particularly high expectations for her recovery. The patient, however, may perceive the therapist's enthusiasm as an empathic failure to appreciate how defeated and awful the patient feels. The therapist may be seen as demanding and sadistic ("She's torturing me"), and the patient as unmotivated and resistant ("She's not working up to her potential").

Alternately, the 60-year-old patient, quadriplegic from a spinal cord malignancy, may evoke despair and helplessness in her therapist ("She'll wind up going to a nursing home and dying anyway"), even though the patient may be well adjusted to her prognosis and may be highly motivated to increase her level of independence during her remaining years of life. The therapist may show little enthusiasm when treating the patient, perhaps becoming emotionally withdrawn or constricted, and the patient may come to feel rejected and abandoned. Another therapist working with this patient may have had a mother who died of a rapidly fatal malignancy. This therapist may view the patient's having several more years of life expectancy and the desire to live those years as fully as possible as positive factors and a cause for enthusiasm and may be able to form an effective therapeutic alliance that has realistic as well as idealized aspects. Moreover, when two therapists have such divergent perceptions of and relationships with a patient, particularly when the differences relate strongly to the intrapsychic needs of one or both therapists, conflict among team members may develop.

Clearly, all health care personnel and all patients have the capacity for distorted perceptions of one another. However, distorted interpersonal perceptions increase as relationships become more intimate, and rehabilitation hospitalizations are measured in months, rather than the days typical of acute care hospitalization. While not directly a part of the rehabilitation team's mission, agenda, or identity, these intense and interdependent relationships, with their capacity to stimulate distortions, are an unavoidable and inevitable part of the day-to-day activities influencing patients' well-being and successful participation in rehabilitation.

The Psychologist's Role

In addition to providing direct clinical services, the psychologist's role on the rehabilitation team is to enhance patient care by integrating psychological knowledge and interventions into the team's approach to treating patients. The ways in which the psychologist does so depend to a large extent on the form of relationship that he or she has with the team. Currently, three different types of relationships are common: psychologists may be fully integrated *members* of the team; psychologists may be *consultants* to the team, seeing only those patients who are identified by the team as needing psychological services; and, on the uptrend in recent years, particularly in brain injury and chronic pain rehabilitation programs, psychologists may be *leaders* or *coleaders* (usually jointly with a physician) of the team.

The model where the psychologist is a fully integrated member of the team has distinct advantages. In this model, the psychologist routinely evaluates all patients treated by the particular team of which he or she is a member, provides direct treatment to patients and families as needed, and serves as a behavioral sciences consultant to the team. Routinely seeing all patients admitted allows the psychologist to serve several quite useful functions. Patients with preexisting mental disorders or with disturbances reactive to their disability can be identified promptly, and the team can be alerted to any special management needs of such patients before such needs exert a disruptive influence on the rehabilitation program and produce an impasse between the patient and the

team. Even when patients show no substantial psychopathology, there may be enormous benefits when the psychologist assists the team to gain a psychological understanding of the patient. This can include information about the patient's personality and coping style, ways of handling threats to independence and autonomy, past history of losses and response to them, unique "meaning" of the injury or illness, expectations for the future, self-concept, cognitive abilities, interpersonal skills, and other descriptions that assist the team to view the patient as a differentiated and unique psychological being. Providing such information *early* during a patient's hospitalization allows the team to incorporate a psychological understanding of the patient into their initial formulation of the treatment program, thus customizing it from the outset to the patient's needs, preferences, and capacities. Such information also reduces the team's potential misperceptions of the patient's behavior, such as interpreting a depressed patient's behavior as reflecting lack of motivation and rejection of what the team has to offer.

Seeing all patients initially on admission also allows the psychologist gradually to sensitize the team to psychological factors as they apply in general to rehabilitation patients, and to develop the team's capacity to create a healthy psychological milieu for patients. Enhancing the psychological sophistication of the team, in turn, increases the psychologist's ability to develop treatment approaches that require the active collaboration of the entire team. It also helps the team to understand and value the contributions and point of view of the psychologist, and to accept psychological goals as legitimate pursuits within the rehabilitation context. Further, as the team becomes psychologically sophisticated, this sophistication is communicated to patients and families in subtle as well as in overt ways, and their acceptance of the psychologist's active involvement in their care is facilitated.

One difficulty in being a fully integrated member of the team relates to how the psychologist defines his or her role and status on the team in relation to the Bachelor's and Master's level therapists who comprise the team. The psychologist may have difficulty accepting complete parity with Bachelor's and Master's level clinicians, particularly given the professional alternatives available outside the rehabilitation setting (e.g., a university or medical school professorship, independent private practice, or the leadership of mental health teams and programs) and the implications that parity may have in areas such as salaries and clinical autonomy. A number of approaches are possible. For example, medical staff membership with either full privileges or more restricted "consulting" privileges allows the psychologist to function as a member of the treatment team, but with a recognition of her or his level of training and expertise, as well as with an organizationally sanctioned and broadly understood basis for an appropriate degree of professional autonomy and for competitive salary levels. Faculty status at a university or a medical school can confer a similar benefit.

The psychologist may also add selected team leadership functions to his or her team membership role. The hospital's policies and procedures can be written to give the psychologist formal leadership over specific clinical issues, such as evaluating suicidal or assaultive patients and supervising the team in instituting appropriate precautions, using seclusion and restraint for behavioral management, or coordinating domains of treatment such as "cognitive retraining." One such formal mechanism that has proved quite useful in the authors' setting has been the use of a "psychology–neuropsychology team conference," led by the psychologist. This conference is the behavioral science parallel of the medically oriented team conference led by the physician, and it allows the psychologist to coordinate the team's activities as they relate to the psychological and neuropsychological needs of patients. The conference has the side benefit of formally demonstrating that although the psychologist is a member of the treatment team, he or she also has a role that includes coordination and leadership over a portion of the team's activities. Leadership may also be undertaken in the areas of research and program development

and evaluation, areas where the psychologist may have substantial expertise and training.

It should be pointed out that although the above methods may be quite helpful in creating a satisfying role for the psychologist, they still leave the psychologist with the difficult task of balancing being a member of the team for some activities and being a leader for others. Tact, diplomacy, and a good sense of boundaries are essential to making this situation productive rather than disruptive and alienating. The psychologist must have a sincere and manifest respect for the contributions of the rest of the team, derived from a thorough knowledge and understanding of the different disciplines. Moreover, psychologists who cannot tolerate being, at least a portion of the time, identified as peers of Bachelor's level professionals are perhaps best advised to avoid this model of psychological practice.

An alternate model of service delivery is for the psychologist to restrict his or her role to consultation to the team. In this model, the psychologist evaluates and treats only those patients formally referred by the physician and the team. Although serving exclusively as a consultant produces a more clearly defined and less awkward role for the psychologist, it has disadvantages. Having nonpsychologists select those patients who will receive psychological services can result in the psychologist becoming involved only after a situation has developed into an irreparable impasse between the patient and the treatment team that would have been preventable with earlier identification and intervention. Referrals may be limited to patients with more obvious psychological disturbances, and patients who have more subtle disturbances or who may benefit from supportive psychotherapy may be overlooked. Moreover, because the psychologist's interactions with the team are more sporadic and generally center on "putting out fires," there is less opportunity for the psychologist to enhance the general psychological sophistication of the treatment team and to work with them to create an overall psychologically sensitive milieu that addresses the needs of all patients, and there is less opportunity for the psychologist to show the team the various but perhaps less obvious ways in which psychological services can assist in rehabilitation.

In summary, although the psychologist as a consultant has a better protected status and a clearer professional role, they are gained at the expense of more limited access to patients and more restricted opportunities for applying psychological services in their broadest extent. Overall, the psychologist as a team member provides the best opportunities for the maximal involvement of psychologists in clinical care, and this role can be augmented with formal and informal leadership responsibilities that recognize psychologists' level of training, expertise, and professional autonomy.

Psychological Services

As is shown below, the rehabilitation hospital offers psychologists the opportunity to apply virtually the full spectrum of clinical psychological, neuropsychological, and behavioral medicine services.

Some patients admitted to the rehabilitation hospital may have preexisting psychological or psychiatric disorders in addition to the illness or injury that is the basis of the rehabilitation hospitalization. These patients need evaluation and management during the rehabilitation hospitalization, with special emphasis on the psychologists assisting a team that often is not prepared to integrate the treatment of traditional physical medicine problems with the management of a significant psychiatric disturbance. A complication is that, sometimes, the team may have great distaste for such integration. It is not uncommon to see patients who have spinal cord injury because of psychosis ("The voices told me to jump out of the window"), traumatic amputation because of self-multilatory behavior, brain injury secondary to depression and failed suicide attempts, and so on.

Another group of patients may have no prior psychiatric history but may develop psychological disturbances in reaction to their injuries or diseases. These may range from adjustment reactions to severe and incapacitating disorders

so severe that the patient is entirely unable to participate in the rehabilitation program. Moreover, even when adjustment reactions are not so severe as to be diagnosable psychopathology, there may be many useful interventions that the psychologist can undertake to optimize the patient's adaptation and well being. Such patients draw the psychologist into the traditional and well-known mental health role. Other needs of rehabilitation patients require the psychologist to go beyond this role. Among the most frequent reasons for rehabilitation hospitalization are cerebrovascular accident, head injury, and other diseases and injuries to the brain. In addition to preexisting and reactive psychological disorders, these patients may have organic cognitive, personality, and behavioral disturbances that require that the psychologist possess expertise in neuropsychological assessment and treatment. In some settings, a distinction is made between a clinical psychologist who provides more traditional assessment and psychotherapy services, and a clinical neuropsychologist who provides assessment and treatment for neurocognitive and neurobehavioral disorders. Although such a distinction can be justified on the basis of the scarcity of psychologists with neuropsychological training, it is more difficult to justify on the basis of how best to serve brain-impaired patients: these patients and their families have clinical psychological and neuropsychological needs that are interwoven and difficult to compartmentalize.

Other patients may have problems with acute or chronic pain, and a whole field has developed around the methods for psychological intervention with such patients (Fey & Williamson-Kirkland, 1987). Still other patients may require behavioral medicine services for lifestyle modification in the areas of smoking cessation, diet, exercise, compliance with medical regimens, and stress management. Substance abuse problems may also be prevalent in rehabilitation populations, in some cases being a direct contributor to the disease or injury that has occasioned the need for rehabilitation, as in the patient who has wrecked a car or fallen while intoxicated and has sustained a brain or spinal cord injury, and who risks doing so again if substance abuse is not identified and treated during rehabilitation. An often overlooked role of the psychologist in the rehabilitation hospital is being an expert in human learning. Patients come to a rehabilitation hospital largely in order to acquire *skills* to enable them to function optimally despite the limitations imposed by an illness or injury. In a study of 20 consecutive admissions to a spinal cord injury unit, Bleiberg and Merbitz (1981) found that 64% of the entries in the medical record related directly to the *acceleration or deceleration of specific behaviors*. Psychologists' extensive expertise in human learning can be applied productively to many rehabilitation issues.

In addition to providing direct clinical services to patients, a variety of consultation and education services may be provided. As discussed previously, the intrapsychic needs of patients and staff inevitably become a part of the therapeutic matrix, and the psychologist provides clarification and insight in order to bring such issues to adaptive resolution. This can be done during team conferences, when the team raises specific concerns about a patient's behavior or mental state. The psychologist addresses such concerns by describing the psychological dynamics underlying the patient's behavior, sometimes (cautiously) introducing, for the team's consideration, thoughts about how aspects of their approach to the patient may be related to how the patient is responding to them. At these and at psychology–neuropsychology conferences, the psychologist also validates the team's tribulations with particularly difficult or overwhelming patients by providing a context where staff can talk about their feelings and by providing mutual understanding and support.

Active problem solving with the team is essential and may occur in the above kinds of conferences or in "hallway consultations." In many cases, these result in the psychologist's assisting a team member to incorporate a psychological intervention within a traditional medical treatment, such as assisting the physical therapist to develop a behavioral contract with a noncompliant patient. In some cases, this can extend to actual cotreatment, as when

the psychologist provides hypnosis for relaxation and pain management while the physical therapist performs range of motion exercises on a painful joint. (Although such treatment is very effective in some cases, consideration should be given beforehand to its possible effects on any concurrent psychotherapy relationship that may exist between the psychologist and the patient.)

Relations with Psychiatry

It is rare for psychiatrists to be members of rehabilitation teams, but it is routine for them to be available as consultants. Moreover, many of the psychological services needed by rehabilitation patients are provided both by psychiatrists and by psychologists (as well as by others, as discussed in the next section), so that competitive tension may develop between the two professions. (See Chapter 6 by Belar for discussion of professional relationships.) As a consultant, the psychologist is exposed to the full force of this competitive tension and to the variations in how it is resolved in different settings. The rehabilitation physician team-leader is free to select the consultant of his or her choice, whether a psychologist or a psychiatrist. The psychologist is vulnerable to the team leader's notions of what psychologists and what psychiatrists should do, and to prejudices about one or the other profession (which, surprisingly, in rehabilitation settings, often favor the psychologist). To be sure, the psychologist can emphasize types of services not typically offered by psychiatrists, such as neuropsychological testing or biofeedback, but there inevitably remains a large area where the two professions can and do overlap and compete.

As a team member, the psychologist has more opportunity to select a collaborative rather than a competitive resolution of this issue. Because the psychologist has regular and frequent contact with the physician and the team, there is a greater opportunity to educate them about the types of situations well managed by psychologists and those where psychiatric consultation is advisable. Moreover, because the psychologist evaluates all patients on admission to the rehabilitation unit, the psychologist can identify those situations where psychiatric consultation is indicated and can participate in specifying the nature and topics of the consultation request to the psychiatrist. In the authors' setting, it is self-imposed policy to request a psychiatric consultation in all cases of major affective disorder and active psychosis; it is optional in other cases, based on the psychologist's clinical judgment. By agreement with the medical staff (of which psychologists are members), psychologists participate in formulating the referral request for psychiatric consultation, in most cases actually making the phone call to the psychiatrist and describing what is desired from the consultation. Our experience has been that this procedure minimizes misunderstandings between psychologists and psychiatrists. Minimizing such misunderstandings increases the extent to which psychologists will initiate requests for psychiatric consultation and fosters collaborative rather than competitive relationships. It also has the substantial benefit of preventing the psychologist from coming to work one morning and finding that his or her patient not only has had a psychiatric evaluation the day before but now is in psychotherapy with both a psychologist and a psychiatrist who have not yet even discussed the case.

Relationships with Other Disciplines

The psychologist new to the rehabilitation setting will be astonished by the amount of "psychological" activity routinely performed by nonpsychologists. Physicians and nurses routinely provide counseling to patients and families. Occupational therapists commonly evaluate visual-spatial, psychomotor, and memory functions and provide intensive treatment in these areas. Speech and language pathologists have a long-standing professional involvement in the evaluation and treatment of the aphasias and, in recent years, have extended this to include cognitive disturbances secondary to various neurological disorders. Social workers include mental health diagnosis and psychotherapy in their training and professional identity. Vocational rehabilitation counselors use tests of mental abilities and apti-

tudes and provide counseling and guidance. Moreover, the increased availability of computerized testing and test interpretation, often backed by very aggressive marketing to nonpsychologists, has brought many procedures previously restricted to psychologists within easy use of other professionals.

In rehabilitation settings that have not previously employed a psychologist, the above disciplines, perhaps with the assistance of a psychiatric consultant, are likely to have incorporated into their established roles many of the functions and activities that the psychologist would wish to perform. The psychologist will feel a legitimate claim to these activities but so will the other professions, and the issue will become how to divide and share roles and activities equitably among several qualified providers. Adaptive, but sometimes maladaptive and outright bizarre, solutions are possible. Among the worst solutions are those in which the parties involved avoid addressing the issue altogether, as though it did not exist. Such solutions can result in patients, simultaneously having *four* individual psychotherapists (a psychiatrist, a psychologist, and a social worker, all engaged in psychotherapy, and a rehabilitation nurse providing sexual and family counseling); the situation may be further entangled because different members of the treatment team and the family have formed alliances and may have consulted with different "psychotherapists." Current reimbursement practices for rehabilitation encourage rather than discourage such practices. Rehabilitation remains exempt from diagnostic-related-group (DRG) prospective reimbursement, so that the more the service given to a patient, the more the revenue for the hospital. There thus is no financial incentive to regulate and monitor duplicative treatments. However, although such situations may not trigger administrative regulation, they can and should be approached from the point of view of sound clinical practices.

An equitable resolution of turf conflicts is essential to prevent divisive and unproductive relationships among the team members. Because these conflicts are so emotionally arousing and can threaten the integrity of the team, there is frequently collusion among the members to keep the conflicts from becoming overt. Our experience has been that although there may be initial resistance to discussing these issues, discussion can be beneficial and usually results in mutually acceptable compromises. These typically are based on the involved parties' first acknowledging the reality that their disciplines overlap, and sharing their mutual distress at having others interfere with what they see as their spheres of authority and expertise. As the involved parties begin to recognize that they all feel a similar sense of having their professional identities encroached on, it often becomes possible to negotiate compromises, and the results may then be formalized as hospital policy and procedure. Moreover, such discussions may also help to identify areas where professional overlap is desirable and productive, further reducing tensions and misunderstandings.

Finances and Productivity

As health care reimbursement becomes increasingly restrictive, and hospitals scrutinize their revenues and expenses ever more closely, increasingly "tough" negotiations about productivity are inevitable between psychologists and hospital administrators.

The unit of commerce in the rehabilitation hospital is the "billable hour." This is a familiar unit to psychologists, who are accustomed to generating income based on the time they invest in providing a service, but it has some special implications in the rehabilitation setting. The most important has to do with the fact that the productivity, or "billable hours per week," of psychologists is viewed by hospital administrators in the context of the productivity of other disciplines, such as occupational or physical therapy, and the psychologist, for several reasons, is at a decided disadvantage when this measure is used. In many settings, these other disciplines generate 30 or more billable hours per week, but, as is shown below, this amount is unrealistic for psychologists.

Patients typically receive 3 hours per *day* of occupational and physical therapy in order to justify their stay in a rehabilitation hospital.

Assuming that a physical therapist provides 1½ of the 3 hours, a caseload of four patients is sufficient to yield 6 hours per day, or 30 hours per week, of billable activity. This means that the physical therapist can generate 30 hours per week while attending team and family conferences, receiving phone calls and making referrals, writing weekly progress notes in the medical chart, and so on, *for only four patients.* The psychologist, on the other hand, is likely to see patients, on average, for 1 hour per week, so that a caseload of 20 or more patients is typical. However, the indirect activities increase dramatically as a result. For example, with a caseload of 20 patients who are team-conferenced for 30 minutes every two weeks, 5 hours per week (not including the additional time for preparation and documentation) are used for this function alone, as contrasted to the physical therapist's caseload of four patients, who require 1 hour per week for this function. Similarly, activities such as phone calls from families, making discharge arrangements, informal "hallway consultations," scheduling, and other seemingly minor indirect activities affect the psychologist at a 5 to 1 ratio compared to most of the other therapists.

The main point of this analysis is to demonstrate that psychologists cannot achieve the level of productivity, as defined by billable hours, of many of the other rehabilitation disciplines. Even maintaining a level of 20 hours per week, which is two thirds of that typical of other disciplines, clearly requires the psychologist to work more than a 40-hour week, and time for research, training, and program development is not even included. Hopefully, psychologists can use analyses such as the one above to document and justify reasonable standards of productivity. Moreover, when psychologists contemplating employment in a rehabilitation setting are offered a full-time position that requires 20 hours per week of direct patient contact, they should be alert that this is not quite the easy task it might be in other employment settings.

Several approaches to psychologists' favorably negotiating productivity issues with hospital administrations are possible, but all require that the unit of productivity measurement, the billable hour, be abandoned. This course is far more difficult than it may appear, as the billable hour is highly entrenched in the rehabilitation hospital culture, just as the 50-minute hour is in the psychological culture. However, if psychologists' productivity can be measured in terms of revenue, then several approaches to enhancing productivity are available. One is to increase the price of services in order to reflect and offset the large amount of indirect service to the patient that underlies each hour of direct service. The experience in the authors' and other settings has been that such practices are well accepted by third-party payers, provided they are not overdone and there is reasonable documentation to justify them. Another approach is to develop charge systems to capture some portion of the indirect services. Whereas psychologists in mental health settings may not be accustomed to charging for consultation services where they are assisting another professional to treat the patient, it may be appropriate to charge, for example, for time spent in developing and monitoring a behavior modification program that is implemented by a physical therapist or a cognitive training program implemented by an occupational therapist. Again, the standard of elementary reasonableness applies, and every hallway chat about the patient should not become subject to a charge. Last, psychologists can enhance productivity through the *judicious* use of lower salaried technicians and psychometrists, particularly for such functions as neuropsychological testing.

The above has been a brief review of the rehabilitation hospital, with an emphasis on aspects of the environment that influence the role of psychologists. In the next two sections, clinical work with two specific rehabilitation patient populations is reviewed.

Psychological Services for Patients with Spinal Cord Injury

Demographic studies of spinal cord injury were reviewed by Trieschmann (1988). Approximately 8,000 individuals in the United States sustain new spinal cord injuries each year. Improving medical technology has resulted in increased survival of the initial injury and the potential for a normal life span following the injury. Of spinal-cord-injured patients, 82% are male, and 61% are between 16 and 30 years old, with a modal age of onset of 19. Although the most common cause is motor vehicle accidents, the etiology varies by race, gender, and age. Males sustain more sports-related injuries than females, whites sustain more sports-related and fewer violence-related injuries than nonwhites, and injuries due to falls increase dramatically with age. The incidence rates of paraplegia and quadriplegia are similar, though quadriplegia is more frequent with advanced age.

Spinal cord injury is devastating and often life-threatening, and its neurological effects involve the renal, urological, reproductive, cardiovascular, and respiratory systems (Ducharme & Ducharme, 1984). Spinal cord injury may occur at any level of the vertebral column as a result of either injury or disease. Damage to the ascending and descending nerve tracts results in sensory and motor losses at and below the level of injury. Complete injuries of the spinal cord leave no function and no recovery of function below the level of damage, though incomplete injuries can show varying degrees of recovery. Cervical lesions affect both the upper and the lower extremities (quadriplegia), and thoracic, lumbar, or sacral lesions affect the lower extremeties (paraplegia). In addition to paralysis and loss of sensation, typical physical changes include loss of bowel and bladder control, disturbances in the regulation of body temperature, sexual and reproductive dysfunction, and respiratory difficulties. Chronic pain and severe muscle spasm in the affected limbs also are common problems.

Psychosocial Issues in Spinal Cord Injury

Focusing rehabilitation on the physical aspects alone (as encouraged by health insurance reimbursement practices) has proved to be far too narrow. Krause and Crewe (1987), for example, showed that good psychosocial adjustment is a critical feature of long-term survival with spinal cord injury. Recovery and adjustment both are long-term processes that may involve major alterations in lifestyle, often with significant financial and social implications (Cook, Bolton, & Taperek, 1981). The reactions of society may engender social isolation (Haney & Rabin, 1984), may inhibit the expression of sexuality and of negative emotional states (Zola, 1982a), and may be overprotective (Zola, 1982b). Moreover, the stress related to adaptation does not necessarily diminish over time (Frank & Elliot, 1987). For example, McGowan and Roth (1987) found spinal-cord-injured patients more than two years postinjury to have greater psychosocial difficulties than did those less than two years postinjury. Thus, physical rehabilitation from the injury is only the first step in what is a lifelong process of adjustment and readjustment.

Successful rehabilitation has been related to many psychological and social factors, including the patient's personality style (Athelstan & Crewe, 1979), marital and family situation, self-concept, and education, as well as the degree to which the patient was responsible for the injury (Dew, Lynch, Ernst, Rosenthal, & Judd, 1985). Trieschmann (1988) identified two critical coping responses associated with successful long-term adaptation: a deep and abiding will to live and a willingness to accept responsibility for one's own life and care. She found that patients with an internal locus of control tend to adapt more successfully, as they are better at taking responsibility for the daily care activities necessary to survival (though she also found that such patients often are perceived by rehabilitation staff as "problem patients" because they demand more control over their environment and are less compliant and obedient). Moreover, despite the widely held belief that psychological recovery requires that patients sequentially traverse a series of fixed stages—typically including periods of denial, protest, depression, and anger—Trieschmann found no evidence to support this claim and found wide variation across patients.

A patient's response to hospitalization for spinal cord injury is a function of his or her personality, the life circumstances and role obligations at the time of injury, and the characteristics of the care setting and the caregivers. For example, the contemplative individual who has prided himself or herself on autonomy, and who withdrew from others during times of stress, may find especially toxic the loss of privacy and the necessity of relying on others to accomplish activities of daily living, such as eating, blowing the nose, or having a bowel movement. The adjustment process is also influenced by such factors as the stability of the patient's marriage, the type of vocation and whether it can be pursued despite physical disability, and the degree to which personal appearance or athletic performance was central to self-esteem. Moreover, the medical care setting can be psychologically traumatic in its own right: helicopter evacuation to a trauma center, immobilizing devices, tracheostomies, catheterization, major surgery, and either the uncertainty of knowing if paralysis is permanent, or the blunt pronouncement "You'll never walk (or use your arms) again." During this period of acute medical care, the patient may show confusion, disorientation, anxiety, or psychotic behavior (Judd & Burrows, 1986). Loss of sensation may cause panic. In the face of these various traumata, the patient is led to believe that his or her sense of wholeness and integrity will return as a result of a period of treatment in the rehabilitation hospital.

Thus, the patient arrives at the rehabilitation hospital seeking to regain the continuity of his or her life—physically, socially, vocationally, interpersonally, and intrapsychically. Instead, the patient finds another medical system with rules, procedures, and values, and this system makes demands and has expectations. In cases where the acute care system has not been explicit with the patient about prognosis, the patient may have wildly unrealistic expectations of the rehabilitation stay: "I'm going to *walk* out of this hospital." Even when the prognosis has been addressed quite directly, the enormity of its implications may not be comprehended fully.

The structure of the rehabilitation environment and the attitudes of the staff influence how the patient will perceive the injury and go about reconstructing his or her identity as a person with a disability. Therefore, a key role for the psychologist is to facilitate the development of a psychologically sophisticated and supportive environment. This may include instructing the staff about psychodynamic issues, helping to structure the rehabilitation program in ways that make the environment as "nontoxic" as possible, and assisting the staff to integrate the principles of learning theory into their interactions with patients. Individual, group, and family psychotherapy allow the psychologist to facilitate the patient's adaptation to the injury, as well as to assist the patient to cope with the rehabilitation setting and its demands and limitations. Knowledge about the patient derived through the psychotherapy relationship, in turn, allows the psychologist to consult with the staff regarding how to work most effectively with the patient, including assisting the staff to understand and manage their reactions to particular patients.

Assessment of Spinal-Cord-Injured Patients

The initial session is used to assess the patient's current emotional adjustment, understanding of the nature of the injury and the prognosis, the availability and quality of family and social support, style of coping and perception of and reaction to the rehabilitation hospital. Given that the patient typically has not asked to see a psychologist, nor did the patient come to the rehabilitation hospital because he or she was seeking a psychological evaluation, it is often best for the psychologist not to pursue assessment too vigorously or rigidly during this initial contact, and rather to emphasize constructing a benign therapeutic environment that can serve as the basis for a future trusting alliance.

Some patients may have preexisting notions and prejudices about talking to a psychologist ("I don't need to talk to you—I'm not crazy"). Other patients may view the psychologist as an adversary, engaged in examining and thereby challenging the integrity of precisely those functions that they believe have been spared by the injury. These patients may need to protect such preserved aspects of their identity in order to tolerate the physical impairment. Because the perceived intactness of cognitive and personality processes contributes significantly to the recovery and reconstruction of self-esteem, the psychologist must weigh the benefits of assessment against the possible costs to the patient and to the future psychotherapy relationship. Information about cognitive ability and personality is useful to the treatment team, but poor (real or imagined) performance may constitute a devastating narcissistic injury to a patient trying to reconstitute his or her identity. Assessment, therefore, should be sensitive to the patient's conscious and unconscious efforts to psychologically structure an adequate set of coping responses, especially during the early phases of rehabilitation hospitalization.

The above is not to say that many patients will not be "ripe" for an in-depth exploration of psychological issues starting at the initial contact. Many will be grateful to have the opportunity to talk at last to someone interested in how they feel emotionally rather than how they function (or do not function) physically. Others may be in crisis. Spinal cord injury often happens in the context of auto accidents, and family members or close friends may have been passengers. It is not uncommon for the patient to have lost a spouse or a child in the accident, a source of crisis that may be compounded when the patient was at fault in the accident. Or the patient may have been in crisis *before* the injury, with a failing marriage or work situation that further deteriorates after the injury. For these as well as the previously discussed reasons, there may be good cause to defer formal assessment procedures until a stable therapeutic alliance has been formed.

Because assessment frequently blends with the initiation of psychotherapy, it is important in the initial session for the psychologist to explain to the patient the purposes of psychological services, particularly in ways that emphasize "normal" adjustment issues and that deemphasize psychopathology and its associated stigmata. The nature of the therapeutic relationship should also be explained, along with the psychologist's role on the treatment team. The psychologist's possible discussions of the patient with other team members, confidentiality and its limits, and the psychologist's procedures for the management of confidential clinical information should be discussed.

The initial sessions emphasize developing an understanding of the patient's family history and current family functioning; academic record, including performance and attitude concerning prior education and level of motivation to pursue further education; vocational history; and personal and family psychiatric and drug history. The availability of support systems and the patient's involvement with them is reviewed. Ancillary stressors predating the injury are identified. The patient's manner of coping, views about depending on others, and prior unresolved family issues will influence the perception of and response to the treatment setting and team. An understanding is developed of the patient's personality structure and possible preinjury psychopathology, as manifested in close or intimate relationships, especially as these may surface in the rehabilitation setting. The patient's perception of himself or herself and of the impact of the injury is explored, particularly disturbances in self-concept and self-esteem and threats to identity and body integrity. Responses to recurrent stressors, dependency, and loss of control are addressed. Also considered are areas of strength, particularly former successful coping with trauma and loss.

In the course of the interview, information regarding loss of consciousness and other alterations in mental status should be obtained. The presence of head injury is not uncommon in spinal-cord-injured patients (Silver, Morris, & Otfinowski, 1980; Wagner, Kopaniky, & Esposito, 1983; Weiss, 1974). Although subtle sequelae of head injury or hypoxic brain damage are typically not searched for in spinal-cord-injured patients (Davidoff, Morris, Roth, & Bleiberg, 1985b), these are more prevalent than was earlier believed (Wilmot, Cope, Hall, & Acker, 1985). Wilmot *et al.* found that 64% of

their sample showed cognitive impairment (though, importantly, 44% had preinjury poor academic functioning), a finding similar to that of Davidoff, Morris, Roth, and Bleiberg (1985a) who found a cognitive impairment incidence 57%. Wilmot *et al.* suggested that the patients most at risk of cognitive deficits are those whose injury was the result of sudden impact and who suffered respiratory complications from the injury. However, as Trieschmann (1988) noted, the above studies very likely overestimated the incidence of head injury because other factors, such as anxiety, inattention, depression, and poor premorbid cognitive abilities may have contributed to the observed prevalence of cognitive impairment. Because much of rehabilitation is based on the patient's ability to learn and retain information, potential cognitive deficits and learning disabilities need to be identified and accommodated in the rehabilitation program, regardless of their etiology.

Significant emotional or cognitive disturbances identified during the interview can be assessed more fully with formal psychological or neuropsychological testing. Although useful in identifying psychopathology, psychological tests are perhaps even more useful and more important with the spinal cord injury patient in identifying areas of strength and sometimes even areas of unexpected talent. Many spinal-cord-injured patients are blue-collar workers who never considered higher education or the pursuit of a profession, and they are unlikely to view themselves as candidates for such goals. However, the discovery of cognitive strengths can lead the patient and the team to consider many educational and vocational options that otherwise may not have seemed appropriate.

Treatment Issues

A variety of treatment goals guide the psychologist working with spinal-cord-injured patients. Some are best achieved through direct psychotherapeutic services to the patient, but others are best accomplished through interventions with the rehabilitation environment and staff.

The primary purpose of psychotherapy during the inpatient hospitalization is to enhance the patient's ability to derive maximum benefit from the rehabilitation program. Although many psychotherapy goals can be identified, it should be kept in mind that because of the expense of rehabilitation hospitalization the patient is likely to have only one chance to acquire skills related to independence. Psychotherapy goals, therefore, should be tailored to maximize the gains derived from the hospitalization, and it should be kept in mind that some psychotherapy goals may be pursued most effectively on an outpatient basis after the patient has completed the rehabilitation hospitalization.

Among the most important of the inpatient goals is to assist the patient to adapt to the demands and limitations of the rehabilitation environment. The precise issues addressed vary from patient to patient, depending on the patient's attitudes toward authority, reactions to dependency, and introspectiveness, among many other factors. However, in general, the psychologist assists the patient to navigate the intense interpersonal relationships that develop between himself or herself and the many members of the treatment team, provides emotional support, promotes increased insight, and addresses psychiatric symptomatology as needed. Assisting the patient to preserve self-esteem is essential (Judd & Brown, 1987). Psychotherapy may focus on body image, perceptions of the self as damaged, response to the losses that have accrued, alterations in relationships with family or friends (or fears of such alterations), anxieties regarding future vocational and social role possibilities, and feelings about reactions from a community largely populated by able-bodied persons. Moreover, psychotherapy may address reactions to mechanical equipment and assistive devices, which many patients find alienating (Ohry, 1987).

The rehabilitation hospital is also a mirroring environment for the patient, and the "rehabilitation mirror" is a critical element in the patient's struggle with the changes in herself or himself following the injury. The rehabilitation mirror reflects images of disability and dependence on others and on mechanical devices, which to the patient make up a blatant disconfiguration of the recently healthy person that he or she was and are in marked contrast to the body image and sense of self that are the product of a lifetime of development. The psychotherapy relationship provides the patient with an opportunity to explore issues stimulated by the rehabilitation mirror. Frequently, refusal of treatment, deficient participation, and "poor motivation" are responses to some aspect of the mirroring environment (although sometimes the patient's reactions to the environment reflect disguised communications about the psychotherapy alliance).

One of the most important psychotherapy goals is to increase the patient's knowledge, and therefore the patient's capacity for mastery, regarding his or her injury. Knowledge about an injury has been shown to be facilitated most effectively through group activities (Miller, Wolfe, & Spiegel, 1975). Such groups may be directed at improving knowledge of the medical aspects of the injury (e.g., skin integrity, bowel and bladder management, and sexuality) as well as the psychosocial consequences. For example, as Wilmuth (1987) noted, sexual counseling for spinal-cord-injured patients must include information regarding the physical as well as the interpersonal aspects of sexual functioning, and the program described by Romano and Lassiter (1972) emphasizes education regarding the effects of injury on sexual performance, instruction on technique and hygiene, and the development of communication skills regarding sexuality.

Social-skills-training groups address community-based issues and provide patients with opportunities to express their views and to practice responses to specific situations. Such groups may focus on responses to a variety of social and interpersonal situations and may be used to promote assertiveness, the use of humor, anger management, the capacity to put others at ease, requesting help, and sharing information with others about the disability. Older patients have reported greater social discomfort than younger patients (Dunn, 1977) and are an important group to target. Romano (1976) emphasized social skills training to improve social competence, and image awareness groups focus on body image and appearance, offering specific training in grooming, hygiene, and dress. Stress management groups assist patients to manage stressful situations (Garrison, 1978).

Many patients come to the rehabilitation hospital with long histories of alcohol or other substance abuse. Addressing substance abuse with such patients is essential, and in the authors' setting a weekly Alcoholics Anonymous meeting is held in the hospital. The psychologist may also become involved with patients who are abusing drugs while in the hospital. It is important not to force too intense a focus on substance abuse during the early phases of rehabilitation, especially if the patient is still reacting to the emotional trauma of recent injury. Attempts to focus on substance abuse during a time that the patient feels emotionally assaulted by the recent injury can result in even more resistance and denial than is typical of substance abusing populations.

Another approach to providing psychological services is based on interventions that promote the psychological healing properties of the rehabilitation environment. Gans (1987) presented a cogent argument that "the most important target for mental health intervention for resolving staff/patient conflict in a rehabilitation setting is the staff, not the patient" (p. 187). Gans argued that the non-mental-health members of the team frequently form the most intimate and transference-laden relationships with patients: they are the ones who teach patients to "walk, talk, eat, dress, and even urinate and defecate" (p. 188). However, such team members are frequently not prepared to manage the psychologically intense, and often primitive, relationships that arise. Gans described a consultation procedure, termed the "team-attended psychological interview" (TAPI), in which the team present a "problem" patient, then are present while the patient undergoes a psychiatric interview, and then, after the patient leaves the room, discuss the patient. Although the TAPI is an educational procedure, it avoids dry and abstract psychological concepts and instead involves the team in the emotional immediacy of a patient with whom they have daily and intimate interactions.

The TAPI is only one of many potential approaches to assisting the team to understand more fully the psychological status of their patients and the nature of the relationships that are formed. Ideally, the pursuit of such understanding will be "institutionalized" within the program through a formal set of activities that may include TAPIs, in-service presentations on psychological topics, and a recognition that discussion of patients' psychological issues is an important and *routine* component of team conferences. This can extend so far as to include psychosocial goals in the mission statement of the hospital or the program and should be included in brochures and other formal descriptions of the program's services and goals. Moreover, it is extremely helpful if the various mental health providers on the team, be they psychologists, social workers, psychiatrists, or psychiatric nurses, resolve their internal "turf" issues so that they can form a working alliance and can provide a coherent psychosocial program.

In the authors setting, the mental health providers on the team lead a psychosocial task force that is a formal part of the spinal cord injury program's governance. The mandate of the task force is to integrate psychosocial treatments into all aspects of the program, and to evaluate and enhance the existing services. One critical area of concern has been the development of adequate procedures to orient new patients and families to the program, before admission, so that they understand the nature of the program and what their role in it will be. The orientation requires that the staff also make explicit its expectations, thus creating more clearly defined "boundaries" regarding the patient–staff relationship. The task force also works with the team to develop policies regarding noncompliance with therapies and procedures for the coherent application of behavior modification, and it constantly reviews the treatment environment to identify and correct psychologically toxic aspects.

Psychological Services for Patients with Brain Injury

As discussed above, the psychologist's task with spinal cord injury is to facilitate adaptation to the broad consequences of physical impairment. Brain injury, however, adds an additional psychological dimension: it can produce physical impairments of comparable severity to spinal cord injury, but it routinely also produces impairments in cognition, affect, personality, and behavior. This additional layer is formidable, in that the psychologist is still faced with the task of promoting adaptation to impairment but must do so with patients who have mental in addition to physical impairments, and whose "organ of adaptation"—the brain—has been compromised.

Psychological Consequences of Brain Injury

Although there are many causes of damage to the brain, the primary cause of admission to brain injury rehabilitation programs is traumatic brain injury (TBI), and the ensuing discussion is limited to this diagnostic group. The incidence of TBI is high: there are 422,000 U.S. inpatient admissions per year (Kalsbeek, McLaurin, Harris, & Miller, 1980), a figure that excludes patients with injuries sufficiently mild not to warrant hospital admission and those with immediately fatal injuries. Of those admitted to the hospital, approximately 80% have mild TBI (Caveness, 1977), though there is substantial controversy about whether any TBI truly is "mild." The incidence is highest for young adults, males outnumber females by 2 to 1, motor vehicle accidents are the most frequent cause (assaults, falls, and industrial and athletic accidents are less frequent causes), alcohol and drug intoxication are often contributing factors, and a large number of patients have a history of prior head injuries (Rimel & Jane, 1983).

Because the severity of injury varies so widely, it is difficult to generalize about the impairments of TBI patients. In cases of milder injury, the impairments may be minimal or subtle, and in more severe cases, the impairments may be devastating and may encompass all aspects of physical and mental functioning. The typical requirement that a patient need inpatient physical and occupational therapy in order to be admitted to a rehabilitation hospital, however, tends to select patients with moderate and

severe injuries. With very rare exception, TBI sufficiently severe to produce physical impairments that justify rehabilitation hospitalization also produces a substantial degree of cognitive and behavioral impairment.

The nature and severity of TBI patients' physical, cognitive, and behavioral impairments may change dramatically during the course of the rehabilitation hospitalization. Coma, obtundation, and decreased alertness are frequently early features during the hospitalization, often giving way to disorientation and agitation later on. During the period when these features are present, many other cognitive impairments may be masked. As the features of the early stages of recovery subside, neurologically based impairments of cognition and personality, as well as psychological reactions to such impairments, become manifest. Because so many of the "psychological" factors in TBI patients are directly neurological, it is essential that the psychologist understand the neuropathology of TBI and its relation to cognitive and behavioral impairment. Several excellent sources are available (Adamovich, Henderson, & Auerbach, 1985; Alexander, 1987; Auerbach, 1986; Gennarelli, 1986; Levin, Benton, & Grossman, 1982), and the discussion below is only a very brief review.

TBI patients generally incur four types of injury to the brain. First, acceleration-deceleration of the brain, as is typical in car accidents and falls, causes "strain" and "shearing" of axons in the midbrain, the brain stem, and the corpus callosum. Damage to these areas is related to impairments in level of alertness, attention, concentration, and speed of information processing. Second, acceleration-deceleration of the brain also causes contusions and lacerations in areas where movement of the brain is likely to be opposed by bone or membrane, primarily the poles of the frontal and temporal lobes, with their rich connections to other cortical and to limbic areas. Such damage is related to impairments in the regulation and activation of intellectual and emotional behavior. Third, the prior two types of injury, as well as other factors, cause swelling of the brain and increased intracranial pressure. Increased intracranial pressure reduces the brain's blood flow, which may already have been compromised in the TBI patient by systemic hypotension and ventilatory insufficiency from other injuries and may result in hypoxic damage to highly oxygen-dependent areas, most prominently the limbic region. Damage to limbic structures is related to impairments in memory and emotionality. Last, there are other, though much less frequent, types of damage that may affect *any* brain area. There may be contusion of the brain in a coup and contrecoup distribution, depressed skull fracture and resulting penetration of the brain by bone, and various vascular complications such as subdural and epidural hematoma, intracerebral bleeding and vasospasm.

Thus, the neuropathology of TBI is apt to affect the anterior frontal and temporal regions and their underlying white matter, the limbic areas, the corpus callosum, and the midbrain and the brain stem. The behavioral implications are that TBI patients are most likely to show impairments in the activation and regulation of behavior, mental and emotional flexibility, speed of information processing, memory, and concentration and attention. Note that these are "general" mental functions, in that they activate, guide, and regulate—provide "executive control" (Stuss, 1987)—of specific cognitive and behavioral functions such as speech, language, arithmetic, visual-spatial analysis, and motor output. Thus, it is not uncommon to find TBI patients who recover to the point of being able to walk and talk and perform various discrete cognitive operations, but who remain substantially disabled because they can not regulate the application of such intact skills to meet social, vocational, interpersonal, and internal psychological needs.

Not only are the above "general" deficits the most prevalent form of impairment following TBI, they also are the most disruptive of successful rehabilitation and positive outcome. Surveys in diverse countries have uniformly designated change in personality, behavioral dyscontrol, loss of motivation and initiative, and impaired judgment as far stronger correlates of poor outcome than are impairments of specific intellectual or sensorimotor abilities

(Lezak, 1983; Oddy, Humphrey, & Uttley, 1978; Panting & Merry, 1972; Prigatano, Fordyce, Zeiner, Roueche, Pepping, & Wood, 1986b; Rosenbaum & Najenson, 1976; Thomsen, 1974). Thus, even when specific cognitive functions are impaired, they may not be the primary factors underlying the patient's disabilities and handicaps. It is much more likely that the TBI patient will have impairments related to the activation and modulation of cognitive and emotional activity, and to the regulation of the interplay between the person and a constantly changing internal and external environment. This is not to say that specific impairments, if present, are not an important focus of evaluation and treatment; rather, they are most productively approached within the context of the more general impairments.

Assessment of Brain-Injured Patients

The assessment of TBI patients is a complex task. The central issues are selecting appropriate assessment procedures and instruments; matching the assessment procedures and instruments to the patient's stage of recovery and level of function; using the assessment data to develop and monitor the rehabilitation treatments; and using the assessment data for prognosis and for planning regarding the patient's long-term needs. It should also be noted that assessment issues cannot be discussed without a consideration of the qualifications and training of the psychologist performing the assessment; the following discussion assumes that the psychologist has a reasonable level of neuropsychological, in addition to clinical-psychological, expertise.

The core methodology for assessing TBI patients is a neuropsychological test battery. (See Chapter 18 by Sweet for an overview of testing.) Many such test batteries are available, and the major ones—such as the Luria-Nebraska (Golden, Hammeke, & Purisch, 1980), the Halstead-Reitan (Reitan & Wolfson, 1985), and the "Boston" or "Portland" (Lezak, 1983)—all have been used effectively with TBI patients. The selection of the battery is based primarily on the training, experience, and preference of the psychologist involved; the specific test battery that is chosen is not nearly as important as the psychologist's ability to use that battery to meet the needs of TBI patients.

There are many reasons for using a formal neuropsychological test battery. The major test batteries have extensive normative and validation data and thus are useful for comparing a given patient with populations of various types and degrees of known pathology, as well as with "normal" populations. The major test batteries are also composed of multiple, overlapping tests that sample different areas of function, thus permitting the analysis of patterns of performance within an individual patient. Thus, the psychologist may identify intact as well as impaired areas of function, knowledge of both being necessary for the development of a rehabilitation plan. An analysis of patterns of performance also assists in an approximation of the patient's preinjury abilities, which has clinical as well as medicolegal utility.

Although a formal neuropsychological test battery serves as a core for the evaluation process, in some clinical situations other assessment procedures and instruments are needed. Formal neuropsychological test batteries are quite time-consuming, the briefer ones requiring about three hours and the lengthier ones often exceeding six hours. Some TBI patients, especially during the early stages of recovery, when impaired alertness, disorientation, and agitation are most likely, simply cannot tolerate or cooperate with such procedures. Other patients may have such severe impairments that the basic intent of many of the tests is defeated, as, for example, in the case of severely inattentive patients; with these patients, despite the intent of the test to measure language, visual-spatial, or memory skills, the likelihood is that attentiveness will be measured instead. Moreover, during the early phases of recovery, it is not unusual for patients to show substantial change from week to week, so that it is not cost-effective to administer a lengthy and expensive test battery, the results of which may be outdated before the report even reaches the patient's medical record. In such cases, briefer procedures, such as the Cognitive Status Ex-

amination (Barrett & Gleser, 1987), or more flexible approaches, such as a neurobehaviorally oriented mental status exam (Strub & Black, 1985), are all that may be possible or needed. Further, in cases of extremely severe confusion and disorientation, or early during recovery from coma, a differentiated cognitive status exam may be impossible, and global behavioral scales specifically developed for TBI patients, such as the Level of Cognitive Functioning Scale (Hagen, 1982) or the Galveston Orientation and Amnesia Test (Levin, O'Donnell, & Grossman, 1979), may be most appropriate.

Another limitation on the use of neuropsychological test batteries relates to the nature of their validation. The primary validation of these tests has been regarding their ability to predict the presence or absence of brain dysfunction, as well as its localization. Although there are many occasions for the use of the tests for this purpose with TBI patients, a far more prevalent use of the tests is for prediction of how a patient will perform in real-life vocational, educational, and other functional activities. Recent studies of the "ecological" validity of these tests have shown statistically significant, though modest-magnitude, correlations (Little, Williams, & Long, 1986; McSweeney, Grant, Heaton, Prigatano, & Adams, 1985). Stuss (1987) suggested that one reason for the discontinuity between test and real-life performance in TBI patients is that much of the handicap in such patients is due to impaired frontal lobe function and that frontal lobe deficits may not be elicited"if the examiner is a good tester, providing a controlled, standardized, structured environment, and, in a sense, becoming the frontal lobes of the patient" (p. 174).

The potential discontinuity between test and real-life performance in TBI patients has led to the supplementation of neuropsychological tests with rating scales that directly assess real-life functioning. Among the more commonly used are the Katz Adjustment Scale—Relative's Form (KAS-R; Katz & Lyerly, 1963) and the Sickness Impact Profile (SIP; Bergner, Bobbitt, Pollard, Martin, & Gilson, 1976) and its recent revision to increase its sensitivity and appropriateness to TBI patients (Temkin, McLean, Dikmen,

Gail, Bergner, & Almes, 1988). On the KAS-R, a relative or close caregiver rates the patient's status across a range of domains of daily-life functioning. The SIP involves similar ratings, except that they are based on self-report by the patient.

Although functional rating scales such as the above are extremely helpful in assessing the day-to-day real-life status of TBI patients, they must be used with caution because of possible response bias. As McKinlay and Brooks (1984) showed, TBI patients substantially underestimate their own level of impairment. Patients' families also underestimate the patient's level of impairment, though this underestimation is restricted to the period when the patient is in the hospital; after a period of living with the patient, family members report more impairment than do the patients. Such response bias can be addressed by using scales that rate a patient's performance on simulations of real-life tasks. For example, the Street Survival Skills procedure of Linkenhoker and McCarron (1983) rates a TBI patient's performance on a series of simulated tasks, such as following a map, managing money and a checkbook, and responding to a job advertisement. In some functional areas, simulations have been found to be the assessment method of choice, such as for predicting the TBI patient's adequacy in driving an automobile, where on-the-road testing and performance in simulators provide better prediction than does neuropsychological testing (Van Zomeren, Brouwer, Rothegatter, & Snoek, 1988).

Another limitation of neuropsychological test batteries is their vulnerability to "practice effects" when administered within closely spaced time intervals. Although practice effects from testings separated by several months have been shown to be negligible in TBI populations (Mandelberg, 1976), some assessment applications in the rehabilitation setting, such as evaluating the effects of cognition-enhancing medications, may require weekly or more frequent measurement. For such applications, procedures from the cognitive psychology laboratory may be the most useful, especially the computer-generated forms of these tests devel-

oped specifically for psychopharmacological studies (Crook & Larrabee, 1988; Thorne, Genser, Sing, & Hegge, 1985).

In addition to assessing neuropsychological status, a comprehensive evaluation of the TBI patient should address factors that mediate the relation between cognitive abilities and their expression as real-life skills, such as mood and psychological adjustment, personality and coping style, capacity for accurate self-appraisal, willingness to use compensatory procedures, and interpersonal and social skills. These factors have a strong influence on outcome, and as Diller and Gordon (1981) noted, they are an important and often productive focus for treatment interventions.

Embedded within all of the above is the need to form some impression of the TBI patient's preinjury cognitive, emotional, and behavioral status. Such an impression takes on clinical value by providing an understanding of how the patient has been changed by the injury. Some traits that may be "impairments" in some patients are simply preinjury characteristics of other patients. One of the present authors examined a Hell's Angel with TBI whose goal following rehabilitation was to "fix my motorcycle, tear flesh, and break bones." This represented a good recovery to premorbid status for this patient but may have been a frontal lobe pseudopsycopathic syndrome in another patient. Similarly, there was an angry and quite hostile borderline patient who had not had a head injury, and who was in traditional long-term psychotherapy with one of the authors. One day, after seeing the patient twice a week for over a year, his psychotherapist was on the neurosurgery floor to do a consultation and saw the patient sitting in the hallway in a wheelchair with his head bandaged, having sustained a TBI in an assault three days earlier. The head nurse immediately called the psychologist over and told him to cancel the scheduled consultation because it was urgent that he see the patient in the hallway, who had the "worst" frontal lobe syndrome she and the neurosurgeon had ever seen. Examination of the patient, however, showed him to be little changed since his last psychotherapy session, though perhaps because of the stress of the situation, he was even more eager to irritate and enrage others than usual.

A careful assessment of preinjury status also has important medicolegal value. TBI occurs quite frequently under circumstances that involve legal liability or entitlement to compensation. Because many of the most handicapping impairments of TBI are cognitive and behavioral, it is quite likely that the psychologist will be asked to provide evidence on the patient's behalf, much of which will consist of descriptions and measurements that define the capacities that the patient has lost as a result of the injury. (The high incidence of forensic involvement related to TBI patients—including litigation for damages from auto and workplace accidents, hearings regarding financial entitlements such as worker's compensation and disability income, and proceedings regarding legal competency and guardianship—should alert the psychologist to the high probability that his or her work will at some time be read by attorneys, often years after it was written. Therefore it is prudent to treat all psychological services to TBI patients as *potentially* entering the legal arena, to keep careful documentation, and to use tests and assessment procedures that will hold up to scrutiny and challenge.)

Treatment Programs

Treatment Settings

There is no single all-purpose program for TBI rehabilitation. Rather, there are many diverse programs that, together, define a continuum of care. The continuum begins with emergency and acute medical and surgical services that stabilize a patient's injuries. Once medically stabilized, however, many patients remain in coma. Those who remain persistently in coma may remain in the acute care hospital or may be transferred to nursing-home-based coma programs. Those who recover from coma, or whose coma is lightening or showing other signs of favorable prognosis, are typically transferred to the rehabilitation hospital.

Although psychologists, for obvious reasons, typically do not provide extensive services to comatose patients, other disciplines on the rehabilitation team do. These services are provided to sustain life, and to maintain the patient's skin, muscles, and joints in optimal condition, and many members of the team invest substantial clinical energy in patients who may never emerge from coma or who may not do so for months and, in some cases, years. This can be a cause of substantial stress to the team, partly because of the day-to-day difficulty of working with an unresponsive or minimally responsive patient, and partly because working with such patients stimulates powerful ambivalence about moral and ethical issues, as well as arouses latent conflict regarding helplessness. The patient's family is also likely to experience great stress. Thus, even though the psychologist may have little clinical involvement with the patient, there is much that can be done to assist the team and the family to cope adaptively with what is always a difficult situation.

As patients become more alert, they are able to participate in an active rehabilitation program that may range in duration from three to nine months. Cognitive and behavioral issues are likely to be addressed throughout this period. However, during the early stages, basic biological and self-care issues such as eating, toileting, personal hygiene, dressing, and ambulation are the predominant focus. Unlike spinal cord injury, where the neurological deficit is fixed and relatively unlikely to improve during the course of the rehabilitation hospitalization, it is almost the rule that TBI patients show substantial neurological recovery during the rehabilitation hospitalization. It is thus likely that a large number will master self-care and other basic tasks relatively quickly. These patients then go on to pursue goals related to the more complex tasks of reintegration into their communities and families, and it is here that cognitive and behavioral impairments become a central focus of treatment.

The middle and latter stages of rehabilitation hospitalization focus on a reintegration of the patient into his or her former lifestyle, responsibilities, and roles. In some cases, these are resumed with little or no modification, but in many other cases, there is a need to extensively redefine former roles and responsibilities in ways that accommodate the patient's altered capacities. Depending on the severity of the residual impairment, it may be possible to accomplish this redefinition during the rehabilitation hospitalization, though in the overwhelming majority of cases, some form of continued treatment following hospital discharge is needed, in the form of a postacute rehabilitation program.

Some postacute programs are residential. These programs, typically called *transitional living programs*, are designed for patients who no longer require the 24-hour-a-day medical and nursing care of the rehabilitation hospital, but who have cognitive and behavioral deficits sufficiently severe to preclude their return to home or to independent living without additional treatment. Most such programs are housed in large, single-family homes in residential neighborhoods. They provide the full range of therapies as does the rehabilitation hospital, and provide 24-hour supervision (though not medical or nursing care). The context of treatment is the community, and the focus is on developing skills for functioning effectively and independently within it.

Following an average of six to nine months of treatment, many patients in transitional living programs no longer require 24-hour supervision and they can enter a semi-independent living program. This is usually a part of the transitional living program or is closely affiliated with it. Here, supervision is gradually faded, and the patient assumes increasing responsibility for managing his or her daily activities and responsibilities, which can include sheltered or competitive employment, returning to school, and social and recreational activities in the community. Fading the amount and type of supervision also permits an assessment of the minimal level of supervision that the patient needs to function effectively and safely, which then is used as the basis for a long-term maintenance plan.

Services similar to those of transitional living can be given through outpatient day-treatment programs. The choice between residential and outpatient treatment is sometimes based on clinical factors, such as the degree of behavioral disturbance and need for a controlled environment in which to treat it. The decision may also be based on the client's and the family's preference or capacity, or on the availability of an outpatient program within commuting distance of the patient's home. Often, however, the decision is based on the types of services that the patient's insurance will cover.

Last, there are specialized programs that treat behavior disorders. Because some degree of behavioral disturbance is routine in TBI, all of the previously mentioned programs provide for its treatment. However, some patients have, for want of a better term, spectacular behavior disorders that require a special treatment environment and an appropriately trained staff. These programs, such as the one described by Wood (1987), use a highly structured and controlled environment, with extensive application of behavioral techniques, sometimes combined with pharmacotherapy.

As can be seen from the above, TBI rehabilitation spans a broad and diverse range of programs and settings. Psychologists interested in TBI have the opportunity to work in acute care hospitals in departments of neurosurgery, in rehabilitation hospitals, in residential postacute treatment settings, and in outpatient clinics and programs. The different settings in which TBI rehabilitation is practiced provide different role opportunities for psychologists. In the acute care medical setting, which generally takes the form of a neurosurgical intensive-care unit, the psychologist is likely to be involved as consultant, performing assessments of selected patients who have recovered from coma sufficiently to participate in such activity, and perhaps meeting with the team and with families to discuss their feelings and frustrations, or participating in research activities. The psychologist in the acute care setting may also develop ongoing treatment relationships with patients who have milder injuries that do not justify rehabilitation hospitalization, but that may nonetheless be substantially disabling. The rehabilitation hospital offers the psychologist the above role of consultant. However, it also offers the psychologist the opportunity to be a member of the team, participating actively in broad aspects of patient care, and there is some opportunity for a concurrent leadership role regarding selected aspects of care and research. Postacute programs, such as transitional-living and day-treatment programs, offer consultation and team membership opportunities, but they are especially rich in leadership opportunities for psychologists. In essence, as the TBI patient recovers, the general trend is for medical needs to decline, and for behavioral and psychosocial needs to predominate.

Neuropsychological Treatment

Although the types of treatment provided by psychologists vary across settings and programs, it is likely that treatments aimed at remediating specific cognitive impairments will be used. For example, Sohlberg and Mateer (1987), Ben-Yishay, Piasetsky, and Rattok (1987), and Wood (1986) have described training procedures for improving the attentiveness of TBI patients and have shown these procedures to be effective, with generalization of improvement to performance on other tasks and in other settings, and with maintenance of improvement at follow-up testing as long as six months later. Craine (1982) described procedures for treating impaired initiation and self-regulation that are based on teaching the patient to use "overt verbal medication"—in essence having the patient learn sets of messages to herself or himself—to cue and organize poorly regulated behaviors, and Sohlberg, Sprunk, and Metzlaar (1988) reported on the successful application of such a procedure.

The above is only a brief illustration of a large group of recent studies showing the successful application of cognitive retraining procedures to many types of cognitive dysfunction. Studies of the remediation of memory deficits, which are pervasive and quite disabling in many TBI patients, have been less promising. Treatment procedures based on drill and rote practice, where "memory" is treated as a muscle that will

perform better if exercised, were reviewed by Glisky and Schacter (1986), who concluded that "there is no evidence that such practice improves memory for any other materials [or] generalizes to other situations, tasks, or stimuli" (p. 55). Alternate approaches to improving memory have been based on the use of "memory prosthetics" such as notepads, diaries, and the more recently available wristwatches that store and retrieve information, and these have shown more promising results (Rimmele & Hester, 1987).

Comprehensive Programs

In addition to interventions that address specific areas of impairment, the psychologist is likely to be involved in complex and multidimensional interventions that simultaneously address a patient's many impairments as they relate to disability in real-life performance. This field is best illustrated by describing a day treatment program based heavily on such interventions (Prigatano et al., 1986a). This program treats six to eight patients at a time, for a six-month period, with a staff of three neuropsychologists, a speech and language pathologist, an occupational therapist, and a physical therapist. Patients are seen six hours per day, four days per week, and the day is broken into seven segments: (1) small-group cognitive retraining, where the focus is on remediating specific cognitive deficits and also emphasizes an increasing self-awareness of strengths and deficits by having clients chart their progress; (2) social skills training, in a group format, where clients practice real-life social tasks such as conflict resolution, asking for a date, and job interviews; (3) group psychotherapy, in which emotional and motivational issues are discussed; (4) individual psychotherapy; (5) physical, occupational, vocational, or speech therapy, based on individual needs; (6) independent therapy, where clients work alone, in order to develop time-management skills and internal regulation; and (7) a "milieu" community discussion group at the end of the day. In this and similar programs, psychologists *combine* clinical psychological and neuropsychological assessment and intervention in order to address the broad issues of diability and handicap. Specific impairments are not remediated as an end in themselves; rather, multiple interventions are applied synergistically with the holistic intent of reducing disability and handicap. Descriptions of comprehensive residential programs are also available (Fryer & Haffey, 1987).

Conclusions

The previous discussion has emphasized two major points. First, the rehabilitation hospital *is not* a mental health setting. Patients do not go to the rehabilitation hospital because they want psychotherapy, their physicians do not refer them to the rehabilitation hospital because of mental health needs, most of the staff view their mission in term of physical function and do not think of themselves as mental health workers, and insurance companies do not authorize rehabilitation hospitalizations based on psychological needs. Second, as has been shown throughout this chapter, the rehabilitation hospital *is*, without question, a mental health setting. Much of psychology's future task will be to assist the rehabilitation community—patients, families, staff, insurance companies, and legislatures—to see and appreciate the enormous behavioral science context in which rehabilitation occurs.

References

Adamovich, B.B., Henderson, J. A., & Auerbach, S. (1985). *Cognitive rehabilitation of closed head injured patients.* San Diego: College-Hill Press.

Alexander, M. P. (1987). The role of neurobehavioral syndromes in the rehabilitation and outcome of closed head injury. In H. S. Levin, J. Grafman, & H. M. Eisenberg (Eds.), *Neurobehavioral recovery from head injury.* New York: Oxford University Press.

Athelstan, G. T., & Crewe, N. M. (1979, April). Psychological adjustment to spinal cord injury as related to manner of onset of disability. *Rehabilitation Counseling Bulletin,* pp. 311–319.

Auerbach, S. H. (1986). Neuroanatomical correlates of at-

tention and memory disorders in traumatic brain injury: An application of neurobehavioral subtypes. *Journal of Head Trauma Rehabilitation, 1,* 1–12.

Barrett, E. I., Jr., & Gleser, G. C. (1987). Development and validation of the Cognitive Status Examination. *Journal of Consulting and Clinical Psychology, 55,* 877–882.

Ben-Yishay, Y., Piasetsky, E. B., & Rattok, J. (1987). A systematic method for ameliorating disorders in basic attention. In M. J. Meier, L. Diller, & A. L. Benton (Eds.), *Neuropsychological rehabilitation.* New York: Guilford Press.

Caveness, W. F. (1977). Incidents of craneocerebral trauma in the United States, 1970–1975. *Annals of Neurology, 1,* 507.

Cook, D. W., Bolton, B., & Taperek, P. (1981, November). Rehabilitation of the spinal cord injured: Life status at follow-up. *Rehabilitation Counseling Bulletin,* pp. 111–122.

Craine, J. (1982). The retraining of frontal lobe dysfunction. In L. E. Trexler (Ed.), *Cognitive rehabilitation: Conceptualization and intervention.* New York: Plenum Press.

Crook, T. H., & Larrabee, G. J. (1988). Interrelations among everyday memory tests: Stability of factor structure with age. *Neuropsychology, 2,* 1–12.

Davidoff, G., Morris, J., Roth, E., & Bleiberg, J. (1985a). Closed head injury in spinal cord injured patients: Retrospective study of loss of consciousness and posttraumatic amnesia. *Archives of Physical Medicine and Rehabilitation, 66,* 41–43.

Davidoff, G., Morris, J., Roth, E., & Bleiberg, J. (1985b). Cognitive dysfunction and mild closed head injury in traumatic spinal cord injury. *Archives of Physical Medicine and Rehabilitation, 66,* 489–491.

Dew, M. A., Lynch, K. A., Ernst, J., Rosenthal, R., & Judd, C. M. (1985). A causal analysis of factors affecting adjustment to spinal cord injury. *Rehabilitation Psychology, 30,* 39–46.

Diller, L., & Ben-Yishay, Y. (1987). Outcomes and evidence in neurological rehabilitation in closed head injury. In H. S. Levin, J. Grafman, & H. M. Eisenberg (Eds.), *Neurobehavioral recovery from head injury.* New York: Oxford University Press.

Diller, L., & Gordon, W. A. (1981). Interventions for cognitive deficits in brain-injured adults. *Journal of Consulting and Clinical Psychology, 49,* 822–834.

Ducharme, S. H., & Ducharme, J. (1984). Psychological adjustment to spinal cord injury. In D. W. Krueger (Ed.), *Emotional rehabilitation of physical trauma and disability.* New York: Spectrum.

Dunn, M. (1977). Social discomfort in the patient with spinal cord injury. *Archives of Physical Medicine and Rehabilitation, 58,* 257–260.

Fey, S. G., & Williamson-Kirkland, T. E. (1987). Chronic pain: Psychology and rehabilitation. In B. Caplan (Ed.), *Rehabilitation psychology desk reference* (pp. 101–130). Rockville, MD: Aspen.

Frank, R. G., & Elliott, T. R. (1987). Life stress and psychological adjustment following spinal cord injury. *Archives of Physical Medicine and Rehabilitation, 68,* 344–347.

Fryer, L. J., & Haffey, W. J. (1987). Cognitive rehabilitation and community readaptation: Outcomes from two program models. *Journal of Head Trauma Rehabilitation, 2,* 51–63.

Gans, J. S. (1983). Hate in the rehabilitation setting. *Archives of Physical Medicine and Rehabilitation, 64,* 176–179.

Gans, J. S. (1987). Facilitating staff patient interaction in rehabilitation. In B. Caplan (Ed.), *Rehabilitation psychology desk reference* (pp. 185–218). Rockville, MD: Aspen.

Garrison, J. (1978). Stress management training for the handicapped. *Archives of Physical Medicine and Rehabilitation, 59,* 580–585.

Gennarelli, T. A. (1986). Mechanisms and pathophysiology of cerebral concussion. *Journal of Head Trauma Rehabilitation, 1,* 23–30.

Glisky, E. L., & Schachter, D. L. (1986). Remediation of organic memory disorders: Current status and future prospects. *Journal of Head Trauma Rehabilitation, 1,* 54–63.

Golden, C. J., Hammeke, T. A., & Purisch, A. D. (1980). *Manual for the Luria-Nebraska Neuropsychological Battery.* Los Angeles: Western Psychological Services.

Gunther, M. (1977). The threatened staff: A psychoanalytic contribution to medical psychology. *Comprehensive Psychiatry, 18,* 385–397.

Hagen, C. (1982). Language cognitive disorganization, following closed head injury: A conceptualization. In L. E. Trexler (Ed.), *Cognitive rehabilitation: Conceptualization and intervention* (pp. 131–151). New York: Plenum Press.

Haney, M., & Rabin, B. (1984). Modifying attitudes toward disabled persons while resocializing spinal cord injured patients. *Archives of Physical Medicine and Rehabilitation, 65,* 431–436.

Judd, F. K., & Brown, D. J. (1987). Psychiatry in the spinal injuries unit. *Paraplegia, 25,* 254–257.

Judd, F. K., & Burrows, G. D. (1986). Liaison psychiatry in a spinal injuries unit. *Paraplegia, 24,* 6–19.

Kalsbeek, W. D., McLaurin, R. L., Harris, B. S. H., & Miller, J. D. (1980). The National Head and Spinal Cord Injury Survey: Major findings. *Journal of Neurosurgery, 53 (Suppl.),* S19–S31.

Katz, M. M., & Lyerly, S. B. (1963). Methods of measuring adjustment and social behavior in the community. *Psychological Reports, 13,* 503–535.

Krause, J., & Crewe, N. (1987). Prediction of long term survival among persons with spinal cord injury: An 11 year prospective study. *Rehabilitation Psychology, 37,* 205–213.

Langs, R. (1988). *A primer of psychotheraphy.* New York: Gardner Press.

Levin, H. S., Benton, A. L., & Grossman, R. G. (1982). *Neurobehavioral consequences of closed head injury.* New York: Oxford University Press.

Levin, H. S., O'Donnell, V. M., & Grossman, R. G. (1979). The Galveston orientation and amnesia test: A practical scale to assess cognition after head injury. *Journal of Nervous and Mental Diseases, 167,* 675–684.

Lezak, M. D. (1983). *Neuropsychological assessment* (2nd ed.). New York: Oxford University Press.

Linkenhoker, D., & McCarron, L. (1983). *Adaptive behavior: The Street Survival Skills Questionnaire.* Dallas: McCarron-Dial Systems.

Little, M. M., Williams, J. M., & Long, C. J. (1986). Clinical memory tests and everyday memory. *Archives of Clinical Neuropsychology, 1,* 323–334.

Mandelberg, I. A. (1976). Cognitive recovery after severe head injury. *Journal of Learning and Motivation, 39,* 1001–1007.

McGowan, M. B., & Roth, S. (1987). Family functioning and functional independence in spinal cord injury adjustment. *Paraplegia, 25,* 357–365.

McKinlay, W. W., & Brooks, D. N. (1984). Methodological problems in assessing psychosocial recovery following severe head injury. *Journal of Clinical Neuropsychology, 6,* 281–291.

McSweeney, A. J., Grant, I., Heaton, R. K., Prigatano, G. P., & Adams, K. M. (1985). Relationship of neuropsychological status to everyday functioning in healthy and chronically ill persons. *Journal of Clinical and Experimental Neuropsychology, 7,* 281–291,

Miller, D. K., Wolfe, M., & Spiegel, M. H. (1975). Therapeutic groups for patients with spinal cord injuries. *Archives of Physical Medicine and Rehabilitation, 56,* 130–135.

Oddy M., Humphrey, M., & Uttley, D. (1978). Subjective impairment and social recovery after closed head injury. *Journal of Neurology, Neurosurgery and Psychiatry, 31,* 299–306.

Ohry, A. (1987). Ethical questions in the treatment of spinal cord injured patients. *Paraplegia, 25,* 293–295.

Panting, A., & Merry, P. H. (1972). The long-term rehabilitation of severe head injuries with particular reference to the need for social and medical support for the patient's family. *Rehabilitation, 38,* 33–37.

Prigatano, G., Fordyce, D., Zeiner, H., Roueche, F., Pepping, M., & Wood, B. C. (1986a). *Neuropsychological rehabilitation after brain injury.* Baltimore: Johns Hopkins University Press.

Prigatano, G., Fordyce, D., Zeiner, H., Roueche, F., Pepping, M., & Wood, B. C. (1986b). The outcome of neuropsychological rehabilitation efforts. In G. P. Prigatano (Ed.), *Neuropsychological rehabilitation after brain injury* (pp. 119–133). Baltimore: Johns Hopkins University Press.

Reitan, R. M., & Wolfson, D. (1985). *Neuroanatomy and neuropathology: A clinical guide for neuropsychologists.* Tucson: Neuropsychology Press.

Rimel, R. W., & Jane, J. A. (1983). Characteristics of the head injured patient. In M. Rosenthal, E. R. Griffith, M. R. Bond & J. D. Miller (Eds.), *Rehabilitation of the head injured adult.* Philadelphia: F. A. Davis.

Rimmele, C. T., & Hester, R. K. (1987). Cognitive rehabilitation after traumatic head injury. *Archives of Clinical Neuropsychology, 2,* 353–384.

Romano, M. (1976). Social skills training with the newly handicapped. *Archives of Physical Medicine and Rehabilitation, 57,* 302–303.

Romano, M., & Lassiter, R. (1972). Sexual counseling with

the spinal cord injured. *Archives of Physical Medicine Rehabilitation, 53,* 568–572.

Rosenbaum, M., & Najenson, T. (1976). Changes in life patterns and symptoms as reported by wives of severely brain-injured soldiers. *Journal of Consulting and Clinical Psychology, 44,* 881–888.

Silver, J. R., Morris, W. R., & Otfinowski, J. S. (1980). Associated injuries in patients with spinal injury. *Injury, 12,* 137–139.

Sohlberg, M. M., & Mateer, C. A. (1987). Effectiveness of an attention training program. *Journal of Clinical and Experimental Neuropsychology, 2,* 117–130.

Sohlberg, M. M., Sprunk H., & Metzlaar, K. (1988). Efficacy of an external cuing system in an individual with severe frontal lobe damage. *Cognitive Rehabilitation, 6,* 36–41.

Strub, R. L., & Black, F. W. (1985). *The mental status of examination in neurology* (2nd ed.). Philadelphia: F. A. Davis.

Stuss, D. T. (1987). Contribution of frontal lobe injury to cognitive impairment after closed head injury: Methods of assessment and recent findings. In H. S. Levin, J. Grafman, & H. M. Eisenberg (Eds.), *Neurobehavioral recovery from head injury.* New York: Oxford University Press.

Temkin, N., McLean, A., Jr., Dikmen, S., Gale, J., Bergner, M., & Almes, M. J. (1988). Development and evaluation of modifications to the sickness impact profile for head injury. *Journal of Clinical Epidemiology, 41,* 47–57.

Thomsen, I. V. (1974). The patient with severe head injury and his family: A follow-up study of 50 patients. *Scandinavian Journal of Rehabilitation Medicine, 6,* 180–183.

Thorne, D. R., Genser, S. G., Sing, H. C., & Hegge, F. W. (1985). The Walter Reed Performance Assessment Battery. *Neurobehavioral Toxicology and Teratology, 7,* 415–418.

Trieschmann, R. E. (1988). *Spinal cord injuries: Psychological, social, and vocational rehabilitation,* (2nd ed.). New York: Demos.

Van Zomeren, A. H., Brouwer, W. H., Rothengatter, J. A., & Snoek, J. W. (1988). Fitness to drive a car after recovery from severe head injury. *Archives of Physical Medicine and Rehabilitation, 69,* 90–96.

Wagner, K. A., Kopaniky, D. R., & Esposito, L. (1983). The head and spinal cord injured patient: Impact of combined sequelae (abstract). *Archives of Physical Medicine and Rehabilitation, 64,* 519.

Weiss, M. H. (1974). Head trauma and spinal cord injuries: Diagnostic and therapeutic criteria. *Critical Care Medicine, 2,* 311–316.

Wilmot, C. B., Cope, D. N., Hall, K. M., & Acker, M. (1985). Occult head injury: Its incidence in spinal cord injury. *Archives of Physical Medicine and Rehabilitation, 66,* 227–231.

Wilmuth, M. E. (1987). Sexuality after spinal cord injury: A critical review. *Clinical Psychology Review, 7,* 389–412.

Wood, R. Ll. (1986). Rehabilitation of patients with disorders of attention. *Journal of Head Trauma Rehabilitation, 1,* 43–53.

Wood, R. Ll. (1987). *Brain injury rehabilitation: A neurobehavioural approach.* Rockville, MD: Aspen.

Zola, K. I. (1982a). Denial of emotional needs to people with handicaps. *Archives of Physical Medicine and Rehabilitation, 63,* 63–67.

Zola, K. I. (1982b). Social and cultural disincentives to independent living. *Archives of Physical Medicine and Rehabilitation, 63,* 394–397.

CHAPTER 23

Chronic Pain

Psychological Assessment and Treatment

Stanley L. Chapman

Introduction

Pain is a natural response to a mechanical, thermal, or chemical stimulus that activates specific structures called *nociceptors* found in the skin, the viscera, and the deep somatic tissues (Crue, 1983). Pain stimuli travel from nociceptors across nerve fibers to the spinal cord and are processed in the central nervous system as the experience of pain. In most cases, the stimulus that activates the nociceptors is terminated and the pain goes away after a short period of time. Such pain can be labeled *acute pain*, which is defined as pain of recent onset, in which biological or tissue damage is usually dominant. In the absence of a residual structural defect or a systemic disease, acute pain usually subsides in less than 30 days and almost always in six months (U.S. Department of Health and Human Services, 1986).

Interdisciplinary rehabilitation programs are seldom needed in cases of acute pain. The underlying medical condition is treated, or perhaps the patient is instructed to avoid certain activities so that proper healing can take place. Medications that relieve pain may be given for brief periods of time to alleviate the suffering without fear of addiction.

Chronic pain provides a different and puzzling story. The term *chronic* is derived from the Green *kronos*, which means "time." Thus, chronic pain can be defined as pain lasting for long periods of time; more than six months is a commonly used duration. Such pain may be associated with a residual persistent structural defect or with the persistence of a disease process such as arthritis. Chronic pain may also continue past the normal expected healing time or may exist in the absence of any medical findings. A variant is recurrent acute pain, in which there are repetitive episodes of pain over long periods of time with pain-free intervals interspersed.

The statistics on the incidence and cost of chronic pain are staggering. The Nuprin Pain Report (Louis Harris & Associates, 1985) surveyed the number of people who have pain for more than 30 days during the year and found that back pain, joint pain, and headache *each* occurred in 13%–14% of people, whereas chronic muscle pain occurred in 10% of people, stomach pains in 4%, and a variety of other types of pain in smaller percentages.

Stanley L. Chapman • Pain Control and Rehabilitation Institute of Georgia, 350 Winn Way, Decatur, Georgia 30030.

Growth of Pain Centers

In light of the above statistics, it might be surprising to the reader that there were just a few small and scattered clinics specializing in the management of chronic pain until 1960 and 1961, when major clinics were established by John Bonica (University of Washington) and Benjamin Crue (City of Hope in Duarte, California), respectively. These clinics followed the 1953 publication of Bonica's *Management of Pain*, which proposed the revolutionary idea that chronic pain is a different entity from acute pain and requires treatments fundamentally different from narcotic pain medications and surgical procedures. Bonica emphasized that chronic pain patients need an interdisciplinary team approach led by a physician with specialized training in the management of chronic pain.

Equally revolutionary was the work on Wilbert Fordyce and his colleagues. Fordyce's widely read *Behavioral Methods for Chronic Pain and Illness* (1976) outlined the role of the behavioral psychologist in the assessment and treatment of pain. Fordyce emphasized that chronic pain can be viewed in terms of *pain behaviors* which include observable verbal and nonverbal expressions of pain, such as complaints, grimaces, remaining inactive, and taking medications. He asserted that these behaviors are not merely expressions of medical pathology; rather they can be influenced highly by basic processes of learning and can be altered through systematic applications of behavioral interventions based primarily on operant conditioning principles. Many pain centers are modeled largely on these behavioral principles. Fordyce's contribution was revolutionary because the role of the environment and the family had been largely ignored before his work. Chronic pain patients who did not respond to medical treatments often were told simply that they had to live with it or, perhaps, were labeled as "functional" and sent to a psychiatrist.

New and revolutionary ideas throughout history have often been applied injudiciously, and such appears to be the case with Fordyce's. Fordyce was very clear in his writings about the need for a very careful analysis of pain behaviors to assess the extent to which they are governed by principles of operant conditioning. He noted that programs based on operant conditioning models need to be formulated individually and used with carefully selected patients. This author has noted that some professionals assume that pain behavior is operantly conditioned solely because medical evidence cannot be found to "justify" the complaint. Furthermore, some pain programs have developed rigid practices and have applied them to patients regardless of diagnosis. These practices have included forbidding any discussion of pain itself, setting quotas for reactivation regardless of medical limitations, and forbidding all medications for pain. In some cases, the patient's experience of pain is regarded as irrelevant because it cannot be measured as behavior.

A third influence on the development of pain centers is the recognition of the interaction between physical, emotional, and cognitive factors in the pain experience. The era in which pain was regarded as x% "organic" and y% "functional" or "psychological" (where $x + y = 100$) mercifully appears to be coming to an end. Such a concept belies modern research findings (such as in the area of biofeedback) that show that psychological processes have clear *physiological* effects (e.g., Brown, 1977). It also neglects the significant role that physical illness itself plays in the development of emotional changes such as depression. Actually, pain in the absence of malingering must be seen as 100% physical and 100% emotional. This concept is well embodied in Melzack and Wall's influential gate control theory of pain (1965), which has given rise to hundreds of research papers. This theory espouses the concept that the central nervous system plays a major role in inhibiting or facilitating pain perception at many levels of neural transmission.

The establishment of interdisciplinary pain centers also fits with other trends and developments of the 1970s and 1980s. In his book *Megatrends* (1982), John Naisbitt described the increasing development of what he called "high tech/high touch," which he defined as the integration of technology with a very personalized concern for the individual. The trend toward more active individual responsibility for health and fitness, which is also embodied in the philosophies of pain centers, is seen in the integration of psychology and medicine in many other areas, including programs for head injury, alcohol and drug misuse, smoking, obesity, and cardiac problems, to name a few.

Indeed, the pain center movement has grown rapidly from the handful of pain centers that existed in the 1960s. In the 1970s, many pain centers saw patients from large areas, and most of the largest and best known consisted of intensive (and expensive) inpatient units, with durations of treatment of three to six weeks. Since the early 1980s, local pain centers have grown everywhere, and interdisciplinary centers are available in almost every large urban area. The models for treatment have diversified to include more outpatient approaches and an integration of pain programs with other services, particularly in the areas of vocational rehabilitation, work hardening, and disability evaluation. Key events in the pain movement include the founding of the International Association for Studies on Pain and its subchapter, the American Pain Society, in 1974 and 1978, respectively, and the publication of the interdisciplinary journal *Pain* beginning in 1975. Numerous journals, such as *Pain Management*, the *Journal of Pain and Symptom Management*, *The Clinical Journal of Pain*, and *The Pain Clinic*, now are directed largely to the study of chronic pain and its issues. In 1983, the Commission on Accreditation of Rehabilitation Facilities (CARF), with the endorsement of the American Pain Society, began the process of evaluating chronic pain programs for accreditation. The number of accredited centers has grown from 38 at the end of 1985 to 120 as of June 1990. The accreditation process appears to be gaining increasing financial clout, as several states have required accreditation as a precondition for payment through their workers compensation systems.

The Psychology of Pain

Psychologists have been playing an increasing role in the assessment and treatment of chronic pain. Indeed, CARF now mandates that a chronic pain management program include a psychologist or psychiatrist as part of its "core team." This increasing role has been paralleled by increasing research in the ways by which psychological processes affect pain experience and behavior.

Operant Conditioning

Several researchers have found that operant factors influence pain behaviors. Reports of pain levels have been found to vary with the solicitousness of the spouse (Anderson & Rehm, 1987; Block, Kremer, & Gaylor, 1980; Flor, Kerns, & Turk, 1987), and it has been demonstrated in controlled studies that variables such as activity level, drug intake, and pain report can be altered systematically through the application of operant reinforcement contingencies (Cairns & Pasino, 1977; Doleys, Crocker, & Patton, 1982; Linton, Melin, & Götestam, 1984; Sanders, 1983). Despite evidence that operant conditioning can influence the frequency of pain-related behavior, there has been little study of its role in the etiology or development of such behavior. Furthermore, there has been an increasing appreciation that clinicians need to avoid equating pain experience with pain behavior (Schmidt, 1987; Turk & Flor, 1987).

Classical Conditioning

Though the classical or respondent conditioning of pain responses has not been well tested in the research literature, there are ample findings that a wide variety of autonomically mediated animal and human reflex responses, such as heart rate, peripheral vasomotor activity, and pupillary contraction and dilation, can be conditioned to occur in the presence of previously neutral environmental stimuli (Sanders, 1985). Several authors have speculated that pain experience or behavior may become contingent on previously neutral stimuli through classical conditioning: the presence of a bottle of pills on the nightstand can become a discriminative stimulus for drug taking (Morse, 1983), or perhaps the situations or environment in which acute pain was experienced can maintain chronic pain experience or behavior (Chapman, 1983b; Sanders, 1985). Classically conditioned fear of pain itself can become a powerful factor in bringing about physiological changes that contribute to the development of pain, resulting in a closed feedback loop in which pain and anxiety reinforce each other (Turk & Flor, 1984).

Social Modeling

The term *social modeling* refers to the acquisition of a response based on observation of the response in others. Indirect evidence of its importance in chronic pain has come from findings of clear cultural differences in pain responses (Zborowski, 1969), findings that pain complaints tend to run in families (Apley, 1975; Gentry, Shows, & Thomas, 1974; Violon & Giurgen, 1984), and findings that pain behaviors in experimental situations can be altered as a result of the behavior of relevant models and the consequences of that behavior (Craig & Prkachin, 1978; Thelen & Frye, 1981).

Cognitive Aspects of Chronic Pain

The 1970s and 1980s saw a reemphasis on research and clinical interventions designed to alter individuals' cognitions—or the way people think or appraise their experience. Indeed, many psychologists working with chronic pain now describe their interventions as fitting into a cognitive-behavioral model, which seeks to modify "covert behaviors" such as thoughts and attitudes regarding the pain problem.

The importance of cognitive factors in acute pain is suggested by studies of volunteers exposed to an acutely painful stimulus. These studies lead one to conclude that subjects show an increased report of pain and/or a decreased ability to tolerate pain when they are induced to expect that a given stimulus will be very painful (Chaves & Barber, 1974); to perceive themselves as having little or no control over the stimulus (Davison & Valins, 1969; Geer, Davison, & Gatchel, 1970; Staub, Tursky, & Schwartz, 1971); to focus attention on the pain stimulus (Kanfer & Goldfoot, 1966); or to feel anxious (Bobey & Davidson, 1970). Turk, Meichenbaum, and Genest (1983) cited several

studies that they performed suggesting that "catastrophization" (i.e., thinking the worst about future pain experience) is a major factor in increasing the experience of pain and in decreasing one's ability to cope with it. The perceived meaning of pain also appears to be a relevant cognitive dimension. Beecher (1946), for example, contrasted the tremendous pain tolerance displayed by injured soldiers, for whom pain represented an honorable escape from the rigors of war, with the responses of civilians to similar pains, which represented bodily harm and disease.

With chronic pain, these cognitive factors are likely to gain more and more predominance as time passes. Patients often learn to expect and fear pain, to see themselves as having no control over it, to see it as robbing their lives of any positive meaning, and to catastrophize about the future effects of pain on their lives. Lefebvre (1981) found that cognitive distortions about the nature of pain are common in chronic pain patients. Smith, Follick, Ahern, and Adams (1986) found that patients who tend to misconstrue the meaning of the pain sensation, particulary those who predict future disability from present difficulties, tend to be more severely disabled than patients who do not do so. Chapman, Brena, and Bradford (1981) noted that simple education often results in significant changes in pain behaviors and that chronic pain patients often act on erroneous beliefs (e.g., that any increase in activity will make pain worse, that certain activities will lead to reinjury, and that getting off narcotic pain medications will result in unbearable pain).

Pain and Depression

The role of depression in the etiology of chronic pain has been a subject of considerable controversy. There is little doubt that individuals with chronic pain report symptoms similar to those reported by people who are clinically depressed, such as a low energy level, difficulty in getting started and in sleeping, and feelings of being demoralized. Blumer and Heilbronn (1982) hypothesized that chronic pain is a variant of depressive disease, with depressive symptomatology often masked as a chronic pain problem. They cited as evidence studies revealing biological markers for depression such as early escape from dexamethasone and abnormal rapid-eye-movement (REM) latency; the frequent presence of bodily aches and pains among depressives; frequent family histories of depression, alcoholism, and other behavioral problems in chronic pain patients; and the frequency of pain relief with antidepressant medication (e.g., Blumer & Heilbronn, 1984; Blumer, Zorick, Heilbronn, & Roth, 1982). Magni (1987) pointed out that "it is clear that both neurotransmitters, serotonin and noradrenaline, seem to be involved in the process of pain perception, and, at the same time, in the probable pathogenic mechanism of depression" (p. 15).

A number of critics have questioned the conceptionalization of chronic pain as a depressive equivalent, citing a variety of conceptual and methodological weaknesses in the argument and some nonconfirmatory data (Getto, 1988; Rosenbaum, 1982; Turk, Rudy, & Steig, 1987; Turk & Salovey, 1984). For example, Getto (1988) and Magni (1987) have noted that the research evidence for the presence of biological markers for depression and for a family history of depression is conflicting and is characterized by studies with small numbers of subjects. Turk *et al.* (1987b) described how the common correlates of physical limitation and illness could account for seven of the eight criteria for major depression found in the third edition of the American Psychiatric Association's *Diagnostic and Statistical Manual of Mental Disorders* (DSM-III; APA, 1980). Furthermore, Dessonville, Reeves, Thompson, and Gallagher (1984) found that the depression manifested by chronic pain patients was qualitatively different from major depression in that pain patients tend to show fewer affective and cognitive aspects. Leoser (1979) suggested that the relief that chronic pain patients get with antidepressant medications is largely attributable to the effect of those medications on restoring sleep and normal circadian rhythms, rather than on their effect on any underlying affective disorder. Though antidepressant medications are frequently used

to help relieve chronic pain symptomatology, the dose level is often far below that required to relieve a major depression.

Pain and Stress

An often-cited theory is that pain and tension exist in a cycle; that is, that repeated muscular hyperactivity leads to ischemia and oxygen depletion at the affected site and subsequently to the release of pain-eliciting substances such as bradykinin (Turk & Flor, 1984). The pain then acts as an additional stressor, causing over time a host of atypical musculoskeletal, visceral, metabolic, and neurological dysfunctions that can further aggravate the experience of pain (Chapman, 1977; Craig, 1984). Although this theory has received indirect support from the success of clinical programs designed to teach muscle relaxation and stress management, there is conflicting evidence about what role stress plays in the development of chronic pain, and research efforts have not addressed clearly the assumptions inherent in the model of a vicious circle of pain, stress, and tension (Turk & Flor, 1984).

Assessment Methods

The purpose of a psychological-behavioral assessment is to ascertain which of the factors of conditioning described above are relevant to the pain problem and, most important, to design treatment.

The Psychological Interview

The best source of data is likely to be a comprehensive psychological interview. Such an interview permits the direct observation of affect and mental and cognitive status and allows an in-depth exploration of the identified problem areas. It also provides a sampling of behavior in an interpersonal situation. Because of the subjectivity of interview data, it is important to correlate such data with objective measures of behavior and also to gather information from multiple sources, which may include previous physicians, employers, and/or the attorney or rehabilitation counselor. Interviewing the spouse and/or significant others separately and comprehensively is often essential; not only does it provide a reliability check for the information provided by the patient, but it also yields important data regarding interpersonal relationships and the way in which the significant other(s) responds to pain behaviors.

Table 1 provides a summary of the information to be sought in a psychological assessment interview. Focusing on questions relating to pain symptomatology first is often wise, as the patient may resist discussing psychopathology until some rapport is established. Pain complaints that are diffuse, and vaguely described

Table 1. Outline for Psychological-Behavioral Interviewing

1. Pain complaints
   A. Location, description, and patterns of occurrence
   B. Onset
   C. History of treatments, including medications
   D. Previous history of pain-relieving illnesses
   E. Nonverbal pain behavior observed
2. Physical, social, and recreational activities
   A. Level and average day's schedule
   B. Changes and restrictions since onset of pain
   C. Attempts to return to restricted activities
   D. Pacing and consistency of activities
3. Work
   A. Employment, disability, and financial status
   B. Work history and performance
   C. Work satisfaction
   D. Attempts to return to work; perceived restrictions
   E. Emotional and cognitive responses regarding work and disability status
4. Sleep disturbances: Nature and history
5. Social relationships
   A. Previous and current social reinforcement patterns
   B. Modeling of illness or pain behavior
   C. Sexual disturbances
   D. Family attitudes and expectancies
6. Emotional status
   A. Previous and current affective or behavioral disturbances (including physical or sexual abuse)
   B. Mental status
   C. Treatment history for psychological difficulties
7. Cognitive-attitudinal factors
   A. Beliefs regarding the nature of the pain problem
   B. Expectancies of outcome
   C. Openness toward rehabilitation approach
   D. Perceived meaning of pain

and that do not fit with expectations based on medical data may reflect the influences of learning. The patterns of occurrence of pain complaints often yield important diagnostic information. For example, pain behavior that varies according to a spouse's attentiveness or support may be operant; increasing complaints after a stressful day of work are likely to be respondent; and patterns that closely match medical data, such as the worsening of arthritic pain in the morning, suggest the importance of organic factors. The interviewer should also note the nonverbal pain behaviors (such as sighing, grimacing, rubbing a painful area, and abnormalities in gait and posture) observed during an interview and should assess the degree to which these fit with expectations based on medical findings or may be learned responses.

The history and evolution of pain complaints in the patient and within a family system can provide important clues to the underlying processes and mechanisms, such as the presence of social modeling or tendencies to develop illness under stress. Pain behaviors can persist despite multiple previous medical treatments because the underlying psychological issues were not addressed (Chapman & Brena, 1982) or because such treatments reinforced such behavior. Medication and alcohol use patterns can be quite revealing. The opportunity to take medications that induce anxiety relief can be a powerful reinforcer for pain behavior. Once addiction or dependency is present, the cessation of medications can be very aversive.

Because increasing meaningful activities is likely to be a vital part of the rehabilitation process, an assessment of the patient's previous and current activity patterns is essential. Pain behavior can be respondent to overactivity, underactivity, inconsistent activity patterns, poor activity selection, and/or difficulty in pacing activities. Social activities are often restricted by pain patients even though they may be only minimally affected by pain-related physical dysfunction. Such restrictions can be both a cause and an effect of depression or other emotional distress.

To ensure maximum reliability, questions regarding work status, attitudes, and history may need to be directed toward a variety of sources, including the employer and the rehabilitation counselor if applicable. When available, objective data regarding work ratings and history, job absenteeism, and financial benefits are important to gather. Work avoidance is much less likely to condition pain behavior if the patient had a stable work history, enjoyed the previous work, saw himself or herself as competent, received good work performance ratings, is receiving significantly reduced income compared to when he or she was working, has made efforts to return to work, and shows clear medical evidence to support changes in job status or responsibilities.

Sleep disorders are frequently reported by patients with chronic pain and can affect their daily ability to function and to cope with pain and stress. Failure to sleep at night can reflect underlying anxiety or difficulty in distracting oneself from pain, and early morning awakening may be a symptom of underlying depression. Valid conclusions regarding the etiology of sleep difficulties can be made only from the integration of evidence regarding the multiple medical, environmental, psychological, and behavioral factors that determine sleep.

Pain and its attendant changes in affect and lifestyle affect the whole family, which then can become a powerful ally or obstacle in the rehabilitation process (e.g., Turk, Flor, & Rudy, 1987a). An analysis of the expectations, attitudes, and behaviors of significant others before and since the onset of pain is thus essential. Whereas some families reinforce illness behavior, others ignore it or reduce communication with the patient. Both patterns can exacerbate depression and dysfunction.

An investigation of present and past emotional dysfunctions and their temporal relationship with pain behaviors is important because the failure to achieve significant rewards through "well" (i.e., nonpain) behavior enhances the likelihood that the patient will find greater rewards through pain behavior. In addition, failure in emotional function can result in physical dysfunctions associated with the experience of pain. Although research regarding the effect of early emotional experiences on the subsequent

development of pain is generally lacking, there is some suggestive evidence that physical or sexual abuse during childhood predisposes an individual to certain chronic pain syndromes (e.g., Haber & Roos, 1984).

A final area of important inquiry involves cognitive factors such as expectations, goals, beliefs, and self-statements regarding the pain problem. Analysis of these covert factors can reveal the underpinnings of emotional dysfunction and of difficulties in coping with pain and stress, as well as the presence of distortions that are likely to affect the future course of therapy.

Standardized Questionnaires

The use of standardized questionnaires allows a more objective comparison of important aspects of the patient's behavior with that of other patients studied and treated. In recent years there has been an explosion of questionnaires designed to measure pain and its effects on functioning. Although it is impractical to review the majority of them, some of the most widely used and/or the most promising are discussed here briefly.

Measures of Pain Intensity

Pain is a subjective experience intimate to the sufferer. The visual analogue scales represent a simple method of quantifying its self-rated intensity. The most commonly used consist of a horizontal line with end points labeled "no pain" and "pain as bad as it could be." In some versions, the letters of the words *light, moderate,* and *severe* are evenly spaced below the line. The patient is instructed merely to make a mark along the line that most accurately corresponds to his or her pain level. These scales show test–retest reliability (Scott & Huskisson, 1976) and are sensitive to changes in pain intensity (Joyce, Zutish, Hrubes, & Mason, 1975). Furthermore, chronic pain patients have been found to be consistent in recording their ratings of different levels of experimental thermal pain on a visual analogue scale and in comparing such levels with their clinical pain (Price, McGrath, Rafii,

& Buckingham, 1983). The potential disadvantages or limitations of the test include its measurement of pain along only one dimension and its susceptibility to influence from the attitudes of the clinician who administers it; it is often unclear whether these ratings are influenced by affective dimensions.

The McGill Pain Questionnaire (MPQ; Melzack, 1975) represents an attempt to measure pain experience across different dimensions. One part of this questionnaire consists of 78 adjectives merged into 20 different categories that represent the different sensory, affective, and evaluative dimensions of chronic pain experiences. Within each category, two to six words are listed in order, from those that represent the least to those that represent the most intense pain experience. The patient is instructed to select only those categories that contain a word descriptive of his or her pain and to circle the most descriptive word within that category. From these selections, one can derive the number of words chosen, the average rank value of the words, and the sum of the ranks represented by each word.

The validity of the division of pain experience into categories of sensory, affective, and evaluative dimensions has been assessed through a variety of correlational and factor-analytic studies, which have yielded mixed results (Turk, Rudy, & Salovey, 1985); however, the test's validity in describing different aspects of pain experience has been demonstrated in over 60 validity studies (Turk *et al.*, 1985), and the MPQ represents a well-researched and well-validated tool for the quantitative assessment of the quality and intensity of the pain experience. Furthermore, a short form has recently been devised that allows ratings of the intensity of 11 sensory and 4 affective words and has been found to correlate significantly with sensory, affective, and total scores on the MPQ (Melzack, 1987).

Measurement of Overt Pain Behaviors

Because the important goals of pain control centers generally include modifying patients' activity levels while decreasing their reliance on habit-forming medications, diaries have been devised to assess these variables; however, the validity of these self-report diaries has been called into question by studies suggesting poor correlation with the data from automated devices (Sanders, 1983) or staff observations (Kremer, Block, & Gaylor, 1981; Ready, Sarkis, & Turner, 1982). Such findings underscore the need to validate self-report data through the use of automated or objective measures when available, through information from significant others in the patient's environment, and through direct observation of behavior.

Keefe and Block's measurement system (1982) which rates five motor behaviors in low-back-pain patients (bracing, guarded movement, rubbing a painful area, grimacing, and sighing), represents a major advance in the objective measurement of pain behaviors. Observers can be trained to quantify these behaviors with 93%–99% accuracy either during a standard medical examination or through time-sampling methods (Keefe & Block, 1982). The validity of this methodology has been demonstrated in studies with back pain patients that have revealed positive correlations of certain of these motor behaviors with subjective pain intensity (Keefe & Crisson, 1988; Keefe, Wilkins, & Cook, 1984) and with measures of medical status, such as surgical history and positive medical test findings (Keefe *et al.*, 1984). The measure is also sensitive to change with pain-relieving modalities such as nerve blocks (Connally & Sanders, in press) and may represent a method of evaluating treatment outcome.

Tests Normed on Psychiatric Samples

The Minnesota Multiphasic Personality Inventory (MMPI) is probably the instrument most commonly used to assess overall psychological functioning in chronic pain patients. Indeed, a survey of multidisciplinary pain clinics published by Hickling, Sison, and Holtz (1985) revealed that the MMPI was used in 94% of the 65 pain clinics that were surveyed. The MMPI has several advantages for use with chronic pain patients. It now has a broad research base with such patients, and cluster analyses have revealed distinct groups of patients whose responses to chronic pain differ (e.g., Armentrout, Moore, Parker, Hewett, & Feltz, 1982; McGill, Lawlis, Selby, Mooney, & McCoy, 1983; Naliboff, McCreary, McArthur, Cohen, & Gottlieb, 1988). The MMPI is useful as a comprehensive screening instrument for psychological problems and for its ability to identify psychologically defensive patients or those with personality disorders whose self-report data may be particularly suspect and for whom long-term management or compliance problems may develop. The inspection of individual items on elevated scales may help to identify areas that need to be addressed in therapy.

Despite these advantages, several limitations regarding the MMPI should be noted. With the possible exception of findings that a very high score on the Hypochondriasis Scale suggests a poor prognosis with a variety of outcome measures in pain rehabilitation (McCreary, Turner, & Dawson, 1979; Pemberton, 1985; Strassberg, Reimherr, Ward, Russell, & Cole, 1981), MMPI scales or clusters have not consistently predicted the outcome of a variety of treatments for patients with chronic pain. Some confusion regarding the interpretation of the MMPI has occurred because of a failure to recognize that the test was normed on psychiatric patients and not on chronic pain patients. Moore, McFall, Kivlahan, and Capestany (1988), for example, described the many items on Scale 8 on the MMPI that are related to correlates of pain such as medication use, the interference of pain with concentration and memory, and decreased physical function. On other scales as well, elevations may not indicate primary psychological disturbances. Thus, clinicians need to avoid computerized interpretations of the MMPI that are based on psychiatric samples.

The Symptom Check List-90 (SCL-90), which includes a list of 90 physical and emotional symptoms that patients rate on a 5-point scale according to the degree to which they are bothered by them, is another example of a test normed on psychiatric patients that has been applied to the chronic pain population. Though the results provide a good summary of self-reported difficulties, the meaning of scale elevations in chronic pain patients has not been well researched, and the test has been described as being insensitive in differentiating personality variables in these patients (Main, 1983).

Tests Designed for Medical Populations

The significant role of medical variables in influencing psychological function suggests the importance of using tests that have been devised specifically for medical populations or to assess medical problems. An example is the Millon Behavioral Health Inventory (MBHI; Millon, Green, & Meagher, 1979), a 150-item true-false questionnaire that provides a measurement of both the overt and the covert behavior patterns important in physical ailments in which behavioral or emotional components play a significant role. The focus is not on general personality but on a variety of specific personality dimensions, attitudes, events, and cognitions (such as cooperativeness, recent stress, pessimism, chronic tension, and somatic anxiety) that affect the development of physical illness and/or compliance or prognosis in medical treatment programs. One scale is designed specifically to predict outcome in an interdisciplinary chronic pain program. The MBHI was designed from theoretical formulations, internal structural analyses, and studies of relationships of the items and scales to external criteria and variables. Few validity studies have been performed with a chronic pain sample; however, Sweet, Breuer, Hazelwood, Toye, and Pawl (1985) found a significant correlation of .35 between the score on the Pain Treatment Responsivity Scale and a summary outcome measure for 52 chronic pain patients; they also noted very high correlations of this scale with other scales of the MBHI. This finding casts doubt on the specificity of this scale in predicting treatment outcome.

The Sickness Impact Profile (Bergner, Bobbitt, Carter, & Gilson, 1981) is a 136-item questionnaire designed specifically to measure physical, psychosocial, work, and recreational dysfunctions associated with chronic illness. Follick, Smith, and Ahern (1985) provided normative data for patients with chronic low back pain and found that selected scale scores correlated substantially with activity level and emotional distress and are sensitive dependent measures of outcome in a pain rehabilitation program.

The Multidimensional Health Locus of Control scale (Wallston & Wallston, 1978) is an 18-item test that measures the extent to which the respondents see their health as being attributable to their own behavior, the behavior of powerful others such as physicians, or fate. This questionnaire may predict behavior in a pain rehabilitation program, as locus of control has been related to compliance in other health care settings (Wallston & Wallston, 1978).

Tests Designed Specifically for Chronic Pain Patients

Several tests have been developed recently specifically for use with chronic pain patients, including the Psychosocial Pain Inventory (Heaton, Getto, Lehman, Fordyce, Brauer, & Groban, 1982), the Pain Disability Index (Tait, Pollard, Margolis, Duckro, & Krause, 1987), the Chronic Pain Battery (Levitt, 1983), and the Vanderbilt Pain Inventory (Brown & Nicassio, 1987). The research is still too cursory to provide a clear assessment of their utility. One test that appears particularly promising is the West Haven-Yale Multidimensional Personality Inventory, devised by Kerns, Turk, and Rudy (1985). It is a 52-item questionnaire divided into three sections. One section measures five important dimensions of the pain experience: pain severity and suffering, interference with function, support from others, perceived life control, and affective distress. Two other sections relate to specific responses to pain by significant others (such as soliciting, punishing, and distracting), and to the nature of the respondent's social, recreational, and physical activities. This battery was developed with the use of very careful factor and item analysis and with considerable psychometric sophistication, and it clearly evaluates many important

areas of psychosocial impact. Its additional strengths lie in its ease of administration and scoring and its sensitivity to treatment effects (Barrios, Niehaus, & Henke, 1986; Kerns *et al.*, 1985).

Assessment of Malingering

An assessment issue that occasionally surfaces with chronic pain patients is whether the pain report may be an example of malingering, defined by Leavitt and Sweet (1986) as the deliberate faking of symptoms for the sole purpose of obtaining an extrinsic goal. Estimates from questionnaires from 99 orthopedists regarding the scope of malingering among the low-back-pain patients seen in their practice revealed a wide range; whereas 38% estimated a 1%–2% incidence of malingering (as defined above), 35% felt that the incidence was 10% or more, and 10% felt that the incidence was 30% or more (Leavitt & Sweet, 1986). These diffuse perceptions may relate to difficulty in distinguishing "malingering" from pain amplification or exaggeration and from psychogenic conditions that precipitate pain. In addition, the lack of any clear biological or objective markers that correlate closely with subjective pain reports makes the assessment of expected pain levels hazardous at best.

To date, most efforts to study malingering have been based on anecdotal evidence and case reports. Leavitt and Sweet's survey of orthopedists (1986) did suggest that there is a considerable consensus among orthopedists regarding the major indicators of malingering. These were categorizable along the two basic dimensions of exaggeration and incongruity, including such items as "weakness to manual testing not seen in other activities" and "overreacting during the examination." Several authors (e.g., Brena & Chapman, 1984; Ellard, 1970) have pointed to factors such as poor compliance with treatment designed to relieve pain and dysfunction, reports of negative outcome from such treatment, and dramatized pain description as contributing a diagnosis of malingering. Brena and Chapman also stressed that malingering can be diagnosed only from a consistent pattern among multiple indicators and stressed that, taken individually, most indicators are readily attributable to other processes besides malingering.

Only a few controlled studies have been performed that are related to the assessment of malingering. In a series of articles, Leavitt (1985, 1987a,b) reported data from large numbers of healthy volunteers asked to choose words to simulate the presence of back pain that would be severe enough to keep them from working. In comparison with patients in treatment with an orthopedist for actual back pain, the volunteers were more likely to choose words representing intense pain and affective distress (e.g., *grueling*, *stabbing*, and *sickening*) and were less likely to choose words representing mild sensory features (e.g., *numb* and *tingling*). Though these results are provocative, Leavitt admitted that a necessary limitation on interpretation lies in the fact that volunteers asked to simulate pain may be different from malingerers in the clinical setting. An apparent limitation in the research is that the volunteers were asked to simulate pain that would disable them from work, whereas many of the pain patients with whom they were compared may have reported nondisabling pain.

Chapman and Brena's study (1990) may provide indirect evidence related to malingering. They compared the responses during treatment in an interdisciplinary pain program from 143 patients showing no inconsistency with 17 patients labeled as "inconsistent." Inclusion in the inconsistency group necessitated unanimous independent agreement by three raters of a consciously produced clear contradiction among statements or behaviors, or between a statement and behavior. The authors found numerous differences among the two groups that were significant beyond the .05 level. Specifically, the inconsistent group showed less evidence of medical pathology and physical impairment, a higher likelihood of assessed submaximum effort at contracting muscles during EMG examination, a greater incidence of pending disability issues, a greater use of affective language in describing their pain, less reported pain relief from sympathetic injections

given with saline, lower activity levels at pretreatment and posttreatment, and staff ratings indicating higher levels of observed pain behavior and lower levels of interest and compliance during treatment. These results were consistent with predictions that inconsistent patients would show a syndrome of characteristics distinct from that of patients showing no such inconsistency; however, it needs to be emphasized that one example of inconsistency is not the same as malingering.

The assessment of malingering is important in the settlement of disability claims and in containing the cost of health care; however, there are currently few clear guidelines for making this diagnosis, and it is unlikely that one "litmus test" of malingering can ever be developed to make the diagnosis easy. Rather, an analysis of many factors, including the medical findings, the psychological and behavioral history of the patient, the presence of reinforcers for malingering, and the nature and consistency of patients' statements and behaviors, is likely to be required. The clinician needs to be very careful to avoid the diagnosis of malingering solely on the grounds that pain behavior exists that does not fit with expectations based on the medical findings.

Psychological Interventions with Chronic Pain Patients

The emotional distress that accompanies pain often requires extensive psychological intervention. Living with chronic pain and restriction in basic everyday activities is very emotionally and physically draining for many individuals, particularly when there is an absence of support or understanding in the environment. Many patients with chronic pain go through emotional changes similar to those described by Kübler-Ross (1969) in her work with terminal cancer victims. Basically, this sequence is an adaptation to loss, which in chronic pain usually represents a loss of feeling good and the ability to function. It may progress through stages of denial, anger "bargaining" (or anxiety), depression, and, if these stages are worked through emotionally, acceptance. Acceptance represents a condition in which the individual still has goals and finds meaning in life. The sequencing and presence of these stages varies from patient to patient. The psychologist is faced with the challenge of helping chronic pain patients work through these stages so as to gain acceptance of limitations and pain in concert with maximum control over such difficulties.

Group Therapy

Many of the emotions and concerns that arise with chronic pain are almost universal. For example, common conflicts arise in dealing with others in society who cannot understand chronic pain and may tend to minimize it, deny it, or suggest that it can be cured. Statements such as "You look so good . . . you cannot be feeling so bad," "You must be better by now after all the treatment you have had," or "Why don't you do x, y, or z, so that your pain can be cured" are difficult for many to deal with. Embarrassment about the pain is also a frequent issue. Many patients overdo to appear to be "normal," whereas many others become socially isolated out of their fear that the pain may build up in a social situation. Misplaced fears, such as that the pain cannot be handled without reliance on medication or that activities cannot be pursued without reinjury or unbearable pain, often build up and prevent a more normal lifestyle. Groups can be helpful in education, the establishment of a support system for patients, reinforcement for progress, and the enhancement of better communication and greater perspective about difficulties. Groups in which positive models for coping are available can be very powerful in helping a new patient to gain faith and to begin needed rehabilitation efforts.

Family Therapy

Involvement of the family in the treatment is almost always critical to promote understanding and proper reinforcement of the patient's efforts. Marital therapy is often needed to help repair broken relationships and harmful communication patterns.

Individual Therapies

Many patients who are suffering from major emotional difficulties may be in need of indepth individual psychotherapy. For many such individuals, merely teaching basic relaxation or stress management skills may be akin to placing a Band-Aid on a massive hemorrhage. The specific therapeutic interventions that are often relevant for chronic pain patients include cognitive therapy and relaxation, self-hypnosis, and biofeedback.

Cognitive Therapies

In the broadest sense, any kind of psychotherapy can be regarded as "cognitive" in that it can alter appraisals of the self or the situation, of thought processes, and of expectations; however, specific strategies have been used to teach patients with pain to alter their self-statements regarding the experience of pain. Success may be measured by a change in cognitions from "This pain is nothing but terrible misery" and "When I hurt, only a pill will help" to self-statements such as "My life can have meaning and happiness despite the presence of pain" and "I can control this pain through my own efforts." The process of cognitive change often involves the identification of important self-statements, an exploration of how these contribute to the painful experience, and a rehearsal of healthier alternative cognitions. Turk et al. (1983) suggested the importance of breaking down the pain experience into the phases of preparation for pain (or related stress), confrontation with pain, critical moments in dealing with it, and self-reflection and appraisal following a painful episode. The patient may be asked to rehearse a variety of skills when facing pain, such as deep breathing, imagery of pleasant scenes, imaginative transformation of pain (perhaps thinking of pain as being numbed by an anesthetic), attention diversion techniques, and giving oneself reassuring messages to alter conditioned fear responses. Many studies reviewed by such authors as Tan (1982), Turk *et al.* (1983b), and Turner and Chapman (1982) have revealed significant reductions in subjective pain intensities with such strategies, but often, the specific mechanism of change has remained unclear.

Relaxation and Self-Hypnosis

Relaxation therapy can be regarded as a form of cognitive behavior modification in that it involves a type of mental focusing that alters the attentional processes and increases the perceived control of pain and stress. Its role in altering physiological functions such as muscle tension, blood flow, and metabolism, and in reducing insomnia and the need for medications, also contributes to its successful use in chronic pain. Its utility is underscored by patients' posttreatment reports of its helpfulness. Chapman (1983a) reported that after *one* 60-minute session of relaxation with the use of a tape for home practice, 60% of a mixed group of 237 chronic pain patients rated it as "very helpful" and 34% as "somewhat helpful." These figures, equivalent to those for physical therapy, exceeded those reported for almost all other modalities of treatment, including pain-relieving modalities such as transcutaneous electrical nerve stimulation and nerve blocks.

Chronic and continuous pain is likely to require a higher level of patient skill than acute or recurrent pain states because patients have to learn to relax while in pain and to integrate relaxation into their lifestyle. Merely using a procedure at pain onset or when trigger events for pain are present is likely to be insufficient. For this reason, the development, through practice, of brief relaxation methods and the use of a host of stress management techniques (e.g., time management, exercise, balanced activities, proper diet, humor, and assertiveness and communication training) are essential. Inasmuch as different patients find different methods of relaxation most helpful, it is important to teach a variety of methods, including progressive relaxation, breathing methods, and imagery.

Some patients with chronic pain with high hypnotic susceptibility may be able to relieve pain more directly through the addition of suggestion to a procedure that permits deep relaxation. Numerous authors, whose work has been summarized in Grzesiak and Ciccone (1988), have reported significant improvement in pain control with self-hypnosis, but the hypnotic control of chronic pain is an area in great need of systematic and controlled research. Hypnotic treatment protocols certainly do provide powerful messages that patients can control their pain, as well as an impressive ritual of procedures.

Biofeedback Therapy

Biofeedback involves the use of equipment to teach patients to change their physiological responses by providing them with immediate feedback from their responses (Brown, 1977). The most frequent pain complaint to which biofeedback has been applied has been headaches. One commonly used type of feedback used primarily for muscle-contraction or mixed headaches is electromyographic (EMG) biofeedback, whose goal is reducing muscle tension in the frontalis muscle or in neck, shoulder, or other facial muscles. Thermal biofeedback, used primarily for vascular headaches, is directed to teaching patients to warm their hands. The successful use of biofeedback modalities in patients with chronic or recurrent muscle-contraction and migraine headaches was summarized by Blanchard, Andrasik, Ahles, Teders, and O'Keefe (1980), who used "meta-analysis" to review treatment outcome studies. In this method, mean changes in headache measures across research studies form the unit of analysis. The authors used the percentage of improvement in headache density as a dependent measure whenever possible. If density data were unreported, they used other headache measures, such as intensity, frequency, and duration, if available; otherwise, they used the gross percentage of subjects who were improved. They found mean posttreatment percentages of improvement ranging from 51.8% to 65.1% for thermal or EMG biofeedback (with or without relaxation training) for migraine and muscle-contraction headache sufferers.

Several reviewers (e.g., Chapman, 1986; Holmes & Burish, 1983; Williamson, 1981) have summarized the results of follow-up studies of one year or longer, which have revealed maintenance of, if not improvement in, posttreatment levels.

These findings certainly are impressive, particularly when one recognizes that the subjects in these studies generally had suffered headaches that lasted for long periods of time and that had been quite refractory to medication; however, doubts about the *mechanism* or the differential utility of biofeedback have come from numerous studies suggesting that (1) biofeedback and relaxation therapies yield similar treatment outcomes; (2) there is no clear evidence that headache sufferers differ in physiological parameters targeted with biofeedback from normals without headache; (3) changes in such parameters are generally poorly correlated with improvements in pain levels; (4) migraine subjects have been found to improve just as much with hand-cooling biofeedback as with hand-warming; and (5) pseudofeedback (i.e., feedback not contingent on the subject's responses) results in nearly as much improvement in muscle-contraction headache subjects as does actual EMG feedback. The vast number of studies on these points are explored comprehensively in Chapman's review (1986). In light of these results, authors have speculated that the critical elements in biofeedback may involve such factors as increases in hope, self-mastery, and perceived control (Frank, 1982; Meichenbaum, 1976; Rickles, Onoda, & Doyle, 1982); an increased awareness of the role of emotions and stress in headaches (Turner & Chapman, 1982a); the application of coping strategies to control stress (Andrasik & Holroyd, 1983; Jessup, Neufeld, & Merskey, 1979); and the motivation of patients to practice relaxation techniques (Silver & Blanchard, 1978).

As the methodology in psychophysiological research improves, some significant relationships between pain and aspects of physiology may emerge; however, based on current knowledge, biofeedback for headaches may best be conceptualized as a tool by which subjects can alter important cognitive, emotional, and behavioral responses. Thus, it is an adjunct to psychotherapy; the use of biofeedback technicians relatively untrained in psychotherapy and the provision of many repetitive biofeedback sessions focused merely on the control of a single response thus seem to be highly questionable clinical practices.

Biofeedback has also been used for many other pain syndromes besides headaches, but its use is poorly documented and researched. The commonly observed practice of using frontalis EMG biofeedback or thermal biofeedback for any pain problem is unwarranted in light of the relatively low correlations of such parameters with pain reports or with other measures of stress (e.g., Carlson, Basilio, & Heaukulan, 1983; Whatmore, Whatmore, & Fisher, 1981). The use of EMG biofeedback from the paraspinal muscles in low-back-pain subjects has been questioned for the same kinds of lack of physiological correlations and pain reports noted above for headache subjects. Furthermore, the EMG levels of the paraspinal muscles may be specific to certain positions or movements (Ahern, Follick, Council, Laser-Wolston, & Litchman, 1988; Wolf, Nacht, & Kelley, 1982) and may have different patterns among different diagnostic groups with low back pain (Arena, 1988).

Interdisciplinary Pain Centers

The Commission on Accreditation of Rehabilitation Facilities (1988) defined a chronic pain management program as "a program designed to reduce pain, improve quality of life, and decrease dependence on the health care system, for persons with pain which interferes with physical, psychosocial, and vocational functioning, through the provision of coordinated, goal-oriented, interdisciplinary team services" (p. 51). CARF (1987) established a committee in 1986 to design a model method of evaluating the effectiveness of such programs. This committee showed remarkable unanimity in defining the major goals of pain programs as including a reduction in the use and misuse of medications; a reduction in the use of subsequent health care services for the presenting pain problem; a return to productive lives; an increase in physical activities; an increase in the ability to manage pain and related problems; a decrease in subjective pain intensity; and the containment of program cost. Indeed, interdisciplinary pain programs are most appropriate for individuals who show many elements of the disease of the D's (dysfunction, disuse, dramatized complaints, depression or demoralization, disability, and/or drug misuse). Individuals who show only a few such symptoms may benefit from individualized programs provided by a pain center but may not require a full, comprehensive program.

Despite the similarity of their goals, pain programs show great diversity in the scope, nature, and type of services they provide. Both inpatient and outpatient treatments are commonly offered by pain centers. The major advantages of inpatient programs include the possibility of the development of a 24-hour-a-day therapeutic community, the opportunity to observe behavior reliably and to control it intensively, the temporary elimination of previous cues for pain behavior, and controlled drug withdrawal. Outpatient approaches, on the other hand, usually cost much less than inpatient programs, which may be as high as $20,000. Outpatient approaches allow the observation of progress and problems in the home environment and allow patients to maintain other productive activities while receiving treatment. They also minimize such potential problems inherent in inpatient programs as the difficult transition from hospital to home, the possibility that being hospitalized will reinforce one's identity as being sick, and the adjustments and inconveniences associated with being away from the home environment.

The pain center most commonly sees patients referred by medical disciplines when pain and dysfunction have persisted for many months. Patients undergo an array of treatments and often spend much of their day at the center. Although CARF now mandates that the core team include at a minimum a physician, a specialized nurse, a physical therapist, and a clinical psychologist or psychiatrist, many include vocational counselors and evaluators, social workers, recreational and occupational therapists, and many other types of coun-

selors. Hickling et al.'s survey of 76 centers (1985) revealed that individual psychological therapy, transcutaneous electrical nerve stimulation, biofeedback, psychotropic medication, couples and family therapy, and operant programs were the most common therapies offered. Though physical therapy was not listed as a treatment modality, the finding that 75% of clinics employed a physical therapist suggests its widespread use.

The role of the psychologist in a pain center is a critical one. The Hickling et al. (1985) survey revealed that psychology was the most frequent discipline participating in a pain clinic (85.5% of the clinics employed at least one psychologist). Psychologists' roles almost always included patient evaluation and therapy, but psychologists were also found to be active in administration, coordination of treatment, supervision, research, teaching, consultation, and public relations. The psychologist's skills in helping other staff respond consistently and appropriately to the behaviors and attitudes of patients so as to maximize compliance and treatment effectiveness are often critical. In addition, psychologists can help provide guidance to prevent or deal with staff burnout. Given the chronic nature of patients' difficulties and the frequently personal and intense nature of interactions with patients, burnout is a significant risk at pain centers.

The diversity of approaches at pain centers suggests the need for controlled research to evaluate the efficacy of different approaches, but little such research has been done, largely because of the lack of comparability in outcome measures and patient selection variables from setting to setting. In addition, interdisciplinary pain programs have not been compared with other treatment approaches in controlled studies; however, scores of studies have suggested that patients with low back pain and other chronic musculoskeletal complaints who go through such programs generally report increased activity levels and a reduced use of medications for pain without increases (and, often, with some decreases) in subjective pain intensities. Generally, the data have revealed that 20%–50% of unemployed workers have returned to work at follow-ups of one to three years. Unfortunately, most of these data have been based on the patient's self-report. The reader is referred to reviews by Aronoff, Evans, and Enders (1983), Enders (1983), and Chapman (1985) for a more comprehensive description and criticism of such studies.

Characteristics of Excellence in Pain Programs

In the absence of firm data, this author wishes to draw on his experience to suggest some critical criteria for excellence in pain centers. Excellent programs provide comprehensive assessment and consistent treatment with a well-structured interdisciplinary team approach. Consistency is critical because so many patients with chronic pain are confused and angered by the inconsistent messages heard previously about the etiology and treatment of their pain. Programs in which treatments from different disciplines are offered on a contemporary rather than a sequential basis often have greater consistency. In addition, treatments across disciplines have great power to reinforce each other. For example, activity increase or drug withdrawal may be made difficult in the absence of any modalities that provide some pain relief; patients may not be fully alert and able to function if drugs are not managed or withdrawn properly; critical physical therapy exercises may not be performed unless depression is managed; and depression may not be relieved without an increase in activity and function. Thus, the team members must share a common philosophy of treatment and must communicate regularly, both informally and through team conferences. This approach can be contrasted with a consultation approach, in which each specialist treats the patient only from the standpoint of his or her own discipline and then refers the patient for other treatments.

The treatment must also be flexible, individualized, and related to specific and realistic behavioral goals. It is a poor practice to provide the same treatment for everyone in a package without regard to individual differences, as there is a greater deal of evidence that there is no such thing as a "typical pain personality."

One unfortunate but commonly observed practice is to provide package treatment and then simply to blame the patient if this treatment appears to fail. Usually, the label put on the patient is that he or she was "simply not motivated" or "didn't want to get better." Based on early work on motivation by Thorndike and others, this author suggests defining "motivation" or the probability of behavior as being proportional to the expectancy of reward, multiplied by the sum of the intrinsic value of the reward and externally applied incentives, divided by the effort required to obtain the reward. Thus, if someone is not increasing her or his activities as instructed, it could be that the person does not expect success, does not find the activities very reinforcing, has no incentive for increasing her or his activities, and/or feels that the effort involved is too great.

Quality programs involve careful planning for the maintenance of behavioral changes over the long term. This planning is particularly important in chronic pain because of the likelihood of flare-ups and the continual long-term effort needed for successful rehabilitation. Planning may need to include a preparation for subsequent problems and a rehearsal of strategies, the careful fading of treatment and maintenance of contact, and, perhaps most important, involvement in the education of many individuals in the environment with whom the patient will be interacting. For example, successfully helping the injured worker return to productivity is likely to involve careful coordination with the family, the rehabilitation counselor, the insurance adjustor, the attorney, the primary-care physician, and the subsequent employer.

One of the areas that is essential in the accreditation of programs through CARF is the development of a program evaluation system that provides for the regular measurement of success versus costs in meeting major goals. From such an evaluation, the reasons for the failure to meet goals must be identified, and corrective action must be taken. Program evaluation data at the Pain Control and Rehabilitation Institute of Georgia have yielded results with major implications for treatment. These have included findings that relaxation and physical therapies are judged more helpful by patients than medical modalities for pain relief; that patients on workers' compensation show treatment outcome similar to that of those with no pending litigation; and that seeing patients as outpatients rather than inpatients yields similar outcomes with a greatly reduced cost.

A final criterion for excellence in chronic pain management programs is the caring, involvement, and enthusiasm of the staff. Many patients have expressed how important these factors are to them in the face of their feelings of despair and helplessness and in their low feelings of self-worth. One only needs to review the literature on the power of the placebo for pain (e.g., Routon, 1983) to recognize the importance of such emotional and cognitive variables in pain control.

Summary

Major advances in pain management have occurred with the recognition that the mind and the body are interdependent. Affective, cognitive, and behavioral processes have major effects on the physiological processes that underlie the experience of pain. In turn, pain, particularly when chronic, affects psychological processes greatly. Increasingly specific and sophisticated methods of psychological assessment and treatment have been devised to help the chronic pain patient and are often most effective when provided in the context of an interdisciplinary team approach. Despite these advances, the understanding of the psychological and physical process and the mechanisms of chronic pain remains in its infancy and awaits advances from future research.

References

Ahern, D. K., Follick, M. J., Council, J. R., Laser-Wolston, N., & Litchman, H. (1988). Comparison of lumbar paravertebral EMG patterns in chronic low back pain patients and non-patient controls. *Pain, 34*, 153–160.

American Psychiatric Association, (1980). *Diagnostic and statistical manual of mental disorders*. Washington, DC: Author.

Anderson, L. P., & Rehm, L. P. (1987). The relationship between strategies of coping and perception of pain in three chronic pain groups. *Journal of Clinical Psychology, 40*, 1170–1177.

Andrasik, F., & Holroyd, K. A. (1983). Specific and nonspecific effects in the biofeedback treatment of tension headaches: 3-year follow-up. *Journal of Clinical and Consulting Psychology, 51*, 634–636.

Apley, J. (1975). *The child with abdominal pains*. Oxford, England: Blackwell.

Arena, J. G. (1988, June). *Psychophysiologic assessment of the pain patient*. Presentation at the Conference on Behavioral Medicine, Medical College of Georgia, Augusta.

Armentrout, D. P., Moore, J. E., Parker, J. C., Hewett, J. E., & Feltz, C. (1982). Pain–patient MMPI subgroups: The psychological dimensions of pain. *Journal of Behavioral Medicine, 5*, 201–211.

Aronoff, G. M., Evans, W. O., & Enders, P. L. (1983). A review of follow-up studies of multidisciplinary pain units. *Pain, 16*, 1–11.

Barrios, F. X., Niehaus, J. C., & Henke, K. J. (1986, March). *A comparison of pre-treatment, discharge, and follow-up status of chronic pain patients*. Paper presented at the annual meeting of the Society of Medicine, San Francisco.

Beecher, H. K. (1946). Pain in men wounded in battle. *Annals of Surgery, 123*, 96–105.

Bergner, M., Bobbitt, R. A., Carter, W. B., & Gilson, B. S. (1981). The Sickness Impact Profile: Development and final revision of a health status measure. *Medical Care, 19*, 787–805.

Blanchard, E. B., Andrasik, F., Ahles, A., Teders, S. J. & O'Keefe, D. (1980). Migraine and tension headaches: A meta-analytic review. *Behavior Therapy, 11*, 613–631.

Block, A. R., Kremer, E., & Gaylor, (1980). Behavioral treatment for chronic pain: The spouse as a discriminative cue for pain behavior. *Pain, 9*, 243–252.

Blumer, D., & Heilbronn, M. (1982). Chronic pain as a variant of depressive disease. *Journal of Nervous and Mental Disease, 170*, 381–394.

Blumer, D., & Heilbronn, M. (1984). "Chronic Pain as a Variant of Depressive Disease:" A Rejoinder. *Journal of Nervous and Mental Disease, 172*, 405–407.

Blumer, B., Zorick, F., Heilbronn, M., & Roth, T. (1982). Biological markers for depression in chronic pain. *Journal of Nervous and Mental Disease, 170*, 425–428.

Bobey, M. J., & Davidson, P. O. (1970). Psychological factors affecting pain tolerance. *Journal of Psychosomatic Research, 14*, 371–376.

Bonica, J. J. (1953). *The management of pain*. Philadelphia: Lea & Febiger.

Brena, S. F., & Chapman, S. L. (1984). Pain and litigation. In P. D. Wall & R. Melzack (Eds.), *Textbook of pain* (pp. 832–839). Edinburgh: Churchill Livingstone.

Brown, B. B. (1977). *Stress and the art of biofeedback*. New York: Harper & Row.

Brown, G. B., & Nicassio, P. M. (1987). Development of a questionnaire for the assessment of active and passive coping strategies in chronic pain patients. *Pain, 31,* 53–64.

Cairns, D., & Pasino, J. (1977). Comparison of verbal reinforcement and feedback in the operant treatment of disability due to chronic low back pain. *Behavior Therapy, 8,* 621–630.

Carlson, J. G., Basilio, C. A, & Heaukulan, J. D. (1983). Transfer of EMG training: Another look at the general relaxation issue. *Psychophysiology, 20,* 530–536.

Chapman, C. R. (1977). Psychological aspects of pain patient treatment. *Archives of Surgery, 112,* 767–772.

Chapman, S. L. (1983a). Behavior modification. In S. F. Brena & S. L. Chapman (Eds.), *Management of patients with chronic pain* (pp. 145–159). New York: SP Medical and Scientific Books.

Chapman, S. L. (1983b). The role of learning in chronic pain. In S. F. Brena & S. L. Chapman (Eds.), *Management of patients with chronic pain* (pp. 55–61). New York: SP Medical and Scientific Books.

Chapman, S. L. (1985). Behaviour modification for chronic pain states. In S. F. Brena & S. L. Chapman (Eds.), *Clinics in anaesthesiology. Chronic pain: Management principles* (pp. 111–142). London: Saunders.

Chapman, S. L. (1986). A review and clinical perspective on the use of EMG and thermal biofeedback for chronic headaches. *Pain, 27,* 1–43.

Chapman, S. L., & Brena, S. F. (1982). Learned helplessness and responses to nerve blocks in chronic low back pain patients. *Pain, 14,* 353–364.

Chapman, S. L., & Brena, S. F. (1990). Patterns of conscious failure to report, accurate self-report data in patients with low back pain. *Clinical Journal of Pain, 6,* 178–190.

Chapman, S. L., Brena, S. F., & Bradford, L. A. (1981). Treatment outcome in a chronic pain rehabilitation program. *Pain, 11,* 255–268.

Chaves, J. F., & Barber, T. X. (1974). Cognitive strategies, experimenter modeling and expectations in the attenuation of pain. *Journal of Abnormal Psychology, 83,* 356–363.

Commission on Accreditation of Rehabilitation Facilities. (1987). *Program evaluation in chronic pain programs.* Tucson: Author.

Commission on Accreditation of Rehabilitation Facilities. (1988). *Standards manual for organizations serving people with disabilities.* Tucson: Author.

Connally, G. H., & Sanders, S. H. (in press). Relationship of overt pain behavior, cognitive coping style and response to lumbar sympathetic nerve blocks. *Pain.*

Craig, K. D. (1984). Emotional aspects of pain. In P. D. Wall & R. Melzack (Eds.), *Textbook of pain* (pp. 153–161). Edinburgh: Churchill Livingstone.

Craig, K. D., & Prkachin, K. M. (1978). Social modeling influences on sensory decision theory and psychophysiological indexes of pain. *Journal of Personality and Social Psychology, 36,* 805–815.

Crue, B. L. (1983). The neurophysiology and taxonomy of pain. In S. F. Brena & S. L. Chapman (Eds.), *Management of patients with chronic pain* (pp. 21–31). New York: SP Medical and Scientific Books.

Davison, G. S., & Valins, S. (1969). Maintenance of self-attributed and drug-attributed behavior change. *Journal of Personality and Social Psychology, 11,* 25–33.

Dessonville, C. L., Reeves, J. L., Thompson, L. W., & Gallagher, D. (1984, September). *The pattern of depressive symptomatology in geriatric normals, depressives, and chronic pain patients.* Presentation at the Fourth World Congress on Pain, Seattle.

Doleys, D. M., Crocker, M., & Patton, D. (1982). Response of patients with chronic pain to exercise quotas. *Physical Therapy, 62,* 1111–1114.

Ellard, J. (1970). Psychological reactions to compensable injury. *The Medical Journal of Australia* (August 22), pp. 349–355.

Flor, H., Kerns, R. D., & Turk, D. C. (1987). The role of the spouse in maintenance of chronic pain. *Journal of Psychosomatic Research, 31,* 251–260.

Follick, M. J., Smith, T. W., & Ahern, D. K. (1985). The Sickness Impact Profile: A global measure of disability in chronic low back pain. *Pain, 21,* 67–76.

Fordyce, W. E. (1976). *Behavioral methods for chronic pain and illness.* St. Louis: Mosby.

Frank, J. P. (1982). Biofeedback and the placebo effect. *Biofeedback and Self-Regulation, 7,* 449–460.

Gentry, W. D., Shows, W. D., & Thomas, M. (1974). Chronic low back pain: A psychological profile. *Psychosomatics, 15,* 174–177.

Getto, C. J. (1988). Depression. In N. T. Lynch & S. V. Vasudevan (Eds.), *Persistent pain: Psychosocial assessment and intervention* (pp. 93–101). Boston: Kluwer.

Geer, J. H., Davison, G. C., & Gatchel, R. I. (1970). Reduction of stress in humans through nonveridical perceived control of aversive stimulation. *Journal of Personality and Social Psychology, 16,* 734–738.

Grzesiak, R., & Ciccone, D. S. (1988). Relaxation, biofeedback, and hypnosis in the management of pain. In N. T. Lynch & S. V. Vasudevan (Eds.), *Persistent pain: Psychosocial assessment and intervention* (pp. 163–188). Boston: Kluwer.

Haber, J., & Roos, C. (1984, September). *Effects of spouse abuse and/or sexual abuse in the development and maintenance of chronic pain in women.* Presentation at the Fourth World Congress on Pain, Seattle.

Louis, Harris & Associates. (1985). *Nuprin pain report.* New York.

Heaton, R. K., Getto, C. J., Lehman, R. A. W., Fordyce, W. E., Brauer, E., & Groban, S. E. (1982). A standardized evaluation of psychosocial factors in chronic pain. *Pain, 12,* 165–174.

Hickling, E. J., Sison, G. F. P., & Holtz, J. L. (1985). Role of psychologists in multidisciplinary pain clinics: A national survey. *Professional Psychology: Research and Practice, 16,* 868–880.

Holmes, D. S., & Burish, T. G. (1983). Effectiveness of biofeedback for treating migraine and tension headaches: A review of the evidence. *Journal of Psychosomatic Research, 27,* 515–532.

Jessup, B. A., Neufeld, R. W. J., & Merskey, H. (1979). Biofeedback therapy for headache and head pain: An evaluative review. *Pain, 7,* 225–270.

Joyce, C. R., Zutish, D. W., Hrubes, V., & Mason, R. M. (1975). Comparison of fixed interval and visual analogue scales for rating chronic pain. *European Journal of Clinical Pharmacology, 8,* 415–420.

Kanfer, F. H., & Goldfoot, D. A. (1966). Self-control and tolerance of noxious stimulation. *Psychological Reports, 18,* 79–85.

Keefe, F. J., & Block, A. R. (1982). Development of an observational method for assessing pain behavior in chronic low back pain patients. *Behavior Therapy, 13,* 363–375.

Keefe, F. J., & Crisson, J. E. (1988). Assessments of behaviors. In N. T. Lynch & S. V. Vasudevan (Eds.), *Persistent pain: Psychosocial assessment and intervention* (pp. 61–73). Boston: Kluwer.

Keefe, F. J., Wilkins, R. H., & Cook, W. A. (1984). Direct observation of pain behavior in low back pain patients during physical examination. *Pain, 20,* 59–68.

Kerns, R. D., Turk, D. C., & Rudy, T. E. (1985). The West Haven-Yale Multidimensional Personality Inventory (WHYMPI). *Pain, 23,* 345–356.

Kremer, E. F., Block, A. J., & Gaylor, M. S. (1981). Behavioral approaches to treatment of chronic pain: The inaccuracy of patient self-report measures. *Archives of Physical Medicine and Rehabilitation, 62,* 188–191.

Kübler-Ross, E. (1969). *On death and dying.* New York: Macmillan.

Leavitt, F. (1985). Pain and deception: Use of verbal pain measurement as a diagnostic aid in differentiating between clinical and simulated low-back pain. *Journal of Psychosomatic Research, 29,* 495–505.

Leavitt, F. (1987a). Detection of simulation among persons instructed to exaggerate symptoms of low-back pain. *Journal of Occupational Medicine, 29,* 229–233.

Leavitt, F. (1987b). A linguistic pain signature in simulated low-back pain. *Journal of Pain and Symptom Management, 2,* 83–88.

Leavitt, F., & Sweet, J. J. (1986). Characteristics and frequency of malingering among patients with low back pain. *Pain, 25,* 357–364.

Lefebvre, M. F. (1981). Cognitive distortion and cognitive errors in depressed psychiatric and low back pain patients. *Journal of Consulting and Clinical Psychology, 49,* 517–525.

Levitt, S. R. (1983). *Chronic pain battery.* Durham, NC: Pain Resource Center.

Linton, S. J., Melin, L., & Götestam, K. G. (1984). Behavioral analysis of chronic pain and its management. In M. Hersen, R. M. Eisler, & P. M. Millers (Eds.), *Progress in behavior modification,* Vol. 18 (pp. 249–291). New York: Academic Press.

Loeser, J. D. (1979). Letter to the editor: Re: "Effects of Doxepin on Perception of Laboratory-induced Pain in Man." *Pain, 6,* 241–242.

Magni, G. (1987). On the relationship between organic pain and depression when there is no organic lesion. *Pain, 31,* 1–21.

Main, C. J. (1983). The Modified Somatic Perception Questionnaire (MSPQ). *Journal of Psychosomatic Research, 27,* 503–514.

McCreary, G., Turner, J., & Dawson, B. (1979). The MMPI as a predictor of response to conservative treatment for low back pain. *Journal of Clinical Psychology, 35,* 278–284.

McGill, J. C., Lawlis, G. F., Selby, D., Mooney, V., & McCoy, C. E. (1983). The relationship of Minnesota Multiphasic Personality Inventory (MMPI) profile clusters to pain behaviors. *Journal of Behavioral Medicine, 6,* 77–92.

Meichenbaum, D. (1976). Cognitive factors in biofeedback therapy. *Biofeedback and Self-Regulation, 1,* 201–216.

Melzack, R. (1975). The McGill Pain Questionnaire: Major properties and scoring methods. *Pain, 1,* 277–299.

Melzack, R. (1987). The short-form McGill Pain Questionnaire. *Pain, 30,* 191–197.

Melzack, R., & Wall, P. D. (1965). Pain mechanism: A new theory. *Science, 150,* 971–979.

Millon, T., Green, C. J., & Meagher, R. B. (1979). The MBHI: A new inventory for the psychodiagnostician in medical settings. *Professional Psychology* (August), pp. 529–539.

Moore, J. E., McFall, M. E., Kivlahan, D. R., & Capestany, F. (1988). Risk of misinterpretation of MMPI Schizophrenia scale elevations in chronic pain patients. *Pain, 32,* 207–213.

Morse, R. H. (1983). Pain and emotions. In S. F. Brena & S. L. Chapman (Eds.), *Management of patients with chronic pain* (pp. 47–54). New York: SP Medical and Scientific Books.

Naisbitt, J. (1982). *Megatrends.* New York: Warner.

Naliboff, B. D., McCreary, J. P., McArthur, D. L., Cohen, M. J., & Gottlieb, H. J. (1988). MMPI changes following behavioral treatment of chronic low back pain. *Pain, 35,* 271–277.

Pemberton, J. S. (1985). *Personality variables and prediction of response to treatment of chronic low back pain.* Unpublished doctoral dissertation, Georgia State University.

Price, D. D., McGrath, P. A., Rafii, A., & Buckingham, B. (1983). The validation of visual analogue scales as ratio scale measures for chronic and experimental pain. *Pain, 17,* 45–56.

Ready, L. B., Sarkis, E., & Turner, J. A. (1982). Self-reported versus actual use of medications in chronic pain patients. *Pain, 12,* 285–294.

Rickles, W. H., Onoda, L., & Doyle, C. C. (1982). Task-force study section report: Biofeedback as an adjunct to psychotherapy. *Biofeedback and Self-Regulation, 7,* 1–34.

Rosenbaum, J. F. (1982). Comments on "Chronic Pain as a Variant of Depressive Disease: The Pain-Prone Disorder." *Journal of Mental and Nervous Disease, 170,* 412–414.

Routon, J. (1983). The placebo response. In S. F. Brena & S. L. Chapman (Eds.), *Management of patients with chronic pain* (pp. 205–210). New York: SP Medical and Scientific Books.

Sanders, S. H. (1983). Component analysis of behavioral

treatment program for chronic low back pain. *Behavior Therapy, 18,* 697–705.

Sanders, S. H. (1985). The role of learning in chronic pain states. In S. F. Brena & S. L. Chapman (Eds.), *Clinics in anaesthesiology: Chronic pain: management principles* (pp. 57–73). London: Saunders.

Schmidt, A. J. M. (1987). The behavioral management of pain: A criticism of a response. *Pain, 30,* 285–291.

Scott, J., & Huskisson, E. C. (1976). Graphic representation of pain. *Pain, 2,* 175–184.

Silver, B. V., & Blanchard, E. B. (1978). Biofeedback and relaxation training in the treatment of psychophysiological disorders, or, are the machines really necessary? *Journal of Behavior Medicine, 1,* 217–239.

Smith, T. W., Follick, M. J., Ahern, N. K., & Adams, A. (1986). Cognitive distortion and disability in chronic low backpain. *Cognitive Therapy and Research, 10,* 201–210.

Staub, E., Tursky, B., & Schwartz, G. E. (1971). Self-control and predictability: Their effects on reactions to aversive stimulation. *Journal of Personality and Social Psychology, 18,* 157–162.

Strassberg, D. S., Reimherr, F., Ward, M., Russell, S., & Cole, A. (1981). The MMPI and chronic pain. *Journal of Consulting and Clinical Psychology, 49,* 220–266.

Sweet, J. J., Breuer, S. R., Hazelwood, L. A. Toye, R., & Pawl, R. P. (1985). The Millon Behavioral Health Inventory: Concurrent and predictive validity in a pain treatment center. *Journal of Behavioral Medicine, 8,* 215–226.

Tait, R. C., Pollard, C. A., Margolis, R. B., Duckro, P. N., & Krause, S. J. (1987). The Pain Disability Index: Psychometric and validity data. *Archives of Physical Medicine and Rehabilitation, 68,* 438–441.

Tan, S-Y. (1982). Cognitive and cognitive-behavioral methods for pain control: A selective review. *Pain, 12,* 201–228.

Thelen, M. H., & Frye, R. A. (1981). The effect of modeling and selective attention on pain tolerance. *Journal of Behavior Therapy and Experimental Psychiatry, 12,* 225–229.

Turk, D. C., & Flor, H. (1984). Etiological theories and treatments for chronic back pain: 2. Psychological models and interventions. *Pain, 19,* 209–233.

Turk, D. C., & Flor, H. (1987). Pain behavior: The utility and limitations of the pain behavior construct. *Pain, 31,* 277–295.

Turk, D. C., & Salovey, P. (1984). "Chronic Pain as a Variant of Depressive Disease": A critical reappraisal. *Journal of Nervous and Mental Disease, 172,* 398–404.

Turk, D. C., Flor, H., & Rudy, T. E. (1987a). Pain and families: 1. Etiology, Maintenance and psychosocial impact. *Pain, 30,* 3–27.

Turk, D. C., Meichenbaum, D., & Genest, M. (1983). *Pain and behavioral medicine: A cognitive-behavioral perspective.* New York: Guilford Press.

Turk, D. C., Rudy, T. E., & Salovey, P. (1985). The McGill Pain Questionnaire reconsidered: Confirming the factor structure and examining appropriate uses. *Pain, 21,* 385–397.

Turk, D. C., Rudy, T. E., & Steig, R. L. (1987b). Chronic pain and depression: 1. "Facts." *Pain Management, 1,* 17–25.

Turner, J. A., & Chapman, C. R. (1982a). Psychological interventions for chronic pain: A critical review: 1. Relaxation training and biofeedback. *Pain, 12,* 1–22.

Turner, J. A., & Chapman, C. R. (1982b). Psychological interventions for chronic pain: A critical review: 2. Operant conditioning, hypnosis, and cognitive-behavioral therapy. *Pain, 12,* 23–46.

U. S. Department of Health and Human Services. (1986). *Report of the Commission on the Evaluation of Pain.* Washington: U.S. Government Printing Office.

Violon, A., & Giurgen, D. (1984). Familial models for chronic pain. *Pain, 18,* 199–203.

Wallston, K. A., & Wallston, B. S. (1978). Development of the Multidimensional Health Locus of Control (MHLC) scales. *Health Education Monographs, 6,* 160–170.

Whatmore, G. B., Whatmore, N. J., & Fisher, L. D. (1981). Is frontalis activity a reliable indicator of the activity in other skeletal muscles? *Biofeedback and Self-Regulation, 6,* 305–314.

Williamson, D. A. (1981). Behavioral treatment of migraine and muscle-contraction headaches: Outcome and theoretical explanations. *Progress in Behavior Modification, 11,* 163–201.

Wolf, S. L., Nacht, M., & Kelley, J. L. (1982). EMG feedback training during dynamic movement for low-back pain patients. *Behavior Therapy, 13,* 395–406.

Zborowski, M. (1969). *People in pain.* San Francisco: Jossey-Bass.

CHAPTER 24

Program for Behavioral and Learning Disorders in Children

Jean C. Elbert and Diane J. Willis

Introduction

The psychologist in a medical setting has unique training and skills that can be invaluable to the physician in the overall medical management of patients. Children hospitalized for a variety of medical problems such as renal disorders, cardiac problems, severe asthma or diabetes, seizures, or other chronic handicapping conditions often require the services of a psychologist because of learning problems or possible accompanying biosocial problems. Within pediatric outpatient clinics, a large percentage of children are referred and evaluated because of learning- and behavior-related disorders. The referral is often made to the medical setting by school personnel or other agencies for evaluation and the distinguishing of organic from nonorganic contributions to the disorder. It is recognized that early intervention in both learning and behavior problems can do much to alleviate and prevent more serious disorders at a later date. Recognizing that the totality of treatment of the patient and the disease must focus on both the mind and the body, a growing

number of medical centers are adding behavioral or pediatric psychologists to their staffs. Early behavioral pediatrics focused on "biosocial development and learning difficulties of children and adolescents" (Russo & Varni, 1982, p. 14), but in recent years, the field has broadened and has become much more comprehensive. With the advances in medicine and technology, we know that children are surviving conditions that, earlier, they might not have survived. Often, however, there are cognitive deficits or behavioral sequalae later in their young lives.

Since the late 1970s, numerous textbooks have been published on the psychological and behavioral management of pediatric problems (Lavigne & Burns, 1981; Magrab, 1978, 1984; Russo & Varni, 1982; Tuma, 1982; Wright, Schaefer, & Solomons, 1979). Additionally, the market is flooded with texts on learning and/or behavior problems in children.

In this chapter, we present a review of behavior and learning disorders. We briefly review the literature on attention-deficit–hyperactivity disorder (ADHD), learning disorders, and oppositional and conduct disorders. Children with learning problems may or may not have ADHD and vice versa. Additionally, many children referred to medical settings are re-

Jean C. Elbert and Diane J. Willis • Department of Pediatrics, University of Oklahoma Health Sciences Center, Oklahoma City, Oklahoma 73117.

ferred because of behavior problems with accompanying overactivity. These children may or may not also have learning problems. In truth, however, learning and behavior disorders co-occur. After briefly reviewing the literature on the above, we focus on methods of assessment, current clinical interventions, and suggestions for developing and maintaining a learning- and behavior-disorders clinic.

Literature Review

Behavior Disorders

An important consideration in the differentiation and the clinical management of behavior disorders in general is that rigorous scientific efforts to understand, classify, and validate childhood behavior disorders are of relatively recent origin (Achenbach, 1981, 1982; Hobbs, 1975; Shaffer, 1980). Although the delineation and validation of discrete categories and subclassifications of childhood behavior disturbance continue, professionals have generally agreed on the delineation of two major dimensions of child psychopathology: those children exhibiting overcontrolled or internalizing disorders and, conversely, those with undercontrolled or externalizing behaviors (Achenbach & Edelbrock, 1978; Quay, 1979). Our focus in this chapter is on developing clinical services for those children who manifest externalizing or undercontrolled behavior.

Attention-Deficit–Hyperactivity Disorder (ADHD)

Children who exhibit disorders in impulsivity, attention, and self-control are among the most frequent types of referrals to child mental health and pediatric outpatient clinics (Pelham, 1982; Stewart, Pitts, Craig, & Dieruf, 1966). Indeed, children with suspected hyperactivity constitute the group most commonly referred to child guidance clinics (Barkley, 1981). Depending on the rigor of the diagnostic criteria, estimates of the prevalence of this disorder have ranged from 3% to 14% of the population

(Lambert & Sandoval, 1980; Sandoval, Lambert, & Sassone, 1980; Trites, Dugas, Lynch, & Ferguson, 1979). However, the most commonly accepted prevalence figures suggest that the disorder constitutes 3%–5% of the school-aged population. Barkley (1981) suggested that a generally accepted gender ratio is 6 to 1 in favor of males. In addition, the early onset of the disorder, its relatively pervasive nature, and its chronicity in terms of lasting well into adolescence have made ADHD the most widely studied childhood disorder of the past decade (Barkley, 1981).

As numerous investigators in the area of childhood psychopathology have agreed, no childhood behavior disorder has been subjected to as many reconceptualizations, redefinitions, and renamings as the current DSM-III-R (American Psychiatric Association, 1987) diagnostic category "Attention Deficit–Hyperactivity Disorder" (Hinshaw, 1987; Lahey, Pelham, Schaugency, Atleins, Murphy, Hynd, Russo, Hartdagen, & Lorys-Vernon, 1988). Although it is beyond this chapter to discuss all aspects of ADHD, there has been considerable debate about whether ADHD represents a clinical syndrome or a cluster of symptoms identifying a biologically distinct entity. (See the review by Conners & Wells, 1986.) Although several prominent child psychiatrists have described ADHD as a syndrome (Cantwell & Baker, 1987; Laufer & Denhoff, 1957), the failure of investigators to agree on rigorous clinical diagnostic criteria has reduced efforts to determine a biological validation of ADHD. In a review of the psychophysiology of attention deficit disorder with hyperactivity, Zametkin and Rapoport (1987) noted a sizable number of neuroanatomically based hypotheses about the etiology of the syndrome: dysfunctional diencephalic structures (Laufer & Denhoff, 1957), decreased reticular-activating-system excitation (Wender, 1974), a deficient forebrain inhibitory system (Dykman, Ackerman, Clements, & Peters, 1971), and, more recently, blood flow research implicating the central frontal lobes (Lou, Henriksen, & Bruhn, 1984). To date, the available evidence from anatomical, genetic, physiological, and pharmacological studies has yielded a weak or nonspecific relationship between biological factors and hyperactivity (Rapoport & Ferguson, 1981). However, such psychophysiological findings have pointed toward a central nervous system mechanism in the development of the ADHD symptoms (Chelune, Ferguson, Koon, & Dickey, 1986; Hastings & Barkley, 1978).

In the context of multivariate research on childhood behavior disorders, ADHD has generally been regarded as a specific subclassification of the externalizing disorders (which also include acting-out, disruptive, and aggressive behavior), though controversy exists with respect to the validity of these narrow-band syndromes as subsets of the broad dimensions (e.g., Boyle & Jones, 1985). Hinshaw (1987) carefully reviewed factor- and cluster-analytic research regarding the distinctiveness of the two major classes of externalizing disorders of childhood: attention-deficit–hyperactivity and conduct-problems–aggression. The earliest studies using the Behavior Problem Checklist (Quay & Peterson, 1975, 1983) yielded four independent factors entitled "Conduct Problem," "Socialized Aggression," "Attention Problem–Immaturity," and "Motor Tension–Excess." Such factor-independence had suggested both the independence of attention-deficit-disorder (ADD) and the possibility of a subtype of ADD with hyperactivity. The landmark studies of Achenbach (1978) and Achenbach and Edelbrock (1979) also demonstrated a replication of hyperactivity and aggression factors across both age and gender.

Hinshaw's review (1987) concluded that the currently available research does provide supportive evidence of the distinctiveness of the hyperactivity–attentional-deficits and the conduct-problems–aggression dimensions. However, in spite of the emergence of relevant factors, the correlations between factor scores were often substantial. Thus, with respect to strict medical criteria, the narrow-band domains of attention-deficit–hyperactivity and conduct-problems–aggression may not constitute valid independent syndromes, as the studies to date have revealed neither distinct biological and etiological precursors nor differential re-

sponses to treatment. Therefore, whether ADHD constitutes a disorder distinct from the diagnostic category *conduct disorder* continues to be researched (Milich, Loney, & Landau, 1982; O'Leary & Steen, 1982; Prinz, Connor, & Wilson, 1981). However, it should be stressed that whether or not these disorders are clinically distinct, they frequently co-occur. The available literature suggests that antisocial parents, family hostility, and low SES are most commonly associated with conduct disorders, whereas ADHD children are most apt to display cognitive and achievement deficits. Behavioral and social outcomes are generally poorer for conduct-disordered children and are poorer still for children displaying a combination of hyperactivity-inattention and conduct-problems–aggression (Hinshaw, 1987). Children in whom these disorders co-occur are more likely to engage in physical aggression and to exhibit increased antisocial behavior than children with a single diagnosis of conduct disorder, despite their being younger at the time of the initial referral (Hynd & Willis, 1988).

Several issues regarding the diagnostic classification of ADHD children have been debated: (1) whether the attention deficit and the hyperactivity are diagnostically distinct; (2) the pervasiveness of symptoms; and (3) the degree of overlap between ADHD and learning disabilities. With regard to the first issue, the most recent revision of the DSM-III-R (American Psychiatric Association, 1987) collapsed the two subtypes of ADD with and without hyperactivity into a single diagnostic classification. However, investigators continue to argue that ADD with hyperactivity (ADD/H) and ADD without hyperactivity (ADD/WO) can be reliably differentiated, (e.g., Lahey *et al.*, 1984, 1987, 1988); consequently controversy as to their separability ensures that there will be further revisions of the attention deficit and hyperactivity symptom complexes (Hynd & Willis, 1988).

A second diagnostic distinction among ADHD children pertains to the pervasiveness of their symptoms. Schacher, Rutter, and Smith (1981) described "true hyperactives" (those rated as hyperactive by both parents and teachers) as distinct from "situational hyperactives" (rated as hyperactive at home but not at school). Campbell and colleagues (1977a,b) likewise described behavioral differences between these two groups, and some investigators would argue against making the ADHD diagnosis in those who are not pervasively (i.e., cross-situationally) hyperactive (Barkley, 1982).

A third diagnostic subtyping issue is the degree to which the learning-disabled (LD) and ADHD categories overlap. On theoretical grounds, arguments have been made for the single identity of attentional and learning disorders. The idea of an attentional deficit as the central syndrome affecting learning-disabled children was popularized by Dykman and colleagues (1971, 1979) and was a natural extension of the original notion of minimal brain dysfunction (Clements, 1966; Wender, 1971). In contrast, Douglas and Peters (1979) and Douglas (1984) have argued for the existence of primary and secondary attentional disorders that may lead to the same behavioral outcome: academic underachievement. In the child with primary ADHD, impulsivity, impaired arousal, weak effort and attention, and a strong need to seek immediate gratification lead to poor academic performance. Conversely, in the learning-disabled child, the basic deficit in linguistic or perceptual information processing results in a weakened tolerance of frustration, reduced effort, failure experience, and subsequently, disturbed attention secondary to the primary learning disorder. Douglas and her colleagues have done extensive research in identifying the psychological and cognitive disturbance in ADHD children (see Douglas, 1983, 1984). It has been well documented that a sizable fraction of hyperactive children have significantly poorer school achievement than their normal peers (e.g., Cantwell & Satterfield, 1978). In Silver's investigation (1981) of the relationship between hyperactivity and learning disorders, 92% of the children referred for hyperactivity were found to have learning disabilities. Conversely, of the learning-disabled group, almost half displayed distractibility and hyperactivity, a finding indicating that there is a considerable overlap between the two disorders.

Oppositional and Conduct Disorders

Oppositional disorder, now referred to as *oppositional defiant disorder* "is a pattern of negativistic, hostile, and defiant behavior without the more serious violations of the rights of others that are seen in conduct disorder" (DSM-III-R, American Psychiatric Association, 1987, p. 56). Although this disorder was distinguished from conduct disorder in the more recent revision of the DSM-III-R, several investigators believe that there is insufficient evidence for separate classifications (McMahon, 1987; McMahon & Forehand, 1987; Wells & Forehand, 1985). Nevertheless, in a hospital setting, the clinician is likely to be referred numerous children who are defiant in public, at home, and with peers and adults, and who are, in general, sufficiently disruptive so that the parents are seeking help for the problem. Disturbances in behavior of at least six months' duration suggest problems, depending on the severity of the disturbance (mild, moderate, or severe). Oppositional disorders, especially in young children aged 2–7 or 8, are more readily treatable in the clinic. The cases that are more difficult to treat include serious conduct disorders and ADHD cases with secondary oppositional or conduct problems.

Evidence has begun to accumulate that suggests that early noncompliance in children may be a hallmark of the development of later conduct-disordered behavior (McMahon, 1987). Further, ongoing work has begun to differentiate the developmental patterns of progression of conduct disorders, including subtypes of overt and covert disorders, and such models have implications for the assessment and intervention with conduct-disordered children (see Baum, 1989; McMahon, 1987; Wells & Forehand, 1985).

Other behavioral symptoms seen in hospital settings include crying and colic, feeding problems, temper tantrums, breath holding, school refusal, anxiety, fears and phobias, and enuresis and encopresis, to name a few. It is not the intent of this chapter to focus on these more common types of disorders, even though they are a source of numerous referrals by other health care professionals. The interested reader is referred to Gabel (1981) and Schaefer and Millman (1981) for suggestions on treating specific behavior problems.

Learning Disorders

Learning Disorders with Known CNS Dysfunction

A primary group of children requiring specialized diagnostic services and intervention are those pediatric patients whose learning, developmental, or behavioral disorders are associated with known medical conditions, either congenital or acquired. In the rapidly developing field of pediatric neuropsychology, considerable attention has been given to the specialized diagnostic assessment of children with known brain dysfunction (Hynd & Obrzut, 1981; Hynd & Willis, 1988; Rourke, 1985; Rourke, Bakker, Fisk, & Strang, 1983; Rourke, Fisk, & Strang, 1986; Rutter, 1983). Children seen in pediatric psychology or neuropsychology clinics include patients with the following types of disorders: genetic or chromosomal abnormalities, metabolic disorders, CNS infections, abnormalities of CNS development, CNS cancers and solid tumors, seizure disorders, neuromuscular disease, and head trauma. In addition to well-studied chromosomal disorders such as Down syndrome and others associated with mental retardation, such sex chromosome abnormalities as Klinefelter syndrome (XXY), Turner syndrome (XO), fragile-X syndrome, XYY, and XXX are associated with specific learning deficits and behavioral disturbances that are most appropriately diagnosed in medical settings, where the psychologist has access to genetic and cytogenetic diagnostic findings.

Turner syndrome, for example, is generally associated with possible right-hemisphere pathology, as cognitive deficits in spatial functioning (Waber, 1979) have been documented, as well as autisticlike behaviors in the child with fragile-X syndrome. In such metabolic disorders as phenylketonuria (PKU), medical treatment compliance may involve the psychol-

ogy unit, as both behavioral disorders (Stevenson, Hawcroft, Lobascher, Smith, Wolff, & Graham, 1979) and conceptual, visuospatial deficits (Pennington, van Doorninck, McCabe, & McCabe, 1985) have been reported to increase in those children who are not compliant with dietary restrictions. The psychological sequelae of meningitis include developmental delays in receptive and expressive language, hearing impairment, intellectual deficit, seizures, and visual and motor impairments (Sell, 1983). Of the many neurodevelopmental malformations, spina bifida is common enough (1–4 per 1,000) so that many tertiary-care medical centers have a meningomyelocele clinic to coordinate the multispeciality care that these children require. Disturbances in brain function occur as a result of hydrocephalus, defects in brain cell formation, and ascending infection (Graham, 1983). Because of the improved surgical shunt procedures used with these children, the mortality rate has steadily dropped, and effort is being devoted to assessing long-term psychological outcomes. Assessment of their associated language and visual-perceptual-motor disorders (Swisher & Pinsker, 1971) is important in terms of documenting the need for specialized school services for these children.

The increasing survival rates of children who have sustained such childhood cancers as acute lymphocytic leukemia and cranial tumors necessitate specialized knowledge in assessing the neuropsychological sequelae of CNS irradiation and neurosurgical treatment. Cognitive deficits, including impaired speed of visuomotor and visuospatial functioning, acquired aphasias, and socially disinhibited behavior, have been described in children sustaining head injuries (Chadwick, Rutter, Shaffer, & Shrout, 1981). Neuropsychological assessment is often required to determine the rehabilitation needs of these children. (See Chapter 22 for additional discussion of rehabilitation services for these patients.)

Specific Learning Disability (SLD)

Recognition of a population of normally intelligent children who exhibit difficulties in specific areas of learning, such as auditory-linguistic and language processing or visual-spatial and perceptual motor learning, has been apparent since the turn of the century. Clinical case studies of impaired reading skills appeared in the medical literature early in the century (Hinshelwood, 1900) and continued with Orton's early description (1928) of congenital word blindness. It was recognized that learning problems in children often mirror the dysfunction of adults with known brain lesions (Goldstein, 1942), and this knowledge was then extended to studies of known or suspected brain damage in children through the work of Strauss and Lehtinen (1947) and later Strauss and Kephart (1955). (See the review of the development of the learning-disabilities field in Hallahan & Cruickshank, 1973.)

Specific learning disability (SLD) was defined and came into popular usage as one of the special education categories included in Public Law 94-142, the Education for All Handicapped Children Act of 1975 (U.S. Office of Education, 1977). Traditionally, definitions of specific learning disability have included varying emphases on the following components: (1) a task failure component (e.g., poor academic achievement in one or more areas); (2) an exclusionary component (e.g., that discriminates (SLD) from skill deficits attributable to sensory impairment, mental retardation, emotional disturbance, and environmental disadvantage); (3) an etiological component (e.g., attribution to CNS dysfunction); (4) a discrepancy component (e.g., attained skill levels below expected performance); and (5) a psychological process component (e.g., disorders in attention, perception, memory, and conceptualization) (National Joint Committee on Learning Disabilities; Hammill, Leigh, McNutt, & Larsen, 1981). Despite general agreement with federal definitional guidelines, however, research investigators, clinicians, individual state departments of education, and local school districts have used widely differing specific criteria for the diagnosis of learning disability. In particular, the operational definition of a "significant discrepancy" has variously been based on age- or grade-level expectancies, or on predicted achievement based on IQ scores. The criteria used in determining the magnitude of this discrepancy between intellectual potential and attained skill levels differ widely, with the result that an individual child might meet LD eligibility criteria in one school district and not in another. Because of such widely varying criteria for diagnosis and classification, the reported prevalence of learning disabilities has varied from 2% to over 20% (Broman, Bien, & Shaughnessy, 1985).

This lack of consensus concerning the diagnostic procedures and criteria that should be used to classify a student as learning-disabled has provided considerable divergence of opinion in the learning-disabilities field. From the early focus on children with neurological difficulties, the field of learning disabilities has gradually shifted to serving a variety of children with other problems whose only similarity is that they are experiencing difficulty in school (Torgesen, 1986). The definitional guidelines designate as exclusionary criteria those instances of poor achievement that are primarily a result of physical handicaps, visual and hearing impairments, and mental retardation. In addition, motivational problems, attentional disorders, emotional disturbance, and environmental, cultural, or economic disadvantage are all factors that must be considered in the diagnosis of learning disabilities. When the discrepancy between intellectual potential and achievement is emphasized as the overriding criterion for a diagnosis of learning disability (as emphasized in the *Federal Register*, 1977), many underachieving children are likely to be inappropriately identified as learning-disabled. The U.S. Department of Education's Sixth Annual Report to Congress (1984) stated that the prevalence of learning disabilities had doubled in the past 10 years, and more than 40% of all school pupils served in special education were classified as learning-disabled, a figure that represented 4% of all schoolchildren (Chalfant, 1989). Such a twofold increase in the identification of these children has obviously raised significant concerns about the funding for special-educational programs, and some have proposed placing a cap of 2% on the proportion of the population of schoolchildren who can be classified as learning-disabled (Chalfant, 1989).

Despite their common diagnostic label, it is well known that children with learning disabilities represent a very heterogeneous group with multiple types of problems and, very likely, differing etiologies. In recent years, research has focused on attempts to classify these children into relatively homogeneous subtypes, so that developmental course, response to intervention, and outcome may be better understood. Such attempts to differentiate more homogeneous patterns of cognitive strengths and weaknesses represent a more differentiated level of diagnosis than simply documenting that a child is underachieving. Formal attempts at subclassification have been based on etiological inferences (e.g., neurological or genetic); patterns of performance on a variety of psychometric measures of perception, memory, language, and cognitive processing; and patterns of performance on measures of academic achievement in reading recognition, reading comprehension, spelling, written language, and mathematics. It is beyond the scope of this chapter to provide a detailed discussion of such studies, and various reviews of the LD subtyping literature are available (e.g., Hynd, Connor, & Nieves, 1987; McKinney, 1984). Some of the earliest attempts at classification were based on a clinical inferential classification system (e.g., Mattis, French, & Rapin, 1975) and gave empirical support to long-standing clinical observations that children with learning disabilities included subgroups with auditory processing and oral language disorders, visual-spatial-perceptual disorders, and articulatory and/or visual-motor impairment, as well as children with mixed disorders who were more seriously handicapped (Boder, 1973; Johnson & Myklebust, 1967). In contrast, Rourke and colleagues have studied patterns of neuropsychological performance in those learning-disabled children grouped on the basis of their respective academic performance in reading, spelling, and math (Petrauskas & Rourke, 1979; Rourke, 1985). Finally, more statistically refined studies of empirically derived subtypes have appeared, beginning with data from the Flor-ida Longitudinal Study (Satz, Taylor, Friel, & Fletcher, 1978), and including numerous Q techniques (e.g., Doehring & Hoshko, 1977) and cluster analyses (e.g., Lyon & Watson, 1981; Spreen & Haaf, 1986).

Public school systems provide the classification of most children as learning-disabled, following formal assessment by a school psychologist and agreement by an eligibility team composed of designated school personnel. However, the law also provides that a child's parent may request an independent evaluation, which the school eligibility team must consider in deciding the educational needs of the child and whether special-education services are warranted. Referrals of children to medical center clinics frequently consist of requests for a second opinion regarding the child's learning difficulty, or for a more refined diagnosis with implications for educational remediation in a child who may be identified as learning-disabled, but who is not making expected progress. Specialized services for the comprehensive assessment of learning disabilities may therefore represent a needed supplement to school evaluations and/or may identify needs for children who may not be eligible for special-education programs in the public schools. Public policy decisions that affect the number of children eligible for special education are likely to increase the numbers of such children being referred to private agencies and/or medical settings.

Establishing a Learning- and Behavior-Disorders Clinic

The clinicians who move rapidly to establish a learning- and behavior-disorders clinic will find themselves inundated with referrals. These types of clinics can be extremely helpful to the hospital, from both a service–public-relations and a financial standpoint. Excellent facilities offering a comprehensive assessment of specific school learning disorders of children, especially when the child's condition may be complicated or exacerbated by health-related problems or by an attention-deficit–hyperactivity disorder, are not routinely found within communities. The assessment of children with complex learning–activity–attention problems requires the service of a physician, a psychologist, and often a speech pathologist and a special educator. Children who continue to be frustrated by school learning and ADHD problems may also develop oppositional or conduct problems.

Establishing the clinic may require the approval of the departmental chair or at least the director of the psychology service. It can be pointed out that intervention in learning and behavioral problems is part of good medical practice in terms of preventing more serious emotional problems. Additionally, the clinic can contribute to the teaching, service, and research goals of the department and the medical center.

Once approval from the department chair or the head of the section is obtained to establish the clinic, the clinician will want to describe clearly the types of referrals accepted by this clinic, the day or days on which the clinic meets, to whom referrals are to be made, and the fee structure.

Methods of Psychological Assessment

Once the clinic for learning and behavioral disorders is established, formal assessment may be streamlined and made more efficient by obtaining background information and previous records before scheduling an appointment with the parent and the child. This information may prevent costly duplications of assessment of which a referring individual may be unaware. Second, clarification of the referral question is important in preventing an assessment that is not appropriately targeted. In hospital settings, many referring physicians, particularly interns and residents unfamiliar with psychological assessment, may simply order psychological testing in much the same way as they order other medical laboratory procedures. Although some referral questions may be straightforward (i.e., ruling out mental retardation), in most referrals for evaluating learning and be-

havior disorders, speaking with the referring individual to clarify the diagnostic question(s) is useful. When a referral is made by another professional, it is also useful to address the parents' questions, which may be quite different.

Depending on the referral question(s), additional information is frequently helpful before scheduling the first appointment. The types of information one can obtain include the following: (1) a parent form giving basic developmental and background information on the child; (2) one or more parent and teacher behavior-rating scales (to be described in this section); (3) reports of previous psychological testing by school or private agency; and (4) reports from mental health professionals who have previously been involved in intervention with the child or the family. (The information from other agencies must, of course, be obtained with the parents' informed consent.) The criteria useful in assessing oppositional defiant disorder, conduct disorder, and ADHD are listed in Figure 1 (pp. 430–431). These criteria, taken from the DSM-III-R (American Psychiatric Association, 1987), can be used during an interview with parents to aid the clinician in diagnosis.

Schroeder and Gordon (1990) developed a framework for assessing child behavior problems that includes sending out questionnaires and checklists before the family's scheduled appointment in the clinic. With these data, they attempt to clarify the referral question with the parents, to determine the social context in which the behavior is occurring, to determine the general and specific areas of assessments that may be needed, and/or to develop an intervention approach (see Table 1, p. 432). Following this initial stage of data gathering, the clinician may be able to triage the referral, depending on the background information and data collected, and subsequently, a more specialized approach can be planned (see Figure 2, p. 433).

Children referred for the diagnosis and management of ADHD should optimally have an evaluation of basic intellectual functioning and current academic achievement in order to rule out basic disorders of learning and/or cognition. The following types of assessment are used in most ADHD clinics: behavior rating scales (for both parents and teachers), structured behavioral observations, parent and child interviews, and other broad-band and narrow-band personality assessment measures. Both general and specific approaches to assessment are discussed in the following sections.

Behavior Disorders: ADHD and Oppositional and Conduct Disorders

Interviewing

For any mental health professional in a medical or a nonmedical setting, the clinical or diagnostic interview is the most frequently used assessment procedure (Matarazzo, 1983). Information gathering during the interview focuses on developmental issues of the target child, a history of the pregnancy, early illnesses, educational history, family and sibling history, and often the parents' marital issues. Rarely is there uniformity in the clinical interview among mental health professionals, as the data obtained from parents often focuses on issues surrounding the presenting problem and how that problem is affecting the family as well as who or what sustains the problem. Willis and Holden (1990) suggested that, when a child is referred for learning or behavior disorders, an interview cover the following: (1) the period of gestation and whether or not there is a history of an abnormal pregnancy, substance abuse, fever, trauma, or accidents during the gestation period; (2) the birth history (the type of delivery, the birth weight, and problems such as anoxia, infections, hypoglycemia, or jaundice); (3) postnatal history (developmental history, accidents or trauma, fevers or infections, and temperament), the caretaker of the child, and the educational history; and (4) family history, marital stresses experienced by the family, the history of the target problem, and factors reinforcing or sustaining the problem. (Much of this information can be obtained before the first appointment from a form developed by the clinician and filled out by the parent.) Additionally, during the initial interview,

the parents can be provided feedback about the clinician's impressions of the problem, as well as what diagnostic procedures and/or treatment appear to be warranted. In a hospital setting, it may be difficult to ensure that parents will return for further appointments, and compliance is greater the more informed the parents are and the less they have to wait before being seen.

Behavioral Observation and Assessment

In the hospital setting, a highly useful behavior and observation tool for screening for parent–child interaction problems of the child aged 2–7 is Eyberg's Dyadic Parent-Child Intervention Coding System (Eyberg, 1974; Eyberg & Robinson, 1982, 1983; Robinson, Eyberg, & Ross, 1980). Within a brief 15- to 30-minute session, the clinician can determine the pattern of parent–child interaction problems that may be creating and/or maintaining the behavior problem. The behavior coding system, paired with a detailed parent interview, pinpoints positive or negative parenting behaviors and the child's response to the parent (e.g., compliance or noncompliance). A description of the various child and parent behaviors coded are represented in Figure 3 (p. 434).

Various behavioral observation coding methods have been used with ADHD children in order to obtain objective measures of actual behaviors and social interactions. Two methods developed specifically for classroom observations are the Stoney Brook Code by Abikoff, Gittelman-Klein, and Klein (1977) and the Hyperactive Behavior Code developed by Jacob, O'Leary, and Rosenblad (1978). Such coding procedures are designed to provide time-sampled recordings of such behaviors as out-of-seat, off-task, fidgeting, verbalizations, compliance or noncompliance, motor movements, and restlessness, which are consistent with the diagnostic criteria of ADHD. Although such observational data are highly valuable in obtaining baseline clinical data, they presume the use of trained observers in the classroom. Training school personnel in coding procedures or making trips from the clinic to the child's school is

DSM-III-R CHECKLIST

Child's Name: _____   Date: _____
Rater's Name: _____

1 = rarely   2 = occasionally   3 = pretty often   4 = very often

Oppositional Defiant Disorder (313.81)
A.   Have the problems been present for at least six months? _____
B.   Compared with other children the same age, are at least five of these
      problems? _____   (either "3" or "4" must be circled)
1 2 3 4   1) often loses temper
1 2 3 4   2) often argues with adults
1 2 3 4   3) often actively defies or refuses adult requests or rules,
                 e.g., refuses to do chores at home
1 2 3 4   4) often deliberately does things that annoy other people,
                 e.g., grabs other children's hats
1 2 3 4   5) often blames others for his or her own mistakes
1 2 3 4   6) is often touchy or easily annoyed by others
1 2 3 4   7) is often angry and resentful
1 2 3 4   8) is often spiteful or vindictive
1 2 3 4   9) often swears or uses obscene language

Check the appropriate level of severity (make this rating at the conclusion
of the intake)
_____   Mild: Few, if any, symptoms in excess of those required to make the
                 diagnosis and only minimal or no impairment in school and social
                 functioning.
_____   Moderate: Symptoms or functional impairment intermediate between mild
                 and severe.
_____   Severe: Many symptoms in excess of those required to make the
                 diagnosis and significant and pervasive impairment in
                 functioning at home and in school and with other adults and
                 peers.

Attention-deficit Hyperactivity Disorder (314.01)
A.   Have the symptoms been present for at least six months? _____
B.   Compared to other children the same age, are at least eight of these
      problems? _____   (either "3" or "4" must be circled)
1 2 3 4   1) often fidgets with hands or feet or squirms in seat
1 2 3 4   2) has difficulty remaining seated when it is required
1 2 3 4   3) is easily distracted by extraneous stimuli
1 2 3 4   4) has difficulty awaiting turn in games or group situations
1 2 3 4   5) often blurts out answers to questions before they have been
                 completed
1 2 3 4   6) has difficulty following through on instructions from others
                 (not due to oppositional behavior or failure of
                 comprehension), e.g., fails to finish chores
1 2 3 4   7) has difficulty sustaining attention to tasks or play
1 2 3 4   8) often shifts from one uncompleted activity to another
1 2 3 4   9) has difficulty playing quietly
1 2 3 4   10) often talks excessively
1 2 3 4   11) often interrupts or intrudes on others, e.g., butts into
                 other children's games

Figure 1. Suggested DSM-III-R checklist for oppositional disorder, ADHD, and conduct disorder (adapted from DSM-III-R criteria).

1 2 3 4 12) often does not seem to listen to what is being said to him or
              her
1 2 3 4 13) often loses things necessary for tasks or activities at
              school or home, e.g., toys, pencils, books, homework
1 2 3 4 14) often engages in physically dangerous activities without
              considering possible consequences (not for the purpose of
              thrill seeking), e.g. runs into street without looking.

Check the appropriate level of severity (make this rating at the conclusion
of the intake)
_____ Mild
_____ Moderate
_____ Severe

Conduct Disorder (group type: 312.20; solitary aggressive type: 312.00;
                  undifferentiated type: 312.90)
A.   Have the symptoms been present for at least six months? _____
B.   Are at least three of the following present? _____ (either "3" or "4"
      must be circled)
1 2 3 4   1) has stolen without confrontation of a victim on more than one
               occasion
1 2 3 4   2) has run away from home overnight at least twice
1 2 3 4   3) often lies
1 2 3 4   4) has deliberately engaged in firesetting
1 2 3 4   5) is often truant from school
1 2 3 4   6) has broken into someone's house, building or car
1 2 3 4   7) has deliberately destroyed other's property
1 2 3 4   8) has been physically cruel to animals
1 2 3 4   9) has used a weapon in more than one fight
1 2 3 4 10) often initiates physical fights
1 2 3 4 11) has stolen with confrontation of a victim
1 2 3 4 12) has been physically cruel to other people

Check the appropriate level of severity (make this rating at the conclusion
of the intake)
_____ Mild: Few, if any, conduct problems in excess of those required to
              make the diagnosis, and conduct problems cause only minor harm
              to others.
_____ Moderate: Between mild and severe.
_____ Severe: Many conduct problems in excess of those required to make the
              diagnosis, or conduct problems cause considerable harm to others
              (e.g. serious physical injury to victims, extensive vandalism or
              theft, prolonged absence from home).

Check the appropriate type
_____ Group: The essential feature is the predominance of conduct problems
              occurring mainly as group activity with peers. Aggressive
              physical behavior may or may not be present.
_____ Solitary Aggressive: The essential feature is the predominance of
              aggressive physical behavior, usually toward both adults and
              peers, initiated by the person (not as a group activity).
_____ Undifferentiated: this subtype represents a mixture of clinical
              features that cannot be classified as either Solitary Aggressive
              type or Group type.

Figure 1. (*Continued*)

Table 1. A Framework for Assessing Child Behavioral Problems[a]

I. *The initial contact.*
    Identify the problem. Send out questionnaires or checklists.
II. *Clarify the referral question.*
    After the parent has told you the problem, make certain that you and the parent are thinking about the same problem. You can do this by simply reflecting what the parent said: "It sounds as if you are concerned about your child's getting up in the night, as well as the different ways you and your husband are handling the situation."
III. *Determine the social context.*
    A child is referred because someone is concerned. This doesn't mean a child needs treatment or that the child's behavior is the problem. Ask, "Who is concerned about the child?" "Why is this person concerned?" "Why is this person concerned now rather than at some other time?" Listen to the parents' affect in describing the problem: Are they overwhelmed, depressed, nonchalant?
IV. *General areas to assess.*
    A. Development
    B. Environment
    C. Consequences of behavior
    D. Physical status
V. *Specific areas to assess.*
    A. Persistence of behavior
    B. Changes in behavior
    C. Severity of behavior
    D. Frequency of behavior
    E. Situation specificity
    F. Type of problem
VI. *How is the behavior affecting the child and others?*
    A. Who is suffering?
    B. Does the behavior interfere with the developmental process?
VII. *Where can you intervene?*
    A. Development
        1. Teach new responses to the child, parent, or school.
        2. Change the behavior by increasing or decreasing it.
    B. Environment
        1. Change the cues that set off the behavior or prevent it from occurring.
        2. Change the emotional atmosphere.
        3. Change parental expectations, attitudes, or beliefs.
    C. Consequences of behavior
        1. Changes parents' responses to the behavior.
        2. Change others' responses to the behavior.
        3. Change the payoff for the child.
    D. Physical status
        1. Intervene in the cause of the problem.
        2. Treat the effect of the problem.

[a]From C. S. Schroeder and B. N. Gordon. (1990). "Assessment of Behavior Problems in Young Children." In D. J. Willis and J. L. Culbertson (Eds.), *Testing Young Children.* Austin, TX: Pro-Ed. Adapted from Schroeder *et al.*, 1983.

often impractical for most clinicians in a medical setting. Similar coding procedures have been designed to code child behaviors in the clinic setting (Milich *et al.*, 1982).

Behavior-Rating Scales

There are numerous published and experimental rating scales and behavioral checklists that clinicians can use in a hospital clinic. Several of these scales and checklists are discussed here, but additional readings can be found in Barkley (1987, 1988a), Mash and Terdal (1988), Magrab (1984), Ollendick and Hersen (1984), and Willis and Culbertson (1990). A behavioral checklist derived from the DSM-III-R diagnostic criteria for "Oppositional Defiant Disorder," "Conduct Disorder," and "Attention-Deficit

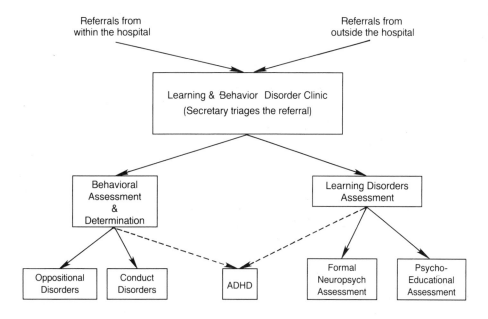

Figure 2. Diagram of proposed learning and behavioral disorders clinic.

Hyperactivity Disorder" is provided in Figure 1 and can be used in conjunction with a parent interview and/or child observation.

Unlike many other childhood disorders, ADHD comprises behaviors that are evident at times and to some degree in all children. Because ADHD is defined not by a few pathognomonic signs, but by deviation in the frequency and severity of behaviors, it has been argued that the diagnosis of ADHD should be dimensional (e.g., on a continuum of normal to severely deviant) as opposed to categorical (e.g., simply present or absent, such as a chromosomal defect) (Barkley, 1987; Kazdin, 1988).

From this model, accurate diagnosis and classification depend heavily on subjective ratings of the child's behavior in natural settings such as school and home. Thus, standardized behavior checklists and rating scales have become basic tools in the assessment of ADHD children. (See the reviews of such scales in Barkley, 1987, 1988a; Edelbrock & Rancurello, 1985.)

For parents' ratings, the most frequently cited and used measures include the Conners Parent Rating Scale—Revised (CPRS-R; Goyette, Conners, & Ulrich, 1978), the Werry-Weiss-Peters Activity Rating Scale (Routh, Schroeder, & O'Tuama, 1974), the Child Behav-

Family Name: _____   Observer: _____
Child's Name: _____   Date: _____
Mother _____   Father _____   CDI _____   PDI _____   Clean-Up _____   Time _____

Baseline	Treatment Sessions	Follow-Up
A	B		1	2	3	4	5	6	7	8	9	10	11	12	13	14	15		A	B	C

Parent Behaviors	Total
Acknowledge	
Irrelevant Verbalization	
Critical Statement	
Physical Negative	
Physical Positive	
Unlabeled Praise	
Labeled Praise	
Desc/Refl Question	
Reflective Statement	
Descriptive Statement	
Indirect Command followed by: No Opportunity	
Compliance	
Noncompliance	
Direct Command followed by: No Opportunity	
Compliance	
Noncompliance	
Other	

Child Behaviors		Total
Changes Activity		
Cry	Ignored	
	Responded to	
Yell	Ignored	
	Responded to	
Whine	Ignored	
	Responded to	
Smart Talk	Ignored	
	Responded to	
Destructive	Ignored	
	Responded to	
Physical Negative	Ignored	
	Responded to	
Other	Ignored	
_____	Responded to	
Response Following Noncompliance		
Chair Warning followed by: Compliance		
Noncompliance		
Chair		
Deviant Behavior on Chair		
Leaves Chair without Permission		
Spank		
Are you Ready? Yes		
No		
Comply following Chair		

Figure 3. Data recording sheet. (From S. Eyberg and E. Robinson, "Dyadic Parent-Child Interaction Coding System: A Manual," *Psychological Documents*, 1983, *13*, Ms. 2582. Reprinted by permission.)

ior Checklist and Profile (CBCL; Achenbach, 1978; Edelbrock & Rancurello, 1985), the Home Situation Questionnaire (HSQ; Barkley, 1981), the SNAP Checklist (Pelham & Murphy, 1981), and the Yale Children's Inventory (Shaywitz, Schnell, Shaywitz, & Towle, 1986). Barkley (1981) suggested that a thorough assessment of hyperactive children should include at least the Conners PRS-R, the Achenbach CBCL, and Barkley's HSQ; consequently, these scales are briefly described.

The Conners Parent Rating Scale—Revised (Goyette *et al.*, 1978) is normed for use with children aged 3–17. Parents rate 48 behaviors on a zero to 3-point Likert scale, and the obtained profile is determined from five scales: conduct, learning, psychosomatic problems, impulsivity-hyperactivity, and anxiety.

The Child Behavior Checklist (CBCL; Achenbach & Edelbrock, 1983) is a rigorously researched instrument that is normed for children aged 4–16, and an additional form is available for preschool children aged 2–3 (Achenbach, Edelbrock, & Howell, 1987). Although the majority of parent rating scales are narrow-band scales designed to document the presence of ADHD symptoms, the CBCL is a broad-band measure that contains individual scales representing both internalizing and externalizing disorders; thus, it allows a more comprehensive measurement of the presence and patterns of the most common psychopathological behaviors. This scale is now widely used in clinic settings and is useful in pinpointing behavioral traits that may then need to be assessed in greater detail. The derived CBCL scales provide a profile of internalizing problems such as depression, social withdrawal, and somatic complaints, and the externalizing problems of hyperactivity, delinquency, and aggression. An additional broad-based rating scale useful as an initial diagnostic tool is the Personality Inventory for Children (PIC; Wirt, Lachar, Klinedinst, & Seat, 1984), which is designed to measure behavior, ability, affect, and family functioning in children and adolescents aged 3–16.

The Home Situations Questionnaire (HSQ; Barkley, 1981) was developed in order to assess situational variation in children's behavior dis-

orders, by obtaining information about specific settings and contexts in which the behaviors occur. Parents rate 16 items for both the presence and the severity of the problem in a particular context, and the profile provides useful information in planning contingency management interventions.

With regard to teacher rating scales, the CBCL has a teacher report form (CBCL-TRF; Edelbrock & Achenbach, 1984), and this and the Conners Teacher Rating Scale (CTRS-R; Goyette *et al.*, 1978) are used widely both in clinical diagnosis and for identifying research samples. Additional rating scales completed by teachers are the ADD/H Comprehensive Teacher Rating Scale (Ullman, Sleator, & Sprague, 1984) and the Self Control Rating Scale (SCRS; Kendall & Wilcox, 1979). All of these scales permit a comparison of the child with behavior and attentional disorders with data from normal age- and sex-matched children to assist in establishing the degree of deviance of the oppositional and/or ADHD symptoms.

In the clinic setting, as the clinician interviews the parents, the DSM-III-R checklist of symptoms characterizing oppositional defiant disorders, ADHD, and conduct disorders can be given to the parents to fill out, or it can be mailed to the parents before the first clinic visit (see Figure 1). This simple checklist can aid the clinician in clarifying the child's problem and in developing appropriate interventions.

Parent Scales

Because parent and family functioning are often critical to assess in addition to the individual child's behavior, several scales are recommended to provide information about parents and family contexts. The Parenting Stress Index (PSI; Abidin, 1983) assesses the level of stress that the parents may be experiencing with their child and can be used to assess stress levels as treatment progresses. The 101 items are rated on a 5-point scale, with items covering such areas as marital relationships, parental attachment to the child, aspects of the child's temperament, how reinforcing the child is to the parent, and parental depression. Additional measures that evaluate possible environmental influences on the child's problem are the Home Environment Questionnaire (Sines, 1983), the Family Environment Scale (Moos & Moos, 1981), and the Life Events Questionnaire (Herzog, Linder, & Samaha, 1981).

Attention, Vigilance, and Impulsivity

In addition to the methods described, specialized objective measures of vigilance and impulsivity and timed tests of immediate memory, perceptual planning, and perceptual speed are frequently used, both in assessing behavioral baselines, and in evaluating response to medical, psychological, and educational management.

Beginning with the research of Douglas and colleagues (see the reviews in Douglas, 1983, 1984), the demonstration of deficits in cognitive processes in ADHD children has led to attempts to document these deficits both as an aid to clinical diagnosis and as an indication of the effectiveness of the intervention procedures. Douglas argued that the ADHD child's most basic attentional problems are related to deficits in the investment, organization, and maintenance of attention and effort, and to deficits in the inhibiting of impulsive responding (Douglas & Peters, 1979). Many direct objective measures of attention and impulsivity have consisted of laboratory tasks that have not been adequately normed for use with children or subjected to stringent psychometric analyses (Ostrom & Jenson, 1988). However, adaptations of some laboratory tasks, as well as child-normed measures have been used in the specialized clinical assessment of various aspects of attention, concentration, and impulsivity (e.g., Conners & Wells, 1986; Kirby & Grimley, 1986), and the use of such measures as an adjunct to parent and teacher behavior-rating scales both contributes to the validity of the diagnostic procedures and provides objective baseline data against which to assess clinical progress and response to medication. Such laboratory measures of vigilance, impulsivity, and activity level are being more widely used clinically, particularly in medical settings and uni-versity-based clinics, where resources are more available.

Vigilance has long been assessed by varying types of the continuous performance test (CPT), originated by Rosvold, Minsky, Sarason, Brasome, and Beck (1956). Although several versions have been described for experimental use, a behavior-based measure of vigilance has been developed for clinical use and is available commercially (Gordon Diagnostic System, or GDS; Gordon, 1983). A microprocessor is designed to administer an eight-minute vigilance task in which the child sees randomly presented numbers on an electronic display and is directed to respond with a lever press only when she or he sees a designated number sequence (e.g., 9 preceded by 1). Normative data for the number of correct and incorrect responses are available on over 1,200 children aged 4–16 (Gordon & Mettelman, 1988). However, much more validation research is needed before the CPT (and the GDS) becomes a reliable diagnostic tool. One such study suggested that the CPT yielded both false positives and false negatives when validated against other neuropsychological measures, and the authors concluded that caution should be exercised in the use of this device for ADHD screening (Trommer, Hoeppner, Lorber, & Armstrong, 1988).

Impulsivity has often been assessed experimentally by various perceptual search measures requiring the inhibition of responding and match-to-sample techniques. The Matching Familiar Figures Test (MFFT; Kagan, 1966) involves 12 match-to-sample pictures and is scored for the mean time taken to make the first response (latency) and the total number of errors (incorrectly identified pictures). A longer (20-item) version has been developed for older children (MFF-20; Cairns & Cammock, 1978), and norms are available for 5- to 12-year-old children (Cairns & Cammock, 1978, 1984). Although the MFFT has been widely used and has been shown to significantly discriminate ADHD from normal children (Campbell, Douglas, & Morgenstern, 1971), its reliability and validity have been criticized (Milich & Kramer, 1984). Another match-to-sample task is the Children's Embedded Figures Test (CEFT;

Karp & Konstadt, 1971), in which the child must discriminate geometric shapes embedded in pictures. A different type of impulse control task is the Delay Task included in the GDS (Gordon, 1986). Table 2 lists sources of the specialized measures for assessing various aspects of attention.

Learning Disorders

In a medical center program, the psychological assessment of children with suspected learning disorders may be undertaken for varying purposes and presumes varying levels of data gathering, ranging from simple to complex assessment. Types of referrals include (1) requests for current cognitive functioning and/or academic achievement levels, implying straightforward psychometric procedures; (2) the evaluation of children with known brain pathology to assess the magnitude of the impairment, the degree of deterioration, the potential for rehabilitation, and the relative importance of neuropsychological and social-emotional factors as influencing current behavioral patterns; (3) the

comprehensive psychoeducational assessment of children with complex learning and/or language disorders for the purpose of determining prescriptive intervention and appropriate remediation; (4) the evaluation of school readiness in preschool children; (5) the developmental assessment of high-risk infants demonstrating delayed development; and (6) the differential diagnosis of psychopathology, emotional or motivational problems, and family pathology that may present initially as school failure.

Because both the determinants and the manifestations of a child's learning disability may vary widely, it is rare that assessment is straightforward and simple. Merely documenting a child's intellectual potential and current achievement levels may be sufficient to determine eligibility for specialized school placement; however, such a level of assessment is available in the public school at no cost to the parent. Brief assessments in clinical settings often stem from practical considerations of heavy clinical caseloads, paired with the time required for a comprehensive assessment (sometimes rang-

Table 2. Specialized ADHD Assessment

Sustained attention or vigilance	
Continuous Performance Test (CPT)	Rosvold *et al.* (1956)
Gordon Diagnostic System	Gordon (1983)
Freedom from Distractibility	Kaufman (1975)
(WISC-R: Arith, Cod, DS)	Wechsler (1974)
Background Interference Procedure (Bender Gestalt)	Canter (1976)
Reaction Time	
Delayed Reaction Time (DRT)	Gordon (1983)
Perceptual search or impulsivity	
Matching Familiar Figures Test (MFFT)	Kagan (1966); Salkind and Nelson (1980)
MFF-20	Cairns and Cammock (1978)
Children's Embedded Figures Test (CEFT)	Karp and Konstadt (1971)
Organization and planning	
Porteus Maze Test	Porteus (1959)
WISC-R Mazes	Wechsler (1974)
Immediate memory	
WISC-R Digit Span	Wechsler (1974)
Detroit Word Order	Hammill (1985)
Detroit Sentence Imitation	Hammill (1985)
McCarthy Memory Scale	McCarthy (1972)
Stanford-Binet Memory Scale	Thorndike, Hagen, and Sattler (1986)
K-ABC Sequential Processing Scale	Kaufman and Kaufman (1983)

ing from 4 to 12 hours, depending on whether a complete neuropsychological battery is required). Various approaches to the diagnosis of learning disabilities have advocated a flow diagram set of procedures with a designated series of steps and recommended tests (e.g., Aaron, 1981). However, it must be cautioned that such approaches may be quite simplistic and may fail to result in an accurate portrayal of the multiple determinants of a child's actual school performance or may be insufficient for prescribing a well-integrated intervention plan.

For example, the Wide Range Achievement Test-R (though widely used in assessment and effective as a screening instrument) cannot provide a realistic appraisal of a child's contextual-reading or written-language skill. Similarly, a more comprehensive theoretical framework (e.g., Johnson & Myklebust, 1967) would dictate that deficiencies in written-language expression reflect underlying oral language deficiencies, which should also be assessed for a determination of the remedial objectives. Thus, although learning deficits may occasionally be circumscribed, frequently they have more wide-ranging consequences. An assessment of the child in isolation (i.e., without input from school personnel) may fail to detect the possibility of unusually high expectations, competition from more intelligent and/or achievement-oriented classmates, and failure to recognize that poor school achievement may be caused by an interaction of child, teacher, and home variables. As described in the section of ADHD characteristics, attentional problems are frequently associated with poor school achievement; consequently, the learning-disabilities assessment must frequently be combined with an assessment of ADHD. Finally, such social-emotional variables as depressed affect, passive style, poor motivation, low tolerance of frustration, and inefficient problem-solving strategies are some of the many factors that must also be considered in the assessment process. For these reasons, a multifactorial approach to the assessment of learning disabilities has been advocated by a number of professionals (Adelman, 1971; Gaddes, 1983; Johnson & Blalock, 1987; Johnson & Myklebust, 1967;

Taylor, 1988). Because basic psychometric assessment is often provided by school systems, it is our opinion that medical center specialty clinics can provide very important clinical services in conducting the more comprehensive learning-disabilities evaluations.

A related issue with regard to the assessment of childhood learning disorders is the frequent need for a multidisciplinary assessment of these children. Over the past several decades, there has been increasing specialization and refinement in the assessment of children's learning and developmental disorders, so that an integrated multidisciplinary team approach may be indicated for the diagnosis and management of many complex learning disorders. Because the behavioral consequences of such disorders are most apparent in the child's school functioning, it is important that a clinic include educational specialists or individuals with cross-speciality training in applied psychology as well as special education, in order to aid in formulating recommendations and providing intervention, which are often primarily educational in nature. The relatively high proportion of learning disorders that reflect problems in processing phonological and linguistic information necessitates consultation with communication disorders specialists in speech and language pathology and audiology. Those children whose disorders reflect complex sensory integration and motor output impairment may require input from professionals in both medicine (developmental and behavioral pediatrics or pediatric neurology) and physical medicine (occupational and physical therapy). The advantage as well as the challenge of such a multidisciplinary team approach is the provision of a careful integration of findings by varying professionals, which will prevent the fragmentation of services and will develop a well-integrated program of intervention and regularly scheduled follow-up.

Psychological and Psychoeducational Assessment

Federal legislation has mandated appropriate educational programs for school-aged children, including formal psychoeducational assessment (U.S. P.L. 94-142, 1977), and in addition, legislation providing an extension of such remedial programming for preschool children from birth to 5 years old has recently been enacted (P.L. 99-457). Thus, the formal assessment of many children with learning problems is provided by public schools and state agencies, and parents applying to medical center clinics should be apprised of their right to such an evaluation at no cost. However, assuming that the parent and/or the referral source may be requesting an independent and/or more comprehensive assessment, the following general areas have been regarded as important in the evaluation of learning disabilities (Barkley, 1981): (1) developmental-cognitive processes (i.e., verbal-linguistic, visual-spatial-constructional, sequential-analytic, and planning processes); (2) academic achievement in reading, spelling, math, and written expression; (3) environmental demands (i.e., demands placed on the child at home and at school); (4) reactions of others, such as family members, peers, and teacher(s), to the child's school difficulties; and (5) interaction effects (i.e., the interaction of the child's specific pattern of strengths and deficits, motivational and social-emotional factors, and environmental responses as they affect the child's overall performance and adjustment).

Because there is no single standard battery for the assessment of learning disabilities, the selection of assessment procedures (including interview, formal standardized test measures, rating scales, and informal or criterion-referenced assessment) should obviously be based on the specific referral question. In many clinics that incorporate learning-disabilities and/or educational specialists, formal assessment using standardized measures may be followed by a period of diagnostic teaching, which incorporates an informal testing of the hypotheses generated during the formal assessment, an assessment of the rate of learning of particular types of content, and a systematic application of the instructional strategies derived from formal and informal assessment. Reviews of specific psychoeducational procedures are available (e.g., Salvia & Ysseldyke, 1981; Sattler, 1988) and Table 3 lists some of the most

widely used formal measures in the area of learning-disabilities assessment. A description of the formal child neuropsychological assessment procedures is not possible in this chapter, and the interested reader is referred to reviews in Hynd and Obrzut, 1981; Hynd and Willis, 1988; and Rourke *et al.*, 1983.

Clinical Interventions

Oppositional and Conduct-Disordered Children

Parent–Child Interaction Methods

There is now a large accumulation of evidence demonstrating that the interactions of parent and child are major determinants of each other's behavior (e.g., Patterson, 1976). Consequently, among the most effective techniques for intervening with oppositional children are *in vivo* methods designed to alter inappropriate parent–child interactions that have served to maintain the child's misbehavior. Forehand and McMahon (1981) and Eyberg and Robinson (1982) have developed treatment programs designed to teach parents to modify their children's noncompliance and related deviant behavior. Both of these programs are based on behavioral principles originally developed by Hanf and Kling (1973) and are designed primarily for young children aged 3–8. Following a period of behavioral observation and assessment, during which baseline behavioral levels are obtained (see Figure 3), parents and child enter into an intervention program. In Forehand and McMahon's two-stage program, for example, Phase I is designed to teach parents differential attention to their child's behavior by demonstrating and practicing instances of attending, rewarding, and ignoring specific behaviors. The practice of these skills occurs at home during a daily 5- to 15-minute session with the child entitled the "child's game" or "child-directed interaction." In the second phase, specific problem behaviors are targeted, and the parent is taught appropriate methods for issuing commands, for reinforcing compliance, and for applying systematic consequences for noncompliance, such as a

Table 3. Learning-Disabilities Assessment

General category	Age range	Selected test
Neuropsychological test batteries	8–12	Luria-Nebraska Neuropsychological Battery—Children's Revision (Golden, 1987)
	5–8	Reitan-Indiana Neuropsychological Test Battery for Children (Reitan *et al.*, 1985)
	9–14	Halstead Neuropsychological Test Battery for Children (Reitan *et al.*, 1985)
Intelligence test batteries	3–7	Wechsler Preschool and Primary Scale of Intelligence—Revised (Wechsler, 1989)
	2–8	McCarthy Scales of Children's Abilities (McCarthy, 1972)
	2–12	Kaufman Assessment Battery for Children (Kaufman *et al.*, 1983)
	2–adult	Stanford-Binet Intelligence Scale—IV (Thorndike *et al.*, 1986)
	6–16	Wechsler Intelligence Scale for Children—Revised (Wechsler, 1974)
Auditory-linguistic batteries		
Language abilities	4–13	Test of Language Development—2 (Primary; Intermediate) (Hammill *et al.*, 1988)
	1–7	Preschool Language Scale (Zimmerman *et al.*, 1969)
	2–10	Illinois Test of Psycholinguistic Abilities (Kirk *et al.*, 1968)
	5–16	Clinical Evaluation of Language Functions (Semel *et al.*, 1980)
	3–7	Northwestern Syntax Screening Test (Lee, 1971)
Receptive abilities	5–8	Auditory Discrimination Test (Reynolds, 1987)
	5–adult	Goldman-Fristoe-Woodcock Test of Auditory Discrimination (Goldman *et al.*, 1974)
	5–adult	Lindamood Auditory Conceptualization Test (Lindamood *et al.*, 1979)
	2–adult	Peabody Picture Vocabulary Test—Revised (Dunn *et al.*, 1981)
	5–6	Boehm Test of Basic Concepts—Revised (Boehm, 1986)
	3–10	Test for Auditory Comprehension of Language—Revised (Carrow-Woolfolk, 1985)
	3–12	Token Test for Children (Part V) (DiSimoni, 1978)

(Continued)

time-out procedure. The programs use modeling, role playing, and behavioral rehearsal with immediate feedback, as well as home practice assignments for the parents. The particular benefit of such techniques as these is that, by the use of an observation room and a bug-in-the-ear device, parents receive immediate feedback regarding their management of the child's behavior while in the clinic setting and can then modify their responses and practice more appropriate responses.

Behavioral Contracting

Parent–child/adolescent problems may be managed through assessing the family interaction patterns and modifying negative or coercive communication patterns. Additionally, altering or changing inappropriate reinforcement patterns can be helpful. Often, there is a lack of reciprocal reinforcing interactions between parent and adolescents (Gross, 1983), so efforts to permit adolescents to negotiate privileges depending on certain behaviors are a means of contracting between parent and adolescent. Alexander and Parsons (1973) used behavioral contracting combined with training in negotiation and communication skills with a group of predelinquent and delinquent adolescents and their families. The recidivism rate was lowered in the experimental population. Other behavioral contracting techniques are reported in Kelley and Stokes (1982, 1984).

Table 3. *(Continued)*

General category	Age range	Selected test
Expressive abilities	2–16	Expressive One-Word Picture Vocabulary Test (Gardner, 1979, 1983)
	5–10	Boston Naming Test (Kaplan *et al.*, 1983)
Visual-motor batteries	5–adult	Bender Gestalt (Koppitz, 1964)
	2–15	Developmental Test of Visual-Motor Integration (Beery, 1982)
	8–adult	Revised Visual Retention Test (Benton, 1963)
Specialized cognitive abilities		
Visual-spatial, perception, memory	6–18	Detroit Tests of Learning Aptitude—2 (Hammill, 1985)
	5–adult	Woodcock-Johnson Psychoeducational Battery—Revised (Tests of Cognitive Ability) (Woodcock & Johnson, 1989)
	4–9	Motor-Free Visual Perception Test (Colarusso & Hammill, 1972)
Academic achievement batteries		
Achievement	5–adult	Wide Range Achievement Test—Revised (Jastak *et al.*, 1984)
	5–18	Peabody Individual Achievement Test—Revised (Markwardt, 1989)
	6–18	Kaufman Test of Educational Achievement (Kaufman *et al.*, 1985)
	5–adult	Woodcock-Johnson Psychoeducational Battery—Revised (Woodcock & Johnson, 1989)
Reading	5–adult	Woodcock Reading Mastery Test—Revised (Woodcock, 1987)
	6–14	Gilmore Oral Reading Test (Gilmore, 1968)
	6–adult	Gray Oral Reading Test—Revised (Wiederholt *et al.*, 1986)
	5–18	Gates-MacGinitie Reading Test (MacGinitie, 1978)
	6–14	Durrell Analysis of Reading Difficulty (Durrell *et al.*, 1980)
	6–14	Decoding Skills Test (Richardson *et al.*, 1985)
Spelling, writing, and written language	5–13	Test of Written Spelling (Larsen *et al.*, 1976)
	7–18	Test of Written Language—2 (Hammill *et al.*, 1988)
	7–17	Picture Story Language Test (Myklebust, 1965)
	5–12	Slingerland Screening Tests (Slingerland, 1970)
Math	5–16	Key Math Revised (Connolly, 1988)
	6–18	Stanford Diagnostic Math Test (Beatty *et al.*, 1976)
Motivational factors	9–18	Piers-Harris Children's Self-Concept Scale (Piers, 1984)
	7–15	Nowicki-Strickland Child Locus of Control Scale (1973)

Classroom Intervention

The clinician can use contingency management and other behavioral approaches described above in modifying a child's behavior in the classroom. Time may not always permit the clinician to travel to schools on behalf of the child, and if such consultation is necessary, charges may need to be discussed with the family.

Management of ADHD Children

Although treatments abound for ADHD children, long-term efficacy and the evidence for proposing one type of intervention over another continue to be investigated (see Barkley, 1981; Ross & Ross, 1976). However, approaches most commonly represent a combination of medical and behavioral management of these children. At present, substantial research has demonstrated the efficacy of stimulant medication in the treatment of ADHD children (Barkley, 1977; Cantwell & Carlson, 1978; Whalen & Henker, 1976), and the criteria for the referral of a child for a medication trial have been discussed by Barkley and his colleagues (1981, 1988a). Thus, if the treatment of this clinical group of children is to be undertaken in a learning- and behavior-disorders clinic, collaboration with a physician is indicated, and protocols for the assessment of drug response are beginning to appear in the literature (e.g., Barkley, Fischer, Newby, & Breen, 1988). It is generally assumed that psychologists will be involved in directing the behavioral treatment of these children, in addition to stimulant drug management. Such a need is demonstrated by studies showing that, although stimulant drugs may enhance attention and decrease hyperactivity, they do not improve school achievement (Barkley & Cunningham, 1978). Similarly, controlled studies have demonstrated that a combination of stimulant medication and behavior therapy is more effective than medication alone (e.g., Gittelman, Klein, Abikoff, Katz, Pollack, & Mattes, 1980). Thus, a number of behavioral approaches have been used with these children: parent training in contingency management (including token economy systems and parent–child interaction training as previously reviewed), self-control training using cognitive-behavioral techniques and self-monitoring (Kendall & Braswell, 1985), and classroom management procedures (involving training teachers in contingency management techniques; Barkley, 1981).

Developing and Maintaining the Program

Referral Source

The development of a specialized clinical service for assessment and intervention with learning- and behavior-disordered children generally presumes the establishment of referral sources from both inside the designated medical setting and the community. The primary medical disciplines to be contacted are ordinarily the primary-care specialities of pediatrics and family medicine, and those specialty areas that may be most apt to have referrals are pediatric neurology and child psychiatry.

If the clinician can handle more referrals than are available in the medical setting and decides to market the learning- and behavior-disorders clinic, letters to physicians in the private sector can be mailed, as well as letters to private schools. Detailed records of the numbers and types of referrals to the clinic must be kept. This type of documentation may enable the clinician to gain hospital approval to hire additional staff, especially as the waiting list begins to develop. The clinician who works closely with the schools and seeks to understand the "politics" of schools and their placement decisions will have greater success in attaining the services required for the child evaluated through the clinic. (See Table 4.)

It is critical that the clinician provide timely feedback to the referring source. Once the referral is made to the clinic, the clinician can send or mail a standard letter stating:

> Thank you for referring—to the learning- and behavioral-disorders clinic. We are in the process of obtaining preliminary information on the child before scheduling him (or her) for an appointment. The results of our assessment will be sent to you after we see this child and his (or her) family. Again, thank you for the referral.

Alternatively, the clinician may want to schedule the clinic on specified days of the week, and referring sources can call and schedule the child and the parents for the next available opening. Because this procedure parallels the format of many medical clinics, physicians often prefer this means of patient scheduling. The drawback is that some patients will be scheduled who are not appropriate for the clinic. If this format is chosen, the secretary to the clinic then mails out forms and questionnaires appropriate to the referral question. The referral of any school-aged child who may have been evaluated at school can be followed by a telephone call to the parents requesting that they provide reports of past school (or other agency) assessment. (See Figure 2.)

Assessing behavioral disorders and what maintains the deviant behavior requires a one-to two-hour time period, so that appointments may be scheduled accordingly, allowing time to see four to six patients per clinic day. Treatment days, of course, are scheduled at another time.

At the conclusion of the overall evaluation a copy of the report should be inserted in the

Table 4. Handling of Referrals (Options)[a]

Option 1	Option 2
1. Secretary accepts referral, schedules and triages to learning- and behavior-disorders clinic day.	1. Secretary accepts referral, clarifies referral question, obtains patient's name, address, and telephone number as well as telephone number of referral source.
2. Questionnaires or behavioral forms may or may not be sent, depending on the time or first appointment.	2. Secretary sends appropriate questionnaire *or* obtains past testing. Clinician determines if assessment is necessary. If assessment is not necessary, the outside school (or other) report and a notation are put in the hospital chart.
3. Secretary checks to see if child has been evaluated at school and obtains report if time permits.	3. If assessment is completed, a letter to the referral source is dictated; the report is sent and a copy is also placed in the hospital chart.
4. Once assessment is complete (behavioral or learning), a letter is sent to the referring source; report (and letter) is placed in the hospital chart.	

[a] A file of all referrals, and referral questions, plus disposition, is kept by the clinician. This type of documentation may be helpful in obtaining support to hire additional staff.

hospital record, and a notation should be made on the hospital progress sheet that is dated and notes that the child has been seen and the report filed. A letter should be sent to the referring physician that provides an overview of the findings and the recommendations. Realistically, the clinician in a medical setting cannot follow all children, so it is critical that the clinician be familiar with the referral services in the community and use them as needed. Because the clinician is being asked for a consultation by the referring source, it is necessary to consult again with that source before sending the child elsewhere. Most physicians, however, leave to the clinician the decision of whom to refer for treatment.

Children referred because of learning disorders require a longer assessment time, and therefore, the scheduling of these children does not conform to a clinic day when several children are scheduled. Again, it is important to learn what services are provided by the schools and which schools provide special-education classrooms or resource programs.

Educating Physicians

The learning- and behavior-disorders clinic can be an excellent teaching clinic for residents and those training to be developmental pediatricians, child psychiatrists, or family medicine practitioners. First, however, to ensure that the clinic will receive appropriate referrals, the clinic must be well defined. As referrals are made to the clinic, referral questions are often ambiguous or ill defined, and part of the task of clinicians is to teach physicians how to make good referrals. A referral form can be developed for the clinic on which physicians must write out what they want to learn as a result of the assessment. Or, if telephone referrals are made, the secretary can be trained to help find out what the physician wants to know and to write this on the referral form. Inappropriate referrals can be deflected, and suggestions about other referral sources can be made. For example, suicidal adolescents or a teenager hallucinating because of drugs may be referred to psychiatry or to another colleague.

Part of the process of educating physicians will occur as they read the clinicians' reports or as clinicians talk to them about individual cases.

Residents or other physicians may observe psychological testing and learn that it is time-consuming and very different from the M.D.'s examination of patients. Teaching physicians about psychological testing and providing them a cursory education in the types of psychological tests can aid their performance on the board exams required of them.

Finally, teaching physicians how to use behavioral scales can be helpful to them in their practice. As a result, they may be better able to distinguish ADHD from other behavioral disorders and to make appropriate referrals early in the child's life. Efforts toward early intervention and the prevention of problems, hopefully, will be the outcome of training in the learning- and behavior-disorders clinic.

Consultation with Health Teams

Children with learning or ADHD problems may require services of other health care professionals. In the planning stages of the clinic the clinician will want to establish a working relationship with people in physical medicine, professionals in communication disorders, neurologists, and pediatricians knowledgeable about ADHD. Educational specialists are extremely helpful and are needed to help with the educational assessment of children. Children with a language-based SLD or a child whose learning and motor activity has been hampered as a result of meningitis may need multiple services. In a learning-disorders clinic, the clinician can combine the findings of the various disciplines in one integrated report. Additionally, the clinician may be able to work out staffing time so that the various health care professionals meet together to integrate their findings on a particular child.

The clinician will also be called on to consult with health professionals about children admitted to the hospital for a variety of reasons, from automobile accidents with head injury to physical or sexual abuse. Compliance issues often require behavioral strategies, and consultations may be requested to address ways of ensuring that a diabetic will adhere to his or her diet.

Clinical Case Report

The following case report includes a clinical description of a child who demonstrated both behavioral and learning problems and for whom methods of assessment and behavioral intervention are discussed.

Referral and Background Information

Mark W., a 7-year-old boy, was referred to the learning- and behavior-disorders clinic of a large children's hospital. The referring pediatrician noted that this child had been followed by both the pediatric neurology and the genetics services in this hospital since infancy, when he was initially evaluated for delayed development. His birth history information included documentation of a small-for-gestational-age newborn infant (5 lb, 9 oz), who was cyanotic at birth, and who was placed in a neonatal unit for several weeks following delivery. An inpatient hospitalization for a diagnostic work-up of this infant at age 9 months resulted in a diagnosis of Prader-Willi syndrome. This is a birth defect whose significant clinical features include muscular hypotonia, feeding problems (poor suck), hypogonadism, mild dysmorphism, delayed motor development, and short stature (Hohn, Sulzbacher, & Pipes, 1981). Although approximately 50% of patients exhibit an interstitial deletion of chromosome 15, cytogenetic studies of this child confirmed a normal karyotype. Of primary concern for cognitive and social development is the documented CNS dysfunction in these children, including mental retardation in 90% of cases (usually within the mild MR range), and moderate to severe behavioral problems. These children are generally described as being affectionate and happy in the early years, but temper tantrums and stubbornness generally appear before age 5 and are believed to be a function of the generalized CNS pathology. Two additional behavioral symptoms are of considerable concern. Many children exhibit diminished satiety between 2 and 3 years, frequently resulting in voracious appetites and a risk of severe obesity.

Assessment

When Mark was referred to the clinic for an evaluation of his learning-disorder and behavioral problems, he was in the first grade and had been placed in a learning-lab resource room for two hours per day. A school psychoeducational evaluation had been completed two months before this referral. A summary of the teacher's observations contained the following remarks:

> Mark is outstanding in verbal fluency and group participation but poor in sustaining attention, motivation, self-discipline, and effort. He does not complete assignments and requires much individualized instruction. He demands attention and will go to great lengths to obtain it, including breaking pencils, breaking his glasses, wetting his pants, and using inappropriate language for shock effect.

Additional information obtained by interview suggested that, at times, Mark's behavior was in control for up to 10 days, but that at other times, he was severely disruptive and was isolated behind a screen in the classroom. Subsequently, his rocking his chair, hollering, and using inappropriate language generally resulted in his being able to continue disrupting the class, and ultimately in his being sent to the principal's office.

The following battery had been administered to Mark by the school psychologist when Mark was 7 years, 3 months of age:

WISC-R: Verbal IQ = 82; Performance IQ = 70; Full Scale IQ = 74
Subtest

Subtest	Scaled score
Information	5
Similarities	8
Arithmetic	7
Vocabulary	9
Comprehension	9
Digit Span	5
Picture Completion	7
Picture Arrangement	2
Block Design	5
Object Assembly	7
Coding	7

Stanford Binet (Form LM): IQ = 85; MA + 6 years, 6 months
Bender Gestalt Test: Koppitz Score 14; Visual Motor Age = 4-10 to 4-11
Woodcock-Johnson Psychoeducational Battery:

Area of achievement	Standard score
Oral Language Cluster	80
Written Language Cluster	87
Reading Cluster	75
Passage Comprehension	<63
Calculation	79
Applied Problems	67

Because the school psychoeducational evaluation had been recently performed, no further formal intellectual assessment was deemed necessary, and the assessment was designed to evaluate more carefully Mark's relative strengths and weaknesses, as well as to focus on evaluating Mark's reported noncompliant, disruptive, and negative attention-getting behaviors. The following additional measures were administered to Mark: perceptual-motor—Development Test of Visual Motor Integration, Motor-Free Visual Perception Test, and Purdue Pegboard Test; academic achievement—Wide Range Achievement Test-R, Gates MacGinitie Reading Test, and Key Math Diagnostic Arithmetic Test; and social-emotional—Sentence Completion Test, Figure Drawings, Roberts Apperception Test, and Michigan Pictures Test-R. The parent and teacher forms of the Achenbach Child Behavior Checklist and the Conners Scale were obtained, together with interviews with both parents and telephone interviews with teachers and the school principal.

Test results suggested that Mark's general mental ability ranged between borderline and low average levels, consistent with the typical clinical manifestations of Prader-Willi syndrome. He demonstrated particular deficits in motor skills, both gross motor coordination (as previously assessed) and graphomotor skills, fine motor planning, and coordination, which were significantly impaired relative to Mark's generalized cognitive delay. Although his single-word identification, spelling, and simple written concrete-number-calculation skills were consistent with his grade placement, Mark demonstrated difficulty with the integration and comprehension of more abstract material. He showed particular difficulty in applying basic knowledge of quantity, manipulating numbers, calculating mentally, recognizing information needed, and following a series of logical sequential steps. In this regard, Mark had considerable difficulty with reasoning that involved the interpretation and integration of meaningful sequential information (e.g., in his very deficient WISC-R Picture Arrangement performance), a cognitive deficiency that was useful in explaining some of his behavior.

Data from the objective behavior-rating scales indicated that his parents and his teachers viewed Mark in a similar manner; there were elevations within the clinical range on the Aggressive and Delinquent scales of the CBCL. However, Mark reportedly did not show physically aggressive behavior toward others; though bossy and demanding with peers, he did not strike out, and, in fact, he showed fearfulness when his provocations resulted in aggression toward him by other children. Mark's teachers rated the Inattentiveness scale of the Conners Scale as significantly elevated for Mark's mental age; however, neither his teachers nor his parents reported an increased activity level or impulsivity. The ratings by the lab teacher and the regular classroom teacher differed significantly, indicating that Mark was much better controlled in the small-group setting. Mark's attentional difficulties were reflected in his inability to persist and complete tasks, and his overwhelming need for individual adult attention generally led to disruptive behavior. Thus, the most salient diagnostic impression appeared to be that of the oppositional-defiant child, whose lowered general intelligence and specific perceptual-motor impairment further limited his behavioral controls.

Interviews with both parents and teachers suggested that Mark had learned to control his environment through inappropriate attention-getting behaviors that had been inadvertently, but consistently, negatively reinforced both at home and at school. Typical examples of such contingency-based behaviors were in Mark's managing to obtain the attention of the entire class and his teacher by yelling and using bad language, and additionally, when he deliberately wet his pants, his mother was called to the school to bring dry clothing, resulting in even further negative attention. Although Mark's behavior was somewhat more appropriate with his father, he appeared to test limits almost constantly with his mother, who acknowledged that she was ambivalent about disciplining

Mark because of his handicap and had given in to Mark's demands at a very early age. Although such methods as charting, stickers, and time-out procedures had been attempted by the parents, nothing had had any lasting effects on Mark's behavior.

Although many behaviorally oriented psychologists may not routinely use other more traditional types of personality assessment, a structured interview with Mark (using a sentence completion format), together with projective storytelling, provided additional insight into Mark's behavior, and this information, in turn, proved helpful in the planning of a behavioral intervention. Mark's responses to pictures depicting interpersonal situations with family, peers, and school indicated that he had difficulty recognizing or comprehending antecedents and consequences, and his responses suggested that his practical social judgment was very likely quite impaired, and that he probably had difficulty predicting his and other's emotional responses. Apparent as well was Mark's acute sensitivity to adult affect; he appeared to view women, in particular, as being frightened by strong emotional reactions in others. This appeared to reflect our observations of his mother's inability to take control of Mark when he was oppositional or tantrumming, thus reinforcing his feelings of omnipotence. Mark himself acknowledged that he often felt out of control when he began inappropriate behavior and could not stop, and his descriptions of himself were of a "mean" and "naughty" child.

Intervention

A behavioral intervention program was initiated, with the goal of modifying Mark's inappropriate behavior at home. Because the evaluation took place toward the end of Mark's first-grade school year, there was insufficient time to initiate a structured behavioral program at school, with the exception of such procedures as having the mother leave a set of dry clothing at school so that wetting his pants would not be rewarded by his mother's coming to school. Firm directives that Mark must change his own clothing and return to his class were effective in eliminating this behavioral symptom. A meeting with school personnel resulted in the plan to place Mark in a small self-contained class of 10 children the following school term. Before our evaluation, they had been doubtful about whether he would require a more restrictive residential or hospital-based program. Although he did not meet the strict guidelines for learning disabilities, the school system had no classroom available for children with behavior disorders, and classes for EMH (educable mentally handicapped) and ED (seriously emotionally disturbed) children were judged to be inappropriate for this child (because Mark's general achievements were well above those of the EMH children, and it was feared that he would further model inappropriate behaviors observed of classmates in an ED class). The school principal, the special-education teacher, and a representative from the state department of special education agreed to cooperate in developing a contingency-based token economy program for rewarding Mark's appropriate academic and social behavior (as in Barkley, 1981). In addition, a time-out room removed from the classroom was designated, so that Mark's outbursts would not be negatively reinforced by capturing the attention of the students and the teacher.

The parents agreed to an intensive behavior management program to be carried out over the summer months, with the goal of seeing whether Mark's behavior could be brought under control at home and could then be generalized to situations outside the home. A program was initiated using Forehand and McMahon's methods (1981) for parent training in behavior management. During an initial session, Mark's parents were provided with an overview of the program and a rationale for the use of contingency-based approaches to noncompliant and inappropriate behavior that is believed to have been learned. The programmatic training was performed during a total of 15 weekly clinic sessions with both parents and Mark, during which behavioral charting was done regularly by the parents at home, and progress was reviewed during the clinic visits.

When Mark entered school in the fall, a contingency management program had been es-tablished that included rewarding the other students for ignoring Mark's inappropriate responses. He began school with a halfday program, which was gradually increased as Mark improved in his behavior. Periodic follow-up visits were scheduled through the clinic, including individual supportive sessions for Mark's mother. At the end of the school year, Mark's behavior had improved sufficiently to allow him to attend a full-day program, with occasional time-out periods. Although the contingency management program was effective in substantially reducing his noncompliant behaviors, this child's social cognition deficits dictated that he also needed intervention in his interactions with peers. He was subsequently referred to a social-skills-training group.

Summary

Children presenting with learning and behavior disorders are a high source of referral in a medical setting both from physicians on staff and from agencies or sources outside the medical setting. This chapter presented an overview of the literature on learning and behavior disorders in children, with a particular emphasis on ADHD and learning disabilities. A discussion of methods of assessing learning or behavior disorders was presented, along with suggestions for clinical intervention. Finally, the authors suggested ways to develop and maintain a learning- and behavior-disorders clinic within a medical setting.

References

Aaron, P. G. (1981). Diagnosis and remediation of learning disabilities in children: A neuropsychological key approach. In G. W. Hynd & J. E. Obrzut (Eds.), *Neuropsychological assessment and the school-age child*. New York: Grune & Stratton.

Abidin, R. R. (1983). *Parenting Stress Index*. Charlottesville, VA: Pediatric Psychology Press.

Abikoff, H., Gittelman-Klein, R., & Klein, D. (1977). Validation of a classroom observation code for hyperactive children. *Journal of Consulting and Clinical Psychology, 45*, 772–783.

Achenbach, T. M. (1978). The Child Behavior Profile: 1. Boys aged 6–11. *Journal of Consulting and Clinical Psychology*, *46*, 478–488.

Achenbach, T. M. (1981). The role of taxonomy in developmental psychopathology. In M. E. Lamb & A. L. Brown (Eds.), *Advances in developmental psychology* (Vol. 1). Hillsdale, NJ: Erlbaum.

Achenbach, T. M. (1982). *Developmental psychopathology* (2nd ed.). New York: Wiley.

Achenbach, T. M., & Edelbrock, C. S. (1978). The classification of child psychopathology: A review and analysis of empirical efforts. *Psychological Bulletin*, *85*, 1275–1301.

Achenbach, T. M., & Edelbrock, C. S. (1979). The Child Behavior Profile: 2. Boys aged 12–16 and girls aged 6–11 and 12–16. *Journal of Consulting and Clinical Psychology*, *47*, 223–233.

Achenbach, T. M., & Edelbrock, C. (1983). *Manual for the Child Behavior Checklist and Revised Child Behavior Profile*. Burlington, VT: Thomas Achenbach.

Achenbach, T. M., Edelbrock, C., & Howell, C. T. (1987). Empirically based assessment of the behavioral/emotional problems of 2- and 3-year old children. *Journal of Abnormal Child Psychology*, *15*, 629–650.

Adelman, H. S. (1971). The not so specific learning disability population. *Exceptional Children*, *37*, 528–533.

Alexander, J. F., & Parsons, B. V. (1973). Short-term behavioral intervention with delinquent families: Impact on family processes and recidivism. *Journal of Abnormal Child Psychology*, *81*, 219–225.

American Psychiatric Association. (1987). *Diagnostic and statistical manual of mental disorders* (3rd ed. rev.; DMS-III-R). Washington, DC: Author.

Barkley, R. A. (1977). A review of stimulant drug research with hyperactive children. *Journal of Child Psychology and Psychiatry*, *18*, 137–165.

Barkley, R. A. (1981). *Hyperactive children: A handbook for diagnosis and treatment*. New York: Guilford Press.

Barkley, R. A. (1982). Guidelines for defining hyperactivity in children: Attention deficit disorder with hyperactivity in children. In B. E. Lahey & A. Kazdin (Eds.), *Advances in clinical psychology* (Vol. 5). New York: Plenum Press.

Barkley, R. A. (1987). The assessment of attention deficit-hyperactivity disorder. *Behavioral assessment*, *9*, 207–233.

Barkley, R. A. (1988a). Attention deficit disorder with hyperactivity. In E. J. Mash & L. G. Terdal (Eds.), *Behavioral assessment of childhood disorders* (2nd ed.). New York: Guilford Press.

Barkley, R. A. (1988b). Child behavior rating scales and checklists. In M. Rutter, A. H. Tuma, & I. S. Lann (Eds.), *Assessment and diagnosis in child psychopathology*. New York: Guilford Press.

Barkley, R. A., & Cunningham, C. E. (1978). Do stimulant drugs enhance the academic performance of hyperactive children? *Clinical Pediatrics*, *17*, 85–92.

Barkley, R. A., Fischer, M., Newby, R. F., & Breen, M. J. (1988). Development of a multimethod clinical protocol for assessing stimulant drug response in children with attention deficit disorder. *Journal of Clinical Child Psychology*, *17*, 14–24.

Baum, C. (1989). Conduct disorders. In T. H. Ollendick & M. Hersen (Eds.), *Handbook of child psychopathology* (2nd ed.). New York: Plenum Press.

Beatty, L. S., Madden, R., Gardner, E. F., & Karlsen, B. (1976). *Stanford Diagnostic Mathematics Test*. New York: Harcourt Brace Jovanovich.

Beery, K. E. (1982). *Revised administration, scoring and teaching manual for the Developmental Test of Visual-Motor Integration*. Cleveland: Modern Curriculum Press.

Benton, A. L. (1963). *Benton Visual Retention Test* (rev. ed.). San Antonio: Psychological Corporation.

Boder, E. (1973). Developmental dyslexia: A diagnostic approach based on three atypical reading-spelling patterns. *Developmental Medicine and Child Neurology*, *15*, 663–687.

Boehm, A. E. (1986). *Boehm Test of Basic Concepts—Revised*. San Antonio: Psychological Corporation.

Boyle, M. H., & Jones, S. C. (1985). Selecting measures of emotional and behavioral disorders of childhood for use in general populations. *Journal of Child Psychology and Psychiatry*, *26*, 137–159.

Broman, S., Bien, E., & Shaughnessy, P. (1985). *Low achieving children: The first seven years*. Hillsdale, NJ: Erlbaum.

Cairns, E., & Cammock, T. (1978). Development of a more reliable version of the Matching Familiar Figures Test. *Developmental Psychology*, *14*, 555–560.

Cairns, E., & Cammock, T. (1984). The development of reflection-impulsivity: Further data. *Personality and Individual Differences*, *5*, 113–115.

Campbell, S. B., Douglas, V. I., & Morganstern, G. (1971). Cognitive styles in hyperactive children and the effect of methylphenidate. *Journal of Child Psychology and Psychiatry*, *12*, 55–67.

Campbell, S. B., Endman, M. W., & Bernfeld, G. (1977a). A three-year follow-up of hyperactive preschoolers into elementary school. *Journal of Child Psychology and Psychiatry*, *18*, 239–249.

Campbell, S. B., Schleifer, M., Weiss, G., & Perlman, T. (1977b). A two-year follow-up of hyperactive preschoolers. *American Journal of Orthopsychiatry*, *47*, 149–162.

Canter, A. (1976). Manual for the Canter Background Interference Procedure (BIP) for the Bender Gestalt Test. Nashville, TN: Counselor Recordings and Tests.

Cantwell, D. P., & Baker, L. (1987). Differential diagnosis of hyperactivity. *Developmental and Behavioral Pediatrics*, *8*, 159–165.

Cantwell, D., & Carlson, G. (1978). Stimulants. In J. Werry (Ed.), *Pediatric pharmacology*. New York: Brunner/Mazel.

Cantwell, D. P. & Satterfield, J. H. (1978). The prevalence of academic underachievement in hyperactive children. *Journal of Pediatric Psychology*, *3*, 168–171.

Carrow-Woolfolk, E. (1985). *Test for Auditory Comprehension of Language* (rev. ed.). Allen, TX: DLM Teaching Resources.

Chadwick, O., Rutter, M., Shaffer, D., & Shrout, P. (1981). A prospective study of children with head injuries: 4. Specific cognitive deficits. *Journal of Clinical Neuropsychology, 3* 101–120.

Chalfant, J. C. (1989). Learning disabilities: Policy issues and promising approaches. *American Psychologist, 44,* 392–398.

Chelune, G. J., Ferguson, W., Koon, R., & Dickey, T. O. (1986). Frontal lobe disinhibition in attention deficit disorder. *Child Psychiatry and Human Development, 16,* 221–232.

Clements, S. D. (1966). *Minimal brain dysfunction in children.* U.S. Government Printing Office (NINBD Monograph No. 3, USPHS Publication No. 1415).

Colarusso, R. P., & Hammill, D. D. (1972). *Motor-Free Visual Perception Test.* Novato, CA: Academic Therapy Publications.

Conners, C. K., & Wells, K. C. (1986). Hyperkinetic children: A neuropsychosocial approach. In A. Kazdin (Ed.), *Developmental clinical psychology and psychiatry* (Vol. 7). Beverly Hills, CA: Sage.

Connolly, A. J. (1988). *Key Math Revised: A Diagnostic Inventory of Essential Mathematics.* Circle Press, MN: American Guidance Service.

DiSimoni, F. G. (1978). *The Token Test for Children.* Boston: Teaching Resources.

Doehring, D. G., & Hoshko, I. M. (1977). Classification of reading problems by the Q-technique of factor analysis. *Cortex, 13,* 281–294.

Douglas, V. I. (1983). Attentional and cognitive problems. In M. Rutter (Ed.), *Developmental neuropsychiatry.* New York: Guilford Press.

Douglas, V. I. (1984). The psychological processes implicated in ADD. In L. M. Bloomingdale (Ed.). *Attention deficit disorder: Diagnostic, cognitive and therapeutic understanding.* New York: Spectrum.

Douglas, V. I., & Peters, K. G. (1979). Toward a clearer definition of the attentional deficit of hyperactive children. In G. A. Hale & M. Lewis (Eds.). *Attention and cognitive development.* New York: Plenum Press.

Dunn, L. M., & Dunn, L. M. (1981). *Peabody Picture Vocabulary Test—Revised.* Circle Pines, MN: American Guidance Service.

Durrell, D. D., & Catterson, J. H. (1980). *Durrell Analysis of Reading Difficulty* (3rd ed.). San Antonio: Psychological Corporation.

Dykman, R. A., Ackerman, P. T., Clements, S. D., & Peters, J. E. (1971). Specific learning disabilities: An attentional deficit syndrome. In H. R. Myklebust (Ed.), *Progress in learning disabilities* (Vol. 2). New York: Grune & Stratton.

Dykman, R. A., Ackerman, P. T., & Oglesby, D. M. (1979). Selective and sustained attention in hyperactive, learning-disabled, and normal boys. *Journal of Nervous and Mental Disease, 16,* 288–297.

Edelbrock, D., & Achenbach, T. W. (1984). The teacher version of the Child Behavior Profile: 1. Boys aged 6–11. *Journal of Consulting and Clinical Psychology, 52,* 207–217.

Edelbrock, C., & Rancurello, M. D. (1985). Childhood hyperactivity: An overview of rating scales and their applications. *Clinical Psychology Review, 5,* 429–445.

Eyberg, S. M. (1974). *Manual for coding dyadic parent-child interactions.* Unpublished manuscript. Portland: Oregon Health Sciences University, Department of Medical Psychology.

Eyberg, S. M., & Robinson, E. A. (1982). Parent child intervention training: Effects on family functioning. *Journal of Clinical Child Psychology, 11,* 130–137.

Eyberg, S. M., & Robinson, E. A. (1983). Dyadic Parent-Child Interaction Coding System: A manual. *Psychological Documents, 13,* Ms. 2582.

Forehand, R. L., & McMahon, R. J. (1981). *Helping the noncompliant child.* New York: Guilford Press.

Gabel, S. (1981). *Behavioral problems in childhood: A primary care approach.* New York: Grune & Stratton.

Gaddes, W. H. (1983). Applied educational neuropsychology: Theories and problems. *Journal of Learning Disabilities, 16,* 511–514.

Gardner, M. F. (1979). *Expressive One-Word Picture Vocabulary Test.* Novato, CA: Academic Therapy.

Gardner, M. F. (1983). *Expressive One-Word Picture Vocabulary Test, Upper Extension.* San Francisco: Academic Therapy.

Gilmore, J. V., & Gilmore, E. C. (1968). *Gilmore Oral Reading Test.* New York: Harcourt Brace Javanovich.

Gittelman, R., Klein, D. F., Abikoff, H., Katz, S., Pollack, E., & Mattes, J. (1980). A controlled trial of behavior modification and methylphenidate in hyperactive children. In C. K. Whelan & B. Henker (Eds.), *Hyperactive children: The social ecology of identification and treatment.* New York: Academic Press.

Golden, C. J. (1987). *Luria-Nebraska Neuropsychological Battery: Children's Revision.* Los Angeles: Western Psychological Services.

Goldman, R., Fristoe, M., & Woodcock, R. (1974). *G-F-W Test of Auditory Discrimination.* Circle Pines, MN: American Guidance Service.

Goldstein, K. (1942). *After effects of brain injuries in war.* New York: Grune & Stratton.

Gordon, M. (1983). *The Gordon Diagnostic System.* DeWitt, NY: Gordon Systems.

Gordon, M. (1986). Microprocessor-based assessment of Attention Deficit Disorders. *Psychopharmacology Bulletin, 22,* 288–290.

Gordon, M., & Mettelman, B. B. (1988). The assessment of attention: 1. Standardization and reliability of a behavior-based measure. *Journal of Clinical Psychology, 44,* 682–690.

Goyette, C. K., Conners, C. K., & Ulrich, R. F. (1978). Normative data on Revised Conners Parent and Teacher Rating Scales. *Journal of Abnormal Child Psychology, 6,* 221–236.

Graham, P. J. (1983). Specific medical syndromes. In M. Rutter (Ed.), *Developmental neuropsychiatry.* New York: Guilford Press.

Gross, A. M. (1983). Conduct disorders. In M. Hersen (Ed.), *Outpatient behavior therapy: A clinical guide.* New York: Grune & Stratton.

Hallahan, D. P., & Cruickshank, W. M. (1973). *Psychoeduca-*

tional foundations of learning disabilities. Englewood Cliffs, NJ: Prentice-Hall.

Hammill. D. D. (1985). *Detroit Test of Learning Aptitude—2.* Austin, TX: Pro-Ed.

Hammill, D. D., & Newcomer, P. L. (1988). *Test of Language Development—2.* Austin, TX: Pro-Ed.

Hammill, D. D., & Larsen, S. C. (1988). *Test of Written Language—2.* Austin, TX: Pro-Ed.

Hammill, D. D., Leigh, J. E., McNutt, G., & Larsen, S. C. (1981). A new definition of learning disabilities. *Learning Disability Quarterly, 4,* 336–342.

Hanf, C., & Kling, J. (1973). *Facilitating parent-child interactions: A two-stage training model.* Unpublished manuscript, University of Oregon Medical Center.

Hastings, J. E., & Barkley, R. A. (1978). A review of psychophysiological research with hyperactive children. *Journal of Abnormal Child Psychology, 7,* 413–447.

Herzog, J., Linder, H., & Samaha, J. (1981). *The measurement of stress: Life events and the interviewer's ratings.* (Project Competence Report No. 1). Minneapolis: University of Minnesota.

Hinshaw, S. P. (1987). On the distinction between attentional deficits/hyperactivity and conduct problems/aggression in child psychopathology. *Psychological Bulletin, 101,* 443–463.

Hinshelwood, J. (1900). Congenital word blindness. *Lancet, 1,* 1506–1508.

Hobbs, N. (1975). *Issues in the classification of children.* San Francisco, CA: Jossey-Bass.

Hohn, V. A., Sulzbacher, S. J., & Pipes, P. L. (Eds.). (1981). *Prader-Willi syndrome.* Baltimore: University Press.

Hynd, G. W., & Obrzut, J. E. (1981). *Neuropsychological assessment and the school-age child: Issues and procedures.* New York: Grune & Stratton.

Hynd, G. W., & Willis, W. G. (1988). *Pediatric neuropsychology.* Orlando, FL. Grune & Stratton.

Hynd, G. W., Connor, R. T., & Nieves, N. (1987). Learning disability subtypes: Perspectives and methodological issues in clinical assessment. In M. G. Tramontana & S. R. Hooper (Eds.), *Assessment issues in child neuropsychology.* New York: Plenum Press.

Jacob, R. G., O'Leary, K. D., & Rosenblad, C. (1978). Formal and informal classroom settings: Effects on hyperactivity. *Journal of Abnormal Child Psychology, 6,* 47–59.

Jastak, S., & Wilkinson, G. S. (1984). *Wide Range Achievement Test—Revised.* Wilmington, DE: Jastak Associates.

Johnson, D. J., & Blalock, J. M. (Eds.). (1987). *Adults with learning disabilities: Clinical studies.* Orlando, FL: Grune & Stratton.

Johnson, D. J., & Myklebust, H. (1967). *Learning disabilities: Educational principles and practices.* New York: Grune & Stratton.

Kagan, J. (1966). Reflection-impulsivity: The generality and dynamics of conceptual tempo. *Journal of Abnormal Psychology, 71,* 17–24.

Kaplan, E., Goodglass, H., & Weintraub, S. (1983). *Boston Naming Test.* Philadelphia: Lea & Febiger.

Karp, S. A., & Konstadt, N. (1971). *Children's Embedded Figures Test.* Palo Alto, CA: Consulting Psychologists Press.

Kaufman, A. S. (1975). Factor analysis of the WISC-R at 11 age levels between 6½ and 16½ years. *Journal of Consulting and Clinical Psychology, 43,* 138–140.

Kaufman, A. S., & Kaufman, N. L. (1983). *K-ABC: Kaufman Assessment Battery for Children.* Circle Pines, MN: American Guidance Service.

Kaufman, A. S., & Kaufman, N. L. (1985). *Kaufman Test of Educational Achievement.* Circle Pines, MN: American Guidance Service.

Kazdin, A. (1988). The diagnosis of childhood disorders: Assessment issues and strategies. *Behavioral assessment, 10,* 67–94.

Kelley, M. L., & Stokes, T. S. (1982). Contingency contracting with disadvantaged youth: Improving classroom performance. *Journal of Applied Behavior Analysis, 15,* 447–454.

Kelley, M. L., & Stokes, T. F. (1984). Student teacher contracting with goal-setting for maintenance. *Behavior Modification, 8,* 223–244.

Kendall, P. C., & Braswell, L. (1985). *Cognitive-behavioral therapy for impulsive children.* New York: Guilford Press.

Kendall, P. C., & Wilcox, L. E. (1979). Self-control in children: Development of a rating scale. *Journal of Consulting and Clinical Psychology, 47,* 1020–1029.

Kirby, E. A., & Grimley, L. K. (1986). *Understanding and treating attention deficit disorder.* New York: Pergamon Press.

Kirk, S. A., McCarthy, J. J., & Kirk, W. D. (1968). *The Illinois Test of Psycholinguistic Abilities.* Urbana: University of Illinois Press.

Koppitz, E. M. (1964). *The Bender-Gestalt Test for Young Children.* New York: Grune & Stratton.

Lahey, B. B., Schaughency, E. A., Strauss, C. C., & Frame, C. L. (1984). Are attention deficit disorders with and without hyperactivity similar or dissimilar disorders? *Journal of the American Academy of Child Psychiatry, 23,* 302–310.

Lahey, B. B., Schaughency, E. A., Hynd, G. W., & Carlson, C. L. (1987). Attention deficit disorder with and without hyperactivity: Comparison of behavioral characteristics of clinic-referred children. *Journal of the American Academy of Child and Adolescent Psychiatry, 26,* 718–723.

Lahey, B. B., Pelham, W. E., Schaugency, E. A., Atkins, M. S., Murphy, H. A., Hynd, G., Russo, M., Hartdagen, S., & Lorys-Vernon, A. (1988). Dimensions and types of attention deficit disorder. *Journal of the American Academy of Child and Adolescent Psychiatry, 27,* 330–335.

Lambert, N. M., & Sandoval, J. (1980). The prevalence of learning disability in a sample of children considered hyperactive. *Journal of Abnormal Child Psychology, 8,* 33–50.

Lambert, N. M., Sandoval, J., & Sassone, D. (1978). Prevalence of hyperactivity in elementary school children as a function of social system definers. *American Journal of Orthopsychiatry, 48,* 446–463.

Larsen, S. C., & Hammill, D. D. (1976). *Test of Written Spelling.* Austin, TX: Pro-Ed.

Laufer, M. W., & Denhoff, E. (1957). Hyperkinetic behavior syndrome in children. *Journal of Pediatrics, 50,* 463–472.

Lavigne, J. V., & Burns, W. J. (1981). *Pediatric psychology: An introduction to pediatricians and psychologists.* New York: Grune & Stratton.

Lee, L. (1971). *Northwestern Syntax Screening Test.* Evanston, IL: Northwestern University Press.

Lindamood, C., & Lindamood, P. (1979). *Lindamood Auditory Conceptualization Test* (rev. ed.). Allen, TX: DLM Teaching Resources.

Lou, H. C., Henrikson, L., & Bruhn, D. (1984). Focal cerebral hyperfusion in children with dysphasia and/or attention deficit disorder. *Archives of Neurology, 41,* 825–829.

Lyon, R., & Watson, B. (1981). Empirically derived subgroups of learning disabled readers: Diagnostic characteristics; *Journal of Learning Disabilities, 14,* 256–261.

MacGinitie, W. H. (1978). *Gates-MacGinitie Reading Tests* (2nd ed.). Boston: Houghton Mifflin.

Magrab, P. (1978). *Psychological management of pediatric problems* (Vols. 1, 2). Baltimore: University Park Press.

Magrab, P. R. (Ed.), (1984). *Psychological and behavioral assessment: Impact on pediatric care.* New York: Plenum Press.

Markwardt, F. C. (1989). *Peabody Individual Achievement Test-Revised.* Circle Pines, MN: American Guidance Service.

Mash, E. J., & Terdal, L. G. (Eds.), (1988). *Behavioral assessment of childhood disorders* (2nd ed.). New York: Guilford Press.

Matarazzo, J. (1983). Computerized psychological testing. *Science, 221,* 323.

Mattis, S., French, J. H., & Rapin, I. (1975). Dyslexia in children and young adults: Three independent neuropsychological syndromes. *Developmental Medicine and Child Neurology, 17,* 150–163.

McCarthy, D. A. (1972). *Manual for the McCarthy Scales of Children's Abilities.* San Antonio: Psychological Corporation.

McKinney, J. D. (1984). The search for subtypes of specific learning disability. *Journal of Learning Disabilities, 17,* 43–50.

McMahon, R. J. (1987). Some current issues in the behavioral assessment of conduct disordered children and their families. *Behavioral Assessment, 9,* 235–252.

McMahon, R. J., & Forehand, R. (1987). Conduct disorders. In E. J. Mash & L. G. Terdal (Eds.), *Behavioral assessment of childhood disorders* (2nd ed.). New York: Guilford Press.

Milich, R., & Kramer, J. (1984). Reflections on impulsivity: An empirical investigation of impulsivity as a construct. In K. Gadow & I. Bialer (Eds.), *Advances in learning and behavioral disabilities* (Vol. 3). Greenwich, CT: JAI Press.

Milich, R., Loney, J., & Landau, S. (1982). Independent dimensions of hyperactivity and aggression: A validation with playroom data. *Journal of Abnormal Psychology, 91,* 183–198.

Moos, R. H., & Moos, B. S. (1981). *Family Environment Scale manual.* Palo Alto, CA: Consulting Psychologists Press.

Myklebust, H. R. (1965). *Picture Story Language Test.* New York: Grune & Stratton.

Nowicki, S., & Strickland, B. R. (1973). A locus of control scale for children. *Journal of Consulting and Clinical Psychology, 40,* 148–154.

O'Leary, S. G., & Steen, P. L. (1982). Subcategorizing hyperactivity: The Stoney Brook Scale. *Journal of Consulting and Clinical Psychology, 50,* 426–432.

Ollendick, T. H., & Hersen, M. (Eds.), 1984). *Child behavioral assessment.* New York: Pergamon Press.

Orton, S. T. (1928). Specific reading disability-strephosymbolia. *Journal of the American Medical Association, 90,* 1095–1099.

Ostrom, N. N., & Jenson, W. R. (1988). Assessment of attention deficits in children. *Professional School Psychology, 3,* 253–269.

Patterson, G. R. (1976). *Living with Children.* Webster, NC: Psytec.

Pelham, W. S. (1982). Childhood hyperactivity: Diagnosis, etiology, nature, and treatment. In R. G. Gatchel, A. Baum, & J. Singer (Eds.), *Handbook of psychology and health: Clinical psychology and behavioral medicine: Overlapping disciplines* (Vol. 1). New York: Erlbaum.

Pelham, W., & Murphy, H. (1981). *The SNAP Checklist: A teacher checklist for identifying children with attention deficit disorders.* Unpublished manuscript, Florida State University, Tallahassee.

Pennington, F., van Doorninck, W. J., McCabe, L. L., & McCabe, E. R. B. (1985). Neuropsychological deficits in early treated phenylketonuric children. *American Journal of Mental Deficiency, 89,* 467–474.

Petrauskas, R., & Rourke, B. P. (1979). Identification of subgroups of retarded readers: A neuropsychological multivariate approach. *Journal of Clinical Neuropsychology, 1,* 17–37.

Piers, E. V. (1984). *Piers-Harris Children's Self-Concept Scale.* Los Angeles: Western Psychological Services.

Porteus, S. D. (1959). *The Maze test and clinical psychology.* Palo Alto, CA: Pacific Books.

Prinz, R. J., Conner, P. A., & Wilson, C. C. (1981). Hyperactive and aggressive behaviors in childhood: Intertwined dimensions. *Journal of Abnormal Child Psychology, 9,* 191–202.

Quay, H. C. (1979). Classification. In H. C. Quay & J. S. Werry (Eds.), *Psychopathological disorders of childhood* (2nd ed.). New York: Wiley.

Quay, H. C., & Peterson, D. R. (1975). *Manual for the Behavior Problems Checklist* (rev. ed.). Unpublished manuscript, University of Miami, Coral Gables, FL.

Quay, H. C. & Peterson, D. R. (1983). *Interim manual for the Revised Behavior Problem Checklist* (1st ed.). Unpublished manuscript University of Miami, Coral Gables, FL.

Rapoport, J. L., & Ferguson, H. B. (1981). Biological validation of the hyperkinetic syndrome. *Developmental Medicine and Child Neurology, 23,* 667–682.

Reitan, R. M., & Wolfson, D. (1985). *The Halstead-Reitan Neuropsychological Test Battery.* Tucson: Neuropsychology Press.

Reynolds, C. M. (1987). *Auditory Discrimination Test.* Los Angeles: Western Psychological Services.

Richardson, E., & DiBenedetto, B. (1985). *Decoding Skills Test*. Parkton, MD: York Press.

Robinson, E. A., Eyberg, S. M., & Ross, A. W. (1980). The standardization of an inventory of child conduct problem behaviors. *Journal of Clinical Child Psychology 9*, 22–28.

Ross, D. M., & Ross, S. A. (1976). *Hyperactivity: Research, theory, and action*. New York: Wiley.

Rosvold, H., Minsky, A., Sarason, I., Brasome, E., & Beck, A. (1956). A continuous performance test of brain damage. *Journal of Consulting Psychology, 20*, 343–352.

Rourke, B. P. (1985). *Learning disabilities in children: Advances in subtype analysis*. New York: Guilford Press.

Rourke, B. P., Bakker, D. J., Fisk, J. L., & Strang, J. D. (1983). *Child neuropsychology: An introduction to theory, research, and clinical practice*. New York: Guilford Press.

Rourke, B. P., Fish, J. L., & Strang, J. D. (1986). *Neuropsychological assessment of children*. New York: Guilford Press.

Routh, D. K., & Roberts, R. D. (1972). Minimal brain dysfunction in children: Failure to find evidence for a behavioral syndrome. *Psychological Reports, 31*, 307–314.

Routh, D. K., Schroeder, C. S., & O'Tuama, L. (1974). Development of activity level in children. *Developmental Psychology, 10*, 163–168.

Russo, D. C., & Varni, J. W. (1982). *Behavioral pediatrics*. New York: Plenum Press.

Rutter, M. (Ed.). (1983). *Developmental neuropsychiatry*. New York: Guilford Press.

Salkind, N. J., & Nelson, C.F. (1980). A note on the developmental nature of reflection-impulsivity. *Developmental Psychology, 6*, 237–238.

Salvia, J., & Ysseldyke, J. E. (1981). *Assessment in special and remedial education* (2nd ed.). Boston: Houghton Mifflin.

Sandoval, J., Lambert, N. M., & Sassone, D. (1980). The identification and labeling of hyperactivity in children: An interactive model. In C. K. Whalen & B. Henker (Eds.), *Hyperactive children: The social ecology of identification and treatment*. New York: Academic Press.

Sattler, J. M. (1988). *Assessment of children* (3rd ed.). San Diego: Jerome M. Sattler.

Satz, P., Taylor, H. G., Friel, J., & Fletcher, J. (1978). Some developmental and predictive precursors of reading disabilities: A six-year follow-up. In A. L. Benton & D. Pearl (Eds.), *Dyslexia: An appraisal of current knowledge*. New York: Oxford University Press.

Schacher, R., Rutter, M., & Smith, A. (1981). The characteristics of situationally and pervasively hyperactive children: Implications for syndrome definition. *Journal of Child Psychology and Psychiatry, 22*, 375–392.

Schaefer, C. E., & Millman, H. L. (1981). *How to help children with common problems*. New York: Van Nostrand Reinhold.

Schroeder, C. S., & Gordon, B. N. (1990). Assessment of behavior problems in young children. In D. J. Willis & J. L. Culbertson (Eds.), *Testing young children*, Austin, TX: Pro-Am Publications.

Sell, S. H. (1983). Long-term sequelae of bacterial meningitis in children. *Pediatric Infectious Disease, 2*, 90–93.

Semel, E. M., & Wiig, E. H. (1980). *CELF: Clinical Evaluation of Language Functions*. Columbus, OH: Charles E. Merrill.

Shaffer, D. (1980). An approach to the validation of clinical syndromes in childhood. In S. Salinger, J. Antrobus, & J. Glick (Eds.), *The ecosystem of the "sick" child*. New York: Academic Press.

Shaywitz, S. E., Schnell, C., Shaywitz, B. A., & Towle, V. R. (1986). Yale Children's Inventory (YCI): An instrument to assess children with attentional deficits and learning disabilities: 1. Scale development and psychometric properties. *Journal of Abnormal Child Psychology, 14*, 347–364.

Silver, L. B. (1981). The relationship between learning disabilities, hyperactivity, distractibility, and behavioral problems: A clinical analysis. *Journal of the American Academy of Child Psychiatry, 20*, 385–397.

Sines, J. O. (1983). *Home Environment Questionnaire manual*. Iowa City: Psychological Assessment Services.

Slingerland, B. H. (1970). *Slingerland Screening Tests for Identifying Children with Specific Learning Disability* (rev. ed.). Cambridge, MA: Educators Publishing Service.

Spreen, O., & Haaf R. G. (1986). Empirically derived learning disability subtypes: A replication attempt and longitudinal patterns over 15 years. *Journal of Learning Disabilities, 19*, 170–180.

Stevenson, J. E., Hawcroft, J., Lobascher, M., Smith, I., Wolff, O. H., & Graham, P. J. (1979). Behavioral deviance in children with early treated phenylketonuria. *Archives of Disease in Childhood, 54*, 14–18.

Stewart, M. A., Pitts, F. N., Craig, A. G., & Dieruf, W. (1966). The hyperactive child syndrome. *American Journal of Orthopsychiatry, 36*, 861–867.

Strauss, A. A., & Kephart, N. C. (1955). *Psychopathology and education of the brain-injured child* (Vol. 2). Orlando, FL: Grune & Stratton.

Strauss, A. A., & Lehtinen, L. E. (1947). *Psychopathology and education of the brain-injured child* (Vol. 1). Orlando, FL: Grune & Stratton.

Swisher, L. P., & Pinsker, E. J. (1971). The language characteristics of hyperverbal hydrocephalic children. *Developmental Medicine and Child Neurology, 13*, 746–755.

Taylor, H. G. (1988). Learning disabilities. In E. J. Mash & L. G. Terdal (Eds.), *Behavioral assessment of childhood disorders*. New York: Guilford Press.

Thorndike, R. L., Hagen, E. P., & Sattler, J. M. (1986). *Guide for administering and scoring the Stanford-Binet Intelligence Scale* (4th ed.). Chicago: Riverside.

Torgesen, J. K. (1986). Learning disabilities theory: Its current state and future prospects. *Journal of Learning Disabilities, 19*, 339–407.

Trites, R. L., Dugas, F., Lynch, G., & Ferguson, H. B. (1979). Prevalence of hyperactivity. *Journal of Pediatric Psychology, 4*, 179–188.

Trommer, B. L., Hoeppner, J. B., Lorber, R., & Armstrong, K. (1988). *Developmental and Behavioral Pediatrics, 9*, 339–345.

Tuma, J. (1982). *Handbook for the practice of pediatric psychology*. New York: Wiley.

Ullman, R. K., Sleator, E. K., & Sprague, R. L. (1984). A new rating scale for diagnosis and monitoring of ADD children (ACTeRS). *Psychopharmacology Bulletin, 29,* 160–164.

U.S. Office of Education. (1977). *Education of Handicapped Children: Implementation of Part B of the Education of the Handicapped Act.* Federal Register 42: 42471–518.

U.S. Office of Education. (1984). Sixth Annual Report to Congress on the Implementation of the Education of the Handicapped Act. Washington, DC: U.S. Department of Health, Education, and Welfare.

Waber, D. P. (1979). Neuropsychological aspects of Turner's syndrome. *Developmental Medicine and Child Neurology, 21,* 58–70.

Wechsler, D. (1974). *Manual for the Wechsler Intelligence Scale for Children—Revised.* San Antonio: Psychological Corporation.

Wechsler, D. (1989). *Wechsler Preschool and Primary Scale of Intelligence—Revised.* San Antonio: Psychological Corporation.

Wells, K. C., & Forehand, R. (1985). Conduct and oppositional disorders. In P. H. Bornstein & A. E. Kazdin (Eds.), *Handbook of clinical behavior therapy with children.* Homewood, IL: Dorsey.

Wender, P. H. (1971). *Minimal brain dysfunction in children.* New York: Wiley Interscience.

Wender, P. (1974). Some speculations concerning a possible biochemical basis of minimal brain dysfunction. *Life Sciences, 14,* 1605–1621.

Whalen, C. K., & Henker, B. (1976). Psychostimulants and children: A review and analysis. *Psychological Bulletin, 83,* 1113–1130.

Wiederholt, J. L., & Bryant, B. R. (1986). *Gray Oral Reading Test—Revised.* Austin, TX: Pro-Ed.

Willis, D. J., & Culbertson, J. (Eds.). (1990). *Testing young children.* Austin, TX: Pro-Ed.

Willis, D. J., & Holden, E. W. (1990). Etiological factors contributing to deviant development. In J. H. Johnson & J. Goldman (Eds.), *Developmental assessment in clinical child psychology.* New York: Pergaman Press.

Wirt, R. D., Lachar, D., Klinedinst, J. K., & Seat, P. D. (1984). *Multidimensional description of child personality: A manual for the Personality Inventory for Children* (1984 revision by David Lachar). Los Angeles: Western Psychological Services.

Woodcock, R. W. (1987). *Woodcock Reading Mastery Tests—Revised.* Circle Pines, MN: American Guidance Service.

Woodcock, R. W., & Johnson, M. B. (1989). *Woodcock-Johnson Psycho-Educational Battery—Revised.* Allen, TX: DLM Teaching Resources.

Wright, L., Schaefer, A., & Solomons, G. (1979). *Encyclopedia of pediatric psychology.* Baltimore: University Park Press.

Zametkin, A. J., & Rapoport, J. L. (1987). Neurobiology of attention deficit disorder with hyperactivity: Where have we come in 50 years? *Journal of the American Academy of Child and Adolescent Psychiatry, 26,* 676–686.

Zimmerman, I. L., Steiner, V. G., & Pond, R. E. (1967). *Preschool Language Scale* (rev. ed.). Columbus, OH: Charles E. Merrill.

CHAPTER 25

Development of a Program for Sleep Disorders

Rosalind D. Cartwright

Introduction: Definition of Clinical Problems and Patients

Difficulties with sleep are very common in our time. Survey results from several large sample studies (Bixler, Kales, Soldatos, Kales, & Healey, 1979; Karacan, Thornby, & Williams, 1983; Welstein, Dement, Redington, Guilleminault, & Mitler, 1983) show that the rate at which persons report having difficulty getting to sleep, or staying asleep long enough to be rested the following day, on a chronic basis, runs between 20% and 30%. Not all of these persons think of themselves as patients, nor do they complain to their physician. Nonetheless, according to a study by the Institute of Medicine (1979), the sale of over-the-counter sleep aids and of health food products claiming to facilitate sleep is so high as to confirm that there are many who attempt to treat themselves. In increasing numbers, these patients are now finding their way to the new specialized sleep centers.

Common as it is, not getting enough sleep is only one of many types of sleep difficulties

recognized in the field of clinical sleep disorders. The problems are divided roughly into three major groups: trouble with sleeping, trouble with staying awake, and troublesome behaviors intruding into sleep. Within these three clusters of disorders are many subclasses associated with different pathological factors underlying similar symptom pictures. Within these categories, there are also differing distributions of patients by age, by sex, and even by personality patterns.

Disorders of sleep, in other words, need to be understood by clinicians as a broad group of disturbances of the 24-hour rest–activity rhythm. Some of these have primary psychopathology, some have secondary psychopathology, and others have no psychopathology associated with them. These disorders have only recently been recognized as needing specialized attention. A diagnostic nosology has been developed to aid clinicians in identifying their patients' sleep–wake problems (Roffwarg, 1979). This manual, currently under revision, outlines the criteria for ruling in and ruling out disorders falling within four major classes labeled A, B, C, and D (see Table 1).

Within the A category are those patients presenting with difficulty in initiating or maintaining sleep, the so-called DIMS, or insomnias.

Rosalind D. Cartwright • Sleep Disorder Service and Research Center, Rush–Presbyterian–St. Luke's Medical Center, 1653 West Congress Parkway, Chicago, Illinois 60612.

**Table 1. Diagnostic Classification
of Sleep and Arousal Disorders
(Examples Only, Not Complete)**

A. DIMS: Disorders of initiating and maintaining sleep
   1. Psychophysiological
      a. Transient and situational
      b. Persistent
   2. Associated with psychiatric disorders
   3. Associated with the use of drugs and alcohol
   4. Associated with sleep-induced respiratory impairment
   5. Associated with sleep-related nocturnal myoclonus and restless legs
B. DOES: Disorders of excessive somnolence
   1. Associated with sleep-induced respiratory impairment
   2. Narcolepsy
   3. Idiopathic CNS hypersomnolence
   4. Intermittent (periodic) DOES
      a. Kline-Levin syndrome
      b. Menstrual-associated syndrome
   5. Insufficient sleep
C. DOSWS: Disorders of sleep–wake schedule
   1. Jet lag syndrome
   2. Work shift change
   3. Delayed sleep phase syndrome
   4. Advanced sleep phase syndrome
   5. Non-24-hour sleep–wake syndrome
D. DYS: Dysfunctions associated with sleep or partial arousals (parasomnias)
   1. Sleep walking
   2. Sleep terrors
   3. Sleep-related enuresis
   4. Sleep-related gastrointestinal reflux
   5. Sleep-related head banging

The B group consists of disorders of excessive somnolence, or the DOES. The Cs are patients for whom timing of sleep is the main difficulty. These are patients with a disorder of their sleep–wake schedule, or DOSWS. Finally, there is the D group, which consists of patients who have some disordered behavior intruding into sleep. These are called *parasomnias*, or *dyssomnias*, known as DYS.

From a survey of 8,000 patients seen at sleep centers (Coleman, 1983), it has been reported that, among the A group of patients, the most common are those whose difficulty in getting enough sleep is associated with personality disorders or affective disorders. These make up about 35% of patients presenting to the sleep disorder centers complaining that they have difficulty sleeping. The next most common are the so-called psychophysiological DIMS. These patients appear to be hyperaroused on a chronic basis, so that sleep initiation and/or maintenance is a problem for them. About 15% of poor sleepers fall into this class. The third most frequent among the DIMS patients are those whose trouble is secondary to alcohol or drug use, including those who have become tolerant of or who are withdrawing from some prescription sleep medication. These account for about 13% of the DIMS. In another 12%, the poor sleep is found to be related to an underlying sleep disorder called *periodic movements of sleep*, or PMS. This is a rhythmic, persistent muscle twitching, usually of the lower extremities, and sometimes as subtle as a flexion of only the big toe. Even though the movement is slight, it is often sufficient to lighten sleep or to fragment it with many microarousals. A few patients also have awakenings at night because of disturbing dreams, sleep-related pain (such as arthritis), a noisy environment, panic attacks, abnormally aroused EEGs (alpha-delta sleep), or respiratory disorders. The underlying causes of short sleep, or sleep of such poor quality that it leaves the patient unrefreshed, are so varied that it is clear that this group of patients needs to be carefully assessed before treatments are undertaken. This evaluation must include psychological assessment for the true understanding of the DIMS patient.

By and large, it is safe to say that at least half the patients complaining of not getting sufficient sleep to carry on productive waking lives also have significant psychopathology that requires a psychological diagnosis and treatment. In these cases, the sleep problem is a surface manifestation of an underlying inability to handle stress and/or affect appropriately, so that tensions are not discharged during waking in a healthy fashion. Both anxiety and depression are enemies of sleep. Anxiety is usually associated with delayed sleep onset and depression with early sleep offset or early-morning awakening. Both types of patients feel their sleep to be inadequate.

In the remaining half of the DIMS patients, some physical pathology is at the base of the disorder, but they may have developed a conditioned anxiety around their inability to sleep or a secondary depression resulting from a long history of chronic daytime fatigue. These, too, need to be addressed, with an emphasis on correcting the basis of the sleep disruption.

DIMS patients can be complicated and always require a careful multidisciplinary assessment of both their waking and sleep functioning before the undertaking of any treatment program. Unfortunately, most patients who complain are usually put directly on some pharmacological agent. This treats the symptom but may mask the problem rather than correct it.

The B category of patients share the symptom of excessive daytime sleepiness, commonly referred to as *EDS*. This definition may not seem to distinguish them clearly from the A patients, who are tired because of having too little nighttime sleep. The difference here is that the Bs usually have no nighttime sleep complaints at all. Many avow that they "sleep like a rock," yet they have overwhelming sleepiness during their waking activities.

Two diagnoses are associated with daytime sleepiness following a night of sleep; they account for most of these patients: narcolepsy and sleep apnea syndrome. These are by far the commonest problems presenting to the specialized sleep disorder centers for help. In many ways their symptom pictures in the office resemble each other. Both types of patients report difficulty in maintaining wakefulness, particularly under sedentary conditions: driving, watching TV, even eating, and especially while sitting in church or at long meetings. A careful history taken in the office soon separates these two (see Table 2).

Further clarification of the differences between these two can be made by interviewing the spouse. In most instances, sleep apneic patients are not self-referred; they come for treatment at the insistence of the employer or of the bed partner. Employers refer patients because they fall asleep at work, and bed partners because their own sleep is disrupted by the raucous snoring of the patient. When noisy breathing is followed by a cessation of respiration (the apneic pause) lasting more than 10 seconds, followed by a gasping, choking noise as respiration begins again, the diagnosis can be presumed but must be confirmed in the sleep laboratory. The patient is often unaware of this behavior, which results in comprised oxygenation and many microarousals throughout the night. Both of these are implicated in their difficulty in maintaining daytime wakefulness, despite the patients' own impression that they are good sleepers who fall asleep as soon as their heads hit the pillow. The fight for

Table 2. Differentiating Sleep Apnea and Narcolepsy in the Interview

	Narcolepsy	Sleep apnea
Age of onset	Early years 12–22	Middle years 40–50
Sex ratio	Equal	85% male
Associated with		
Structural abnormality of the upper airway	No	Yes
Weight gain	No	Yes
Nocturnal snoring	No	Yes
Hypertension	No	Yes
Cataplexy	Yes	No
Sleep paralysis	Yes	No
Automatic behaviors	Yes	No
Daytime sleep attacks	Short	Long
Psychological factors	Absent	Often present
Marital situation	Often single	Marriage disrupted
Work history	Often in jeopardy	Depends on severity

breath can be described by partners only if they have not already left the shared bed to sleep in peace elsewhere. The narcoleptic, on the other hand, sleeps quietly, although many have sleep interruptions throughout the night.

According to Coleman (1983), narcolepsy makes up 25% and sleep apnea 50% of those who present with symptoms of difficulty in maintaining wakefulness. The remaining patients are a mixture of bipolar patients in the depressive phase of their illness, natural long sleepers trying to fit their lives into an eight-hour sleep schedule, patients with chronic obstructive pulmonary disease, and assorted others.

In the C group are patients suffering from sleep schedule difficulties. These are less common than either the As or the Bs. They fall into two main subgroups: phase-advanced and the phase-delayed sleepers. The phase-advanced are sleepy too early in the evening and wake too early in the morning; the so-called larks or early birds. The phase-delayed patients are the opposite. They are not sleepy until very late at night and are not ready to wake up until very late in the day; these are the so-called sleepyheads or nightowls. Two other diagnoses appear often enough to be mentioned as part of this group: those with non-24-hour sleep–wake patterns and those with jet lag or shift-work-related sleep problems. Both phase-advanced and phase-delayed patients may present as DIMS patients, complaining of inadequate amounts of sleep. The phase-advanced person who is sleepy at about 8:00 p.m. but fights off sleep until a more socially acceptable bedtime of 10:30 or 11:00 p.m. may wake naturally at 4:00 a.m. Such patients need their eight hours to feel rested but get only 5 or 5½ hours. The phase-delayed person who is not sleepy but goes to bed at 10:30 or 11:00 because of a need to be up at 7:00 a.m. and who is unable to fall asleep until the early morning hours also suffers from an inadequate amount of sleep. These two types of patients are easily identified if they allowed to follow their own internal body clock schedule for a week or two while they keep a log of their sleep–wake hours. If they also log their temperatures are regular inter-

vals, the phase-advanced person may show an early-evening temperature drop, and the phase delayed sleeper a late reduction in temperature. Because the sleep–wake rhythm of most people is closer to 25 than 24 hours, young people who habitually stay up later and sleep later in the morning during holiday periods may find that their sleep–wake rhythm has drifted. When school starts, these youngsters find they cannot fall asleep earlier and wake up on time. These individuals need some specialized treatment to help them get back on track.

The last class of disorders, the D group, or the parasomnias, are more common in the very young and in the elderly than they are during the adult years. Young children may have nocturnal enuresis, sleep terrors, sleep-walking episodes, or stereotyped behaviors such as sleep rocking or head banging, which are very upsetting to the household. They themselves usually have no recollection of these episodes and show no evidence of daytime disturbance. These are developmental disorders that rarely persist beyond the teenage years but may recur later in life if the person is under stress. Commonly, these sleep-related behaviors occur during the first third of the night, usually at times of transition: at sleep onset in the case of stereotyped movements, or at the transition from the first deep sleep of the night to more active rapid-eye-movement (REM) sleep for those with night terrors, sleep walking, and some types of enuresis.

In the elderly, there are confusional states that are related to the sleep–wake cycle and are called sundowning and a newly recognized type of REM-sleep-related acting out of violent behaviors in some who have lost the protective muscle paralysis that usually accompanies dreaming sleep (Schneck, Bundlie, Ettinger, & Mahowald, 1986). Parasomnias can be dangerous. The child sleepwalker may get out of the house. The night terror person may crash through a glass window in a panic state. Elderly REM-behavior-disorder patients may attack the sleeping spouse or break their own bones defending themselves against a dreamed intruder. All of these individuals require help.

As this brief introduction shows, sleep disorders can occur in persons of all ages. The first complaint may be of a physical or psychological difficulty rather than of a sleep disturbance. No wonder these patients are not always recognized by either physicians or mental health workers as suffering from a disorder of sleep requiring specialized attention. Many go from physician to physician seeking some help, only to be told to "take a vacation" or "take a sleeping pill" before finding someone with the expertise to address the diagnostic questions appropriately. Such experts are usually to be found in the sleep disorder service of a major medical center, although such services are now being initiated in many small hospitals and even in some free-standing clinics. The important issue for patients seeking help is not the size of the service but whether there is both psychological expertise and medical competence available, and whether they will be fully evaluated before a treatment program is undertaken. Not only is a team approach important, it is even more important that at least one member of the team be accredited to diagnose and treat sleep disorders (i.e., hold a degree as an accredited clinical polysomnographer, or ACP) and that the service itself be accredited by the Association of Sleep Disorder Centers. This accreditation ensures that patients with sleep–wake difficulties will be fully evaluated, that the sleep laboratory test will be properly conducted, and the results will be accurately interpreted.

A Review of the Literature

The literature in this field is divided into two types: (1) the basic studies of normative sleep, including its development from infancy through old age, and the various individual differences that are within normal limits and (2) the literature of clinical sleep disorders. This literature is now vast, and no short review can do it justice. In fact, a growing number of new general textbooks is now available (Anch, Browman, Mitler, & Walsh, 1988; Kryger, Roth, & Dement, 1989; Williams, Karacan, & Moore, 1988) as well as some excellent more specialized works, such

as those of children's sleep disorders (Ferber, 1985; Guilleminault, 1987; Weissbluth, 1984), on sleep in aging (Esmer, Kurtz, & Webb, 1985), on insomnia (Kales & Kales, 1984), and on the epidemiology of these disorders (Guilleminault & Lugarasi, 1983). There is a specialized journal devoted to this literature called, appropriately, *Sleep*, as well as an annual volume of abstracts, *Sleep Research*. What is not well represented in this literature are the more behavioral and psychological aspects involved in the genesis of these problems, their impact on the patient, and their management. In fact, as the field expands clinically to meet the high demand for services from patients, it is becoming increasingly "medicalized." With the influx of medically trained specialists seeking new patients, pulmonologists have become involved with the respiratory disorders of sleep (apnea), neurologists with the sleep seizures and narcoleptic patients, psychiatrists with the insomnias, specialized biological rhythm experts with patients presenting with phase-shifted sleep, and pediatricians with parasomnias. Psychologists have increasingly given ground to these practitioners and have devoted themselves more to research than to practice endeavors. This does not serve the patients well. A most important aspect of this field has been its recognition of the complexity and the interactive impact of sleep disorders on waking functioning and, conversely, of how waking behaviors impact sleep. This approach requires psychological input and an interdisciplinary approach.

The most important literature in this field starts with the Rechtschaffen and Kales (1968) manual for the laboratory recording and scoring of normal human sleep. This manual has now been supplemented by a volume of laboratory procedures edited by Guilleminault (1982) and by the important classification system of these disorders referred to earlier, edited by Roffwarg (1979). These three are the basic technical texts defining laboratory recording and procedures for the reduction of the data needed to diagnose a wide variety of sleep disorders.

The treatment literature is more scattered. For the psychophysiological insomnias, several excellent behavioral programs are available, including the stimulus control program developed by Bootzin and Nicassio (1978) and the sleep restriction method developed by Spielman, Saskin, and Thorpy (1987). Both are aimed at reducing wake time in bed. Both require sensitive management and continued patient contact. For the anxiety- and depression-related insomnias, psychotherapy is often indicated (Kales & Kales, 1984). Here, the cognitive therapy approach may be useful (Beck, Rush, Shaw, & Emery, 1979). For the "conditioned" insomnias, some forms of relaxation training can be effective if the patients are carefully selected (Hauri, 1978).

Among the large group of patients falling into the DOES diagnosis, the sleep apneic patients are the most prevalent. These patients are clearly suffering from a physical problem: an airway that is compromised by the relaxation of sleep. These have now been well described in several review papers (Kurtz & Krieger, 1987; Mishoe, 1987; Strohl, Cherniak, & Gothe, 1986). Their treatment options are many: surgery (Fujita, Conway, Zorick, & Roth, 1981), mechanical devices (Cartwright & Samelson, 1982; Sullivan, Berthon-Jones, & Issa, 1981), and behavioral programs (Cartwright, Lloyd, Lilie, & Kravitz, 1985). The marital problems of apneic patients have only recently been noted (Cartwright & Knight, 1987). There has not yet been enough work done to define the criteria for assigning patients to specific treatment options, although this work is clearly needed (Cartwright, Stefoski, Caldarelli, Kravitz, Knight, Lloyd, & Samelson, 1988).

The research on narcolepsy has established this disorder as a specific disturbance of the REM sleep system (Rechtschaffen & Dement, 1969; Richardson, Carskadon, Flagg, Van Den Hoed, Dement, & Mitler, 1978). The importance of genetic counseling for these patients, the need for spouse support, and the importance of patients' own understanding of good health habits for the control of their daytime symptoms are often noted in this literature. In general, follow-up psychological counseling to supplement the usual pharmacological treatments has been neglected.

Important to the understanding of daytime sleepiness that is not caused by either sleep apnea or narcolepsy is the work of Carskadon and Dement (1981) and Carskadon (1982), who have studied the sleep of adolescents, young adults, and the elderly under natural and experimentally controlled sleep restriction schedules. They have mapped the effects of shortened sleep on napping behavior the following day. This work shows clearly that voluntary restriction of noctural sleep increases daytime sleepiness markedly. This has not been well recognized by patients. The restriction of sleep hours is better tolerated by adolescents than it is later in life. Daytime sleepiness resulting from voluntary sleep restriction is diagnosed as insufficient sleep syndrome, a funny title for a serious and widespread problem.

The literature on sleep schedule difficulties is not extensive. Perhaps the best of this work is that by Weitzman, Czeisler, Coleman, Spielman, Zimmerman, and Dement (1981), which defines the problem of phase delay as one of chronobiology and a treatment program of chronotherapy to correct it.

In the area of the parasomnias, night terrors, sleep walking, and noctural enuresis have been modeled as disorders of arousal (Broughton, 1968). These are related to an unusually deep sleep in the first or second cycle, leading to difficulty in accomplishing a normal shift to the more aroused REM sleep without some behavioral arousal being triggered. This model has now been supplemented by one in which abnormal behaviors do not abort REM but actually occur in REM sleep itself (Schneck *et al.*, 1986; Schneck, Bundlie, Patterson, & Mahowald, 1987) as true enactments of dreaming.

Methods of Psychological Assessment

By now, it is clear that a good psychological assessment is essential for patients presenting to sleep disorder services. There is considerable pressure for such services to act as laboratories to which the patient is referred for a night of sleep recording (polysomnography) without any prior assessment by a sleep expert. This is seen as saving money for the HMOs and private insurance companies, but in fact it may prove more expensive in the long run. A good assessment before a night study may indicate that the more expensive laboratory test is not necessary, or that some additional monitor should be added at the time of the recording to clarify the diagnosis.

A sleep disorder assessment requires some data generated by the patient at home before an office visit and a good diagnostic interview by a well-trained interviewer. The recommended assessment data include a sleep log for at least two weeks, in which patients record their pre-sleep mood, all medications taken before sleep, lights-out time, time to fall asleep, number of awakenings, wake time, and postsleep mood. Many laboratories use a 24-hour log covering daytime nap episodes as well as the nocturnal sleep period. These are usually a little more complicated for the patient to understand and so are best given in the office, where the instructions can be spelled out. The logs are then returned at the next visit or by mail. In addition to the log of sleep–wake times, most services ask patients to fill out a questionnaire describing their sleep–wake behavior, covering such items as the type of events that disturb their sleep (e.g., the need to urinate, pain, coughing or choking, heartburn, or reflux, noise, light, heat, snoring by partner or by self, nightmares, or cramping or twitching of muscles). These data help to focus the clinical interview on factors that the patient has noted.

To fill out the picture of the patient, it is also important to administer a short psychological test battery. Sometimes, this is sent to the patient in advance, along with the sleep log and the questionnaire. At other times, it is completed at the time of the first appointment. The usual battery consists of the Minnesota Multiphasic Personality Inventory (MMPI), the Trait/State Anxiety Scale, a sentence completion test, and a schedule of recent events. This battery gives an overview of how depressed or anxious the patient may be, whether psychotherapy appears to be indicated, and how well it may be accepted. If the patient's profile shows high levels of cognitive confusion or a propensity for angry acting-out behaviors, it is better to know it before designing a treatment program.

The most important part of the prelaboratory assessment is the clinical interview, which should follow up on any of the items identified in the self-report data. Is the patient a Sunday-night-only insomniac, or is the insomnia consistent across all nights? Does the patient report restless legs before sleep and many sleep interruptions in the log? Does the patient report no trouble with nocturnal sleep, but many episodes of daytime drowsiness along with a weight gain and dry mouth on awakening? The clinical interviewer must be knowledgeable about all the signs of the various sleep disorders and must be alert in pursuing the symptoms of each that he or she suspects may account for the symptom picture. For DIMS patients, this includes a history of the difficulty; any precipitating events associated with the original episode, as well as subsequent exacerbations; what treatments have been tried in the past; the current health status; the use of alcohol, caffeine, nicotine, and other known sleep disrupters; the present stressors in the patient's life, including her or his own health; and the status of the patient's relationships, employment, and finances. The attempt here is to separate those DIMS patients who are unable to relax into sleep because of high levels of psychological tension, or an affective disorder, from those who are phase-delayed; have an organic problem underlying their sleep disturbance, such as periodic movements in sleep (PMS); have an occult sleep apnea that causes them to arouse frequently because of inadequate respiration; or use or withdrawal from substances that disrupt sleep.

In general, the knowledge about sleep is so poor both among in the public and among internists that misdiagnoses are frequent. A narcoleptic who suffers disturbed nocturnal sleep can be confused with an insomniac and may be treated inappropriately if not seen by a knowledgeable person who traces the early history of daytime sleepiness and the other symptoms related to this disorder.

The best training in these diagnostic skills

comes from a thorough familiarity with the roster of sleep disorders, which is usually gained only by working with such patients on a sleep disorder service. The best test of the level of competency in this task is passing the board examination and being awarded the ACP degree.

Once the initial assessment is done in the office, the next step is usually a laboratory evaluation. This consists of one or two nights of recording for seven hours, during the patient's normal sleep time. The specific parameters measured depend on the presumptive diagnosis, although some leads are standard for all patients: at least two channels of EEG—one for eye movements (EOG), and one for chin muscle (EMG)—and one EKG are always recorded. The other parameters—nasal and oral airflow, thoracic and abdominal impedance, and ear oximetry—are needed to diagnose sleep apnea. The EMG of the anterior tibialis muscles is recorded to look for periodic movements of sleep. The EMG of the masseter muscles is recorded for bruxism and so on. The all-night recording may be followed by a series of daytime naps. This test is always done if narcolepsy is the suspected diagnosis, and it may be done to estimate the degree of daytime sleepiness following a night of sleep for those patients who report difficulty in remaining awake while driving.

The montage for diagnosing each sleep disorder has been set out in the laboratory procedures text edited by Guilleminault (1982). It goes without saying that getting a good recording requires good technical assistants who not only are capable of producing artifact-free records but can interact well with patients, explain the procedures clearly, put patients at their ease, and be responsive to their needs for assistance at night. These assistants must also be alert to major medical problems, must be trained in CPR, and must be aware when to call for medical assistance. Sleep patients may have cardiac arrhythmias, sleep seizures, confusional episodes, or other problems that are sleep-induced. Technicians who are in charge must be very knowledgeable about patient management during the long recording period. Each 30-second epoch of the resulting polysomnogram, or PSG, must then be scored both for sleep stages and for clinical events before a diagnosis is made and its severity estimated. Only then should an intervention be planned.

Clinical Interventions

Treating the DIMS

For the DIMS patient, the intervention depends on the diagnosis of the underlying cause. These treatments are usually behavioral or pharmacological or both. For those patients whose poor sleep is associated with a stressful event that is transient, the intervention may consist of support only, with reassurance that good sleep will return. If daytime functioning is severely impaired by lack of sleep, some medication may be recommended on a short-term basis only. The rule of thumb for sleep medications is to pick one that is short-acting, such as Halcion or Restoril, and that will not interfere with alertness during waking hours, and to phase it out after no more than three weeks.

For patients who have long-standing difficulty in falling asleep, but who can sleep through the night once sleep has been initiated, it is important first to rule out an affective or anxiety disorder, or a phase delay of sleep. The first can be estimated from the MMPI and the clinical interview, and it can also be identified from a sleep recording showing an abnormally early first REM period (see Figures 1 and 2). If the patient appears to be suffering from a major affective disorder, this, of course, should be treated either by an antidepressant or psychotherapy or both. If the patient has a heightened level of anxiety, it, too, may be treated with an antianxiety medication, psychotherapy, relaxation training, or all three.

If the patient's sleep is phase-delayed, and sleep onset is prolonged, but a normal length of sleep can be sustained on weekends or holidays when the usual wake-up time can be extended, then the onset of sleep can be adjusted

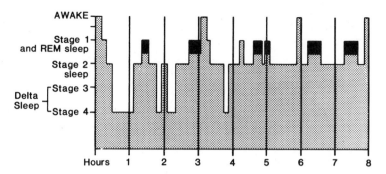

Figure 1. Typical sleep pattern of a young human adult. Stage 1 sleep and REM sleep are graphed on the same level because their EEG patterns are very similar.

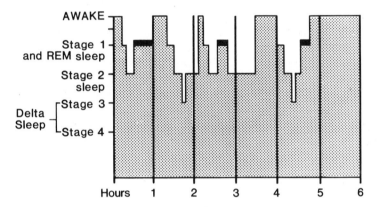

Figure 2. Typical sleep in depression.

to occur earlier. There are two main treatment approaches. The first is recycling the sleep itself through a program of delaying bedtime by two hours each night until the desired onset time has been reached. This program takes strong motivation and about 10 days to accomplish (see Figure 3).

The second method is to use a standard wake-up time regardless of how much sleep has been obtained and to have patients expose themselves to very bright, full-spectrum light or sunlight for at least an hour at this time. This early-morning light treatment helps to reset the circadian clock to an earlier awake time. Special lights are available commercially on a rental basis for this purpose. Those patients who are neither phase-delayed nor suffering from an affective or anxiety disorder, but who have dif-

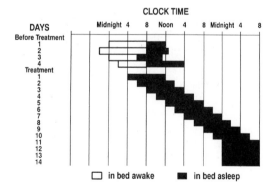

Figure 3. An example of chronotherapy.

ficulty obtaining enough sleep, may be helped by any one of a variety of other interventions (e.g., taking a 20-minute hot bath two hours before bedtime) (Horne & Shackell, 1987; Sewitch, 1987). This bath artificially raises the core body temperature, and the hour-and-a-half interval after it allows time for the temperature to fall before sleep time. This reduction in temperature helps to deepen and prolong natural sleep. Although it is controversial in the literature, L-tryptophan, a natural amino acid, taken in amounts of 1–4 grams before bedtime, has proved effective in shortening sleep onset time for some patients (Hartmann, Lindsley, & Spinweber, 1983). Currently this has been withdrawn from the market while the Center for Disease Control investigates the source of a contaminant introduced in the supply from one manufacturer.

Strict sleep scheduling (always maintaining the same bedtime and rise time) also helps the sleep–wake rhythm to become more clearly established. Relaxation training, especially of the type using muscle stretching and deep breathing, has proved useful for those insomnia patients with racing thoughts that interfere with sleep initiation. Given that many DIMS patients are found to have elevations in the 2, 4, and 7 scales of the MMPI (Kales & Kales, 1984) and may be described as internalizers of tensions, many profit from a psychotherapeutic approach. This should be aimed at helping the patient to learn better techniques for dealing with tensions and at increasing their sense of self-efficacy. Many need to learn that they can survive if they do express their feelings.

Sleep maintenance is a somewhat more difficult problem to manage without medication. First, it must be established whether there is any underlying cause of the nocturnal arousals, such as periodic movements in sleep (PMS), or nightmaring. This can be done only with a sleep recording. For PMS, this recording must include an EMG of both right and left legs. If this diagnosis is sustained by a recording showing rhythmic bursts of muscle activity that disrupt sleep at a rate of more than five per hour, this activity may be treated either by a muscle relaxant such as Klonopin or by an electrical

stimulator (patterned after a TENS unit, which fatigues these muscles prior to sleep). For nightmares, REM sleep interruptions to obtain dream reports should be followed by psychotherapy to address the issues uncovered.

If no underlying pathology is found, frequent awakenings can be approached by the stimulus control method of Bootzin and Nicassio (1978) or by Spielman, Saskin, and Thorpy's sleep restriction method (1987). The first requires the patient to get out of bed if not asleep within 20 minutes but to stay up no longer than 20 minutes before trying again. This program ensures that the bed will not be associated with lying awake for long periods of time. The problem is that many patients lose patience with getting up and down as often as they must before gaining enough control over their sleep to benefit. If they work at it, patients can become sufficiently fatigued to sleep better. The sleep restriction method is easier for most patients to accomplish successfully, This requires that they set their wake-up time at a desired hour and limit their bedtime to the average number of hours actually slept for the past two weeks. This may be very few for some severe insomniacs. If, for example, the sleep log shows an average of only four hours of sleep and if the rise time must be 6:00 A.M., the patient may not go to bed until 2:00 A.M. and must stick to this schedule for as long as it takes her or him to sleep through without interruption. The patient is then allowed to increase the time spent in bed very gradually, 15 minutes a week, but only as long as she or he continues to sleep through. This program can take a few months to accomplish, but it is often successful in giving patients a normal sleep span again.

Another major cause of sleep disturbance is alcohol or drug addiction or the withdrawal from these substances. These patients may need a carefully planned reduction of their medication and the use of some more benign substance for an interim, along with a good deal of support from their sleep clinician.

In general, DIMS patients need careful psychological diagnosis and treatment, as the greatest proportion of their problems are associated with some psychopathology, poor sleep habits, or behavioral problems, such as substance abuse.

Treating the DOES

The problems in this diagnostic group, as a whole, are often due to basic sleep pathology, psychological difficulties being secondary. The most common DOES diagnosis is sleep apnea syndrome. The interventions available for this sleep-related breathing disorder vary widely. There are various types of surgery, such as submucous resection (SMR) which opens the nasal passages surgically; uvulopalatopharyngoplasty (UPPP), which widens the opening of the oropharynx; tonsillectomy; advancement of the mandible; and tracheostomy. All are performed in an effort to eliminate this problem. The difficulty with these surgeries, other than tracheostomy, is that they are often unsuccessful in preventing the continued collapse of the upper airway. More often, mechanical devices are used to hold the airway open. Nasal continuous positive airway pressure, or nCPAP (Sul-

livan *et al.*, 1981), has become a common intervention for those with severe apnea (see Figure 4). The nose mask and hose are attached to a compressor. The pressure level is adjusted in the sleep laboratory so that it is high enough to prevent airway collapse. Although this is a successful treatment for many patients, compliance with it as a long-term treatment is difficult to maintain. Patients need to be encouraged during this treatment to undertake weight loss through dietary changes and exercise so that they may reduce the severity of their sleep disorder or even eliminate the problem altogether.

Another device that can be helpful for some apneic patients is called the tongue-retaining device, or TRD (Cartwright & Samelson, 1982). This device, worn during sleep, holds the tongue in a forward position by suction. It prevents the retrolapse of the tongue and the resulting blockage of the airway that commonly occurs when the patient sleeps in the supine position. This device is much smaller and easier to tolerate on a long-term basis than the nCPAP. However, it is appropriate only for those with a patent na-

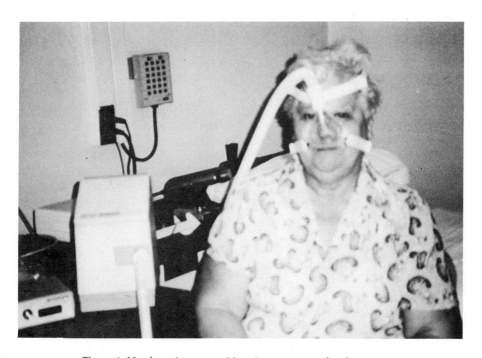

Figure 4. Nasal continuous positive airway pressure for sleep apnea.

sal airway and a clear increase in the rate of apneic pauses when they sleep in the supine position.

The observation that 60% of all sleep apneic patients have significantly more snoring and apneic events when they are supine than when they sleep in the lateral position has led to the development of other mechanical devices. One simple behavioral intervention of this kind is a position alarm worn on the chest, which sounds whenever the patient remains in the supine position for more than 15 seconds (see Figure 5). Another is a T-shirt with a pocket on the back to hold tennis balls (see Figure 6). These sleep position trainers are especially useful for those patients whose rate of apnea during lateral sleep time is at or near normal limits. They need to be coupled with patient education in good health habits to prevent any progressive worsening of the apnea. These habits include a weight loss program for those who are significantly above ideal body weight, exercise, and alcohol restriction.

Aside from direct attention to the apnea, these patients often need marital counseling. They are so noisy at night and sleepy during the day that many spouses become resentful. They feel that they have no real partner during the day or peaceful night sleep for themselves. A support society for these patients (the Sleep Apnea Support Society) has recently been established and can be of great benefit to both partners.

The second major cause of daytime sleepiness, narcolepsy, is most often treated with a daytime stimulant medication or a combination of such an agent (usually Cylert or Ritalin) along with a medication to control the cataplexy or loss of muscle tone associated with the expression of strong emotion experienced by many of these patients. Here, the clinical psychologist has a role in helping the patient learn to live with his or her disorder. Most narcoleptics can have their most troublesome symptoms controlled with medication, but often, there is a long delay before the correct diagnosis is given and the appropriate treatment is started. Because this disorder usually appears in teenage

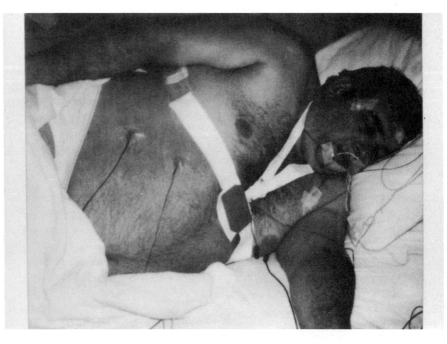

Figure 5. Position monitor and alarm.

Figure 6. T-shirt with tennis balls for sleep apnea.

years, a youngster's life can be severely affected by his or her inability to maintain wakefulness in school and/or failure to maintain body tonus when excited. Many narcoleptic children have been accused of being on drugs, of being lazy, or even of being mentally retarded. They are often discriminated against by peers and disciplined by authorities for behavior which is not under their control. The potential damage to self-image and social relations is real and needs psychological attention.

Treating the DYS

The sleep-related behavior disorders nocturnal enuresis, bruxism, sleep walking, night terrors, head banging or body rocking, and nightmaring are all more common in young children than in adults. For the most part, these behaviors take care of themselves as the child matures. Some of those that are severe and persistent can be approached behaviorally. The bell-and-blanket routine for enuresis is well known. A similar approach has been devel-

oped for bruxism, or tooth grinding. This involves monitoring the masseter muscles during sleep. When grinding begins, an alarm sounds. Some patients with sleep walking, night terrors, and stereotyped movements in sleep have been helped by hypnosis. Nightmaring in children and adults often yields to dream therapy, in which the patient works out an understanding of the fear and gets practice in handling it more directly during waking.

Developing and Maintaining a Sleep Service

Sleep disorder services draw their patients from many sources: primary-care physicians who are baffled by a patient's sleep symptoms; psychiatrists who have been treating patients with benzodizepines for sleep and find them now tolerant of these medications; otolaryngologists who suspect a sleep apnea due to some obstruction in the upper airway; neurologists whose patient's spouse or parents report that the patient makes strange movements in sleep; and pediatricians when a family is disturbed by a child's night terrors or his or her inability to wake in time for school. But most of all, the referrals stem from the patients themselves once they learn that a sleep disorder service is now available. It is important for professionals in this field to reach out to inform their community of their availability by accepting the many offers to appear on local radio stations or TV news spots to discuss the signs and symptoms of sleep troubles. Physicians also need to be educated. Psychologists at medical centers should take the opportunity to lecture to medical students, to offer to give grand rounds, and to open up their service for medical students to do elective clerkship rotations—in short, to use all avenues to communicate about these disorders and their services.

It is also important for a new service to "give," as well as to "get," referrals in order to maintain the flow of patients. Those who come directly to the service will need consultation with other health care professionals, such as dieticians for the obese apneic; pulmonary function testing is also appropriate for these patients. Many require the help of a cardiologist to manage the hypertension and cardiac arrhythmias associated with severe sleep apnea. An examination by a good ear, nose, and throat (ENT) specialist is also recommended. Patients who are referred by a physician need to have their doctor kept up to date and consulted about their further evaluation at each point. Nothing cuts off patient flow faster than "patient stealing." The patient's own physician needs to be included on the health care team. Copies of all results from the laboratory should be sent to her or him along with an explanation of these results and the suggested treatments discussed.

Within a sleep disorder service, there needs to be a multidisciplinary group that meets regularly to discuss the incoming patients and their evaluation and care. There should be on the team at least one mental health expert and one physical health expert who are well versed in sleep disorders. Others who should be invited to become consultants are pharmacologists, pediatricians, pulmonary physicians, psychiatrists, neurologists, ENT specialists, and nursing personnel. In a teaching hospital, there may also be psychology interns rotating through the service and participating fully.

Follow-up of patients is also important. Patients should be given full feedback on their laboratory test findings. For those who are sent home on a mechanical device such as the nasal CPAP, inquiry must be made into whether the pressure needs to be raised or lowered to maintain control of the apnea or whether they can be withdrawn from this treatment altogether when sufficient weight has been lost. Patients put on medication may become tolerant and need to be followed for this tolerance and problems with side effects.

Because sleep disorder services are outpatient in nature, patients need to be informed of the importance of clearing with their health care carrier concerning their coverage for this type of service. Most third-party payers cover 80% of the charges associated with a sleep evaluation but may not cover psychotherapy, for example. Running a laboratory is expensive. The acquisition and maintenance of equipment, supplies, and personnel make this a high-cost service. Starting a new service requires an investment of approximately $35,000 per bed in equipment alone. Personnel costs are high because the testing and sleep record scoring are very labor-intensive. Computerization is coming to this field but is not yet reliable.

Clinical Case Examples

G.B.: A Sleep Apnea Patient

Presenting complaints: G.B., a 38-year-old married man, consulted the sleep disorder service in March 1987, complaining of nocturnal snoring, which had become progressively louder; disrupted sleep; flailing of the arms; and sleep talking. He was also falling asleep at dinner parties, on airplanes, at the wheel of a car, and as a passenger. He had no exercise activity and had had a steady weight gain over the last 10 years from 200 to 269 pounds. He came to the interview with a very concerned wife, who reported that the patient's father had recently died and had weighed over 500 pounds. The patient felt that he was genetically doomed to be obese and that his business entertaining at lunches and dinners required a diet of rich foods and alcohol.

Assessment: Psychologically, G.B. was functioning well. His MMPI showed him to have strong defenses, to be rather uninsightful, and to be socially poised, as befitted his profession of investment counselor. Only his Scales 3 and 5 peaked above 70. His marriage was a good one, and he drew strength from it. In the sleep laboratory, he slept only 41% of his 7 hours of bedtime. He had a total of 46 awakenings. Ninty-four percent of his sleep consisted of EEG Stages 1 and 2 (the lightest sleep). During this period, he had 263 episodes of obstructive sleep apnea. The average duration of these was 23 seconds. His rate of respiratory distress was 180 events per hour. Clearly, he was sleepy in the daytime because of his nightly struggle to breathe.

Treatment: Following his follow-up interview, when the treatment options were discussed, G.B. returned to the laboratory for a trial of nasal continuous positive airway pressure (nCPAP). At a pressure of 10cm/H_2O, his apnea rate dropped from 180 to 25 per hour. This equipment was ordered for him on a rental basis on the contingency that he begin a weight loss program.

After three months, his wife called to report that he had not yet started the recommended program. He was contacted and reminded that the nCPAP was a crutch, not a cure. The cure was up to him. Eight months later, he returned to the laboratory and weighed in at 218, a 50-pound loss. He had taken up a nightly walking program with his wife and a change in food habits.

When G.B. was reevaluated without the nCPAP to see if his apnea count was now reduced, he had no apnea. He slept 70% of the seven-hour night, and for the first time in the laboratory, he had delta (deep sleep). He was given feedback that he could now give up the nCPAP equipment. He was skeptical and decided to come back six months later to be sure he had remained apnea-free. In June 1988, he had another sleep test. He had continued to lose weight in the interim, a total of 81 pounds. He remained apnea-free, with normal sleep architecture, no sleep flailing, no sleep talking, no daytime sleepiness, and no snoring. The nCPAP had served its purpose. It kept him breathing at night and active during the day while he changed his health habits: no business dinners, and salads only for lunch. His wife was

very proud of him and continued to support his changed lifestyle.

M.A.: An Insomnia Patient

Presenting complaints: M.A. is a 58-year-old married man with a long history of poor sleep, which he considered psychogenic and for which he had undertaken five years of psychoanalysis. He explained his poor sleep as being due to his nightly excessive ruminations concerning the health problems of his family. His first wife had died young, and the second had had critical illnesses. He also admitted to dissatisfactions with himself at midlife.

Assessment: The patient's MMPI showed a chronic depressive picture. He appeared to fear that he was in a rut in his career and to be dissatisfied with a poor sex life in the marriage. He felt lonely and used repression, denial, and somatization as defenses.

Treatment: The patient was given relaxation training, thought-stopping techniques, and the Bootzin stimulus control instructions (no more than 20 minutes in bed and, if not sleeping, getting up for no more than 20 minutes before returning to bed). In addition, the patient was seen every other week for five sessions of marital therapy, during which he expressed to his wife his needs for more sexual closeness. He continued to log his sleep, which improved steadily. Originally, he would take three to four hours to fall asleep. This time was reduced to one half or one hour without the use of medication. After nine weeks, he declared he was now satisfied with his sleep and terminated treatment. The marriage and sexual relationship improved, the wife having learned to be more responsive to him and the patient having learned how to help her feel this way.

Summary

Patients with disorders of sleep represent a new and growing opportunity for patient care. Their problems often involve both psychological and medical components. They require a team approach to their evaluation and management. Programs can be based in any one of a number of hospital departments but require a psychologist as a full partner. Sleep is an area of health influenced so heavily by emotional stress, and by the patient's health behaviors, that many kinds of psychological knowledge need to be brought to bear on the diagnosis and management of these problems. The psychologist brings his or her special viewpoint of seeing symptoms in the context of the patient's whole life and uses this viewpoint to help clarify the issues and design treatments that patients can use productively to gain more restorative sleep.

References

Anch, M., Browman, C., Mitler, M., & Walsh, J. (1988). *Sleep: A scientific perspective.* Englewood Cliffs, NJ: Prentice-Hall.

Beck, A., Rush, J., Shaw, B., & Emery, G. (1979). *Cognitive therapy of depression.* New York: Guilford Press.

Bixler, E., Kales, A., Soldatos, C., Kales, A., & Healey, S. (1979). Prevalence of sleep disorders in Los Angeles metropolitan area. *American Journal of Psychiatry, 136,* 1257–1262.

Bootzin, R., & Nicassio, P. (1978). Behavioral treatments for insomnia. In M. Hersen, R. Eissler, & P. Miller (Eds.), *Progress in behavior modification, Vol. 6,* (pp. 1–45). New York: Academic Press.

Broughton, R. (1968). Sleep disorders: Disorders of arousal? *Science, 159,* 1070–1078.

Carskadon, M. (1982). The second decade. In C. Guilleminault (Ed.), *Sleeping and waking disorders: Indications and techniques* (pp. 99–125). Menlo Park, CA: Addison-Wesley.

Carskadon, M., & Dement, W. (1981). Cumulative effects of sleep restriction on daytime sleepiness. *Psychophysiology, 18,* 107–113.

Cartwright, R., & Knight, S. (1987). Silent partners: The wives of sleep apneic patients. *Sleep, 10,* 244–248.

Cartwright, R., & Samelson, C. (1982). The effect of a nonsurgical treatment for sleep apnea. *Journal of the American Medical Association, 248,* 705–709.

Cartwright, R., Lloyd, S., Lilie, J., & Kravitz, H. (1985). Sleep position training as treatment for sleep apnea syndrome. *Sleep, 8,* 87–94.

Cartwright, R., Stefoski, D., Caldarelli, D., Kravitz, H., Knight, S., Lloyd, S., & Samelson, C. (1988). Toward a treatment logic for sleep apnea: The place of the tongue retaining device. *Behavior Research and Therapy, 26,* 121–126.

Coleman, R. (Chairman). (1983). Diagnosis, treatment, and follow-up of about 8,000 sleep/wake disorder patients. In C. Guilleminault & E. Lugaresi (Eds.), *Sleep/wake disorders: Natural history, epidemology, and long term evolution.* New York: Raven Press.

Esmer, W., Kurtz, D., & Webb, W. (Eds.). (1985). *Sleep, aging, and related disorders.* Basel: Karger.

Ferber, R. (1985). *Solve your child's sleep problems.* New York: Simon & Schuster.

Fujita, S., Conway, W., Zorick, F., & Roth, T. (1981). Surgical correction of anatomic abnormalities in obstructive sleep apnea: uvulopalatopharyngoplasty. *Otolaryngology and Head Neck Surgery, 89,* 923–934.

Guilleminault, C. (Ed.). (1982). *Sleeping and waking disorders: Indications and techniques.* Menlo Park, CA: Addison-Wesley.

Guilleminault, C. (Ed.). (1987). *Sleep and its disorders in children.* New York: Raven Press.

Guilleminault, C., & Lugaresi, E. (Eds.). (1983). *Sleep/wake disorders: Natural history, epidemiology, and long-term evolution.* New York: Raven Press.

Hartmann, E., Lindsley, J., & Spinweber, C. (1983). Chronic insomnia: Effects of tryptophan, flurazepam, secobarbital and placebo. *Psychopharmacology, 80,* 138–142.

Hauri, P. (1978). Biofeedback techniques in the treatment of chronic insomnia. In R. Williams & I. Karacan (Eds.). *Sleep disorders: Diagnosis and treatment* (pp. 145–159). New York: Wiley.

Horne, J., & Shackell, B. (1987). Slow wave sleep elevations after body heating: Proximity to sleep and effects of aspirin. *Sleep, 10,* 383–392.

Institute of Medicine. (1979). *Sleeping pills, insomnia, and medical practice.* Washington, DC, National Academy of Sciences.

Kales, A., & Kales, J. (1984). *Evaluation and treatment of insomnia.* New York: Oxford University Press.

Karacan, I., Thornby, J., & Williams, R. (1983). Sleep disturbances: A community survey. In C. Guilleminault & E. Lugaresi (Eds.), *Sleep/wake disorders: Natural history, epidemiology, and long-term evolution* (pp. 37–59). New York: Raven Press.

Kryger, M., Roth, T., & Dement, W. (Eds.). (1989). *Principles and practice of sleep medicine.* Philadelphia: Saunders.

Kurtz, D., & Krieger, J. (1987). Review of the therapeutic approaches to sleep apnea. In J. Peter, T. Podszus, & P. von Wichert (Eds.), *Sleep related disorders and internal diseases* (pp. 346–359). New York: Springer-Verlag.

Mishoe, S. (1987). The diagnosis and treatment of sleep apnea syndrome, *Respiratory Care, 32,* 183–200.

Rechtschaffen, A., & Dement, W. (1969). Narcolepsy and hypersomnia. In A. Kales (Ed.), *Sleep: Physiology and pathology* (pp. 119–130). Philadelphia: J. B. Lippincott.

Rechtschaffen, A., & Kales, A. (Eds.). (1968). *A manual of standardized terminology, techniques and scoring system for sleep stages of human subjects.* Los Angeles: UCLA Brain Information Service/Brain Research Institute.

Richardson, G., Carskadon, M., Flagg, W. Van den Hoed, J., Dement, W., & Mitler, M. (1978). Excessive daytime sleepiness in man: multiple sleep latency measurement in narcoleptic and control subjects. *Electroencephalography and Clinical Neurophysiology, 45,* 621–627.

Roffwarg, H. (Ed.). (1979). Diagnostic classification of sleep and arousal disorders. *Sleep, 2,* 1–137.

Schneck, C., Bundlie, S., Ettinger, M., & Mahowald, M. (1986). Chronic behavioral disorders of human REM sleep: A new category of parasomnia. *Sleep, 9,* 293–308.

Schneck, C., Bundlie, S., Patterson, A., & Mahowald, M. (1987). Rapid eye movement sleep behavior disorder. *Journal of the American Medical Association, 257,* 1786–1789.

Sewitch, D. (1987) Slow wave sleep deficiency insomnia: A problem in thermo-down regulation at sleep onset. *Psychophysiology, 24*, 200–215.

Spielman, A., Saskin, P., & Thorpy, M. (1987). Treatment of chronic insomnia by restriction of time in bed. *Sleep, 10*, 45–56.

Strohl, K., Cherniack, N., & Gothe, B. (1986). Physiologic basis of therapy for sleep apnea. *Review of Respiratory Diseases, 134*, 791–802.

Sullivan, C., Berthon-Jones, M., & Issa, F. (1981). Nocturnal nasal airway pressure for sleep apnea. *New England Journal of Medicine, 309*, 112.

Weissbluth, M. (1984). *Crybabies: Coping with colic*. New York: Berkley Books.

Weitzman, E., Czeisler, C., Coleman, R., Spielman, A., Zimmerman, J., & Dement, W. (1981). Delayed sleep phase syndrome. *Archives of General Psychiatry, 38*, 737–746.

Welstein, L., Dement, W., Redington, D., Guilleminault, C., & Mitler, M. (1983). Insomnia in the San Francisco Bay Area: A telephone survey. In C. Guilleminault & E. Lugaresi (Eds.), *Sleep/wake disorders: Natural history, epidemiology, and long-term evaluation* (pp. 73–85). New York: Raven Press.

Williams, R., Karacan, I., & Moore, C. (Eds.). (1988). *Sleep disorders: Diagnosis and treatment* (2nd. ed.), New York: Wiley.

CHAPTER 26

Diabetes

Clinical Issues and Management

Daniel J. Cox, Linda Gonder-Frederick, and J. Terry Saunders

Introduction to Diabetes and Program Development

Mechanisms and Etiology

The term *diabetes* actually embraces a number of different disease processes characterized by chronically high blood glucose (BG) levels (Cox, Gonder-Frederick, Pohl, & Pennebaker, 1986). In the nondiabetic person, glucose metabolism is largely controlled by a feedback system between BG concentration and the beta cells of the pancreas that produce and release the hormone insulin. Insulin secretion lowers BG concentration by (1) allowing glucose to enter and be utilized by cells; (2) promoting glucose storage; and (3) inhibiting the release of stored glucose. This finely tuned system prevents BG levels from becoming abnormally high (hyperglycemia) or abnormally low (hypoglycemia), maintaining a relatively narrow homeostatic range between approximately 60 and 150 mg/dl.

Daniel J. Cox, Linda Gonder-Frederick, and J. Terry Saunders • Behavioral Diabetes Research, Department of Behavioral Medicine and Psychiatry, University of Virginia Health Sciences Center, Charlottesville, Virginia 22901.

Diabetic hyperglycemia is caused by abnormalities in insulin production or utilization. Table 1 summarizes diabetic abnormalities in BG as well as underlying mechanisms. Because glucose is not properly metabolized, it accumulates in the bloodstream. There are two major types of diabetes: Type I and Type II (Drash, 1984; Sperling, 1988). Type I diabetes, also called *insulin-dependent diabetes*, can occur at any age but typically begins before the age of 30, the onset being most likely between ages 5–6 and 10–12. Genetic, autoimmune, and infection factors appear to be involved in the development of Type I diabetes. The most popular formulation (Craighead, 1978) is that a genetically predisposed individual encounters a stressful condition, like a viral infection that triggers an autoimmune response that destroys pancreatic beta cells, and the result is an inability to produce adequate insulin. After the onset of Type I diabetes, patients must take insulin injections in order to survive.

At the time of diagnosis, most Type I patients are severely hyperglycemic and suffering from dehydration and ketoacidosis. Dehydration and the classic symptoms of frequent urination and increased thirst are a result of the kidneys' attempts to eliminate excess glucose through

Table 1. BG Fluctuations in Insulin-Dependent Diabetes Mellitus

BG range	BE level (mg/dl)	Causes
Hyperglycemia		
Diabetic ketoacidosis (coma)	1200	Insufficient insulin
		Insulin resistance
Severe hyperglycemia	400	Excess food intake
		Fever or infection
Moderate hyperglycemia	250	Stress or emotional changes
		Unknown factors
Mild hyperglycemia	160	
Euglycemia		
Normal postprandial	140	Normal glucose metabolism
Normal fasting	80	
Hypoglycemia		
Mild hypoglycemia	70	Excessive insulin
		Insufficient food intake
Moderate hypoglycemia	60	Strenuous exercise
		Stress or emotional changes
Severe hypoglycemia	50	Unknown factors
Insulin shock (coma)	20	

the urine. Ketoacidotic comas are caused by decreased blood pH, which is a result of increased levels of ketones, a by-product of body fat breakdown when glucose is not available to the cells.

Type II diabetes, known as *non-insulin-dependent diabetes* (NIDDM), is also characterized by chronic hyperglycemia, but the onset typically occurs after age 40 (Drash, 1984). It is the most common chronic disease in the U.S. Unlike in Type I, Type II symptoms typically emerge very gradually; therefore, the illness is often undetected. As in Type I diabetes, abnormalities in insulin action are involved, but insulin resistance rather than insufficient insulin production is the basic mechanism (Olefsky, 1976). There is a strong genetic component, with nearly a 100% concordance rate between identical twins. Obesity and minimal exercise also appear to contribute to insulin resistance. Approximately 90% of Type II patients are obese, and it is estimated that 90% of them could be treated with diet and exercise alone (Newburgh & Conn, 1979). However, most take oral medications or insulin injections to lower BG.

The personal and societal impact of diabetes is reflected in the number of people affected and the effects of the disease. In the U.S., there are 5–6 million diagnosed cases of diabetes involving 5% of the population. The vast majority (90%) of these are Type II cases. It is estimated that another 4–5 million undiagnosed cases exist. Because of the long-term complications associated with diabetes, life expectancy after diagnosis is thought to be reduced by an average of 30%.

Long-Term Complications

Both Type I and Type II diabetes are associated with long-term complications. Diabetes can adversely affect any part of the body, but the most common effects are seen in the eyes, the kidneys, the feet, the blood vessels, and the nervous system. There is growing evidence that chronic hyperglycemia plays a causal role in the development of long-term diabetic complications; however, genetic factors may also contribute (Cahill, Etzwiler, & Freinkel, 1971; Cox *et al.*, 1986). Both the rate of development and the severity of complications vary greatly across individual patients.

Visual Impairment. The most common complication is retinopathy, a degenerative disease of the small blood vessels in the retina that can cause visual impairment and blindness. By 15 years following diagnosis, 97% of Type I and 80% of Type II patients show evidence of retinopathy. Approximately 12% of the people who have had diabetes for more than 30 years are blind, and in the U.S., diabetes is the leading cause of blindness occurring between ages 20 and 74.

Nephropathy. In the U.S., diabetes is also responsible for 20–30% of all kidney disease. By the 15th year after diagnosis, approximately 33% of Type I and 20% of Type II patients have developed some degree of kidney disease. Of Type I patients, 30–40% eventually develop end-stage renal disease.

Neuropathy. Diabetes can result in damage to any part of the nervous system, but the peripheral and autonomic nervous systems are most commonly affected. Peripheral neuropathy typically begins with abnormal sensations and/or severe pain in the lower limbs and feet. Eventually, loss of sensation occurs, which results in a decreased awareness of injury to the feet and the legs. This insensitivity to pain, in combination with peripheral vascular disease, which impairs healing processes, contributes to lower limb amputations. Autonomic neuropathy contributes to cardiovascular problems, gastrointestinal problems, and sexual impotence.

Macrovascular Disease. Both Type I and Type II diabetes greatly increase the risk of coronary heart disease and cerebrovascular disease. All macrovascular diseases occur at an earlier age and are more extensive and severe in the diabetic population. There is increased risk of myocardial infarction and a two to six times higher rate of stroke. Hypertension is quite common, occurring in 50% of cases.

Treatment

In response to the growing evidence that chronic hyperglycemia is a major contributor to the above complications, the medical management of diabetes has increased its emphasis on a tighter control of glucose levels. The goals of treatment are to normalize glucose metabolism as much as possible, while avoiding overtreatment, in order to reduce the risk of long-term complications. In essence, the person with diabetes must attempt to duplicate behaviorally the normal and automatic metabolic functions via a complex self-treatment plan that involves balancing medication or injections, dietary modification, exercise, and self-testing of BG or urine. These four major components of the diabetic regimen are described in more detail in the following sections.

It is important to point out that there is no one diabetic regimen. Self-treatment regimens vary greatly across types of diabetes, individual patients, and prescribing physicians. To appreciate the psychological and behavioral demands imposed by diabetes, it is useful to remember that the regimen must be performed on a daily basis for the remainder of the patient's life.

Insulin and Other Medications

Insulin injections are the keystone of treatment for Type I patients. Although insulin therapy is used in Type II diabetes, oral medications, which lower BG by increasing insulin secretion, are more common. At least one injection must be taken daily, but as many as 3 to 5 daily injections may be prescribed, depending on the complexity of the regimen. Insulin is classified according to the time course of its action (short, intermediate, or long-acting) as well as its source (beef, pork, or synthetic human).

The most common and simplest regimen involves a combination of a short-acting and an intermediate or long-acting insulin taken in one daily injection before breakfast. In the more traditional (and still most common) model of treatment, a fixed insulin dose is prescribed by the physician, and the patient attempts to keep food intake (amount and schedule) and activity levels constant from day to day. More intensive and flexible insulin regimens are necessary for the optimal control of Type I diabetes, which requires multiple daily injections. The ideal regimens involve the use of multiple injections, with self-determined doses adjusted on a day-to-day basis according to BG test results.

Insulin therapy also has complications, the most important being hypoglycemia. Hypoglycemic episodes occur when BG levels become too low to provide sufficient metabolic fuel for the maintenance of normal bodily function, primarily brain function. At 50–70 mg/dl, symptoms begin to occur, including epinephrine-induced symptoms caused by hormonal counterregulatory mechanisms (e.g., trembling, heart rate increase, and lightheadedness) and neuroglycopenic symptoms caused by inadequate brain glucose (e.g., confusion, slurred speech, and motor discoordination). As BG levels decrease, the symptoms become more severe, and seizures or coma can occur. Hypoglycemic episodes are common with insulin therapy and increase in frequency when patients attempt to maintain BG levels closer to a normal range. From the patient's perspective, hypoglycemic episodes are physically unpleasant, potentially embarrassing, frightening, and dangerous.

Diet and Exercise

The concept of a *diabetic diet* has gone through many transformations since the early 1980s. Current recommendations include limited intake of simple carbohydrate (sugar), increased consumption of complex carbohydrates, and reduced fat intake (Arky, 1983; National Institutes of Health, 1987). Although regimens vary greatly, almost all persons with diabetes follow some sort of dietary plan. For patients using insulin therapy, the timing of food intake is as important as food amount and type. Diet is the treatment of choice for the Type II patient, who typically needs to reduce caloric intake and increase energy expenditure. Diet and aerobic exercise in combination appear to produce greater therapeutic benefits that either alone, so regular and frequent aerobic exercise is often prescribed.

Self-Monitoring

The purpose of the self-testing of blood or urine is to (1) prevent unacceptable BG levels; (2) monitor overall diabetes control; and (3) evaluate the effectiveness of self-treatment (Schiffrin & Belmonte, 1982; Skyler, 1981b). At the most minimal level, daily urine tests are needed to detect the presence of ketones, signifying extreme and prolonged hyperglycemia. Optimally, self-testing of blood samples is done, using a drop of blood obtained via finger prick that is placed on reagent strips to be either visually interpreted or read by meters. The frequency of self-testing varies greatly; some patients perform only a few tests per week, and others perform several tests each day. Three to four daily tests appear to be needed for Type I patients before any improvements in diabetes control occur. With complex regimens involving multiple daily injections or insulin pumps and adjusted insulin dose, more frequent testing is required (American Diabetes Association, 1987).

Education

The self-treatment of diabetes requires the person to acquire a large amount of information and to master a number of skills. For this reason, education and training play a major role in diabetes management. At diagnosis, the person needs to quickly acquire basic survival skills, such as insulin administration and treatment of hypoglycemia. Continuing education is also important, as the individual's needs change over the course of the disease and as new advances are made in treatment. Unfortunately, the majority of patients receive inadequate education, which is one of the greatest barriers to optimal diabetes care (Cox *et al.*, 1986; Miller, Goldstein, & Nicolaison, 1978).

Adopting and Maintaining Self-Care Behavior

From the patient's point of view, diabetes is an extremely challenging disease. It is a life-threatening, lifelong condition whose treatment requires an uncommon degree of active involvement on the part of the patient:

> People who live with diabetes always live with un-relieved stress. Throughout each day, no matter what else they are doing, they must be tuned-in to their own bodies in order to anticipate the need for more insulin, more food (quick sugar), or even exercise. For the person with diabetes a simple headache must be analyzed. He needs to know why he has the feeling? What caused it? Is it potentially threatening? He may then have to take immediate and appropriate action. As well as anticipating insulin reaction, infections, and illness, and making necessary adjustments, there are smaller, daily frustrations. He must resist the temptations of sweets, alcohol, and too much food, as well as cope with the awkward social situations such actions create. From the moment a person develops diabetes, for 24 hours a day, 365 days a year, for the rest of his life, he is responsible for managing the unmanageable, controlling the uncontrollable, and coping with the incurable—his diabetes. (Hoover, 1983, p. 41)

It is hardly surprising, therefore, that one of the most common reasons for psychological referral of people with diabetes is failure to adopt and/or to maintain some aspect of self-care behavior, that is, "noncompliance." In order to evaluate and treat these problems adequately, the psychologist should consider at least three areas of interpersonal and psychological functioning: the patient–practitioner relationship, the patient's values and health beliefs, and the patient's perceived barriers to self-care.

Patient–Practitioner Relationship

In the literature on health psychology, the problem of adopting and maintaining self-care behavior is often discussed as an issue in *compliance* (e.g., Feuerstein, Labbe, & Kuczmierczyk, 1986; Gatchel & Baum, 1983; Stone, Cohen, & Adler, 1979). The use of this term suggests a therapeutic relationship in which the practitioner prescribes a course of treatment that the patient is expected to follow. As Anderson (1985) pointed out, the inequality and lack of reciprocity in such a relationship can create the problem of noncompliance by removing patients, their values, and their concerns from the goal-setting aspects of the treatment process. This problem is difficult to address either individually or systemically because its roots are deeply embedded in standard medical practice and cultural expectations.

To some extent, the problem can be addressed in individual cases by educating patients and encouraging them to become more active participants in the decision-making process. In the professional education literature on diabetes, there is now a strong emphasis on patient activation (Hiss, 1986; Rost, 1989; Roter, 1977), on building educational partnerships with patients (Powers, 1989), and on reconceptualizing the problem of compliance as one of adherence. The advantage of the latter term is that it implies "a more active, voluntary collaborative involvement of the patient in a mutually acceptable course of behavior to produce a desired preventative or therapeutic result" (Meichenbaum & Turk, 1987, p. 20). (Also see Chapter 15, by Turk & Meichenbaum, in this book.)

Although these are steps in the right direction, the standard values of medical training and the context of the medical setting are powerful influences that act to maintain traditional patterns of care provision. For example, in a recent study of goal setting by physicians providing diabetes care (Dobson, Nord, & Haire-Joshu, 1989), none of the six physicians studied used statements pertaining to goal acceptance by the patient. The psychologist should be aware of these dynamics in the patient–practitioner relationships when assessing the underlying causes of a referral for failure to comply with medical treatment.

Patients' Values and Health Beliefs

There is evidence that the health beliefs of persons with diabetes are related to their adoption and maintenance of self-care behaviors (Bloom, Cerkoney, & Hart, 1980). Becker and

Maiman (1975) described a health belief model that contains several major elements: perceived severity of the health threat and susceptibility to it, perceived cost versus benefits of treatment, and cues to action (i.e., internal or external stimuli that initiate health behaviors). Each of these elements represents an area of belief held by patients that can influence their decision to adopt or not to adopt a recommended health care behavior. For example, the person with diabetes who believes that he or she has "just a touch of sugar" (low severity of health threat) is less likely to adopt a self-care behavior aimed at treating this condition than a person who believes that he or she has serious diabetes. In treating problems of adopting or maintaining self-care behaviors, the psychologist should assess the health beliefs of the person with diabetes to determine if this is a contributing factor. Harris, Linn, Skyler, and Sandifer (1987) developed and validated a 38-item Diabetes Health Belief Scale based on the health belief model that may be useful for this purpose (see "Assessment").

Perceived Barriers to Self-Care

Another important factor in the patient's ability to adopt and maintain self-care behaviors is perceived environmental barriers. Timely and accurate administration of medication, regular exercise, controlled eating, and regular and accurate testing of blood glucose are all behaviors that can be made easier or more difficult by the patient's environment. To the degree that the environment presents barriers to these behaviors (e.g., lack of time, lack of an appropriate setting, lack of necessary resources or supplies), the patient may be expected to experience difficulty in adopting or maintaining self-care behaviors. Scales for measuring environmental barriers to adherence in diabetes have been developed by Glasgow, McCaul, and Schafer (1986) and Irvine, Saunders, Blank, and Carter (1989). These scales are clinically useful in that they reveal patients' perceptions and attributions regarding the effect of the environment on their self-care. Thus, the scales provide a useful point of departure for more in-depth discussions of patients' everyday experiences and feelings about coping with diabetes and diabetes self-care.

Clinical Issues in Diabetes Mellitus

Overview

The following eight sections describe clinical problems with unique relevance to patients with diabetes, concluding with the clinical implications in each problem area. This is not an exhaustive listing of either clinical problems or possible interventions. Additionally, all individuals with diabetes do not necessarily have such problems. However, clinicians should be alerted to these problems and their implications.

Adjustment

Problem

Developing and managing diabetes are riddled with adjustments. Though there is sparse research on the subject, we conceptualize such adjustments as being analogous to grieving a loss, passing through the five stages (Sperling, 1988) of denial, anger, negotiation, depression, and accommodation. Although this adjustment process may be a natural course, the sequence of stages may differ across individuals, and some individuals may either be delayed in reaching or never achieve accommodation.

The first major adjustment is to the diagnosis, and to the loss of a healthy body and all of the spontaneous behaviors that such a body permits. We speculate that Type I diabetes demands a more dramatic adjustment, given that a failure to make the most basic accommodations, such as insulin injections, immediately threaten survival. Unfortunately, the person with Type II diabetes can frequently tolerate, in the short run, the perspective, "I've only got a little sugar in my blood, so I'll stop eating candy and everything will be OK." Failure to come to terms with the diagnosis of diabetes is likely to lead to inadequacy in or a total absence of self-treatment and eventual long-term complications (see "Patients' Values and Health Beliefs").

In addition to adjusting to the diagnosis and the treatment regimen, the inevitable complication demands adjustment. Losing one's leg, sexual functioning, kidney functioning, bowel control, or vision or being condemned to chronic pain so that even pulling the sheet over the feet is excruciating, represents a loss of major proportions. If the patient fails to accept and accommodate to such losses, then life can be frozen at the time when such complications develop. The impact of these complications is even more damning because of their gradual development. For example, visual loss may have an on-again–off-again course, so that a person may awaken one morning with blood in the vitreous totally clouding vision and two weeks later may awaken with clear vision. Individuals may be uncertain about whether they are going blind. In fact, the transitional process has been shown to be more distressing than terminal blindness (Bernbaum, Albert, & Duckro, 1988). A similar transitional process may occur with male sexual functioning; a man may interpret a sexual failure as reflecting neuropathy and may subsequently avoid sexual function with his partner.

The adjustment to such traumatic events is not a simple phenomenon. Adjustment is accomplished in degrees. As in the field of plastic surgery, where there is no correlation between degree of physical disfigurement and subsequent psychosocial adjustment (Edgerton & Langman, 1982), there is probably no correlation between the nature and the type of loss and the diabetic patient's subsequent quality of adjustment.

Implications and Interventions

There are few specific guidelines available to aid the clinician either in prophylatically helping patients to adjust to complications or in correcting an ongoing poor adjustment process. Although participation in diabetic support groups may be most efficacious because individuals burdened by the same issues can support, confront, and aid one another, those individuals who need such groups the most are probably the least likely to participate.

As clinicians, we should be alert to these issues, and we should consider two levels of intervention: First, realizing that adjustment to loss is a normal process, we can educate and aid patients and their families as losses are experienced, thereby possibly aiding in the adjustment process and avoiding any secondary problems. Second, when the adjustment to losses continues to be dysfunctional beyond six months (American Psychiatric Association, 1987, "Adjustment Disorders"), direct clinical intervention is indicated. One problem is the definition of *dysfunctional* and the typical unwillingness of individuals with dysfunction to accept help. *Dysfunction* means that daily function and quality of life are significantly disrupted beyond their premorbid levels, in excess of what is justified by the actual loss. For example, it would not be considered dysfunctional if a recent amputee discontinues his previous hobby of skiiing, but it would be dysfunctional if that same individual gave up socializing with friends. When patients recognize this dysfunction and find it ego-dystonic, they will probably be receptive to regrief work (Zisook, 1987). However, for individuals who deny either such dsyfunctions or realistic limitations (e.g., who continue to drive when their vision has deteriorated), access to the patient may be best gained through couples or family therapy.

Depression

Problem

Depression has been conceptualized as learned helplessness, anger turned inward, and adjustment to loss. We have already discussed the numerous losses that are part of the diabetes process, as well as depression as a part of the grieving process. Depression may become permanent if the grieving process is not completed.

Although there are strong speculations that complications result from poor metabolic control, we do not know how good this control needs to be nor whether there is a critical phase in the life course of the disease during which this control needs to be optimized. There is also evidence that genetic makeup significantly influences who will and will not develop various complications and when these complications will surface. What we do know is that it is virtually impossible to have diabetes for a long time and to avoid all complications. This is fertile ground for learned helplessness: "I've been trying to control my blood glucose for the past five years, exercise regularly, and even quit smoking, but that didn't prevent my heart attack." Alternatively, patients may experience such losses with the thoughts, "If I had have only worked harder," "It is all my fault that I am losing my sight," or, "Why didn't I take my diabetes more seriously?" Given these factors, it is easy to imagine how anger can be turned inward when complications begin to develop.

These may be reasons why repeated studies have demonstrated that individuals with diabetes, as a group, are significantly more depressed than their nondiabetic counterparts (Lustman, Griffith, & Clouse, 1988). This finding is supported whether one uses simple self-report measures, such as the Hopkins Symptoms Checklist 90, or more elaborate interview schedules. However, this is not to say that, if one has diabetes, a concurrent diagnosis should also be depression.

Implications and Interventions

If the depression can be linked to the onset of complications, then regrief work (Zisook, 1987) should be considered. When depression appears to be related to unrelenting self-care demands, cognitive therapies should be considered (Beck, 1976).

Although this discussion suggests that diabetic depression is secondary, it is also possible that diabetic patients have a concurrent diagnosis of primary depression, which needs to be considered along with pharmocotherapy or traditional psychotherapy.

Family Adjustment

Overview

The impact of diabetes on the family cannot be overemphasized. The impact of the family on diabetes management is also enormous. The family unit has been described as the "environment in which diabetes management" and coping occur (Newbrough, Simpkins, & Maurer, 1985). The coping resources and responses of family members, both adaptive and maladaptive, will influence the adjustment to and the incorporation of diabetes management. The onset of diabetes in a family member is best viewed as a source of significant and chronic stress on the family structure. Immediately on diagnosis, family members need to begin acquiring new information and skills for basic diabetes management, which often necessitates unwelcome changes in family routines and lifestyle. Other diabetes-related stressors include financial strain, anxieties about complications and reduced life expectancy, and frequent medical crises, such as hypoglycemic episodes.

For all of the above reasons, it is not surprising that families living with diabetes often experience problems that require psychological or behavioral intervention. The difficulties commonly experienced by families are determined to a large extent by which family member has diabetes. Many of the problems encountered by the parents of a child with diabetes are unique and are different from those encountered by the spouse of a person with diabetes. For this reason, these two types of family units are discussed separately here.

Problems in the Child or Adolescent with Diabetes. The degree and frequency of maladjustment and poor outcome in children with diabetes are a controversial question, and studies have yielded mixed results. Some researchers have found social dysfunction and other symptoms more prevalent in diabetic children (Jacobson, Hauser, Wertlieb, Wolfsdorf, Orleans, & Vieryra, 1986), and others have found no increased risk of psychosocial difficulties (Kovacs, Finkelstein, Feinberg, Crouse-Novak, Paulauskas, & Pollack, 1985). At present, the most reasonable interpretation appears to be that, although diabetes may contribute to psychosocial problems in children, they do not necessarily occur.

Two common patterns of maladjustment have been described: overdependence and overindependence (Mattsson, 1979). Overdependence is usually associated with increased fearfulness, reduced activity levels, and reduced social interaction. There is some evidence that this pattern is most common in cases of early onset (Holmes, 1986). Parental anxiety has been cited as the primary contributor to the child's sense of vulnerability and the resulting overdependence. It is not uncommon for the parents of a child with diabetes to refuse to leave the child alone, even long after being left alone is developmentally appropriate. Many of these children are not allowed to be away from their parents for long periods, such as sleepovers, or to participate in sports and other group activities. Although one can empathize with the overanxious parents who have witnessed their 2-year-old having hypoglycemic seizures, overprotection will significantly interfere with normal development.

Problems with overdependence are often noticed when transitions toward self-care in the child are difficult. By age 10, and often much earlier, children should become responsible for at least part of their own self-treatment (Drash, 1979; Wysocki, Meinhold, Cox, & Clarke, 1990). This should include performing self-tests of blood or urine, giving their own insulin injections, and knowing how to treat hypoglycemia. Between the ages of 12 and 14, the adolescent child should be capable of complete self-treatment.

The other common maladaptive response, overindependence, occurs when the child will not accept parental advice and support and may even indulge in risk-taking behaviors (Goldstein & Hoeper, 1987; Mattsson, 1979). Not surprisingly, this pattern is most common in adolescents, but it can occur in younger children. Diabetes-related issues are extremely important and anxiety-provoking to parents, as well as constantly present, so that they are almost perfect vehicles for the expression of rebellion and control in the adolescent. Covert mismanagement of insulin, meals, or exercising (frequently resulting in medical crises) is not uncommon in adolescent patients. Typically, the adolescent expression of conflicts over independence and the testing of limits via diabetes management result in parents' remaining overinvolved in the child's treatment.

The most positive prognosis for good adjustment and diabetes management occurs in families characterized by a stable structure, supportive relationships, and an ability to cope well with adversity before the diabetes diagnosis (Johnson, 1980; Simonds, Goldstein, Walker, & Rawlings, 1981). In contrast, problems are most common in families characterized by marital conflict, unstable structures, limited resources, and social isolation (Koski & Kumento, 1977; Orr, Golden, Myers, & Marrero, 1983). Families already under a great deal of stress may find their coping capacities overwhelmed by the added demands of diabetes management. All children with chronic diseases, including diabetes, are at increased risk of parental neglect, which most often takes the form of medical care neglect or failure to comply with treatment recommendations (Horan, Gwynn, & Renzi, 1986).

Implications and Interventions

Obviously, family therapy is often the intervention of choice for family adjustment problems and crisis situations. For parents having marital difficulties that are exacerbated by diabetes-related issues, couples therapy is usually effective. For the very young child, play therapy has been recommended to provide an outlet for emotional distress and the development of coping skills. A relatively new treatment innovation for older children involves peer support and counseling, involving work in group settings where children can share diabetes-related issues and learn coping skills from each other. Another alternative is a "buddy system," in which newly diagnosed children are paired with other children who have successfully adjusted, offering an opportunity for modeling and learning good coping skills. This may be accomplished in summer diabetic camps.

For the adolescent child, the presenting problems are often issues surrounding non-adherence to treatment requirements. In such cases, family therapy may be helpful to improve communication and methods of coping. Behavioral interventions, such as token economy techniques, have proved useful in such cases (Epstein, Figueroa, Farkas, & Beck, 1981), and even without contingency management, techniques such as goal setting and contracting may improve self-care behaviors (Schafer, Glasgow, & McCaul, 1982). Parents may be included in behavioral interventions as providers of reinforcement or participants in contracting. However, in cases where the diabetes regimen has become a vehicle for power struggles, individual work with the adolescent may help to make self-treatment an issue of adolescent responsibility and to improve self-efficacy and self-esteem.

Problems for the Married Couple

Unfortunately, the impact of diabetes on the quality of marital relationships has received virtually no attention by researchers. We can probably assume, however, that, for adults as well as children, supportive, well-functioning couples with adequate emotional, financial, and social resources have the best prognosis for successful coping, whereas dysfunctional and nonsupportive couples will have more difficulties.

Balancing responsibilities for diabetes management becomes an issue for many couples. Individuals who are overdependent on their spouses continue these patterns in diabetes-related issues, increasing the risk of psychological burden on the spouse and potential conflict. Other individuals exhibit overindependence, refusing to accept positive, beneficial forms of support. There is some evidence that diabetes control is better in married than in unmarried men but is not better in married than in unmarried women. Although speculative, this evidence may reflect a tendency for nondiabetic female spouses to instigate changes in the family's eating and other routines, whereas female diabetic spouses fail to instigate similar changes on their own behalf.

A particularly difficult situation arises when a diabetic spouse denies the seriousness and the possible implications of his or her illness. In our experience, these cases have most often involved nondiabetic wives who are frightened and angry, and who do not know how to deal with a diabetic husband who denies his illness; the result is frequent hypoglycemia or complications.

Surveys of young adults suggest that the presence of diabetes has a significant impact on marriage and family planning (Ahlfield, Soler, & Marcus, 1985). Diabetes was viewed as interfering with family activities and as a "source of friction" by more than one third of the people surveyed; the men were more likely to endorse diabetes as a marital stressor. More diabetic than nondiabetic adults decide not to have children; the decision may be related to concerns about decreased health or life expectancy, fear of genetic transmission, or fear of unsuccessful pregnancies.

The presence of diabetes can be expected to interact with, and perhaps to exacerbate, any previously existing problems in a marriage. Couples with poor communication skills and an inability to resolve conflicts are likely to have difficulties dealing effectively with diabetes-related issues, which require a high level of cooperative behavior. For these couples, the added stress of diabetes management is likely to cause marital problems that require intervention.

Implications and Interventions

The young diabetic adult who is considering marriage may have significant anxieties concerning childbearing, future health status, and reduced life expectancy. In addition, the partner who is considering marrying a person with diabetes may have similar, but often unexpressed, concerns about the future. Premarital counseling is highly recommended in such cases, as it provides an opportunity for both individuals to express and work through diabetes-related issues, including feelings about future uncertainties. This early intervention may also be beneficial in helping the young couple to develop agreements concerning their attitudes toward and responsibilities in diabetes management.

For the married couple, the presence or onset of diabetes may create psychological distress or may exacerbate previously existing problems in the relationship. Assessments of marital difficulties in couples with diabetes should always directly address such possible interactions. Intervention may involve modifications in traditional couples therapy, for example, focusing on marital conflict surrounding diabetes-related issues. However, therapeutic benefits may also require a multimodal approach in which separate interventions are included to address diabetes-specific problems. The case study section describes such a case, which required conventional marital techniques to improve communication and conflict resolution skills, as well as individualized blood glucose awareness training.

Obesity

Problem

Obesity is strongly associated with NIDDM. Approximately 80% of individuals with NIDDM are obese at the time of diagnosis, and the incidence of NIDDM increases exponentially with an increasing degree of obesity (Salans, Knittle, & Hirsch, 1983). Although there is no clear evidence of a causal connection between obesity and NIDDM, weight reduction is indicated for persons with NIDDM for at least two reasons. First, even small amounts of weight loss are often followed by significant reductions in hyperglycemia. Second, weight loss may lead to reductions in hypertension, which is one of the major complications of diabetes (see "Hypertension"). In this sense, weight reduction can be viewed as a tool for controlling diabetes and for controlling or preventing the development of hypertension.

Although weight loss and weight maintenance are extremely important aspects of dia-

betes treatment, they present formidable challenges to the patient and the practitioner. Studies have shown that persons with diabetes may have more difficulty losing weight than those who do not have the disease (Henry, Wiest-Kent, Schaeffer, Kolterman, & Olefsky, 1986; Wing, Marcus, Epstein, & Salata, 1987). It is not clear whether this difficulty is due to behavioral or metabolic factors. However, obese persons with NIDDM are often referred for psychological treatment because they are perceived or perceive themselves to be unable to comply with prescribed weight-loss regimens. In addressing this issue, it is important to determine whose goal weight loss is. It is essential that the patient be personally motivated and take personal responsibility for efforts directed at weight loss, regardless of how desirable these efforts are from a medical standpoint. The consequence of attempting to supply extrinsic motivation for difficult and permanent behavior changes is likely to be failure to lose weight or weight loss followed by regain to an even higher weight. The latter outcome is a major problem among lifelong dieters. There is evidence to indicate that this "yo-yo" pattern of weight loss and regain leads to a lowering of metabolic rate, making weight loss or maintenance even more difficult (Buckmaster & Brownell, 1988).

Implications and Interventions

In terms of treating obesity, there are at present no indications for differential treatment of persons with diabetes. It is important to realize that a variety of psychological, metabolic, social, and genetic factors influence the development and maintenance of obesity. Psychological treatment addresses cognitive and behavioral factors in obesity to a limited extent and with limited success. In general, behavior modification programs for weight loss, either by themselves in the case of mild obesity, or in combination with very low-calorie diets in the case of moderate obesity, have been shown to be effective in reducing weight. Long-term weight-loss maintenance is much more difficult to achieve. Treatment components that improve weight maintenance include behavioral screening, ongoing self-monitoring, a structured eating plan, group support, and structured exercise (Green & Saunders, 1988). Because of the multifaceted nature of effective obesity treatment, a programmatic approach with multidisciplinary support is essential.

Hypertension

Problem

The prevalence of hypertension among patients with diabetes, particularly NIDDM, is extremely high (Hall, 1986). Estimates run between 30% and 60%, increasing with age and the duration of the diabetes (Runyan, 1988). The pathophysiology of hypertension in diabetes is multifactorial and not fully understood. Elevated levels of plasma insulin, a characteristic of many patients with NIDDM who are "insulin-resistant," may cause sodium retention by the kidney, leading to increases in blood volume that result in hypertension (Hall, 1986; Peterson & Jovanovic-Peterson, 1988). Alternatively, insulin-deficient patients may be unable to eliminate salt and water because of renal insufficiency; the result is, again, hypertension (Hall, 1986).

Hypertension is viewed as the most important factor in accelerating the development of renal disease and retinopathy (American Diabetes Association, 1988). Diabetic patients with hypertension are also at increased risk of the full range of cardiovascular complications of diabetes, including coronary artery disease, myocardial infarction, stroke, and retinopathy. In upper age groups, over 70% of the deaths of diabetics are due to renal and cardiovascular complications (Runyan, 1988). Obesity, another cardiovascular risk factor, is present in about 80% of patients with both diabetes and hypertension. Thus, hypertension, diabetes, and obesity often present clinically as a triad.

Implications and Interventions

The treatment of hypertension is complicated because of the need to avoid pharmacological interventions that will adversely affect glucose control or lipid levels. Beta blockers are generally avoided because they raise lipid levels and decrease insulin secretion and may cause hypoglycemia while masking many of its symptoms. There are other antihypertensive drugs that do not have these side effects, including angiotensin-converting enzyme (ACE) inhibitors, alpha 1 and alpha 2 blockers, and calcium channel blockers (Peterson & Jovanovic-Peterson, 1988). However, nonpharmacological interventions should be the treatment of first choice in approaching hypertension among persons with diabetes.

Peterson and Jovanovic-Peterson (1988) advocated a multifaceted approach directed toward normalizing blood glucose and weight, increasing exercise, eliminating smoking and alcohol, and reducing stress and sodium intake. Weight reduction is the primary consideration for the large number of patients who have the triad of diabetes, hypertension, and obesity because it reduces the insulin resistance common to all three diseases, as well as blood glucose and blood pressure levels. Behavioral techniques are useful in modifying eating behavior (see "Obesity"), alcohol consumption, smoking, and exercise. Biofeedback and relaxation techniques are useful in teaching patients to cope better with stress and may help to reduce hypertension. Patients with hypertension and diabetes should be assessed with respect to their potential to benefit from weight loss, lifestyle change, and stress reduction.

Sexual Dysfunction

Problem

Sexual dysfunction, particularly impotence, is a serious and well-documented problem among diabetic males; it can occur any time after adolescence. The prevalence of impotence among diabetic men has been found to be as high as 50–60%, far higher than in the normal male population (Ellenberg, 1983). Although up to 90% of impotence in the general population is due to psychological causes, 80–90% of the impotence among men with diabetes is at least partly organic or neurological (Clarke, 1988). The latter sources of impotence include congenital abnormalities, trauma, systemic disease, vascular disease, endocrine deficiency, and central, spinal, and peripheral neuropathy. Impotence can also be induced by drugs, including certain antihypertensives and antidepressants. It is common for impotence to have multiple determinants and to be exacerbated by problems of psychological adaptation. For example, Clarke (1988) discussed a case of impotence in a male diabetic that originated in metabolic control but was exacerbated by alcohol consumption and performance anxiety.

The organic and neurological sources of impotence in diabetic men are relatively well understood, and there are fairly clear bases for distinguishing these from psychological etiologies. Autonomic neuropathy, one of the major complications of diabetes, is also the major cause of organic impotence (Ellenberg, 1983). Diabetic autonomic neuropathy involving the pelvic parasympathetic nerves prevents any possibility of an erection by blocking the pathway for the neurological reflex that dilates the penile arteries. This blocking of the reflex has important implications for differential diagnosis.

Impotence due to autonomic neuropathy has a gradual and progressive onset, developing over a period of months or years. Erections become less firm and less frequent over this period, eventually fading away altogether. Testicular sensitivity is also lost, but libido and ejaculation remain intact (Clarke, 1988). By contrast, impotence of psychological origin often has a more sudden onset and may be associated with specific circumstances or partners. Masturbation may continue to be effective in bringing about erections, and nocturnal erections still occur. Testicular sensitivity remains, but libido is often diminished.

In women, the role of organic factors is not as clear or well understood. There is evidence that women with Type II diabetes have an increased prevalence of sexual dysfunctions, but this difference does not hold up among those with Type I diabetes (Schreiner-Engel, 1988). If organic or neurological mechanisms were the primary cause of sexual dysfunction in diabetic women, then one would expect the prevalence of these problems to be at least as great among Type I women as among Type II women. Psychological factors may offer a better explanation. For example, Schreiner-Engel (1988) suggested that the relatively early onset of Type I diabetes may enable women to incorporate the presence of the disease into their developing sexual identities and relationships. By contrast, the later onset of Type II diabetes may have a much more detrimental effect on preexisting female sexual identity, self-image, and marital relationships.

Implications and Interventions

Although there is no hope of restoring a normal erection in men who suffer from diabetic autonomic neuropathy, aids are available that allow the individual to engage in vaginal intercourse. These include penile implants, suction–constriction devices, and external mechanical aids (Witherington, 1989). Because the use of these devices involves major adaptation and commitment by both sexual partners, psychological counseling is an important component of this treatment approach. In treating diabetic patients, the practitioner must be aware of the increased probability of sexual dysfunction, and whether organic factors are clearly implicated in the etiology of sexual dysfunction or not, psychological factors must always be addressed. Treatment includes a variety of approaches, including individual counseling or psychotherapy, marital counseling, and/or sexual therapy.

Incontinence

Problem

One of the more psychologically and socially disturbing consequences of neuropathy can be fecal incontinence. Neuropathy can lead to loss of feeling in the rectum. Rectal distension sensations are the typical signal that a bowel movement is pending and that one needs to find a bathroom or to tighten the external anal sphincter muscle to prevent fecal leakage. Additionally, neuropathy can lead to loss of control of the anal sphincter muscles, which results in failure to prevent a bowel movement. Although a low-frequency complication, incontinence may have major ramifications: damage to one's self-concept and restrictions of one's social activities in fear of having a humiliating fecal accident in public. Like sexual dysfunction, this is a complication that is frequently not discussed by either patient or health care provider.

Implications and Interventions

Interventions occur at two levels: First, if the accidents are uncontrollable, individuals can be helped to accept and cope with the problem. The patient can be assisted in developing hygienic behaviors such as frequent toileting, using enemas, wearing adult diapers, and using deodorants that allow the resumption of social activities. Handicapped parking stickers can alleviate the concern that "If I have an accident, I won't be able to get to my car to leave the situation and care for myself." Second, anal EMG (Sims, Remler, & Cox, 1988) and manometric biofeedback procedures (MacLeod, 1983) have been demonstrated to treat incontinence effectively. Manometric biofeedback involves placing a balloon in the rectum, inflating that balloon to simulate a stool, and having pressure sensors from the balloon display the rectal distension on a polygraph. This feedback is designed to train individuals to discriminate the sensations of a full rectum. An additional balloon is inflated in the anal canal, with pressure sensors displaying external anal sphincter strength on the polygraph. This procedure allows patients to see the activity of the sphincter, and permits training and strengthening of this muscle. Surface EMG biofeedback involves the use of routine biofeedback devices using pediatric disposable EKG electrodes taped bilaterally to the anal opening to monitor and subsequently train striated external anal sphincter EMG.

Symptom Recognition and Interpretation

Problem

Diabetic hypoglycemia and hyperglycemia are associated with physical symptoms, although the specific symptomatology differs from individual to individual. At first glance, the relationship between diabetic blood glucose symptoms and clinical problems may not be obvious. However, our experience in both treatment and research has repeatedly shown that the failure to accurately interpret blood glucose symptoms is a significant contributor to psychological distress and problems with self-treatment (Cox et al., 1986; Gonder-Frederick & Cox, 1990). Hypoglycemic symptoms appear to present the most difficulties for persons with diabetes, partly because the physiological effects of hypoglycemia, such as epinephrine secretion and neuroglycopenia, result in mood lability and feelings of anxiety, irritability, and confusion (Gonder-Frederick et al., 1989).

For some individuals, it is difficult to distinguish hypoglycemia-related mood changes and genuine emotional responses. They frequently misattribute their mood lability to some other source. This misattribution may cause problems in self-perception and interpersonal relationships. For example, the anxiety-like symptoms associated with hypoglycemia may be misinterpreted as a panic attack, or genuine anxiety symptoms may be misattributed to low BG. The negative emotional feelings that occur with hypoglycemia may also be mistakenly attributed to people in the environment, such as co-workers, friends, spouses, and children, with obvious negative implications for social interactions and relationships. In addition to the misattribution of emotions, the inability to recognize hypoglycemic symptoms may cause other forms of interpersonal stress. Bizarre responses and behaviors may occur with hypoglycemia, as well as extreme mood change, resulting in significant social and occupational problems, and even in job loss for some individuals.

Failure to recognize glucose symptoms, both hypoglycemic and hyperglycemic, interferes with optimal self-treatment and diabetes control (Gonder-Frederick & Cox, 1990; O'Connell, Hamera, Knapp, Cassmeyer, Eaks, & Fox, 1984). A person cannot treat and correct unacceptably low or high BG levels that are undetected. Our own research, as well as others', has repeatedly demonstrated that persons with diabetes frequently fail to detect glucose symptoms (Cox, Clarke, Pohl, Gonder-Frederick, Hoover, Zimbelman, & Pennebaker, 1985; Gonder-Frederick, Cox, Pennebaker, & Bobbitt, 1986). For example, it is very common for individuals to "miss" the early, more subtle hypoglycemic symptoms, which if untreated may result in severe hypoglycemia. Because of autonomic neuropathy and other physiological changes, symptoms may also disappear or alter as diabetes progresses, and individuals may be unable to rely on cues that previously served as indicators of hypo- and hyperglycemia.

Implications and Interventions

Because of the frequency and significance of problems in symptom recognition, we have developed an intervention specifically designed to improve BG detection and estimation in diabetic patients. Blood glucose awareness training (BGAT) involves systematic instruction in two major areas: awareness of the symptoms, moods, and cognitive changes that signal BG extremes and awareness of the environmental and behavioral factors that influence BG. Sessions are conducted weekly in a group setting for seven weeks. Throughout BGAT, the participants complete a BG diary that requires them to estimate their own glucose level several times each day. To complete a diary entry, they record all of the information (symptoms, food, and time of day) relevant to their estimate, estimate their BG, and then measure their actual BG level. The accuracy of the estimate is then evaluated and graphed. The positive effects of this systematic instruction and feedback have now been documented in several studies (Cox, Carter, Gonder-Frederick, Clarke, & Pohl, 1988; Cox, Gonder-Frederick, Julian, Carter, & Clarke, 1989); BGAT produces im-

provements in estimation accuracy and metabolic control.

If formal BGAT is not feasible, the clinician should still consider and assess the influence of BG symptomatology on presenting problems, especially with self-reported episodes of mood lability or anxiety attacks. This can be done by having the patient monitor and record both daily moods and physical symptoms, along with concurrent BG levels that are analyzed for patterns and relationships. If BG fluctuations are a contributing factor, then either normalization of glucose levels to reduce symptoms or improved recognition of glucose symptoms may offer some relief. Although BG fluctuations are typically ignored as a significant factor, it has been our experience that practitioners are likely to encounter situations where glucose symptomatology exacerbates other psychological and social difficulties (see Case Study 2).

Phobias

Problem

From a simplistic perspective, the theoretical model of *in vivo* exposure would suggest that individuals with diabetes should not be afraid of injections or finger pricks for self-monitoring of BG (SMBG), because they experience both regularly. However, it has been our experience that a minority of patients have a significant fear of such routine behaviors that functions as a barrier to their frequent routine performance. For example, instead of using a multiple injection regimen or doing SMBG four times a day, patients use a single injection or measure their blood glucose only when absolutely necessary.

During these routine behaviors, such patients frequently engage in a lot of muscle bracing, hypoventilation, and either anxiety-provoking ("It is going to hurt; I can't do it") or dissociative ("Its not really happening to me") cognitions. Patients with such difficulties are often embarrassed by their fears and may try to hide them. However, these fears are typically detectable both by reviewing the low frequency of such self-care behaviors and by watching their performance.

Hypoglycemia can lead to aversive physical symptoms, mood disruptions, cognitive dysfunctions, accidents, and even death. As a result of such personal or vicarious traumatic experiences, psychologically predisposed individuals may develop a phobia to hypoglycemia. This can lead the patient either to avoidance behaviors that result in elevated blood glucose, that provides a cushion between their regular blood glucose level and the dreaded hypoglycemia, or to agoraphobiclike symptoms, where the patient fears being left alone, which represents the possibility of no one being available to treat their hypoglycemia (Cox et al., 1987).

It is also possible that people lacking a healthy fear of hypoglycemia may be more vulnerable to severe hypoglycemic episodes and would benefit from emotional reconditioning to enhance their respect for the seriousness of hypoglycemia. This reconditioning may be extremely important, as approximately one diabetic patient in eight has a severe hypoglycemic episode yearly, that may not only result in life- and property-threatening accidents but also lead to permanent cognitive deficits (Wredling, Lins, Adamson, Theorell, & Levander, 1988).

Implications and Interventions

Once such fears are detected and evaluated, they may resolve with typical fear-reducing behavioral interventions, such as reciprocally inhibiting relaxation training and cognitive restructuring, hypnosis, or, in rare instances, more in-depth psychotherapy.

Fear of hypoglycemia may be a little more difficult, as it is difficult to expose an individual systematically to in vivo graded hypoglycemia, and hypoglycemia should be reasonably feared. Although it has not yet been tested, theoretically individuals with such fears and elevated glycosylated hemoglobins can be identified and put through a fear reduction process of education, in vivo exposure to hypoglycemia, cognitive restructuring, and blood glucose awareness training. Individuals are typically hypoglycemic just before meals, when insulin action is peaking, or sometimes following intensive exercise. These naturally occurring events may be a primary source of in vivo exposure to hypoglycemia.

Assessment

When a patient sees a health care professional, the typical agenda is to assess current blood glucose control and the development and progression of any physical complications. Consequently, the focus is on diaries of blood glucose and insulin records, and on a physical exam of the eyes, heart, and feet. Because of this limited focus, the press for time, and the limited treatment options, frequently little attention is given to a psychological examination.

The routine screening of psychological symptoms is important not only because individuals with diabetes are more vulnerable to psychological distress, but also because psychological distress may significantly disrupt patients' self-care behaviors and may thus put them at risk of acute and chronic complications.

To avoid confusion, it must be said that we are not implying that all or even most individuals with diabetes have significant psychological disturbances. What we are saying is that psychological symptoms are common and may have serious ramifications; therefore, they justify routine screening, similar to the routine retina exams given because of the increased risk of retinopathy. In assessing psychological functioning, one must consider two dimensions: the premorbid-postmorbid dimension and the psychological-physical dimension (see Figure 1).

Individuals with diabetes have a premorbid personality on which the disease diagnosis and management is superimposed. A passive-dependent personality may respond very differently to the demands of self-injection and self-measurement of blood glucose from an obsessive-compulsive personality (Jacobson & Hauser, 1983). These factors should be assessed with traditional psychological assessment tools. Because the focus here is on a static preexisting personality, such an assessment need be done only once.

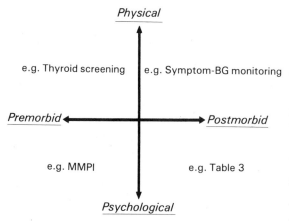

Figure 1. A conceptual model of considerations in the psychological assessment of patients with diabetes.

The postmorbid factors are diabetes-specific behaviors, adjustment reactions to the diabetes itself, and general psychological changes that occur following diagnosis. Skyler (1981a) called for the development of diabetes-specific instruments that would more specifically address the unique needs of the patient with and the demands of diabetes. Although there has been a groundswell of support for this idea it is important to realize that the traditional psychological tools have their place, such as the assessment of depression and anxiety. Like retinopathy, this type of general screening should be routinely done on at least an annual basis. Table 2 illustrates such a screening instrument.

Psychological symptoms may have either a psychological or a physiological basis (see Case Study 2). For example, lethargy and depression can be caused by chronic hyperglycemia secondary to caloric loss and increased acidosis and blood–urea nitrogen (BUN). Chronic pain syndromes may be influenced, in part, by hyperglycemia, which may lower the pain threshold (Herrman-Lee, 1989). Anxiety, irritability, atypically increased sexual arousal, and apparent irrational behaviors may be due to hypoglycemia (Gonder-Frederick, Cox, Bobbitt, & Pennebaker, 1989). Acute cognitive dysfunction both during an assessment session and elsewhere may be due to hypoglycemic neuroglycopenia (Ryan, 1988).

Although traditionally it has been difficult to separate endogenous from exogenous depression in individuals with chronic illness, the potential etiological role of blood glucose can be assessed in two ways: First, in a behavioral analysis, blood glucose may be monitored during times when the psychological symptom is present and when it is not, for a determination of any blood glucose-symptom patterns (see Case Study 2). Second, the relation of euglycemia and the abatement of symptoms can be established. If the symptoms are associated with hypoglycemia, the symptoms should resolve in a matters of minutes, whereas if the symptoms are due to chronic hyperglycemia, they may not resolve until after a week of good metabolic control.

Health care providers are generally not going to look for psychological morbidity; and consequently, they infrequently refer their patients for either psychological evaluation or intervention. Therefore, it may be useful to have the diabetic unit routinely administer a very brief psychological screening. When patients endorse certain items in certain ways, further inquiry is indicated. This could result in a psychological consultation, an informal call to a staff psychologist, or further questioning by the medical staff (see Table 2).

In addition to standard psychological and neuropsychological test instruments, new diabetes-specific instruments for psychological-postmorbid conditions should be considered. Examples of these appear in Table 3.

Case Studies: Clinical Problems Associated with Diabetes

Case Study 1

The patient was a 46-year-old while female with a four-year history of NIDDM and morbid obesity (5'4", 254 pounds) of 25 years duration. The treatment of her diabetes had included attempted weight loss, exercise, and oral hypoglycemic agents. Many options had been tried to achieve weight loss, including supplemented fasting, "fad" diets, hypnosis, diet pills, diet clubs, and having her jaws wired for three months. Some months before, a ran-

Table 2. Diabetes Psychological Screening Questionnaire[a]

Name _____   Date _____

1. How do you feel about having diabetes? (check one)
    ___ I find it hard to believe that I really have diabetes.
    ___ It makes me angry. "Why me?"
    ___ If I do the right things, it will go away.
    ___ It means that life and my body will never be the same.
    ___ It is one of life's problems and I just try to handle it.
2. How often would you say arguments or conflicts in your family happen? Place an X in the parentheses ( ) above the phrase that best describes how often arguments or conflicts occur.

( )	( )	( )	( )	( )	( )	( )
Almost never	Much less often than typical families	Less often than typical families	About average	More often than typical families	Much more often than typical families	Almost constantly

3. Do you have any difficulty with your sexual functioning? ___ Yes   ___ No

	Not at all				Extremely
4. How serious do you think your diabetes is?	1	2	3	4	5
5. How likely do you think it is that you will develop any complications of diabetes or that existing complications will worsen?	1	2	3	4	5
6. How effective do you think your diabetes treatment is at preventing serious long-term complications?	1	2	3	4	5
7. How much of a hardship or burden is your diabetes care?	1	2	3	4	5
8. How difficult is it for you to take care of your diabetes?	1	2	3	4	5
9. How satisfied are you with your involvement in making decisions about your diabetes care?	1	2	3	4	5
10. To what extent do you feel sad, little interest in life, and low energy levels?	1	2	3	4	5
11. How much do your family and friends help care for your diabetes?	1	2	3	4	5
12. To what extent do you worry about and keep your blood sugar high to avoid severe low-blood-sugar reactions (an insulin reaction)?	1	2	3	4	5

13. Do you smoke? ___ Yes   ___ No
14. How many alcoholic drinks do you have each day? ___ a day
15. All things considered, how satisfied would you say you are with your life with your life right now. Place an X in the parentheses ( ) above the word or phrase that best describes how satisfied you are.

( )	( )	( )	( )	( )	( )	( )
Very satisfied	Satisfied	Somewhat satisfied	Neither satisfied nor dissatisfied	Somewhat dissatisfied	Dissatisfied	Very dissatisfied

16. Is there anything not asked in this questionnaire that you would like to tell your health care provider?

---

[a]Item content: 1 = adjustment; 2 = family stress; 3 = sexual dysfunction; 4–7 = health beliefs; 8 = barriers; 9 = patient–practitioner relationship; 10 = depression; 11 = social support; 12 = fear of hypoglycemia; 13–14 = health behavior; 15 = quality of life; 16 = other information.

Table 3. Diabetes-Specific Assessment Instruments

Instrument	Reference	Number of items	Assessment objective
ATT-39	Dunn *et al.* (1986)	39	General psychological adjustment to diabetes
Barriers Scale	Irvine *et al.* (1989)	54	Environmental factors that block or interfere with execution of self-care behaviors
Diabetes Locus of Control Scale	Bradley *et al.* (1984)	42	Locus of control scale that identifies patients who are poor risks for pump therapy
Diabetes Family Behavior Checklist	Schafer *et al.* (1986)	16	Family factors that enhance appropriate self-care behaviors
Hypoglycemic Fear Survey	Cox *et al.* (1987)	27	Identification of patients with significant fear of future hypoglycemic episodes that may lead to poor metabolic control
Diabetes Knowledge Scale	Garrard *et al.* (1987)	15	A generic clinical diabetes knowledge scale
Diabetes Health Belief Scale	Harris *et al.* (1987)	38	Assessment of severity or vulnerability and cost–benefit dimensions of the health belief model
Diabetes Quality of Life Scale	DCCT (1988)	46	Quality-of-life scale measuring satisfaction, impact, and social and diabetes worry
Male Sexual Dysfunction Questionnaire	Abel *et al.* (1982)	7	Differentiates psychogenic from organic impotence

dom blood glucose test yielded a value of 296 mg/dl, and her glycosylated hemoglobin was 20.1%; both were extremely high values. At that time, the options available to her for diabetes and weight control were presented, including insulin therapy, the gastric bubble, and gastric stapling. She refused all of these. It was explained that she was at acute risk of developing complications of diabetes. She stated that she understood but was helpless to do anything about her problems. She was willing to see a psychologist, and a referral was made.

The patient was seen for about 25 sessions over the course of a year. The treatment focused on (1) helping the patient to understand the reasons for her inability to cope more effectively with diabetes self-care, particularly weight loss, and (2) working with the patient to develop behavioral strategies for controlling her eating behavior. The outcomes of treatment were limited improvements in eating behavior (decreased bingeing and the development of a 1,500-calorie meal plan) and glycosylated hemoglobin (17.1%), as well as a better understanding by the patient of how her values and health-related beliefs influenced her efforts to control her eating. With respect to the latter, the patient came to realize that dating and past sexual experiences had made her extremely ambivalent about the possibility of losing weight and becoming more physically attractive. Also, as long as the threat of complications from diabetes remained theoretical, the patient was un-

willing to sacrifice the pleasure she derived from her lifestyle for potential improvements in health. The patient left therapy with a clearer understanding and acceptance of the risks she was running, and why she was doing so.

The value of this case in the present context is twofold. First, it illustrates the difficulties that psychologists can expect to encounter in the area of diabetes self-care. Second, it highlights some of the complexities that arise when psychological treatment is used as a tool to achieve medically desirable lifestyle changes. Because of the importance of the patient's own values and goals, it may be that a result that would be seen as a "failure" from a medical point of view is at least acceptable from a psychological perspective.

Case Study 2

A couple in their mid-40s, married for 18 years, presented to the behavioral medicine clinic complaining of problems with the husband's hypoglycemia. He had had diabetes for 23 years and used insulin pump therapy to control his blood glucose. The wife reported that he was having frequent hypoglycemic episodes, which were quite frightening to her, and that he often became very irritable when these episodes occurred. The husband admitted to recent mood lability but denied that it was related to his diabetes. Further assessment revealed a long history of frequent arguments, difficulty in expressing emotions, and an inability to resolve conflicts.

A treatment plan was designed to address both the diabetes-specific and the more traditional marital problems. The husband first underwent glucose counterregulation testing (constant intravenous insulin infusion until hypoglycemia occurs to allow an assessment of the ability to hormonally stabilize or reverse blood glucose). The test revealed (1) defective counterregulation and (2) an inability to detect hypoglycemia despite neuroglycopenic symptoms.

Throughout his 11 weekly therapy sessions, the husband kept a blood glucose diary in which he recorded (1) his symptoms; (2) his estimated current glucose level; and (3) his actual glucose level as determined by SMBG four times daily. An analysis of these data showed that the husband was very frequently hypoglycemic although he was unaware of it (e.g., estimating his glucose to be normal when it was actually quite low). Based on these findings, the husband's diabetes treatment plan was revised to include more frequent self-tests, snacks before physical activity, and increased trust in his wife's suggestion that he test his blood glucose.

Traditional marital therapy focused on improving communication and negotiation skills. A behavioral approach was used in which the couple learned and practiced more appropriate reactions in conflict situations, as well as the expression of positive emotions. The homework assignments included communication and listening exercises, increased physical contact, and the recording of arguments. On termination, the couple had significantly decreased the frequency of their arguments and had improved their conflict resolution skills. They indulged in more frequent positive interactions, including taking walks and planning one weekly pleasure outing together. The husband's awareness of his inability to detect hypoglycemia and its effect on his moods convinced him to use SMBG more frequently and to increase his efforts to avoid low blood glucose and to reduce his hypoglycemic episodes.

This case illustrates the need for specific assessment and intervention that focuses on the diabetes-related aspects of the presenting problems. In this particular case, the wife's desire to improve the quality of the relationship was greatly increased by her husband's willingness to change his diabetes management. The positive outcome demonstrates the potential benefits of combining diabetes-specific and traditional therapeutic approaches.

Summary

The management of diabetes is clearly a complex behavioral process, influenced by a multitude of disease, individual, social, and situational variables. It is clear that behavior affects diabetes and that diabetes affects behavior. Consequently, the understanding and management of diabetes will progress with the mutual respect and collaboration of medical and behavioral science professionals at both a clinical and a research level (Cox et al., 1986). Major clinical problems with unique relevance to patients with diabetes have been described in this chapter. Guidelines for psychological assessment and interventions have also been presented in an effort to integrate clinical psychology into the treatment of diabetes.

ACKNOWLEDGMENTS

The authors wish to express their deepest appreciation to Diana Julian for her skillful efforts in both coordinating the work of the three coauthors and assuming all responsibility for the preparation of the manuscript. Additionally, this chapter was supported in part by Virginia State funds supporting the Clinical Diabetes Research Center and NIH grant RO1 AM28288.

References

Abel, G. G., Becker, J. V., Cunningham-Rathner, J., Mittelman, M., & Primack, M. (1982). Differential diagnosis of impotence in diabetics: The validity of sexual symptomatology. *Neurourology and Urodynamics, 1,* 57–69.

Ahlfield, J. E., Soler, N. G., & Marcus, S. P. (1985). The young adult with diabetes: Impact of the disease on marriage and having children. *Diabetes Care, 8,* 52–56.

American Diabetes Association. (1987). Consensus statement on self-monitoring of blood glucose. *Diabetes Care, 10,* 95–99.

American Diabetes Association. (1988). *Physician's guide to insuling-dependent (Type I) diabetes: Diagnosis and treatment.* Alexandria, VA: Author.

American Psychiatric Association. (1987). *DSM-III-R: Diagnostic and Statistical Manual of Mental Disorders.* Washington, DC: American Psychiatric Association.

Anderson, R. M. (1985). Is the problem of noncompliance all in our heads? *The Diabetes Educator, 11*(1), 31–34.

Arky, R. A. (1983). Nutritional managment of the diabetic. In M. Ellenberg & H. Rifkin (Eds.), *Diabetes mellitus: Theory and practice.* New Hyde Park, NY: Medical Examination Publishing.

Beck, A. T. (1976). *Cognitive therapy and the emotional disorders.* New York: International Universities Press.

Becker, M. H., & Maiman, L. A. (1975). Sociobehavioral determinants of compliance with health and medical care recommendations. *Medical Care, 13*, 10–24.

Bernbaum, M., Albert, S. G., & Duckro, P. N. (1988). Psychosocial profiles in patients with visual impairment due to diabetic retinopathy. *Diabetes Care, 11*, 551–557.

Bloom Cerkoney, K. A., & Hart, L. K. (1980). The relationship between the health belief model and compliance of persons with diabetes mellitus. *Diabetes Care, 3*(5), 594–598.

Bradley, C., Brewing, C., Gamsu, D. S., & Moses, J. L. (1984). Development of scales to measure perceived control of diabetes mellitus and diabetes related health beliefs. *Diabetic Medicine, 1*, 213–218.

Buckmaster, L., & Brownell, K. D. (1988). Behavior modification: The state of the art. In R. T. Frankle & M-U. Yang (Eds.), *Obesity and weight control.* Rockville, MD: Aspen.

Cahill, G. F., Etzwiler, D. D., & Freinkel, N. (1971). Blood glucose control in diabetes. *Diabetes, 20*, 785–799.

Clarke, D. H. (1988). Erectile dysfunction in diabetes. *Clinical Diabetes, 6*(5), 103–105.

Cox, D. J., Clarke, W., Pohl, S., Gonder-Frederick, L., Hoover, C., Zimbelman, L., & Pennebaker, J. W. (1985). Accuracy of perceiving blood glucose in IDDM patients. *Diabetes Care, 8*(6), 529–536.

Cox, D. J., Gonder-Frederick, L., Pohl, S., & Pennebaker, J. W. (1986). Diabetes. In K. A. Holroyd & T. L. Creer (Eds.), *Self-management of chronic diseases: Handbook of clinical interventions and research.* New York: Academic Press, (pp. 305–346).

Cox, D. J., Irvine, A., Gonder-Frederick, L., Nowacek, G., & Butterfield, J. (1987). Fear of hypoglycemia: Quantification, validation, and utilization. *Diabetes Care, 10*, 617–621.

Cox, D. J., Carter, W. R., Gonder-Frederick, L. A., Clarke, W. L., & Pohl, S. (1988). Training awareness of blood glucose in IDDM patients. *Biofeedback and Self-Regulation, 13*(3), 201–217.

Cox, D. J., Gonder-Frederick, L. A., Julian, D. M., Carter, W. R., & Clarke, W. L. (1989). Blood glucose awareness training among IDDM patients: Effects and correlates. *Diabetes Care, 13*, 313–318.

Craighead, J. E. (1978). Current views on the etiology of insulin-dependent diabetes mellitus. *New England Journal of Medicine, 299*, 1439–1445.

DCCT Research Group. (1988). Reliability and validity of a diabetes quality-of-life measure for the Diabetes Control and Complications Trial (DCCT). *Diabetes Care, 11*, 725–732.

Dobson, J. J., Nord, W. R., & Haire-Joshu, D. (1989). The use of goal setting by physicians in the treatment of diabetes. *The Diabetes Educator, 15*(1), 62–65.

Drash, A. L. (1979). The child with diabetes mellitus. In B. A. Hamburg, L. F. Lipsett, G. E. Inoff, & A. L. Drash (Eds.), *Behavioral and psychosocial issues in diabetes: Proceedings of the National Conference.* National Institute of Health, Washington, DC.

Drash, A. L. (Ed.). (1984). *The physicians guide to Type II diabetes (NIDDM): Diagnosis and treatment.* New York: American Diabetes Association.

Dunn, S. M., Smartt, H. H., Beeney, L. J., & Turtle, J. R. (1986). Measurement of emotional adjustment in diabetic patients: Validity and reliability of ATT39. *Diabetes Care, 9*, 480–489.

Edgerton, M. T., & Langman, M. (1982). Psychiatric considerations. In E. H. Courtias (Ed.), *Male esthetic surgery.* St. Louis: Mosby.

Ellenberg, M. (1983). Diabetic neuropathy. In M. Ellenberg & H. Rifkin (Eds.), *Diabetes mellitus theory and practice,* (3rd Ed., Vol. 2, pp. 777–801). New Hyde Park, NY: Medical Examination Publishing.

Epstein, L. H., Figueroa, J., Farkas, G. M., & Beck, S. (1981). The short-term effects of feedback on accuracy of urine glucose determination in insulin-dependent diabetic children. *Behavior Therapy, 12*, 560–564.

Feuerstein, M., Labbe, E. E., & Kuczmierczyk, A. R. (1986). *Health psychology.* New York: Plenum Press.

Garrard, J., Joynes, J. O., Mullen, L., McNeil, L., Mensin, C., Feste, C., & Etwiler, B. D. (1987). Psychometric of patient knowledge test. *Diabetes Care, 10*(4), 500–509.

Gatchel, R. J., & Baum, A. (1983). *An introduction to health psychology.* New York: Newbery Award Records.

Glasgow, R. E., McCaul, K. D., & Schafer, L. (1986). Barriers to regimen adherence among persons with insulin-dependent diabetes. *Journal of Behavioral Medicine, 9*, 65–77.

Goldstein, D. E., & Hoeper, M. (1987). Management of diabetes during adolescence: Mission impossible? *Clinical Diabetes, 5*, 1–9.

Gonder-Frederick, L. A., & Cox, D. J. (1990). Symptom perception and blood glucose feedback in the self-treatment of IDDM. In C. S. Holmes (Ed.), *Neuropsychological and behavioral aspects of diabetes.* New York: Springer Verlag.

Gonder-Frederick, L. A., Cox, D. J., Pennebaker, J. W., & Bobbitt, S. A. (1986). Blood glucose symtom beliefs of diabetic patients: Accuracy and implications. *Health Psychology, 5*(4), 327–342.

Gonder-Frederick, L. A., Cox, D. J., Bobbitt, S. A., & Pennebaker, J. W. (1989). Changes in mood state associated with blood glucose fluctuations in insulin dependent diabetes mellitus. *Health Psychology, 8*(1), 45–59.

Green, J. G., & Saunders, J. T. (1988). Strategies for long-term weight maintenance. *On the cutting edge* (Newsletter of the Diabetes Care and Education Practice Group of The American Dietetic Association), pp. 13–16.

Hall, W. D. (1986). Hypertension in the patient with diabetes. In J. K. Davidson (Ed.), *Clinical diabetes mellitus*, pp. 454–459. New York: Thieme.

Harris, R., Linn, M. W., Skyler, J. S., & Sandifer, R. (1987). Development of the Diabetes Health Belief Scale. *The Diabetes Educator, 13*(3), 292–297.

Henry, R. R., Wiest-Kent, T. A., Schaeffer, L., Kolterman, O. G., & Olefsky, J. M. (1986). Metabolic consequences of very low-calorie diet therapy in obese non-insulin-dependent diabetic and nondiabetic subjects. *Diabetes, 35,* 155–164.

Herrman-Lee, J. (1989). *Glycemic control and analgesic response in diabetic rats*. Dissertation, University of Virginia.

Hiss, R. G. (1986). The activated patient: A force for change in diabetes health care and education. *The Diabetes Educator, 12,* 225–231.

Holmes, D. M. (1986). The person and diabetes in psychosocial context. *Diabetes Care, 9,* 194–206.

Hoover, J. W. (1983). Patient burnout, and other reasons for noncompliance. *The Diabetes Educator, 9*(3), 41–43.

Horan, P. F., Gwynn, C., & Renzi, D. (1986). Insulin-dependent diabetes mellitus and child abuse: Is there a relationship? *Diabetes Care, 9,* 302–307.

Irvine, A. A., Saunders, J. T., Blank, M., & Carter, W. (1990). Environmental barriers to diabetic regiment adherence: A scale validation. *Diabetes Care, 13*(7).

Jacobson, A. M., & Hauser, S. T. (1983). Behavioral and psychological aspects of diabetes. In M. Ellenberg & H. Rifkin (Eds.), *Diabetes mellitus: Theory and practice*. New Hyde Park, NY: Medical Examination Publishing.

Jacobson, A. M., Hauser, S. T., Wertlieb, D., Wolfsdorf, J. I., Orleans, J., & Vieryra, M. (1986). Psychological adjustment of children with recently diagnosed diabetes mellitus. *Diabetes Care, 9,* 323–329.

Johnson, S. B. (1980). Psychosocial factors in juvenile diabetes: A review. *Journal of Behavioral Medicine, 3,* 95–115.

Koski, M. L., & Kumento, A. (1977). The interrelationship between diabetic control and family life. *Pediatric and Adolescent Endocrinology, 3,* 41–45.

Kovacs, M., Finkelstein, R., Feinberg, T. L., Crouse-Novak, M., Paulauskas, S., & Pollack, M. (1985). Initial psychologic responses of parents to the diagnosis of insulin-dependent diabetes mellitus in their children. *Diabetes Care, 8,* 568–575.

Lustman, P. J., Griffith, L. S., & Clouse, R. E. (1988). Depression in adults with diabetes: Results of 5-yr follow-up study. *Diabetes Care, 11,* 605–612.

MacLeod, J. H. (1983). Biofeedback in the management of partial and incontinence. *Diseases of the Colon and Rectum, 26,* 244–246.

Mattsson, A. (1979). Juvenile diabetes: Impacts on life stages and systems. In B. A. Hamburg, L. F. Lipsett, G. E. Inoff, & A. L. Drash (Eds.), *Behavioral and psychosocial issues in diabetes: Proceedings of the National Conference*. National Institute of Health, Washington, DC.

Meichenbaum, D., & Turk, D. C. (1987). *Facilitating treatment adherence*. New York: Plenum Press.

Miller, L. V., Goldstein, J., & Nicolaison, G. (1978). Evalua-

tion of patient's knowledge of diabetes self-care. *Diabetes Care, 1,* 275–280.

National Institutes of Health. (1987). Consensus development conference on diet and exercise in non-insulin-dependent diabetes mellitus. *Diabetes Care, 10,* 639–644.

Newbrough, J. R., Simpkins, C. G., & Maurer, H. (1985). A family development approach to studying factors in the management and control of childhood diabetes. *Diabetes Care, 8,* 83–92.

Newburgh, L. H., & Conn, J. W. (1979). A new interpretation of hyperglycemia in obese, middle-aged persons. *Journal of the American Medical Association, 112,* 7–11.

O'Connell, K. A., Hamera, E. K., Knapp, T. M., Cassmeyer, V. L., Eaks, G. A., & Fox, M. A. (1984). Symptom use and self-regulation in Type II diabetes. *Advances in Nursing Science, 6,* 19–28.

Olefsky, J. M. (1976). The insulin receptor: Its role in insulin resistance of obesity and diabetes. *Diabetes, 25,* 1154–1164.

Orr, D. P., Golden, M. P., Myers, G., & Marrero, D. G. (1983). Characteristics of adolescents with poorly controlled diabetes referred to a tertiary care center. *Diabetes Care, 6,* 170–175.

Pennebaker, J. W., Cox, D. J., Gonder-Frederick, L., Wunsch, M. G., Evans, W. S., & Pohl, S. (1981). Physical symptoms related to blood glucose in insulin-dependent diabetics. *Psychosomatic Medicine, 43*(6), 489–500.

Peterson, C. M., & Jovanovic-Peterson, L. (1988). First-line treatment of hypertension with diabetes. *Diabetes Professional,* pp. 12–14.

Powers, M. A. (1989). Building the educational partnership. *The Diabetes Educator, 15*(2), 155–156.

Rost, K. (1989). Patient activation interventions: Research issues in dissemination. *The Diabetes Educator, 15*(1), 80–82.

Roter, D. L. (1977). Patient participation in patient-provider interactions: The effects of patient question asking on the quality of interaction, satisfaction, and compliance. *Health Education Monographs, 5,* 281–315.

Runyan, J. W. (1988). Coexisting hypertension and diabetes. *Practical Diabetology, 7*(3), 1–7.

Ryan, C. M. (1988). Neurobehavioral complications of Type I diabetes: Examination of possible risk factors. *Diabetes Care, 11,* 86–93.

Salans, L. B., Knittle, J. L., & Hirsch, J. (1983). Obesity, glucose intolerance, and diabetes mellitus. In M. Ellenberg & H. Rifkin (Eds.), *Diabetes mellitus theory and practice* (3rd ed., Vol. 2, pp. 269–479). New Hyde Park, NY: Medical Examination Publishing.

Schafer, L. C., Glasgow, R. E., & McCaul, K. D. (1982). Increasing the adherence of diabetic adolescents. *Journal of Behavioral Medicine, 5,* 353–362.

Schafer, L. C., McCaul, K. D., & Glasgow, R. E. (1986). Supportive and nonsupportive family behaviors: Relationships to adherence and metabolic control in persons with Type I diabetes. *Diabetes Care, 9,* 179–185.

Schiffrin, A., & Belmonte, M. M. (1982). Multiple daily self-glucose monitoring: Its essential role in long-term glucose control in insulin-dependent diabetic patients treated with pump and multiple subcutaneous injections. *Diabetes Care, 5,* 479–484.

Schreiner-Engel, P. (1988). Diagnosing and treating the sexual problems of diabetic women. *Clinical Diabetes, 6*(6), 1261–136.

Simonds, J. R., Goldstein, P., Walker, B., & Rawlings, S. S. (1981). The relationship between psychological factors and blood glucose regulation in insulin-dependent diabetic adolescents. *Diabetes Care, 4,* 610–615.

Sims, C. G., Remler, H., & Cox., D. J. (1988). Biofeedback and behavioral treatment of elimination disorders. *Clinical Biofeedback and Health, 10,* 128–135.

Skyler, J. (1981a). Psychological issues in diabetes. *Diabetes Care, 4,* 656–657.

Skyler, J. (Ed.). (1981b). Symposium on blood glucose self-monitoring. *Diabetes Care, 4,* 392–426.

Sperling, M. A. (Ed.). (1988). *The physician's guide to insulin-dependent (Type I) diabetes: Diagnosis and treatment.* Alexandria, VA: American Diabetes Association.

Stone, G. C., Cohen, F., & Adler, N. E. (1979). *Health psychology.* San Francisco: Jossey-Bass.

Wing, R. R., Marcus, M. D., Epstein, L. H., & Salata, R. (1987). Type II diabetic subjects lose less weight than their overweight nondiabetic spouses. *Diabetes Care, 10*(5), 563–566.

Witherington, R. (1989). Mechanical aids for treatment of impotence. *Clinical Diabetes, 7*(1), 1–14.

Wisocki, T., Meinhold, P., Cox, D. J., & Clarke, W. (1990). Survey of diabetes professionals regarding developmental changes in diabetes selfcare. *Diabetes Care, 13*(1), 65–68.

Zisook, S. (1987). *Biopsychosical aspects of bereavement.* New York: American Psychiatric Press.

CHAPTER 27

Psychological Theory, Assessment, and Interventions for Adult and Childhood Asthma

Thomas L. Creer, Russ V. Reynolds, and Harry Kotses

Introduction: Description of the Disorder

Although the symptoms that comprise asthma were described by many ancient writers, a precise definition of asthma has eluded scholars and scientists alike. As Sol Permut observed, "It's like love—we all know what it is, but who would trust anybody else's definition?" (cited in Gross, 1980, p. 203). Gross (1980) suggested that Permut undoubtedly made his remarks because

> the etiology or etiologies are obscure, the clinical picture is diverse, and the pathophysiological mechanisms are seemingly multiple, yet it borders or overlaps other conditions to the extent that it makes it necessary to decide "what is asthma and what is not," if we wish to communicate what is being referred to. (p. 203)

The problem of defining asthma is illustrated by comparing three definitions. Pearlman (1984) emphasized that asthma is a disorder of the bronchial tree in which there is recurrent and at least partially reversible generalized obstruction to airflow. Williams (1982) stressed the hyperirritability of the airway to various stimuli and the extreme variability of the disorder. Finally, a definition by a committee of the American Thoracic Society (1987) features the characteristic that attacks can reverse either with treatment or spontaneously. Considering all these descriptions, the following definition might be proposed: Asthma is a disorder characterized by increased hyperreactivity of the airways to various stimuli, including (1) allergens; (2) nonspecific irritants, such as exercise and cold air; and (3) infections. Several responses may occur, including (1) constriction of the smooth muscle in the bronchial wall; (2) swelling of the bronchial walls; (3) increased mucus secretion; (4) infiltration of the inflammatory cells; or (5) a combination of these factors. The occurrence of these responses is commonly referred to as an asthma attack; these attacks, episodes, or flare-ups occur intermittently, vary in severity, and may reverse either spontaneously or as a result of treatment. This definition emphasizes four major characteristics of asthma: the hyperreac-

Thomas L. Creer, Russ V. Reynolds, and Harry Kotses • Department of Psychology, Ohio University, Athens, Ohio 45701.

tivity of the airways and the intermittency, variability, and reversibility of attacks.

Hyperreactivity of the Airways

Stimuli that have no effect when inhaled by normal individuals can trigger bronchoconstriction in patients with asthma. An increasingly large number of stimuli have been identified as producing attacks, although the specific stimuli that trigger episodes vary both from patient to patient and from attack to attack in the same patient. There are patients, mainly children, who suffer seasonal asthma. They experience asthma when allergen counts are high but are often asthma-free during the remainder of the year. They are sometimes said to have extrinsic asthma because their attack triggers are outside the body. A second category of patients, often adults with late-onset asthma, experience flare-ups on a perennial basis. A gamut of stimuli may trigger their attacks, including nonspecific irritants and infections. Their asthma is often referred to as intrinsic or ideopathic with precipitating stimuli that are hard to identify. (To further complicate this picture, most patients have mixed asthma; this means that both extrinsic and intrinsic stimuli trigger their attacks.)

Intermittency of Attacks

The frequency of attacks varies from individual to individual and, for any given patient, from time to time. A patient may suffer several attacks during a day or may go months, even years, between attacks. The frequency of attacks suffered by a particular patient is a function of the number and diversity of the stimuli that trigger his or her asthma. As noted, some patients suffer attacks only during certain seasons when pollens are present in their environment; for others, however, any number and variety of stimuli can precipitate flare-ups.

Variability of Attacks

Variability refers to the severity both of discrete attacks and of the overall condition of a patient's asthma. This characteristic was sug-

gested by Williams (1982) to be the reason why asthma has escaped a precise definition. Asthma severity presents two major concerns to medical and behavioral scientists. First, there is no standard way of classifying either a given attack or the overall course of a patient's disorder as mild, moderate, or severe asthma. Second, the lack of operational definitions concerning the severity of attacks complicates the already intricate matter of assessing a condition that may change over time.

Reversibility of Asthma

Reversibility is the *sine qua non* of asthma according to McFadden (1980); it distinguishes the condition from other types of respiratory conditions, particularly emphysema, where there is no reversibility. This characteristic presents two major concerns to the asthma treatment team. First, reversibility is a relative condition. Although the majority of patients show complete reversibility of airway obstruction, others do not, even with intensive therapy. Second, the ability of attacks to remit spontaneously makes it difficult to establish, with certainty, which intervention produced the change (Creer, 1982).

Psychological Factors: A Systems Perspective

Although psychological variables are important in the assessment and treatment of many cases of adult and childhood asthma, they are not causative in a primary sense; psychological events precipitate or exacerbate asthma only for those individuals with hypersensitive airways (Sadler, 1982). Because the relationship between psychological variables and illness is reciprocally multidirectional, a concise summary of the connections between thoughts, emotions, actions, environment, and physiological asthmatic processes is not possible (see Engel, 1986; Tobin, Reynolds, Holroyd, & Creer, 1986). Therefore, any discussion of psychological factors and asthma is likely to oversimplify the clinical reality of a single case.

Two useful schemes assist our thinking about the relationship of psychological events, psychological interventions, and asthma. Matus (1981) outlined three levels of psychological factors that operate with asthma: psychological precipitants, such as emotional arousal, that have a direct effect on lung function; psychological exacerbants, such as panic, that contribute to a worsening of asthma because of their inhibiting effects on adequate self-care actions; and psychological factors, such as secondary gain, low self-esteem, or inadequate asthma knowledge, that interfere with preventive actions. Similarly, Creer (1982) discussed three levels of psychological interventions: psychotherapy aimed at resolving the conflicts that hypothetically cause asthma (an approach now in disfavor); the institution of behavioral procedures, such as relaxation, for their direct effects on lung function; and the use of behavioral and psychotherapeutic methods to modify behavior (e.g., medication noncompliance) or to resolve conflict (e.g., ambivalence about authority figures such as physicians) that has led indirectly to the mismanagement of the patient's asthma.

The next section of this chapter focuses on the hypothesized direct effects of emotion, stress, and relaxation on lung function. The subsequent section provides an overview of behavioral and psychotherapeutic interventions targeted for psychological factors that contribute indirectly to asthma.

Direct Psychological Effects: Stress and Relaxation

A direct link between asthma and psychological factors has been hypothesized for over two millennia. In the Middle Ages, Maimonides reiterated some of the teachings of Hippocrates when he wrote that mental anguish or distress in asthma patients prevented normal respiration. In the modern era, a similar theme was echoed first by Mackenzie (1886), who told how a patient with "rose asthma" experienced constriction of the nasal passages, respiratory discomfort, tightening of the chest, and nasal secretion when brought into contact with an artificial rose. The ideas embodied both in the ancient teachings and in the more recent clinical observations culminated, in the 20th century, in the psychosomatic formulation of asthma.

Despite the long association between asthma and psychology, the systematic study of the role of psychological factors in asthma began only in recent years. The research in this area has focused on how psychological factors exacerbate or reduce asthma. Psychological factors associated with the exacerbation of asthma involve stress; factors associated with the reduction of asthma have to do with relaxation. Kotses, Hindi-Alexander, and Creer (1989) recently reviewed research that showed that stress and relaxation, respectively, exacerbated and reduced asthma. Although the findings from this work are both positive and unambiguous, the authors' interpretation is subject to debate.

Experimental Evidence for Direct Psychological Factors in Asthma

Three approaches have been used experimentally to bring about increases in airflow resistance in asthma patients: threat of aversive stimulation, imagination of aversive events, and performance of aversive tasks (see Kotses *et al.*, 1989). The most popular of the three, threat of aversive stimulation, involves the use of a suggestion procedure. Patients have been told they would experience symptoms of asthma after inhaling a substance that was identified as a bronchoconstrictor. Usually, a neutral substance was administered. In a few cases, however, a bronchoconstrictor was administered to determine how the response to the drug plus the warning differed from the response to only the drug (see Kotses *et al.*, 1989).

When a neutral substance is presented as a bronchoconstrictor, many asthma patients exhibit increases in airflow resistance. When a bronchoconstrictor is administered and identified, asthma patients exhibit a greater increase in airflow resistance than they do when the same drug is administered without identification. These findings parallel those from other experiments, in which asthma patients either imagined aversive events or performed aversive tasks. Adult asthma patients experienced increased airflow resistance when they imagined asthma, cough, fear, and anger; pediatric asthma patients evidenced increased airflow resistance when they imagined fear and anger. Both adult and pediatric asthma patients increased airflow resistance when they watched stressful movies, and asthmatic adults did so when they performed mental arithmetic. Clearly, a variety of procedures have resulted in airflow resistance increases in asthma patients. What these procedures have in common is that they all incorporate elements of stress. On this basis, it may be concluded that stress is responsible for airflow resistance increases in asthma patients (see Kotses *et al.*, 1989).

Just as stress exacerbates asthma, relaxation leads to improvement in asthma. Two approaches to relaxation have been used: general relaxation and limited relaxation. The most popular general relaxation procedure used with asthma patients has been a form of progressive relaxation training (Jacobsen, 1938). It has been effective in producing decreases in airflow resistance when used by itself (Alexander, 1972; Alexander, Miklich, & Hershkoff, 1972), and when combined with biofeedback (Davis, Saunders, Creer, & Chai, 1973; Scherr, Crawford, Sergent, & Scherr, 1975), autogenic training (Alexander, Cropp, & Chai, 1979; Moore, 1965), guided imagery (Hock, Rodgers, Reddi, & Kennard, 1978), or reciprocal inhibition (Miklich, Renne, Creer, Alexander, Chai, Davis, Hoffman, & Danker-Brown, 1977). Usually, the combination of progressive relaxation training plus another procedure results in greater improvement in airflow than progressive relaxation training alone. Other general relaxation procedures that have resulted in asthma improvement include hypnosis plus relaxation instructions (Ben-zvi, Spohn, Young, & Kattan, 1982; Maher-Loughnan & Kinsley, 1968; Smith, Colebatch, & Clarke, 1970), suggestion of easier breathing (Luparello, Leist, Lourie, & Sweet, 1970), quasi-hypnotic instructions (Tal & Micklich, 1976), and imagining relaxation plus reciprocal inhibition (Yorkston, Eckert, McHugh, Philander, & Blumenthal, 1979; Yorkston, McHugh, Brady, Serber, & Sergeant, 1974).

Limited relaxation also effects improvements in asthma patients. This form of relaxation is achieved through electromyographic biofeedback training. It is called limited because its relaxation effects are restricted to the trained response (Glaus & Kotses, 1979; Kotses & Glaus, 1982); biofeedback-induced relaxation of one set of muscles is not reflected in other muscles. Relaxation limited to the facial musculature led to decreases in airflow resistance in children with asthma immediately after training (Kotses, Glaus, Bricel, Edwards, & Crawford, 1978), a few hours after training (Kotses, Glaus, Crawford, Edwards, & Scherr, 1976), and several months after training (Kotses, Harver, Segreto, Glaus, Creer, & Young, 1989). In the latter case, biofeedback-induced facial relaxation was combined with instructions to practice facial relaxation at home without the aid of electronic feedback.

The research described above clearly demonstrates that stress and relaxation, respectively, increase and decrease airflow resistance in asthma patients and thereby influence symptoms of asthma. These findings have been taken to mean that emotional or psychological factors modulate asthma. Contributing to the conclusion is the traditional idea that increased airflow resistance and asthma are equivalent. In all likelihood, however, the conclusion is not warranted. Asthma and increased airflow resistance are not equivalent, and the effects of psychological factors on airflow resistance do not occur exclusively in asthma patients. Recent work has shown that some of the procedures producing airflow resistance increases in asthma patients also produce increases in healthy individuals. The procedures investigated include mental arithmetic (Kotses, Westlund, & Creer, 1987) and threat of aversive stimulation (Kotses, Rawson, Wigal, & Creer, 1987).

Psychological Assessment and Clinical Interventions

General Considerations

Physicians and other members of the medical treatment team usually refer an asthmatic patient for behavioral medicine or psychiatric consultation when the individual's asthma is not controlled, despite the use of usual and appropriate medical care. In addition, a patient may be referred to a clinical psychologist because it is suspected that psychological factors contribute to or in some way complicate the management of his or her asthma. In some cases, despite adequate management of the patient's asthma, a physician may judge that a psychological intervention could further improve the overall asthma management and/or quality of life for the asthmatic. Alternately, some patients are referred simply because they are uncooperative with treatment or present some other behavior problem.

The complexity of asthma as a medical disorder and the host of psychological factors that are associated with adaptation to asthma make it a challenging clinical problem to assess and treat. Two types of preliminary psychological assessment strategies have been developed for asthmatic patients. The first approach is the use of a checklist or structured interview such as the Asthma Problem Behavior Checklist (APBC; Creer, Marion, & Creer, 1983). This instrument provides self-report information concerning a broad range of asthma-related behavioral, emotional, and lifestyle problems. The content areas surveyed with the APBC include medication compliance, addictive behaviors, asthma triggers, attack early-warning signs, panic, assertiveness skills, psychological symptoms (e.g., depression), medication side effects, family conflict, limits in daily living, and self-management skills of the patient and the family.

An alternative approach to the psychological assessment of an asthmatic patient is the use of the Battery of Asthma Illness Behavior (e.g., Dirks & Kinsman, 1982; Dirks & Kreischer, 1982). This battery consists of three paper-and-pencil instruments: the Minnesota Multiphasic Personality Inventory (MMPI), the Respiratory Illness Opinion Survey (RIOS), and the Asthma Symptoms Checklist (ASC). The MMPI assesses general personality characteristics as interpreted from the standard clinical scales and experimental scales (Barron, 1953; Dirks, Schraa, Brown, & Kinsman, 1980; Lachar, 1974; Navran, 1954). The RIOS assesses attitudes toward asthma and its treatment such as optimism about coping with the illness, regard for medical staff, locus of control over asthma manage-

ment, and the degree of stigma patients associate with their illness (Staudenmayer, Kinsman, & Jones, 1978). The ASC measures the subjective symptomatology that patients experience during acute asthma episodes, including somatic symptoms (e.g., hyperventilation and airway obstruction), affective symptoms (e.g., panic-fear and irritability), and fatigue (Kinsman, Dahlem, Spector, & Staudenmayer, 1977).

The structured interview and the psychological testing strategies have different inherent strengths and weaknesses. However, both provide a comprehensive preliminary assessment of asthmatic referrals. In most hospital settings, there is limited time to conduct an assessment, to formulate a treatment plan, and to carry out an intervention; for example, in 1984, the average hospital stay for a child admitted for asthma was 3.6 days (Halfon & Newacheck, 1986). Therefore, an efficient yet comprehensive assessment, as afforded by the two methods discussed above, is a necessity. (See Matus, 1981, for a comprehensive overview of the assessment of psychological factors that contribute to precipitating, exacerbating, or maintaining asthma.)

Specific Problems

Panic and Phobic Responses

Symptoms of panic are often associated with attack onset in adults and children with asthma. In some cases these symptoms are anticipatory because they occur with the mildest symptoms of asthma, whereas for other individuals the panic occurs only with moderate to severe asthma episodes. Asthmatic patients can also develop phobic reactions to medical equipment and procedures, or to the hospital itself (Creer, 1979). In any case, panic and other fear reactions can contribute to the onset or worsening of an asthma episode and can directly interfere with a patient's ability to make decisions and effectively manage an attack. Further, panic is implicated as contributing to asthma deaths in many cases (Friedman, 1984).

Symptoms of panic include sudden apprehension or fear, shortness of breath, dizziness, choking, increased heart rate, shaking, sweating, nausea, numbness or tingling, and chest pain (American Psychiatric Association, 1987). Several of these symptoms, such as shortness of breath, are common signs of asthma and therefore may be a simple indication of attack onset and not panic *per se*. In assessing for panic reactions in the context of asthma, it is important to document the presence of other symptoms, such as sudden apprehension or fear, that are less likely to be confounded with primary asthma symptoms.

Panic assessment strategies can include the use of the Asthma Problem Behavior Checklist (Creer *et al.*, 1983), The Panic-Fear scale on the MMPI (Dirks *et al.*, 1980), the Panic-Fear scale on the Asthma Symptom Checklist (Kinsman *et al.*, 1977), or a simple review of the acknowledged symptoms of panic (American Psychiatric Association, 1987).

The treatment of choice for panic or phobic reactions is systematic desensitization (Wolpe, 1958) or some other self-directed exposure method involving relaxation. There are generally three steps in these exposure methods. First, the patient is taught to relax using progressive deep-muscle (Jacobson, 1938) or some other form of relaxation. Second, situations or experiences associated with panic are arranged by the patient in a hierarchy from least fearful to most fearful. In the case of asthma-induced panic, such a list might include the experience of asthma early-warning signs or the sight of an emergency room. Last, the experience of relaxation, as induced by the chosen relaxation method, is paired with imagining or actual exposure to the feared situation(s). Several other sources provide a more detailed introduction to using these methods with asthmatics (see Creer, Renne, & Chai, 1982; Creer & Reynolds, 1990a; Reynolds, Kotses, Creer, Bruss, & Joyner, 1989).

The desired impact of relaxation-based exposure treatments are at least threefold. First, they are designed to decrease anticipatory fear and panic before attacks. In this regard, at least one study has shown a long-term decrease in chronic anxiety following biofeedback-induced facial relaxation (Kotses *et al.*, 1989). Second, different forms of relaxation have been found

to have a direct effect on asthma by inducing a modest improvement in lung function (e.g., Alexander *et al.*, 1972; Kotses *et al.*, 1978), although these effects appear to be short-term in most cases (Kotses & Glaus, 1981). (Of course, most medication effects are time-limited following ingestion as well.) Third, teaching patients an active, self-initiated coping strategy such as relaxation can enhance their confidence in their ability to manage their asthma. Theoretically, such a belief is thought to be related to the initiation and persistence of self-management efforts (Tobin, Wigal, Winder, Holroyd, & Creer, 1987).

Systematic desensitization was originally developed as a procedure to be used in the clinician's office; however, relaxation is more commonly introduced today as a coping skill that patients use primarily in their own daily living. Research with a variety of lifestyle-related problems has demonstrated the importance of transferring these skills to daily living. For example, the only biofeedback–relaxation study that has demonstrated long-term effects on asthma included a home practice component (Kotses *et al.*, 1989). The use of such a coping-skills approach should help provide for the generalization of treatment effects to the home environment after discharge from the hospital.

Medication Noncompliance

There is little doubt that poor medication compliance is a significant problem in the treatment of asthma. Estimates of the degree of medication compliance in children have been as low as 2% (Sublett, Pollard, Kadlec, & Karibo, 1979). Although the course of asthma is variable, and the direct relationship between interventions and asthma outcomes are difficult to determine, one study found that children who adhere poorly to the proper use of theophylline, a commonly prescribed bronchodilator, experience more wheezing and decreased lung function compared to children who are compliant (Cluss, Epstein, Galvis, Fireman, & Friday, 1984). Also, the increase in pediatric hospitalization for asthma, despite

503

the development of more effective medications, has been attributed to poor medication compliance (Jerome, Wigal, & Creer, 1987).

There are several direct and indirect methods of assessing compliance to asthma medication (Jerome et al., 1987). Such methods include blood and serum assays (Weinberger & Cuskey, 1985), biological markers such as riboflavin (Cluss et al., 1984), observation (Jerome et al., 1987), self-report via the use of a daily asthma diary (Creer et al., 1988), and pill or liquid counts (Rapoff & Christophersen, 1982). It's important to note that, unfortunately, significant others, including parents, may not provide more accurate medication reports than the identified patient (Parish, 1986). The characteristics of a particular clinical setting and patient population dictate the most appropriate method of assessment.

Several interventions are suggested for improving medication compliance in adult and child asthmatics. For example, education about asthma medications should be a part of any compliance intervention (Haynes, Taylor, & Sackett, 1979; Rapoff & Christophersen, 1982). Topics to be covered include medication mechanisms of action, medication benefits, and probable side effects. (A discussion of asthma education programs appears below.) Two behavioral techniques, shaping and contracting, have also been found to be effective methods of promoting medication compliance. Shaping involves the systematic positive reinforcement of gradual approximations of the timely and appropriate use of asthma medications (Creer & Reynolds, 1990b). Behavioral contracting is a common method used to promote adherence to behavioral or pharmacological prescriptions. This method requires that a written contract be developed that clearly specifies the behavioral goals (e.g., taking theophylline before breakfast and dinner) and the consequences of compliance and/or noncompliance (see Jerome et al., 1987).

Improving asthma medication compliance sometimes necessitates that the patient negotiate with the physician about changes in the medication regimen. Minimizing a regimen's complexity and tailoring medication use to the patient's lifestyle are accepted features of an individualized medication regimen (Garfield, 1982; Haynes et al., 1979). A detailed review of medication compliance assessment and intervention methods can be found elsewhere (Jerome et al., 1987).

Misperception of Asthma Severity

Poor awareness of changes in asthma severity has been attributed to psychological factors, such as coping or defense styles, and to physiological mechanisms, including low hypoxic drive (e.g., Rubinfeld & Pain, 1976; Steiner, Higgs, Fritz, Laszlo, & Harvey, 1987). Whatever the process, the misperception of changes in lung function can adversely affect the management of asthma. For example, individuals who misperceive the severity of their symptoms are less likely to comply with their prescribed medication regimen (Becker, Radius, Rosenstock, Drachman, Schuberth, & Teets, 1978). Also, an inaccurate perception of asthma may contribute to the rising mortality rate for asthmatics, especially the sudden-onset deaths (Barger, Vollmer, Felt, & Buist, 1988).

There are two complementary approaches to assessing the presence of problems caused by symptom misperception which may be used in conjunction with one another. First, the clinician may inquire about a prior history of sudden-onset attacks that required emergency medical treatment. A lack of general awareness of one's asthma symptoms and trouble in identifying asthma triggers may also indicate that there are problems with accurate symptom perception. The Asthma Problem Behavior Checklist, discussed above, provides a structured interview format for eliciting this information (Creer et al., 1983). Alternately, one can assess symptom discrimination objectively through the use of lung function measures. Other sources provide an adequate discussion of the lung function tests used to identify patients with perceptual difficulties (Gottfried, Altose, Kelsen, & Cherniack, 1981; Rubinfeld & Pain, 1976).

The recommended intervention for patients with symptom discrimination problems (daily lung function assessment at home with the use of a portable peak flow meter; see Williams, 1982) is important for at least two reasons. First, daily lung function assessment may help patients learn to discriminate changes in airway obstruction (i.e., asthma severity), although research has produced contradictory findings in this regard (Higgs, Richardson, Lea, Lewis, & Laszlo, 1986; Silverman, Mayer, Sabinsky, Williams-Akita, Feldman, Schneider, & Chiaramonte, 1987; Sly, Landau, & Weymouth, 1985). Second, lung function assessment can be used to confirm the onset or worsening of an attack. Lung function values can help patients decide when to implement attack prevention steps (e.g., drinking water and resting), to take additional as-needed medication (e.g., inhaled bronchodilators), or to seek emergency medical treatment (Reynolds *et al.*, 1989; Williams, 1982).

In addition, research suggests that lung function values, through the use of Bayesean or other statistical methods, can be used to predict individual attacks for certain patients. The initial research suggests that patients need to self-monitor asthma symptoms and lung function values for at least four weeks before prediction is possible (Harm, Kotses, & Creer, 1985). For some patients, such a sustained data collection effort is worthwhile. Probability estimates based on peak expiratory flow rate (PEFR) data, collected twice daily, have produced a three- to fivefold increase in the predictability of asthma episodes over base rate for asthmatic children (Harm *et al.*, 1985; Taplin & Creer, 1978).

Clearly, daily peak flow assessment provides the technology for increasing the predictability of asthma episodes. However, research has yet to demonstrate that lung function data will be used by patients to improve self-management of their asthma. For example, the only treatment study to include asthma prediction training with other self-management training yielded no additional improvement in the patient's ability to manage their asthma over the effects of a multicomponent self-management program alone (Marion, 1987).

Several technological and practical problems exist that may limit the effective use of lung function and probability data in asthma decision making and self-management for individual patients. For example, when asthma episode base rates are very low or very high, there will be little or no improvement and, in some cases, a decrement in asthma-episode-prediction accuracy (Harm *et al.*, 1985). In addition, it is not clear that patients will comply with self-management recommendations even in the face of increased probability of asthma; they may disregard lung function data as they do other symptom feedback. Making the prediction procedure as simple and clear as possible is likely to increase compliance. Also, no study has demonstrated that the improved rates are stable and therefore hold up over time (Taplin & Creer, 1978). Finally, the Bayesean model lends itself to the establishment of a single lung-function cutoff score, below which a patient would be urged to take specific asthma management actions. Alternatively, using daily PEFR values in a logistic regression model, which can yield daily attack-probability estimates for any lung function value, may provide more useful information to the patient and the health care team (Bruss, 1989). (A discussion of Bayesean and logistic regression prediction models is beyond the scope of this chapter; see Harm and colleagues, 1985, and SPSS, 1988.)

Despite the empirical and practical questions that remain to be studied, anecdotal observations of adults and children with asthma indicate that some patients make effective use of lung function data in the same way that they make effective use of observed daily changes in other asthma symptoms (e.g., wheezing and fatigue). Even if lung function values do not prove to be useful in the daily prediction of asthma, patients can use them as powerful aids in the asthma-management decision-making process.

Medication Side Effects

Five primary drug classes are used to control asthma: xanthine, beta-adrenergic, corticosteroid, cromolyn sodium, and anticholinergic medications (Matts, 1984; Scanlon, 1984). Any one of these medications, with the possible exception of cromolyn sodium, can produce physical, emotional, or cognitive side effects that are easily misinterpreted as psychological symptoms. Further, the common need for polypharmacy in the treatment of asthma increases the likelihood and variety of such side effects.

Primary in the assessment of medication side effects is a working knowledge of the most common side effects produced by the major classes of asthma medications. With this knowledge, the clinician can choose among three approaches to detecting medication side effects: structured interviews or survey questionnaires, personality inventories, and/or neuropsychological instruments. Structured interviews or surveys that may be useful include the Asthma Problem Behavior Checklist (Creer *et al.*, 1983) or the Child Behavior Checklist (Achenbach & Edelbrock, 1979). Standardized personality inventories such as the MMPI or the Personality Inventory for Children have also been used (e.g., McLoughlin, Nall, Isaacs, Petrosko, Karibo, & Lindsey, 1983). Neuropsychological assessment has included the Wechsler intelligence scales and components of the Halstead-Reitan or Luria-Nebraska batteries (e.g., Furakawa, Shapiro, DuHamel, Weiner, Pierson, & Bierman, 1984; Kasenberg & Bloom, 1987). Regardless of what assessment method is used, it is useful to collaborate with physicians and to use an ABA manipulation of medication use to determine which drug is causing which side effect (Murphy, Dillon, & Fitzgerald, 1980).

Xanthine medications, which have a chemical structure similar to caffeine, can cause a number of side effects that may be brought to the attention of a psychologist. The most commonly prescribed xanthine medications are formulations of theophylline. Theophylline can cause acute and potentially life-threatening side effects, which are due to drug overdose or a unique theophylline sensitivity in individual patients (Weinberger, Lindgren, Bender, Lerner, & Szefler, 1987). (See Ellis, 1985, for a discussion of the side effects associated with theophylline toxicity.) It is unlikely that a psychologist will be confronted with symptoms of theophylline toxicity (e.g., vomiting blood), as these symptoms are likely to bring a patient to the emergency room. A consulting psycholo-

gist is more likely to be faced with more subtle and transient theophylline side effects that may present as a psychological symptom or syndrome. For example, case reports and clinical trials have identified side effects suggestive of anxiety and depression such as irritability, depressed mood, crying, or suicidal ideation, as well as nausea, nervousness, sleep disturbance, decreased appetite, attention deficits, irritability, restlessness, and decreased fine-motor coordination (e.g., Joad, Ahrens, Lindgren, & Weinberger, 1986; Kasenberg & Bloom, 1987). It appears that children are more likely to have idiosyncratic reactions to theophylline, although a small percentage of adults ($< 10\%$) also have theophylline sensitivity (Nelson & Schwartz, 1987). (See Hill & Szefler, 1987, or Weinberger and colleagues, 1987, for a discussion of the medical controversies that surround the use of theophylline.)

Beta-adrenergic bronchodilators, another commonly used antiasthmatic medication, stimulate the central nervous system (CNS). Common symptoms of CNS overstimulation include dizziness, headache, increased blood pressure, nausea and vomiting, nervousness, disturbed sleep, increased sweating, and weakness (Fernandez, 1987; Galant, 1983). These problems, if they do occur, are usually short-lived. Patients suffering more serious physical side effects, such as chest pain or irregular heartbeat (Fernandez, 1987; Galant, 1983), are likely to contact their physician as soon as possible. Fortunately, newer adrenergic bronchodilators, such as albuterol, produce fewer cardiovascular symptoms and other side effects of CNS excitation (Wolfe, Yamate, Biedermann, & Chu, 1985).

One problem associated with the use of inhaled beta-adrenergic medications is their overuse. This is a serious problem because overuse of inhaled albuterol or other adrenergic bronchodilators can lead to the constriction of the bronchial tubes and a worsening of an asthma attack (Galant, 1983). Asthma that is out of control, requiring frequent emergency room visits, may be due to the overuse of these medications.

Corticosteroids are the most powerful anti-asthma drugs available (Milner, 1982). Unfortunately, when corticosteroids are used in pill or injectable forms they have the greatest potential for causing serious physical and emotional side effects, so that their *long-term* use is limited except in the most severe cases (Hollister & Bowyer, 1987; King & Chang, 1987). The availability of inhaled corticosteroids in recent years has changed the way these medications are used (Clark, 1985). Whereas the effects of inhaled steroids are largely restricted to the lungs, oral steroids, because they rely on digestion and entry into the blood system, affect many organs and produce undesirable effects throughout the body (Hollister & Bowyer, 1987). Therefore, oral steroids affect the body more globally and cause more serious side effects than inhaled steroids.

The potential for side effects from oral steroid treatment depends on the dose level and the length of the treatment (King & Chang, 1987). Major physical side effects, such as growth suppression in children, take months to develop. Treatment for a few days or a few weeks to help an individual recover from an asthma flare-up is relatively safe (Milner, 1982). Long-term daily use at high dose levels is more likely to produce serious side effects. The possible physical side effects of long-term use of oral corticosteroids include cataracts, weakness and fatigue, osteoporosis, changes in body shape, unusual weight gain or weight loss, adrenal suppression, and, in children and adolescents, growth suppression. Psychologists are sometimes called on to help patients cope with these life-changing side effects. Psychological side effects that have been reported include "steroid euphoria," psychosis, and mood swings (Hollister & Bowyer, 1987; Medical Economics Company, 1988).

Because anticholinergic bronchodilators are inhaled, their primary effect is in the lungs. They are poorly absorbed by the body and are excreted rapidly, so that the likelihood of side effects is minimized. The most common side effect is coughing, found in about 6% of the patients who use the drug. Less common side effects include nervousness, nausea, gastrointestinal problems, and dry mouth, all found in less than 3% of the patients who use the drug (Medical Economics Company, 1988).

Asthma Knowledge

Basic knowledge about asthma and its care has been found to be poor among asthma patients and family members (Bucknall, Robertson, Moran, & Stevenson, 1988; Martin, Landau, & Phelan, 1982; Spykerboer, Donnelly, & Thong, 1986). Deficits in asthma knowledge have been thought to contribute to medication noncompliance (Martin *et al.*, 1982), poor symptom discrimination (Creer, 1983), inadequate self-care and a reliance on emergency care (Bucknall *et al.*, 1988), and overall asthma morbidity and mortality (Hindi-Alexander, 1987). We also speculate that limited asthma knowledge is related to family conflict around asthma and asthma-precipitated anxiety and panic. In addition, variability in physician knowledge, due to differences in specialization, has been associated with adequacy of care for asthma (Greenwald, Peterson, Garrison, Hart, Moscovice, Hall, & Perrin, 1984). Thus, knowledge about asthma is essential to the patient, the family, and the health care provider.

Asthma education efforts have taken many forms. Asthma education programs have been developed for children (Rubin, Leventhal, Sadock, Letovsky, Schottland, Clemente, & McCarthy, 1986), families (Creer, Backial, Burns, Leung, Marion, Miklich, Morrill, Taplin, & Ullman, 1988; Hindi-Alexander, 1987), and adults (Creer, Kotses, & Reynolds, 1989; Snyder, Winder, & Creer, 1987). In addition to traditional asthma education (e.g., pathophysiology, asthma triggers and early-warning signs), several of these programs also provide training in specific asthma self-management skills such as relaxation, daily monitoring of lung function, problem solving, and assertion training (e.g., Creer *et al.*, 1988).

Several excellent reference books are available for patients, families, and nonmedical clinicians working with asthmatics (Gershwin & Klingelhofer, 1986; Plaut, 1988; Weinstein, 1987). Also, other sources provide patient and therapist materials for the development of outpatient asthma education and self-management programs (Creer, Backial, Ullman, & Leung, 1986; Creer, Kotses, & Reynolds, 1989; Reynolds *et al.*, 1989).

Family Dysfunction

Dysfunctional family relationships, particularly between an asthmatic child and his or her mother, have long been a focus in the asthma literature. For example, cross-sectional studies have documented a high degree of dependency, enmeshment, overprotectiveness, and parental anxiety in some families with an asthmatic child (e.g., Rubenstein, King, & London, 1979; Staudenmayer, 1981). Contemporary wisdom, which recognizes asthma as a physical disorder, highlights the reciprocal interaction between the management of a patient's asthma and the larger context of family relationships (Matus, 1981; Sadler, 1982).

Ultimately, dysfunctional family systems are associated with the mismanagement of chronic illness (e.g., Masterson, 1985; Staudenmayer, 1981). Structural family theory provides a useful model for conceptualizing the communication and relationship problems of dysfunctional families (Liebman, Minuchin, & Baker, 1974; Masterson, 1985). The most common structural problems found in families with a chronically ill member include the overinvolvement of one parent with a sick child, referred to as a *coalition* (Penn, 1983), or a family exclusively focused on a chronically ill member as a way to bind the family together and to cope with other conflicts such as a failing marriage (Masterson, 1985). The goals of structural family therapy include creating flexible family boundaries, where closeness and distance are both tolerated depending on the situational demands; involving both parents in the management of a child's illness; improving family conflict resolution; addressing concrete problems that involve other family members, which are not being dealt with because of the patient's asthma; and encouraging the asthmatic patient to assume age-appropriate responsibilities and freedoms (Liebman *et al.*, 1974; Masterson, 1985). Considerable improvement in family relationships and asthma management have resulted from the use of structural family therapy (e.g., Liebman *et al.*, 1974). An adequate treatment of the theory, assessment, and intervention strategies associated with structural family therapy is beyond the scope of this chapter (see Liebman *et al.*, 1974; Masterson, 1985).

Developing and Maintaining the Program

Introducing an intervention program, including the teaching of self-management skills, has enhanced our approach to developing and maintaining programs for both inpatients and outpatients of hospitals and treatment facilities. This experience will serve as the basis for the remainder of the discussion.

Inpatient Referrals

Most clinics and hospitals have procedures for referring patients to psychologists for evaluation or treatment. These psychologists are often members of behavioral medicine or consultation–liaison psychiatry units (Brown & Waterhouse, 1987; McKegney & Schwartz, 1986; Sadler, 1982). Referrals range from a formal written request to an informal comment made during morning rounds attended by psychologists and physicians. Most of these requests reflect legitimate concerns, although Creer, Renne, and Christian (1978) noted four recurring referral problems.

Consensual Ambiguity

Consensual ambiguity refers to physicians and nurses both agreeing that a certain patient exhibits a particular problem behavior. The difficulty is that the same term may be used by different professional staff with little agreement on how the terms are to be defined behaviorally. Thus, instead of having a known target behavior, psychologists often spend considerable time in attempting to determine the exact behavior they are to assess and, if possible, change.

New-Patient Syndrome

Staff often focus more on the behaviors of newly admitted patients than on their asthma. As a result, referrals are often made for an evaluation of a patient early in his or her stay even though, with the passage of only a few days, these behavioral patterns disappear.

Big-Name Phenomenon

Certain patients are referred to psychologists more than others. It is not always clear why these patients require greater attention by the psychological staff. Sometimes, these patients are credited with exhibiting far more inappropriate behaviors than they could ever have performed. These patients are not necessarily more troublesome than others, but their names are simply better known to the attending medical staff.

Behaviors Involving Only One Patient

If a patient behaves in a way that disturbs other patients and the medical staff, he or she is almost immediately referred to a psychologist. However, behaviors such as head banging, enuresis, and even encopresis are not always referred to the psychological staff, as these behaviors do not usually affect other patients.

Outpatient Referrals

In many cases, the psychologist is concerned only with inpatients. However, with increased competition among hospitals, many psychologists are being asked to develop and market programs for smoking cessation, weight reduction, eating disorders, and other types of physical disorders. Having had a number of years' experience with outpatient programs, we will briefly highlight the techniques used to contact and maintain referral sources. Generally ineffective techniques include personal letters (they are ignored even if they reach the physician), telephone calls (despite promises by medical personnel, they often prove unproduc-

tive), and talks to local medical societies (the physicians who attend these meetings often promise assistance, but they tend to forget such promises). Visits to health maintenance organizations are not effective because these groups usually have their own programs, and although interest is shown by personnel at city hospitals and neighborhood health centers, referred patients are often unreliable. More successful strategies for developing a referral base include the use of public service announcements, brochures describing the program that can be left with physicians and in clinic waiting rooms, and most important, continual personal contacts with physicians. A single phone call or letter, as noted above, is usually not effective. This last tactic mimics that used by successful pharmaceutical sales personnel. The psychologist should keep in contact with physicians to see if she or he can encourage them to enroll their patients in the program(s) that the psychologist is involved in. A large number of patient referrals from only a few physicians or groups of physicians can ensure the success of most psychological programs. Finally, the word of mouth of those patients who have been successfully worked with also ensures a constant source of patients for the program.

Collaborating with Physicians

We could paraphrase Tennessee Williams in *A Streetcar Named Desire* by saying that we "have always depended upon the kindness of strangers" because, overall, physicians have cooperated with us to help achieve the successes we have attained together. Setting up any inpatient or outpatient program to work with asthma patients requires some level of collaboration with physicians. There are a few practical tips we would like to offer with respect to working with physicians. First, it is impossible to work with all physicians; the psychologist must be selective. Physicians with the better reputations are often more receptive to psychological approaches. They know that they do not have all the answers for managing a complex problem such as asthma, and they welcome any assistance. Also, most physicians are initially skeptical of working with psychologists. However, once they observe that the psychologist has made progress with a patient, they are not only cooperative but willing to read educational materials. We have heard physicians casually describe what occurred with a patient in terms that indicate that they have become very familiar with the psychological aspects of asthma. Finally, we would suggest that psychologists publish some of their articles in medical journals, both to educate physicians and to become known to the medical community.

Consultation with Health Team

There are a few stylistic considerations in regard to serving as effective consultants to health care teams. First, the psychologist must be prepared to answer questions and must know patient charts, as well as any changes that may have occurred in a patient's behavior. A form of daily monitoring contributes immeasurably to this process. Second, psychologists should avoid treading on medical turf. They should know something about the medical progress of a patient, but it is up to medical personnel to discuss these problems. Third, the psychologist should read medical journals. Important articles relevant to the psychosocial management of asthma appear regularly. The purpose of keeping abreast of these journals is not to turn psychologists into physicians, but to permit them to understand what is said during rounds and other meetings of the treatment team. They also enhance the psychologist's ability to use behavioral procedures with patients afflicted by asthma. Fourth, it is helpful to attend annual meetings of the American Academy of Allergy and Immunology and the American Thoracic Society/American Lung Association. Generally, there is more relevant information available at these meetings than at those designed for psychology or behavioral medicine. Finally, psychologists should know their role on the treatment team. They are the psychologists on the treatment team; others should recognize this and listen to whatever advice they have to offer.

Patient Flow through System

The system designed for asthmatic patients will be a function of whether they are inpatients or outpatients. Patient flow through the system encompasses preentry, intake evaluation, ongoing monitoring, termination or discharge evaluation, and follow-up evaluation. Each of these components is briefly summarized here.

Preentry

It is useful to collect information from patients before they enter a clinic or hospital. This preliminary assessment information can be used to gauge the changes that occur as a consequence of hospitalization. In outpatient self-management programs, it is useful to have patients begin collecting diary and report-of-attack data before they receive self-management training. This information provides patients and clinicians with objective information about asthma episodes that can be used to make early treatment decisions.

Intake Evaluation

When a patient is admitted to a hospital or an outpatient program, the psychologist wants to detect the psychological contributions to the patient's asthma as soon as possible. This evaluation should be comprehensive, using a structured interview and/or a battery of psychological measures (see above).

Treatment and Ongoing Monitoring

The use of structured interviews, behavior checklists, and psychological assessment batteries is discussed earlier in this chapter. Other instruments that may be useful in an intake or ongoing evaluation include the Asthma Self-Efficacy Scale (Tobin et al., 1987) and the Piers-Harris Self-Concept Scale (Piers & Harris, 1964). These measures assess attitudes and beliefs relevant to the management of asthma. In addition, with the Child Behavior Checklist (Achenbach & Edelbrock, 1979), one can obtain invaluable information from the parents of children with asthma. The ongoing use of behavioral checklists, often completed by nurses, can be invaluable in monitoring an inpatient's progress. With outpatients, we ask that a Report of Attack/Episode Form (Creer *et al.*, 1986) be completed by the patient after each attack. This procedure has proved useful with both children and adults in helping them learn what stimuli trigger attacks and how they respond to such episodes.

Termination or Discharge Evaluation

Before a patient is terminated from an outpatient program or is discharged, it is useful to evaluate him or her to determine if progress was made in specific problem areas and if further intervention is necessary. This evaluation usually involves a readministration of the instruments used at intake.

Follow-Up

Whether the patient was treated as an inpatient or an outpatient, it is wise to assess his or her progress 6–12 months after discharge or completion of a program. This assessment not only provides useful information on the patient's progress but permits the psychologist to evaluate the effectiveness of the program and any aftercare interventions.

Clinical Case Example

The case we have selected conveys the type of problems faced by both adults and children with asthma. The patient, Bob, is a 39-year-old man who was diagnosed as having asthma in early childhood. He is married and successful in his chosen profession. He was enrolled in a course to acquire self-management skills to manage his asthma.

The patient was diagnosed as asthmatic when he was 5 years old. Although his asthma was considered perennial, in that he could expect to suffer attacks throughout the year, he was particularly prone to asthma flare-ups during the summer months. In an attempt to prevent attacks, Bob's parents restricted him to his room. There was no air-conditioning unit or other air-filtering system in the room; instead, Bob was forced to stay in a hot, hu-

mid room with closed windows. He was not allowed to participate in activities with his peers, nor, judging from his descriptions, did he receive adequate medical treatment (although his treatment was probably standard for that time). Bob believes that his parents, especially his father, did what they thought was correct in their attempt to manage Bob's asthma. At the same time, however, he expressed feelings of loss and bitterness about his childhood. Furthermore, he noted that his parents thought that the asthma would disappear of its own accord as Bob grew older. Although it is true that some asthmatic children do experience fewer symptoms as they grow into adulthood, there is no way to predict which children will be fortunate in this respect. In addition, more recent evidence suggests that children really do not "outgrow" their asthma: although there may be a decrease in frequency over a period of time, asthma attacks can unexpectedly return and may often be of greater severity in adulthood. In summary, Bob believes that he was denied not only a normal childhood, but the opportunity to select a more appealing profession because of lifestyle limitations placed on him by his parents because he suffered asthma.

Bob has continued to experience attacks as he has grown older. His asthma is such that he requires several medications, prescribed on both a maintenance and an as-needed schedule. He is not compliant with his medication instructions and prefers to follow his own judgment. Although his noncompliance has not resulted in a serious asthma attack, such as status asthmaticus or steadily worsening asthma, there is always the potential for such an event to occur. The noncompliance is perhaps a reflection of Bob's attitudes toward physicians. His negativism toward medicine developed when he admitted himself to an emergency room with acute asthma. The physician in charge did not recognize the severity of Bob's attack but nevertheless hospitalized him for observation. Bob wanted more intensive care because he believed he was about to lose consciousness. It turned out that Bob was correct and the physician was wrong: if it had not been for the alert monitoring of a respiratory technician, he could have succumbed to his attack. As it was, the status asthmaticus he experienced was reversed, and Bob survived. However, he does not trust physicians to any great extent, although he has carefully shopped around to find the most reputable physician from whom to receive treatment.

There is another problem that repeatedly plagues Bob: He is concerned about whether he and his wife should have children because he fears they will not only suffer asthma but experience the same emotional scars he bears from his childhood. Although he has been married for 10 years, he and his wife have postponed having children.

To solve the above problems, several strategies have been taken. First, Bob has been an active participant in a self-management program for asthma. This has helped him realize that many of the problems he has experienced are not unique to him; if anything, he has discovered that he has not suffered either the severity of attacks or the number of behavioral consequences of other members of his group. He likes to share his experiences at these meetings and has requested occasional group meetings after the formal aspects of training were completed. Second, Bob has learned that, although there is a possibility that he and his wife may have a child who has asthma, the chances are that any asthma would be better controlled than was possible when Bob was a child. Thus, he and his wife have decided to have a family. Third, in an attempt to overcome the bitterness he feels because of his childhood, Bob has started individual psychotherapy. This has assisted him to come to grips with some of the feelings he has about his asthma and the consequences of the disorder. Finally, despite his increased knowledge about and skill in managing his asthma, Bob is still noncompliant with medication instructions, particularly with respect to maintenance medications. A specific program and method for monitoring Bob's compliance will need to be established to correct this problem. His improved respect for his physician should also help. Thus, any success attained with Bob in improving his compliance will rely on a synthesis of behavioral and medical expertise.

Summary

A variety of methods for psychological assessment of and intervention with asthma have been described. It is likely that psychologists will increasingly be called on to use these and other methods to assist asthmatic patients. At the turn of the century, a leading English physician, Sir William Osler, said that asthmatic patients were of little medical concern as, he quipped, they "panted towards old age." As we charge toward the end of this century, this statement is increasingly fallacious. Survey after survey reports that childhood asthma is the leading cause for the hospitalization of children. This situation is not confined to the United States but depicts a worldwide trend with respect to childhood asthma. More and more adults also experience what is referred to as *late-onset asthma*. These cases often reflect what physicians view as severe and difficult-to-manage asthma. Asthma can appear suddenly in adults who have heretofore experienced excellent health. These individuals are often unprepared to cope with the lifestyle limitations that asthma can bring. Further, the number of stimuli that trigger asthma in the work environment—occupational asthma—is soaring. At one time, only a few stimuli were thought to trigger asthma; now, a host of different triggers are being identified. With the continued deterioration of our environment, this trend will continue.

The increasing rate of deaths due to asthma is particularly troublesome in an age when improved medical procedures and medications are available. Although asthma still accounts for a small proportion of mortality compared to such conditions as heart disease and cancer, the rise of asthma-related deaths recently resulted in experts gathering from across the globe to discuss how the trend could be controlled (Sheffer & Buist, 1987). A reversal of these trends in asthma mortality and morbidity will require an interdisciplinary treatment approach. As Selner (1988) recently told his medical colleagues in the Bela Schick Lecture on the future of allergy, "The allergist must establish a relationship with a psychologist. There are 30,000 of them out there, find one" (p. 76). Selner's advice could not come at a more opportune time for the millions who suffer from asthma.

ACKNOWLEDGMENT

Preparation of this chapter was supported, in part, by Grant No. 32538 from the National Heart, Lung, and Blood Institute.

References

Achenbach, T. M., & Edelbrock, C. S. (1979). The child behavior profile: 2. Boys aged 12–16 and girls aged 6–11 and 12–16. *Journal of Consulting and Clinical Psychology, 47*, 223–233.

Alexander, A. B. (1972). Systematic relaxation and flow rates in asthmatic children: Relationship to emotional precipitants and anxiety. *Journal of Psychosomatic Research, 16,* 405–410.

Alexander, A. B., Cropp, G. J. A., & Chai, H. (1979). Effects of relaxation training on pulmonary mechanics in children with asthma. *Journal of Applied Behavior Analysis, 12,* 27–35.

Alexander, A. B., Miklich, D. R., & Hershkoff, H. (1972). The immediate effects of systematic relaxation training on peak expiratory flow rates in asthmatic children. *Psychosomatic Medicine, 34,* 338–394.

American Psychiatric Association. (1987). *Diagnostic and statistical manual of mental disorders* (3rd ed. rev.). Washington, DC: Author.

American Thoracic Society. (1987). Standards for the diagnosis and care of patients with chronic obstructive pulmonary disease (COPD) and asthma. *American Review of Respiratory Disease, 136*(1), 225–244.

Barger, L. W., Vollmer, W. M., Felt, R. W., & Buist, A. S. (1988). Further investigation into the recent increase in asthma death rates: A review of 41 asthma deaths in Oregon in 1982. *Annals of Allergy, 60,* 31–39.

Barron, F. (1953). An ego-strength scale which predicts response to psychotherapy. *Journal of Consulting Psychology, 17*(5), 327–333.

Becker, M. H., Radius, S. M., Rosenstock, I. M., Drachman, R. H., Schuberth, K. C., & Teets, K. C. (1978). Compliance with a medical regimen for asthma: A test of the health belief model. *Public Health Reports, 93,* 268–277.

Ben-zvi, Z., Spohn, W. A., Young, S. H., & Kattan, M. (1982). Hypnosis for exercise-induced asthma. *American Review of Respiratory Disease, 125*(4), 392–395.

Brown, T. M., & Waterhouse, J. (1987). A psychiatric liaison service in a general hospital—Eighteen years on. *Health Bulletin, 15,* 190–196.

Bruss, G. (1989). *The prediction of asthma episodes using peak respiratory flow rates in a logistic regression model.* Unpublished doctoral dissertation proposal, Ohio University, Athens.

Bucknall, C. E., Robertson, C., Moran, F., & Stevenson, R. D. (1988). Management of asthma in hospital: A prospective audit. *British Medical Journal, 296,* 1637–1639.

Clark, T. J. H. (1985). Inhaled corticosteroid therapy: A substitute for theophylline as well as prednisolone? *Journal of Allergy and Clinical Immunology, 75,* 330–334.

Cluss, P. A., Epstein, L. H., Galvis, S. A. Fireman, P., & Friday, G. (1984). Effect of compliance for chronic asthmatic children. *Journal of Consulting and Clinical Psychology, 52*(5), 909–910.

Creer, T. L. (1979). *Asthma therapy: A behavioral health-care system for respiratory disorders.* New York: Springer.

Creer, T. L. (1982). Asthma. *Journal of Consulting and Clinical Psychology, 50*(6), 912–921.

Creer, T. L. (1983). Response: Self-management psychology and the treatment of childhood asthma. *Journal of Allergy and Clinical Immunology, 72*(5), 607–610.

Creer, T. L., Marion, R. J., & Creer, P. P. (1983). Asthma problem behavior checklist: Parental perceptions of the behavior of asthmatic children. *Journal of Asthma, 20*(2), 97–104.

Creer, T. L., Renne, C. M., & Chai, H. (1982). The application of behavioral techniques to childhood asthma. In D. C. Russo & J. W. Varni (Eds.), *Behavioral pediatrics: Research and practice.* New York: Plenum Press.

Creer, T. L., Renne, C. M. & Christian, W. P. (1978). Unpredictable problems in applying social learning principals in a child care facility. *Child Care Quarterly, 7,* 142–155.

Creer, T. L., & Reynolds, R. V. (1990a). Asthma. In M. Hersen & V. B. Van Hasselt (Eds.), *Psychological aspects of developmental and physical disabilities: A casebook.* Newbury Park, CA: Sage.

Creer, T. L., & Reynolds, R. V. (1990b). Asthma. In A. M. Gross & R. S. Drabman (Eds.), *Handbook of clinical behavioral pediatrics.* New York: Plenum Press.

Creer, T. L., Backial, M., Ullman, S., & Leung, P. (1986). *Living with asthma: Part 1. Manual for teaching parents the self-management of childhood asthma; Part 2. Manual for teaching children the self-management of asthma* (NIH Publication No. 86-2364), Washington DC: U.S. Government Printing Office.

Creer, T. L., Kotses, H., & Reynolds, R. V. (1989). *Living with Asthma: Help for adults with asthma; A group leader's guide to self-management.* Unpublished manual, Ohio University, Athens.

Creer, T. L., Backial, M., Burns, K. L., Leung, P., Marion, R. J., Miklich, D. R., Morrill, C. Taplin, P. S., & Ullman, S. (1988). Living with Asthma: Part I. Genesis and development of a self-management program for childhood asthma. *Journal of Asthma, 25,* 335–362.

Creer, T. L., Kotses, H., & Reynolds, R. V. (1989). Living with Asthma: Part 2. Beyond CARIH. *Journal of Asthma, 26,* 53–63.

Davis, M. H., Saunders, D. R., Creer, T. L., & Chai, H. (1973). Relaxation training facilitated by biofeedback apparatus as a supplemental treatment in bronchial asthma. *Journal of Psychosomatic Research, 17,* 121–128.

Dirks, J. F. & Kinsman, R. A. (1982). Bayesian prediction of noncompliance: As-needed (PRN) medication usage patterns and the battery of asthma illness behavior. *Journal of Asthma, 19*(1), 25–31.

Dirks, J. F., & Kreischer, H. (1982). The battery of asthma illness behavior: 1. Independence from age of asthma onset. *Journal of Asthma, 19*(2), 75–78.

Dirks, J. F., Schraa, J. C., Brown, E. L., & Kinsman, R. A. (1980). Psycho-maintenance in asthma: Hospitalization rates and financial impact. *British Journal of Medical Psychology, 53,* 349–354.

Ellis, E. F. (1985). Theophylline toxicity. *Journal of Allergy and Clinical Immunology, 76,* 297–301.

Engel, B. T. (1986). Psychosomatic medicine, behavioral medicine, just plain medicine. *Psychosomatic Medicine, 48*(7), 466–479.

Fernandez, E. (1987). Beta-adrenergic agonists. *Seminars in Respiratory Medicine, 8,* 353–365.

Friedman, M. S. (1984). Psychological factors associated

with pediatric asthma death: A review. *Journal of Asthma,* 21(2), 97–117.

Furakawa, C. T., Shapiro, G. G., DuHamel, T., Weiner, L., Pierson, W. E., & Bierman, C. W. (1984). Learning and behavior problems associated with theophylline therapy. *The Lancet,* 1, 621.

Galant, S. P. (1983). Current status of beta-adrenergic agonists in bronchial asthma. *Pediatric Clinics of North America,* 30, 931–942.

Garfield, E. (1982). Patient compliance: A multifaceted problem with no easy solution. *Current Comments,* 37, 5–14.

Gershwin, M. E., & Klingelhofer, E. L. (1986). *Asthma: Stop suffering, start living.* Reading, MA: Addison-Wesley.

Glaus, K. D., & Kotses, H. (1979). Generalization of conditioned muscle tension: A closer look. *Psychophysiology,* 16, 513–519.

Gottfried, S. B., Altose, M. D., Kelsen, S. G., & Cherniak, N. S. (1981). Perception of changes in airflow resistance in obstructive pulmonary disorders. *American Review of Respiratory Disease,* 124, 566–570.

Greenwald, H. P., Peterson, M. L., Garrison, L. P., Hart, L. G., Moscovice, I. S., Hall, T. L., & Perrin, E. B. (1984). Interspecialty variation in office-based care. *Medical Care,* 22(1), 14–29.

Gross, N. J. (1980). What is this thing called love?—or defining asthma. *American Review of Respiratory Diseases,* 121, 203–204.

Halfon, N., & Newacheck, P. W. (1986). Trends in hospitalization for acute childhood asthma, *1970–84. American Journal for Public Health,* 76, 1308–1311.

Harm, D. L., Kotses, H., & Creer, T. L. (1985). Improving the ability of peak expiratory flow rates to predict asthma. *The Journal of Allergy and Clinical Immunology,* 76(5), 688–694.

Haynes, R. B., Taylor, D. W., & Sackett, D. L. (Eds.) (1979). *Compliance with therapeutic regimens.* Baltimore, MD: Johns Hopkins University Press.

Higgs, C. M. B., Richardson, R. B., Lea, D. A., Lewis, G. T. R., & Laszlo, G. (1986). Influence of knowledge of peak flow on self assessment of asthma: Studies with a coded peak flow meter. *Thorax,* 41, 671–675.

Hill, M. R., & Szefler, S. J. (1987). Theophylline update: Current controversies. *Seminars in Respiratory Medicine,* 8, 372–380.

Hindi-Alexander, M. C. (1987). Asthma education programs: Their role in asthma morbidity and mortality. *Journal of Allergy and Clinical Immunology,* 80(3), 492–494.

Hock, R. A., Rodgers, C. H., Reddi, C., & Kennard, D. W. (1978). Medico-psychological interventions in male asthmatic children: An evaluation of physiological change. *Psychosomatic Medicine,* 40, 210–215.

Hollister, J. R., & Bowyer, S. L. (1987). Adverse side-effects of corticosteroids. *Seminars in Respiratory Medicine,* 8, 400–405.

Jacobson, E. (1938). *Progressive relaxation.* Chicago: Universit of Chicago Press.

Jerome, A., Wigal, J. K., & Creer, T. L. (1987). A review of

medication compliance in children with asthma. *Pediatric Asthma, Allergy and Immunology,* 1(4), 193–211.

Joad, J. P., Ahrens, R. C., Lindgren, S. D., & Weinberger, M. M. (1986). Extrapulmonary effects of maintenance therapy with theophylline and inhaled albuterol in patients with chronic asthma. *Journal of Allergy and Clinical Immunology,* 78(6), 1147–1153.

Kasenberg, D. F., & Bloom, L. (1987). Potential neuropsychological side effects of theophylline in asthmatic children. *Pediatric Asthma, Allergy and Immunology,* 1(3), 165–173.

King, T. E., & Chang, S. W. (1987). Corticosteroid therapy in the management of asthma. *Seminars in Respiratory Medicine,* 8, 387–399.

Kinsman, R. A., Dahlem, N. W., Spector, S., & Staudenmayer, H. (1977). Observations on subjective symptomatology, coping behavior, and medical decisions in asthma. *Psychosomatic Medicine,* 39, 102–119.

Kotses, H., & Glaus, K. D. (1981). Applications of biofeedback to the treatment of asthma: A critical review. *Biofeedback and Self-Regulation,* 6(4), 573–593.

Kotses, H., & Glaus, K. D. (1982). Generalization of conditioned muscle tension: Sharpening the focus. *Psychophysiology,* 19, 498–500.

Kotses, H., Glaus, K. D., Bricel, S. K., Edwards, J. E., & Crawford, P. L. (1978). Operant muscular reduction and peak expiratory flow rate in asthmatic children. *Journal of Psychosomatic Research,* 22, 19–23.

Kotses, H., Glaus, K. D., Crawford, P. L., Edwards, J. E., & Scherr, M. (1976). Operant reduction of frontalis EMG activity in the treatment of asthma in children. *Journal of Psychosomatic Research,* 20, 453–459.

Kotses, H., Harver, A., Segretto, J., Glaus, K. D., Creer, T. L., & Young, G. A. (1989). Long-term effects of biofeedback induced facial relaxation on measures of asthma severity in children. *Biofeedback and Self-Regulation,* in press.

Kotses, H., Hindi-Alexander, M., & Creer, T. L. (1989). A reinterpretation of psychologically-induced airway changes. *Journal of Asthma,* 26, 53–63.

Kotses, H., Rawson, J. C., Wigal, J. K., & Creer, T. L. (1987). Respiratory airway changes in response to suggestion in normal individuals. *Psychosomatic Medicine,* 49, 536–541.

Kotses, H., Westlund, R., & Creer, T. L. (1987). Performing mental arithmetic increases total respiratory resistance in individuals with normal respiration. *Psychophysiology,* 24, 678–682.

Lachar, D. (1974). *The MMPI: Clinical assessment and automated interpretation.* Los Angeles, CA: Western Psychological Services.

Liebman, R., Minuchin, S., & Baker, L. (1974). The use of structural family therapy in the treatment of intractable asthma. *American Journal of Psychiatry,* 131(5), 535–540.

Luparello, T., Leist, N., Lourie, C. H., & Sweet, P. (1970). The interaction of psychologic stimuli and pharmacologic agents on airway reactivity in asthmatic subjects. *Psychosomatic Medicine,* 32, 509–513.

Mackenzie, J. N. (1886). The production of the so-called

"rose cold" by means of an artificial rose. *American Journal of the Medical Sciences, 91*, 45–57.

Maher-Loughnan, G. P., & Kinsley, B. J. (1968). Hypnosis for asthma—a controlled trial. *British Medical Journal, 4*, 71–76.

Marion, R. J. (1987). *Teaching children to predict asthma using an in-home pulmometer.* Unpublished doctoral dissertation, Ohio University, Athens.

Martin, A. J., Landau, L. I., & Phelan, P. D. (1982). Asthma from childhood at age 21: The patient and his disease. *British Medical Journal, 284*, 380–382.

Masterson, J. (1985). Family assessment of the child with intractable asthma. *Developmental and Behavioral Pediatrics, 6*(5), 244–251.

Matts, S. G. F. (1984). Current concepts in the overall management of asthma. *The British Journal of Clinical Practice, 38*, 205–212.

Matus, I. (1981). Assessing the nature and clinical significance of psychological contributions to childhood asthma. *American Journal of Orthopsychiatry, 51*(2), 327–341.

McFadden, E. R., Jr. (1980). Asthma: Pathophysiology. *Seminars in Respiratory Medicine, 1*, 297–303.

McKegney, F. P. & Schwartz, C. E. (1986). Behavioral medicine: Treatment and organizational issues. *General Hospital Psychiatry, 8*, 330–339.

McLoughlin, J., Nall, M., Isaacs, B., Petrosko, J., Karibo, J., & Lindsey, B. (1983). The relationship of allergies and allergy treatment to school performance and student behavior. *Annals of Allergy, 51*, 506–510.

Medical Economics Company. (1988). *Physician's desk reference.* Oradell, NJ: Author.

Miklich, D. R., Renne, C. M. Creer, T. L., Alexander, A. B., Chai, H., Davis, M. H., Hoffman, A., & Danker-Brown, P. (1977). The clinical utility of behavior therapy as an adjunctive treatment for asthma. *Journal of Allergy and Clinical Immunology, 60*, 285–294.

Milner, A. D. (1982). Steroids and asthma. *Pharmacology Therapy, 17*, 229–238.

Moore, N. (1965). Behaviour therapy in bronchial asthma: A controlled study. *Journal of Psychosomatic Research, 9*, 257–276.

Murphy, M. B., Dillon, A., & Fitzgerald, M. X. (1980). Theophylline and depression. *British Medical Journal, 281*, 1322.

Navran, L. (1954). A rationally derived MMPI scale to measure dependence. *Journal of Consulting Psychology, 18*(3), 192.

Nelson, L. A., & Schwartz, J. I. (1987). Theophylline-induced age-related CNS stimulation. *Pediatric Asthma, Allergy and Immunology, 1*(3), 175–183.

Parish, U. M. (1986). Parent compliance with medical and behavioral recommendations. In N. A. Krasnegor, J. D. Arastan, & M. F. Cataldo (Eds.), *Child health behavior: A behavioral pediatrics perspective.* New York: Wiley.

Pearlman, D. S. (1984). Bronchial asthma: A perspective from childhood to adulthood. *American Journal of Diseases of Children, 138*, 459–466.

Penn, P. (1983). Coalitions and binding interactions in families with chronic illness. *Family Systems and Medicine, 1*, 16–25.

Piers, E. V., & Harris, D. B. (1964). Age and other correlates of self-concept in children. *Journal of Educational Psychology, 55*, 91–95.

Plaut, T. F. (1988). *Children with asthma: A manual for parents* (2nd ed.). Amherst, MA: Pedipress.

Rapoff, M. A., & Christophersen, E. R. (1982). Improving compliance in pediatric practice. *Pediatric Clinics of North America, 29*, 339–357.

Reynolds, R. V., Kotses, H., Creer, T. L., Bruss, G. & Joyner, C. A. (1989). *Living with asthma: Help for adults with asthma. A client's guide to self-managment.* Unpublished manual, Ohio University, Athens.

Rubenstein, H. S., King, S. H., & London, E. L. (1979). Adolescent and postadolescent asthmatics' perception of their mothers as overcontrolling in childhood. *Adolescence, 14*(53), 1–18.

Rubin, D. H., Leventhal, J. M., Sadock, R. T., Letovsky, E., Schottland, P., Clemente, I., & McCarthy, P. (1986). Educational intervention by computer in childhood asthma: A randomized clinical trial testing the use of a new teaching intervention in childhood asthma. *Pediatrics, 77*(1), 1–10.

Rubinfield, A. R., & Pain, M. C. F. (1976). Perception of asthma. *The Lancet, 1*, 882–884.

Sadler, J. E. (1982). Childhood asthma from the point of view of the liaison child psychiatrist. *Psychiatric Clinics of North America, 5*(2), 333–343.

Scanlon, R. T. (1984). Asthma: A panoramic view and a hypothesis. *Annals of Allergy, 53*, 203–212.

Scherr, M. S., Crawford, P. L., Sergent, C. B., & Scherr, C. A. (1975). Effect of bio-feedback techniques on chronic asthma in a summer camp environment. *Annals of Allergy, 35*, 289–295.

Selner, J. C. (1988). Bela Schick Lecture: The future of allergy. *Annals of Allergy, 60*, 73–79.

Sheffer, A. L., & Buist, A. S. (1987). Proceedings of the asthma mortality task force. *Journal of Allergy and Clinical Immunology, 80*, 361–514.

Silverman, B. A., Mayer, D., Sabinsky, R., Williams-Akita, A., Feldman, J. Schneider, A. T., & Chiaramonte, L. T. (1987). Training perception and airflow obstruction in asthmatics. *Annals of Allergy, 59*, 350–354.

Sly, P. D., Landau, L. I., & Weymouth, R., (1985). Home recording of peak expiratory flow rates and perception of asthma. *American Journal of Diseases in Children, 139*, 479–482.

Smith, M. M., Colebatch, H. J. H., & Clarke, P. S. (1970). Increase and decrease in pulmonary resistance with hypnotic suggestion in asthma. *American Review of Respiratory Disease, 102*(2), 236–242.

Snyder, S. E., Winder, J. A., & Creer, T. L. (1987). Development and evaluation of an adult asthma self-mangement program: Wheezers Anonymous. *Journal of Asthma, 24*(3), 153–158.

SPSS. (1988). *SPSS-X user's guide* (3rd ed.). Chicago: Author.

Spykerboer, J. E., Donnelly, W. J., & Thong, Y. H. (1986). Parental knowledge and misconceptions about asthma: A controlled study. *Social Science Medicine, 22*(5), 553–558.

Staudenmayer, H. (1981). Parental anxiety and other psychosocial factors associated with childhood asthma. *Journal of Chronic Disease, 34*, 627–636.

Staudenmayer, H., Kinsman, R. A., & Jones, N. F. (1978). Attitudes toward respiratory illness and hospitalization in asthma. *Journal of Nervous and Mental Disease, 166*, 624–634.

Steiner, H., Higgs, C. M. B., Fritz, G. K., Laszlo, G., & Harvey, J. E., (1987). Defense style and perception of asthma. *Psychosomatic Medicine, 49*(1), 35–44.

Sublett, J. L., Pollard, S. J., Kadlec, G. J., & Karibo, J. M. (1979). Non-compliance in asthmatic children: A study of theophylline levels in pediatric emergency room population. *Annals of Allergy, 43*, 95–97.

Tal, A., & Miklich, D. P. (1976). Emotionally-induced decreases in pulmonary flow rates in asthmatic children. *Psychosomatic Medicine, 38*, 190–200.

Taplin, P. S., & Creer, T. L. (1978). A procedure for using peak expiratory flow-rate data to increase the predictability of asthma episodes. *The Journal of Asthma Research, 16*(1), 15–19.

Tobin, D. L., Reynolds, R. V., Holroyd, K. A., & Creer, T. L. (1986). Self-management and social learning theory. In K. A. Holroyd, & T. L. Creer (Eds.), *Self-management of chronic disease: Handbook of clinical interventions and research* (pp. 29–55). Orlando, FL: Academic Press.

Tobin, D. L., Wigal, J. K., Winder, J. A., Holroyd, K. A., & Creer, T. L. (1987). The "asthma self-efficacy scale". *Annals of Allergy, 59*(10), 273–277.

Weinberger, A. G., & Cuskey, W., (1985). Theophylline compliance in asthmatic children. *Annals of Allergy, 54*, 19–24.

Weinberger, M., Lindgren, S., Bender, B., Lerner, J. A., & Szefler, S. (1987). Effects of theophylline on learning and behavior: Reason for concern or concern without reason? *The Journal of Pediatrics, 111*(3), 471–474.

Weinstein, A. M. (1987). *Asthma: The complete guide to self-management of asthma and allergies for patients and their families.* New York: McGraw-Hill.

Williams, M. H., Jr. (1982). Expiratory flow rates: Their role in asthma therapy. *Hospital Practice, 17*(10), 95–110.

Wolfe, J. D., Yamate, M., Biedermann, A. A., & Chu, T. J. (1985). Comparison of the acute cardiopulmonary effects of oral albuterol, metaproterenol, and terbutaline in asthmatics. *Journal of the American Medical Association, 253*(14), 2068–2072.

Wolpe, J. (1958). *Psychotherapy by reciprocal inhibition.* Stanford, CA: Stanford University Press.

Yorkston, N. J., McHugh, R. B., Brady, R., Serber, M., & Sergeant, H. G. S. (1974). Verbal desensitization in bronchial asthma. *Journal of Psychosomatic Research, 18*, 371–376.

Yorkston, N. J., Eckert, E., McHugh, R. B., Philander, D. A., & Blumenthal, M. N. (1979). Bronchial asthma: Improved lung function after behavior modification. *Psychosomatics, 20*, 325–331.

CHAPTER 28

Psychological Characteristics and Treatment of Patients with Gastrointestinal Disorders

Harry S. Shabsin and William E. Whitehead

Introduction

The gastrointestinal tract has long been known to be among the most reactive of the organ systems in humans (Beaumont, 1833; Cannon, 1929; Pavlov, 1910). The esophagus, the stomach, and the small and large intestines make up the major organs of this system, and the general anatomical structure of these organs is similar in that they are composed of an outer layer of longitudinally oriented smooth muscle covering an inner circular smooth muscle layer (Vander, Sherman, & Luciano, 1975). It is the contractions of these outer longitudinal and inner circular smooth muscle layers that produce the propulsive or peristaltic activity of the gastrointestinal tract. Beneath these muscle layers lie the muscularis mucosa, composed of both circular and longitudinal muscle fibers, and the submucosal and mucosal linings of the lumen, which contain exocrine gland cells se-

creting protective fluids and, in the intestine, digestive enzymes.

Medical disorders affecting each of the major components of the gastrointestinal (GI) tract have at one time or another been hypothesized to be related to stress or emotional trauma. These disorders include functional illnesses such as the irritable bowel syndrome (IBS) and globus hystericus, as well as psychosomatic illnesses with more well-defined tissue pathology, such as peptic ulcer disease (PUD) and inflammatory bowel disease (IBD). Diagnostically, the major difference between functional and psychosomatic gastrointestinal disorders lies in the occurrence of verifiable tissue lesions in the cell wall of the lumen of the GI tract. When GI symptoms occur in the absence of tissue pathology or a known disease entity, they are diagnosed as being of a functional origin. In such instances, complaints of pain or other symptoms are thought to be related to abnormal functioning of an otherwise healthy organ rather than to specific organic lesions or tissue pathology. When GI symptoms are seen in the presence of tissue lesions of unclear or unknown etiology, they are often thought of as being psychosomatic in origin. This is espe-

Harry S. Shabsin and William E. Whitehead • Department of Psychiatry and Behavioral Sciences, Johns Hopkins University School of Medicine, and Division of Digestive Diseases, Francis Scott Key Medical Center, Baltimore, Maryland 21224.

cially true if patients with these types of disorders appear to exhibit personality characteristics that differentiate them from individuals who do not suffer such illnesses.

From a medical perspective, functional disorders are often considered related to abnormal motor or myoelectrical activity of the GI tract (Almy & Tulin, 1947; Christensen, 1975), whereas illnesses considered psychosomatic have been attributed to inappropriate immune responses or to a breakdown in mucosal barriers (Kirsner & Shorter, 1982; Schrager & Oates, 1978). However, regardless of the medical diagnosis, psychological factors appear to play a part in exacerbating or bringing on symptoms in many functional and psychosomatic gastrointestinal disorders by affecting physiological activity. As a result, functional and psychosomatic GI illness can also be thought as being psychophysiological in nature (Whitehead & Schuster, 1985).

Functional Disorders of the GI Tract

Irritable Bowel Syndrome

Although the etiology of functional GI disorders is unclear, they account for a majority of the patients presenting with gastrointestinal complaints. One of the most prevalent functional gastrointestinal disorders is the irritable bowel syndrome. It is the most commonly occurring diagnosis made by gastroenterologists, accounting for as many as 50–70% of all office visits made by patients with abdominal complaints (Ferguson, Sircus, & Eastwood, 1977; Kirsner & Palmer, 1958). Although not all of these patients experience IBS on a regular basis, it is experienced as a chronic condition by a smaller, but still sizably significant, proportion of the general public. Studies suggest as many as 8–22% of the population in Western countries experience this disorder on a recurring basis (Drossman, Sandler, McKee, & Lovitz, 1982; Thompson & Heaton, 1980; Whitehead, Winget, Fedoravicius, Wooley, & Blackwell, 1982).

As is typical of functional disorders, IBS is a difficult disorder to diagnose because its symptoms are similar to those of a variety of organic diseases, such as ulcerative colitis or diverticular disease, that affect the GI system. The most frequently reported symptoms associated with IBS are abdominal pain, constipation, diarrhea, abdominal distension, intestinal gas, or alternating periods of constipation and diarrhea in the absence of any organic disease or pathology (Schuster, 1984; Thompson, 1979). Because of the difficulty in diagnosing IBS based on the presenting complaints, several studies have attempted to define the features of this disorder that separate it from organic diseases presenting with similar symptoms. Manning, Thompson, Heaton, and Morris (1978) found increased bowel movements associated with the onset of abdominal pain, looser stools associated with the onset of abdominal pain, abdominal pain often eased after a bowel movement, and visible abdominal distension to be the best discriminators of IBS. Statistical analysis showed IBS patients to be significantly more likely to report each of these four symptoms than patients with other GI disorders. In their patients with IBS, Manning *et al.* (1978) found 91% to have two or more of these symptoms, compared with only 30% with other diseases who reported more than two of these symptoms. These investigators also found feelings of abdominal distension, rectal mucus, and feelings of incomplete evacuation to occur more often in IBS, although these symptoms did not occur significantly more frequently than in patients with organic diseases. Kruis, Thieme, Weinzierl, Schussler, Holl, and Paulus (1984) found alternating constipation and diarrhea or a combination of abdominal pain and irregularities of bowel habits to occur significantly more often in IBS than in other GI disorders. However, unlike the Manning *et al.* (1978) study, these symptoms were predictive only when organic disease had been ruled out.

Motility and Myoelectric Activity

Many of the symptoms associated with IBS have been attributed to abnormal intestinal motility (Connell, 1974; Kellow & Philips, 1987; Shabsin & Whitehead, 1988; Thompson, Laidlow, & Wingate, 1979). Differences in the various indices of colonic motility have been found between normals and IBS patients during rest (Dinoso, Goldstein, & Rosner, 1983; Latimer, Sarna, Campbell, Latimer, Waterfall, & Daniel, 1981), after a meal (Harvey & Read, 1973; Sullivan, Cohen, & Snape, 1978), and in response to distensions of the colon (Ritchie, 1973; Whitehead, Engel, & Schuster, 1980). Abnormal myoelectric slow-wave activity of the colon has also been reported to be associated with IBS (Snape, Carlson, & Cohen, 1976; Taylor, Darby, & Hammond, 1978; Welgan, Meshkinpour, & Hoehler, 1985). However, other investigators (Latimer *et al.*, 1981) have failed to support such findings, suggesting that further research is needed to clarify the relationship between IBS and intestinal myoelectric activity.

Psychological Considerations

Medical clinic patients with IBS symptoms have been found to be significantly more neurotic than other medical groups (Esler & Goulston, 1973; Palmer, Crisp, Stonehill, & Walker, 1974). In a well-controlled study, Ryan, Kelly, and Fielding (1983) found IBS patients to exhibit significantly more somatic and free-floating anxiety than matched controls, and Hislop (1971) found anxiety to occur three times more frequently in an IBS group than in age- and sex-matched controls. Hysteria, the development of physical symptoms in response to stress or psychological conflict (Graham, 1977), in conjunction with anxiety or depression, has also been found to be a common psychological trait in IBS patients (Liss, Alpers, & Woodruff, 1973; Young, Alpers, Norland, & Woodruff, 1976). Patients with IBS have also been found to have more somatic complaints than normal control groups, to visit their physicians more often for minor complaints, to rate their colds as more serious than do others, and to visit their doctors and have more hospitalizations for acute physical illness than PUD patients, even through PUD is a more serious illness than IBS (Sandler, Drossman, Nathan, & McKee, 1984; Whitehead *et al.*, 1982). Latimer, Campbell, Latimer, Sarna, Daniel, and Waterfall (1979) suggested that IBS patients differ from other groups in their preoccupation with bowel symptoms and in their health-seeking behavior, and Whitehead *et al.* (1982) hypothesized that the increased somatic concerns exhibited by IBS patients are an expression of learned illness behavior acquired in childhood through rewards and attention provided by parents for somatic symptoms.

Support for the involvement of illness behavior or a somatization disorder in IBS also comes from studies that have investigated the symptoms of this disorder in the general population, that is, individuals who have not sought medical assistance for IBS-like symptoms. Overall, only about 20–50% of the general population with IBS symptoms seeks medical assistance (Drossman, Sandler, McKee, & Lovitz, 1982; Thompson & Heaton, 1980; Whitehead, Bosmajian, Zonderman, Costa, & Schuster, 1988), and those who seek medical advice have been differentiated from those who do not based on neurotic psychiatric traits, that is, elevated levels of hypochondriasis, depression, hysteria, and general neuroticism rather than the occurrence of symptoms (Drossman, McKee, Sandler, Mitchell, Cramer, Lowman, & Burger, 1988; Greenbaum, Abitz, VanEgeren, Mayle, & Greenbaum, 1984; Whitehead *et al.*, 1988). Whitehead *et al.* (1988) and Drossman *et al.* (1988) also found that non-medical-consulters with IBS symptoms were no different psychologically from asymptomatic controls, suggesting that psychological characteristics are an important factor in determining the decision to seek medical consultation for the symptoms of IBS but are not necessarily a prerequisite to the development of these symptoms.

Esophageal Disorders

Functional GI disorders have also been described in the esophagus. The primary purpose of the esophagus is the transportation of food from the mouth to the stomach (Hightower, 1974). This transportation is accomplished by distally migrating bands of contracting circular muscle activity. This motility is initiated primarily by the act of swallowing and secondarily by distension of the esophagus. Once begun, the contractions proceed in an involuntary fashion from the upper boundary, the pharyngoesophageal sphincter, or *upper esophageal sphincter* (UES), of the esophagus to its lower boundary, or gastroesophageal sphincter, often termed the *lower esophageal sphincter* (LES). The esophagus, it should also be noted, is slightly different from other GI muscle tissue in that its upper third consists of striated muscle rather than smooth muscle. This gradually changes to smooth muscle in the middle third of esophagus, and the lower third of this organ consists entirely of smooth muscle tissue similar to the remainder of the GI tract.

Two common disorders of the esophagus that appear to be functional are diffuse esophageal spasm and nutcracker esophagus. Diffuse esophageal spasm (DES) is a frequently described motility disorder of the esophagus and is characterized by repetitive, nonpropulsive contractions that occur either spontaneously or following normal contractions initiated by swallowing (Castell, 1982). These contractions are of greater amplitude and longer duration than the esophageal contractions seen in normals and do not appear to be related to esophageal disease or tissue damage (Earlam, 1975; Henderson, 1980; Whitehead & Schuster, 1985). The typical symptoms associated with DES are chest pain and dysphagia (difficulty in swallowing).

Nutcracker esophagus is also a frequently occurring disorder in patients complaining of chest pain. In this disorder, peristalsis is normal but the amplitude of esophageal contractions is approximately double that of normal levels. Together, DES and nutcracker esophagus account for over 55% of the motility findings in patients complaining of noncardiac chest pain (Katz, Dalton, Richter, Wallace, Wu, & Castell, 1987).

Globus hystericus, the feeling of a lump in the throat, is another frequently reported esophageal symptom occurring in between 18% and 46.5% of the general population and is found more often in women and in patients with IBS (Thompson & Heaton, 1982; Watson, Sullivan, Corke, & Rush, 1978). Globus occurs most frequently between meals and has been suggested to be related to physical disorders such as hiatal hernia, heightened cricopharyngeal

pressures, reflux esophagitis, peptic ulcer disease, and cervical spinal disorders (Malcomson, 1966; Watson & Sullivan, 1974). However, it appears that these disorders do not account for the symptoms of globus in many patients complaining of these sensations (Caldarelli, Andrews, & Derbyshire, 1970; Mair, Schroder, Modalsli, & Maurer, 1974).

Psychological Considerations

The psychological profile of patients with noncardiac chest pain is remarkably similar to that of IBS patients, a finding perhaps suggesting a similar emotional etiology in these disorders. In general, patients have heightened levels of anxiety, depression, or hypochondriasis. Using the Million Behavioral Health Inventory (MBHI), Richter, Obrecht, Bradley, Young, and Anderson (1986b) compared patients with nutcracker esophagus to IBS patients, patients with benign structural abnormalities of the esophagus, and a normal control group. Both patients with nutcracker esophageus and those with IBS had significantly higher scores on measures of somatic anxiety and general gastrointestinal susceptibility, a measure of the likelihood that an individual will respond to stress with GI symptoms. Clouse and Lustman (1983), using DSM-III criteria (American Psychiatric Association, 1980), found patients with DES to exhibit somatization disorders, anxiety disorders, or depression. In this study, 50 consecutive referrals for chest pain were blindly rated on the occurrence of specific distal esophageal motility disorders. Of 25 patients with such disorders, 21 had psychiatric diagnoses, compared to only 4 of 13 without motility disorders.

Both of these studies are interesting in that they compared groups of patients with similar esophageal symptoms, only some of whom had motility disorders. It is often argued that psychiatric diagnoses are the result of symptoms rather than being a factor in the etiology of complaints. If this were the case, one should expect all patients with symptoms to display elevated psychological traits associated with neuroticism. However, Richter *et al.* (1986) and Clouse and Lustman (1983) found only patients with specific motility disorders to exhibit psychiatrically elevated scores. Patients with symptoms related to structural abnormalities of the esophagus or with nonspecific motility abnormalities were not found to be psychologically different from normal controls or the general population. This finding suggests that symptoms alone are not enough to cause psychopathology and that elevated levels of somatic anxiety or depression may be a precipitating factor in the onset of chest symptoms related to esophageal motility disturbances. Support for this position also comes from reports that have found the symptoms of DES to occur more often when patients are anxious or tired (Earlam, 1975; Henderson, 1980).

Globus sensations have also been associated with emotional states and have been attributed to suppressed crying by psychoanalytic theorists (Glaser & Engel, 1977). They are reported to occur frequently during periods of strong emotions and to abate with crying (Thompson & Heaton, 1982). The sensations of globus have also been attributed to the voluntary act of repeated swallowing, which may occur during time of stress or tension (Schatzki, 1964). This results in a reduction of saliva, and thus brings about the feeling of a lump in the throat, which may become worse as a person focuses on this feeling or increases swallowing in an attempt to get rid of the feeling. In general, although globus symptoms appear to occur frequently in the general population, they do not appear to be considered serious by those experiencing these sensations, as they account for only 1–3% of ear, nose, and throat medical consultations (Freeland, Ardran, & Emrys-Roberts, 1974; Malcomson, 1966).

Psychosomatic Disorders of the GI Tract

Inflammatory Bowel Disease

Ulcerative colitis and Crohn's disease are considered the major inflammatory diseases of the bowel. Ulcerative colitis involves recurrent inflammation and ulceration of the mucosal portion of the colon and the rectum and occurs with an incidence of approximately 50 to 100 cases per 100,000 in the general population, although this figure varies among different cultural and ethnic groups (Mendeloff, 1975; Stonnington, Phillips, Melton, & Zinsmeister, 1987). Crohn's disease is a more generalized subacute and chronic inflammation that may involve any portion of the gastrointestinal tract, but it is most commonly found in the distal ileum, the colon, and the anorectal area (Kirsner & Shorter, 1982). Its occurrence in the general population appears to be between 20 and 40 cases per 100,000 (Binder, Both, Hansen, Hendriksen, Kreiner, & Torp-Pedersen, 1982; Mendeloff, 1975). The most common symptoms of IBD are diarrhea, rectal bleeding, weight loss, and abdominal pain.

Psychological Considerations

Inflammatory bowel disease has historically been considered a psychosomatic disorder related to obsessive-compulsive, passive-dependent, and anxious or depressive personality characteristics because of the high incidence of these traits found in patients with this disorder, many of whom have been studied intensively by psychoanalytically oriented psychotherapists (Bellini & Tansella, 1976; Engel, 1955; Ford, Glober, & Castelnuovo-Tedesco, 1969; Mahoney, Ingram, Hundley, & Yaskin, 1949). However, many of the earlier studies associating IBD with psychological traits were retrospective or were not well controlled, and not all studies have supported the finding of psychopathology or the uniqueness of specific psychological traits in IBD (Esler & Goulson, 1973; Feldman, Cantor, Soll, & Bachrach, 1967a,b; Goldberg, 1970). A number of more recent controlled studies, however, have found IBD patients to be significantly more neurotic, depressed, hypochondriacal, hysterical, obsessive-compulsive, and alexithymic than normal or medical control groups, although not necessarily more so than non-IBD patients with psychosomatic illnesses (Fava & Pavan, 1976–1977b; Hezler, Stillings, Chammas, Norland, & Alpers, 1982; McMahon, Schmitt, Patterson, & Rothman, 1973; Sheffield & Carney, 1976; West, 1970).

In addition, McKegney, Gordon, and Levine (1970) found the level of psychopathology as measured on the Cornell Medical Index, a health questionnaire that has been found to be sensitive in detecting emotional disturbances in medical populations (Abramson, 1966), to be significantly correlated with the duration and severity of IBD. These authors also reported that 68% of the patients with Crohn's disease and 86% of the patients with ulcerative colitis had had a well-defined and serious life crisis in the six months before the onset of their physical illness. This relationship between emotional stress and the onset of IBD has also been reported by others (Grace, 1953; Hislop, 1974; Whybrow, Kane, & Lipton, 1968), and the relationship between specific stressors and the onset of symptoms has been shown to be quite similar (Fava & Pavan, 1976–1977a) for patients with IBD and those with IBS, a disorder considered primarily psychophysiological in origin. These findings suggest that the interaction between emotional stress and psychological characteristics can affect the onset of exacerbation of the symptoms associated with IBD.

On the other hand, although the theory that psychological traits contribute to the development of IBD is intriguing, the mechanism by which personality traits may result in the gastrointestinal pathophysiology resulting in IBD remains to be elucidated. Consequently, more emphasis has been placed recently on organic or environmental explanations, in which inappropriate immune system responses, bacterial contagions, or diet is hypothesized to account for the development of IBD (Kirsner & Shorter, 1982). However, efforts in this direction have been no more successful in conclusively determining the cause of IBD than have psychologically oriented theories.

Peptic Ulcer Disease

Peptic ulcer disease involves the development of ulcerating lesions in the lining of the stomach or the proximal portion of the small intestine. Peptic ulcers are related to a failure of the mucosal lining to protect the stomach or the intestine from gastric acid or pepsin, a digestive agent that breaks down protein. The factors that are thought to be involved in the development of PUD include abnormalities in the neutralization of acid by the secretion of bicarbonate, physical abnormalities in the mucosal lining, or the hypersecretion of gastric acid or pepsinogen, the precursor to pepsin (Card & Marx, 1960; Donaldson, MacCrae, & Parks, 1983; Flemstrom, 1981; Schrager & Oates, 1978; Wormsley, 1969).

Approximately 10% of the population experiences PUD (Pflanz, 1971), and duodenal (proximal small intestine) ulcers occur almost three times as frequently in Western cultures as do gastric (stomach) ulcers (Bonnevie, 1975a,b). Studies of Oriental cultures, on the other hand, show gastric ulcers to occur almost twice as often as duodenal ulcers (Kawai, Shirakawa, Misaki, Hayahsi, & Watanabe, 1989). The reason for this discrepancy is unclear, although it is possible that diet, smoking, or other cultural factors are involved. The primary diagnostic symptom of PUD is a gnawing hungerlike pain that is relieved by eating, although abdominal pain is not always reported in patients with peptic ulcers and may occur in the absence of documented ulcerations (Peterson, Sturdevant, Frankl, Richardson, Isenberg, Elashoff, Sones, Gross, McCallum, & Fordtran, 1977). The serious complications of PUD include vomiting, internal bleeding, and perforation of the stomach or intestine, resulting in infection of the peritoneal cavity.

Psychological Considerations

In addition to physiological abnormalities, personality characteristics and emotional stress have also been implicated in the development of PUD. Alexander (1950) proposed that unconscious psychological conflicts result in autonomic arousal or alterations in endocrine responses that may initiate the disease processes. Such autonomic and endocrine alterations are the result of emotional states such as anger, depression, or anxiety produced by unconscious psychological conflict. Frustrated oral and dependency needs have also been considered essential emotional conflicts in the development of PUD (Alexander, French, & Pollack, 1968). Graham and his colleagues (Grace & Graham, 1952; Graham, Lundy, Benjamin, Kabler, Lewis, Kunish, & Graham, 1962) interviewed patients with a variety of illnesses thought to be psychosomatic in origin and found that patients with pud often described feeling that they had not been given what they were due by others and expressed a desire to avenge what they considered to be wrongful acts against them by others. There is also evidence that peptic ulcer patients tend to be more introverted, anxious, and depressed than individuals without PUD (Ecklenberger, Overbeck, & Biebel, 1976; Whitehead et al., 1982).

Although many of the studies associating personality characteristics with PUD are inferential (i.e., psychological characteristics separating PUD patients from other groups are assumed to be related to the onset of peptic ulcerations, but underlying disease processes are not studied), there are some studies that have investigated the relationship between emotional conflict and alterations in GI physiology that may be related to the onset of illness. Weiner, Thaler, Reiser, and Mirsky (1957) investigated personality characteristics in army recruits with high and how pepsinogen levels. By the use of blindly scored projective tests, 71% of the pepsinogen hypersecreters were identified based on their level of immaturity, dependency, and oral needs in relation to their level of masculine defenses and paranoid ideations. In the study, increased pepsinogen levels were correlated with increased gastric acid output, and 7 of the 10 subjects with the greatest dependency needs went on to develop peptic ulcerations. Also using projective measures, Cohen, Silverman, Waddell, and Zuidema (1961) found ulcer patients to exhibit significantly more dependency needs and less aggression and anger than control patients with other illnesses.

Stress and Abnormal GI Functioning

A feature common to many psychophysiological disorders is the occurrence of stressful events preceding symptom onset or exacerbation. A recent study (Sternback, 1986) found gastrointestinal symptoms to be the second most often reported response to stress in America, and it is estimated that more than 50% of all medical visits to gastroenterologists are for functional disorders known to be affected by stress (Drossman, Powell, & Sessions, 1977; Kirsner & Palmer, 1958). Studies suggest that stressful environmental or emotional events may interact with the psychological characteristics described above to precipitate psychophysiological gastrointestinal disorders.

The stressors frequently associated with psychophysiological illnesses in the GI tract include worry over money, health, family, relationships, work, and everyday mundane problems (Chaudhary & Truelove, 1962; Hill & Blendis, 1967; Pflanz, 1971; Sternback, 1986; Wolf & Almy, 1949). Emotional and environmental stresses have been found to be a factor in the expression of symptoms associated with IBD and DES (Clouse & Lustman, 1983; Faulkner, 1940; Hislop, 1974; Monk, Mendeloff, Siegel, & Lilienfeld, 1970) as well as in the exacerbation or onset of symptoms related to PUD and IBS (Chaudhary & Truelove, 1962; Davies & Wilson, 1937; Grayson, 1972; Hill & Blendis, 1967; Hislop, 1971).

However, the type of stress experienced may play a role in the development of gastrointestinal symptoms. For instance, using Paykel's scale for the qualitative differentiation of life events (Paykel, Prusoff, & Uhlenhuth, 1971) and the Holmes and Rahe (1967) scale, Fava and Pavan (1976–1977a) compared 20 IBS, 20 IBD, and 20 appendicitis patients for levels of stress in the six months preceding their illness. Both the IBD and the IBS patients had similar numbers of stressful undesirable events involving losses and exits from the social field preceding their illness, as measured on the Paykel scale, which uses modified subsets of the Holmes and Rahe scale. However, the IBS patients reported more stressful events from the entire Holmes and Rahe scale preceding their illness than did the IBD patients, although the IBS and the IBD patients reported more stressful events preceding their illness on both scales than did the appendicitis patients. Mendeloff, Monk, Siegel, and Lilienfeld (1970), using the Holmes and Rahe scale, also found that IBS patients reported more stressful events in the six months before hospitalization for their illness than did IBD patients. There findings suggest that IBD patients may be more sensitive to social stresses and that IBS patients respond to and experience a greater variety of overall stress preceding their illness.

Stressful environmental events also appear to be an important factor in the onset of PUD. Duodenal ulcers occur more often in urban than in rural settings and are found more often in populations under duress of war than in times of peace (Pflanz, 1971). In addition, professions with high levels of inherent stress, such as police officer or air traffic controller, also have greater incidences of PUD (Cobb & Rose, 1973; Grayson, 1972; Richard & Fell, 1975). This epidemiological evidence relating environmental stress to PUD is also supported by controlled studies that have measured stressful events before the onset of illness or that have compared stressful events in PUD patients to those in matched controls (Alp, Court, & Grant, 1970; Sapira & Cross, 1982), although there is some evidence that duodenal ulcers may be more related to stress than gastric ulcers (Piper, Greig, Shinners, Thomas, & Crawford, 1978).

Although the physiological process by which personality traits might interact with stress to produce the symptoms in psychosomatic GI disorders remains to be pinpointed, it appears that alterations in motility produced by stress may be responsible for many of the physical complaints reported by patients with functional GI disorders.

Almy and his associates (Almy, 1951; Almy & Tulin, 1947) were the first modern investigators to document alterations in colonic motility in subjects under stress. In a series of observational studies using proctoscopic visualization of the colon, Almy found that stressors such as cold-pressor or experimenter-evoked anxiety altered motility in both IBS and normal subjects and that these alterations ceased after the termination of the stressful period. However,

such reversals took longer in IBS patients, and alterations in colonic activity appeared to be related to coping strategies. Changes in colon motility were not observed in subjects experiencing physical stress until they expressed an emotional reaction to their situation, and the type of colon activity was reported as being related to the attitude that the subjects adopted. Subjects expressing hostility, defensiveness, or active coping strategies showed increased motility, while subjects expressing helplessness, defeat, or a subdued acceptance showed decreased motility. Similar findings were reported by Wolf and Wolff (1947), and changes in intestinal motility were also found to be related to personality characteristics by Lechin, Van Der Dijs, Gomez, Lechin, and Arocha (1983), who reported that decreased motility was associated with somatization disorders, hypochondriasis, and obsessive-compulsive traits, whereas increased motility was found to be related to guilt, anxiety, agitation, and depersonalization.

Using modern electrical and pressure recording equipment, Welgan et al. (1985) investigated colonic motility in IBS patients and normal controls to stressors involving cold pressor, mental arithmetic, and fear. Although IBS patients had significantly more motor activity at rest than did normals, both groups significantly increased their colon motility during stress. The IBS group also showed a significant increase in slow-wave (2–4 cpm) myoelectric activity during stress compared to controls. In a separate study, Welgan, Meshkinpour, and Beeler (1987) found IBS patients to have significantly more colonic motility during periods of anger than a control group. No difference was found between IBS and normals during periods of rest. In both of these studies, IBS patients displayed significantly greater amounts of hypochondriasis, hysteria, and depression than did normals as measured on the Minnesota Multiphasic Personality Inventory (MMPI).

Studies such as Welgan's suggest that stress alone is not enough to produce IBS. Because both normals and IBS patients increase colonic motility in response to a variety of stressors, it would appear that it is the combination of personality traits and the experience of emotional or environmental stress that leads to the development of IBS. Support for this conclusion comes from studies that have investigated IBS symptoms in individuals who either do or do not seek medical consultation for their abdominal complaints (Drossman *et al.*, 1982; Sandler *et al.*, 1984; Whitehead *et al.*, 1988). In general, these studies have found patients more likely to report stress influencing their abdominal symptoms and to have significantly higher levels of neuroticism than nonpatients.

Emotional or environmental stress has also been shown to cause increased esophageal contractions or esophageal spasms in both normal subjects and patients complaining of noncardiac chest pain (Faulkner, 1940; Rubin, Nagler, Spiro, & Pilot, 1962; Stacher, Steinringer, Blau, & Landgraf, 1979; Wolf & Almy, 1949; Young, Richter, Anderson, Bradley, Katz, McElveen, Obrecht, Dalton, & Snyder, 1987). Anderson, Dalton, Bradley, and Richter (1989) compared 10 noncardiac chest pain patients with esophageal contraction abnormalities, 9 noncardiac chest pain patients without esophageal contraction abnormalities, and 20 normal controls during baseline and laboratory-induced stress. The stressors in this study included unpredictable bursts of 100-db white noise or difficult cognitive discrimination tasks. All subjects showed significant increases in esophageal contraction amplitudes during stress, and patients initially diagnosed with esophageal contraction abnormalities showed the greatest increases. Correlated with these changes in esophageal function were significant increases in both observed and self-rated reports of anxiety for each of the three groups during the stressful conditions.

Studies such as that by Anderson *et al.* (1989) appear to delineate a direct relationship between stress, motility, and symptoms of noncardiac chest pain (NCP). However, the relationship between NCP and abnormalities of esophageal motility is not clear cut. A number of researchers have found little correlation between periods of NCP and the occurrence of contraction abnormalities (Anderson *et al.*, 1989; Clouse, Staiano, Landau, & Schlacter,

1983; Katz, Dalton, Richter, Wu, & Castell, 1987; Peters, Maas, Petty, Dalton, Penner, Wu, Castell, & Richter, 1988). This dissociation of motility disorders from NCP led Shabsin, Katz, and Schuster (1988) to suggest that esophageal motor abnormalities may simply be physiological markers for individuals more apt to experience NCP symptoms in response to various environmental stimuli, including stress.

Evidence indicating that patients report their NCP symptoms more frequently during periods of fatigue, anxiety, or emotional arousal (Benjamin, Richter, Cordova, Knuff, & Castell, 1983; Henderson, 1980; Jacobson, 1927), taken together with studies showing that contraction abnormalities of the esophagus are significantly associated with elevations of anxiety, depression, and somatic concerns (i.e., Clouse & Lustman, 1983) suggests that NCP, like IBS, has a strong psychological component in the etiology or exacerbation of symptoms. Further evidence for a psychological component in both IBS and NCP comes from studies that have found patients with symptoms associated with both of these disorders to report pain at significantly lower distension pressures in the esophagus or colon than do normals (Richter, Barish, & Castell, 1986a; Ritchie, 1973; Whitehead *et al.*, 1980) and to have significant elevations in somatic anxiety and gastrointestinal susceptibility, a measure of the likelihood of an individual's responding to stress with GI symptoms (Richter *et al.*, 1986b). It thus appears at present that emotional or environmental stress, combined with psychological characteristics, are important considerations in the symptoms associated with NCP and IBS.

Treatment

Since Jacobson's early descriptions (1927) of treating gastrointestinal symptoms with progressive muscle relaxation (PMR), a number of investigators have reported successfully treating a variety of GI disorders with behavioral, cognitive, and psychotherapeutic procedures. Many of these studies are descriptive case reports, and others have used small numbers of subjects, usually without adequate controls. Thus, further research is needed on the effectiveness of psychotherapy in treating GI disorders. However, there are a few large and well-controlled studies (e.g., Svedlund, Sjodin, Ottoson, & Dotevall, 1983) that have produced positive results and that lend support to many of the case reports finding psychotherapy to be an effective means of treating chronic GI symptoms.

Irritable Bowel Syndrome

Svedlund *et al.* (1983) divided 101 IBS patients into two matched groups and followed them at 3 and 15 months. Both groups received medical management consisting of bulk-forming agents and, when appropriate, anticholinergic agents, antacids, and minor tranquilizers. The control group received only medical management, while the experimental group also received 10 one-hour psychotherapy sessions spread over the first three months of the study. Svedlund and his associates used a variety of techniques, including behavior modification, problem solving, and stress management, to help patients modify maladaptive behavior and to find solutions to problems. The therapy was focused on ways of coping with stress and emotional problems, and both physical and psychological symptoms were used to assess improvements during this study. The physical symptoms consisted of ratings of somatic complaints, abdominal pain, and bowel dysfunction. The psychological outcome measures were based on scores obtained for total psychological symptoms, an asthenic-depressive syndrome, and an anxiety syndrome. The assessments took place before the subjects were assigned to groups and at 3- and 15-month follow-ups.

Both groups studied by Svedlund *et al.* (1983) showed improvement in somatic and psychological scores after three months of treatment. The experimental group showed significantly greater improvement, however, on total somatization and abdominal pain than the controls at first follow-up, and all three somatic measures were significantly improved over those in controls at 15 months. In addition, the

patients receiving psychotherapy continued to show improvement between 3 and 15 months, whereas those receiving only medication regressed toward the initial severity of their symptoms at the 15-month follow-up. In contrast to the physical scores, no changes in psychological symptoms occurred after 3 months, and the differences in the psychological symptoms between groups was not significant at either the 3- or the 15-month follow-up. In addition to supporting the use of psychotherapy in the treatment of IBS, this study also validates reports showing a poor prognosis for the long-term treatment of IBS patients with medications alone (e.g., Waller & Misiewicz, 1969).

Other therapies based on the work of Ellis (1962) directed at changing irrational cognitions concerning both symptoms and stressful life events or the work of Fordyce (1976) using behavioral therapy to decrease reinforcers of chronic illnesses have also been found effective for treating the symptoms of IBS. The majority of these studies have used singly or in combination, a variety of strategies, including PMR, cognitive restructuring, stress management, behavior modification, or desensitization procedures (Cohen & Reed, 1968; Harrell & Beiman, 1978; Hedberg, 1973; Jacobson, 1927; Khatami & Rush, 1978; Mitchell, 1978; Youell & McCullough, 1975). Hypnotically induced relaxation and aversive behavioral therapy have also been reported to be successfully used in the treatment of IBS (Dolezalova, Cerny, & Jirak, 1978; Legalos, 1977; Miller & Kratochwill, 1979; Whorwell, Prior, & Colgan, 1987).

Biofeedback has also been reported as an effective treatment for IBS symptoms. In a group study using a waiting list control population, Giles (1978) found relaxation training using biofeedback to be an effective procedure for treating symptoms arising from functional gastrointestinal disorders. However, the best results were obtained from a combination of biofeedback plus psychotherapy. Other studies have also reported successfully treating IBS using biofeedback (Blanchard, Schwartz, & Neff, 1988; Weinstock, 1976). The above studies used frontalis EMG biofeedback or finger temperature training, and patients achieved the greatest improvements when biofeedback was given in combination with psychotherapy involving behavioral or cognitive strategies.

Specific biofeedback based on bowel sounds has also been attempted as a treatment for IBS. Using electronic stethoscopes, Furman (1973) and Radnitz and Blanchard (1988) were able to reduce symptoms in patients by providing feedback based on bowel sounds. The patients were taught to alternatively increase and decrease bowel sounds in order to gain control over bowel activity. However, others have not been able to replicate these results (Weinstock, 1976), and only half of the patients treated by Radnitz and Blanchard (1988) showed improvement in their symptoms by using bowel sound feedback.

In general, it appears that nonspecific biofeedback aimed at increasing overall relaxation may be more effective than specific biofeedback based on bowel activity. This conclusion is supported by an interesting study conducted by Whitehead (1985), in which two groups of IBS patients were given either biofeedback based on colon motility or stress management based on PMR and desensitization training. The patients provided with stress management substantially reduced their complaints of abdominal pain but showed no alteration in colonic activity. The group provided with motility biofeedback, on the other hand, showed substantial decreases in abnormal motility patterns but no change in abdominal pain complaints. Half the biofeedback patients were then crossed over to the stress management group. These patients responded to stress management with decreased symptoms but no further changes in motility. Although the total number of subjects in this study was small, the efficacy of stress management for the crossed-over patients adds increased significance to these findings and suggests that psychological factors may be more important in the treatment of IBS symptoms than are abnormalities in physiological activity.

Few studies have reported on the use of psychodynamic or nondirective psychotherapy for the treatment of IBS. Hislop (1980) described a series of 52 IBS patients whom he provided with brief psychodynamically oriented therapy designed to explore emotional states and to help in the expression of repressed emotions. Recent life events were also explored, and reassurance was provided. For the majority of the patients, therapy was made available in the form of three or fewer one-hour sessions spaced two to four weeks apart. Following treatment, 46% of the patients described their symptoms as absent or improved on self-report questionnaires, but no statistical analysis was performed and no control group was used. As a result, these findings must be accepted with caution, especially as others (Waller & Misiewicz, 1969) have indicated that follow-up contact alone seems to help patients with IBS, and psychodynamically oriented therapy typically requires more than three hours to be effective. Overall, then, relaxation procedures, cognitive psychotherapy, and behavior modification strategies appear to be the only documented psychological therapies effective in helping patients with IBS and would seem to be the treatments of choice for this disorder.

Esophageal Disorders

As with IBS, one of the more useful therapeutic procedures for treating psychophysiological disorders of the esophagus appears to be relaxation training. Jacobson (1927) was the first to use relaxation procedures in the treatment of gastrointestinal disorders and reported several cases of esophageal spasm treated successfully with progressive muscle relaxation. Latimer (1981) reported on a single case of esophageal spasm treated successfully with frontalis EMG feedback and PMR. Latimer's patient was able to reduce her total perceived spasm time from an initial 10 hours a week to less than 1 hour a week by the end of a 15-week treatment period. Latimer also used a double swallowing technique that proved helpful to this patient. These gains were maintained at a six-month follow-up after the treatment was terminated.

Shabsin et al. (1988) reported using PMR, frontalis EMG, and index finger temperature biofeedback in conjunction with behavior modification and cognitive restructuring in the successful treatment of chest pain due to vigorous achalasia, a neurological disorder causing portions of the esophagus to become aperistaltic. The typical symptoms of this disorder include chest pain and dysphagia, and manometric findings often show DES as well as aperistaltic activity. The patient in this report had proved refractory to both medical and surgical management for control of her symptoms over a 10-year period. After completion of the treatment, the patient was followed at 3 and 12 months without a return of her symptoms, although her abnormal motility patterns did not change from their pretreatment state. This dissociation of esophageal motility from symptoms was also reported by Latimer (1981) and, as with IBS, suggests the importance of psychological characteristics in chronic symptoms related to esophageal dysfunction.

Although there are no reports of the behavioral management of globus hystericus, several case histories have been published describing the use of biofeedback, relaxation strategies, or behavior modification in the treatment of patients complaining of difficulty swallowing or upper esophageal dysphagia. Using biofeedback and psychotherapy, Shabsin (1988) treated a 62-year-old woman complaining of swallowing difficulties, nausea, vomiting, and abdominal pain. The patient's symptoms had begun eight months before treatment, at a time when her husband was undergoing cardiac bypass surgery, from which he successfully recovered, and they had a fairly stereotypical pattern, beginning with the patient's focusing on her difficulty in swallowing. This was followed by attempts to clear her throat. When these failed to alleviate her swallowing difficulty, she began to clear her throat more vigorously and to increase the contractions of her abdominis rectus (abdominal) muscles. This process resulted in her symptoms of abdominal pain, nausea, and retching, which the patient termed vomiting.

The treatment of this patient initially consisted of PMR, combined with EMG biofeedback from electrodes placed symmetrically on the lateral aspects of the cricopharyngeal (throat) muscles. It has been suggested that patients with swallowing difficulties experience spasms of the cricopharyngeal muscles (Watson & Sullivan, 1974), and it was hypothesized that biofeedback would enable this patient to voluntarily decrease any inappropriate activity of these striated muscles. The patient was also provided with cognitive stress management training involving positive statements about her health and recovery, and behavior modification was used by withdrawing attention that the patient was receiving from her family for her symptoms. Over the 12-week treatment program provided to this patient, her symptoms decreased from their initial level of four to six times a day to less than once a week. Frontalis EMG biofeedback, relaxation training, and aversive behavior therapy have also been reported as being used successfully in the treatment of patients with swallowing difficulties (Haynes, 1976; Solyom & Sookman, 1980).

Inflammatory Bowel Disease

Several studies have reported that supportive psychotherapy (i.e., non-insight-oriented, non-anxiety-provoking therapy in which the therapist supports the patient with approval and help in dealing with problems that the patient finds difficult) produces both symptomatic and psychological improvement in patients with IBD (Grace, Pinsky, & Wolff, 1954; Groen & Bastiaans, 1951; Karush, Daniels, Flood, O'Connor, Druss, & Sweeting, 1977; Karush, Daniels, O'Connor, & Stern, 1968). Typically, 60%–70% of IBD patients treated with supportive psychotherapy show improvements in their illness ranging from total remission to a decrease in symptoms. In the best controlled of these studies, Grace et al. (1954) compared 34 IBD patients treated with psychotherapy with 34 age, sex, and severity-of-disease matched IBD controls. The patients treated with psychotherapy differed from those not receiving psychotherapy in terms of number of patients experiencing improvements in their illness (65% vs. 32%), hospitalizations for relapses (53% vs. 29%), the average number of weeks spent in the hospital (9.5 vs. 11.5), and the number of patients needling ileostomy surgery (3 vs. 10).

In general, the psychological improvements found in IBD patients treated with supportive psychotherapy involve decreases in anxiety and an improvement in their abilities to interact with others in social or work-related situations (Karush *et al.*, 1968). Personality characteristics, on the other hand, were generally not altered to any significant degrees in IBD patients responding to psychotherapy. This finding would suggest that long-term insight-oriented psychotherapy directed toward altering personality is not necessary in treating IBD patients.

Relaxation exercises have also been found to be helpful in treating patients with IBD. Shaw and Ehrlich (1987) reported randomly assigning 40 ulcerative colitis patients to either a PMR treatment group or a waiting control group contacted by phone once per week. The groups were of equal number and were sex matched but not age matched, and none of the patients had a history of psychosis or psychiatric hospitalization. Following treatment, the PMR group reported significantly less distress from their symptoms and significantly less intense and less frequent abdominal pain related to their IBD than did the controls. In addition, after treatment, but not before, significantly fewer patients treated with PMR were using anti-inflammatory drugs than were controls. This reduction in medication is quite encouraging, as many anti-inflammatory agents, such as prednisone, used to treat IBD have serious side effects. Other forms of relaxation training involving autogenic training (Luthe & Schultz, 1969) have also been reported in uncontrolled studies to be effective in treating patients with ulcerative colitis (DeGossely, Koninckx, & Lenfant, 1975; Schaeffer, 1966).

Overall, the studies conducted so far indicate that decreases in psychological anxiety and physical arousal can be helpful in treating the patient with IBD. They also suggest that behavioral and cognitive strategies may be helpful for the patient with IBD, although these studies remain to be done.

Using behavior modification and cognitive restructuring, we have recently treated a patient with IBD at the Gastrointestinal Pain Center of the Francis Scott Key Medical Center.

The patient was a 46-year-old male with a 20-year history of Crohn's disease. In the six years before the patient was seen at the Pain Center, his illness had become much more severe, and he was taking large amounts of steroids with only moderate alleviation of his symptoms.

When first seen, this patient was experiencing a great deal of social and family disability related to his symptoms, although he continued to work. He was also very angry and bitter about his illness and felt it had ruined his life. The patient spent most of his free time lying down or complaining to his wife about his illness. The behavior modification involved decreasing the attention that the patient's wife gave him as a result of his symptoms. It also included increasing the recreational and social activities engaged in by the patient, both on his own and with his wife. The cognitive aspects of the treatment involved altering the patient's perceptions that he was too sick to engage in nonwork activities and helping him accept his illness without anger or despair. After six months of weekly psychotherapy, the patient's abdominal symptoms from his illness were greatly improved (i.e., two to three or fewer episodes of abdominal pain per month compared to the daily and almost constant pain he had experienced before therapy). The patient was followed over the next year and maintained the improvements he had obtained with psychotherapy. In addition, his anti-inflammatory medication was reduced by 50% during this time by his gastroenterologist without any increase in his symptoms.

Although a case study, this report is similar to other larger studies in that the patient improved using non-insight-oriented forms of psychotherapy, and it suggests the usefulness of behavior modification and cognitive psychotherapy in the treatment of symptoms arising from IBD.

Peptic Ulcer Disease

In several controlled studies (Chappell, Stefano, Rogerson, & Pike, 1936; Svedlund & Sjodin, 1985), dynamically oriented psychotherapy using cognitive and behavioral strategies aimed at reducing stress, anxiety, and emotional problems has been shown to be effective in treating symptoms arising from PUD. Svedlund and Sjodin randomly assigned 103 peptic ulcer pa-

tients to either a control group or a psychotherapeutic group. Both groups received medical
management as needed, consisting of antacids, anticholinergics, H_2-receptor antagonists, or minor tranquilizers, and the experimental group received three months of weekly
one-hour psychotherapy in the form of stress
management and problem solving for issues of
emotional importance to the patient. Supportive therapy was primarily used, but insight
and anxiety-provoking therapy was also employed with those patients who could tolerate
it. At three months following therapy, both
groups showed improvement in their symptoms, although the patients given psychotherapy
showed greater decreases in somatic symptoms than did the controls. However, these differences between groups were not statistically
significant. At 15-month follow-up, however,
the psychotherapy group showed significantly
fewer somatic symptoms than did the control
group, which had begun to regress back toward the original level of symptoms before
medical treatment.

Chappell *et al.* (1936) treated 32 peptic ulcer
patients with behavior modification aimed at
decreasing the social rewards for having PUD
symptoms. Cognitive psychotherapy involving the patients' thinking of themselves as recovering or being healthy, as well as distraction
to combat anxiety and worry, was also used.
The patients were also informed of the relationship between emotional arousal, worry, and
physiological changes in their GI tract that
might lead to PUD. The patients received psychotherapy daily in small groups for six weeks.
After six weeks, 30 of the 32 experimental patients were free of symptoms and were eating a
normal diet compared to only 2 of the 20 PUD
control patients. These gains had been maintained at three-year follow-up, with 26 of the 28
experimental patients reporting no symptoms
or only mild symptoms over this time period.

Brooks and Richardson (1980) reported similar results using intensive but briefer psychotherapy involving rational emotive therapy (e.g.,
Ellis, 1962) and assertiveness training. The patients were seen for eight sessions over a two-
week period and were then followed at 60 days

and 42 months. At both follow-ups, the patients receiving psychotherapy were experiencing fewer days of ulcer pain and were consuming fewer antacids than the control patients
provided with three sessions of supportive
therapy. The experimental patients also had
fewer hospitalizations following treatment than
did controls, and after 42 months, only one of
the nine experimental patients had experienced
an illness relapse, compared to five of the eight
controls. Frontalis EMG biofeedback and PMR
have also been reported in several case studies
as being effective in relieving PUD symptoms
and in healing peptic ulceration as shown by
radiological evidence (Aleo & Nicassio, 1978;
Beaty, 1976).

As with other psychophysiological GI illnesses, brief forms of behavioral and cognitive
psychotherapy appear to be effective in treating PUD. However, of all the disorders discussed in this chapter, the medical management of PUD is the most effective. The use of
histamine H_2 antagonists and antacids has
been scientifically well documented as healing
peptic ulcers and as preventing relapses (Dronfield *et al.*, 1979; Winship, 1978). This finding
seems to have led to a decreased interest in the
use of psychotherapy in the treatment of PUD,
as medical management is more cost-effective
in terms of both time and money. On the other
hand, for the patient who continues to experience frequent episodes of PUD and who has
the psychological characteristics associated
with this illness, psychotherapy may provide
the most beneficial treatment over an extended
period of time.

Program Development
and Implementation

In spite of evidence suggesting that psychological traits and stress can influence the course
of certain gastrointestinal disorders, the medical community, with notable exceptions, has
not readily sought psychotherapy as a form of
treatment for GI illnesses. A portion of the explanation lies in the fact that physicians generally receive little training in the psychological

aspects of illness and are not oriented by their education to see illness from this perspective. On the other hand, neither are psychologists typically given much exposure to the medical aspects of GI illnesses in their training, so that they may misinterpret a patient's complaints or refuse medical referrals that are appropriate.

Developing a psychological practice in a medical setting, therefore, requires, first, the ability to communicate effectively with the medical community. Often, this necessitates developing an understanding of both the medical and the psychological aspects of a particular illness. Lack of medical knowledge makes it difficult to discuss physical illness from a psychological perspective or to provide effective information to the medical community about ways in which psychotherapy can be used to help patients. Physicians are concerned about their patients and are more likely to trust a patient to psychotherapy if they feel that the therapist has an adequate understanding of the medical considerations involved in a patient's symptoms. In addition, without such information, it is difficult to treat patients psychologically who may be experiencing complicated and sometimes health-threatening medical symptoms.

Information about the interaction between psychology and medicine can be conveyed by giving talks at hospital grand rounds or other medical meetings and conferences, by maintaining an active correspondence about a patient's care with the referring physician, by offering to provide in-service talks to medical support teams, or by personal communications. Such activities increase the credibility of using psychotherapy for medical patients and help to promote a collegial relationship between the professionals involved in a patient's health care.

Establishing independence is also an important element in providing health care in a medical setting. Psychologists should request hospital privileges in order to see inpatients and make notes in medical charts. Obtaining staff privileges facilitates being perceived as part of the medical community and helps to promote patient referrals. Establishing independence also means participating in referring patients to other specialists when appropriate for further medical or psychiatric evaluation. Hospitals or medical institutions can best be thought of as groups of specialized consultants who refer back and forth to each other as needed in order to provide health care to a patient. Participating in this network helps to establish psychological evaluation or treatment as part of the patient care process.

Our experience at the Francis Scott Key Medical Center has been mostly with outpatients, although we do provide inpatient evaluation and short-term treatment programs for hospitalized patients. We do not provide treatment for terminally ill patients, and most other inpatients are discharged before much psychotherapy for treating disorders related to long-term personality characteristics can be accomplished. Often, however, we continue to treat patients once they have been released from the hospital. For the most part, we find that treating stress-related GI disorders is more effective in an outpatient program, where patients can learn to cope with stressful situations on a daily basis.

We also believe that evaluation is an important aspect of any treatment program. Each of our patients has a nonstructured interview as part of the intake procedure. This portion of the evaluation is designed to provide information on drug dependency (prescription or nonprescription), secondary-gain issues, family problems, emotional status, and other factors that may be pertinent to the patient's symptoms. Personality characteristics are obtained by means of the MMPI (Hathaway & McKinley, 1942) or the Symptom Check List-90 (SCL-90) (Derogatis, Lipman, & Covi, 1973), and a depression inventory such as the Beck (Beck, Ward, Mendelson, Mock, & Erbaugh, 1961) or Hamilton (Hamilton, 1967) is administered as part of the standard intake evaluation. Measures of anxiety, assertiveness, mental status, and other personality characteristics are also obtained as deemed appropriate. Such information is then used in communicating the patient's needs and emotional state to the referring physician, as well as in planning the treatment protocol for the patient.

A brief outline for the psychological management of patients with gastrointestinal disorders is given below:

- Medical evaluation to rule out or treat organic disorders.
- Psychological evaluation including personality, behavioral, and cognitive assessments.
- A treatment program capable of providing some combination of: Psychotherapy for depression, anxiety, and somatization disorders. Behavior modification. Family therapy. Stress management. Cognitive restructuring strategies. Biofeedback and relaxation training.

Summary

This chapter has detailed the psychological characteristics associated with psychophysiological disorders or the gastrointestinal tract and the psychological treatment strategies that have been shown useful in alleviating or eliminating the symptoms accompanying these disorders. The most effective procedures have involved behavioral, cognitive, and psychophysiological (i.e., biofeedback and relaxation training) forms of psychotherapy. In general, many of these techniques may be seen as involving various forms of stress management, and their effectiveness would appear to be associated with the central role that stress has been shown to play in the onset or exacerbation of functional and psychosomatic gastrointestinal disorders. At present, psychological and behavioral management programs, in conjunction with medical screening, appear to provide the best approach to treating the patient with psychophysiological disorders of the gastrointestinal tract.

References

Abramson, J. H. (1966). The Cornell Medical Index as an epidemiological tool. *American Journal of Public Health, 56*, 287–298.

Aleo, S., & Nicassio, P. (1978). Auto-regulation of duodenal ulcer disease: A preliminary report of four cases. *Proceedings of the Biofeedback Society of American (Ninth Annual Meeting)* (pp. 278–281). Denver, CO: Biofeedback Society of America.

Alexander, F. (1950). *Psychosomatic medicine*. New York: W. W. Norton.

Alexander, F., French, T. M., & Pollack, G. H. (1968). *Psychosomatic specificity* (Vol. 1). Chicago: University of Chicago Press.

Almy, T. P. (1951). Experimental studies on the irritable colon. *American Journal of Medicine, 9*, 60–67.

Almy, T. P., & Tulin, M. (1947). Alterations in colonic function in man under stress: Experimental production of changes simulating the "irritable colon." *Gastroenterology, 8*, 616–626.

Alp, M. H., Court, J. H., & Grant, A. K. (1970). Personality pattern and emotional stress in genesis of gastric ulcer. *Gut, 11*, 773–777.

American Psychiatric Association. (1980). *Diagnostic and statistical manual of mental disorders* (3rd. ed.). Washington, DC: Author.

Anderson, K. O., Dalton, C. B., Bradley, L. A., & Richter, J. E. (1989). Stress: A modulator of esophageal pressures in healthy volunteers and non-cardiac chest pain patients. *Digestive Diseases and Sciences, 34*, 83–91.

Beaty, E. T. (1976). Feedback assisted relaxation training as a treatment for peptic ulcers. *Biofeedback and Self Regulation, 1*, 323–324.

Beaumont, W. (1833). *Experiments and observations on the gastric juice and the physiology of digestion*. Plattsburgh, NY: F. P. Allen.

Beck, A. T., Ward, C. H., Mendelson, M., Mock, J. E., & Erbaugh, J. K. (1961). An inventory for measuring depression. *Archives of General Psychiatry, 4*, 561–571.

Bellini, M., & Tansella, M. (1976). Obsessional scores and subjective general psychiatric complaints of patients with duodenal ulcer or ulcerative colitis. *Psychological Medicine, 6*, 461–467.

Benjamin, S. B., Richter, J. E., Cordova, C. M., Knuff, T. E., & Castell, D. O. (1983). Prospective manometric evaluation with pharmacologic provocation of patients with suspected esophageal motility dysfunction. *Gastroenterology, 84*, 893–901.

Binder, V., Both, H., Hansen, P. K., Hendrikson, C., Kreiner, S., & Torp-Pederson, K. (1982). Incidence and prevalence of ulcerative colitis and Crohn's disease in country of Copenhagen, 1962–1978. *Gastroenterology, 83*, 563–568.

Blanchard, E. B., Schwartz, S. P., & Neff, D. F. (1988). Two-year follow-up of behavioral treatment of irritable bowel syndrome. *Behavior Therapy, 19*, 67–73.

Bonnevie, O. (1975a). The incidence of duodenal ulcer in Copenhagen County. *Scandinavian Journal of Gastroenterology, 10*, 385–393.

Bonnevie, O. (1975b). The incidence of gastric ulcer in Copenhagen County. *Scandinavian Journal of Gastroenterology, 10*, 231–239.

Brooks, G. R., & Richardson, F. C. (1980). Emotional skills

training: A treatment program for duodenal ulcer. *Behavior Therapy, 11,* 198–207.

Caldarelli, D. D., Andrews, A. H., & Derbyshire, A. J. (1970). Esophageal motility studies in globus sensation. *Annals of Otolaryngology, 79,* 1098–100.

Cannon, W. B. (1929). *Bodily changes in pain, hunger, fear, and rage* (2nd. ed.). New York: Appleton-Century-Crofts.

Card, W. I., & Marx, I. N. (1960). The relationship between acid output of the stomach following "maximal" histamine stimulation of the parietal cell mass. *Clinical Sciences, 19,* 147–163.

Castell, D. O. (1982). Pathophysiology and spectrum of clinical syndromes of esophageal motility disorders. *Journal of the Society of Gastrointestinal Assistants, 4,* 17–23.

Chappell, M. N., Stefano, J. J., Rogerson, J. S., & Pike, F. H. (1936). The value of group psychological procedures in the treatment of peptic ulcer. *American Journal of Digestive Diseases and Nutrition, 3,* 813–817.

Chaudhary, N. A. & Truelove, S. C. (1962). The irritable colon syndrome. *Quarterly Journal of Medicine, 31,* 307–322.

Christensen, J. (1975). Myoelectric control of the colon. *Gastroenterology, 70,* 601–609.

Clouse, R. E., & Lustman, P. J. (1983). Psychiatric illness and contraction abnormalities of the esophagus. *The New England Journal of Medicine, 42,* 337–342.

Clouse, R. E., Staiano, A., Landau, D. W., & Schlachter, J. L. (1983). Manometric findings during spontaneous chest pain in patients with presumed esophageal "spasm." *Gastroenterology, 85,* 395–402.

Cobb, S., & Rose, R. M. (1973). Hypertension, peptic ulcer, and diabetes in air traffic controllers. *Journal of the American Medical Association, 224,* 489–492.

Cohen, S. I., & Reed, J. L. (1968). The treatment of "nervous diarrhea" and other conditioned autonomic disorders by desensitization. *British Journal of Psychiatry, 114,* 1275–1280.

Cohen, S. I., Silverman, A. J., Waddell, W., & Zuidema, G. D. (1961). Urinary catecholamine levels, gastric secretion and specific psychological factors in ulcer and non-ulcer patients. *Journal of Psychosomatic Research, 5,* 90–115.

Connell, A. M. (1974). Clinical aspects of motility. *Medical Clinics of North America, 58,* 1201–1216.

Davies, D. T., & Wilson, A. T. (1937). Observations on the life-history of chronic peptic ulcer. *Lancet, 2,* 1350–1360.

DeGossely, M., Koninckx, N., & Lenfant, H. (1975). La rectocolite hémorragique: Training autogene. A propos de quelques cas graves. *Acta Gastro-Enterologica Belgica, 38,* 454–462.

Derogatis, L. H., Lipman, R. S., & Covi, L. (1973). SCL-90: An outpatient psychiatric rating scale—Preliminary report. *Psychopharmacology Bulletin, 9*(1), 13–27.

Dinoso, V., Goldstein, J., & Rosner, B. (1983). Basal motor activity of the distal colon: A reappraisal. *Gastroenterology, 85,* 637–642.

Dolezalova, V., Cerny, M., & Jirak, R. (1978). Relaxation and EMG activity in neurotics and patients with psychosomatic gastrointestinal disorders. *Activa Nervosa, 20* (Suppl. 1), 35–36.

Donaldson, J. D., MacCrae, K. D., & Parks, T. G. (1983). Comparison of mucus substances in gastric juice of normal subjects, duodenal ulcer and dyspeptic patients. *European Journal of Surgical Research, 15,* 11–17.

Dronfield, M. W., Bachelor, A. I., Lackworth, W., & Langman, M. J. S. (1979). Controlled trial and maintenance cimetidine treatment in healed duodenal ulcer: Short and long-term effects. *Gut, 20,* 526–530.

Drossman, D. A., Powell, D. W., & Sessions, J. T. (1977). The irritable bowel syndrome. *Gastroenterology, 73,* 811–822.

Drossman, D. A., Sandler, R. S., McKee, D. C., & Lovitz, A. J. (1982). Bowel patterns among subjects not seeking health care. *Gastroenterology, 83,* 529–534.

Drossman, D. A., McKee, D. C., Sandler, R. S., Mitchell, M., Cramer, E. M., Lowman, B. C., & Burger, A. L. (1988). Psychological factors in the irritable bowel syndrome. *Gastroenterology, 95,* 701–708.

Earlam, R. (1975). *Clinical tests of oesophageal function.* New York: Grune & Stratton.

Ecklenberger, D., Overbeck, G., & Biebel, W. (1976). Subgroups of peptic ulcer patients. *Journal of Psychosomatic Medicine, 20,* 489–499.

Ellis, A. (1962). *Reason and emotion in psychotherapy.* New York: Lyle Stuart.

Engel, G. L. (1955). Studies of ulcerative colitis: 3. The nature of the psychologic process. *American Journal of Medicine, 17,* 231–256.

Esler, M. D., & Goulston, K. J. (1973). Levels of anxiety in colonic disorders. *The New England Journal of Medicine, 288,* 16–20.

Faulkner, W. B. (1940). Severe esophageal spasm. *Psychosomatic Medicine, 2,* 139–140.

Fava, G. A., & Pavan, L. (1976–77a). Large bowel disorders: 1. Illness configuration and life events. *Psychotherapy and Psychosomatics, 27,* 93–99.

Fava, G. A., & Pavan, L. (1976–1977b). Large bowel disorders: 2. Psychopathology and alexithymia. *Psychotherapy and Psychosomatics, 27,* 100–105.

Feldman, F., Cantor, D., Soll, S., & Bachrach, W. (1967a). Psychiatric study of a consecutive series of 19 patients with regional ileitis. *British Medical Journal, 4,* 711–714.

Feldman, F., Cantor, D., Soll, S., & Bachrach, W. (1967b). Psychiatric study of a consecutive series of 34 patients with ulcerative colitis. *British Medical Journal, 3,* 14–17.

Ferguson, A., Sircus, W., & Eastwood, M. A. (1977). Frequency of "functional" gastrointestinal disorders. *Lancet, 2,* 613–614.

Flemstrom, G. (1981). Gastric secretion of bicarbonate. In L. R. Johnson (Ed.), *Physiology of the gastrointestinal tract* (pp. 603–616). New York: Raven Press.

Ford, C. V., Glober, G. A., & Castelnuovo-Tedesco, P. (1969). A psychiatric study of patients with regional enteritis. *Journal of the American Medical Association, 208,* 311–315.

Fordyce, W. E. (1976). Behavioral concepts in chronic pain. In P. O. Davidson (Ed.), *The behavioral management of anxiety, depression and pain.* New York: Brunner/Mazel.

Freeland, A. P., Ardran, G. M., & Emrys-Roberts, E.

(1974). Globus hystericus and reflux oesophagitis. *Journal of Laryngology, 88*, 1025–1031.

Furman, S. (1973). Intestinal biofeedback in functional diarrhea: A preliminary report. *Journal of Behavior Therapy and Experimental Psychiatry, 4*, 317–321.

Giles, S. L. (1978). Separate and combined effects of biofeedback training and brief individual psychotherapy in the treatment of gastrointestinal disorders. *Dissertation Abstracts International, Part B*, 2495.

Glaser, J. P., & Engel, G. L. (1977). Psychodynamics, psychophysiology and gastrointestinal symptomatology. In T. P. Almy & J. F. Fielding (Eds.), *Clinics in gastroenterology*. London: W. B. Saunders.

Goldberg, D. (1970). A psychiatric study of patients with diseases of the small intestine. *Gut, 11*, 459–465.

Grace, W. J. (1953). Life stress and regional enteritis. *Gastroenterology, 23*, 542–553.

Grace, W. J., & Graham, D. T. (1952). Relationship of specific attitudes and emotions to certain bodily diseases. *Psychosomatic Medicine, 14*, 243–251.

Grace, W. J., Pinsky, R. H., & Wolff, H. G. (1954). The treatment of ulcerative colitis. *Gastroenterology, 26*, 462–468.

Graham, D. T., Lundy, R. M., Benjamin, L. S., Kabler, J. D., Lewis, W. C., Kunish, N. O., & Graham, F. K. (1962). Specific attitudes in initial interviews with patients having different "psychosomatic" diseases. *Psychosomatic Medicine, 24*, 257–266.

Graham, J. R. (1977). *The MMPI: A practical guide.* New York: Oxford University Press.

Grayson, R. R. (1972). Air controller syndrome: Peptic ulcer in air traffic controllers. *Illinois Medical Journal, 142*, 11–115.

Greenbaum, D., Abitz, L., Van Egeren, L., Mayle, J., & Greenbaum, R. (1984). Irritable bowel symptom prevalence, rectosigmoid motility and psychometrics in symptomatic subjects not seeing physicians. *Gastroenterology, 84*(5 Pt. 2), 1174.

Groen, J., & Bastiaans, J. (1951). Psychotherapy of ulcerative colitis. *Gastroenterology, 17*, 344–352.

Hamilton, M. (1967). Development of a rating scale for primary depressive illness. *British Journal of Social and Clinical Psychology, 6*, 278–296.

Harrell, T. H., & Beiman, I. (1978). Cognitive-behavioral treatment of the irritable colon syndrome. *Cognitive Therapy and Research, 2*, 371–375.

Harvey, R., & Read, A. (1973). Effects of cholecystokinin on colonic motility and symptoms in patients with irritable bowel syndrome. *Lancet, 1*, 1–3.

Hathaway, S. R., & McKinley, J. C. (1942). *Minnesota Multiphasic Personality Inventory.* Minneapolis: University of Minnesota Press.

Haynes, S. N. (1976). Electromyographic biofeedback treatment of a woman with chronic dysphagia. *Biofeedback and Self Regulation, 1*, 121–126.

Hedberg, A. G. (1973). The treatment of chronic diarrhea by systematic desensitization: A case report. *Journal of Behavior Therapy and Experimental Psychiatry, 4*, 67–68.

Henderson, R. D. (1980). *Motor disorders of the esophagus* (2nd ed.). Baltimore: Williams & Wilkens.

Hezler, J. E., Stillings, W. A., Chammas, S., Norland, C. C. & Alpers, D. H. (1982). A controlled study of the association between ulcerative colitis and psychiatric diagnoses. *Digestive Disease and Sciences, 27*, 513–518.

Hightower, N. C. (1974). Applied anatomy and physiology of the esophagus. In H. L. Bockus (Ed.), *Gastroenterology* (3rd ed., pp. 124–142). Philadelphia: W. B. Saunders.

Hill, O. W., & Blendis, L. (1967). Physical and psychological evaluation of "non-organic" abdominal pain. *Gut, 8*, 221–229.

Hislop, I. G. (1971). Psychological significance of the irritable colon syndrome. *Gut, 12*, 452–457.

Hislop, I. G. (1974). Onset setting in inflammatory bowel disease. *The Medical Journal of Australia, 1*, 981–984.

Hislop, I. G. (1980). Effect of very brief psychotherapy on the irritable bowel syndrome. *The Medical Journal of Australia, 2*, 620–623.

Holmes, T. H., & Rahe, R. H. (1967). The social readjustment rating scale. *Journal of Psychosomatic Research, 11*, 213–218.

Jacobson, E. (1927). Spastic esophagus and mucous colitis. *Archives of Internal Medicine, 39*, 433–455.

Karush, A., Daniels, G. E., O'Connor, J. R., & Stern, L. O. (1968). The response to psychotherapy in chronic ulcerative colitis: 1. Pretreatment factors. *Psychosomatic Medicine, 30*, 255–276.

Karush, A., Daniels, G., Flood, C., O'Connor, J., Druss, R., & Sweeting, J. (1977). *Psychotherapy in chronic ulcerative colitis.* Philadelphia: W. B. Saunders.

Katz, P. O., Dalton, C. B., Richter, J. E., Wu, W. C., & Castell, D. O. (1987). Esophageal testing of patients with noncardiac chest pain or dysphagia. *Annals of Internal Medicine, 106*, 593–597.

Kawai, K., Shirakawa, K., Misaki, F., Hayashi, K., & Watanabe, Y. (1989). Natural history and epidemiologic studies of peptic ulcer disease in Japan. *Gastroenterology, 96*, 581–585.

Kellow, J. E., & Phillips, S. F. (1987). Altered small bowel motility in irritable bowel syndrome is correlated with symptoms. *Gastroenterology, 92*, 1885–1893.

Khatami, M., & Rush, J. (1978). A pilot study of the treatment of outpatients with chronic pain: Symptom control, stimulus control, and social system intervention. *Pain, 5*, 163–172.

Kirsner, J. B., & Palmer, W. L. (1958). The irritable colon. *Gastroenterology, 34*, 490–493.

Kirsner, J. B., & Shorter, R. G. (1982). Recent developments in "nonspecific" inflammatory bowel disease. *New England Journal of Medicine, 306*, 775–785, 837–848.

Kruis, W., Thieme, C., Weinzierl, M., Schussler, P., Holl, J., & Paulus, W. (1984). A diagnostic score for the irritable bowel syndrome. *Gastroenterology, 87*, 1–7.

Latimer, P. (1981). Biofeedback and self-regulation in the treatment of diffuse esophageal spasm: A single-case study. *Biofeedback and Self Regulation, 6*, 181–189.

Latimer, P., Campbell, D., Latimer, M., Sarna, S., Daniel,

E., & Waterfall, W. (1979). Irritable bowel syndrome: A test of the colonic hyperalgesia hypothesis. *Journal of Behavioral Medicine, 2,* 285–295.

Latimer, P., Sarna, S., Campbell, D., Latimer, M., Waterfall, W., & Daniel, E. (1981). Colonic motor and myoelectric activity: A comparitive study of normal subjects, psychoneurotic patients, and patients with irritable bowel syndrome. *Gastroenterology, 80,* 893–901.

Lechin, R., Van der Dijs, B., Gomez, R., Lechin, E., & Arocha, L. (1983). Distal colonic motility and clinical parameters in depression. *Journal of Affective Disorders, 5,* 19–26.

Legalos, C. N. (1977). Aversive behavior therapy for chronic stomach pain: A case study. *Pain, 4,* 67–72.

Liss, J. L., Alpers, D., & Woodruff, R. A. (1973). The irritable colon syndrome and psychiatric illness. *Diseases of the Nervous System, 34,* 151–157.

Luthe, W., & Schultz, J. H. (1969). *Autogenic therapy* (Vol. 1) New York: Grune & Stratten.

Mahoney, U. P., Bockus, H. L., Ingram, M., Hundley, J. W., & Yaskin, J. C. (1949). Studies in ulcerative colitis: 1. A study of personality in relation to ulcerative colitis. *Gastroenterology, 13,* 547–563.

Mair, W. S., Schroder, K. E., Modalsli, B., & Maurer, H. J. (1974). Aetiological aspects of the globus symptom. *Journal of Laryngology, 88,* 1033–1040.

Malcomson, K. G. (1966). Radiological findings in globus hystericus. *British Journal of Radiology, 39,* 583–586.

Manning, A. P., Thompson, W. G., Heaton, K. W., & Morris, A. F. (1978). Towards positive diagnosis of the irritable bowel. *British Medical Journal, 2,* 653–654.

McKegney, F. P., Gordon, R. O., & Levine, S. M. (1970). A psychosomatic comparison of patients with ulcerative colitis and Crohn's disease. *Psychosomatic Medicine, 32,* 153–166.

McMahon, A. W., Schmitt, P., Patterson, J. F., & Rothman, E. (1973). Personality differences between inflammatory bowel disease patients and their healthy siblings. *Psychosomatic Medicine, 35,* 91–103.

Mendeloff, A. I. (1975). The epidemiology of idiopathic inflammatory bowel disease. In J. B. Kirsner & R. G. Shorter (Eds.), *Inflammatory bowel disease* (pp. 3–19). Philadelphia: Lea & Febiger.

Mendeloff, A. I., Monk, M., Siegel, C. I., & Lilienfeld, A. (1970). Illness experience and life stresses in patients with irritable colon syndrome and with ulcerative colitis. *The New England Journal of Medicine, 282,* 14–17.

Miller, A. J., & Kratochwill, T. R. (1979). Reduction of frequent stomachache complaints by time out. *Behavior Therapy, 10,* 211–218.

Mitchell, K. (1978). Self-management of spastic colitis. *Journal of Behavior Therapy and Experimental Psychiatry, 9,* 269–272.

Monk, M., Mendeloff, A. I., Siegel, C. I., & Lilienfeld, A. (1970). An epidemiological study of ulcerative colitis and regional enteritis among adults in Baltimore: 3. Psychological and possible stress-precipitating factors. *Journal of Chronic Disease, 22,* 565–578.

Palmer, R. L., Crisp, A. H., Stonehill, E., & Waller, S. L. (1974). Psychological characteristics of patients with the irritable bowel syndrome. *Postgraduate Medical Journal, 50,* 416–419.

Pavlov, I. (1910) *The work of the digestive glands,* trans. W. H. Thompson. London: Griffin.

Paykel, E. S., Prusoff, B. A., & Uhlenbuth, E. H. (1971). Scaling of life events. *Archives of General Psychiatry, 25,* 240–347.

Peters, L. J., Maas, L. C., Petty, D., Dalton, C. B., Penner, D., Wu, M. B., Castell, D. O., & Richter, J. E. (1988). Spontaneous non-cardiac chest pain: Evaluation by 24 hour ambulatory esophageal motility and pH monitoring. *Gastroenterology, 94,* 878–886.

Peterson, W. L., Sturdevant, R. A. L., Frankl, H. D., Richardson, C. T., Isenberg, J. I., Elashoff, J. D., Sones, J. Q., Gross, R. A., McCallum, R. W., & Fordtran, J. S. (1977). Healing of duodenal ulcer with antacid regimen. *New England Journal of Medicine, 297,* 341–345.

Pflanz, M. (1971). Epidemiological and sociocultural factors in the etiology of duodenal ulcer. *Advances in Psychosomatic Medicine, 6,* 121–151.

Piper, D. W., Greig, M., Shinners, J., Thomas, J., & Crawford, J. (1978). Chronic gastric ulcer and stress. *Digestion, 18,* 303–309.

Radnitz, C. L., & Blanchard, E. B. (1988). Bowel sound biofeedback as a treatment for irritable bowel syndrome. *Biofeedback and Self Regulation, 13,* 169–179.

Richard, W. C., & Fell, R. D. (1975). Health factors in police job stress. In W. H. Kroes & J. J. Hurrell (Eds.), *Job stress and the police officer: Identifying stress reduction techniques* (pp. 73–84). (HEW Publication No. NIOSH 76–187). Washington, DC: U.S. Government Printing Office.

Richter, J. E., Barish, C. F., & Castell, D. O. (1986a). Abnormal sensory perception in patients with esophageal chest pain. *Gastroenterology, 91,* 845–852.

Richter, J. E., Obrecht, W. F., Bradley, L. A., Young, L. D., & Anderson, K. O. (1986b). Psychological comparison of patients with nutcracker esophagus and irritable bowel syndrome. *Digestive Diseases and Sciences, 31,* 131–138.

Ritchie, J. (1973). Pain from distention of the pelvic colon by inflating a balloon in the irritable colon syndrome. *Gut, 14,* 125–132.

Ryan, W. A., Kelly, M. G., & Fielding, J. F. (1983). Personality and the irritable bowel syndrome. *Irish Medical Journal, 76,* 140–141.

Rubin, J., Nagler, R., Spiro, H., & Pilot, M. (1962). Measuring the effect of emotions on esophageal motility. *Psychosomatic Medicine, 24,* 170–176.

Sandler, R. S., Drossman, D. A., Nathan, H. P. & McKee, D. C. (1984). Symptom complaints and health care seeking behavior in subjects with bowel dysfunction. *Gastroenterology, 87,* 314–318.

Sapira, J. D., & Cross, M. R. (1982). Pre-hospitalization life change in gastric ulcer (GU) versus duodenal ulcer (DU). *Psychosomatic Medicine, 44,* 121.

Schaeffer, G. (1966). Ergebnisse des autogenen trainings bei der colitis ulcerosa. In J. J. Lopez Ibor (Ed.), *IV World*

Congress of Psychiatry, Madrid, 5–11, IX. Amsterdam: Excerpta Medica Foundation, International Congress Series No. 117.48.

Schatzki, R. S. (1964). Globus hystericus (globus sensation). *The New England Journal of Medicine, 270,* 676.

Schrager, J., & Oates, M. D. G. (1978). Human gastrointestinal mucus in disease states. *British Medical Bulletin, 34,* 79–82.

Schuster, M. M. (1984). Irritable bowel syndrome: Applications of psychophysiological methods of treatment. In R. Hoelzl & W. E. Whitehead (Eds.), *Psychophysiology of the gastrointestinal tract: Experimental and clinical applications.* New York: Plenum Press.

Shabsin, H. S. (1988). Behavioral considerations in evaluating and treating chronic gastrointestinal pain. *Endoscopy Review, 5*(3), 67–72.

Shabsin, H. S., & Whitehead, W. E. (1988). Psychophysiological disorders of the gastrointestinal tract. In W. Linden (Ed.), *Biological barriers in behavioral medicine.* New York: Plenum Press.

Shabsin, H. S., Katz, P. O. & Schuster, M. M. (1988). Behavioral treatment of intractable chest pain in a patient with vigorous achalasia. *The American Journal of Gastroenterology, 83,* 970–973.

Shaw, L., & Ehrlich, A. (1987). Relaxation training as a treatment for chronic pain caused by ulcerative colitis. *Pain, 29,* 287–294.

Sheffield, B. F., & Carney, M. W. P. (1976). Crohn's disease: A psychosomatic illness? *British Journal of Psychiatry, 128,* 446–450.

Snape, W., Carlson, G., & Cohen, S. (1976). Colonic myoelectric activity in the irritable bowel syndrome. *Gastroenterology, 70,* 326–330.

Solyom, L., & Sookman, D. (1980). Fear of choking and its treatment. *Canadian Journal of Psychiatry, 25,* 3–34.

Stacher, G., Steinringer, H., Blau, A., & Landgraf, M. (1979). Acoustically evoked esophageal contraction and defense reaction. *Psychophysiology, 16,* 234–241.

Sternback, H. A. (1986). Pain and hassles in the U.S.: Nuprin Pain Report. *Pain, 24,* 69–80.

Stonnington, C. M., Philips, S. F., Melton, L. J., & Zinsmeister, A. R. (1987). Chronic ulcerative colitis: incidence and prevalence in a community. *Gut, 28,* 402–409.

Sullivan, M., Cohen, S., & Snape, W. (1978). Colonic myoelectrical activity in irritable-bowel syndrome. *The New England Journal of Medicine, 298,* 878–883.

Svedlund, J., & Sjodin, I. (1985). A psychosomatic approach to treatment in the irritable bowel syndrome and peptic ulcer disease with aspects of the design of clinical trials. *Scandanavian Journal of Gastroenterology, 20*(suppl. 109), 147–151.

Svedlund, J., Sjodin, I., Ottoson, J., & Dotevall, G. (1983). Controlled study of psychotherapy in irritable bowel syndrome. *Lancet, 2,* 589–592.

Taylor, I., Darby, C., & Hammond, P. (1978). Comparison of rectosigmoid myoelectrical activity in irritable colon syndrome during relapse and remission. *Gut, 15,* 559–607.

Thompson, D. J., Laidlow, J. M., & Wingate, D. L. (1979). Abnormal small bowel motility demonstrated by radio-telemetry in a patient with irritable colon. *Lancet, 2,* 1321–1323.

Thompson, W. G. (1979), *The irritable gut.* Baltimore: University Park Press.

Thompson, W. G., & Heaton, K. W. (1980). Functional bowel disorders in apparently healthy people. *Gastroenterology, 79,* 283–288.

Thompson, W. G., & Heaton, K. W. (1982). Heartburn and globus in apparently healthy people. *Journal of the Canadian Medical Association, 126,* 46–48.

Vander, A. J., Sherman, J. H., & Luciano, D. S. (1975). *Human physiology: The mechanisms of body function.* New York: McGraw-Hill.

Waller, S. L., & Misiewicz, J. J. (1969). Prognosis in the irritable bowel syndrome. *Lancet, 2,* 753–756.

Watson, W. C., & Sullivan, S. N. (1974). Hypertonicity of the cricopharyngeal sphincter: A cause of globus sensation. *Lancet, 2,* 676.

Watson, W. C., Sullivan, S. N., Corke, M., & Rush, D. (1978). Globus and headache: Common symptoms of the irritable bowel syndrome. *Journal of the Canadian Medical Association, 118,* 387–388.

Weiner, H., Thaler, M., Reiser, M. F., & Mirsky, I. A. (1957). Etiology of duodenal ulcer: 1. Relation of specific psychological characteristics to rate of gastric secretion (serum pepsinogen). *Psychosomatic Medicine, 19,* 1–10.

Weinstock, S. A. (1976). The reestablishment of intestinal control in functional colitis. *Biofeedback and Self Regulation, 1,* 324.

Welgan, P., Meshkinpour, H., & Hoehler, F. (1985). The effect of stress on colon motor and electrical activity in irritable bowel syndrome. *Psychosomatic Medicine, 47,* 139–149.

Welgan, P., Meshkinpour, H., & Beeler, M. (1987). Effect of anger on colon motor and myoelectric activity in irritable bowel syndrome. *Gastroenterology, 94,* 1150–1156.

West, K. L. (1970). MMPI correlates of ulcerative colitis. *Journal of Clinical Psychology, 26,* 214–229.

Whitehead, W. E. (1985). Psychotherapy and biofeedback in the treatment of irritable bowel syndrome. In N. W. Read (Ed.), *Irritable bowel syndrome* (pp. 245–256). London: Grune & Stratton.

Whitehead, W. E., & Schuster, M. M. (1985). *Common gastrointestinal disorders: Physiological and behavioral basis for treatment.* New York: Academic Press.

Whitehead, W. E., Engel, B., & Schuster, M. M. (1980). Irritable bowel syndrome: Physiological and psychological differences between diarrhea-predominant and constipation-predominant patients. *Digestive Diseases and Sciences, 25,* 404–413.

Whitehead, W. E., Winget, C., Fedoravicius, S. S., Wooley, S., & Blackwell, B. (1982). Learned illness behavior in patients with irritable bowel syndrome and peptic ulcer. *Digestive Diseases and Sciences, 27,* 202–208.

Whitehead, W. E., Bosmajian, L., Zonderman, A. B., Costa, P. T., & Schuster, M. M. (1988). Symptoms of psychological distress associated with irritable bowel syndrome. *Gastroenterology, 95,* 709–714.

Whorwell, P. J., Prior, A., & Colgan, S. M. (1987). Hypnotherapy in severe irritable bowel syndrome: further experience. *Gut, 28,* 423–425.

Whybrow, P. C., Kane, F. J., & Lipton, M. A. (1968). Regional ileitis and psychiatric disorder. *Psychosomatic Medicine, 30,* 209–221.

Winship, D. H. (1978). Cimetidine in the treatment of duodenal ulcer: Review and commentary. *Gastroenterology, 74,* 402–406.

Wolf, S., & Almy, A. P. (1949). Experimental observations of cardiospasm in man. *Gastroenterology, 13,* 401–421.

Wolf, S., & Wolff, H. G. (1947). *Human gastric function: An experimental study of man and his stomach* (2nd ed.). New York: Oxford University Press.

Wormsley, K. G. (1969). Response to duodenal acidification in man: 1. Electrolyte changes in duodenal aspirate. *Scandinavian Journal of Gastroenterology, 4,* 717–726.

Youell, K. J., & McCullough, J. P. (1975). Behavioral treatment of mucous colitis. *Journal of Consulting and Clinical Psychology, 43,* 740–745.

Young, L. D., Richter, J. C., Anderson, K. D., Bradley, L. A., Katz, P. O., McElveen, L., Obrecht, W. F., Dalton, G. B., & Snyder, R. M. (1987). The effects of psychological and environmental stressors on peristaltic esophageal contractions in healthy volunteers. *Psychophysiology, 24,* 132–141.

Young, S. J., Alpers, D. H., Norland, C. C., & Woodruff, R. A. (1976). Psychiatric illness and the irritable bowel syndrome: Practical implications for the primary physician. *Gastroenterology, 70,* 162–166.

CHAPTER 29

Programs for the Treatment of Dental Disorders

Dental Anxiety and Temporomandibular Disorders

Barbara G. Melamed and David J. Williamson

Introduction

Surveys indicate that 4%–10% of the adult population avoid dental treatment because of extreme fears, and probably an even larger percentage (up to 25%) avoid dental treatment except when they are symptomatic (Ayer & Corah, 1984). Extreme dental fears were reported in 23% of 14- to 21-year-olds, and there is a continuing reduction in fear with age. Dentists are generally inconsistent in their approach to dealing with anxious patients. Corah (1988) reported that three quarters of the dentists surveyed felt that patient anxiety was the greatest barrier to regular dental care. Very few dentists, however, ask the patient whether he or she is anxious. They are afraid that the question will suggest that there is something to be anxious about. Many dentists admit that they themselves become anxious if they deal with anxious patients. Dentists need to know both how to recognize patient anxiety and what to do about it. The first part of this chapter reviews the etiology of dental anxiety and indicates how it can be measured in child and adult patients. Strategies for reducing patient anxiety are provided in terms of the treatment of patients by psychologists and the education of dentists in behavior management strategies for preventing dental phobia. The second part of the chapter deals with temporomandibular disorder (TMD), a pain syndrome increasingly recognized as being influenced by psychological factors. The term *temporomandibular disorder* has been adopted instead of the more commonly *temporomandibular joint disorder* (TMJ) in order to encompass a wider range of pathogenic processes both within and outside the temporomandibular joint that contribute to the symptoms reported by patients suffering from this syndrome. Etiological theories of TMD as well as behavioral interventions aimed at alleviating the problem are discussed.

Barbara G. Melamed • Ferkauf Graduate School of Psychology and Albert Einstein College of Medicine, Yeshiva University, 1300 Morris Park Avenue, Bronx, New York 10461. **David J. Williamson** • Department of Clinical and Health Psychology, University of Florida Health Science Center, Gainesville, Florida 32610.

Dental Anxiety

Anxiety is a complex variable. The prospect of dental treatment may evoke fears of many different things. A questionnaire survey found that 12 million American people avoid dental treatment because of psychological concerns (Friedson & Feldman, 1958). Kleinknecht, Klepac, and Alexander (1973) found that many aspects of the dental situation contribute to anxiety, including a fear of

1. Criticism for poor oral hygiene
2. Loss of control
3. Pain
4. The anesthetic injection
5. The sound and sensation of the drill

These fears may generalize, and the result is that the dentist and associated stimuli, such as offices, support personnel, and treatment instruments, become sources of discomfort.

In our survey of children (Cuthbert & Melamed, 1982), those between 6 and 7 years of age reported the greatest fear. Across an age range of 4–14 years, a fear of choking was ranked the highest, followed by injections and drilling. These findings are consistent with an Australian survey (Herbertt & Innes, 1979) that found that third- and fourth-grade children (8–9 years of age) were reporting the least cooperation and the most anxiety during dental treatment.

Children learn to fear the dentist vicariously through the media or by hearing from peers or parents about the dread of dental treatment. Children who have not previously experienced dental examinations may have anticipatory fears that are quickly alleviated by the actual experiences that they undergo in the dental operatory. The avoidance of dental treatment by adults and their children continues to be a major public health concern, despite the initiative of pedodontists in seeking early visits with the child to employ active preventive measures and minimize dental problems.

The discomfort experienced during dental visits may sometimes be unpleasant, but not objectively harmful. Unfortunately, like other health services, dental care is often neglected because of immediate discomfort and the long-term nature of deleterious consequences. Patients who are severely dental-phobic do not go for routine work. The vast majority of people eventually do seek dental treatment. The few who are true dental phobics are characterized by avoiding dental appointments, forgetting appointments, or requiring extreme sedation, even general anesthesia, in order to cope with routine dental treatment. Many seek dental care only when pain is already involved. This provides an ideal situation for the classical conditioning of emotionally aversive responses and avoidance behavior. The true dental phobic may tolerate discomfort and deterioration of oral health for many years. Fearful patients of this type have been found to score higher on Eysenck's Neuroticism Scale and lower on the Extraversion scale than nonfearful controls (Lautch, 1971). Adult dental patients who are highly anxious in general may require a different psychological approach from those who may be nervous about the dentist, but who are low in general anxiety. These individuals need to be referred to clinical psychologists.

A second important role for the clinical psychologist is the education of dental practitioners. The dentist's handling of the patient in the operatory not only affects the quality of care that can be delivered but influences the patient's future behavior with regard to dental care. Thus, the iatrogenic effects of an adverse dentist–patient interaction is yet another source of dental phobia. Studies with children (Melamed, Hawes, Heiby, & Glick, 1975a; Melamed, Weinstein, Hawes, & Katin-Borland, 1975b) have provided evidence that a high level of fear in children during a dental examination predicted the degree of disruption during actual restorative treatment. Pedodontists have used the tell-show-do techniques for many years in managing the normal child patient. These techniques provide information to the patient about what to expect and how to react. The literature on imparting information about impending stressful events, such as medical procedures, suggests that different types of individuals need different types and amounts of information (Melamed, 1982). Yet, in the literature on dentist–patient communication, such undefined terms as *empathy, friendliness,* and *a calm, competent image* are thought to be sufficient for the treatment of most patients.

Many dentists are encouraged to use positive reinforcement to obtain cooperation with the dental procedures. Unfortunately, they often delay the reinforcement (i.e., the child receives a general treat at the end of the session). Comments of praise are often too general for the child to learn much about the expected behavior. Although a few studies exist in which positive tangible reinforcement improved the dental chair behaviors of mentally retarded children (Kohlenberg, Greenberg, Reymore, & Mass, 1972) or psychotic adults (Klinge, 1979), there is little definitive research on reinforcement strategies consistently applied with normal dental patients.

With the difficult child dental patient, dentists are taught to be firmer and to use a loud voice and restraints if necessary to control disruptive behavior. Yet, retrospective reports suggest that the development of dental phobias involved previous noxious experience in the dental setting and negative family attitudes (Forgione & Clark, 1975; Shoben & Borland, 1954). In spite of the general cautions in the application of punishment procedures in the parent training and the animal behavior literature, 95% of pedodontists polled sometimes use restraints or hand-over-mouth techniques in the control of their patients (Levy & Domoto, 1979). Although the suppression of uncooperative behavior may be an immediate result of the use of criticism, punishment, or ridicule, the effect may be short-lived. In fact, the patient's return for further dental treatment, as well as future avoidance behaviors in other medical situations, may be negatively affected by the anxiety generated by this experience.

This chapter includes documented case illustrations and controlled research investigations that address many of the same issues that confront psychologists who treat fear and anxiety disorders. The practical issues of getting the phobic patient into the dental office includes applications of behavioral approaches already known to reduce irrational fear behavior. Many methods of presenting preparatory

information for both adults and children in improving their cooperation during dental treatment have been adopted by dentists, but there is a lack of empirical validations regarding which intervention is most likely to succeed with a given patient.

Assessment of Psychological Factors: A Three-Systems Approach

Psychologists have been using the dental operatory as a laboratory for studying fear behaviors and for comparing the effectiveness of different behavioral strategies. Melamed (1979) provided evidence that there exists a wide variety of instruments sensitive to the dental anxiety of adult and child patients. The need for a multidimensional approach to the study of fear-related behaviors is based on the lack of high correlations between subjective, behavioral, and somatic indices (Kleinknecht, Bernstein, & Alexander, 1977).

Separate measures for adults and children are used, as developmental level and social desirability affect which behaviors are likely to be observed. Although self-reported fears of adult patients have been found useful in discriminating between high- and low-anxiety dental patients (Corah, 1969; Gale, 1972; Kleinknecht et al., 1973; Kleinknecht & Bernstein, 1979), the children's self-reported ratings are not as reliable an indication of their fear. General anxiety measures such as the Taylor Manifest Anxiety Scale and the State-Trait Anxiety Inventory (Spielberger, Gorsuch, & Lushene, 1970) have been adapted and found valid in reflecting change of anxiety in adults in the dental situation (Lamb & Plant, 1972).

On the other hand, children's *behavior* in the dental setting has been a more reliable and valid indication of their discomfort when measured by observations of movement, attempts to dislodge instruments, protests, and cries (Melamed, 1979) than the measures of self-report. In preschool children, movement appears to be related to anxiety (Glennon & Weisz, 1978). Kleinknecht et al. (1977) found, on the other hand, that the most frightened adult patients were those who moved the least during actual treatment. Perhaps adults have learned more subtle cues for signaling anxiety (e.g., pleading eye movements and grunting).

Patients' anticipatory arousal is often higher than their actual response during treatment, and patients' failure to sit still or cooperate may endanger proceeding without taking special care to help patients control their affective discharge. Physiological measures of arousal, particularly heart rate and palmar sweating, have been found to discriminate between high- and low-anxiety dental patients. These measures are also sensitive to which dental procedures elicit the most anxiety in a given person.

Children's Fear Measures

Self-Reports

Statements such as "When will it be over?" "Will it hurt?" "I want my mommy," or "I've got to go to the potty" may indicate fear. The Children's Fear Survey Schedule, modified for dental items (Figure 1; Cuthbert & Melamed, 1982; Melamed *et al.*, 1975b) has been found to be predictive of the degree of disruption of the child in the dental chair. It has also been validated as an index in change of dental fear following therapeutic modeling. However, its use is limited to children above the age of 4. It also needs to be individually administered orally to younger children.

CHILDREN'S FEAR SURVEY SCHEDULE

Name _____

Age _____ Sex _____

Session _____

Child was:

Cooperative	yes	no
Anxious about questions	yes	no
Answered questions	yes	no
Understood questions	yes	no

Items	Not afraid at all 1	A little afraid 2	A fair amount 3	Pretty much afraid 4	Very afraid 5
1. dentists	____	____	____	____	____
2. doctors	____	____	____	____	____
3. injections (shots)	____	____	____	____	____
4. having somebody examine your mouth	____	____	____	____	____
5. having to open your mouth	____	____	____	____	____
6. having a stranger touch you	____	____	____	____	____
7. having somebody look at you	____	____	____	____	____
8. The dentist drilling	____	____	____	____	____
9. the sight of the dentist drilling	____	____	____	____	____
10. the noise of the dentist drilling	____	____	____	____	____
11. having somebody put instruments in your mouth	____	____	____	____	____
12. choking	____	____	____	____	____
13. having to go to the hospital	____	____	____	____	____
14. people in white uniforms	____	____	____	____	____
15. having the nurse clean your teeth	____	____	____	____	____

Figure 1. Dental subscale of the Children's Fear Survey Schedule.

Behavioral Indices

The practicing dentist would prefer a quick screening device to call attention to the children with a high probability of disruptiveness. The decision to use local anesthesia is sometimes made out of convenience than because of patients' needs. However, this injection procedure itself elicits one of the strongest fears in children as measured by heart rate increase, enhanced anxiety reports, and movement. Frankel (1980) suggested the use of a tickle test to help the dentist decide which children to use local anesthesia with. In 100 children between the ages of 6 and 14 who were inexperienced in dental treatment, a significant difference was found between those children who required anesthetic treatment and those who did not by using a prior response to being tickled. Children, especially boys, who laughed were more likely to require an anesthetic during treatment. Unfortunately, children who were considered management problems were excluded from the study.

The use of mothers' expectations of how their children will behave, when based on previous observation of their children during dental treatment, is an excellent way to predict children who may require more management (Melamed, 1979).

Treatment

Those investigators using behavioral rating scales (Klorman, Hilpert, Michael, LaGana, & Sveen, 1980; Koenigsberg & Johnson, 1975; Melamed, Yurcheson, Hawes, Hutcherson, & Fleece, 1978; Venham & Quatrocelli, 1977) have been very successful in showing the consistency of behavioral disruption across treatment sessions. This allows a standard behavioral device for predicting disruptiveness and evaluating the effects and influences of dentists' management strategies. Different stages of dental treatment have also been identified as producing varying amounts of stress. Venham and his colleagues (Venham, Bengston, & Cipes, 1978) found that, although patients' responses to cavity preparation remained fairly consistent across treatment sessions (for 3- to 5-year-olds), there

was an increase in clinically rated anxiety and heart rate and a decrease in cooperation during the injection of local anesthetic. Observational rating scales have been specifically developed for younger toddlers (Chambers, Fields, & Machen, 1981; Fields, Machen, Chambers, & Pfefferle, 1981).

Physiological Measures

The research on psychophysiological recordings on children (Lewis & Law, 1958) indicates that these are useful and valid measures of transient stress during dental treatment. Heart rate has been found to be a particularly sensitive measure of fear during dental treatment. Children inexperienced in dental procedures showed a greater change in heart rate and more variability than those experienced in the procedures. Pulse rates increased even more during the anticipation of the procedure than during the actual implementation (Duperson, Burdick, Koltek, Chebib, & Goldberg, 1978).

It is only by doing a profile analysis of concordance and desynchrony between these indicants of anxiety that we can more fully understand the process of fear change. The more intense the phobia, the more likely these systems are to positively covary and to call for a broad-based treatment program. There is also a need to be practical. Except for research purposes, it would probably suffice to ask the parent how the child behaved during early treatments and actually to observe the child's disruptive behaviors during less critical intrusive procedures than dental injections (i.e., during prophylaxis or dental x-ray procedures).

Adult Fear Measures

Self-Report

The most popular scale used to assess dental fear is the Corah (1988) Dental Anxiety Scale. It is a four-item multiple choice test that takes less than five minutes to administer. It has been used for over two decades and is reliable and has predictive validity. Unfortunately, not all dentally phobic patients obtain high scores on this scale. A small percentage of patients showing extremely marked physiological responses indicative of anxiety during dental treatment do not necessarily score high on this scale. Early and Kleinknecht (1978) reported that personality factors may also show different responses regarding fear based on their tendency toward the extreme on repression-sensitization. They found that sensitizers were more physiologically aroused than repressors as measured by the palmar sweat index during auditory tapes of the dentist's drill. There was a positive relationship between the degree of fear of dentistry and the tendency toward sensitization.

Clinical Interventions

One would expect procedures directed at arousal reduction to have a more specific effect on the physiological system, whereas behavior change (opening the mouth wide) may induce cooperative behavior while still leaving the patient autonomically aroused. Self-reported fear reduction often lags behind other changes (Hodgson & Rachman, 1974). A broad-range treatment directed at alleviating discomfort in all three systems would probably have the most likelihood of resulting in long-term effectiveness. After an analysis is made of the intensity and the disruptiveness of behaviors within the three systems (subjective, behavioral, and physiological), treatment should be focused first on the system most susceptible to change. It is often useful to pinpoint the origin of the fear and to specify the current antecedents of arousal and avoidance behaviors.

Treating Dental Fears in Children

The specialty of pedodontics recognizes that children present unique problems in the dental operatory. The behavioral management of the child cannot be separated from the quality of the dentist's work. The emotional state of the child and what he or she learns about the pleasantness or unpleasantness of the early dental appointment will influence his or her future

oral health behaviors. Fear has been identified as an important factor in the disruptive behavior of the school-aged child in the dental office. Practicing dentists consider the fearful, disruptive child to be among the most troublesome of problems in their clinical work (Levy & Domoto, 1979). Yet, Brockhouse and Pinkham (1980) demonstrated that dentists without years of experience had a poor ability to predict how anxiously a child will behave.

The child must cooperate or must at least passively comply with the dentist's procedures in order to have the technical procedures completed. Although some dentists prefer to premedicate children so as to reduce the possibility of disruption, others prefer to involve the child as an active participant. From the point of view of child development (Murphy, 1952), a child who copes successfully with stress acquires feelings of mastery that contribute to a healthy self-concept. Therefore, dentists may be viewed as socializing agents, teaching children from very early how to handle discomfort appropriately.

The treatment approach in the area of children's dental behavior focuses on the issues of (1) identifying the variables that will predict disruptive behavior in the dental office; (2) examining parental (usually maternal) influences on the child's behavior; (3) preventing dental phobias through preparatory information; (4) replacing maladaptive anxiety with good coping skills; and (5) management procedures used to promote cooperation during treatment.

Predictors of Disruptiveness

Psychologists can help by identifying which types of children are likely to present management problems. A generally nervous or high-strung disposition was described in 54% of dentally anxious children, but in none of the nonanxious patients (Lautch, 1971). Children who are already identified as difficult by the school or the parent should be seen by pedodontist.

Preschool children (3–5 years) may not have the coping skills to deal with the new experience of receiving dental treatment. Developmentally delayed children exhibit a more anxious response to dental visits. In research by Venham and his colleagues (1978, 1979), anxiety among normal preschoolers was related to scores indicative of lower self-adequacy, higher guilt, self-distrust, and self-rejection. Children who are generally more anxious or insecure exhibited greater situational stress. Higher heart rates or anxiety ratings were correlated with personality disorders of passive withdrawal, cautiousness, pessimism, and avoidance of failure.

The child's previous experience in the dental and medical settings also predicted cooperativeness. Children with prior experience were rated as more cooperative. The Behavior Profile Rating Scale during prophylaxis was a good predictor of treatment disruptiveness (Melamed *et al.*, 1975a,b).

Presence of Others: Maternal Variables

A great deal of controversy has been generated about whether the mother or father should be allowed in the dental operatory during the child's treatment. Although the mother's presence with younger children is assumed to be anxiety-inhibiting, it is less clear what to expect with an older child. Separation anxiety is the younger child's natural reaction to the removal of the mother from the operatory. Therefore, for children under 3, it is widely recommended that the mothers be permitted to remain with the child. Preschool children derive the most benefit from their mother's presence if the mother is properly instructed and motivated (Frankl, Shiere, & Fogels, 1962). Venham, Bengston, and Cipes (1977, 1978) systematically evaluated the effect of the mother's presence on the child's reaction to dental stress. In one study, they varied whether the mother was present or absent over two independent visits. A combination of physiological, self-report, and behavioral measures failed to produce a significant effect. However, the mother's presence on Visit 1 and absence on Visit 2 produced lower anxiety and greater cooperation than the reverse order. Younger children were more anxious on the second visit regardless of the mother's pres-

ence. In a more naturalistic design, in which the decision regarding the parent–child separation was left up to the parent and the child, the presence or absence of the parent did not produce any significant differences in behavior (Venham *et al.*, 1978). The reaction depended in part on what the parents did during the visit. If they were calm and responded with confidence when their child was moderately anxious, no adverse reaction was produced.

In an attempt to help mothers become less anxious, Pinkham and Fields (1976) investigated the effects of preappointment procedures on maternal anxiety. Previous research (Wright, Alpern, & Leake, 1973) had demonstrated that, when the dentist provided pretreatment information to mothers, their anxiety was reduced. The 3- to 5-year-old children and their mothers either (1) were given a preappointment visit one week before the dental appointment; (2) were given the preappointment visit during which the child watched a preparatory videotape separately; or (3) did not have a preappointment visit. There was a significant reduction in the maternal scores on the Taylor Manifest Anxiety Scale for the combined treatment–film group. However, there was no correlation between maternal anxiety and the child's behavior during dental treatment.

Preparatory Information

The tell-show-do technique (Addelston, 1959) has been widely used in preparing children for dental treatment. The procedure has some components of familiarization and some of modeling, in that the child is told what will happen and observes a demonstration by the dentist. Early management techniques for introducing fearful children to the dental operatory involved having them observe a successful peer (Adelson & Goldfried, 1970) or sibling (Ghose, Giddon, Shiere, & Fogels, 1969), which added the ingredient of providing the child with a response to compete with anxiety. Even a pleasant interaction with the dental assistant outside the operatory is just as effective as these behavior techniques (Sawtell, Simon, & Simeonsson, 1974). These behavioral techniques all have in common providing the child with information regarding what will occur.

Melamed and her colleagues (1975a,b) found clear evidence that children without prior experience showed reduced fear-related disruptive behavior in the operatory after viewing a preparatory videotape immediately before the impending event. This approach involves a child scheduled to undergo dental treatment viewing another person, usually a peer, receiving the same or similar treatment. The model who undergoes treatment experiences no adverse consequences and usually receives reinforcement for "good" behaviors (i.e., sitting quietly in the chair, and keeping the head still and the mouth open). Thus, both information about the procedure and a template about how to behave are provided. This procedure is most effective for children who have not had previous dental treatments.

The fact is that some children do not code the information presented by videotape or imagery instruction (Chertok & Bornstein, 1979). One third of children between 5 and 13 years were unable to respond to covert modeling experiences suggested by instructions to visualize other children in dental situations. This study found that the covert modeling groups did not reduce fear to any greater extent than imagining the fear-relevant stimuli alone. There is some evidence of a curvilinear relationship between the amount of information provided and the anxiety of children (Herbertt & Innes, 1979; Melamed et al., 1983b). The stress produced by too much information may heighten their awareness, particularly if they have had a negative previous experience. For younger children, play with dentistry-related toys may be more effective than imagery in giving them a sense of control.

Coping-Skills Training

In focusing on preventive procedures for preschool children (ages 44–71 months), Siegel and Peterson (1980) demonstrated the effective use of both sensory information and coping instructions (relaxation, controlled breathing, and distraction) as compared to no treatment

for control groups subjects. There was a significantly lowered heart rate response in the coping-group subjects after the treatment and immediately preceding dental work, although both treatment groups behaved well.

Melamed, Yurcheson, Hawes, Hutcherson, and Fleece (1978) clearly found that previous experience attenuated the effect of modeling. Children with previous experience did not benefit from either a demonstration of the local anesthetic or peer modeling as compared with viewing an unrelated film. They probably had the information about what would occur. However, children with no previous experience had reduced disruptiveness following a modeling videotape of a youngster adequately handling his dental experience. Klorman et al. (1980) found similar effects.

In a study of experienced dental-phobic children, Klingman, Melamed, Cuthbert, and Hermecz (1984) found that the opportunity to practice along with the peer models depicted in the videotape had a greater effect on fear reduction, both subjectively and behaviorally, than merely observing these coping strategies used by the peer models. This behavioral rehearsal is thought to work either by strengthening an already-existing coping strategy (i.e., controlled breathing) or by teaching the child a new way of handling fear, such as distraction or relaxing imagery. "Stress inoculation" is identifying when to use these strategies and practicing under simulated conditions so that the child feels more confident about his or her success of using them when needed (Meichenbaum, 1977).

The complexity of these findings for treatment recommendations suggests that different types of preparation procedures work better with different children. The therapist needs to find out what children already use to cope with stress and to remind them to use this during the dental visit.

Patient Management

The dental literature has an abundance of articles (Barenie & Ripa, 1977; Chambers, 1977; Weinstein & Getz, 1978) promoting the use of behavior modification techniques in managing the child dental patient. However, with the widespread use of nitrous oxide and analgesics, there is less need to deal with the child's anxiety-related disruptive behaviors. Weinstein and Nathan (1988) cautioned against the unselective use of nitrous oxide, as limited research exists and they found that younger children (under 6) did not benefit as much as did older children. Their data (Weinstein, Domoto, & Holleman, 1986) suggested that children who have been exposed to N_2O in stressful settings may be more difficult to manage during future administrations of nitrous oxide. There is a need to wean children so that they can handle treatment without the drug. Chambers (1977) argued that some anxiety or uncertainty about future events is adaptive and that children should not routinely be deprived of the experience of coping with and mastering their fears.

In a recent survey of 223 pedodontic diplomates (Levy & Domoto, 1979), 95% reported that, in most situations, undesirable behavior could be managed through personal interaction alone. The use of local anesthesia when performing restorative procedures was deemed desirable by 96%. This is the aspect of treatment most consistently found to evoke anxiety. The mildly fearful child is most likely to resist the dentist at this phase of treatment.

The management of the child patient is one of the most difficult skills to teach dental students. There is agreement that the degree of cooperation of the child during treatment influences the quality of the work and the speed with which it can be accomplished. Dentists need to know that taking some time to familiarize the child with the procedures will actually shorten the treatment time. Children should be told that, if they sit still, they will be done quicker. Johnson, Pinkham, and Kerber (1979) showed, by use of the Psychological Stress Evaluator (a voice indicator of stress), that dentists, regardless of their level of experience, show stress when faced with difficult child management situations. Yet, very little coursework is devoted to identifying potential behavioral problems and how they should be han-

dled. The negative effects of some behavior management strategies currently taught (e.g., hand-over-the-mouth with airway restriction), are not made clear during dental training.

There is an assumption by some that a loud voice and, if necessary, either physical or mechanical restraints will solve all problems. Although surveys show that only a small percentage of child patients disrupt treatment to the extent that they fail to have the necessary work, Levy and Domoto (1979) estimated that 6% of children may learn through poor experiences to avoid dental treatment later on. Hand-over-mouth management had been used by 87.8% of those dentists surveyed to manage temper tantrums, hysteria, aggression, or resistance in their 2- to 9-year-old patients and in handicapped children. Some feel that these restraints are necessary for some normal children to protect them and the dentist from injury and to make them aware that their undesirable coping strategy is neither necessary or useful. Learning theory in psychology suggests that punishment procedures may have a short-lived suppressive effect, and that, in the long run, it may result in such deleterious behaviors as increased anxiety or aggression, or avoidance of the punishing agent.

Melamed and her colleagues (1983a) systematically investigated the effects of dentists' use of instruction alone; instructions plus praise for cooperative behavior (positive reinforcement); instructions plus criticism, restraint, or loud voice for inappropriate behavior (punishment); or the use of instructions with both contingent reward and contingent punishment. The data on two samples of urban and rural lower- to middle-socioeconomic-class children aged 4–12 years consistently revealed that instructions alone facilitated cooperation during subsequent treatment sessions. The use of contingent praise and punishment also reduced behavior disruptiveness. These treatment management strategies had different effects depending on the child's level of fear and previous experience in dental treatment. Children high in disruptive behaviors during the examination did not show any significant tendency to improve with repeated experience alone. They did best with the contingent use of both reward and punishment. Those children who were older than 7½ years of age, who were low in dental fear, or who had had previous treatment experience were made worse if the dentist used punishment as a treatment strategy. Weinstein and his colleagues (1982) also found that the dentist is able to obtain compliance in about 85% of children by directive guidance alone. In fact, the use of criticism is more likely than any other response style to be followed by negative child behaviors.

Intervention Strategies with Adults

Systematic Desensitization

Growing out of Pavlovian respondent conditioning notions, systematic desensitization represents an attempt to countercondition the fear that a patient has learned to associate with the sights, smells, and sensations involved in dental treatment. Adult patients, like children, are also sensitive to criticism. Systematic desensitization involves working with the patient to construct a hierarchy of the specific things that patients fear.

Table 1 represents one patient's hierarchy. First, the individual is taught progressive relaxation. The hierarchy is then presented from the lowest to the highest fear-provoking situation, while the patient is instructed to imagine each scene vividly while remaining relaxed. If at any point the patient feels anxious, the scene is aborted, and the patient again assumes a relaxed state. Repetitions occur until the complete hierarchy can be imagined without anxiety. Then, exposure to real-life events is prompted. This therapy usually takes 11 sessions and is accomplished outside the dental office.

Graduated Exposure

This technique involves gradual exposure to the feared stimuli in imagination, in videotape simulation, or in real life. The therapist involvement is less extensive, and no hierarchy of items is constructed. Instead, the material is organized by chronological steps. By having

Table 1. Tentative Dental Hierarchy for Patient C.P.
(Arranged from Low- to High-Anxiety Items)

1. You are picking up the phone and calling the dentist for an appointment.
2. You are getting out of the car and looking at the outside of the dentist's offices.
3. You are walking into the receptionist's area of the dentist's office, and you tell one of the receptionists that you're here for your appointment.
4. You have been sitting in the reception area of the dentist's office for about 5 minutes.
5. You have been waiting in the office for your turn for about 20 minutes.
6. You are sitting in the dental chair in an upright position.
7. You are getting a novocaine shot for numbness.
8. The saliva ejector is placed in your mouth.
9. You have your mouth open while the dental assistant stands over you and places her hand inside your mouth.
10. You have been in the dental chair for about 15 minutes.
11. You hear the noise of the dentist's drill (both fast and slow speed).
12. You are sitting in the dental chair in a full reclining position.
13. You are having stitches removed from your gum.
14. You have been in the dental chair for about 45 minutes.
15. You feel the pain after having a tooth pulled.
16. You feel the numbness of the novocaine wearing off and the dull, tingly feeling as it is happening.
17. Dentist is putting moist material in your mouth to make a mold, and you keep your mouth closed for one minute.
18. You have your mouth open wide while the dentist *and* the assistant stand over your and place their hands inside your mouth, forcing you to breathe through your nose.

the patient watch simulated dental settings (e.g., watch a film), the results have been some excellent reductions in fear by self-report (Corah, Gale, & Illig, 1979) and on behavioral measures (Kleinknecht *et al.*, 1977). Exposure allows the patient to anticipate events.

Imaginal Flooding

This procedure also involves exposure to frightening stimuli, but there is an attempt to elicit maximal anxiety instead of presenting the hierarchy gradually. This procedure does not require learning relaxation responses and must be conducted by a clinician who is skilled in the technique, as high levels of anxiety are elicited and the less experienced therapist may cause the patient to drop out of treatment.

Modeling

This procedure involves exposure to the elements of the dental situation with the demonstration of appropriate behaviors. It is rarely used with adults. The most common use of this procedure, as it is applied to dental treatment, involves both *in vivo* exposure and symbolic modeling.

Stress Inoculation

This procedure involves modeling, information, and coping-skills provision and involves three steps. The patients are shown how pain and fear contribute to their experience of discomfort. The second phase helps them to select three or four coping strategies to use to tolerate dental treatment. Then, they integrate these skills into an active coping plan and practice it under simulated conditions. Then, they are asked to use it during an actual treatment session.

Biofeedback

This is an attempt to directly condition the arousal response by asking the patient to reduce autonomic activity in a particular system (e.g., muscle tension, or heart rate) by means of controlling a display that represents their own variation. Both electromyographic and heart-rate or breathing biofeedback procedures have been used with and without systematic desensitization and have resulted in the successful reduction of self-rated anxiety (Hirschman, Young, & Nelson, 1979; Miller, Murphy, & Miller, 1978). It appears that the perceived sense of control is the effective ingredient.

Hypnosis

There have been studies that show a decrease in bleeding and discomfort during ex-

traction in patients treated under hypnosis (Forgione, 1988). However, much of the same control has been accomplished with the autogenic-training, imaginal, and cognitive-control procedures already discussed. Because hypnosis must be learned through professional training, it is unlikely to be used directly by many practitioners.

Distraction and Self-Control

Corah and his colleagues (1978, 1979) examined the effects of distraction (playing a videogame), relaxation (via audiotape), perceived control (pressing a button to stop the dentist), and a no-treatment control condition on physiological arousal level (heart rate and skin resistance) and self-reports of discomfort and anxiety. Both relaxation and active distraction reduced anxiety ratings, particularly of high-fear patients, from the first visit to the second visit. In another study examining perceived control over aversive events, patients who were given a device to signal the dentist regarding their discomfort showed significantly less discomfort than patients who felt that they had no control over what the dentist did (Thrash, Marr, & Box, 1982).

In summary, despite these readily available treatments, there is little in the research literature to suggest which patients may benefit from which procedures. It is clear that the patient who avoids the dentist must first be approached outside the dental operatory and encouraged to discuss her or his fears and concerns.

Program Development

This chapter provides the rationale for considering dental anxiety a multifaceted problem that has different causes and different manifestations. A graduated approach involving parents, teachers, dentists, and psychologists is recommended. The three-systems approach is used to evaluate patients' fears, specific avoidance behaviors, and physiological responses to dental treatment before a strategy is selected.

Dental Fear Clinic
within the Psychology Service

For many years, the Department of Behavioral Sciences in the University of Florida Dental School had a fear clinic. Despite massive efforts to educate the public and to invite the patients to receive treatment, this facility was vastly underutilized. In fact, the investment of money in building a mock dental operatory in which patients could be desensitized for their concerns seemed cost-ineffective. The dental setting itself elicited too many fears to allow the dentally anxious person even to begin treatment. The dentists were encouraged to handle their own patient management problems; however, an identified psychologist in the dental school was on call to assist with disturbed patients. The child dental clinic had its own way of handling the anxious child. For very young or retarded children, a papoose board or mother was used to provide restraint. In cases of loud crying or disruptive activity, a child was removed from the clinic and isolated in a separate quiet area. In this location, dentists often instilled behavioral control by loud voice or punishment-oriented techniques. Therefore, we established a separate clinic for the treatment of fear and anxiety disorders within the Clinical and Health Psychology Division. The fact that it was located within the Dental Towers building was not highlighted. Patients were welcomed by a receptionist and interviewed in a quiet, dimly lit office that had the comfortable appearance of a living room. No instruments, sounds, or smells of dentistry were apparent. The task of the clinical psychologist consists of identifying the factors in the medical, dental, and family history of the patient that may have elicited the anxiety.

Primary Prevention

The Number 1 rule is to educate the patient and the health care team about the importance of preventive approaches. This means attempting to get the young patient in for treatment before there is any painful occurrence. Parents should never trick children or threaten them with dental visits; instead, the dentist should be presented as their health advocate, just like the pediatrician. With the advent of fluoridation programs that successfully compete with carious development, more attention needs to be focused on the child patients and procedures that can generate proper daily dental care. Elaborate school programs have been developed for elementary and secondary schools (Lund & Kegeles, 1979; Martens, Frazier, Hirt, Meskin, & Proshek, 1973). These focus on teaching compliance with self-care by providing contingent rewards. The problem of long-term maintenance of these habits still exists. If schools are used to fill the educational role of providing preventive care instructions about good diet, oral hygiene, and preventive visits, much will be accomplished toward decreasing dental anxiety in the general population. Community dentistry approaches (Bailey & Reiss, 1979; Reiss, Piotrowski, & Bailey, 1976) have encouraged the use of appointment reminders and financial rewards to parents of lower economic classes for bringing their children to the dentist when finances as well as education may hinder early attention to dental needs.

The most critical feature in good patient cooperation is the quality of the relationship between the dentist and the patient. The skill that the practitioner has in putting the patient at ease is much more important than the time he or she takes to accomplish treatment. Patients report that dentist behaviors that would reduce their anxiety included explaining the procedures before starting; giving specific information during the procedures; instructing the patient to be calm; giving warnings about the possibility of pain; verbally supporting the patient with reassurance; helping the patient to redefine the experience; giving the patient some control over the procedures; teaching the patient to cope with distress; and providing distraction. These behaviors would help the patient build trust in the dentist. Patients who perceive the dentist as warm and empathic as a result of these types of interactions are more likely to report satisfaction with their treatment (Corah, 1988). Baron, Logan, and Kao (1990) found that dentists were accurate in detecting patient pain, but less so about patient's anxiety, especially if the procedure being performed was stressful to the dentist.

Treatment of the Dentally Avoidant Adult

Part of what anxious patients fear is being ridiculed for their fear. Given the high number of patients who avoid regular treatment because of their concerns about pain or their fear of loss of control, the practitioner should listen carefully to their concerns. A recent study of memory of dental pain (Kent, 1985) indicated that patients who had greater expectation of pain had more consistent pain reports regardless of the actual amount of pain experienced. Sometimes, merely voicing their concerns allowed them to get past their initial hesitancy. Verbal and nonverbal communication skills are important for dental students, and psychologists can play an integral role in the behavioral science curriculum by teaching dental practitioners to recognize fear and anxiety in themselves and in their patients.

The answer is not to put dentally anxious patients "out" for their dental treatment. Although pharmacological approaches are sometimes indicated for extensive treatment or dental surgery, the use of premedication and intravenous conscious sedation must be clearly warranted as they often deny patients an opportunity to learn how to cope. The goal of the practitioner should be to quickly wean the patient off all but local anesthetics because we undergo dental treatment throughout our life span.

The "stepped approach" to treating anxious patients is adopted from Klepac's recommendation (1988). The lowest step is education. This may come by allowing patients to view educational materials about the treatment they will undertake in the dental setting. Gatchel (1986) reported that, by itself, the group administration of a desensitization type of film to community members avoiding dental treatment brought many patients back into treatment. There also exist self-administered stress inoculation manuals from which the patient can obtain self-help with minimal therapist contact. There are many books and audiotapes now on the market designed to be used by the fearful dental patient (Horowitz, 1987; Kroeger, 1988; Milgrom, Weinstein, Kleinknecht, & Getz, 1985). An entire issue of *The Dental Clinics of North America* (October 1988) was devoted to dental phobia and anxiety.

The next step may involve some professional time: if the patient is not able to progress on his own, the potentially disturbing steps need to be organized so that systematic desensitization or graduated exposure can be accomplished. Many commercially available videotapes can be used with monitoring by a professional.

If the dentist invites the patient in for a familiarization or nontreatment session, just the opportunity to see, hear, and smell the environment may decrease some concerns. The patient also has an opportunity to become comfortable with the practitioner. The opportunity to share concerns and provide suggestions to the practitioner about how they can interact to assist the dentist during treatment may do a lot to instill a sense of control in the patient.

The highest level of involvement is the most costly and should be called into use only if the other alternatives have failed. This level requires a therapist-administered behavioral intervention, such as the use of stress inoculation, systematic desensitization, *in vivo* and participant modeling, biofeedback and hypnosis. Even in the worst of cases, the therapist needs finally to eliminate herself or himself from the actual treatment process so that the patient will accomplish a feeling of self-control.

Referral

In certain instances, patients should be referred to a specialist in behavioral modification. Children who have failed to receive treatment continuously because they present management problems should be seen by pedodontists rather than chancing a further buildup of noxious experiences.

Adults who have been unable to receive dental treatment without medication should be considered for day surgery, in which general anesthesia can be used to begin the restorative process by quickly alleviating pain that may have developed from neglect of their oral needs.

In summary, the most important approach to dental anxiety is the treatment of the patient as an individual with a problem that he or she neither created nor desires. An evaluation of what has led to patients' discomfort as well as of what coping skills they already have for handling other distressful experiences will guide the practitioner, be he or she dentist, psychologist, parent, teacher, or peer.

Case Illustrations

In order to illustrate the need to individualize assessment and treatment, four patients who were treated at the University of Florida Fear and Anxiety Disorders Clinic are described. We have selected four patients who avoided dental treatment. All of them had a dental injection phobia, but they required different approaches based on the nature of the presenting complaint. Our lack of success with one of the patients illustrates what happens when coordination between the dentist and the psychotherapist fails. We discuss each case in terms of the presenting problem, the assessment used, the major treatment approach, and the criteria for improved behavior. Traditional psychological assessment, including the Minnesota Multiphasic Personality Inventory (MMPI) and the Anxiety Disorders Interview Scale, was undertaken in the cases where underlying personality problems or severe psychopathology needed to to be ruled out.

Case 1

Reason for Referral

Louise was a 16-year-old female referred to the clinic by her general practitioner because of a long-standing fear of injections. She had always fainted at the sight of needles. She did not have any traumatic events associated with her fear, although as an adolescent she had had back surgery and had had to wear a brace for scoliosis for three years. As a child she is

thought to have had a seizure. Her father also had vasovagal syncope, and her mother got faint and rigid in anticipation of receiving an injection. The immediate reason for seeking treatment was that Louise would be unable to continue at school unless she received vaccinations, as the family had lived abroad and she was not up-to-date on the required inoculations. She had begun to realize that this phobia was restricting her life because she was reluctant to seek a medical career, couldn't travel to countries that required shots, and was concerned about blood tests for marriage and other screenings.

Assessment

Louise received a battery of self-report questionnaires and was interviewed at length. The results of the test battery, including the Child Diagnostic Screening Scale, the Fear Survey Schedule, the Dental Fear Questionnaire, and the Anxiety Disorders Interview Schedule, indicated that hers was a well-circumscribed fear of injections, needles, and other blood and mutilation items. Although her fear of dentistry was not above the normal range, her complaints of nausea and fainting made her a somewhat undesirable patient. Her main fears did not keep her from going to the dentist, but if she received advance warning about an injection or an operative procedure, she would be extremely weary and exceedingly apprehensive to the degree that it interfered with her normal school day.

Treatment

The first phase of Louise's treatment involved having her construct scenes for use in an assessment of her imagery ability and the specific circumstances surrounding her injection fears. She was then asked to imagine herself in these situations, as if they were actually happening, and then to rate how much anxiety they made her feel. This systematic desensitization hierarchy was also used *in vivo* while she received treatment by a dentist with the chair tilted back to prevent fainting. Her blood pressure was continuously monitored. Her most fear-provoking scene was a memory of when she went to the doctor's office to get her gamma globulin shot at the age of 9. She described this in the following script:

> You are in the doctor's office to get your gamma globulin shot. Everyone is in a large white room. Your heart beats faster as one by one your mom, sister, and brother go first and you are next. Your palms are sweaty and you feel nauseous. Your dad and the doctor and you enter a small, dim room with a flat black table in the center. When your turn comes, you are forced bodily onto the table, as you need to have the shot in your rear end. You feel tense all over. You are screaming, and your fists are clenched so tight that you break a Barbie doll you are holding. You remember the air conditioning even though you are feeling hot all over, and you can smell the alcohol.

Physiological recordings made during four sessions as she imagined this scene (imaginal flooding) revealed that, although she experienced a moderate degree of vividness, the anxiety as revealed by heart rate and galvanic sweat response was not elevated.

A behavior test was constructed in which she was confronted by a hierarchy of situations by means of videotape or actual life exposure. These included sitting on a doctor's examining table, watching an injection into an orange, and sitting in the dental chair in view of the syringes, alcohol, and instruments. She also watched blood drawn in the laboratory, watched children self-inject and have a finger prick on the diabetes unit. She also was taken by car to the doctor's office, and had her arm swabbed by the nurse in preparation for a shot. Because the patient actually fainted during her observation of the self-injection, we consulted with the neurologist in order to rule out organic seizure activity. The dentist was willing to work with her to help her become more accepting of injections. We assisted the dentist by continuously recording Louise's blood pressure reactions in order to anticipate any possible rapid drop in blood pressure so the dental chair could be reclined. She was given local anesthetic injection and dental examination with no problems. The presentation of other videotaped scenes caused no marked drops in blood pressure.

Termination Criterion

The patient was accompanied to the Board of Health. We arranged the situation to match her imaginal fear scene. The patient had no adverse reaction to watching her brother receive his shots. She remarked that the small room, the smells of alcohol, and the air conditioning no longer led to an increase in her discomfort. She received her injections and returned to school. There was no further need for our assistance.

Case 2

Reason for Referral

"Doc" was a 52-year-old white male referred to the Fear Clinic by his local dentist, who was unable to treat him because of his reaction to local anesthetic injections. This reaction involved feeling faint, clammy, and nauseous; having a rapid heart beat; and getting spots before his eyes when any "caine" anesthetic was used. The local allergist refused to test Doc, as he believed that the idiosyncratic reaction was potentially fatal. Other than these reactions, Doc felt that he was in good health. He had a similar reaction when viewing any sort of invasion into the body, through an accident or a medical procedure, and he wanted to find a medical reason for his reaction. He had difficulty watching television and had not been able to visit his fiancee following a surgical procedure. He had much anxiety related to dental work. Although he required extensive treatment, he was unable to complete this work because of his refusal to take local anesthetics. He was obsessed with information about health and played the role of doctor adviser to his coworkers, on whom he performed blood pressure checkups. He brought to the first therapy appointment a folder full of medical articles on the subject of injection allergies and kept copious notes on his progress.

Assessment

Doc's assessment consisted of self-report and physiological and behavioral assessment.

He reported extreme fear of dentists, open wounds, witnessing surgical operations, human blood, and fainting. As part of the behavioral assessment, he was monitored in a mock dental operatory, in which he observed and handled needles. He reported little anxiety except during a videotaped demonstration of a venipuncture and color slides depicting oral injections. He had a marked physiological reaction specific to his fear scenes. His mother was a dental assistant and had told him how she lost her teeth following childbirth because the patient and his brother had used up her calcium. She had had the patient's nonincisor first teeth extracted under local anesthetic at age 11. As a teenager he had had gingivitis that required penicillin tooth powder, which he thought may have led to the initial allergy. He did not have a high pain threshold. He was reportedly susceptible to certain types of motion sickness. He had had reactions that he called "shock reactions" after his son's operation on an Achilles tendon and his fiancee's lumpectomy. The patient reported that 15 years before, he had self-injected for sinusitis and could give injections to others without any problem. His MMPI was within normal ranges, with elevations on the masculinity and hysteria scales.

Treatment

The patient constructed two personal fear scenes related to his dental concerns.

Fear Scene 1. You are in the dental chair as the dentist gets the needle ready. You breathe deeply. He sprays a topical in your mouth. After receiving the novocaine injection, you feel a rush in your whole body. As it starts to take effect you feel faint, nauseous, clammy, and have snowy vision. You need to breathe deeply. Your dentist tells you that you need to get a filling taken care of right away. Your feelings return and you feel tense all over. You begin to worry about the lengthy procedure. You heart beats faster as you breathe deeply. Time passes slowly as you hear the cracking of the tooth as the dentist asks for the scalpel. He tells you he's beginning to cut into your gums.

Fear Scene 2. You are having a surgical procedure for periodontal disease, which is a lengthy procedure. You need to remain conscious. You feel terrified as the dentist injects various places in your mouth. You feel trapped as he is laying the gums open and scraping the bone. You breathe deeply and think, "Get this over with." It feels as if it will never end. You feel fatigued from the tension. Your hands feel clammy as the dentist says, "I hope we can save that tooth."

The physiological reaction showing increased heart rate and sweating was specific to Doc's fear items. During the imaginal flooding sessions that followed, he showed a greater ability to tolerate the imaginal stimuli without any autonomic reactivity. He reported that he was now going out of his way to watch TV shows showing operations.

He was also seen by a dentist who agreed to conduct an experimental condition to evaluate whether his "caine reaction" was really of an allergic nature. We saw the patient in the emergency dental operatory. The patient was fully informed about the procedures, in which he would receive two injections: one would contain an active anesthetic, lidocaine, and the other would be a saline solution. He was not told in which order the injections would be administered. He showed a larger anticipatory anxiety reaction, including an increased breathing rate, a flushed face, and an increased heartbeat, during the first injection with the inert substance than during the second injection. In fact, the greatest anxiety symptoms occurred as he was awaiting the procedure.

Termination Criterion

After this session, Doc was convinced that he really did not have an allergic reaction and could resume dental treatment. The TV shows that had made him anxious if they showed operations no longer affected him, and he was able to hear other patients' accident stories without becoming ill or overinvolved. He returned to his local dentist, who carried out periodontal surgery, with and without novocaine. The patient wrote us a letter sharing his success in having gall bladder surgery without any problems and in making an adjustment to his dentist's newest partner, who had been able to complete the necessary extractions and dental restorations.

Case 3

Reason for Referral

Ms. Clem was a 20-year-old white female who was self-referred to the Fear Clinic for the treatment of a long-standing dental phobia. She had been unable to tolerate dental procedures since age 8, when she had been traumatized in her dentist's office. Physical abuse, involving physical restraint and dental extractions in the absence of anesthesia, was accompanied by foul language. There had been an attempt by her parents to reinstate dental treatment when the patient was 11 and 13 years of age, with further failure. The patient recalled that she had been held against her will by the dental assistants and that neither time had the restorations actually taken place. As she recalled these incidents, the patient became overwhelmed, cried, became flushed, and perspired heavily, and her hands shook. She had not sought dental treatment for the past seven years, although her teeth were irregularly spaced and she reported pain in the rear teeth. Nevertheless, the patient was attractive and wanted to pursue a career as an airline stewardess or a model.

Assessment

The patient's psychological history had several psychosomatic features: she had had recurrent abdominal pain and possible ulcers, and she had missed much school during adolescence. Her MMPI was within normal range. She was an above-average high school student and had worked part time since graduation as a receptionist at a beauty salon. She was married and had no children. The initial evaluation consisted of self-report measures of personality. Her fear scores showed that she had a fairly circumscribed area of concern, including dentists, blood, and injections. Her scores on the Dental Fear Scale were very high.

The behavioral assessment involved asking her to observe a film of an intravenous (IV) injection, and it was stated that a simulated injection would take place. During further observational assessment, the patient progressed

through a series of tasks that involved her moving progressively closer to and touching a set of syringes and needles. She refused to allow a needle to be placed next to her skin.

Treatment

Our oral surgeon saw Ms. Clem to determine the extent of her extraction, orthodontic, and restorative needs. She was cooperative during the dental history but refused initially to open her mouth; there was much hysteria and crying as she told about her early dental experiences. After a brief examination, it was found that she had moderately severe gingivitis, dental caries in several teeth, and a severe Class I malocclusion with a maxillary mandibular arch discrepancy. Her needs included complete dental X-rays; dental prophylaxis; caries control; removal of third molars; and possible extraction of Tooth 37. The surgeon recommended general anesthesia for the extractions but felt that the patient needed to learn to accept dental treatment while awake in order to undertake the orthodontics treatment.

She was seen for 16 sessions of behavioral treatment over the next nine months. The initial phase of treatment involved imaginal flooding to scenes used by the patient to describe her traumatic experiences.

Fear Scene 1. You are 11 years old and are in the dentist's office for extractions. Your heart races and you feel nauseous. Your palms and face are sweaty. Your whole body is shaking, you're crying, and you can't catch your breath. Strong hands are holding you down. You are gasping for air; your fists are clenched as the hands force a black mask with terrible-smelling gas on your face. You feel suffocated. You have no control over the situation. You scream into the mask, but the doctor and the nurse keep telling you to quit being ridiculous. They try to put an IV in your arm. It's a sharp stinging pain. They keep saying your veins roll.

By the end of the flooding, the patient was able to talk about these events without becoming anxious and depressed. The next phase of treatment involved assisting Ms. Clem in breaking down her fear into components that could be more easily addressed: (1) sounds and sights of the dental operatory; (2) injections; (3) loss of control; and (4) confinement. Each of these components was addressed with exposure-based methods. The patient's fearful reactions to the sights and sounds of dental instruments were addressed by gradually approaching the actual setting—first by listening to audiotapes, and then by giving her a set of dental instruments and asking her to handle them and put them in her mouth. The final phase was allowing the therapist to put the instruments in her mouth. A stranger then conducted a mock examination. The patient then met our dental hygienist, who did not perform any prophylaxis for two sessions, during which the patient listened to nice music through headphones.

The injection component was treated by systematic desensitization and film modeling. She learned cue-controlled relaxation and imagery distraction techniques to deal with her loss of control. She practiced these coping skills while allowing herself to be in a small, enclosed office without any way out. She also learned to replace catastrophic thoughts with the positive consequences of dental treatment, including good health and improved appearance. A female dentist in operative dentistry worked with her restorative and extraction needs. Ms. Clem then transferred without difficulty to a male orthodontist. One year later, she had completed orthodontic and dental treatment and had moved to another city. She had been able to continue to have routine dental work without problems.

Case 4

Reason for Referral

Mrs. S. was a 45-year-old married woman who was referred to our Fear and Anxiety Disorders Clinic because of agoraphobia. She had a nine-year history of panic attacks. These had begun to occur when her husband returned from service. The symptoms appeared to be related to marital strife and feelings that she had little control over her life. She could not drive alone and often accompanied her husband to his job so as not to be alone at home. She had avoided dentists for over 20 years be-

cause of a fear of dying. As a result, her teeth were in poor condition. There were many missing teeth. She had difficulty chewing and was in constant pain. She had a long history of psychiatric problems. The clinical diagnosis was agoraphobia with panic attacks. She could not be left alone, refused to drive, and had difficulty in any situation of confinement, such as a beauty shop or a dental office. The family was being manipulated by her need to return to a distant city so that she could live closer to her mother.

Assessment

Testing revealed that her basic problem was agoraphobia, not dental anxiety. She could tolerate injections but would not allow anything to be placed in her mouth. She had avoided dentists, she claimed, because of the expense and the travel required to maintain consistent treatment. The MMPI revealed much psychopathology, with extreme elevations in depression and psychasthenia. Her physiological assessment revealed an intense autonomic response to situations in which she believed she might die.

As part of her assessment, we had her seen by an oral surgeon who agreed to do her dental work under general anesthesia, so that she could avoid many of the concerns she had about experiencing unusual bodily sensations. She also expressed concern about her appearance and did not feel that she could go in for the extraction of any teeth if she would have to be toothless for any length of time.

Treatment

Because of her inability to eat and her serious anorectic condition, we felt that her dental health needed to be explored. The visit with the oral surgeon went poorly. Although we had discussed the patient's fears, he ignored her wish to have all dental work delayed until after the Thanksgiving holidays. He told her she needed to have the remaining lower teeth extracted immediately. The therapist attended the oral surgery with her and held her hand

and coached her in controlled breathing until she was anesthetized. Further work was needed in preparation for dentures. She kept putting her appointments off. After several attempts to have her complete the dental work failed, the patient also became more resistant to resuming the therapy contacts to deal with her ongoing agoraphobia.

Termination Criterion

The patient failed to make progress. Her dental needs seemed to take precedence, but when the therapist attempted to coordinate this effort for her, the patient, who did not like the oral surgeon, withdrew from treatment. The miscalculation of the oral surgeon about the patient's readiness to undertake the oral surgery scared her out of treatment. Our coordination of the treatment efforts failed to accomplish the aims of the therapy. The patient was fitted for lower dentures six months later by a private dentist of her choosing. Filewich (1988) wrote about the difficulty of providing treatment for the agoraphobic dentally phobic patient. In this case, we failed to establish the two primary steps that needed to be given priority: (1) providing the patient with a sense of being in control of the situation and her anxiety symptoms and (2) training in relabeling or reinterpreting her symptoms as indicating nondisaster. In our eagerness to alleviate her poor dental condition, we went along with the oral surgeon's suggestion of general anesthesia. Perhaps a combination of systematic densensitization and cognitive restructuring would have been more effective. The dentist's style is also particularly important in treating the agoraphobic. The dentist should be friendly, kind, supportive, understanding, flexible, and uncontrolling, but firm. We failed to find a dentist who would encourage her to undertake the extractions at a pace that she could tolerate.

In each of these cases, dental treatment was avoided because of feared aspects of the treatment. It was important in the development of the individual treatment programs to discover what aspect or situations evoked the fears and to address treatment in those areas. Thus, both

Patient 1 and Patient 2 needed to deal with their fears of a catastrophic event, death, or fainting in reaction to the injection. Patient 3 needed to unlearn her conditioned aversive reaction to many more aspects of the treatment setting, given her history of mistreatment by dentists. Patient 4 was not given any help with her overwhelming panic regarding the activation of her own bodily symptoms. The direct approach was overwhelming.

Temporomandibular Disorders

Pain-Related Dental Problems

In addition to the psychologist's role in dealing with dental fears, a great deal of progress has been made since the early 1970s in applying cognitive-behavioral strategies to the treatment of a number of pain disorders. As more psychologists are being included on interdisciplinary pain management teams, the role of psychological factors in influencing the etiology and progress of a variety of pain syndromes is being examined more closely than ever before. In a dental setting, one of the pain disorders most commonly seen is temporomandibular disorder (TMD), in which the patient presents with a nagging pain (usually unilateral) in the region of the temporomandibular joint and the masseter muscle. This disorder is also marked by occasional "popping" and "clicking" of the joint, limitations in jaw mobility, headache, and facial tenderness. Epidemiological studies indicate that a large percentage of the population (up to 60%) experiences at least one symptom; therefore, we refer to this syndrome as TMD to include all of the disorders. The disorder affects men and women in equal numbers; however, in clinic samples of TMD patients, women usually outnumber men by a ratio of at least 2 to 1 (Clarke, 1982; Kleinknecht, Mahoney, & Alexander, 1987). In addition, the vast majority of TMD patients in clinics are within the 20–40 age range, although epidemiological data do not show this relationship between TMD symptoms and age (Kleinknecht et al., 1987). In those rare cases where TMD occurs in children under the age of 16, the available literature suggests that it may be more likely to be accompanied by psychopathology (most commonly depression) than in the adult population, although the paucity of data on younger TMD patients makes this conclusion somewhat tenuous (Pillemer, Masek, & Kaban, 1987).

The etiology of TMD remains somewhat elusive, although a number of factors have been implicated. The chief factor that is believed to play a role in perpetuating the pain is bruxism, or the grinding and clenching of teeth. In turn, there is great uncertainty about why bruxing occurs, and why many people who brux have no problems with persistent facial pain (Clarke, 1982; Greene, Olson, & Laskin, 1982). A wide variety of mechanisms have been cited as being responsible for causing the pain; unfortunately, many researchers support the veracity of their theories by attributing cause to cure, or by reasoning about the etiology of the disorder based on successful treatment approaches. As TMD patients generally respond very well to a number of diverse treatment approaches, theories have arisen that attribute the problem to various anatomical difficulties (most notably occlusal problems), personality variables, and psychosocial stressors (Clarke, 1982; Granger, 1958; Moulten, 1955). It is now generally agreed that occlusal problems play a minimal role, if any, in the initiation and exacerbation of TMD and that psychological factors such as stress and anxiety play an influential role in the manifestation of the disorder, although the nature of this role is unclear (Clarke, 1982; Glaros & Rao, 1977; Greene et al., 1982).

The two etiological views that seem to best account for the available data are a diathesis–stress model and a neurophysiological model. In the diathesis–stress model, the TMD patient is seen as having a genetic predisposition to responding to environmental stressors by engaging in bruxismlike activity. This activity is carried over and exaggerated during sleep, and the damage to the connective tissue (myofascia) in the jaw area resulting from the excessive strain being put upon this area causes the TMD symptoms. Such a model would account for the findings that the pain that TMD patients suffer tends to get worse in times of stress, and that bruxers in general are more likely than nonbruxers to respond to a stressful situation by bruxing activity (Greene et al., 1982; Rugh & Solberg, 1976). A more detailed theory offered by Kreisburg (1982) suggests that the nocturnal bruxing activity believed to be central in the exacerbation of TMD is intimately intertwined with the human sleep cycle, as it has been observed that, although bruxism occurs in all stages of sleep, it is most likely to occur during periods of transition from the deeper to the lighter stages (Satoh & Harada, 1973). According to this theory, nocturnal bruxism is an arousal reaction driven by the nigrostriatal system's interaction with the limbic and cortical areas responsible for jaw movement. In this framework, TMD's reactivity to psychosocial stressors is primarily a reflection of the stressors' more general effect on the sleep cycle. Although both of these theories are of value heuristically, neither of them has been extensively examined experimentally, so their status is still largely conjectural.

Clinical Interventions

A large number of treatment approaches proved successful with TMD patients, the success rates being between 60% and 95%, reported in most studies; these rates seem to be largely independent of the treatment used (Brooke, Stenn, & Mothersill, 1977; Salter, Brooke, & Merskey, 1986). Two subgroups of TMD patients whose prospects for successful treatment appear less optimistic are those with a high degree of psychopathology and those whose TMD symptoms began after some sort of facial injury (Salter et al., 1986). From a dental standpoint, occlusal splints that keep the patient from bruxing at night are probably the most common treatment. Other forms of dental treatment include occlusal realignment and a variety of surgical procedures designed to alter the relationships between various pieces of the masticatory system (Von Korff, Howard, Truelove, Sommers, Wagner, & Dworkin, 1988).

Research has shown that the psychologist

also has a valuable contribution to make in the treatment of these patients. EMG biofeedback has met with great success in the treatment of TMD, especially when combined with various forms of muscle relaxation (Cassisi, McGlynn, & Belles, 1987; Clarke, 1982). Although the specifics of these approaches vary, the basic idea is to make the patient more aware of her or his maladaptive response to stress (bruxing) and to provide strategies to alter this response. In addition, nocturnal EMG devices have been used successfully in a bell-and-pad-type procedure, in which the patient is awakened by an alarm when EMG ratings in the masseter muscle reach a criterion value (Cassisi *et al.*, 1987). As with the bell-and-pad procedure, it is important that the patient awaken when the alarm sounds rather than merely stir long enough to turn the alarm off, either manually or by relaxation (depending on the device used). It is useful to have the patient enter the time of awakening in a diary placed some distance from the bed. This serves the dual purpose of awakening the patient and yielding information that will prove valuable in keeping track of patient progress. Daily pain ratings (evening and morning) are also valuable pieces of information that the patient can record in this diary. These strategies often produce a rapid treatment effect, which must be closely monitored, as bruxing activity and pain levels often rebound back to pretreatment levels in these patients when treatment is abruptly stopped (Cassisi *et al.*, 1987).

In addition to applying the strict behavioral strategies, the psychologist can also contribute to the treatment effort by being especially cognizant of the behavioral contingencies that may play a role in maintaining the patient's pain. As explained by Fordyce (1976), pain behaviors are subject to the same learning principles as are other behaviors, and as in any chronic pain syndrome, these patterns of contingent reinforcement become more entrenched over time. Thus, the psychologist's experience in elucidating such relationships and working to change them when necessary can prove to be a valuable addition to the treatment plan. In addition, the psychologist is experienced in the diagnosis and treatment of other types of psychopathol-

ogy that may affect the patient's treatment progress in some way. This skill may also be of value in assessing the psychological status of facial pain patients who are candidates for surgery, as there is a subgroup of TMD patients whom a number of investigators cite as very poor risks for surgery (Marciani, Haley, Moody, & Roth, 1987). Characteristics that may alert one to a patient's belonging in this subgroup are (1) a history of pain-related surgeries, especially from similar pains; (2) a family history of multiple surgical procedures; (3) lack of pain relief from large doses of analgesic drugs; (4) a conviction that there must be an organic cause for the pain and denial of any other problems; (5) suggestion or demand of analgesic medication; and (6) current involvement in seeking job or accident compensation for pain caused by an injury or an occupationally related disease. Although none of these characteristics represents a clear diagnostic criterion, the presence of one or more of them should alert the practitioner to proceed very carefully and deliberately in considering treatment alternatives.

Program Development

The psychologist dealing with other health professionals often has another challenge in educating those around her or him about the role that various psychological states (such as anxiety and depression) play in influencing pain disorders in general and TMD in particular. Additionally, many professionals are unaware that treatment strategies other than occlusal splints, muscle relaxants, and surgery exist, so the psychologist may serve an educational role in this area as well. In interdisciplinary efforts aimed at pain management, professional cooperation and mutual education are vital components, and both patients and professionals stand to gain from an awareness of the available treatment strategies and from a willingness to refer patients to other specialists; for instance, just as the psychologist has a valuable contribution to make in the treatment of TMD, a patient complaining of facial pain referred to a psychologist without having first undergone an extensive evaluation by a dentist or other appropriate professional should be immediately referred for a determination of any obvious organic pathology. In some cases, the radiographic evidence clearly indicates those factors that are causing the pain, and these cases are usually handled efficiently by dental professionals with minimal involvement by psychologists in terms of pain management. The only obvious statement emerging from the literature concerning the treatment of TMD is that, because no differences are usually observed between the success rates of conservative approaches and surgeries, which are inherently riskier and more costly, the more conservative treatments must be attempted before surgery is contemplated in cases without clear-cut radiographic evidence of anatomical pathology (Salter *et al.*, 1986; Von Korff *et al.*, 1988).

Thus, a truly interdisciplinary treatment approach to TMD would begin with an extensive evaluation by a dental professional, with further referral contingent on the clarity of the findings of this evaluation. Psychological evaluation should then occur of those patients without clear physiological pathology to examine the behavioral, emotional, and environmental factors that may be influencing the patient's pain. It should be stressed to patients during such evaluations that the veracity of the pain complaint is not being questioned, and that no one thinks they are "crazy," as most patients with pain who are seen by the psychologist have already seen a number of health professionals with little success and may conclude that these professionals have decided that their pain is not real. A simple explanation of the stress–bruxism–pain relationship often helps the patient to get a better conceptualization of what the evaluation and treatment strategies are all about and makes the possible involvement of "psychological factors" somewhat less threatening. At this point, both dental and behavioral approaches may be pursued, singly or in combination, depending on the results of the respective examination.

In order to illustrate the previously discussed assessment and treatment strategies, a TMD patient who was treated at the psychology clinic will be described.

Case 5

Reason for Referral

Linda was a 23-year-old female referred to the clinic by the Department of Oral and Maxillofacial Surgery of Shands Hospital at the University of Florida for evaluation and treatment of myofascial pain secondary to stress. She presented with pain in her jaw that was present when she was chewing and on her awakening in the morning. This pain would often spread to severe headaches. She had first begun to notice the pain during her first year in high school, and it had become steadily worse over time. She was eating virtually nothing of firmer consistency than mashed potatoes, and the pain in her jaw was constantly present to some degree. She was employed in a high-pressure secretarial job that she enjoyed, and she was living at home with her parents. She had been thoroughly evaluated by the referring physician, and no radiographic evidence of any anatomical difficulties had emerged. She had previously attempted to use an occlusal splint, but the force of her nocturnal bruxism broke the splint before it had any positive effect.

Assessment

Assessment consisted of a lengthy interview followed by a battery of questionnaires, which included the Minnesota Multiphasic Personality Inventory, the West Haven-Yale Multidimensional Pain Inventory, the McGill Pain Questionnaire, the Sickness Impact Profile, the State-Trait Anxiety Inventory, and the Zung Depression Scale. The results indicated that there was no gross psychopathology and that her pain significantly interfered with her diet and her personal comfort. Her pain had never kept her from working, and she did not like taking pain medication at any time other than before she went to bed at night because she felt that these medications impeded her performance at work. She was not interested in surgery except as a last resort after every other recourse.

Treatment

The first phase of Linda's treatment involved teaching her some relaxation skills via biofeedback and a 20-minute progressive-muscle-relaxation (PMR) tape. Her initial masseter EMG levels were fairly high, but she rapidly gained control of these during the day and reduced them to normal levels. She was instructed to practice PMR twice per day, and she stated that this usually made her jaw feel better for a little while. She was also instructed to take her scheduled breaks at work, as she would usually work right through them.

After she became more adept at relaxation and became more conscious of the feelings of tension in her jaw, a few new components were added to her treatment. First, she was given a nocturnal EMG monitor to use every night, and she was instructed to write the times when she was awakened by the alarm in a diary across her bedroom. The criterion EMG level at which the alarm would sound was set and adjusted weekly by the therapist in the session. It was lowered across time based on her progress. She was also instructed to record her pain levels (on a 1-to-10 scale) before and after each relaxation session, immediately before going to bed, and immediately on awakening in the morning. She was also trained in cued relaxation and was told to check her jaw tension level each hour on the hour at work and to relax her jaw as needed. In addition to the strictly behavioral techniques, environmental stressors were discussed, and strategies for reducing the stress associated with these areas were implemented as well.

Termination Criterion

The patient reported a significant drop of pain across virtually all situations. She still had some pain after eating very firm foods such as steak, but her diet was gradually expanding.

She reported very minimal pain the mornings, no pain during the day, and complete cessation of the headaches from which she had earlier suffered.

Summary

This chapter has defined two widely experienced phenomena that are troublesome for dentists, namely, dental phobia and temporomandibular disorders. The brief literature review of etiology focuses on behavioral and physiological indices in addition to patient's own experience of the problem. The role of the psychologist is defined in terms of identifying patients with these problems and describing the subjective, behavioral, and physiological manifestations of the disorders. The treatment role of the psychologist with the dental-avoidant adult or child should include prevention by educating the dentist about the sources of these concerns. When efforts to secure cooperation with treatment have failed, the dentist, with the psychologist's assistance, may provide preparatory information, coping-skills training, and patient management techniques. Premedication has also been used with both adults and children to allay anxiety and secure cooperation. In treating children, the presence of the mother may be enough to provide control over behavioral disruption. The extremely fearful child or adult may initially need graduated exposure to the dental setting in the absence of extensive treatment. Techniques such as progressive relaxation, systematic desensitization, imaginal flooding, stress inoculation, biofeedback, film modeling, and controlled breathing are briefly described as adjuncts to treatment. Clinical cases are described to illustrate these procedures with patients exhibiting dental injection phobias that range from circumscribed simple phobia to agoraphobia.

Temporomandibular disorder (TMD) is one of the pain syndromes most commonly seen in the dental setting. The brief literature review identifies bruxism as playing a significant role. Psychosocial factors have been shown to play an important role in influencing the manifestation and progress of the disorder. Dental interventions frequently address the problem by occlusal realignment via splints or surgical procedures. Patients who are poor surgery risks include those with a personal or family history of pain-related surgery, a lack of relief from large doses of analgesics, and involvement in proceedings to gain compensation for their pain. The psychological interventions have been successful with a majority of patients complaining of TMD. EMG biofeedback and stress management procedures are illustrated with a patient with TMD pain.

ACKNOWLEDGMENTS

This chapter was made possible through the support of the National Institutes of Dental Research Training Grant T32-DE07133-07, which was directed by Dr. Barbara G. Melamed. David J. Williamson is receiving predoctoral fellowship support in this program. The authors wish to thank Dr. Carroll Bennett, whose efforts on an earlier chapter were useful in this work. Michael E. Robinson's assistance in the management of the TMD patient was valuable. We would also acknowledge Drs. Dwight and Nery Clark, and Dr. Frank Dolwick, who were the dentists handling the phobic patients.

References

Addelston, H. K. (1959). Child patient training. *Fortnightly Review of the Chicago Dental Society, 38*, 17.

Adelson, D., & Goldfried, M. (1970). Modeling and the fearful child patient. *Journal of Dentistry for Children, 37*, 476–488.

Ayer, W. A., & Corah, N. L. (1984). Behavioral factors influencing dental treatment. In L. K. Cohen & P. S. Bryant (Eds.), *Social sciences and dentistry: A clinical bibliography* (Vol. 2, pp. 267–322). Kingston-upon-Thames, England: Quintessence Publishing Co. for Federation Dentaire Internationale.

Ayer, W. A., & Levin, M. P. (1973). Elimination of tooth grinding habits by massed practice therapy. *Journal of Periodontology, 44*, 569.

Ayer, W. A., & Levin, M. P. (1975). Theoretical bases and application of massed practice exercises for the elimination of tooth grinding habits. *Journal of Periodontology, 46*, 306.

Bailey, J. S., & Reiss, M. L. (1979). Behavior technology advances in community dentistry. In D. J. Oborne, M. M. Grunelberg, & J. R. Eiser (Eds.), *Research in psychology and medicine.* New York: Academic Press.

Barenie, J., & Ripa. L. (1977). The use of behavior modification techniques to successfully manage the child dental patient. *Journal of the American Dental Association, 94*, 329–334.

Baron, R. S., Logan, H., & Kao, C. F. (1990). Some variables affecting dentists' assessment of patients' distress. *Health Psychology, 9*, 143–153.

Bernstein, D. A., Kleinknecht, R. A., & Alexander, L. D. (1979). Antecedents of dental fear. *Journal of Public Health Dentistry, 39*, 113–124.

Brockhouse, R. T., & Pinkham, J. R. (1980). Assessment of nonverbal communication in children. *Journal of Dentistry for Children, 42*–45.

Brooke, R. I., Stenn, P., & Mothersill, K. (1977). The diagnosis and conservative treatment of myofascial pain dysfunction syndrome. *Oral Surgery, 44*, 844–852.

Cassisi, J. E., McGlynn, F. D., & Belles, D. R. (1987). EMG-activated feedback alarms for the treatment of nocturnal bruxism: Current status and future directions. *Biofeedback and Self-Regulation, 12*, 13–30.

Chambers, D. (1977). Behavioral management techniques for pediatric dentists; An embarrassment of riches. *Journal of Dentistry for Children, 44*, 30–35.

Chambers, W. L., Fields, H. W. & Machen, J. B. (1981). Measuring selected disruptive behaviors of the 36- to 60-month-old patient: 1. Development and assessment of a rating scale. *Pediatric Dentistry, 3*, 25–56.

Chertok, S. L., & Bornstein, P. H. (1979). Covert modeling treatment of children's dental fears. *Child Behavior Therapy, 1*, 249–255.

Clarke, N. G. (1982). Occlusion and myofacial pain dysfunction: Is there a relationship? *Journal of the American Dental Association, 115*, 565–571.

Corah, N. (1969). Development of a dental anxiety scale. *Journal of Dental Research, 48*, 596.

Corah, N. (1988). Dental anxiety: Assessment, reduction and increasing patient satisfaction. *The Dental Clinics of North America, 32*(4), 779.

Corah, N., Bissell, G. D., & Illig, S. (1978). Effects of perceived control in stress reduction in adult dental patients. *Journal of Dental Research, 57*, 74–76.

Corah, N., Gale, E. N., & Illig, S. (1979). Psychological stress reduction during dental procedures. *Journal of Dental Research, 58*, 1347–1351.

Cuthbert, M. I., & Melamed, B. G. (1982). A screening device: Children at risk for dental fears and management problems. *Journal of Dentistry for Children, 49*, 432–436.

Duperson, D. F., Burdick, J. A., Koltek, W. T., Chebib, F. S., & Goldberg, S. (1978, Spring). Cardiac activity of children in a dental situation. *Journal of Pedodontics.*

Early, C., & Kleinknecht, R. (1978). The palmer sweat index as a function of repression sensitization and fear of dentistry. *Journal of Consulting and Clinical Psychology, 46*, 184–185.

Fields, H., Machen, J. B., Chambers, W. L., & Pfefferle, J. C. (1981). Measuring selective disruptive behavior of the 36- to 60-month-old dental patient: 2. Quantification of observed behaviors. *Pediatric Dentistry, 3,* 257–266.

Filewich, R. J. (1988). Treatment of the agoraphobic dental patient. *The Dental Clinics of North America, 32*(4), 723.

Fordyce, W. E. (1976). *Behavioral methods for chronic pain and illness,* St. Louis: Mosby.

Forgione, A. G. (1988). Hypnosis in the treatment of dental fear and phobia. *The Dental Clinics of North America, 32* (4), 745.

Forgione, A., & Clark, E. (1975). Comments on an empirical study of the cause of dental fears. *Journal of Dental Research, 53,* 496.

Frankel, R. I. (1980). The ticklish child and local anesthetic need. *The Journal of Pedodontics,* 139–144.

Frankl, S., Shiere, F., & Fogels, H. (1962). Should the parent remain with the child in the dental operatory? *Journal of Dentistry for Children, 29,* 150–163.

Friedson, E., & Feldman, J. J. (1958). The public looks at dental care. *Journal of the American Dental Association, 57,* 325–335.

Gale, E. (1972). Fears of the dental situation. *Journal of Dental Research, 51,* 964–966.

Gatchel, R. J. (1986). Impact of a videotaped dental fear-reduction program on people who avoid dental treatment. *Journal of American Dental Association, 112,* 218–221.

Ghose, L., Giddon, D., Shiere, F., & Fogels, H. (1969). Evaluation of sibling support. *Journal of Dentistry for Children, 36,* 35–49.

Glaros, A. G., & Rao, S. M. (1977). Bruxism: A critical review. *Psychological Bulletin, 84,* 767–782.

Glennon, B., & Weisz, J. R. (1978). An observational approach to the assessment of anxiety in young children. *Journal of Consulting and Clinical Psychology, 46,* 1246–1257.

Granger, E. R. (1958). Occlusion in temporomandibular joint pain. *Journal of the American Dental Association, 56*(5), 659–664.

Greene, C. S., Olson, R. E., & Laskin, D. M. (1982). Psychological factors in the etiology, progression, and treatment of MPD syndrome. *Journal of the American Dental Association, 105,* 443–448.

Heller, R. F., & Strang, H. R. (1973). Controlling bruxism through automated aversive conditioning. *Behavior Research and Therapy, 11,* 327.

Herbertt, R. M., & Innes, J. M. (1979). Familiarization and preparatory information in the reduction of anxiety in child dental patients. *Journal of Dentistry for Children.*

Hirschman, R., Young, D., & Nelson, C. (1979). Physiologically based techniques for stress reduction. In B. D. Ingersoll & W. R. McCutcheon (Eds.), *Clinical research in behavioral dentistry.* Morgantown, WV: University Foundation.

Hodgson, R., & Rachman, S. J. (1974). Desynchrony in measures of fear. *Behavior Research and Therapy, 12,* 319–326.

Horowitz, L. G. (1987). Overcoming your fear of the dentist. *Massachusetts Tetrahedron, Inc.*

Hutcherson, S. (1976). *The occurrence of modeling in children viewing peer model and familiarization videotapes prior to undergoing dental treatment.* Unpublished master's thesis, Case Western Reserve University, Cleveland, Ohio.

Kent, G. (1985). Memory of dental pain. *Pain, 21,* 187–194.

Keys, J. (1978). Detecting and treatment dental phobic children: 1. Detection. *Journal of Dentistry for Children, 4,* 296–300.

Kleinknecht, R. A., & Bernstein, D. A. (1979). Short term treatment of dental avoidance. *Journal Behavior Therapy Exp. Psychiatry, 10,* 311–315.

Kleinknecht, R., Klepac, R., & Alexander, L. (1973). Origins and characteristics of fear of dentistry. *Journal of the American Dental Association, 86,* 842–848.

Kleinknecht, R., Bernstein, D., & Alexander, L. (1977). *Assessment of fear of dentistry.* Paper presented at the annual meeting of the American Association for Dental Research, Las Vegas.

Kleinknecht, R. A., Mahoney, E. R., & Alexander, L. D. (1987). Psychosocial and demographic correlates of temporomandibular disorders and related symptoms: An assessment of community and clinical findings. *Pain, 29,* 313–324.

Klepac, R. (1975). Successful treatment of avoidance of dentistry by desensitization or by increasing pain tolerance. *Journal of Behavior Therapy and Experimental Psychiatry, 6,* 307–310.

Klepac, R. (1988). Behavioral treatment for adult dental avoidance: A stepped-care approach. *The Dental Clinics of North America, 32*(4), 507.

Klinge, V. (1979). Facilitating oral hygiene in patients with chronic schizophrenia. *Journal of the American Dental Association, 99,* 644–645.

Klingman, A., Melamed, B. G., Cuthbert, M. I. & Hermecz, D. A. (1984). Effects of participant modeling on information acquisition and skill utilization. *Journal of Consulting and Clinical Psychology, 52,* 414–422.

Klorman, R., Ratner, J., Arata, C., King, J., & Sveen, D. (1978). Predicting the child's uncooperativeness in dental treatment from maternal trait, state, and dental anxiety. *Journal of Dentistry for Children, 45,* 62–67.

Klorman, R., Hilpert, P., Michael, R., LaGana, C., & Sveen, D. (1980). Effects of coping and mastery modeling on experienced and inexperienced pedodontic patients' disruptiveness. *Behavior Therapy, 11,* 156–168.

Koenigsberg, S. R., & Johnson, R. (1975). Child behavior during three dental visits. *Journal of Dentistry for Children, 42,* 195–196.

Kohlenberg, R., Greenberg, D., Reymore, L., & Mass, G. (1972). Behavior modification and management of mentally retarded dental patients. *Journal of Dentistry for Children, 39,* 61–67.

Kreisburg, M. K. (1982, March–April). Alternative view of the bruxism phenomenon. *General Dentistry,* 121–123.

Kroeger, R. F. (1988). *How to overcome fear of dentistry.* Cincinnati: Heritage Communications.

Lamb, D., & Plant, R. (1972). Patient anxiety in the dentist's office. *Journal of Dental Research, 51,* 986–989.

Lautch, H. (1971). Dental phobia. *British Journal of Psychiatry*, *229*, 151–158.

Levy, R. L., & Domoto, P. K. (1979). Current techniques for behavior management: A survey. *Pediatric Dentistry*, *1*, 160–164.

Lewis, T. M., & Law, D. B. (1958). Investigation of certain autonomic responses of children to specific dental stress. *Journal of the American Dental Association*, *57*, 769–777.

Lund, A. K., & Kegeles, S. (1979). Cognitive and behavioral strategies for children's preventive dental behavior. *Journal of Dental Research*, *58*, 132.

Marciani, R. D., Haley, J. V., Moody, P. M., & Roth, G. I. (1987). Identification of patients at risk for unnecessary or excessive TMJ surgery. *Oral Surgery, Oral Medicine and Oral Pathology*, *64*, 533–535.

Martens, L., Frazier, P., Hirt, K., Meskin, L., & Proshek, J. (1973). Developing brushing performance in second graders through behavior modification. *Health Services Report*, *88*, 818–823.

Meichenbaum, D. (1977). *Cognitive behavior modification: An integrative approach*. New York: Plenum Press.

Melamed, B. G. (1979). Behavioral approaches to fear in dental settings. In M. Hersen, R. M. Eisler, & P. M. Miller (Eds.), *Progress in behavior modification* (Vol. 7). New York: Academic Press.

Melamed, B. G. (1982). Reduction of medical fears: An information processing analysis. In J. Boulougouris (Ed.), *Learning theory approaches to psychiatry*. New York: Wiley.

Melamed, B. G., Hawes, R., Heiby, E., & Glick, J. (1975a). The use of film modeling to reduce uncooperative behavior of children during dental treatment. *Journal of Dental Research*, *54*, 797–801.

Melamed, B. G. Weinstein, D., Hawes, R., & Katin-Borland, M. (1975b). Reduction of fear-related dental management using filming modeling. *Journal of the American Dental Association*, *90*, 822–826.

Melamed, B. G. Yurcheson, R., Hawes, R., Hutcherson, S., & Fleece, E. L. (1978). Effects of film modeling on the reduction of anxiety-related behaviors in individuals varying in level of previous experience in the stress situation. *Journal of Consulting and Clinical Psychology*, *46*, 1357–1367.

Melamed, B. G., & Mealiea, W. L., Jr. (1980). Behavioral intervention in pain-related problems in dentistry. In J. M. Ferguson & C. B. Taylor (Eds.), *A comprehensive handbook of behavioral medicine*. Jamaica, NY: Spectrum.

Melamed, B. G., Bennett, C. G., Jerrell, G., Ross, S. L., Bush, J. P., Courts, F., & Ronk, S. (1983a). Dentists' behavior management as it affects compliance and fear in pediatric patients. *Journal of the American Dental Association*, *106*(3), 323–330.

Melamed, B. G., Dearborn, M., & Hermecz, D. A. (1983b). Necessary considerations for surgery preparation: Age and previous experience with the stressors. *Psychosomatic Medicine*, *45*, 517–525.

Messer, J. G. (1977). Stress in dental patients undergoing routine procedures. *Journal of Dental Research*, *56*, 362–367.

Milgrom, P., Weinstein, P., Kleinknecht, R., & Getz, T.

(1985). *Treating fearful dental patients: A patient management handbook*. Virginia: Reston.

Miller, M. P., Murphy, P. J., & Miller, T. P. (1978). Comparison of electromyographic feedback and progressive relaxation training in treating circumscribed anxiety stress reduction. *Journal of Consulting and Clinical Psychology*, *46*, 1291–1298.

Moulten, R. E. (1955). Psychiatric consideration in maxillofacial pain. *Journal of the American Dental Association*, *51*(4), 408–416.

Murphy, L. (1952). *The widening world of childhood*. New York: Basic Books.

Pillemer, F. G., Masek, B. J., & Kaban, L. B. (1987). Temporomandibular joint dysfunction and facial pain in children: An approach to diagnosis and treatment. *Pediatrics*, *80*(4), 565–570.

Pinkham, J. R., & Fields, H. W. (1976). The effects of preappointment procedures on maternal manifest anxiety. *Journal of Dentistry for Children*, *43*, 180–183.

Reiss, M. L., Piotrowski, W. D., & Bailey, J. S. (1976). Behavioral community psychology: Encouraging low-income parents to seek dental care for their children. *Journal of Applied Behavior Analysis*, *9*, 387–397.

Rugh, J. D., & Solberg, W. K. (1976). Psychological implications in temporamandibular pain and dysfunction. *Oral Sciences Review*, *7*, 3–30.

Salter, M. W., Brooke, R. I., & Merskey, H. (1986). Temporomandibular pain and dysfunction syndrome: The relationship of clinical and psychological data to outcome. *Journal of Behavioral Medicine*, *9*(1), 97–109.

Satoh, T., & Harada, Y. (1973). Electrophysiological study on tooth grinding during sleep. *Electroencephalography and Clinical Neurophysiology*, *35*, 267–275.

Sawtell, R., Simon, J., & Simeonsson, R. (1974). The effects of five preparatory methods upon child behavior during the first dental visit. *Journal of Dentistry for Children*, *41*, 367–375.

Sermet, O. (1975). Emotional and medical factors in child dental anxiety. *Journal of Child Psychology and Psychiatric and Allied Disciplines*, *15*, 313–321.

Shaw, D., & Thoreson, C. (1974). Dental anxiety in children. *British Dental Journal*, *139*, 134–139.

Shoben, E., & Borland, L. (1954). An empirical study of the etiology of dental fears. *Journal of Clinical Psychology*, *10*, 171–174.

Siegel, L. J., & Peterson, L. (1980). Stress reduction in young dental patients through coping skills and sensory information. *Journal of Consulting and Clinical Psychology*, *48*, 785–787.

Simpson, W. J., Ruzicka, R. L., & Thomas, N. R. (1974). Physiologic responses of children to initial dental experience. *Journal of Dentistry for Children*, 465–470.

Solberg, W. K., & Rugh, J. D. (1972). The use of biofeedback devices in the treatment of bruxism. *Journal of the Southern California Dental Association*, *40*, 852–853.

Spielberger, C., Gorsuch, R., & Lushene, R. (1970). *The State-Trait Anxiety Inventory (STAI) Manual*. Palo Alto, CA: Consulting Psychologists Press.

Srp, L., & Kominek, J. (1963). The reaction of children to

dental treatment: An experimental study. *Odontological Review, 14,* 178–186.

Stoffelmayr, B. (1970). The treatment of retching responses to dentures by counteractive reading aloud. *Journal of Behavior Therapy and Experimental Psychiatry, 1,* 163–164.

Thrash, W. J., Marr, J. N., & Box, T. G. (1982). Effects of continuous patient information in the dental environment. *Journal of Dental Research, 61,* 1063–1065.

Venham, L. L., & Quatrocelli, S. (1977). The young child's response to repeated dental procedures. *Journal of Dental Research, 7,* 734–738.

Venham, L., Bengston, D., & Cipes, M. (1977). Children's response to sequential dental visits. *Journal of Dental Research, 56,* 454–459.

Venham, L., Bengston, D., & Cipes, M. (1978). Parents' presence and the child's response to dental stress. *Journal of Dentistry in Children.*

Venham, L. L., Murray, P., & Gaulin-Kremer, E. (1979). Personality factors affecting the preschool child's response to dental stress. *Journal of Dental Research, 58,* 2046–2051.

Von Korff, M. R., Howard, J. A., Truelove, E. L., Sommers, E., Wagner, E. H., & Dworkin, S. (1988). Temporomandibular disorders: Variation in clinical practice. *Medical Care, 63*(3), 1–8.

Weinstein, P., & Getz, T. (1978). Preclinical laboratory course in dental behavioral science: Changing human behavior. *Journal of Dental Education, 42,* 147–149.

Weinstein, P., & Nathan, J. E. (1988). The challenge of fearful and phobic children. *The Dental Clinics of North America, 32,* 667.

Weinstein, P., et al. (1982). The effects of dentists' behaviors on fear-related behaviors in children. *Journal of American Dental Association, 104,* 32–38.

Weinstein, P., Domoto, P., & Holleman, E. (1986). The use of nitrous oxide in the treatment of children: Results of a controlled study. *Journal of American Dental Association, 112* 325–331.

Wright, G. Z., Alpern, G. P., & Leake, J. L. (1973). The modifiability of maternal anxiety as it relates to children's cooperative dental behavior. *Journal of Dentistry for Children.*

CHAPTER 30

Integration of Clinical Psychology into Hemodialysis Programs

Daniel S. Kirschenbaum

Introduction: Problems Associated with End-Stage Renal Disease (ESRD)

What would happen to your life if your physician told you that you had end-stage renal disease (ESRD) or, in more common parlance, complete kidney failure? Your doctor would explain that, in order for you to survive, your blood must be cleansed of waste products by a mechanical device of some sort. The vast majority of the 100,000 or so ESRD patients use hemodialysis, a process in which the blood is circulated through an artificial kidney and then returned to the patient's bloodstream via a fistula (a connection between an artery and a vein created surgically in the arm). Unfortunately, hemodialysis requires 4 to 7 hours per day (including transportation time if it is done outside the home) three times per week. The use of hemodialysis brings with it the constant threat of death and a substantial dependency on medical personnel and machinery for survival. Also, this process is far from equivalent to the

manner in which people with functional kidneys have their blood cleansed. Bodily impurities and poisons remain in the bloodstream (uremia) at levels that can often produce severe reductions in physical energy, strength, and motivation. Dialysis also produces substantial reductions in sexual activities and a variety of more minor, but still rather troublesome, difficulties, such as headaches, nausea, cramping in the legs, sleep disturbances, skin irritations, and problems with access to the blood supply or other technical difficulties that pose considerable personal threat (Nichols & Springford, 1984).

So, what *would* happen to your life if you developed end-stage renal disease? The quality of your life would clearly suffer, and you and your family would have to cope with one of the most difficult chronic stressors that people face. Clinical psychologists and other mental health professionals have been studying the effects of ESRD on patients and their families since the early 1970s. We are beginning to learn what it is like to live with and adjust to ESRD and how psychologists in particular can help patients manage this formidable life change. This chapter reviews the key findings from this

Daniel S. Kirschenbaum • Department of Psychiatry and Behavioral Sciences, Northwestern University Medical School, Chicago, Illinois 60611.

emerging literature, with an emphasis on the clinical implications of the work.

Overview of the Literature: Clinical Problems and Intervention Strategies

"Illness intrusiveness" has been defined as the degree to which an illness and its associated treatments interfere with normal life activities (Binik, Devins, & Chowanec, in press; Devins, Binik, Gorman, Dattel, McCloskey, Oscar, & Briggs, 1982). ESRD is clearly and objectively an extremely intrusive illness. As such, we might expect the illness to produce substantial and adverse effects on both patients and their families. The literature consistently documents precisely that kind of impact (e.g., Blodgett, 1981; Kaplan DeNour, 1981; Nichols & Springford, 1984).

Most dialysands and their families experience a great deal of distress, resulting in moderate to severe levels of anxiety and depression, which are experienced far more frequently than in normative groups. The most dramatic illustration of this point is that ESRD patients are between 100% and 400% more likely to commit suicide than the average person (Abram, Moore, & Westerbelt, 1971). In addition, approximately 75% of dialysands struggle and fail to manage effectively the major restrictions that ESRD places on dietary and fluid intake. This problem only compounds the difficulties inherent in fluctuating levels of uremia; extraordinary demands on one's family; excessive dependency on medical facilities, personnel, and machines; and decreased control of one's life in general. Nichols and Springford (1984) found that the frustrations and the adverse emotional impact of this intrusive illness were most substantial during the first (or training) year of using hemodialysis. However, consider the following percentages of post-training year dialysands who agreed with the following statements ($n = 18$):

> I feel anxious before going on the machine: 27%.

> I feel jumpy and restless while I'm on the machine: 39%.
> I am angry that I can't do the things I used to do: 39%.
> I feel I'm spoiling my partner's life: 48%.
> I feel depressed much of the time: 28%.
> I sometimes want to take my life: 22%.

Clearly, people vary in their ability to cope with the stress imposed by ESRD. Perceptions of its intrusiveness can predict disruptions in life better than some objective indices of its intrusiveness (Binik *et al.*, in press). Nonetheless, the overall impact of ESRD increases the probability of substantial emotional problems in dialysands and their partners and families by a factor of 10 in most respects. The ESRD patient, therefore, has a very high risk of psychological difficulties and is a prime target for intervention by clinicians.

The clinical literature includes an abundance of advice and a substantial body of empirical evidence on how to help people cope with chronic stressors, to decrease generalized and specific anxieties, and to decrease depression. The behavioral medicine or health psychology literature takes many of these concepts and findings and applies them to health problems and difficulties. The writer and his colleagues have developed a specific set of procedures derived from a cognitive-behavioral approach to clinical problems, and we have applied them to hemodialysis patients in particular (Kirschenbaum, Sherman, & Penrod, 1987; Penrod & Kirschenbaum, 1986). The procedures focus on increasing the motivation to take control of as many aspects of the disease as possible, actually exerting maximum control on the treatment process, and improving the self-regulation of fluid intake. Accordingly, techniques such as decision balance sheets, behavioral contracting, self-monitoring, and the use of social support and problem solving are emphasized. The remainder of this chapter examines the specifics of the procedures; how they can be applied by nursing personnel under the supervision of a psychologist; and establishing, maintaining, and evaluating this approach.

Methods of Psychological Assessment

Increasing numbers of psychologists in medical settings will be working with increasing numbers of dialysands and their families. This is likely because accelerating numbers of psychologists and training programs are oriented toward health psychology, and because there is an obvious need for clinical intervention with this population. The usual methods of assessing depression and anxiety seem appropriate in this case, ranging from clinical interviews to the use of such paper-and-pencil tools as the Symptom Checklist 90—Revised (Derogatis, 1977). An assessment of coping skills would add another important dimension to the initial assessment in view of the challenge to such skills imposed by ESRD and the potential to help patients improve such skills (see Tobin, Holroyd, Reynolds, & Wigel, 1989).

Two specialized assessment devices are now available for use with ESRD patients. Nichols and Springford (1984) developed the first assessment tool for dialysis patients, the Dialysis Problem Checklist (DPCL), and a modified version for the partners of dialysands (Dialysis Problem Checklist B). The Dialysis Problem Checklist consists of 126 problems sometimes experienced by hemodialysis patients; the partner version includes 70 problems sometimes experienced by the partners of such patients. The list of problems was derived from the referrals sent by a dialysis unit to a psychological services department over a three-year period. The problems pertained to communication about the dialysis process, the machinery involved and feelings during dialysis, staff and family relationships, general psychosocial concerns (e.g., anger, depression, and anxiety), and physical discomforts and symptoms.

Nichols and Springford (1984) found that a sample of 34 dialysands reported 3–71 problems (maximum = 126), with 50 items endorsed by at least 50% of the patients. Twenty-eight partners reported 2–31 problems (maximum = 70), with 26 identified by at least 20% of the group. The most commonly acknowledged problems reported by dialysands in their first year of dialysis were worries about having trouble with needles or fistulas (67%), feeling no good as a parent (60%), feeling that there was too much strain on their partners (57%), finding that things were not explained well enough (56%), and constant tiredness (56%). After the first year of treatment, the number of problems endorsed declined, but many similar concerns were reported by almost half of the more experienced dialysands (e.g., feeling that there was too much strain on their partners and anger about being unable to do the things they used to do). The partners also checked items indicating increased anxiety, depression, and frustration.

The psychometric characteristics of these interesting checklists have not yet been well delineated. For example, test–retest and inter-item correlations were not reported, and it seems likely that the items vary in clarity and repetitiveness. However, the scales have demonstrated some construct validity by showing more distress in the first year than in subsequent years of treatment; patients who scored at the extremes on the checklists differed from each other in predictable ways on other variables (e.g., experience level, fluid restriction levels, and anxiety and depression ratings); clinicians judged that a few patients had more problems than they indicated on the checklists but that all of the patients who checked a lot of problems really did seem to experience quite a few problems.

Nichols and Springford's scales (1984), although not well refined as psychometric instruments, definitely appear clinically very useful at this juncture. The article summarizing their work is readily available, and it contains the most commonly acknowledged problems experienced by a relatively modest sample of people with ESRD and by their partners. A version of the checklist with these commonly endorsed items can be provided to patients and their partners in any ESRD treatment center on a regular basis (e.g., every six months). Frequent or severe problems can thus be identified, and appropriate referrals that specify the problems of particular concern can then be made. This procedure would provide the staff, the patients, and the partners with regular feedback, and the process may also enhance communication among all of those involved with the ESRD patients. Taking this step alone may improve patients' sense of control and involvement and may thus facilitate their overall adjustment and perceived competence (see Kirschenbaum *et al.*, 1987).

A second measure that was specifically designed for dialysis patients was developed by the writer and his associates, the Self-Care Dialysis Checklist (SCDC; Kirschenbaum *et al.*, 1987). The SCDC (see Table 1) is a behavioral observation checklist designed to assess the degree to which hemodialysis patients direct and manage their own dialysis treatment in hospital settings. Approximately 98% of hemodialysis patients who are treated in hemodialysis units (versus home dialysis or peritoneal dialysis) have nursing-staff members manage and direct most and sometimes all aspects of their treatments (HCFA, 1984). This is the most expensive approach to hemodialysis treatment, and it is also probably the least beneficial psychologically (Kirschenbaum *et al.*, 1987). Millions of dollars spent on health care in this country would be saved if greater numbers of patients managed most of their own treatments in hemodialysis units. For example, the cost of self-directed treatment at Methodist Hospital in Madison, Wisconsin, 1983 was estimated as being $20 less per treatment than staff-directed treatment (Penrod & Kirschenbaum, 1986). The cost reduction resulted from the decreased amount of staff time required for self-directed treatments. If we assume that these calculations apply to other hemodialysis units, the savings generated by getting only 20% of this country's 60,000 plus staff-directed patients to switch to self-directed treatment would equal approximately $34,000,000 per year. An increasing degree of self-directedness would also yield considerable psychological and perhaps medical advantages (e.g., a decreased perception of pain and decreased depression and anxiety) due to the increased control and perceived control associated with such a change (Binik *et*

Table 1. Items from the Self-Care Dialysis Checklist (SCDC)

"Making nest"
1. Picked up supplies box
2. Piced up blanket
3. Obtained ice
4. Set up recreational materials (e.g., TV, radio, books, or phone)
5. Put call light on chair
Pretreatment check of vital signs
6. Recorded weight
7. Recorded blood pressure
8. Recorded pulse
9. Recorded temperature
Machine and equipment preparation
10. Set up machine without need of nurse intervention
11. Prepared needles (drew up heparin, xylocaine, or ethyl chloride and brought needles to chair)
12. Tore tapes for needles[a]
13. Achieved dialysis "setup" on time (no more than 5 minutes late)
Monitoring treatment
14. Recorded TMP[b] and UFR[c] values within 30 minutes after "put-on"
15. Intervened appropriately at first sign of hypotension (reclined chair, took blood pressure, and called nurse if no improvement)
16. Monitored blood presure at least once an hour
17. Stayed awake during treatment
Posttreatment check of vital signs
18. Recorded blood pressure
19. Recorded pulse
20. Recorded weight
Cleanup
21. Removed clutter from station (cups, wrappers, food, etc.)
22. Put supplies box on appropriate shelf
23. Returned blanket to cubby or wash hamper
24. Put chart in basket
25. Achieved scheduled take-off time (±5 minutes)

---

[a]Tore tapes = cut 10 pieces of tape used to secure needles to access site: 6 pieces, 6″ long; 2 pieces, 2″ long; 2 pieces, 10″ long.
[b]TMP = transmembrane pressure.
[c]URF = ultrafiltration rate.

al., 1982; Devins, Binik, Hollomby, Barre, & Gutman, 1981; Kirschenbaum *et al.*, 1987).

The Self-Care Dialysis Checklist was developed because of the potential advantages of helping patients become more self-directed during their dialysis treatments and because we needed an objective definition and measure of self-directedness. Dozens of tasks and activities are associated with hemodialysis treatment. No clear recommendation of how to determine whether the completion of some or all of these activities should be used to define self-directedness had been offered or accepted in the literature at the time of our work in this area. Because we must be able to define and measure what we hope to improve, the SCDC was created.

The format of the checklist is a one-page presentation that includes blanks for the patient's name. The instructions are:

> For each treatment date, please circle whether the patient completed the following tasks. Please answer question 1–12 within 5 minutes after "put-on." Please answer questions 18–25 within 5 minutes after "take-off."

The items were listed as questions which had Y N underneath them, such as:

1. Picked up the supplies box
   Y N

To assess the reliability of the Self-Care Dialysis Checklist, Kirschenbaum *et al.* (1987) had eight different pairs of nurses complete a total of 10 reliability checks. Four patients were observed as unobtrusively and independently as possible by the pairs of reliability checkers during the same treatment (i.e., Patient A was observed and the checklists were completed independently by both nurses who were in a pair). The reliability checkers received one hour of instructions plus some hands-on experience with feedback (one hour or less) before completing their observations. When agreements ÷ agreements + disagreements were used as the measure, reliabilities were quite good, ranging from 84% to 100% ($M = 92\%$). The "exact agreement" method was used to determine disagreements (and, therefore, reliability) most conservatively (i.e., if Nurse A checked a *Yes* for Item 6, but Nurse B checked *No* for that item, two disagreements were recorded for Item 6; Repp, Deitz, Boles, Deitz, & Repp, 1976).

The SCDC appears to be a reliable, easily learned, and easily used means of assessing 25 behaviors that are crucial components of hemo-

dialysis treatment. We now have available a means of operationalizing the degree of self-directedness used during in-center hemodialysis treatment. Self-directedness could be defined quite simply as the extent to which patients complete the 25 items on the SCDC. Of course, some may argue that certain items should be weighted more substantially then others (e.g., "machine and equipment preparation" items are much more involved than the "making nest" items). This weighting approach may prove useful under some circumstances, although it did not affect the data reported by Kirschenbaum *et al.* (1987). In addition, some tasks incorporated into existing items could easily be elaborated to form additional items (with the potential for reduced reliability in assessment, however). Regardless of the changes and modifications desired in the SCDC, as it currently exists it is readily available (e.g., in this chapter and in the original paper), and it can serve as a baseline assessment of self-directedness in work with any hemodialysis unit or patient in that unit. If a patient currently completes only 1 or 2 tasks on the SCDC, that patient may benefit from working toward completing 3 or 4 tasks, depending on the tasks and the patient involved. Similarly, patients who currently complete 19 or 20 items may find some benefits associated with completing 23 or 24 items. In a related vein, hemodialysis units that average only 3 or 4 items per patient may benefit from improving that average by several items. The clinical psychologist who consults with such units can also help the unit increase its emphasis on self-directedness and to improve its communication about it by using the SCDC as a definitional tool.

Clinical Interventions

The preceding analysis indicates that the stressors associated with end-stage renal disease tend to create or exacerbate such common problems as marital distress and generalized anxiety and depression. These difficulties are readily measured by means of the usual clinical assessments, perhaps supplemented with the questionnaires developed specifically for this population by Nichols and Springford (1984). The usual means of treatment could then be applied. There are, however, several specific difficulties that are more directly linked to ESRD. These include attempting to help patients and staffs encourage greater degrees of self-directedness in the hemodialysis process, as was discussed in the previous section of this chapter. An additional problem area is generally described in the health psychology literature as the "compliance" or "treatment adherence" problem. The vast majority of ESRD patients do not achieve "good" compliance with the extremely demanding dietary and fluid restrictions associated with ESRD (e.g., Procci, 1981).

Penrod and Kirschenbaum (1986) emphasized that a cognitive-behavioral therapy approach seems particularly well suited to managing the dialysis-related problems of increasing self-directedness during treatment and improving adherence to dietary and fluid restrictions. If we view the degree of self-directedness as a problem of choice (choices made by patients, staff, and administrators), we can then apply cognitive-behavioral techniques known to improve decision making (see e.g., Janis & Mann, 1977).

Patients and staff confronting the option of a more independent treatment modality face a tough choice. The expense of the setup and maintenance of a self-directed or a home care dialysis training program can seem prohibitive to short-sighted administrators. Additionally, less staff-dependent modes of delivery can be perceived as threatening to health care providers (Barnes, 1980). It can be threatening when more decision-making control and responsibility shift from staff to the well-educated patient. Traditional caregiving roles are challenged. However, cognitive-behavioral techniques can increase patient and staff involvement in treatment modality decisions and treatment management. Additionally, these techniques can reduce patient fears about more self-directed treatment modalities. Finally, cognitive-behavioral interventions may help staff

and patients master new and complex tasks efficiently and effectively.

Compliance can also be viewed as a cognitive-behavioral problem. It is an easier ESRD problem to conceptualize within this model because it is similar to problems that are more traditionally treated via cognitive-behavioral techniques (e.g., weight loss and smoking cessation). The techniques of goal setting, planning, and self-monitoring that can be used successfully with these more traditional problems may help ESRD patients tolerate the thirst and pursue the challenging goal of minimal fluid intake and a limited diet. Finally, cognitive-behavioral research indicates that compliance with dietary restrictions is one of the most refractory clinical problems encountered (Stuart, 1980). It would help to consider the notorious problem of renal patient noncompliance in light of what is known about the difficulties encountered by healthy people who want to restrict calories; both problems require massive behavior changes, great persistence, and involve formidable physiological barriers to success (Kirschenbaum, 1987). Clearly, intensive systematic efforts to reduce these problems are required. Cognitive-behavior therapy can facilitate such efforts.

Increasing Self-Directedness

The writer and his colleagues (Penrod & Kirschenbaum, 1986) developed and implemented the techniques to be described in the following sections of this chapter in the Hemodialysis Unit of Methodist Hospital (Madison, Wisconsin). The unit was firmly committed to the value of self-directed care. The arguments about the emotional and physical benefits of perceived control over one's body (see, e.g., Devins et al., 1981) were clearly supported by the leaders of the unit. Given that social-environmental climate, it was feasible to develop a plan for intervening with individual patients. The plan was to identify a number of patients who could do more for themselves than they were currently doing. Once these patients were identified, a group of nurses began meeting regularly with the author (a clinical psychologist consultant) and a staff medical social worker. For the first month or two of these weekly meetings, cognitive-behavioral principles were introduced and the specific techniques to be used were developed and refined. The authors made the refinements in the techniques based, in part, on input received from the group. Once the procedures were finalized, each nurse in the group was assigned a case with whom she worked. The subsequent several months of meetings were essentially group supervisions of the cases.

The techniques developed to increase self-directedness in treatment were oriented to the operational definition of self-directed treatment reviewed in the assessment section of this chapter, based on the 25 tasks contained in the Self-Care Dialysis Checklist. Essentially, the use of the SCDC defined *self-directed* in operational terms. For example, someone who completes at least 22 of the 25 tasks could be considered a "self-directed patient" (a higher or lower criterion can also be used, of course). Whether a patient had completed each of the 25 tasks in our operational definition was determined reliably by the nursing staff.

Given the above definition of self-directedness, how does one go about getting a patient to consider becoming more self-directed? We developed a five-step approach to this problem:

1. *Introduction of the rationale to the patient.* The staff person who conducts the intervention (e.g., nurse, social worker, or psychologist) explains why the patient is being approached and what the patient can gain from participating. We used the opening lines. "A number of us have been working on how to improve patient care. I'd like to talk with you about your own treatment." This introduction can be followed up with the potential advantages (emotionally and perhaps medically) of taking care of one's own body as much as possible. This rationale makes it clear that a new procedure or treatment modality is not going to be forced on the person. The emphasis is on the choices the patient is making and on assisting the person in making thorough and maximally healthful decisions. The motivational benefits of enhancing

choice and perceived choice (Kirschenbaum, 1985) and of thorough decision making (Janis & Mann, 1977) have been well established empirically.

2. *Promotion of effective decision making.* Janis and Mann (1977) provided a careful review of the qualities of effective (and ineffective) decision making. They also presented procedures that can promote high-quality decision making. We used two of the approaches they presented to help our patients make effective decisions about the degree of self-directed care they wanted to use. The two procedures we used were the "decision balance sheet" (Janis, 1959) and the "awareness of rationalizations" method (Reed & Janis, 1974). Table 2 presents the instructions used by the nurses in our program to complete a standard self-care balance sheet, which was given to patients for review during the third step of completing their own balance sheets.

As can be seen by reviewing Tables 2 and 3, completing a decision balance sheet can help patients review very carefully all of the advantages and disadvantages of completing as many self-care (self-directed) tasks as possible. This type of systematic review can result in increased work toward a goal (i.e., motivation; see Hoyt & Janis, 1975). However, people sometimes make the judgment that a few of the potentially negative aspects of a decision like this one are prohibitive. Therefore, it helps to consider whether there is substantive merit in some of the negative aspects of pursuing goals. The "awareness of rationalizations" approach does this essentially by providing counterarguments for each of the negative points in a decision balance sheet. Sometimes, of course, the negative aspects of pursuing a goal are very germane and realistic. In this case, however, the writer and his associates did not find many of the proposed negative points particularly valid or convincing. Therefore, after an initial balance sheet was completed, the nurse-intervenor reviewed the "counterrationalizations" that appear in the appendix to this chapter (counterarguments for each of the negative anticipations mentioned in Table 3). If this procedure is done in a nonpatronizing, helpful fash-

Table 2. Balance Sheet Procedural Directions for Use by Staff with Patients Considering Self-Care[a]

Step 1. Open-ended interview.

Questions are posed regarding the most salient reasons for accepting or rejecting self-care training. For example:

What consequences would be likely to follow from self-care training?

What are the positive and negative effects of accepting or rejecting self-care training?

Now suppose you committed yourself to self-care training today. What would you think about? I would like to talk about all of your thoughts about this goal from the perspective of committing yourself to it.

Step 2. Introducing the balance sheet.

To stimulate thinking about nonsalient alternatives and their consequences, the patient is asked to complete a balance sheet. The contractor may say, "You've talked about some of the pros and cons. What I'd like you to do now is to go through the possible considerations in a more systematic way." Using the information obtained in Step 1, the contractor then helps the patient complete a grid, such as the already-prepared Self-Care Balance Sheet (Table 3). After the grid is completed, the patient is asked to review all of them together to see if anything was omitted.

Step 3. Using a list of pertinent considerations.

The patient is presented with additional pros and cons taken from the Self-Care Balance Sheet (see Table 3). These serve as a further stimulus for considering additional consequences and alternatives.

Step 4. Identifying the most important considerations.

To rate the relative importance of each alternative, the patient is directed to review each consequence on the balance sheet and to rate its importance. A 3-point scale (from 1 "hardly important" to 3 "extremely important") can be used to quantify these ratings, or the patient can simply indicate which are the three or four most important consequences.

Step 5. Letting the balance sheet show the way.

Sum all negative rankings (e.g., tangible losses for self and others) and subtract this sum from the sum of all the positive rankings (e.g., tangible gains for self and others). The "tentative" choice should be evidence by a large positive score. However, if the scores are close, ask the patient to interpret the results of the balance sheet. Viewing the choice as tentative prevents a premature commitment by allowing time for second thoughts and possible revised rankings.

[a]This material is reprinted with permission of the publishers of *Perspectives: The Journal of the Council of Nephrology Social Workers,* 1985–1986, 7, 65–90, from an article by Joan D. Penrod and Daniel S. Kirschenbaum entitled "Cognitive-Behavioral Techniques."

ion (i.e., in the role of a consultant, not a parent), it, too, can lead to significant positive behavior change (e.g., Reed & Janis, 1974).

3. *Establishment of a behavioral contract.* Once a patient decides to pursue a greater degree of self-directed treatment, he or she will need assistance to keep motivated toward that goal. Behavioral contracting has clearly established itself as a useful technique for enhancing motivation (Kirschenbaum & Flanery, 1983, 1984). We may define a behavioral contract as "an explicit agreement specifying expectations, plans, and/or contingencies for the behavior(s) to be changed" (Kirschenbaum & Flanery, 1983, p. 224). Contracts that are negotiated, that focus on process (not just outcome goals), and that have consequences administered by the client or significant others tend to produce very good results (Kirschenbaum & Flanery, 1983).

We used these principles in devising the contract form shown in Figure 1. The definitions provided in Table 4 help explain the elements used in the contract. Use of the "treatment agreement" form can help the staff and the patient negotiate most of the key aspects of treatment (e.g., take-off and put-on times, degree of self-directedness in treatment). Each week or month a new contract can be renegotiated. This procedure can clarify expectations and consequences (e.g., for failure to show up on time), can increase the uniformity with which patients are treated on a unit, and so on. In Methodist Hospital, we did not emphasize the consequences of, for example, failing to reach goals established for self-directedness. The use of consequences may not be crucial for behavioral contracts to have beneficial effects (Kirschenbaum & Flanery, 1984). On the other hand, some aspects of the contract did have explicit consequences (e.g., lateness by more than 10 minutes resulted in at least a 1-hour delay in start-up). From a clinical perspective, it seemed clear to us that the primary benefit of this procedure was increasing the uniformity of the approach to treatment in the unit and clarifying expectations with each patient.

4. *Promotion of self-monitoring by patients.* It has been well established that people who systematically gather information about, or keep

Table 3. Self-Care Balance Sheet[a]

GOAL: To complete as many self-care tasks as possible

Positive anticipation	Negative anticipation

Tangible gains and losses

To self:
1. Get on first
2. Scheduling flexibility
3. Ad lib ice
4. More immediate nurse attention (via call light)
5. More control over treatment
6. Easier transition to home care
7. Possibly more favorable mortality rates
8. Quicker recognition of machine problems
9. Learning meaning of symptoms, ensuring a safer treatment
10. More awareness of effects of diet
11. Being a partner in medical treatment
12. Saving time in setup (especially for morning people)

1. Extra time to set up
2. More likely to make mistakes during training
3. Not watched by nurses carefully
4. Can't sleep during treatment
5. Increased responsibilities
6. Increased number of tasks
7. Time needed to study procedures
8. Will think of disease more often
9. Greater energy commitment

To others:
1. Help contain institutional costs
2. Teach others self-care
3. Look healthier
4. Less nurse time to setup
5. Apply greater health knowledge at home (blood pressure, temperature, sterile technique, first-aid skills)

1. Away from family longer because of increased setup
2. May look stupid while learning
3. Partner feels less needed
4. Partner feeling scared about patient's new responsibilities

Self-approval and disapproval
1. Increased sense of control over body
2. Increased control over environment
3. Increased security
4. Discovering that I can still learn (despite age or education)
5. Development of greater problem-solving skills in other areas
6. Belonging to a new group
7. Feeling of pride
8. Self-image less of a patient
9. More self-confidence
10. More knowledge about treatment

1. Decreased sense of security
2. Forces me to feel like a "real dialysis patient"
3. Not fair—others don't do it even though they are physically able

Social approval and disapproval
1. Get more attention during training
2. Get to know a couple of nurses really well
3. Demonstrate personal competence to friends, nurses, family
4. Approval by other self-care patients
5. Belonging to a good group
6. Family perceives patient as less sick
7. Decreased fear of visitors

1. Full-care patients may resent my enthusiasm
2. People will see I learn slowly
3. People will see I perform slowly

[a]This material is reprinted with permission of the publishers of *Perspectives: The Journal of the Council of Nephrology Social Workers*, 1985–1986, 7, 65–90, from an article by Joan D. Penrod and Daniel S. Kirschenbaum entitled "Cognitive-Behavioral Techniques."

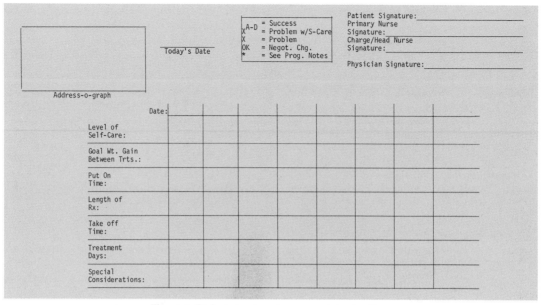

Figure 1. Dialysis treatment agreement (contract).

track of, target problem behaviors (i.e., self-monitor) tend to maintain changes in self-regulated behavior better than those who do not self-monitor (Kirschenbaum, 1987). In other words, it seems to be necessary to keep track of many problem behaviors in a systematic way in order to change those behaviors (i.e., keeping track of the calories consumed facilitates weight reduction). We used this principle when attempting to encourage maximum self-directedness during hemodialysis treatment. We developed a version of the Self-Care Dialysis Checklist that could be used by patients to self-monitor their degree of self-directedness (i.e., 25 behaviors completed = complete self-directedness). They simply checked off which behaviors of the 25 they completed during each dialysis treatment. Self-monitoring provides immediate feedback to patients about the extent to which they are meeting their goals, and it cues other vital self-regulatory processes (e.g., self-consequation, self-evaluation, and planning; see Karoly & Kanfer, 1982).

5. *Sustained use of staff support and problem solving.* Regular weekly review of progress, support, and assistance with problem solving can facilitate a variety of self-regulated behav-

ior changes (Colletti & Brownell, 1982). It seems especially important to maintain such contacts for a relatively lengthy period of time to promote long-term maintenance of change (Kirschenbaum, 1987). Therefore, we included weekly meetings between the patients and their nurse-intervenors in order to help patients become more self-directed. The meetings lasted for 4 to 6 months and varied in length from 10 to 60 minutes. In these meetings, the intervenors reviewed progress, modified the treatment agreements as needed, provided support, and helped the patients resolve treatment-related problems.

Improving Treatment Adherence

The writer and his colleagues also developed a variation of the five-step procedure for improving self-directed care to promote treatment adherence (Penrod & Kirschenbaum, 1986). Specifically, six cognitive-behavioral steps were taken to help patients improve their abilities to regulate (restrict) their intake of fluids:

1. *Provision of a rationale.* The medical risks associated with excess weight gain due to fluid

Table 4. Dialysis Treatment Agreement Definitions[a]

	Success	Met treatment agreement criteria.
X	Problem	Treatment agreement criteria not met.
		Make a note in the box with the X or in the progress note.
XA-D	Problem	Level of self-care criteria not met.
		Note the letter corresponding to the area(s) of self-care not met.
OK	Change negotiated	Treatment agreement criteria not met. Patient requested a change in the agreement on a day before this treatment, and the change was mutually agreed on by the patient and the nurse.
*	See progress note	Write a note in the progress note portion of the patient record when there isn't enough room in the box or on the agreement sheet or when the information is such that it needs to go in the permanent record.

Level of self-care
1. Complete self-care
   A. Does entire setup without assistance
   B. Prepares dialysis station
      Gets ice, food, diversion activity such as books and blankets, and has call light, head phones, and blood pressure cuff on the chair
   C. Runs treatment so that dry weight is achieved without cramping and hypotension that requires nursing intervention except for giving up two doses of saline
   D. Charts on flow sheet and returns to nurses' station
2. Partial self-care
   Any degree of complete self-care less than 100% must specify here:
Goal weight gain between treatments
   Appropriate range of weight gain to stay within as agreed upon by the patient and the interdisciplinary team.
Put-on time
   Span of time (e.g., 6:10 to 6:45 A.M.) during which we put patient on. If late put on, state when the patient went on.
Length of treatment
   Prescribed length of treatment as agreed on by the patient and the team. If less or more than prescription, write the length of treatment in the space.
Take-off time
   Either the latest possible time that the patient can come off the machine or a span of time during which we can take the patient off.
Treatment days
   The days the patient is routinely scheduled for treatment.
Special considerations
   Special agreements that have been made with the patient regarding his or her treatment. A check in this area means that the special considerations were needed. An OK means that the special considerations were not given to the patient because he or she did not require them. Specify special considerations.

[a]This material is reprinted with permission of the publishers of *Perspectives: The Journal of the Council of Nephrology Social Workers*, 1985–1986, *7*, 65–90, from an article by Joan D. Penrod and Daniel S. Kirschenbaum entitled "Cognitive-Behavioral Techniques."

consumption were discussed, including an analysis of the more immediate aspect of excess fluid consumption (e.g., more difficult treatments and aftereffects of treatments, possible shortness of breath).

2. *Promotion of effective decision making*. A decision balance sheet and counterrationalization technique were used, as described in the self-care section above. The goal, however, changed from increasing the degree of self-directedness to decreasing fluid intake between treatments.

3. *Establishment of a behavioral contract*. A contract was created that was oriented to the outcome goal of decreased weight gain between treatments. This contract included techniques such as the self-monitoring of fluid intake and the use of self-instructions and stoppers (as described below). As there is a reasonably clear relationship between fluid consumption and weight gain between treatments, it may be effective to include specific weight gains as targets for change in the contract.

4. *Promotion of self-monitoring.* Figure 2 shows the form we developed to facilitate the self-monitoring of fluid consumption by patients. We used an 8½″ × 11″ version of the form that included seven days at a time, and we also created a small pocket-sized version that included only two days at a time. The patient simply put a line through the chart during the day as he or she drank some liquid (e.g., 1 cup of water led to a line through the 250-cc point). Weekly goals were also written on the seven-day form. Alternatively, three day goals could have been used. These relatively "distal" goals

should be associated with improved self-regulation as compared to the use of more "proximal" daily goals (Kirschenbaum, 1985). We encouraged patients to carry with them at all times the smaller two-day forms and to post on their refrigerators their seven-day forms. They were encouraged to enter their daily data onto the larger forms each day.

5. *Use of self-instructional, stimulus-control, and related techniques.* Self-instructional, stimulus-control, and related procedures have demonstrated their efficacy with a variety of self-regulatory problems (see Karoly & Kanfer, 1982;

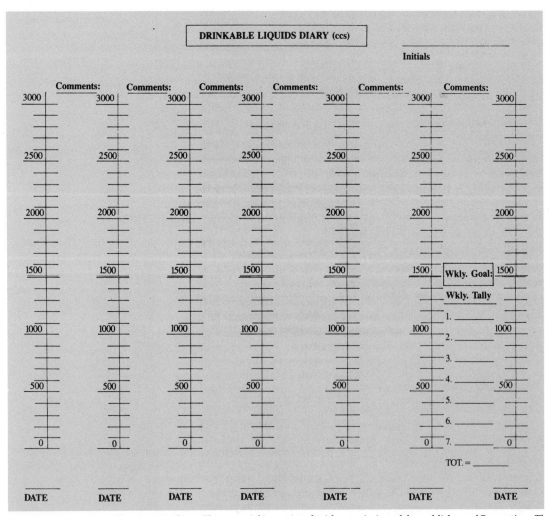

Figure 2. Seven-day self-monitoring form. This material is reprinted with permission of the publishers of *Perspectives: The Journal of the Council of Nephrology Social Workers*, 1985–1986, *7*, 65–90, from an article by Joan D. Penrod and Daniel S. Kirschenbaum entitled "Cognitive-Behavioral Techniques."

Miller & Berman, 1983). Intervenors could help patients modify the way they talked to themselves (i.e., the self-instructions they provided themselves) so that they prompted themselves to work at fluid regulation more diligently, to maintain their self-monitoring more religiously, and otherwise to improve their self-regulation. For example, nurse intervenors were given the following instructions for modifying self-statements at Methodist Hospital. The name of the patient in the example is Sam:

Sam's nurse suggested that Sam might find it easier to reach his goals if his goals became a more obvious part of his everyday life. She asked Sam to help her generate a list of reasons for trying to reduce fluids. Working together (from the old balance sheets), they listed such things in the following ad-slogan format:

DRINK LESS, LIVE LONGER!
DRINK LESS, BREATHE BETTER!
GAIN LESS, EASIER TREATMENT!
A LITTLE WORRY ABOUT DRINKING NOW,
LESS WORRY ABOUT TREATMENT
AND HEALTH LATER!

(Of course, similar sayings would be constructed for any problem in self-control, including increasing some aspects of self-care.) After Sam and his nurse developed this list together, Sam's nurse asked Sam to rewrite the list on an index card. (Better yet, she could have written them out with Sam while they met.) Then, Sam put the index card next to his plate during meals, reviewed it each time he turned on a kitchen light, and otherwise made engaging in a common everyday behavior (e.g., calling someone on the phone, reading a magazine or book, or turning a TV on or off) contingent on a quick review of self-instructions.

The self-instructions should be redone in a week or two just to make them more interesting. it is helpful if the nurse actually rewrites them with the patient a couple of times (over several weeks). In a similar vein, patients can be encouraged to put up signs in their houses with their self-instructional slogans on them. The nurse can again help by writing out such slogans using a magic marker and a large index card (one slogan per 5″ × 7″ card). Putting these signs in closets and kitchens helps patients to remember their goals—and, most important, to increase their chances of successfully reaching them.

Stimulus control techniques are procedures that limit the number of stimuli that reliably elicit undesirable behaviors. Hemodialysis patients who wish to restrict fluid intake find ways to decrease the number of situations in which they consume liquids. For example, the patient may reduce fluid intake by allowing himself or herself a drink only while seated at a table (i.e., eliminating drinking beverages while in a car or while in a living room or bedroom). The use of only one particular small glass may also help, as may keeping a pitcher of water in the refrigerator with a limited amount of water in it from which the patient drinks during the day. Other ideas include a concept related to stimulus control known as *chaining*. It involves elongating the chain of behaviors between the desire to drink and the actual behavioral step of drinking. The use of habits that are essentially incompatible with drinking may decrease the chance of drinking excessively by placing behavioral roadblocks between the desire to drink and the act of drinking (i.e., elongating the "drinking chain"). For example, patients can chew on toothpicks or Stim-u-dents frequently, chew on celery sticks, suck on sour candies (nonsugared), and require themselves to self-monitor their fluid intake before they drink.

6. *Sustained use of staff support and problem solving.* Regular weekly or more frequent meetings are advisable for at least several months (see Kirschenbaum, 1987).

Developing and Maintaining the Program

The services of clinical psychologists on hemodialysis units can emerge from several avenues. Initial referrals for help with particular patients may be sent to psychology services for testing; psychology services for consultation about a particularly difficult case (e.g., anxiety or depression, sexual dysfunction, or family conflict); psychiatry services for consultation about medication or problem cases; and consultation–liaison services for related help. In addition, particular staff members in a hemodialysis program may recognize some problems on the unit, such as conflict between staff members about patient management issues and disagreements about how to encourage or even how to discourage self-directedness of care. An astute social worker or head nurse or physician who acknowledges such problems may ask for help from a psychologist within or outside a hospital system. These various avenues of entree will clearly affect how to approach and maintain involvement with the unit.

An organizational development dilemma such as how the form of entree affects the development of an effective clinical service is beyond the scope of this chapter. However, a few points along these lines may be worth emphasizing. For example, it is clear that, if a psychologist is brought into the system at the request of a key professional who is established and respected in that system, it will be easier to make more substantial unitwide interventions more readily and efficiently (e.g., instruction in the use of behavioral contracts throughout the unit). When the entree comes in the form of a request to help an individual patient, the psychologist may be working from the ground up; in other words, the psychologist must establish his or her usefulness to a key staff member or two at first and then only gradually find opportunities for broader interventions (e.g., conducting workshops with the staff and participating in case conferences). Whatever form the entree takes, from this writer's perspective, the head nurse is usually the single most important person in a hemodialysis unit, followed by one or two physicians who are well liked by the nursing staff. These are the individuals who must appreciate, value, and participate in the clinical interventions. Therefore, whatever the psychologist does in the unit, it would be worthwhile to keep such key people informed of the activity and to obtain their input about the work and the manner in which such services are used in the system.

It is worth adding that nurses can learn to use the cognitive-behavioral techniques outlined in the preceding section. In fact, the writer and his colleagues developed those procedures for the explicit purpose of having selected nurses learn how to use them (Penrod & Kirschenbaum, 1986). The nurses chosen to administer such a program need superior interpersonal skills and a clear willingness to accept and value the approach. They also require a considerable amount of supervision, at least once per week for many months (either individual or small-group supervision). The evidence about the efficacy of relatively structured interventions applied by paraprofessionals (nurses are not usually professional mental-health–behavioral workers) clearly supports this idea (Durlak, 1979).

Clinical Case Examples

The writer and his colleagues selected four patients whom the nursing staff of a hemodialysis unit believed could not become more self-directed in their treatments. The cognitive-behavioral procedures described in the intervention section of this chapter were then tested (for a technical presentation of this work, see Kirschenbaum *et al.*, 1987).

The selection process involved asking nurses to identify patients who might be capable of becoming more self-directed but whom they believed would not be willing or able to work effectively toward that goal. The patients selected were perceived as being too old (ages were 63, 68, 75, and 76) or they were considered "not motivated." We then assigned each patient to a nurse who was in our training group. The training group met for several months before clinical trials and thereafter weekly for supervision for six months. Baseline assessments of self-directedness were made by having the nurse on duty complete the SCDC 25-item measure of self-directedness on each of these patients for four to six weeks (two to three times per week). These measurements were continued throughout the intervention phase as well.

Figure 3 presents the results of the pilot study with the four elderly hemodialysis pa-

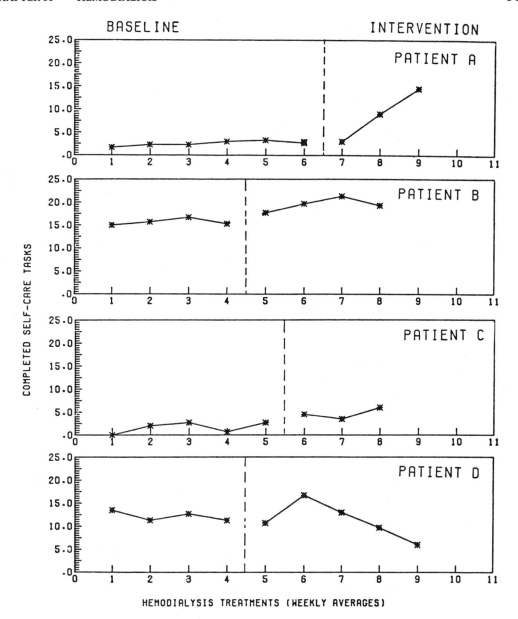

Figure 3. Mean number of self-care hemodialysis tasks completed per week from baseline through intervention phases.

tients. The length of the baselines was varied intentionally to make this a multiple-baseline across-subjects design. Thus, patients were expected to show improvements in the first week or so after the intervention was initiated in the average number of self-care tasks they completed per treatment (two to three treatments per week). The measures were taken by nurses who were not involved in the interventions and who did not know when the interventions had begun.

A review of the data in Figure 3 suggests that the interventions were effective for three of the four patients. Almost as soon as the interventions began, but not beforehand, each of the four patients improved in the number of self-care tasks completed compared to their respective average baseline rates. Improvements were apparent for all patients through the second week of intervention as well. Based on all of the assessments completed during the intervention phase (three to five weeks), the patients changed their average numbers of self-care tasks to the following extent: Patient A = +238%; Patient B = +24%; Patient C = +155%; Patient D = −8%. Although Patient B improved "only" 24%, it must be noted that she was averaging almost 16 of 25 tasks per treatment at baseline. The resulting ceiling effect meant that she could improve only a maximum of 56%. Patient D, who had improved initially but then declined to a level 8% below baseline, became very sick during the latter part of the intervention and died within one month after termination of the intervention phase. The chronic nature of this patient's complex medical problems were such that they clearly were not adversely affected by the intervention.

The efficacy of the cognitive-behavioral intervention was encouraging. Improvements were substantial and relatively immediate in the three other patients, who remained medically stable throughout the study. It must be emphasized that these effects were obtained even though the patients were on a unit that had previously encouraged self-directed treatment and even though the patients themselves were elderly and were considered unlikely to change by the unit's staff. Furthermore, the agents of change were not experienced cognitive-behavioral therapists, but nurses who were quite unfamiliar with the cognitive-behavioral principles used.

The reactions of the nursing staff and the physicians on the unit in which this study was implemented also encourage additional optimism. The patients who participated in the intervention phase seemed to benefit enormously in terms of mood and perceived well-being. Medically, no adverse effects were attributed to the interventions, and some improvements (e.g., in fluid regulation) were reported. However, it must also be added that the improvements shown did require a substantial amount of time from the nurses for training and supervision. Implementing a program like this may also create tensions among the staff and the patients who are not involved. However, once all staff are trained in the approach, these problems should diminish.

Summary

This chapter described end-stage renal disease (ESRD) as a very intrusive chronic illness that places enormous stresses on patients and their families. The physical sequelae of ESRD are chronic fatigue, sexual dysfunctioning, and a host of other very troublesome problems (e.g., frequent nausea, headaches, and muscle cramps). Most ESRD patients must use some version of hemodialysis, which adds to the intrusiveness of the illness (e.g., requiring four to seven hours of treatment three times per week in most cases). These stressors, in turn, result in substantial levels of anxiety, depression, and other psychological difficulties for dialysands and their families.

Clinical psychologists can provide many needed services to ESRD patients and their families, as well as to the medical personnel who treat them. Two assessment devices were described in this chapter that can assist psychologists in determining the specific types of problems experienced by patients and their families: the Dialysis Problem Checklists (Nichols & Springford, 1984) and the Self-Care Dialysis Checklist (Kirschenbaum et al., 1987). Psychologists can use these ESRD-specific measures as well as a host of more traditional assessment devices to examine anxiety, depression, suicidal tendencies, familial conflict, and other common difficulties experienced by this population at unusually high frequencies.

Treatment of the problems associated with ESRD may include a full range of clinical interventions, such as marital counseling, organizational development consultation with staff to improve communication and cooperation, and individual therapy to decrease chronic depression and anxiety. This chapter focused more specifically on cognitive-behavioral techniques that may help dialysands to manage their own treatments more completely and to improve their adherence to the restrictive fluid and dietary requirements associated with hemodialysis. The techniques described included introducing rationales for change, promoting effective decision making, establishing behavioral contracts, facilitating self-monitoring, using staff support and problem solving, and using self-instructional, stimulus-control, and related techniques. These procedures can be implemented by nurses and other non-mental-health professionals under the supervision of psychologists. In fact, a set of cases was presented in this chapter that showed the promise of this approach using a paraprofessional model. Of course, the use of such a systematic approach within a hemodialysis treatment unit has implications for the entire unit. Great care must be taken to involve key staff members in these types of systemwide interventions, and the maintenance of these approaches requires extensive consultation with the leaders of the units.

Psychologists can increase the probability of the continued growth of the field of health psychology (behavioral medicine and medical psychology) by involving themselves with ESRD patients and the medical units that treat them. This is a population that clearly needs and appreciates the kind of resources provided by clinical psychologists. Furthermore, many of the specific problems faced by ESRD patients lend themselves very well to measurement (e.g., self-care and management of fluid intake). Thus, a healthy and productive symbiosis can continue to develop between clinicians, researchers, and the medical staffs involved with ESRD patients.

Appendix: Counterrationalizations Pertaining to the Self-Care Decision*

Instructions: The intervenor (e.g., a nurse) is advised to review each of these potential "negative anticipations" with patients who are con-

*This material is reprinted with permission of the publishers of *Perspective: The Journal of the Council of Nephrology Social Workers*, 1985–1986, 7, 65–90, from an article by Joan D. Penrod and Daniel S. Kirschenbaum entitled, "Cognitive-Behavioral Techniques."

sidering increasing their degree of self-directed treatment. The examples below are taken from the negative anticipations from Table 1 and are written as if they were spoken by the intervenor to the patient.

1. *"Extra time to set up"*
"Setup" can be accomplished quickly, after a patient has experience with the task, sometimes within 20 minutes. Also, because a self-care patient will be put on first, the setup time can be completely made up.

2. *"More likely to make mistakes during training"*
Mistakes during training are part of the normal learning experience. Mistakes are anticipated by the training staff and special attention during this time ensures that such mistakes will be corrected. The training is designed so that the patient masters small steps slowly. Full independence is given only after the patient repeatedly and consistently demonstrates errorless performance.

3. *"Not watched by nurses carefully"*
The nurses do pay special attention to their self-care patients. In fact, it's like having a second pair of eyes because your nurse will be watching, and you will also be watching with "educated" eyes.

4. *"Can't sleep during treatment"*
Important to self-administering a safe treatment is an awareness of both your dialysis machine and your body. Sleep prevents such an awareness and impedes a good treatment. However, patients may relax more confidently when they know that they are receiving a safe treatment and that relaxation is an important component of treatment. Finally, staff-directed patients can't sleep either.

5. *"Increased responsibilities"*
Part of acquiring competence in any area of life means accepting responsibility. Accepting and meeting responsibility can give a patient a sense of control and satisfaction that can increase the quality of life. Doctors tell us that a sense of control has been shown to be associated with increased health. More important, these responsibilities can easily be met. Patients in their 60s and 70s and with no formal education have learned self-care, and these pa-

tients seem to have a renewed sense of confidence. When patients set up their own machine, they can arrange blood lines, clamps, and other things they need during dialysis just the way they want to (within safe guidelines).

6. *"Increased tasks" (see Item 5)*

7. *"Time needed to learn procedure"*

The mastery of any new task requires time. Something must be ventured if one is to gain. Many patients report that they enjoy learning about the machines that sustain their lives. Such learning gives a greater understanding of renal failure and its consequences. Also, during this time, the nursing staff provide much attention and support for this effort. They are eager to answer your questions and to have you become a partner in your treatment. Learning something new can give you a renewed sense of confidence in yourself.

8. *"Will think of disease more often"*

This may be true. But your attitude toward the disease will change. You will gain a greater understanding of its effects and its treatment, as well as a greater sense of control of your disease. Many report that they feel less frightened about their disease, and that their thoughts about the disease are more specific than the vague fears of the unknown they used to have.

9. *"Greater energy commitment" (see Item 5)*

10. *"Away from others because of increased setup time" (see Item 1)*

11. *"May look stupid while learning"*

Part of learning is making errors. Because the training staff understand that errors are part of the learning process, no one is evaluated for "looking stupid." No one has ever learned self-care without making a mistake or misunderstanding something. Mistakes and misunderstandings are expected. So, please don't worry about "looking stupid." In fact, it is our philosophy that making errors makes you eventually "look smart." Everyone learns at a different rate, and the staff appreciate this fact. The rate at which you learn self-care does not reflect your intelligence.

If other patients who do not understand the learning process comment that you are not learning fast enough, or that you look like a klutz, it must be recognized that they do not understand how people really learn. You can choose to ignore what others may say if you understand that learning is a highly individual process. Our only goal is that you learn. How you look and how slow or fast you learn or perform are irrelevant. What you learn and your confidence in what you have learned are our major concerns. In any case, others on the unit are really too busy to monitor closely your performance.

12. *"Partner feels less needed"*

Your partner will still be needed. You will need support and understanding from your partner, and these are needs that will not change. Your partner is invited to learn the self-care procedures with you so that she or he can share in your learning if you wish. Also, if your partner and you learn self-care, you may easily transfer to home care if you want to do that. In that case, your partner will be essential to your treatment. You can be assured that your partner will still fulfill important needs and perhaps some new ones as well, as you become a self-care patient.

13. *"Partner feels scared about patient's ability to handle new responsibilities"*

You can reassure your partner that your training will not be complete until you are an expert in self-care. The staff have trained many patients in self-care procedures and know when a patient is competent in all the ways necessary to ensure a safe and effective dialysis treatment. In fact, your partner will develop new respect for your abilities by seeing you take care of yourself in self-care. And the nursing staff still pay close attention to you during treatment to assure you a good treatment.

14. *"Decreased sense of security"*

Most self-care patients have an increased, not a decreased, sense of security. Knowing your machine, knowing how to trouble-shoot, and knowing how to respond to emergencies build confidence. You also learn how to read your vital signs, bodily signals that may indicate possible problems and what to do about them. Security is definitely enhanced. Remember, two people will be watching your treatment with expert eyes: you and your nurse. Two heads are better than one.

15. "Forces me to feel like a 'real dialysis' patient"

Of course, learning self-care is another reminder that you have kidney disease. Accepting your disease and exercising control in the treatment of it can positively influence your attitude about yourself and can ultimately improve the quality of your life. Not only will you feel that you are a "real dialysis" patient, but you will also feel that you are a "real" partner in the treatment of your disease.

16. "Not fair"

Our goal is to have everyone involved in self-care. You are being selected now because you seem like a good candidate for self-care, that is someone who can learn the task and set an example for others. Besides worrying about what others are doing, what are the personal advantages of self-care to you? That is the really important issue to deal with.

Other patients may have reasons—medical, physical, and/or mental—for not being self-care patients. These reasons are confidential information that we cannot share. Again, your concern should emphasize the advantages of self-care to you.

17. "Full-care patients may resent my enthusiasm"

In fact, most full-care patients respect the self-care patient. They admire your courage and willingness to accept control over your own treatment. The real concern is to evaluate any other persons's attitude in the context of the personal advantages you will experience.

18 and 19. "People will see I learn (perform) slowly" (see Item 11).

References

Abram, H. S., Moore, G. I., & Westervelt, F. B. (1971). Suicidal behavioral in chronic dialysis patients. *American Journal of Psychiatry, 127,* 1199–1204.

Barnes, K. T. (1980). Using transactional analysis to reduce staff stress in dialysis. *Dialysis and Transplantation, 9,* 9.

Binik, Y. M., Chowanec, G. D. & Devins, G. M. (in press). Marital role strain, illness intrusiveness, and their impact on marital and individual adjustment in end-stage renal disease. *Psychology and Health.*

Blodgett, C. (1981). A selected review of the literature of adjustment to hemodialysis. *International Journal of Psychiatry in Medicine, 11,* 97–124.

Colletti, G., & Brownell, K. D. (1982). The physical and emotional benefits of social support: Application to obesity, smoking, and alcoholism. In M. Hersen, R. M. Eisler, & P. M. Miller (Eds.), *Progress in behavior modification* (Vol. 13). New York: Academic Press.

Derogatis, L. R. (1977). *SCL-90 administration, scoring, and procedures manual.* Baltimore, MD: John Hopkins University Press.

Devins, G. M., Binik, Y. M., Hollomy, D. J., Barre, P. E., Binik, Y. M., & Guttmann, R. D. (1981). Helplessness and depression in end-stage renal disease. *Journal of Abnormal Psychology, 90,* 531–545.

Devins, G. M., Binik, Y. M., Gorman, P., Dattel, P. G., McCloskey, B., Oscar, G., & Briggs, J. (1982). Perceived self-efficacy, outcome expectancies, and negative mood states in end-stage renal disease. *Journal of Abnormal Psychology, 91,* 241–244.

Durlak, J. A. (1979). Comparative effectiveness of paraprofessional and professional helpers. *Psychological Bulletin, 86,* 80–92.

HCFA. (1984). *End-stage renal disease program medical information system facility survey tables: 1983.* HCFA Publication No. 03178, Bureau of Data Management, Baltimore.

Hoyt, M. F., & Janis, I. L. (1975). Increasing adherence to a stressful decision via a motivational balance-sheet procedure: A field experiment. *Journal of Personality and Social Psychology, 31,* 833–839.

Janis, L. L. (1959). Motivational factors in the resolution of decisional conflicts. In M. R. Jones (Ed.), *Nebraska Symposium on Motivation,* (Vol. 7). Lincoln: University of Nebraska Press.

Janis, I. L., & Mann, L. (1977). *Decision making: A psychological analysis of conflict, choice, and commitment.* New York: Free Press.

Kaplan DeNour, A. (1981). Prediction of adjustment to chronic hemodialysis. In N. B. Levy (Ed.) *Psychonephrology* (Vol. 1). New York: Plenum Press.

Karoly, P., & Kanfer, F. H. (Eds.). (1982). *Self-management and behavior change: From theory to practice.* Elmsford, NY: Pergamon Press.

Kirschenbaum, D. S. (1985). Proximity and specificity of planning: A position paper. *Cognitive Therapy and Research, 9,* 489–506.

Kirschenbaum, D. S. (1987). Self-regulatory failure: A review with clinical implications. *Clinical Psychology Review, 7,* 77–104.

Kirschenbaum, D. S., & Flanery, R. C. (1983). Behavioral contracts: Outcomes and elements. In M. Hersen, R. Eisler, & P. M. Miller (Eds.), *Progress in behavior modification* (Vol. 15). New York: Academic Press.

Kirschenbaum, D. S., & Flanery, R. C. (1984). Toward a psychology of behavioral contracting. *Clinical Psychology Review, 4,* 597–618.

Kirschenbaum, D. S., Sherman, J., & Penrod, J. D. (1987). Promoting self-directed hemodialysis: Measurement and cognitive-behavioral intervention. *Health Psychology*, *6*, 373–385.

Miller, R. C., & Berman, J. G. (1983). The efficacy of cognitive-behavior therapies: A quantitative review of the research evidence. *Psychological Bulletin*, *94*, 30–53.

Nichols, K. A., & Springford, V. (1984). The psycho-social stressors associated with survival by dialysis. *Behavior Research and Therapy*, *22*, 563–574.

Penrod, J. D., & Kirschenbaum, D. S. (1986). Cognitive-behavioral techniques. *Perspectives: The Journal of the Council of Nephrology Social Workers*, *7*, 65–90.

Procci, W. R. (1981). Psychological factors associated with severe abuse of the hemodialysis diet. *General Hospital Psychiatry*, *3*, 111–118.

Reed, H. D., & Janis, I. L. (1974). Effects of a new type of psychological treatment on smokers' resistance to warnings about health hazards. *Journal of Consulting and Clinical Psychology*, 42, 748.

Repp, A. C., Deitz, D. E. D., Boles, S. M., Deitz, S. M., & Repp, C. F. (1976). Differences among common methods for calculating interobserver agreement. *Journal of Applied Behavior Analysis*, *9*, 109–113.

Stuart, R. B. (Ed.). (1980). *Adherence, compliance, and generalization in behavioral medicine*. New York: Brunner/Mazel.

Tobin, D. L., Holroyd, K. A., Reynolds, R. V. C., & Wigel, J. K. (1989). The hierarchical structure of the Coping Strategies Inventory. *Cognitive Therapy and Research*,*13*, 343–361.

CHAPTER 31

Psychosocial Services for Persons with Human Immunodeficiency Virus Disease

Kathleen Sheridan

Introduction

The acquired immune deficiency syndrome (AIDS) is a vast, rapidly evolving phenomenon. An inherent difficulty in a book chapter on AIDS, no matter what its focus, is maintaining the timeliness and relevance of its content. Our understanding of the basic and clinical science components of AIDS is ever-expanding. Similarly, the complexities involved in developing vaccines and cures and, indeed, in providing health care for patients are sobering.

A second difficulty in dealing with this subject is balancing comprehensiveness and specificity. Clearly, a primary goal is to describe the role and impact of the clinical psychologist, and the contributions of the behavioral sciences and of psychology as disciplines and professions in providing mental health services to people with AIDS. However, that role defies narrow definition. Its interactive boundaries with the psychologist as researcher and the psychologist as educator are fuzzy. The multidimensional role for the clinical psychologist in AIDS is rich in its preparation and far-ranging in its reach.

As a researcher, the clinical psychologist has much to offer in developing prevention models against infection and strategies for measuring their efficacy (Baum & Nesselhof, 1988). The clinical neuropsychologist studies neuropsychological implications during infection, their assessment, and the impact of such deficits on functioning (Tross & Hirsch, 1988). Although the behavioral sciences have been active since the mid-1980s in these areas, only recently have they begun to empirically explore analyses of mood states and their interactions with the disease process (Atkinson *et al.*, 1988). And although calls for health services research have been clear and consistent (Intragovernmental Task Force, 1988), and predictions of the burdens of care on the provider systems and consumers frightening (Fox & Thomas, 1987–1988), thorough evaluations of programs for health care delivery are currently missing.

As an educator, the clinical psychologist possesses the skills in program development and evaluation to propose and test education models for groups at risk of infection, for particular communities that appear to be especially at

Kathleen Sheridan • Department of Psychology, University of Central Florida, Orlando, Florida 32816.

risk, such as young people of color and those traditionally underserved by the health care system, and for the public in general (Temoshok, Sweet, & Zich, 1987). The psychologist seems particularly poised to offer education to health care colleagues and community leaders (Lyons, Sheridan, & Larson, 1989).

As a practitioner, the psychologist has the requisite background in diagnostic assessment and intervention skills. Whether he or she wishes to concentrate at the one-to-one clinical level or at the program development level, the clinician can exert administrative leadership and multidisciplinary collaboration.

Finally, the author presumes that the reader has some current knowledge about AIDS issues. Thus, the chapter is not intended as a primary educational resource on AIDS. The history of AIDS as a disease entity has been brief, intense, and continually revised. This fact alone constitutes a significant challenge for the clinical psychologist. Thus, he or she must not only be up to date on applied psychology literature, but must also remain current with regard to the rapid diagnostic and treatment developments in the medical literature. To compound these difficulties, such powerful political, legal, and social climates surround AIDS that both patients and health care professionals feel acutely challenged.

The nomenclature and definitions concerning AIDS have changed frequently since its recognition in the United States in 1980–1981, and since the shared discoveries of human immunodeficiency virus (HIV) as its causative agent in 1983 ("New CDC Definition," 1987). In *Confronting AIDS: Update 1988*, the Institute of Medicine (1988) recommended a longitudinal or progressive view of infection. Thus, the current preference is to look at the entire spectrum of infectivity as HIV disease. HIV disease ranges from initial detection and confirmation of antibodies to infection, via enzyme-linked immunosorbent assays (ELISA) plus techniques such as the Western Blot,[1] through more-or-

less lengthy stages of infection without symptoms, to stages of infectivity with clinical symptoms that constitute diagnoses of AIDS (Institute of Medicine, 1988). Patients who are HIV-infected can be described as *seropositive*, *asymptomatic*, *symptomatic*, or *persons with AIDS*, depending on their medical status in the progression of HIV disease.

HIV disease has been accurately described as an infectious disease or as a sexually transmitted disease (STD) (Sheridan, 1989). Although it has often been referred to as a behavioral disease, this labeling has generated some controversy because, to some in the nonscientific and the activist communities, it conveys an implication of choice and blame for becoming infected. This implication in turn, they argue, exacerbates nonacceptance and lack of compassion from society in general. For the clinical psychologist, however, the relevance and appreciation of the behavioral or psychosocial components of HIV disease cannot be overstated. Probably no other disease entity carries with it such an essential need to understand lifestyles, interpersonal behaviors, adaptive and maladaptive coping styles, personality profiles, sociodemographic characteristics, and developmental, familial, and psychiatric (if any) histories.

At present, the geographical distribution of HIV disease in the United States is concentrated within four high-incidence clusters. In New York–New Jersey, the largest number of infected persons have contracted the disease via intravenous (IV) drug abuse. Florida's high-incidence population includes an early identified infected group of Haitian immigrants, IV drug users, and homosexual-bisexual persons. In California, San Francisco's infected population is mainly comprised of homosexual and bisexual males. Los Angeles, meanwhile, has seen a predominantly IV-drug-abusing group.

Epidemiologists debate whether the disease will spread from these and other intensely populated areas to smaller communities over the

---

[1]Currently, methods for the direct detection of HIV itself (as opposed to methods that detect the presence of an *antibody* to the virus) are under experimental test. One such promising test is the polymerase chain reaction (PCR) technique, which uses small blood samples and

multiplies their viral genetic components even before the body's manufacture of antibodies (Farzadegan, Polis, Wolinsky, Rinaldo, Sninsky, Kwok, Griffith, Kaslow, Phair, Polk, & Saah, 1988).

next decade (Morbidity and Mortality Weekly Report, 1987).[2] The shift in the rates of new infection from homosexual-bisexual males, where the rates of increase have leveled off, to IV drug communities, where the rates of infection have sharply increased, suggests that urban areas will remain the most critically affected in the near future. HIV also exerts an inordinate impact on persons of color (Mays & Cochran, 1987); this impact is compounded because many minority men and women infected tend also to be economically disadvantaged. As has been well documented, economically disadvantaged minority members of urban communities have insufficient access to and are underserved by the health care system in the United States.

A Model HIV Disease Services Program

Responses to the AIDS epidemic by health care communities grew from crises rather than from planning (Shilts, 1987). Since the epidemic's recognition in 1980–1981 in San Francisco and New York City, health care delivery systems have spun off each city's attempts to respond. The homosexual-bisexual community and health department officials in San Francisco aggressively responded to AIDS in the early years of crisis (though not without painful politics and turmoil; see Shilts, 1987). One of the renowned outgrowths of that activism was an integrated case-management health-care model known as the *San Francisco General model*.

New York City, on the other hand, was slower in its recognition of AIDS as a public health crisis and became quickly overwhelmed by IV drug and related pediatric infections. Its hospitals responded by offering traditional inpatient resources. Cost-effective studies began to demonstrate that any health care for patients would be costly; would heavily burden the already stretched delivery system resources; and would challenge the very foundations of health care

financing, private insurance, Medicare, and Medicaid (Fox & Thomas, 1987–1988). AIDS would represent an unplanned influx of consumers between the ages of 18 and 40 who normally would not seek access to health care. AIDS would also represent a demand for expertise, treatment, and cure that health care providers simply did not possess.

Preliminary (and early) cost-effectiveness estimates suggested that the San Francisco General (or hybrid) model was less expensive than traditional inpatient models. However, some recent factors have clouded the horizon considerably. The latency between initial HIV infection and the progression to symptomatology has lengthened to estimates of between 7 and 11 years (Institute of Medicine, 1988). Similarly, persons at risk of infection seem to be coming forth earlier and more willingly for antibody testing and counseling. The earlier detection of HIV status and better informed diagnoses are leading to longer life expectancies. Although few approved therapies are currently available—zidovudine (popularly, AZT), an antirival agent; aerosolized pentamidine to prevent *Pneumocystis carinii* pneumonia; and alpha interferon, shown to be effective against Kaposi's sarcoma—other approved therapies should regularly appear. On the other hand, if rates of infection are increasing among IV drug users, one could predict that such persons, unless enrolled in chemical dependence programs, would not routinely come forward for HIV antibody testing and counseling. Consequently, an IV drug user may not initially present to the health care system until he or she is symptomatic and, perhaps, seriously ill. His or her health care needs may be intense, whereas his or her ability to afford and sustain care may be very limited.

In effect, these potential and unpredictable factors force the health care planner to speculate cautiously, armed with precious few stable data to predict trends and changes. In view of our limited years of experience and the myriads of unanswerable questions, this section presents the components of a prototype service model. Such a model is likely to be implemented in medical centers or in smaller community hospitals.

[2]Another ongoing debate centers on the validity of the mathematical models being used in HIV prevalence projections for the next decade.

The model, ideally, is multidisciplinary in philosophy and approach, comprehensive in the services and referrals available, and ambulatory. Its bias is toward promoting home- and community-based care and resources over hospital-centered care. Thus, the model emphasizes significant liaisons with community agencies and citizen interest groups. These include chemical dependency programs, church groups, black and Hispanic coalitions, gay and lesbian groups, and school systems, for instance.

Although local and regional needs should ultimately shape what services are to be offered, a model program includes medical, psychological, nursing, social, dental, and, when possible, legal services. Such programs themselves ideally offer a full range of services or provide close links with programs that offer HIV antibody testing and counseling; primary medical care through a team management design; psychological evaluations; psychological interventions; and home-based and end-stage case-management programs. Table 1 presents the health, social, and legal services that can be included or provided through system links and referrals, as well as various liaisons with community agencies and interest groups.

In such a program, depending on the number of professional staff and resources available, and on whether the program is part of a teaching and clinical training facility, the clinical psychologist can choose various roles. The senior psychologist can be administratively responsible for the program, can act as clinical supervisor, or can function as a treatment coordinator. The clinical psychologist who is a direct service provider can best function as a regular member of the primary-care team. Typically, that team consists of the psychologist, an infectious disease physician, a clinical nurse specialist, and a social worker. Other members or consultants include a neurologist, an oncologist, a dentist, a lawyer, and a psychiatrist. In less elaborate systems with fewer resources, the clinical psychologist may act as a regular consultant and referral resource for, say, a primary-care physician's individual or group practice. Obviously, the psychologist may join such a group practice.

The clinical psychologist as a member or consultant within this model not only provides or supervises patient services, he or she directly assists colleagues in other participating disciplines. At the outset, the clinical psychologist should be a part of the interview team for anyone who seeks employment in the program or (unless an organized volunteer program exists) who volunteers to work with HIV-infected patients. Given the very sensitive, complex psychological and social issues surrounding AIDS, a very careful screening of the motivation and the general psychological status of health care professionals and allied staff who work in this area is essential.

During the implementation of program procedures and goals, the psychologist can act as a resource to help colleagues deal with personal issues of stress, burnout, and other reactions that occur when working with seriously ill and

Table 1. Components of a Model Biopsychosocial Services Program for HIV Disease

Services offered	Participating disciplines	Consulting disciplines	Liaisons
HIV antibody testing and counseling	Infectious disease	Neurology	Chemical dependence
	Clinical psychology	Pulmonary medicine	Home-based case management
Primary medical care	Oncology	Psychiatry	
Psychological services	Nursing	Dermatology	Hospice
Social services	Phlebotomy	Hematology	Inpatient units
Dental services or referral	Laboratory technology	Pharmacy	Hemophilia program
Legal services or referral	Social work	Obstetrics/gynecology	Health departments
	Dentistry	Pediatrics	Community agencies
	Law		Community action groups

potentially terminally ill patients. Psychologists can help colleagues anticipate their emotional reactions to patients who are routinely younger than most adult medical patients (and many providers) and who do not always enjoy "unconditional positive regard" from family, friends, health care providers, and society in general. However, when the psychologist is an integral member of the program, as administrator, supervisor, and/or health care provider, he or she cannot play the dual role of team member and outside consultant to the team. When the latter is the case, at least the psychologist can be relied on to recognize staff issues early; to encourage and participate in staff support groups, retreats, or other initiatives that would help improve morale; to prevent debilitating stress; and to maintain quality service delivery to patients.

Ideally, the program will serve as a clinical training site. Whether part of or affiliated with a teaching hospital, any model program at this point will still be regarded as innovative, if not pioneering. One of the true deficits in HIV service delivery is the availability of knowledgeable, motivated health care professionals. Such deficiencies result from the newness of the disease and continuing scientific uncertainty regarding vaccines and cures.

Equally important are the psychological and social implications that are raised for persons wishing to specialize in AIDS. Colleagues in basic sciences and clinical medicine may caution that such specialization creates suspicion about one's sexual orientation or, perhaps, involvement with alcohol or drugs. It is not unusual for colleagues to assume that one who works with HIV-infected individuals has a relative or close friend who has AIDS. Specializing in AIDS service delivery also promises daily contact with primarily young adult patients with very truncated life expectancies. Such career prospects may be daunting to even the most idealistic trainee.

The behavioral sciences and, in particular, clinical psychology have a great challenge and responsibility to prepare practitioners to work with these patients. Some within medicine and basic science currently welcome behavioral science input because they acknowledge that prevention and education are now our only effective weapons. They become extremely shortsighted in their acceptance, however. Even if, for example, a vaccine and cure were developed tomorrow, the psychosocial and neuropsychological problems of persons who are now infected or who have AIDS would not disappear. At the same time, however, psychologists, as well as their psychiatry and social work colleagues, contemporaries, have not yet led the vanguard in clinical training.

Underwriting clinical services is an exercise in creativity and ingenuity that, in every instance, demands an intimate familiarity with exploiting local systems and resources. In general, program support is an amalgam of institutional start-up funds, fees for services, and private and public monies, plus community input and volunteerism.

Because this chapter focuses on the special abilities that the clinical psychologist brings, it also highlights the ways in which he or she may attract research and training funds to support programming. Most important and most needed are cost-effectiveness studies to compare the outcomes of traditional inpatient services and hospice, ambulatory, and home-based case-management systems. Both the Health Resources and Services Administration (HRSA) and the National Institute of Mental Health (NIMH) are potential funding resources for such proposals. Some private foundations—for instance, the Robert Wood Johnson Foundation—have also supported these efforts.

Other opportunities for specific studies include clinical drug trials, supportable by pharmaceutical companies or through the AIDS clinical treatment units under National Institute of Allergy and Infectious Disease (NIAID) auspices; psychological intervention studies (NIMH); special population studies, for instance, of chemical abusers (National Institute for Drug Addictions—NIDA); and special education and service target groups, such as minority teenagers (CDC).

If the program is university-based or -affiliated, NIMH and NIAID offer support for post-doctoral training for behavioral and clinical scientists in multidisciplinary settings for HIV disease. Clinical psychologists can work with departmental directors of clinical training to develop practicum placements for graduate students and predoctoral residency rotations. Many systems funnel back to the program service fees generated by training activities.

Clinical Issues

Exactly who is the consumer of HIV-related services poses questions. HIV *affects* many more people than it *infects*. And the clinical psychologist encounters all of them. He or she works with persons whose behaviors put them at risk for infection; families, friends, and lovers of those infected; the "worried well," including those who work with infected persons, those who are worried that loved ones may be engaging in risk behaviors, those who have children in school with infected youngsters, and those whose own behaviors (although not in themselves risks for infection) cause enough guilt or anxiety to bring them in contact with a mental health professional; and colleagues in the health care professions who work with infected patients. Finally, the psychologist works with HIV-infected men, women, and children.

Persons whose behaviors may put them at risk for HIV infection include homosexual and bisexual males, particularly those who practice unprotected anal intercourse regularly and with anonymous partners; intravenous drug abusers who share, rent, or reuse unsterilized syringes; and persons who had blood transfusions before 1985 (when routine screening of the nation's blood supply for HIV antibodies was introduced), such as hemophiliacs and surgery patients. Others at risk include women who engage in unprotected heterosexual intercourse with partners who are bisexual and/or who use IV drugs, as well as infants whose mothers are HIV-infected when they give birth.

Infected persons or those at risk may approach a clinical psychologist at several times along the HIV disease continuum. At the threshold of the continuum, a person may be concerned about the reliability and safety of his or her sexual[3] or drug practices. Someone may want to talk about whether to undergo HIV antibody testing, trying to weigh the benefits of knowing that information versus "not knowing." This balancing is often accompanied by worries about "If I get tested now and I'm negative, how can I really be sure," particularly if the individual engages in risky behaviors. He or she also needs assurances about confidentiality and any disclosure requirements in the event tests are positive.

In large-scale programs, it is unlikely that the psychologist will directly provide pre- and post-antibody-test counseling. More often, such counseling is provided by nurses, social workers, clinical trainees, or specially designated staff. It is the psychologist's job to provide professional oversight and ongoing supervision to ensure that these staff members will be attentive to and will deal with the emotional variables that affect a patient's ability to take in the information presented and to make informed decisions based on that information.

Once a person's positive serostatus is confirmed, he or she may remain asymptomatic for long periods. Currently, the average incubation period between HIV infection and symptom development ranges from 7 to 11 years. No definitive data suggest what cofactors predict who will go on to develop AIDS versus those who will remain asymptomatic. At the same time, it is unclear whether all those who are HIV-infected will eventually progress to AIDS (Institute of Medicine, 1988). During this asymptomatic period, however, the patient should be encouraged to remain in regular contact with a primary-care physician (in smaller communities or where independent practitioners predominate) or with a primary-care team (more

---

[3]Because this chapter is not an AIDS primer, it does not detail the unprotected sexual practices that enhance the efficiency of HIV transmission. However, the author no longer assumes that clinical psychologists routinely take sexual histories from either heterosexual or homosexual clients during evaluation. Such history taking is critically important to assessing at-risk status and to counseling on safer sex practices and behavior changes when indicated. This observed omission seems to be a prevalent failure in psychological assessment as a general practice.

typical of the model program presented above and large medical centers).

Patients are encouraged to maintain this regular link for several reasons. The effect of HIV infection on the body's immune system, even in the absence of clinical symptomatology, can be monitored by measuring the relative number of T-helper lymphocytes (the cells primarily attacked by the virus) present in the bloodstream. For instance, the number of T-helper cells present in the healthy individual ranges from 800 to 1200. In the HIV-infected person, levels commonly fall below 800, sometimes falling below 200 and 300. In fact, in prospective studies, T-helper-cell levels below 300 have been prognostic of AIDS diagnoses (Polk, Fox, Brookmeyer, Kanchanaraksa, Kaslow, Visscher, Rinaldo, & Phair, 1987). T-helper-cell levels are also used as central inclusion criteria for participation in clinical treatment drug trials sponsored by the National Institute of Allergy and Infectious Disease's 35 AIDS Clinical Trials Units (ACTU) (J. P. Phair, personal communication, 1988). Thus, asymptomatic people can choose to participate in carefully controlled drug studies designed to provide more effective prophylaxis and treatment. Not incidentally, however, asymptomatic people, when faced with the opportunity to participate in drug trials, often benefit from discussing their motivations and options with a psychologist.

This asymptomatic period, although potentially lengthy, can be a period of heightened anxiety and depression. In fact, as HIV-infected persons are living longer without developing symptoms, this latent period has the potential for spawning a variety of psychological problems. The infected person may feel himself or herself to be under a "slow sentence," unable to hope and plan, vigilant and fearful about infecting sexual partners, continually attentive to (if not preoccupied by) physical health, and very helpless at times.

It also represents a period during which some individuals, using their HIV serostatus as precipitants, attempt to deal with chronic family estrangements over choices of sexual or drug-using lifestyles; make decisions about which persons in their lives they choose to inform of their serostatus; struggle with changes they have made or are making in risk behaviors; reconsider future plans (in light of their health status) for life situations, including education, jobs, geographic location; and prepare for death, including estate planning, drafting living wills, and conveying durable powers of attorney.

When a person begins to experience physical symptoms that affect school or job performance, the symptoms may warrant changes in lifestyle from more to less independent functioning. This onset of symptomatology is often paralleled by depressive feelings over loss of autonomy, fears about the future, acute bouts of death anxiety, concerns over changes in body image, and changes in self-concept.

Constitutional symptoms such as vomiting, diarrhea, night sweats, and weight loss tend to be transitory but may be psychologically debilitating. Patients view these symptoms (often accurately) as presages of more serious illness, so that anxiety and stress are heightened. The first signs of illness often become "tests" for a lover or spouse and others in the patient's support network to "pass" as evidence that they really care. Thus, an initial illness may be accompanied by angry interchanges in relationships, plus needs for support from uninfected, affected people.

Hospitalizations with *Pneumocystis carinii* pneumonia (PCP), which represents approximately 60% of all conditions diagnostic of AIDS ("Complete Coverage of HIV," 1988), are frightening, exhausting experiences. Although advances in clinical medicine have resulted in shorter hospital stays and recovery from PCP bouts, the realization of the seriousness of one's condition, the dependency and helplessness engendered by hospitalization, and intensified worry about the future cannot be minimized.

Kaposi's sarcoma (KS) affects some 12% of those diagnosed with AIDS ("Complete Coverage of HIV," 1988). It is a cancer of the capillaries that leaves purplish blotches on the skin, and its course tends to be longer and without the acute episodes of PCP. At the same time, progressive discoloration and lesions on the skin create serious concerns about body image and attractiveness that may lead to the social withdrawal and isolation of the patient. On the other hand, friends and colleagues may flinch or recoil at the sight of cancers and may worry (inaccurately) that touching the patient will lead to infection.

AIDS dementia complex (ADC) is also an AIDS-defining illness. The HIV virus directly infects monocytes and macrophages, including those in the central nervous system (CNS). Early autopsy investigations (Navia, Cho, & Petito, 1986) documented a high percentage of CNS damage in persons diagnosed with AIDS. These early reports led to increased concerns about the nature of CNS impairment in AIDS, its frequency, and its onset. ADC was tentatively defined as a subcortical dementia affecting concentration, attention span, short-term memory, and some motoric responses (Navia, Jordan, & Price, 1986). Next, it was suggested that subtle signs of cognitive decline may precede other symptoms in the progression of HIV disease. Frequencies of ADC were reported as ranging anywhere from 8% to 66% of all AIDS cases (J. A. McArthur, personal communication, 1988). The need for careful differential diagnosis between cognitive deficit and depression, especially in the early stages of HIV infection, has been noted (Holland & Tross, 1985).

Since this early rush of conclusions, behavioral scientists have taken the opportunity and time to stand back and assess the data and methodologies on which these statements were based. Autopsy reports, small study samples,[4] lack of prospective designs, conflicting results with different neuropsychological assessment batteries—all combined to generate preliminary results, results that must be subjected to replication and continued, rigorous empirical test.

[4]Most seropositive samples are also composed of homosexual volunteers. In many studies, these volunteers represent a skewed sample in terms of higher than mean educational, occupational, and salary status. Arguably, they may also be subjects who compensate effectively for subtle cognitive deficit, if it exists. Whether the results would be similar with, say, seropositive IV drug abusers is an open question.

That AIDS patients suffer episodes of dementia is clear. What are the parameters of ADC? How can it be assessed? What is the frequency of cognitive impairment in HIV disease? When does it begin? We cannot say with certainty, although studies are progressing. In preliminary results from the Multicenter AIDS Cohort Study (MACS), a natural history study, comparisons on neuropsychological tests of 836 seronegative homosexual males and 819 seropositive asymptomatic homosexual males matched on education and age showed no differences in cognitive functioning (Selnes, Miller, Becker, Cohen, McArthur, Visscher, Gordon, Satz, Ginzburg, & Polk, 1988).

End-stage illness in HIV disease can occasion some recurrence and reworking of psychological conflicts that heretofore had reached a level of satisfactory resolution. Problems with autonomy and dependency heighten, including further relinquishment of independent activities. More people in the patient's social and work spheres become aware of his or her serious illness. More time, energy, and financial resources are focused on medical care.

As the concept of death anxiety is extremely complex, the reader is referred to the psychological literature on terminal illness for a thorough analysis. For our purposes, it is extremely important to note that the dynamics of death anxiety often show themselves not only in the HIV-infected patient, but in close friends and family. The psychotherapist should recognize that his or her own unresolved issues regarding death may influence countertransference. Such conflict may show itself in a subtle, but definite, withdrawal from the patient, particularly as he or she becomes more feeble, loses weight, and slows in speech, manner, and gait. If unrecognized, such behavior by the therapist can actually be unwittingly supported by the service delivery system. For example, if the patient is ambulatory but very weak, he or she may be encouraged to keep medical appointments, but it may be "understandable" if the patient skips group or individual psychotherapy. Or if the patient is hospitalized, the psychotherapist may give in to the intrusive bustle of inpatient unit scheduling and may not arrange for regularly scheduled psychotherapy sessions.

A patient's death brings a multitude of awarenesses to those affected by it. Ideally, a patient's significant others have had opportunities to prepare themselves emotionally for the loss and its enormity. In more conflictual relationships and when death rapidly follows diagnosis, reactions tend to be more tumultuous. Grief and bereavement processes extend well beyond the point of death, and the psychologist can be helpful to family, lovers, and friends during this period.

Once more, the psychologist who as psychotherapist has known and grown with the patient must acknowledge the effects of the death on himself or herself. Loss, mortality, and inevitability become striking. At another level, the psychologist may begin to question whether the agony and struggle of working through emotional difficulties are not trivial when the patient is facing a not-so-remote death. In confronting the loss of life, then, the psychologist faces the value of his or her own life.

Some synthesizing of clinical issues seems to be in order here. Clinical medicine continues to improve in the earlier detection and diagnosis of HIV disease, and more therapies for symptomatology are becoming available. The clinical psychologist, as a service provider, will encounter increasing numbers of HIV-infected, asymptomatic people who will seek assistance. Their predominant issues will focus on ways to relate to and to cope with, and to integrate the fact of a latent, potentially very serious medical status into a productive, satisfying life. Finally, Table 2 presents *some* psychological issues that may appear during various stages of HIV dis-

Table 2. Some Possible Psychosocial Issues and Resolutions during HIV Progression

Stages of HIV infection	Psychological issues	Adaptive techniques
Pre-HIV-antibody testing	Anxiety	Psychological consultation
	Guilt	
Confirmation of positive serostatus		
Immediate	Shock	Post-HIV-test counseling
	Denial	Learning about HIV
	Depression	Planning
	Suicide potential	Support group attendance
Asymptomatic period	Who to tell	Psychological assessment
	Anxiety	Psychological intervention
	Depression	
	Anger	
	Relationship conflicts	
Symptomatic period (early)	Fear	Neuropsychological evaluation
	Enhanced anxiety	Relationship resolutions
	Depression	Estate planning
	Suicide potential	Living will
	Neuropsychological symptoms	Durable power of attorney
	Death anxiety	
Symptomatic period (advanced)	Body image changes	Primary caregiver
	Self-concept changes	Support manager
	Dependency/independence	Rapprochement with family
	Relationship conflicts	
	Cognitive decline	
Death	Acceptance by patient	Dying with dignity
	Loss	Grief counseling
	Anger	Bereavement group
	Mourning	
	Acceptance by survivors	

ease. However, the reader is cautioned not to regard this listing as definitive or comprehensive. It is intended only as an example. Indeed, persons with HIV disease are individual and unique in their personalities, their emotional lives, and their attempts to deal with stress and illness.

Special discussion of psychological assessment and intervention appear below. Before those sections, however, are two case studies. The value in their presentation is both for their representativeness and for their uniqueness within HIV disease. Although neither case is discussed at length, each presents the reader with several clinical dilemmas for consideration.[5]

Case 1

A 28-year-old gay white male was referred for suicide potential assessment. This young man had been a regular participant in an epidemiological study of HIV disease for over five years. Early in his participation, he had opted not to be told his serostatus. More recently, as the research policy encouraged participants to find out their serostatus, he had asked for his results and had expressed no surprise that he was HIV-positive. He mentioned to the research staff member who discussed the antibody test results with him that, if he became ill, he had made plans to end his own life.

As part of the ongoing research, this man had undergone a series of psychological and neuropsychological tests, including the Centers for the Epidemiological Study of Depression (CES-D) and Beck Depression scales; the Spielberger State-Trait Anxiety scale; the WAIS digit symbol, block design, and digits forward-backward scales; Trials A and B; and Verbal Fluency. At six-month intervals, he had tested within normal ranges on all measures. On interview he was alert, appropriate in affect, and responsive in conversation. He spoke animatedly about his middle-management position, his love relationship of two years, and the support he felt from his brothers, sister, and father. His mother was deceased. He reported no psychiatric history and no drug or alcohol problems before learning his HIV status.

---

[5]The entire and alarming area of pediatric AIDS and its effects on maternal-infant interaction and child development are beyond both the scope of this chapter and the author's expertise.

When asked about how he was coping with his health status, he readily distinguished between being asymptomatic and being sick. If he became so ill, he explained, that he could no longer function independently and make his own decisions, or, if he experienced continuous, intense pain, he would end his life. He had already prepared a living will and a durable power of attorney. He spoke of his future plans with quavering voice and a fearful expression.

With no other counterindicative details, the determination was that the patient was not imminently suicidal. His ongoing commitment to the research project and the description of his support system buttressed this determination. The research team and other members of the primary-care staff were alerted to his psychological state, particularly with a view to the time when his physical condition deteriorated and symptoms appeared. This young man was also encouraged to join a group of HIV-infected people who met weekly to discuss their feelings, fears of becoming ill, and ways of coping with the disease process.

Case 2

A group of five seropositive men and two cotherapists was entering its 12th month. Over the year, one or two members had dropped out and had been replaced by new members, but the group had remained largely intact and generally physically healthy. After a lengthy initial period of establishing trust, of feeling safe enough to attend regularly, and of recognizing the value of discussing all their feelings (positive and negative), the group settled into meaningful discussions of their concerns.

At the end of 10 months, one cotherapist announced that he would be leaving at the end of the year (and that a new cotherapist would be joining them shortly after). Next, at about 10½ months, a member announced he was leaving, satisfied that he had got from the group as much as he could. The next weekend, another member, without any warning, developed PCP, was hospitalized, and died within five days, but not before being visited by all the group members. The group therapists were enlisted "on the spot" by his family members, who had come from out of town, unprepared for the crisis, to help them deal with the patient's illness and death. During the next two weeks, two additional members died. One death was again sudden; the other member's death seemed less surprising because he had come into the group as the person who had been infected longest.

By the end of the year, one cotherapist and one member remained. Complex issues on many levels needed attention, emotional processing, thought, and time. At the programmatic level, the entire staff was left with a variety of losses and reactions: the leaving of a staff member (the cotherapist), the unexpected deaths of two patients, and the death of the third patient—and all within a very short time. In supervision, the remaining cotherapist struggled to deal with the juxtaposition of a successful group process and the loss of group members. Sorting out the reality and meaning of one termination, three deaths, and a cotherapist's leaving, and "surviving" all that, meant long and intense work for him and the supervisor.

At the clinical level, how to carry on the integrity of the the group and how to prepare for new members, including a new cotherapist, were issues of importance. Not the least important was the novelty of this clinical dilemma. At least within the system, no clinician had had such an experience, even those who worked with other chronic illnesses.

Psychological Assessment

Individuals with HIV disease may present to the clinical psychologist in the following ways: referral from another member of the primary-care team; referral from an inpatient attending physician; referral from social service department; referral from chemical dependence programs or other community agencies; or referral by self, family, or close friends. Presenting problems include differential diagnosis (e.g., depression vs. neuropsychological impairment), suicide potential, stress and anxiety levels, impulsivity and aggressiveness, and chemical dependency. Other reasons for consulting a psychologist include concerns about family response to the patient's illness, relationship problems between patients and spouses or lovers, and patients' concerns about important decisions they want to make during the progress of HIV disease.

Depending on the nature of the referral and the presenting problems, the psychologist should be equipped (either personally or via further referral) to conduct a thorough clinical interview, including a mental status examination, suicide assessment, and chemical dependency assessment. In addition, the examiner should have available the resources to conduct intelligence test(s) and a preliminary neuropsychological evaluation. The psychologist should gather information on intimate social and family relationships, and on coping styles. The psychologist should be thoroughly familiar with the person's current medical status and history. The psychologist should also be prepared to recommend psychotropic medication evaluation, for instance. When indicated, he or she should be ready to refer for more intensive neuropsychological evaluation. With the infected person's consent, the psychologist should also be willing to interview a lover, spouse, or other family member who may be integral to planning.

To date, psychologists experienced in providing services to HIV-infected persons have worked primarily with homosexual and bisexual men, and with people infected via blood transfusion, for instance, hemophiliacs. Fewer are experienced with people infected via IV drug abuse. Many psychologists have never worked with people of color who are infected.

It is difficult to estimate, then, the frequency of infected people's pre-AIDS contacts with the mental health system. Certainly, some generalizations seem justified. Among many persons infected via transfusion or intimate sexual practices, dealing with the HIV virus represents the first time they have encountered a clinical interview and assessment. Some persons, particularly homosexual and bisexual men, have had prior psychotherapy histories, typically for reasons of personal growth and to make decisions about alternate lifestyles. Fewer present histories of psychiatric hospitalization or frequent use of psychotropic medication.

Among IV drug abusers, many are referred by chemical dependency programs and have prior psychiatric histories. IV drug abusers who are not in methadone maintenance or treatment programs often do not come to the attention of HIV-related services until they are acutely medically ill; with such patients, psychological assessment and chemical dependency evaluation are delayed until the medical crisis is past.

Many people of color turn to family, neighbors, ministers, and priests for counsel about HIV infection. Thus, referral to a psychologist may be limited to someone who shows striking signs of depression or tension, symptoms that would seriously concern or frighten loved ones. Some homosexual and bisexual men of color experience emotional battles between their sexual behaviors (often closeted) and the cultural expectations of their community. Sadly, such struggles often simmer untended until the infected man becomes physically ill.

It is further difficult to estimate the suicide rate among HIV-infected people. Statistics are sparse and probably difficult to interpret (Flavin, Franklin, & Frances, 1986). In addition, evaluating suicide potential is not a onetime event for this population. The problem is not static. The likelihood of suicide evolves over time in tandem with changes in an individual's physical health, coping styles, and personal relationships. Although people typically experience periods of shock and denial when positive HIV antibody statuses are confirmed (Sheridan & Sheridan, 1988), others do talk about and, indeed, attempt suicide. Still another may become terribly bereft and suicidal over an important person's negative response to the revelation of his or her HIV status. Yet others may become despondent over losing a job because of physical disability or over failing to tolerate zidovudine or during a steady physical decline.

Finally, many infected people see issuing do-not-resuscitate instructions as planning some measure of control over when and how their lives will end. Although they do not prepare to activate plans for suicide at a certain future point (as an alternative to being hospitalized, for instance), they instead clearly provide that, after a certain point or limit, they wish no further extraordinary medical measures to prolong their lives.

Psychological Interventions

No modern-day clinical psychologist can feel expert in providing the varieties of therapeutic interventions that may be useful. At best,

he or she can be adept at matching the needs of those infected and affected by HIV with well-suited interventions. The psychologist should not rule out any modality, including individual, group, couples, and family psychotherapy; stress management; progressive relaxation; biofeedback; and hypnosis. Bereavement counseling for survivors is also important.

There are some guidelines, however. First, many patients, being familiar with the medical atmosphere, initially view psychological interventions as time-limited and issue-oriented. Relatedly, HIV-infected people who opt for multidisciplinary services programs, such as those described above, are very active participants in every phase of their care. Just as they seek second and third medical opinions and become conversant with alternative drug therapies, they will variously seek out medication, acupuncture, massage, aerobic exercise, and other options.

That a psychotherapist must be flexible, no matter what his or her preferred modality, cannot be overstated. A couples psychotherapist often works with homosexual and heterosexual partners. The "family" in family therapy is often uniquely defined; it may include, for instance, the patient's lover, the patient's sister and her husband, plus the patient's aunt. These may constitute the "family" who accept the homosexual relationship or who know the patient is ill.

In dramatic contrast to a classically trained psychoanalyst, for instance, the psychotherapist for an HIV-infected person has a much more complicated role. At times, the therapist gives advice or direct suggestions: "I think you should be tested" or "You and your wife must use condoms." Or he or she indicates the material to be dealt with in therapy: "Have you made out you will?" or "Why are you not reporting your weight loss to the physician?" And, more traditionally, the therapist helps the person to reach his or her decisions or to deal better with anxiety and stress.[6]

---

[6]For a discussion of the psychotherapist's "duty to warn" when an infected person reveals an intent to transmit the virus to an unknowing third party(ies), see Peter and Sanchez (1987).

The popularity of group interventions with HIV-infected people emanates from two sources. Generally, group interventions have proved successful when oriented toward certain medical illnesses (Roback, 1984). Groups that are prescriptively designed and well focused seem to enhance the development of illness-adaptive coping styles and quality social supports among members.

Specifically, the use of volunteer-led, community-based support groups for HIV-infected people was an immediate success in San Francisco and New York. Initiated by homosexual and bisexual communities in the early 1980s, groups of prominence included the Shanti Project in Berkeley, California, and those run by the Gay Men's Health Crisis in New York and by the Howard Brown Memorial Clinic in Chicago, Illinois.

These groups have been led over the years by volunteers (some professionally trained) and paraprofessionals. Although infected persons have appreciated the optimistic, activist, self-help aspects of such groups as being very valuable, other observers (members, leaders, and sponsoring organizations alike) have worried about the difficulties that such groups have in surviving more complex dynamics, such as the deaths of members and the raising of "negative" affects, like depression and anger.

Clearly, then, the clinical psychologist who is expert in group dynamics can provide assistance. First, psychologists can help answer what are the parameters of support groups, who would benefit, who will lead them, and how leaders should be trained. Then, they can define the parameters of professionally led intervention groups, the criteria for membership, and leadership training.

Summary and Conclusions

HIV disease represents a continuum ranging from the initial detection of infection, through a long incubatory and asymptomatic period, to symptomatic states and a diagnosis of AIDS. Currently, there are no cures or vaccines. Education and prevention efforts have shown some success in decreasing the rates of infectivity among homosexuals and bisexuals. The rates are increasing among drug-abusing populations. Early detection and diagnosis plus rapid treatment developments have combined to prolong the lives of infected persons. The psychosocial components of HIV disease have been acknowledged. The extent of the neuropsychological implications of infection is the topic of enormous research activity.

Health care services for infected persons have received only recent attention. Attention to the role of clinical psychology and its practitioners in the provision of comprehensive services is even more recent. Demonstration projects, case management, and services evaluation studies are under way.

The clinical psychologist in a medical setting is poised to play a major role in service delivery, whether at the level of program design, administration and evaluation, clinical training and supervision, or clinical encounters.

Most elements considered important to the delivery of psychological services to HIV-infected people are based on clinical observation and are ripe for empirical verification. Much about the elements relevant to the delivery of such services to underserved communities, to people of color, and to IV-drug-using patients remains to be developed. Some elements in psychological service delivery remain dependent on scientific and clinical medicine advances in HIV disease. The foci and the concentration of psychological services research will need to adapt and change.

The urgency of behavioral science research in HIV-disease service-delivery cannot be overstated. Case-management cost-effectiveness studies, neuropsychological epidemiology, work with chemical abusers, and outcome research on psychological interventions are all essential and can be effective only at vital clinical sites for HIV disease.

The opportunity, the motivation, the challenge, and the reward await the clinical psychologist who wishes to specialize in HIV disease. Planning, providing, and participating in services for infected people are humbling, and trying and, above all, are expressions of faith in the celebration of life.

References

Atkinson, J. H., Grant, I., Kennedy, C. J., Richman, D. D., Spector, S. A., & McCutchan, J. A. (1988). Prevalence of psychiatric disorders among men infected with human immunodeficiency virus: A controlled study. *Archives of General Psychiatry, 45,* 859–864.

Baum, A., & Nesselhof, S. E. A. (1988). Psychological research and the prevention, etiology, and treatment of AIDS. *American Psychologist, 43,* 900–906.

Complete coverage of the human immunodeficiency virus. (1988). Washington, DC: AIDS/HIV record.

Farzadegan, H., Polis, M. A., Wolinsky, S. M., Rinaldo, C. R., Sninsky, J. J., Kwok, S., Griffith, R. L. Kaslow, R. A., Phair, J. P., Polk, B. F., & Saah, A. J. (1988). Loss of human immunodeficiency virus type 1 (HIV-1) antibodies with evidence of viral infection in asymptomatic homosexual men. *Annals of Internal Medicine, 108,* 785–790.

Flavin, D. K., Franklin, J. E., & Frances, R. S. (1986). The acquired immune deficiency syndrome (AIDS) and suicidal behavior in alcohol-dependent homosexual men. *American Journal of Psychiatry, 143,* 1440–1442.

Fox, D. M., & Thomas, E. H. (1987–1988). AIDS cost analysis and social policy. *Law, Medicine & Health Care, 15,* 186–211.

Holland, J. C., & Tross, S. (1985). The psychosocial and neuropsychiatric sequelae of the acquired immune deficiency syndrome and related disorders. *Annals of Internal Medicine, 103,* 760–764.

Institute of Medicine. (1988). *Confronting AIDS: Update 1988.* Washington, DC: National Academy Press.

Intragovernmental Task Force. (1988). *AIDS health care delivery.* Washington, DC: Health Resources and Services Administration.

Lyons, J. S., Sheridan, K., & Larson, D. B. (1989). An AIDS educational model for health care professionals. *Health Education, 19,* 12–15.

Mays, V. M., & Cochran, S. D. (1987). Acquired immune deficiency and Black Americans: Special psychosocial issues. *Public Health Reports, 102,* 224–231.

Morbidity and mortality weekly report. (1987, December 18 Suppl.). *Human immunodeficiency virus infection in the United States: A review of current knowledge.* Atlanta, GA: Centers for Disease Control.

Navia, B., Cho, E., & Petito, C. K. (1986). AIDS dementia complex: II. Neuropathology. *Annals of Neurology, 19,* 525–535.

Navia, B., Jordan, B. D., & Price, R. W. (1986). AIDS dementia complex: I. Clinical features. *Annals of Neurology, 19,* 517–524.

New CDC definition for AIDS. (1987, October). *Medical aspects of human sexuality, 21,* 35–39.

Peter, A. P., & Sanchez, H. (1987). The therapist's duty to disclose communicable diseases. *Western State University Law Review, 14,* 465–478.

Polk, B. F., Fox, R., Brookmeyer, R., Kanchanaraksa, S., Kaslow, R., Visscher, B., Rinaldo, C., & Phair, J. P. (1987). Predictors of the acquired immune deficiency syndrome developing in a cohort of seropositive homosexual men. *New England Journal of Medicine, 316,* 61–66.

Price, R. W., Sidtis, J., & Rosenblum, M. (1988). The AIDS complex: Some current questions. *Annals of Neurology, 23* (Suppl.), 27–33.

Roback, H. B. (Ed.). (1984). *Helping patients and their families cope with medical problems: A guide to therapeutic group work in clinical settings.* San Francisco: Jossey-Bass.

Selnes, O. A., Miller, E. N., Becker, J. T., Cohen, B. A., McArthur, J. C., Visscher, B., Gordon, B., Satz, P., Ginzburg, H. M., & Polk, B. F. (1988). Normal neuropsychological performance in healthy HIV-1 infected homosexual men: The Multicenter AIDS Cohort Study (MACS) (Abstract). *Proceedings of the IVth International Conference on AIDS* (Book 2, p. 399). Stockholm, Sweden.

Sheridan, K. (1989). Mental health practitioners and their roles in the AIDS crisis. In V. M. Mays, G. W. Albee, and S. F. Schneider (Eds.), *Primary prevention of AIDS: Psychological approaches.* Beverly Hills, CA: Sage.

Sheridan, K., & Sheridan, E. P. (1988). Psychological consultation to persons with AIDS. *Professional Psychology: Research and Practice, 19,* 532–535.

Shilts, R. (1987). *And the band played on.* New York: St. Martin's Press.

Temoshok, L., Sweet, D. M., & Zich, J. (1987). A three city comparison of the public's knowledge and attitudes about AIDS. *Psychology and Health, 1,* 43–60.

Tross, S., & Hirsch, D. A. (1988). Psychological distress and neuropsychological complications of HIV infection and AIDS. *American Psychologist, 43,* 929–934.

PART VII

Future Directions

> The empires of the future are empires of the mind.
> —Winston Churchill, Harvard University, 1943

As the future continually impinges on us and becomes our present, for better or worse, clinical psychologists working in medical settings can expect constant change. Whether this change will facilitate growth in the form of extending our roles and our arenas of practice or will lead to retrenchment in the form of scientific stagnation and professional complacence cannot be foreseen now. Surely, these two very different possibilities warrant attention and careful planning. Hopefully, the scientific curiosity that has been so much a part of our history will continue to serve clinical psychologists well in the future. In the final chapter of this handbook, Logan Wright and Alice G. Friedman (Chapter 32) discuss the current issues and scientific and practice trends of our field pertaining to medical settings.

Challenge of the Future
Psychologists in Medical Settings

Logan Wright and Alice G. Friedman

Introduction

In recent years, there has been remarkable growth in the number of psychologists in medical settings and in their contribution to research, teaching, and the clinical care of patients (Thompson, 1987). In 30 years, the percentage of members of the American Psychological Association (APA) who are identified as medical school faculty increased fivefold (Clayson & Mensh, 1987; Pion & Bramblett, 1985). Approximately 3,000 psychologists in North America are currently employed in medical schools (Clayson & Mensh, 1987), and more than 7,000 are employed in hospitals and clinics (National Science Foundation, 1988). The number of psychologists newly employed in medical schools increased from an average of 46 yearly in the mid 1950s to 130 during 1968–1976 (Gentry, Street, Masur, & Asken, 1981). Growth during the 1980s, though not yet documented, appears to have been even greater.

As we embark, on a new century, psychologists in medical settings are faced with challenges from within psychology, from the field of medicine, and from the changing nature of health service delivery. Continued growth and development of the field of medical psychology will depend on an adequate resolution of the resulting political and professional problems. In this chapter, we discuss what we view as the major challenges confronting psychologists in medical settings, and what we anticipate will be adequate solutions to these challenges. First, we discuss the challenges from within the field of psychology. Some of these challenges have prompted recent attempts at reorganization of the American Psychological Association. These center on defining the discipline of psychology. Crucial to this issue are (1) more clearly defining the boundaries of psychology; (2) highlighting what is unique about the discipline; (3) identifying the core curriculum in doctoral-level training programs; (4) defining the minimal standards in experiential areas such as practicum and internship; and (5) establishing standards for entry into the field. These concerns are not specific to psychologists practicing in medical settings. However, ongoing struggles with these issues interfere with psychology's relationship to other disciplines, particularly to medicine, and may inter-

Logan Wright • Department of Psychology, University of Oklahoma, Norman, Oklahoma 73072.   Alice G. Friedman • Department of Psychology, State University of New York at Binghamton, Binghamton, New York 13901.

fere with psychologists' ability to successfully provide their most unique services. Second, we discuss the relationship between psychology and medicine and how political and economic factors influence the functioning of psychologists in medical and other primary-care settings. Last, we discuss why we are optimistic about the future growth of psychology in medical settings.

Current Challenges

Challenges from within the Field

Psychology is currently struggling with major changes within the discipline that appear to threaten the academic and research emphasis of psychology and the commitment of psychology to science. During the past 40 years, there has been a growth in the proportion of Ph.D.'s awarded in the health-service-provider areas (clinical, counseling, school, health, forensic, and applied neuropsychology), accompanied by a marked decrease in the number of new doctorates awarded in such areas as general experimental and comparative psychology. Accompanying the shift toward training in applied psychology has been a rapid growth in the proportion of doctorates awarded by practitioner training programs that do not endorse a scientist-practitioner model of training. In a four-year period, between 1979 and 1983, the number of Psy.D.'s awarded increased fourfold (Howard, Pion, Gottfredson, Flattau, Oskamp, Pfafflin, Bray, & Burstein, 1986). The increase in the number of Ph.D.'s awarded has not been accomplished by an acceleration of the number of graduates from high-ranking academic departments. Rather, poorly ranked programs have doubled and tripled the number of degrees awarded (Howard *et al.*, 1986). Last, although the size of APA membership has increased over the past few years, an increasingly disproportionate number of psychologists in academic fields, rather than private-practicing health-service providers, are either failing to join the APA or are resigning. The APA has been losing its scientists (Howard *et al.*, 1986). This trend of dissatisfaction over the erosion of

psychology's science base eventually culminated in 1988 with the creation of the American Psychological Society (APS). This new alternative national organization espouses a set of values ranging from basic science to Boulder model approaches to health service delivery. The magnitude of this split is dramatized by the fact that such notables as B. F. Skinner, Herbert Simon (psychology's only Nobel Prize winner), and 23 of the 25 living past presidents of the APA are founding members of the APS.

These recent developments have generated controversy over which concepts, principles, experiences, or content areas are essential to training in psychology and how the training of health care providers can be accomplished in a manner that endorses psychology's ties to the science of behavior. The schism between (especially independent) practitioners and basic researchers is not just a contemporary problem (Altman, 1987). In fact, psychologist-practitioners were dissatisfied with the APA in 1917 and split from the organization to form the American Association of Clinical Psychologists. This early schism prompted efforts by the APA to remain unified; the bylaws of the APA were changed to recognize psychology as a science *and a profession*. Twenty years later, tension between practitioners and researchers resulted in the formation of the American Associated for Applied Psychology, which was reunited with the APA four years later (Altman, 1987).

More recently, the schism between the science and the profession of psychology has been played out in the training arena. Training in applied areas of psychology, particularly in clinical psychology, has been the focus of efforts to develop a core curriculum that balances the need for proficiency in research with adequate training in the application of psychological principles to a broad range of settings, problems, and populations. Debates about training issues have been among the most persistent struggles in the development of clinical psychology. Historically, national conferences have provided a forum for resolving the controversies surrounding the requirements for training and entry into the field.

Before the 1940s, the predominant model for

the training of psychologists was that of the scientist (McConnell, 1984). The Conference on Training in Clinical Psychology, held in Boulder, Colorado, in 1949, was the first attempt to develop a graduate training model for university-based clinical psychology programs (Bickman, 1987). The Boulder conference established the scientist-practitioner model as the dominant model for training clinical psychologists. Subsequent conferences (the Institute on Education and Training for Psychological Contributions to Mental Health at Stanford University, in 1955; the Miami Beach Conference on Graduate Education in Psychology, in 1958; and the Chicago conference, in 1965) upheld the recommendations of the Boulder conference. However, at the Chicago conference, an alternative to the scientist-practitioner model of training was introduced. This model, emphasizing clinical experience and a practice-oriented curriculum (the doctorate of psychology), was suggested as an alternative to the scientist-practitioner model in an attempt to diversify doctoral education in psychology. However, the model did not present a major challenge to the Boulder model until the Vail, Colorado, conference in 1973. Many of the participants in this conference, influenced by demands for greater access to mental health services for the poor, endorsed subdoctoral training as a means of confronting societal problems. The Vail conference officially endorsed professionally oriented training programs (Bickman, 1987).

Perhaps inadvertently, the Vail conference encouraged the development of "freestanding schools," which are now perceived by many as a major threat to the field of psychology. More recently, participants at the Morgantown Planning Conference in 1985 and the 1987 Conference at the University of Utah, sponsored by the APA, convened to review the growing concern over the quality of professional and scientific training in doctoral-level programs in psychology and to develop recommendations in an effort to resolve these difficulties (Bickman, 1987).

Questions about the desirability of standardized curricula and about the minimal standards of credentialing and entering the field of psychology have been discussed for years (Herb-

sleb, Sales, & Overcast, 1985). and as noted earlier, the field continues to be divided between those who believe that psychology has neglected training in the applied aspects of the field and those who believe that psychology has neglected training in scientific mentality and methodology (McConnell, 1984). In recent years, the debate has been fueled by the growth of professional schools that do not conform to the usual standards set by traditional academic psychology.

The first Psy.D. degree was awarded at the University of Montreal in the 1940s. However, the Psy.D. remained in "hibernation" for decades (McConnell, 1984). The early programs failed and disbanded, and the concept of was not fully recognized until the 1970s (McConnell, 1984). By the early 1980s, however, there were 44 programs devoted primarily to educating independent practitioners in psychology, 27 awarding Psy.D. degrees and 17 awarding Ph.D. degrees. There has been an accompanying growth of non-university-based professional schools. Although questions of quality control and assurance are relevant to all training programs, freestanding schools more often lack the controls that operate in university-based programs (Fox, 1979). Freestanding schools also frequently lack such essential resources as adequate libraries, acceptable ratios of students and faculty, and well-trained faculty members. Major universities are disinclined to establish practitioner-oriented programs because of the emphasis by such universities on research productivity and scholarship (Peterson, 1985). Thus practitioner-oriented training programs are left to less well-ranked universities and freestanding schools. Unfortunately, although Psy.D.'s awarded by autonomous departments of clinical psychology are frowned on by a large percentage of academic clinical psychologists, they are supported by a majority of applied psychologists (Thelen & Rodriguez, 1987).

The need for adequate standards of training and a minimal core curriculum is pertinent to training, accreditation, and credentialing in psychology for psychologists trained in traditional psychology programs as well as those trained in practitioner-oriented programs (Fox,

1979). Lack of consistency in the level and scope of training compromises psychology's position as a profession among other professions and undermines attempts to promote psychology's future in medical settings. Associated with unaccredited training programs is a twofold risk to the quality assurance of training and experience among graduates of doctoral programs in psychology. The first risk is that such training may produce a glut of poorly skilled practitioners who do not serve the public well and who lower the prestige of the field. The second is that, if training is sufficiently heterogeneous, it is not possible to specify the differences between psychologists and representatives of other fields, such as social work (Wright, 1986).

Development of Psychology as a Profession

Professions appear to follow similar developmental trends. These usually begin with haphazard, inconsistent credentialing and proceed through the loose organization of practitioners in an effort to establish the identifying and unique features of the field. The latter stages involve increasingly stringent and consistent entry requirements and, finally, formal accreditation of the content of training programs and certification or licensure of graduates of the program (Matarazzo, 1977). Psychology, a relative newcomer as an applied science, lags behind other fields. In contrast, medicine undertook its first major reorganization in 1910 with Abraham Flexner's Report. Before the Flexner Report, medical education was accomplished primarily through proprietary schools that were formed whenever and wherever a group of interested practitioners chose to form one. The laissez-faire attitude encouraged a proliferation of medical schools. In St. Louis alone, there were eight such schools (Fox, 1979). Many students entered medical school without a high school diploma; the qualifying exams eliminated only those who were blatantly illiterate (Norwood, 1970).

The Flexner Report was followed by standardization of training. Today, training in medical schools is guided by a common curriculum that is governed by a national body of representatives of the American Medical Association and the Association of American Medical Colleges. In dentistry, before the Gies Report in 1926, training in dentistry was accomplished by private proprietary schools and apprenticeships. Like medicine, the field of dentistry experienced uncontrolled growth. The Gies Report provided specific recommendations about duration of training, established minimal educational standards, and exerted professional control over training. Other professions, such as the practice of pharmacy, have undergone similar shifts from loosely formed guilds to more structured standardization of requirements for entering and remaining in the field.

In psychology, there have been several attempts to develop standardized criteria for entry into the fields. In 1977, Missouri became the 50th state and the 51st jurisdiction to enact statutory controls over the professional practice of psychology (Koocher, 1979). Unfortunately, the education and experience required by the statutes vary greatly from state to state. These inconsistencies are attributable to disagreement about the definition of the term *psychologist*. There is no uniformly agreed-upon definition of the profession and its scope, educational requirements, or standardization of credentials by which a psychologist can be identified (Fox, Barclay, & Rodgers, 1982). In 1983, the APA appointed the Task Force on the Future of Professional Psychology to develop minimum standards for entry into the field (Wright, 1986). In 1987, the APA unanimously approved a model act for the state licensure of psychologists. This act recommends that psychologists be required to graduate from an accredited training program in order to be allowed to sit for a licensing exam. It remains to be seen whether all states will adopt this model act.

In summary, delay in quality control through requiring graduation from an accredited training program before entering the field, along with the proliferation of training programs that do not endorse a scientist- or scholar-practitioner model for health service providers, presents the major threat to the future of psychology, especially within medical settings. In hospitals and other primary-health-care settings, the representatives of all other health service professionals have graduated from accredited training programs. These professionals often look to psychology as a discipline capable of providing a scientific mentality in the interdisciplinary setting.

The movement toward the abandonment of the scientific foundation of psychology represents the major threat to that field for the future. It is psychology's commitment to providing practitioners with a scientific mentality that allows for psychology's unique contribution and that sets it apart from practitioners representing other professions in the health service arena.

The relevant scientific mentality referred to above has been described by the aforementioned Task Force on the Future of Professional Psychology (Wright, 1986) as involving (1) a critical, data-based (rather than purely intuitive) approach to patient care, both diagnostic and therapeutic; (2) a sense of intrinsic obligation to remain abreast of the scientific literature; and (3) a sense of stewardship in generating and communicating new knowledge as an essential part of one's professional practice activities. The latter usually involves structuring one's practice activities so as to generate quantifiable and otherwise meaningful research data.

The Relationship between Psychology and Medicine

The future of medical psychology is inextricably tied to the evolving relationship between psychology and medicine, which has alternated historically between cooperative and antagonistic. Within medicine, psychology has traditionally been linked most closely with psychiatry. During the decades following World War II, psychiatry moved away from its close alliance with medicine (Dana & May, 1986), and psychology gained a greater right to independent practice. In recent years, psychiatry has redirected its emphasis toward a stronger empirical basis for practice. Dana and May (1986) believe that this "remedicalization" of psychiatry will facilitate the development of psychol-

ogy in other medical settings and will facilitate the relationship between psychology and medicine. As psychiatry, for the most part, defines its turf as involving the most severely debilitating forms of psychopathology, it leaves to psychology basic and applied research, *primary* mental health care, and the provision of services to nonpsychiatric medical patients.

Historically, psychology's relationship with primary-care medicine has been more amicable than with psychiatric medicine. Thus, although over half of the psychologists in medical schools have appointments within departments of psychiatry, the practice of psychology within other areas of medical specialty offer opportunities for future growth in clinical and scientific roles (Clayson & Mensh, 1987).

The Joint Commission for Accreditation of Hospitals (JCAH) has had a profound impact on the relationship between psychology and medicine during the past 20 years. Zaro, Batchelor, Ginsberg, and Pallak (1982) described the historical relationship between psychology and the JCAH as being characterized by "persistence, resistance, slow progress, entrenched opposition, and interdisciplinary warfare" (p. 1342). Before the mid-1960s, the JCAH was a loose collective of medical associations that developed minimal standards for hospitals and guidelines for the voluntary accreditation of general hospitals. The organization was relatively unknown to people outside the medical profession (Porterfield, in Matarazzo, Lubin, & Nathan, 1978). In 1965, Medicare and Medicaid legislation was enacted. Congress required that, to be approved for Medicare reimbursement, a hospital must meet the standards of the JCAH, and the JCAH's importance to the fiscal survival of the hospital system took a quantum leap. Hospitals had great financial incentive for complying with JCAH guidelines. Until that time, many psychologists had been full members of the medical staff at university-affiliated medical centers. The JCAH guidelines recommended that membership on the medical staff of hospitals be restricted to physicians and dentists. Many hospital changed their bylaws to comply with the JCAH guidelines and excluded psychologists from staff membership. Hundreds of psychologists lost their privileges (Silver, 1987).

Interestingly, however, psychologists were often unaware of their membership status on the hospital staff. In fact, despite complaints by psychologists about their lack of autonomy, psychologists are often unaware of whether the bylaws of their own medical school enabled psychologists to be full voting members of the medical staff. In 1975, Matarazzo, Lubin, and Nathan (1978) surveyed 113 medical school psychologists, from the faculty of the 115 medical schools in this country, about their membership status on the staff of their university's hospital. Over 30% responded that they were full members of the staff with voting privileges. However, when these psychologists were asked to verify their status, 86% of those who had assumed that they were voting members of the staff turned out to be wrong! Much of this confusion was attributable to very recent changes in JCAH accreditation. However, some of it was attributable to the lack of involvement by psychologists in their hospital's governance.

Psychologists often complain about lack of autonomy in hospitals and other primary-health-care settings. However, they have been slow to confront the opposition to such autonomy and to become actively involved in their future in medical settings. Carr (1987) described the traditional stance of psychologists in medical settings as that of "expatriates," assuming their responsibilities in the medical setting while remaining detached from the political and sociocultural aspects of the setting.

The reluctance of psychologists to be involved in the administrative and fiscal activities of the medical settings in which they practice extends to their understanding of the impact of Medicare legislation on the professional practice of psychology. For a number of reasons, Medicare has an enormous impact on research, training, and the practice of psychology. Medicare and Medicaid, the state-federal health insurance program for the elderly and disabled population of this country, account for 90% of the federal health budget (Uyeda & Moldawsky, 1986). Additionally, the programs serve as models for other health insurance programs, dictating what other insurance policies will reimburse. A large proportion of the patient populations in medical school teaching hospitals are Medicare-eligible. Further, a large percentage of psychologists who teach, practice, and supervise interns are employed by teaching hospitals. Consequently, policies governing reimbursement are central to the survival of psychology in hospital settings (Carr, 1987). Psychology faculty and their students who are unable to be reimbursed for their services become liabilities to their department (Carr, 1987). As noted by Uyeda and Moldawsky (1986), limitations on reimbursement that have been created by the Medicare prospective payment system have interfered directly with the development of behavioral programs in areas such as geriatrics. The elderly are a growing population with considerable health needs, but they are dependent almost entirely on Medicare for coverage.

Many psychologists continue to be uninformed about how they can practice autonomously and bill for their services through Medicare (Uyeda & Moldawsky, 1986). Psychologists are often uninformed about the impact of Medicaid on the practice of psychology. Further, few psychologists know whether their hospital actually reimburses their department for their services (Carr, 1987). This ignorance again points to the need for psychologists to become more involved in the governance of the medical schools and in public policy regarding the future of health care in this country.

Cause for Optimism

Despite some of the difficulties faced in medical settings, psychologists continue to find employment in medical settings in increasing numbers, and psychologists working in medical settings continue to report satisfaction with their jobs (Nathan, Lubin, & Matarazzo, 1981). In fact, many of the challenges faced by psychologists are due, in part, to the positive trends in the field that are discussed in this book. The increase in scientific productivity among psychologists in medical settings, the increasing breadth of research and clinical practice in areas such as clinical neuropsychology and rehabilitation, the increasing demand for psychologists, and the development of relatively new areas such a psychoneuroimmunology and neurotoxicology are only a few examples of the tremendous progress made by psychologists in medical settings since the early 1980s. Changes in the health care system, as well as advances in the field of psychology, provide extremely favorable conditions for psychologists in medical settings. The following discussion highlights some of the major causes for optimism that have been discussed in greater detail in earlier chapters in this book.

Health Promotion

The field of psychology now has an important role in promoting and ensuring the future health of our nation's citizens. Advances in immunology have greatly reduced, and in some cases eliminated, mortality due to infectious diseases. A generation ago, tuberculosis, poliomyelitis, influenza, and pneumonia were leading causes of premature death. Because of remarkable medical achievements over the past several decades, such diseases are rarely the cause of premature death today (Matarazzo, 1982). Instead, most of the major health threats today, such as cancer, cardiovascular disease, substance abuse, and vehicle-related accidents, are closely related to behavior and lifestyle (Matarazzo, 1982). Significant decreases in the incidence of these disorders will depend on developing better strategies to prevent the acquisition of unhealthy habits and to eliminate behaviors that are associated with disease and premature death.

Acquired immunodeficiency syndrome (AIDS) is the major new health problem in this country, and perhaps in the world (Pelosi, 1988). There is currently no vaccine to immunize people against the disease. The major defense against the spread of the disease is the information available about the prevention of AIDS through an avoidance of exposure to the virus. Modification of the *behaviors* associated with the transmittal of the virus is the key to slowing the spread of the disease. Psychology, as the science of behavior, is ideally suited to being in the forefront of a movement to alter the

behaviors associated with the risk of developing the illness. In fact, psychology's greatest successes in health-related areas appear to be in the area of prevention. Although the results of smoking-cessation programs are typically short-lived, programs designed to prevent the acquisition of the behavior in the first place appear to be more successful (Evans, 1984). Psychologists employed in medical settings are in an ideal position (1) to conduct research aimed at identifying behaviors associated with increased risk of physical disability; (2) to develop effective preventive strategies and to demonstrate their efficacy; and (3) to provide consultation to medical staff regarding strategies for increasing patients' adherence to health-enhancing lifestyles.

Current trends suggest that, in the next decade, psychologists will become more involved in the prevention of diseases we have previously assumed to be unrelated to behavior. Advances in relatively new areas of study, such as psychoneuroimmunology and neurotoxicology are providing important information about the complex relationships among behavior, environmental events, and neurochemical functioning. It is increasingly apparent that understanding the links between psychological factors and diseases will facilitate the development of strategies that will decrease premature mortality among high-risk populations. To benefit from these scientific advances, psychologists working in medical settings will find that their research and clinical practices involve increased collaboration with other disciplines, such as biology, endocrinology, the neurosciences, and immunology. This collaboration will require a knowledge of the basic concepts in these related fields to enable conversation across disciplines.

Dispersal of Psychological Services

In 1986, Wright and Burns described primary mental health care as a "find" for psychology. They pointed out that the number of individuals available for psychological interventions in settings other than traditional mental health settings (i.e., in pediatric, internal-medicine, or family-practice settings) far exceed the number who might access psychological services in traditional mental health settings. Whereas the incidence of traditional forms of psychopathology in the general population is relatively small, nearly everyone is physically ill at some point in her or his life. Many individuals must cope with chronic physical problems; some of which negatively affect the individual's quality of life; and many of which have their etiology in unhealthy lifestyles. Only 39% of families in the United States have contact with a mental health practitioner (Regier, Golberg, & Taube, 1978), and only a small percentage of individuals with psychiatric disorders who are provided services by health professionals ever see a mental health provider (Shapiro, Skinner, Kessler, Von Korff, German, Tischler, Leaf, Benham, Cottler, & Regier, 1984). Individuals with psychological problems are more likely to come to a primary medical setting for treatment. Rosen, Locke, Goldberg, and Babigian (1972) reported that, in one general outpatient clinic, 22% of the patients had a primary psychological disturbance. Other studies (see Rosen & Wiens, 1979) have documented a high prevalence of psychological difficulties among patients who present to primary-health-care settings. A survey of the Academy of Pediatrics revealed that the number of patients and their families seen by pediatricians for consultation about psychological problems exceeded the number seen by psychologists and psychiatrists combined (Task Force on Pediatric Education, American Academy of Pediatrics, 1978).

Another study, by Duff, Rowe, and Anderson (1972), describes the large psychological component in pediatric practice. All patients matriculating through a large pediatric outpatient clinic were rated on whether they presented with (1) purely physical problems; (2) purely psychological problems; or (3) some combination of the two. Only 12% of the pediatric outpatients presented with purely physical problems; 36% were rated as having purely psychological problems; and 52% were thought to have some combination of the two.

Last, the incidence of psychological difficulties is elevated among individuals with a history of acute or chronic medical problems

(see, for example, Mulhern, Wasserman, Friedman, & Fairclough, 1989). Thus, if psychologists are to practice in the settings in which there is the greatest need for behavioral interventions and that hold the greatest promise for optimal use of behavioral strategies, they must focus on primary-health-care settings rather than, or in addition to, traditional mental health settings.

The presence of psychologists in medical settings appears to have a duel advantage. First, as noted above, a number of individuals who would not otherwise seek mental health services can be identified and serviced. Second, the provision of psychological services appears to affect positively the use of medical services. An early study (Follette & Cummings, 1967) suggested that individuals in psychological distress had a higher than expected incidence of utilization of both inpatient and outpatient medical services. Psychotherapy resulted in a reduced use of general medical services that persisted for five years. Subsequent studies (Mumford, Schlesinger, & Glass, 1981; Rosen & Wiens, 1979), have supported the notion that even brief mental health interventions can reduce the use of general medical services between 5% and 85% (Budman, Demby, & Feldstein, 1984).

Today over 20 million individuals receive comprehensive health care from one of the approximately 400 health maintenance organizations (HMOs) in this country (Austad, DeStefano, & Kisch, 1988). Members of an HMO pay an annual fee to cover all necessary medical care, and the HMO is required to provide the member with all necessary health care services during the year. The HMO profits from members who use the health care services less frequently. Thus, the HMO movement has greatly increased the value to the field of medicine of psychological intervention and its ability to reduce the inappropriate usage of medical service.

Expanding Roles

Advances in behavioral psychology have greatly expanded the role of the psychologist in primary-health-care settings. Until the mid-1960s, one of psychology's major roles was assessing the mental health status of hospitalized patients. Today, the role of the psychologist in medical settings has broadened considerably. Although assessment continues to be an important function, psychologists more typically divide their time among numerous activities, including assessment, consultation, direct intervention, research, and training medical students and interns.

Assessment

A survey of psychologists in medical settings revealed that between 1955 and 1977, the percentage of time devoted to psychodiagnosis by full-time Ph.D.-level psychologists in medial settings declined from nearly 20% to less than 10% (Nathan, Lubin, Matarazzo, & Persely, 1979). However, psychological assessment continues to be valued in assisting with medical decisions about patients (Elfant, 1985). In a recent survey (Stabler & Mesibov, 1984) of psychologists practicing in health care settings, pediatric psychologists, in particular, reported spending 50% of their time doing assessment (although it was not limited to traditional psychodiagnostic testing), and health psychologists report devoting slightly less time to assessment. Clearly, doing diagnostic work remains an important role for psychologists in medical settings. However, future assessment will extend beyond traditional psychodiagnostics to include aid in decision making about the course of medical treatment. For example, assessment may include behavioral observation of pain behaviors to determine the effectiveness of medical interventions for decreasing pain, the monitoring of nausea and vomiting to assess the effectiveness of behavioral interventions for the reduction of anticipatory anxiety and learned aversions, and observation and self-report to measure patient adherence to medical regimes.

Consultation

The positive impact of psychology on the care of patients, as well as psychology's role in medicine, is apparent from the increasing role of the psychologist as a consultant in primary-care facilities. The importance of consulting in psychology is underscored by the fact that 21 of the 32 chapters in this book deal with this role. This role has been well accepted by the medical staff in most hospital settings, perhaps because it closely matches the model used by most medical specialty areas. In a survey of residents and staff physicians in four medical specialties (family practice, internal medicine, pediatrics, and psychiatry), Liese (1986) found that physicians were typically interested in attending to the psychological concomitants of medical illness and would seek consultation from psychology, psychiatry, and social work. In a recent survey conducted in a pediatric hospital (Olson, Holden, Friedman, Faust, Kenning, & Mason, 1988), physicians, nurses, and social workers reported satisfaction with the provision of psychological consultation services in the hospital and indicated that they were likely to use these services.

The positive growth of psychologists' consultation practices in medical settings has been discussed by many of the authors in this book. However, as noted by Stabler and Mesibov (1984), physicians and other medical practitioners are often unaware of how they can use psychological services. This lack of communication may be attributable, in part, to psychology's tendency to remain uninvolved in many aspects of the medical environment and thus to limit medical practitioners' exposure to psychology. Continued growth will require that psychologists become involved in the sociocultural aspects of their setting. Educating medical colleagues about the services that psychology can provide will become increasingly important as psychology is able to offer more diverse consultation services.

Intervention

The role of the psychologist in treated medically ill patients has increased dramatically since the early 1980s. As clinical practices have become more specialized, they have expanded into new areas. For example, neuropsychology, as a specialty, is relatively new, although clinical psychologists have been evaluating patients for neuropsychological deficits for some time. Neuropsychologists are now involved in cognitive rehabilitation; an area that was hardly a concept in the early 1980s. Such practices are likely to continue to expand as advances in medical care decrease mortality among injured individuals and increase longevity.

Psychologists have traditionally had a limited role in the treatment of medical problems. Interventions typically focus on lessening the behavioral or developmental problems caused by disease, or on treating behavioral problems that negatively affect the management or control of physical disease. However, psychologists now have effective tools for eliminating the medical difficulties caused by maladaptive learning and/or behavior. It is in this area that psychology has seen major breakthroughs. In 1969, Wright, Nunnery, Eichel, and Scott first reported the successful application of behavioral procedures to the treatment of chronic tracheostomy addiction in young children. This was one of the earliest reports of the elimination of a physical problem caused by maladaptive learning and provided one of the first demonstrations of how behavioral psychologists can effectively influence the physical status of medical patients.

Since the early 1970s, there has been tremendous progress in the breadth of impact and in the acceptance of behavioral interventions in medical settings. Today, there are few medical specialty areas on which psychological principles have not had a significant impact. In pediatrics, encopresis has been treated with programs using combinations of medical approaches, reinforcement, and compliance techniques, and the most successful method of curing nocturnal enuresis, the "pad and bell," is based on operant conditioning (see Walker, Milling, & Bonner, 1988). Psychologists are also involved in treating eating disorders such as bulimia, anorexia, and obesity. Drotar (1988) described the psychological treatment of infants suffering from failure to thrive. Psychology has also had a major impact in cardiology, both in the identification of behavioral patterns associated with an increased risk for cardio-

vascular disease and in decreasing the risk of heart attack recurrence by increasing compliance with a medical regime. In ophthalmology and optometry, behavioral principles such as reinforcement, response shaping, and stimulus control have been used to improve visual acuity among myopes (Collins, Epstein, & Gil, 1982). In dermatology, reciprocal inhibition and operant techniques have been used to treat dermatitis successfully (see Wright, Schaefer, & Solomons, 1979). In neurology, aversive conditioning has been effective in reducing self-induced seizures (Wright, 1973).

In oncology, conditioned aversions (anticipatory nausea and vomiting, as well as conditioned food aversions) in patients undergoing chemotherapy can be among the most distressing side effects of treatment (Andrykowski, Redd, & Hatfield, 1985). Severe conditioned aversions can influence the patient's ability and willingness to continue with treatment. Behavioral interventions such as those discussed, including such strategies as relaxation training, systematic desensitization, and biofeedback, have been effective in reducing the negative side effects caused by maladaptive learning.

Behavioral procedures have been successful in increasing compliance and in decreasing distress among adult and child burn victims undergoing physical therapy (Elliott & Olson, 1983; Hegel, Ayllon, Vanderplate, & Spiro-Hawkins, 1986). Psychologists have also been major contributors to behavioral solutions to many other physical problems, including hypertension (Shapiro, Schwartz, Ferguson, Redmond, & Weiss, 1977); cardiac arrhythmias (Hatch, Gatchel, & Harrington, 1982); headaches (Cox & Thomas, 1981); and sexual dysfunction (Geer & Messe, 1982). The preceding clinical program chapters in this book present a wide range of medical problems addressed diagnostically and therapeutically by clinical psychologists.

Conclusion

Future psychologists practicing in medical settings will be faced with the consequences of the challenges that now face the field. Changes in the proportion of psychologists grounded in science, the lack of standardized criteria for entering the field, and the impact of the economy on the health delivery system threaten to affect the field adversely. However, most of these difficulties stem from the rapid and, for the most part, positive growth of the field of psychology since the early 1970s. The opportunities for psychologists to have a positive impact on the health of the nation have never been greater. Hospitals and other health care settings continue to seek increasing numbers of psychologists because of the potential for valuable research, training, and clinical practice. Scientifically oriented psychology continues to develop improved methods both for primary mental health care and for physical problems with lifestyle or other behavioral concomitants. Expanding knowledge and understanding about the interrelationship between health and behavior will bring increasing opportunities for psychologists to contribute in medical settings.

References

Altman, I. (1987). Centripetal and centrifugal trends in psychology. *American Psychologist, 42*, 1058–1069.

American Academy of Pediatrics. (1978). The Task Force on Pediatric Education.

American Psychological Association. (1987a). Model act for state licensure of psychologists. *American Psychologist, 42*, 696–703.

American Psychological Association. (1987b). Resolutions approved by the National Conference on Graduate Education in Psychology. *American Psychologist. 42*, 1070–1084.

Andrykowski, M. A., Redd, W. H., & Hatfield, A. K. (1985). Development of anticipatory nausea: A prospective analysis. *Journal of Consulting and Clinical Psychology, 53*, 447–454.

Austad, C. S., DeStefano, L., & Kisch, J. (1988). The health maintenance organization: 2. Implications for psychotherapy. *Psychotherapy, 25*, 449–454.

Bickman, L. (1987). Graduate education in psychology. *American Psychologist, 42*(12), 1041–1047.

Budman, S. H., Demby, A., & Feldstein, M. L. (1984). Insight into reduced use of medical services after psychotherapy. *Professional Psychology, 15*, 353–361.

Carr, J. E. (1987). Federal impact on psychology in medical schools. *American Psychologist, 42*, 869–872.

Clayson, D., & Mensh, I. N. (1987). Psychologists in medical schools: The trials of emerging political activism. *American Psychologist, 42*, 859–862.

Collins, F. L., Epstein, L. H., & Gil, K. M. (1982). Behavioral factors in the etiology and treatment of myopia. In M. Hersen, R. M. Eisler, & P. M. Miller (Eds.), *Progress in behavior modification* (Vol. 13). New York: Academic Press.

Cox, D. J., & Hobbs, W. (1982). Biofeedback as a treatment for tension headaches. In L. White & B. Tursky (Eds.), *Clinical biofeedback efficacy and mechanisms*. New York: Guilford Press.

Cox, D. J., & Thomas, D. (1981). Relationship between headaches and depression. *Headaches, 21*, 261–263.

Dana, R. H., & May, W. T. (1986). Health care megatrends and health psychology. *Professional Psychology, 17*, 251–255.

Drotar, D. (1988). Failure to thrive. In D. Routh (Ed.), *Handbook of pediatric psychology* (pp. 71–107). New York: Guilford Press.

Duff, R. S., Rowe, D. S., & Anderson, F. P. (1972). Patient care and student learning in a pediatric clinic. *Pediatrics, 50*, 839–846.

Elfant, A. B. (1985). Psychotherapy and assessment in hospital settings: ideological and professional conflicts. *Professional Psychology, 16*, 55–63.

Elliott, C. H., & Olson, R. A. (1983). The management of children's distress in response to painful medical treatment for burn injuries. *Behaviour Research and Therapy, 21*, 675–683.

Evans, R. I. (1984). A social inoculation strategy to deter smoking in adolescents. In J. D. Matarazzo, S. M. Weiss, J. A. Herd, & N. E. Miller (Eds.), *Behavioral health: A handbook of health enhancement and disease prevention* (pp. 765–774). New York: J. Wiley.

Follette, W. T., & Cummings, N. A. (1967). Psychiatric services and medical utilization in a prepaid health plan setting. *Medical Care, 5*, 25–35.

Fox, R. E. (1979). Response in models, modes, and standards of professional training. *American Psychologist, 34*, 339–349.

Fox, R. E., Barclay, A. G., & Rodgers, D. A. (1982). The foundation of professional psychology. *American Psychologist, 37*, 306–312.

Geer, J. H., & Messe, M. (1982). Sexual dysfunction. In R. J. Gatchel, A. Baum, & J. E. Singer (Eds.), *Behavioral medicine and clinical psychology: Overlapping areas*. Hillsdale, NJ: Erlbaum.

Gentry, W. D., Street, W. J., Masur, F. T., & Asken, M. S. (1981). Training in medical psychology: A survey of graduate and internship training programs. *Professional Psychology, 12*, 224–228.

Hatch, J. P., Gatchel, R. J., & Harrington, R. (1982). Biofeedback: Clinical applications in medicine. In R. J. Gatchel, A. Baum, & J. E. Singer (Eds.), *Behavioral medicine and clinical psychology: Overlapping areas*. Hillsdale, NJ: Erlbaum.

Hegel, M. T., Ayllon, T., VanderPlate, C., & Spiro-Hawkins, H. (1986). A behavioral procedure for increasing compliance with self-exercise regimens in severely burn-injured patients. *Behavior, Research and Therapy, 24*, 521–528.

Herbsleb, J. D., Sales, B. D., & Overcast, T. D. (1985). Challenging licensure and certification. *American Psychologist, 40*, 1165–1178.

Howard, A., Pion, G. M., Gottfredson, G. D., Flattau, P. E.,

Oskamp, S., Pfafflin, S. M., Bray, D. W., & Burstein, A. G. (1986). The changing face of American psychology. *American Psychologist, 41*, 1311–1327.

Koocher, G. P. (1979). Credentialing in psychology. *American Psychologist, 34*, 696–702.

Liese, B. S. (1986). Physicians' perceptions of the role of psychology in medicine. *Professional Psychology, 17*, 276–277.

Matarazzo, J. D. (1977). Higher education, professional accreditation, and licensure. *American Psychologist, 32*, 856–859.

Matarazzo, J. D. (1982). Behavioral health's challenge to academic, scientific and professional psychology. *American Psychologist, 37*, 1–14.

Matarazzo, J. D., Lubin, B. L., & Nathan, R. G. (1978). Psychologists' membership on the medical staffs of university teaching hospitals. *American Psychologist, 33*, 23–29.

McConnell, S. C. (1984). Doctor of psychology degree: From hibernation to reality. *Professional Psychology, 15*, 362–370.

Mulhern, R. K., Wasserman, A., Friedman, A. G., & Fairclough, D. (1989). Social competence and behavioral adjustment of children who are long-term survivors of cancer. *Pediatrics, 83*, 18–25.

Mumford, E., Schlesinger, H. J. & Glass, G. V. (1981). Mental health treatment and medical care utilization in fee-for-service system: Outpatient mental health treatment following the onset of a chronic disease. *American Journal of Public Health, 73*, 422–429.

Nathan, R. G., Lubin, B., Matarazzo, J. D., & Persely, G. W. (1979). Psychologists in schools of medicine: 1954, 1964, and 1977. *American Psychologist, 34*, 622–627.

Nathan, R. G., Lubin, B., & Matarazzo, J. D. (1981). Salaries and satisfactions of medical school psychologists. *Professional Psychology, 12*, 420–423.

National Science Foundation. (1988). Profiles of psychology: Human resources and funding. (NSF 88-325). Washington, DC: U.S. Government Printing Office.

Norwood, W. M. F. (1970). Medical education in the United States before 1900. In C. D. O'Malley (Ed.), *The history of medical education* (pp. 463–500). Los Angeles: University of California Press.

Olson, R. A., Holden, E. W., Friedman, A. G., Faust, J., Kenning, M., & Mason, P. (1988). Psychological consultation in a children's hospital: An evaluation of services. *Journal of Pediatric Psychology, 13*, 479–492.

Pelosi, N. (1988). AIDS and public policy: A legislative view. *American Psychologist, 43*, 843–845.

Peterson, D. R. (1985). Twenty years of practitioner training in psychology. *American Psychologist, 40*, 441–451.

Pion, G., & Bramblett, P. (1985). *Salaries in psychology, 1985: Report of the 1985 APA salary survey*. Washington, DC: American Psychological Association.

Regier, D. A., Goldberg, I. D., & Taube, C. A. (1978). The de facto U.S. mental health services system. *Archives of General Psychiatry, 35*, 685–693.

Rosen, B. M., Locke, B. Z., Goldberg, I. D., & Babigian, H. M. (1972). Identification of emotional disturbance in patients seen in general medical clinics. *Hospital and Community Psychiatry, 23*, 364–370.

Rosen, J. C., & Wiens, A. N. (1979). Changes in medical problems and use of medical services following psychological intervention. *American Psychologist, 34*, 420–431.

Shapiro, A. P., Schwartz, G. E., Ferguson, D. C. E., Redmond, D. P., & Weiss, S. M. (1977). Behavioral methods in the treatment of hypertension: A review of their clinical status. *Annals of Internal Medicine, 86*, 626–636.

Shapiro, S., Skinner, E. A., Kessler, L. G., Von Korff, M., German, P. S., Tischler, G. L., Leaf, P. J., Benham, L., Cottler, L., & Regier, D. A. (1984). Utilization of health and mental health services. *Archives of General Psychiatry, 41*, 971–978.

Silver, R. J. (1987). New York State: A case study in organizing psychology. *American Psychologist, 42*, 863–865.

Stabler, B., & Mesibov, G. B. (1984). Role functions of pediatric and health psychologists in health care settings. *Professional Psychology, 15*, 142–151.

Thelen, M. H., & Rodriguez, M. D. (1987). Attitudes of academic and applied clinical psychologists toward training issues: 1969–1984. *American Psychologist, 42*, 412–415.

Thompson, R. J. (1987). Psychologists in medical schools. *American Psychologist, 42*, 866–868.

Thompson, R. J., & Matarazzo, J. D. (1984). Psychology in United States medical schools: 1983. *American Psychologist, 39*, 988–995.

Uyeda, A. K., & Moldawsky, S. (1986). Prospective payment and psychological services. *American Psychologist, 41*, 60–63.

Walker, C. E., Milling, L. S., & Bonner, B. L. (1988). Incontinence disorders: Enuresis and encopresis. In D. Routh (Ed.), *Handbook of pediatric psychology* (pp. 363–397). New York: Guilford Press.

Wright, L. (1973). Aversive conditioning of self-induced seizures. *Behavior Therapy, 4*, 712–713.

Wright, L. (1986). The changing face of clinical psychology: Impact of service delivery changes. *Journal of Clinical Psychology, 42*, 841–844.

Wright, L., & Burns, B. J. (1986). Primary mental health care: A "find" for psychology. *Professional Psychology, 17*, 560–564.

Wright, L., Nunnery, A., Eichel, B., & Scott, R. (1969). Application of conditioning principles to problems of tracheostomy addiction in children. *Journal of Consulting and Clinical Psychology, 32*, 603–606.

Wright, L., Schaefer, A. B., & Solomons, G. (1979). *Encyclopedia of pediatric psychology*. Baltimore: University Park Press.

Zaro, J. S., Batchelor, W. F., Ginsberg, M. R., & Pallak, M. S. (1982). Psychology and the JCAH. *American Psychologist, 37*, 1342–1349.

Afterword

Patrick H. DeLeon

During the decade and a half that I have had the opportunity of observing (and participating in) the development of our nation's health care system, from the vantage point of serving on Capitol Hill, I have come increasingly confident that the field of clinical psychology has much to offer society and to the quality of life of those who need the services of our nation's health care practitioners (DeLeon, 1988). Above all else, ours is a behavioral science profession—one that stresses the fundamental importance of the scientist-practitioner model, both in our day-to-day functioning and in our continuing efforts to expand the traditional boundaries of our profession's scope of practice. At our best, we represent data-driven theories and practices, and we have the professional maturity necessary to modify expectations, relying on measurable outcomes.

In many ways, and especially from a public policy frame of reference, the public at large could (and perhaps even should) view psychology as one of the newest of the specialities. Back in 1974, Marc Lalonde, Minister of National Health and Welfare of Canada, called for formal recognition—at the health policy level—of the importance of disease prevention and

health promotion activities, while expressly noting the health consequences of lifestyle and individual behavior. This was the first time that "behavioral medicine" (or "health psychology") had received such status and had been able to contribute to the highest level of policy debates within the governments of the industrialized world.

Five years later, U.S. President Jimmy Carter released *Healthy People: The Surgeon General's Report on Health Promotion and Disease Prevention* (U.S. Department of Health, Education, and Welfare, 1979). Although very few of our nation's elected officials—or our nation's health policy experts, for that matter—had had formal psychological training, the underlying tenets of psychology's knowledge base and practice orientation were evident throughout this important document. Since the report's release, each of our nation's subsequent Surgeons General has endorsed a similar policy position. In fact, in late 1989, the U.S. Preventive Services Task Force concluded that substituting counseling in methods of maintaining good health was more likely to prevent death and disease than anything else that could be done; that is, it was more likely to reduce morbidity and mortality in this country than any other category of clinical intervention. From our professional orientation, this latest policy recommendation from the administration could

Patrick H. DeLeon • U.S. Senate Staff, Washington, DC 20510.

615

readily be translated into calling for the effective utilization of the clinical/behavioral expertise of clinical psychology.

And yet, notwithstanding the critical importance of behavioral factors to our health and well-being, it is fair to say that, in the public's mind, and particularly in the popular media, "health" is still unequivocally equated with medicine and with physicians, in particular. Weekly, summaries of articles that appear in the *New England Journal of Medicine* are incorporated into our newspapers and news broadcasts. There are, of course, notable examples of the public's (and elected officials') voluntarily using the services of nonphysicians, such as when an individual goes to an optometrist or a pharmacist for advice or has a child delivered by a nurse midwife. Nevertheless, unless one presses for specificity, in our society "health" and "illness" are equated with "medicine." And when we review our nation's current health care priorities and the mechanisms that have been established for paying for services rendered (or needed), the allocation of resources available for important clinical training and research activities, and the professional backgrounds of those selected for health policy administrative positions, the domination of medicine becomes even more evident. Thus, as the chapter authors throughout this handbook emphasize, psychology's survival (and astonishing growth) within *medical settings* has very significant ramifications for all elements of the profession and shows very good promise for our probable future.

During the period that I have served on Capitol Hill, I have also had the opportunity of being actively involved in the governance of the American Psychological Association (APA). Again, with the significant exception of the establishment (and the subsequent extraordinary growth) of the Division of Health Psychology, much of the knowledge base and the health-related orientation described in this handbook has unfortunately escaped the collective consciousness of APA's elected leadership. If one listens to the debates on the floor of our Council of Representatives or within our numerous policy boards and committees, one must reluctantly conclude that, even among our profession's elected leadership, there is very little understanding of psychology's true clinical or scientific potential. As an evolving profession, we represent considerably more than a traditional "mental health speciality." The locus of practice of our clinicians must not be limited conceptually to community mental health centers, state and veterans hospitals, or private psychotherapy practices. Our clinicians and scientists must become active participants in our nation's health delivery system, and we must ensure that they will be accepted by the other relevant disciplines as true professional colleagues. Clinical psychologists are not paraprofessionals or allied health specialists. Psychology's knowledge base and inherent potential for significant contributions to many elements of our nation's health delivery system are too important to accept such a demeaning status.

However, as I have suggested, collectively our profession does not seem yet to understand the bigger picture. Our educational institutions, in particular, do not seem to understand their unique societal and professional responsibilities (DeLeon, 1989). As one listens year after year to the testimony presented by the administration and by numerous public witnesses before the various congressional subcommittees that have jurisdiction over our nation's health care system, one keeps hearing about the progress and tolls of the "medical system." Until very recently, one almost never heard from organized psychology about anything other than the relatively narrow issue of "mental health" training and research. The voices of our training institutions, heralding innovative advancements within the health delivery system due to clinical psychology's unique behavioral expertise, were notably absent. As a profession, we do not have comprehensive training and treatment facilities of our own, and so, we must continue to rely on those of medicine. And perhaps even more important, our training institutions, which could take the lead and establish such facilities, really do not understand the public policy–political process. In light of this political and policy reality, the advances presented in this handbook are truly impressive.

We are making considerable progress both clinically and legislatively. Since the early 1980s, psychology has quietly been making steady inroads in obtaining recognition under the various training and service provisions of Title VII of the U.S. Public Health Service Act, the Health Professions Education Authority. Title VII (commonly known as MODVOPPP) authorizes federal support of training and special-service delivery programs in the fields of medicine, osteopathy, dentistry, veterinary medicine, optometry, podiatry, pharmacy, and public health. When this legislation was first established in the early 1960s, the congressional intent was to increase enrollments at the various health professions' schools (i.e., to address a perceived shortage in medical personnel). During the 1970s, the focus changed to speciality requirements and geographic maldistribution problems (i.e., concern about access). Psychology, with its almost exclusive "mental health" focus, has historically been only tangentially involved, and where any support was received, it would have been under one of the various allied health initiatives. However, a provision of the Orphan Drug Act (P.L. 97-414) modified the underlying public health code to authorize psychologists to serve in the U.S. Public Health Service Regular Corps, which, in a health policy frame of reference, is the prime authority for the federal government's service delivery responsibility and interestingly, appointment by the President as our nation's Surgeon General. Further, the 1985 Health Profession Training Assistance Act (P.L. 99-129) expressly deemed that psychology was no longer to be considered an "allied health profession."

As we have indicated, our evolution into a bona fide health profession has been progressing gradually, although, admittedly, it is not appreciated (or even supported) by many of our colleagues. During the closing hours of the 100th Congress, the Health Omnibus Programs Extension Act of 1988 was enacted into public law (P.L. 100-607). This legislation may have very significant ramifications for the future of our profession. The various institutes of the National Institutes of Health were legislatively mandated to appoint behavioral scientists to their national advisory panels, and further, a psychologist was directed to be appointed to the all-important National Advisory Council on Health Professions Education, which oversees the various MODVOPPP initiatives. Further, clinical psychology *per se* was expressly written into Section 701(4) of the act, which is generally considered the definition section of MODVOPPP. With psychology's inclusion on this advisory council and with our current eligibility for nearly all of the Title VII initiatives, the foundation has finally been established for our training institutions to understand and appreciate the significance of the Title VII (Health Professions) training resources, or, stated another way, that psychology has legislatively become a "health care profession." In a public policy frame of reference, and in the eyes of the other health care professions, the enactment of this legislation provided formal congressional acknowledgment that we are considerably more than a "mental health" speciality; that is, our "health psychology" expertise has become considerably broader in its policy implications than most of us realize.

Consistent with this growing federal recognition and the gradual maturing of our profession, we must now expect that, over time, efforts will be made by the various state associations to explicitly expand their state scope-of-practice acts to more accurately reflect the broadest nature of psychology's clinical involvement and expertise within medical settings. In the early years of our professionalism, we rarely worried about the exact phraseology of our scope-of-practice acts, keeping them as broad and general as possible. However, as newer generations of clinical psychologists become more sophisticated in their clinical interventions and as the evidence of their cost-effectiveness continues to grow, we must develop the additional policy and legislative expertise required to ensure that artificial limitations will not be imposed on their capabilities. Similarly, as we collectively become more involved in programs that are directly related to our nation's changing demography—such as the increased interest in geriatrics—we must expect that those of our profession who are interested in administration and programmatic evaluations will be exposed to the wide range of underlying public policy issues that surround these areas, including how to pay for necessary services and how to allocate Medicare clinical training funds across the range of health professions. As the breadth of expertise demonstrated in this handbook suggests, there is every reason to expect that psychology's field of influence will continue to expand, clinically and programmatically, for many generations to come (DeLeon & VandenBos, 1988).

References

DeLeon, P. H. (1988). Public policy and public service: Our professional duty. *American Psychologist, 43*, 309–315.

DeLeon, P. H. (1989). New roles for "old" psychologists. *The Clinical Psychologist, 42*(1), 8–11.

DeLeon, P. H., & VandenBos, G. R. (1988). Behavioral science: The new health care evolution. *Asia-Pacific Journal of Public Health, 2*(2), 88–89.

Lalonde, M. (1974). *A new perspective on the health of Canadians: A working document.* Ottawa: Government of Canada.

U.S. Department of Health, Education and Welfare (HEW). (1979). *Healthy people: The surgeon general's report on health promotion and disease prevention.* DHEW Pub. No. (PHS) 79-55071. Washington, DC: U.S. Government Printing Office.

U.S. Preventive Services Task Force. (1989). *Guide to clinical preventive services.* Washington, DC: U.S. Government Printing Office.

About the Contributors

Roger T. Anderson is currently a doctoral candidate at the Johns Hopkins University School of Public Health, in the Department of Behavioral Sciences. His primary interest is the influence of the social environment on coronary heart disease morbidity and mortality.

Cynthia D. Belar received her Ph.D. from Ohio University in 1974. From 1974 to 1984, she was on the faculty of the Department of Clinical and Health Psychology at the University of Florida's academic medical center. Since 1984, she has served as Chief Psychologist and Clinical Director of Behavioral Medicine for the Kaiser Permanente Medical Care Program in Los Angeles. She has also served as Chair, Education and Training Committee, Division of Health Psychology; Chair, Graduate Education Committee of the American Psychological Association; Chair, Executive Committee, Association of Psychology Internship Centers; and Chair, Arthritis Health Professions Research Committee of the Arthritis Foundation. She is the author of a recent book, *The Practice of Clinical Health Psychology*, and coeditor of *Health Psychology: A Discipline and a Profession*. Dr. Belar is a diplomate in clinical psychology, American Board of Professional Psychology.

Linas A. Bieliauskas, Ph.D., is an associate professor in the Departments of Psychiatry and Psychology at the University of Michigan and staff psychologist at the Ann Arbor VA Medical Center. He received his Ph.D. in clinical psychology from Ohio University in 1976. From 1976 to 1989, Dr. Bieliauskas was on the faculty and staff of Rush-Presbyterian-St. Luke's Medical Center in Chicago, where he also served as Director of Clinical Training in the Department of Psychology and Social Sciences. He is a diplomate of the American Board of Professional Psychology in both clinical psychology and clinical neuropsychology.

Joseph Bleiberg, Ph.D., is the Director of Psychology at the National Rehabilitation Hospital, where he is also Co-Director of Brain Injury Rehabilitation. He serves as a clinical associate professor in the Department of Neurology at Georgetown University Medical School. He was previously the Director of Psychology at the Rehabilitation Institute of Chicago and an assistant professor at Northwestern University Medical School. He has published primarily in the areas of neuropsychology and rehabilitation.

Bruce Bonecutter received his Ph.D. from the Illinois Institute of Technology in clinical psychology in 1980. He works part time at Cook County Hospital and has a private practice in clinical and health psychology. He has been president of the Illinois Psychological Association, 1989–1990, has held other offices within the state association, and was the Illinois Federal Legislative Activity State Coordinator for the APA's Office of Professional Practice, 1988–1989.

Rosalind D. Cartwright, Ph.D., A.C.P., is professor and chairman of the Department of Psychology and Director of the Sleep Disorder Service and Research Center at Rush-Presbyterian-St. Luke's Medical Center, Chicago, Illinois. She is also a member of the Advisory Council of NIMH and author of *Night Life: Explorations in Dreaming* and a *Primer on Sleep and Dreaming* and coauthored with Carl R. Rogers *Psychotherapy and Personality Change*. She has published over 100 journal articles on psychotherapy, interpersonal relations, sleep disorders, and dream function.

Stanley L. Chapman, Ph.D., is the Director of Psychology at the Pain Control and Rehabilitation Institute of Georgia and currently is a clinical associate professor in the Department of Rehabilitation Medicine at Emory University. He has authored numerous articles and chapters relating to chronic pain and has coedited two books in the area. He is a surveyor for the Commission on Accreditation of Rehabilitation Facilities (CARF) and has served on CARF's National Advisory Committee for developing standards for chronic pain management programs and on its committee for developing a model program evaluation system for such programs. He is on the editorial board of *Topics in Pain Management* and has served as a grant reviewer for the National Institute of Disability and Rehabilitation Research.

James P. Choca, Ph.D., is Chief of the Psychology Service of Veterans Administration Lakeside Hospital at Northwestern University Medical School. Dr. Choca has authored a book, now in its second edition, designed as a practical manual for training psychologists at practicum or internship facilities.

Robert Ciulla, Ph.D., is a clinical psychologist at the Islip Mental Health Center in Islandia, New York. He formerly worked as a clinical psychologist in the Spinal Cord Injury Program at the National Rehabilitation Hospital.

Before that, he was a clinical psychologist at Sunny View Hospital and Rehabilitation Center.

Daniel J. Cox, Ph.D., is professor of Behavioral Medicine and Psychiatry and the Director of the Behavioral Medicine Center at the University of Virginia Health Services Center. He received his Ph.D. in clinical psychology from the University of Louisville in 1976. Dr. Cox has previously authored over 90 book chapters and journal articles on the psychophysiology of stress-related medical disorders.

Thomas L. Creer, Ph.D., is professor of psychology at Ohio University. Before assuming his current position, Dr. Creer served as the Director of the Behavioral Sciences Division at the Children's Asthma Research Institute and Hospital (CARIH), a division of the National Asthma Center, in Denver. For over two decades, his research interests have centered on the behavioral and psychological aspects of chronic respiratory disorders, including asthma, emphysema, and bronchitis.

Nicholas A. Cummings, Ph.D., is the founder and chief executive officer of American Biodyne, Inc., a national capitated mental health provider group in eight states. He is a former president of the American Psychological Association and the founder of the four campuses of the California School of Professional Psychology, the nation's first professional school in psychology. For 25 years he was the chief psychologist for Kaiser-Permanente in northern California, where he wrote the nation's first comprehensive, prepaid mental health plan. He also served as the executive director of the Mental Research Institute, Palo Alto.

Patrick H. DeLeon, Ph.D., J.D., is a member of the Board of Directors of APA and past president of the Divisions of Clinical Psychology, Psychotherapy, and Psychology and the Law. He has been active in the governance of the APA and has served on numerous boards and committees, including as chair of the Board of Professional Affairs and the ad hoc Committee on Legal Issues. He is a fellow and diplomate in clinical and forensic psychology. He has been the recipient of three of the APA's Distinguished Contributions awards. He obtained his Ph.D.

in clinical psychology from Purdue University in 1969, an M.P.H. from the University of Hawaii in 1973, and a J.D. from Catholic University in 1980. He has served on Capitol Hill for nearly 17 years.

Jean C. Elbert is an associate professor in the Department of Pediatrics, University of Oklahoma Health Services Center, and Director of the Communications (Learning Disabilities) Program and the DHD Clinic at the Child Study Center. Dr. Elbert is treasurer of the Clinical Child Psychology section of Division 12, APA, and is on the editorial boards of the *Journal of Clinical Child Psychology* and *Comprehensive Mental Health Care*.

Kevin Franke, Ph.D., received his doctorate in clinical psychology from Loyola University and is currently a staff psychologist in the Eating Disorder Program and a clinical instructor in the Department of Psychiatry and Behavioral Sciences at Northwestern University Medical School.

Alice G. Friedman is an assistant professor of psychology at the State University of New York at Binghamton. She received her master's and Ph.D. degrees in clinical psychology at Virginia Polytechnic Institute and State University. She completed a postdoctoral fellowship in pediatric psychology at the University of Oklahoma Health Sciences Center and was a member of the faculty of St. Jude Children's Research Hospital.

Linda Gonder-Frederick, Ph.D., is an assistant professor of behavioral medicine and psychiatry at the University of Virginia Health Sciences Center. She received her Ph.D. in health psychology from the University of Virginia in 1985 and completed a postdoctoral fellowship in the Department of Internal Medicine at the University of Virginia Health Sciences Center. Dr. Gonder-Frederickson has previously authored more than 50 book chapters and journal articles in the area of diabetes and behavioral issues.

Martin Harrow, Ph.D., is the Director of Psychology at Michael Reese Hospital and Medical Center in Chicago and is a professor in the Departments of Psychiatry and Psychology at the University of Chicago. He has engaged in extensive research on thought disorders, psychosis and long-term adjustment in schizophrenia, and other psychotic disorders and affective disorders. Dr. Harrow has published over 150 scientific papers and two books in these and related areas. Dr. Harrow received his B.A. from the City University of New York in 1955 and his Ph.D. from Indiana University in 1961. He was on the faculty at Yale University for over 11 years before going to Chicago in 1973 to assume positions at Michael Reese Medical Center and the University of Chicago.

David E. Hartman, Ph.D., is a clinical psychologist in hospital and private practice in Illinois. Dr. Hartman is Director of Neuropsychology and Adult Coordinator of the clinical psychology internship program at Cook County Hospital in Chicago. Dr. Hartman is an adjunct assistant professor of psychology in the Department of Psychiatry at the University of Illinois College of Medicine at Chicago. He has published on neuropsychology, psychotherapy, and the ethics of computer use in psychology. Dr. Hartman recently published a book entitled *Neuropsychological Toxicology*. Dr. Hartman is a past chairman of the clinical section of the Illinois Psychological Association.

Heather C. Huszti, Ph.D., is currently a postdoctoral fellow in pediatric psychology with a specialization in hematology-oncology at the University of Oklahoma Health Sciences Center. She has published several articles on pediatric AIDS and AIDS prevention education for adolescents. Her research interests include parental and child coping with chronic illnesses, pediatric pain control, and the evaluation of AIDS prevention programs.

Craig Johnson, Ph.D., received his doctorate in clinical psychology from Oklahoma State University and completed a two-year postdoctoral fellowship in the Department of Psychiatry at Yale. He was Director of the Eating Disorders Program at Northwestern University Medical School for five years. He is currently at the Laureate Psychiatric Clinic and Hospital in Tulsa, Oklahoma. Dr. Johnson has published numerous research articles and books on the treatment of anorexia and bulimia. He founded the *International Journal of Eating Disorders* and serves on several editorial boards.

Michael Jospe, Ph.D., is an associate professor and coordinator of the health psychology track in clinical psychology at the California School of Professional Psychology, Los Angeles. He received his undergraduate education in South Africa and England and his Ph.D. at the University of Minnesota in 1974. He served as Chief Psychologist at Newington (Connecticut) Children's Hospital and, from 1979 to 1987, was coordinator of pediatric consultation–Liaison psychiatry and coordinator of psychosocial oncology at Kaiser-Permanente Medical Center in Los Angeles.

Bonnie L. Katz, Ph.D., is a clinical psychologist in the Spinal Cord Injury Program at the National Rehabilitation Hospital, where she has chaired the program's Psychosocial Task Force and has implemented a comprehensive set of psychological services. She has also conducted research and published on the consequences of and recovery from traumatic victimization.

Daniel S. Kirschenbaum, Ph.D., received his degree in clinical psychology from the University of Cincinnati in 1975. He is currently an associate professor of psychiatry and behavioral sciences at Northwestern University Medical School and the Director of the People at Risk (PAR) Weight Control Program at Northwestern Memorial Hospital (Chicago). Dr. Kirschenbaum is a fellow of the American Psychological Association and of the Association for the Advancement of Applied Sport Psychology. He has coauthored four books and 70 articles, primarily on behavioral self-control with applications to health psychology (particularly obesity) and sport psychology.

Benjamin Kleinmuntz, Ph.D., has been, since 1973, professor of psychology at the University of Illinois at Chicago, where he served as Director of Graduate Clinical Training until 1978. Before that, he was a professor at Carnegie-Mellon University. Dr. Kleinmuntz is a 1958

clinical psychology Ph.D. from the University of Minnesota. He has made numerous contributions in the area of clinical decision making, computerized psychological assessment, and the uses and abuses of polygraphs. He has written numerous articles and eight books on a variety of problem-solving and assessment topics, including the text *Personality and Psychological Assessment*.

Mary P. Koss, Ph.D., is a professor of psychiatry (with a joint appointment in the Department of Psychology) at the University of Arizona College of Medicine. Under contract from the National Institute of Mental Health, she prepared a mental health research agenda in the area of violence against women. Dr. Koss is coauthor of the book *The Rape Victim: Clinical and Community Approaches to Treatment*. She is associate editor of the *Psychology of Women Quarterly* and *Violence and Victims* and is a member of the editorial boards of the *Journal of Consulting and Clinical Psychology* and the *Journal of Interpersonal Violence*. She is the 1989 recipient of the John D. Schaefer Award for Empirical Contributions to Victimization Research given by the National Organization for Victims' Assistance.

Harry Kotses, Ph.D., is a professor of psychology at Ohio University. He has conducted research on the classical and operant conditioning of physiological responses, on the effects of electromyographic biofeedback on asthma, and on the effect of emotion on respiration. He has published scholarly articles on psychophysiology, biofeedback, and asthma. He provides editorial consulting for journals in psychophysiology, health psychology, and medicine.

Asenath La Rue, Ph.D., is an associate professor in the Department of Psychiatry and Biobehavioral Sciences at the University of California–Los Angeles. She is coordinator of the postdoctoral fellowship program in geropsychology at the Neuropsychiatric Institute, UCLA; a member of the Board of Directors, Alzheimer's Disease Association, Los Angeles; and a reviewer for numerous journals in the field of clinical aging. Her primary area of research is cognitive function in normal and pathological aging.

Kris R. Ludwigsen, Ph.D., is in private practice in Walnut Creek, California. She is the editor and contributing author of *Hospital Practice in California: A Manual for Psychologists*. She chairs the Division 42 (Psychologists in Independent Practice) Committee on Hospital Practice, coordinating with the APA Practice Directorate. She has published several articles on hospital practice and has developed a training-certification model to prepare psychologists for hospital practice.

James Malec, Ph.D., works as a consultant in psychology for the Mayo Clinic and Foundation in Rochester, Minnesota, and is an assistant professor in the Mayo Medical School. He is a diplomate in clinical neuropsychology through the American Board of Professional Psychology. Dr. Malec is Program Director of the Brain Injury Outpatient Program at Mayo Medical Center. Additionally, he has published and continues to pursue research questions pertaining to traumatic brain injury, psychological aspects of stroke, spinal cord injury, cancer, and normal aging.

Charles McCreary, Ph.D., is clinical professor of medical psychology, Department of Psychiatry and Biobehavioral Sciences, UCLA School of Medicine. His teaching and research activities focus on the problems that patients face in coping with chronic pain. He has studied the characteristics of patients at risk of a poor response to standard dental and medical treatments for their pain problems, and he has performed research designed to develop self-management skills in patients suffering from chronic pain.

Donald Meichenbaum, Ph.D., is one of the founders of cognitive behavior modification (CBM), and his 1977 book *Cognitive Behavior Modification: An Integrative Approach* is considered a classic in the field. He has also authored *Coping with Stress* and *Stress Inoculation Training*, coauthored *Pain and Behavioral Medicine* and *Facilitating Treatment Adherence: A Practitioner's Guidebook*, and coedited *Stress Reduction and Prevention* and *The Unconscious Reconsidered*.

He is Associate Editor of *Cognitive Therapy and Research* and is on the editorial boards of a dozen journals. He is also editor of the Plenum Press series on stress and coping. He is currently professor of psychology at the University of Waterloo, Waterloo, Ontario, Canada, and a clinical psychologist in private practice. In a recent survey reported in the *American Psychologist*, North American clinicians voted Dr. Meichenbaum one of the 10 most influential psychotherapists of the century. A recent survey of academic psychologists in Canada indicated that Dr. Meichenbaum was the most cited psychology researcher at a Canadian university.

Barbara G. Melamed, Ph.D., is a dean and a professor of psychology at Ferkauf Graduate School of Psychology, Albert Einstein College of Medicine–Yeshiva University. She holds an appointment in the College of Medicine. Dr. Melamed was Program Director of a National Institute of Dental Research Training Program. She is director of the Yeshiva University Fear and Anxiety Disorders Clinic. Dr. Melamed was coauthor of a textbook, *Behavioral Medicine: Practical Applications in Health Care*, awarded Best Behavioral Science Book in 1983. She is a Fellow of the American Psychological Association and was an APA master lecturer in 1983. Her areas of research expertise include preparation for surgery, patient–doctor communication, patient anxiety, and parenting behaviors. She is on the editorial board of *Health Psychology*. She is a member of the Academy of Behavioral Medicine Research and has presented her research internationally.

Russ V. Reynolds, Ph.D., is currently an adjunct assistant professor of psychology at Ohio University. His primary areas of interest are addictive behaviors, asthma, chronic pain, computer-assisted self-management training, and other minimal-therapist-contact treatment modalities.

Ronald H. Rozensky received his Ph.D. from the University of Pittsburgh in 1974. He is the Associate Chairman of the Department of Psychiatry, the Evanston Hospital, Evanston, Illinois, where he is also the Chief of the Psychology Section. In addition, he is an associate professor of clinical psychiatry and behavioral sciences at Northwestern University Medical School and an adjunct associate professor of psychology at Northwestern University. He is a diplomate in clinical psychology from the American Board of Professional Psychology. He has published articles and book chapters in the areas of self-control, obesity, depression, biofeedback, alcoholism, behavioral medicine, and professional issues.

J. Terry Saunders, Ph.D., is assistant professor of medical education and Director for the Diabetes Clinical Research Institute Outreach Program at the University of Virginia Health Sciences Center. He received his Ph.D. in clinical and community psychology from Yale University. Dr. Saunders has also helped to develop the WeLL (Weight Loss for Life) Program at the UVA and codirects the weight maintenance phases of that program. His research interests are in psychological and behavioral adjustment to diabetes, the effectiveness of diabetes patient and professional education, and psychological behavioral factors in weight loss and weight maintenance.

Harry S. Shabsin, Ph.D., is assistant professor of psychiatry and behavioral sciences at the Johns Hopkins University School of Medicine and the Director of the Gastrointestinal Pain Center at the Francis Scott Key Medical Center. He received his Ph.D. in physiological psychology from the University of Tennessee in 1982, after which he was a fellow in behavioral medicine at the Johns Hopkins School of Medicine. His clinical and research interests include the physiology of stress-related gastrointestinal disorders and the psychological characteristics and management of chronic pain syndromes. He has authored book chapters and journal articles on the psychological management of gastrointestinal disorders.

Edward P. Sheridan, Ph.D., is professor and chairman of the Division of Psychology at Northwestern University Medical School. He serves on the executive committees of the Association of Professors of Psychology in Medical Schools, the Council of Graduate Departments of Psychology, and the Council of Directors of Training in Health Psychology. He is a member of the APA Accreditation Committee. In addition, he chaired sections of the 1987 National Conferences on Graduate Education and on Internship Training. He has published widely on graduate education, internship, and postdoctoral training.

Kathleen Sheridan, Ph.D., received her degree in clinical psychology from Fordham University in 1968. In 1983, she received a law degree from Northwestern University. She has been an associate professor in clinical psychiatry and behavioral sciences since 1978. In 1987, she created and became a director of the AIDS Biopsychosocial Services unit at Northwestern. She is also Deputy Director of the Comprehensive AIDS Center at Northwestern University Medical School.

Sharon A. Shueman, Ph.D., received her doctorate in counseling psychology from the University of Maryland in 1977. She is currently employed as an independent consultant in the Los Angeles area, specializing in the development and implementation of quality assurance programs for mental health and other social service programs. From 1978 to 1984, she was on the staff of the American Psychological Association as administrator of the APA's quality assurance and peer review programs. Since 1984, she has also served as a technical consultant to the State of California, working on programs for improving the quality of services in community care facilities for developmentally disabled persons.

George C. Stone, Ph.D., is professor of medical psychology at the University of Calfornia–San Francisco, where he is also Director of Graduate Academic Programs in Social and Behavioral Science of the Department of Psychiatry. He was senior author of *Health Psychology: A Handbook*, was first editor of the journal *Health Psychology* from 1980 to 1984, and was elected president of the Division of Health Psychology, American Psychological Association, serving in 1985–1986. Dr. Stone was senior editor of

Health Psychology: A Discipline and a Profession.
In 1988, he was named official liaison for health
psychology to the Japan Academy of Health
Behavior Science. He received his B.A. degree
(1948), M.A. (1951), and Ph.D. (1954) from the
University of California at Berkeley.

Jerry J. Sweet, Ph.D., is the Director of the
Psychological Evaluation and Testing Service at
Evanston Hospital, Evanston, Illinois, and Associate Director of Clinical Training for the
Clinical Psychology Doctoral Program at Northwestern University. He is an associate professor of clinical psychiatry and behavioral sciences at Northwestern University Medical
School, an adjunct associate professor of psychology at Northwestern University, and a lecturer at Loyola University of Chicago. He is a
diplomate in clinical neuropsychology from
the American Board of Professional Psychology. He is on the editorial boards of the *Journal
of Consulting and Clinical Psychology* and the *International Journal of Clinical Neuropsychology*.
He has published in the areas of clinical neuropsychology, chronic pain, psychological assessment, and professional issues.

Robert J. Thompson, Jr., Ph.D., is professor
and head of the Division of Medical Psychology of the Department of Psychiatry of Duke
University Medical Center. He obtained his
doctorate in clinical psychology from the University of North Dakota in 1971. Subsequently,
he was on the faculties of Georgetown University Medical School and Catholic University of
America. He was awarded the diplomate in
clinical psychology by the American Board of
Professional Psychology in 1976 and is a Fellow
of the Division of Clinical Psychology of the
American Psychological Association and a Fellow of the International Academy of Research
in Learning Disabilities. From 1986 to 1988, he
was president of the Association of Medical
School Professors of Psychology. He is a consulting editor of the *Journal of Pediatric Psychology*. He is the coauthor of two books and has
numerous journal publications.

David L. Tobin, Ph.D., received his doctorate
in clinical psychology from Ohio University

and completed a two-year postdoctoral fellowship in eating disorders at Northwestern University Medical School. He is currently in the
Physicians Group, University of Chicago. Dr.
Tobin has published numerous papers in the
area of eating disorders, as well as in several
other areas of psychological practice in medical
settings.

Steven M. Tovian, Ph.D., is currently Director
of Health Psychology in the Department of
Psychiatry at the Evanston Hospital, Evanston, Illinois. He is an assistant clinical professor
of psychiatry and behavioral sciences at the
Northwestern University Medical School and
an adjunct assistant professor of psychology at
Northwestern University. For several years, he
was Director of Psychosocial Counseling Services at the Kellogg Cancer Care Center at
Evanston Hospital. He has published articles
and book chapters in oncology, adult incontinence, neuropsychology, and rehabilitation
medicine.

Warwick G. Troy, Ph.D., is the Director of Programs in Clinical Psychology at the California
School of Professional Psychology, Los Angeles. He received his Ph.D. from the University of Maryland in 1976. He has served on the
Faculty of Medicine at the University of New
South Wales and as a consultant in medical
education and program evaluation to the World
Health Organization. He has chaired the State
of Maryland's Psychology Training Advisory
Board and an American Psychological Association task force on the accreditation of psychological services settings, He is currently
completing a degree in public health at UCLA.

Dennis C. Turk, Ph.D., is professor of psychiatry and anesthesiology and Director of the Pain
Evaluation and Treatment Institute at the University of Pittsburgh School of Medicine. From
1977 to 1985, he was on the faculty of the Department of Psychology at Yale University. Dr.
Turk is internationally recognized for his work
on the assessment and treatment of chronic
pain and on cognitive-behavioral treatment.
He has published over 120 papers in books and
scholarly journals and has written or edited six

books, including *Pain and Behavioral Medicine: A Cognitive-Behavioral Perspective; Pain Management: A Handbook of Psychological Treatment Approaches;* and *Facilitating Treatment Adherence: A Practitioner's Guidebook.* Dr. Turk is a founding member of both the International Association for the Study of Pain and the American Pain Society and is a fellow of the Academy of Behavioral Medicine Research and the Society of Behavioral Medicine.

C. Eugene Walker, Ph.D., is currently professor of psychology and Director of Training in Pediatric Psychology at the University of Oklahoma Medical School. He is also Associate Chief of Mental Health Services and Director of the Outpatient Pediatric Psychology Clinic at the Oklahoma Children's Memorial Hospital. He has authored over 100 articles on clinical psychology, psychopathology, and psychotherapy in the professional literature and has prepared cassette tape programs on a variety of subjects. He has written numerous invited chapters for psychological and medical textbooks, and for approximately ten books. Included in his latest publications are *Health Psychology: Treatment and Research Issues* (with Arthur Zeiner, Ph.D., and Debra Bendell, Ph.D.); *The Physically and Sexually Abused Child: Evaluation and Treatment* (with Barbara Bonner, Ph.D., and Keith Kaufman, Ph.D.); and *Casebook in Pediatric and Clinical Child Psychology* (with Michael Roberts, Ph.D.).

Sharlene M. Weiss, R.N., Ph.D., has been actively involved in nursing education and health behavior activities for the last 30 years. She is president of Health People, a company that specializes in the design and implementation of health promotion and disease prevention programs. Dr. Weiss currently works in the Medical Illness Counseling Center.

Stephen M. Weiss, Ph.D., is Chief of Behavioral Medicine at the National Heart, Lung, and Blood Institute, National Institutes of Health. He has been actively interested in the relationship of psychosocial factors to cardiovascular health and illness since the early 1960s. He

joined the National Heart, Lung, and Blood Institute in 1974. In addition to his position as Chief of the Behavioral Medicine Branch at the NHLBI, Dr. Weiss also holds academic appointments at the Uniformed Services University School of Medicine, the Johns Hopkins University School of Hygiene and Public Health, and in the Department of Medicine and Physiology of Health Psychology of the American Psychological Association (1979–1980), the Society of Behavioral Medicine (1984–1985), and the Academy of Behavioral Medicine Research (1986–1987). He is currently president-elect of the International Society of Behavioral Medicine. Dr. Weiss has published 10 edited books and over 60 scientific articles, book chapters, and monographs. He is an associated editor for the *Journal of Behavioral Medicine* and serves on the editorial boards of four scientific journals. His international activities include consultation with the U.S. Agency for International Development, the World Health Organization, and the Peace Corps. He serves as Chair of the International Liaison Committee of the Society of Behavioral Medicine and is Co-Chair of the Steering Committee of the International Society of Behavioral Medicine.

William E. Whitehead, Ph.D., is associate professor of medical psychology (Department of Psychiatry) at the Johns Hopkins University School of Medicine and Chief of the Gastrointestinal Physiology Laboratory at the Francis Scott Key Medical Center. He received his Ph.D. in clinical psychology from the University of Chicago in 1973 and worked on the Psychosomatic Service of the Department of Psychiatry at the University of Cincinnati before coming to Johns Hopkins in 1979. His principal research interests are visceral perception and the psychophysiology of gastrointestinal disorders. He is coauthor of a book, *Gastrointestinal Disorders: Behavioral and Physiological Basis for Treatment*, and has authored 80 articles and book chapters.

Jack G. Wiggins, Ph.D., of the Psychological Development Center in Cleveland, Ohio, helped to establish the original APA Peer Review Standards, negotiated the recognition of psychology in CHAMPUS, and founded CAPPS, psychology's original lobbying organization. He chaired an APA Task Force on Medicare, which established APA priorities and allocation of resources on diagnosis and treatment under Medicare and Medicaid. He is a past president of Divisions #42 (Independent Practice) and #29 (Psychotherapy) in the APA, as well as past president of the Association for the Advancement of Psychology (AAP). He is currently an editorial consultant to several journals and has published in the areas of marketing psychological services and guidelines for practice in psychology.

David J. Williamson, B.A., is a predoctoral research fellow at the University of Florida. Mr. Williamson has involved himself in research and clinical work with a number of chronic populations, including those having rheumatoid arthritis, chronic low back pain, and TMJ.

Diane J. Willis, Ph.D., is professor of medical psychology in the Department of Pediatrics at the University of Oklahoma Health Sciences Center and Director of Psychological Services at the Child Study Center. She is a past editor of the *Journal of Pediatric Psychology* and the *Journal of Clinical Child Psychology* and is a past president of the Society of Pediatric Psychology and the Clinical Child Psychology sections of the American Psychological Association.

W. Joy Woodruff, M.S., is a doctoral student in clinical psychology in the Department of Psychology at the University of Arizona. She is completing her psychology internship in behavioral medicine at the Sepulveda VA Hospital in Los Angeles.

Logan Wright, Ph.D., is clinical professor of psychology at the University of Oklahoma and its Health Sciences Center. He received his Ph.D. degree from Peabody College of Vanderbilt University. He was an NIHM Career Development Research Fellow. Dr. Wright is an ABPP diplomate in clinical psychology and a member of both the National Academy of Practice and the Academy of Behavioral Medicine Research. He has served on APA's Board of Directors and as president of Divisions #12 and #31, as well as of the Southwestern Psychological Association. He was also Chair of Operations for AAP. In 1984, Dr. Wright was elected the 96th president of APA. At present, Dr. Wright continues to serve as a professor of psychology at the University of Oklahoma, and as the chief operating officer of the American Psychological Society. Dr. Wright is the author of four books and approximately 100 articles on medical and child psychology, and he is the recipient of several research grant awards. His first book (*Parent Power*) received the Distinguished Contribution Citation at the 1978 APA Media Awards. He coauthored the *Encyclopedia of Pediatric Psychology*.

Index